RELATIVISTIC EFFECTS IN CHEMISTRY

Part B

RELATIVISTIC EFFECTS IN CHEMISTRY

Part B
Applications

Krishnan Balasubramanian

Department of Chemistry
Arizona State University
Tempe, Arizona

A WILEY-INTERSCIENCE PUBLICATION

JOHN WILEY & SONS, INC.

New York / Chichester / Weinheim / Brisbane / Singapore / Toronto

This text is printed on acid-free paper.

Library of Congress Cataloging in Publication Data:

Balasubramanian, Krishnan, 1956–
Relativistic effects in chemistry / Krishnan Balasubramanian)
p. cm.
"A Wiley-Interscience publication."
Includes bibliographical references.
ISBN 0-471-30400-X (pt. A : cloth : alk. paper). — ISBN 0-471-17991-4 (pt. B : cloth : alk. paper)
1. Quantum chemistry. 2. Relativistic quantum theory. I. Title.
QD462.6.R42B35 1997
540′.1′53011—dc20 96-36641

Printed in the United States of America
10 9 8 7 6 5 4 3 2 1

CONTENTS

PREFACE

Why should any chemist be interested in Einstein's theory of relativity proposed in 1905? Answers to several intriguing and yet fundamental mind-boggling chemical questions are hidden in the theory of relativity. Examples of such questions are: what makes gold glitter with yellow color, and what is responsible for the liquid nature of mercury? Why is it that we have stable mercurous compounds but there are no "cadmious" or "zincus" compounds? Why is it that the mercury ion in its monovalent state exists as Hg_2^{2+} rather than as Hg^+? Why is it that the ground state electronic configuration of the Platinum atom is [core]$5d^9 6s^1$, while the corresponding configuration of the Pd atom, which is in the same column of the periodic table, is [core]$4d^{10}$? Likewise, why is it that the ground state electronic configurations of the third row transition metal atoms such as Re, Os, Ir, and Pt differ from their periodic analogs in the second row? Why is it that the sixth row elements thallium through astatine exhibit unusually lower valencies compared to their lighter analogs in the periodic table? What is the real origin of "lanthanide contraction" or "actinide contraction"? Why does radon fluoride exhibit unusual stability and ionic character? Why is it that certain spectroscopic transitions become strong and intense for heavier molecules while for their lighter analogs in the periodic table the same transition is not observed or is very feeble? Why does the lifetime of an electronic state change so much for a heavier molecule compared to its lighter analog in the periodic table? This book is intended for those who wish to seek answers to such intriguing and fundamental questions.

Although the theory of relativity proposed by Albert Einstein in 1905 has been used extensively in several applications in modern physics, it was not until the 1970s that its relevance became apparent in chemistry. Until this time the theory of relativity had a special place mainly in physics, although quantum mechanics had made a substantial impact in chemistry. By this time, applications of quantum mechanics to chemical problems and even biological problems became quite mature, and with the advent of modest computers there was a proliferation of quantum computations of small molecules. Of course these days, with supercomputing power on the desktop, quantum computations of moderate to large molecules are becoming routine.

While quantum mechanics became a natural choice to address the electronic structure of molecules, it was not apparent that relativity should also become an integral part of chemical applications. The discovery of super heavy elements could be attributed as one possible motivation since for such heavy elements the inner electrons are subjected to a large nuclear charge due to the presence of a large number of protons in the nucleus. Thus, in order to maintain balance with the strong electrostatic field, the inner electrons must acquire very large speeds comparable to the speed of light. The importance of relativity was certainly recognized for describing the properties of very heavy atoms many years earlier. In fact, Dirac's equation for a relativistic description of the hydrogen atom is quite old, and is well-known, even though for the hydrogen atom relativistic effects are rather small. Desclaux in 1973 compiled the Dirac–Fock results of several atoms, and it thus became clear that relativistic effects are going to be very important for very heavy atoms.

In 1979 two pioneering *Accounts of Chemical Research* articles, one by Professor Kenneth S. Pitzer at Berkeley [*Acc. Chem. Res.* **12**, 271 (1979)] and the other by P. Pyykko and J. P. Desclaux [*Acc. Chem. Res.* **12**, 276 (1979)] were published. These two stimulating papers explicitly recognize the importance of relativity in chemical applications and how relativity could potentially influence the properties of molecules containing very heavy atoms.

Although by 1979 it became apparent that relativity should play a key role in the properties of molecules containing heavy atoms, relativistic computations could be carried out primarily for diatomics and that too mostly for the diatomic hydrides. Thanks to the relativistic effective core potential techniques, one could carry out relativistic quantum computations of molecules containing very heavy atoms due to the replacement of chemically unimportant core electrons by the frozen relativistic effective core potentials. In the early 1980s relativistic quantum computations of molecules such as Au_2, Tl_2, TlH, AuH, as well as others, were made. The origin of different kinds of relativistic effects could be understood. However, such computations should be considered as very preliminary and "modest" in the current standard. In 1982 the first relativistic configuration interaction computation was carried out on the ground and excited electronic states of the diatomic TlH molecule at Berkeley by P. A. Christiansen, myself, and K. S. Pitzer. Subsequently, this procedure was extended to polyatomics in 1988 by myself, Professor Russel M. Pitzer, and N. W. Winter, independently.

Subsequent to the 1980s numerous relativistic quantum computations have been reported in the literature on both diatomics and polyatomics. In fact, Pyykkö has compiled two volumes containing a list of references to relativistic quantum computations from 1986–1992. A cursory look at these volumes would reveal that there is an exponential growth of relativistic computations made possible both with powerful computers and the development of software.

The current books I have written naturally evolved through the relativistic quantum chemical computations and the results on the electronic structure and properties of molecules containing very heavy atoms. I have been actively involved in this research since 1980. The techniques developed to-date, and the results accumulated up to now on both diatomics and polyatomics of very heavy molecules, need to be combined into a single source which I hope will serve as a valuable reference for any chemist (theoretical or experimental) interested in the properties of molecules containing very heavy atoms.

While there are many books on the applications of quantum mechanics to chemistry, the subject of relativity appears not to have been treated adequately in the chemical context. An interesting feature of this book is that it does not assume that the reader has had adequate exposure to relativity. In fact, after the introductory chapter, the entire second chapter is devoted to the special theory of relativity starting from scratch. No assumptions are made, and the subject matter is built with as many derivations as possible. The topic is developed to the level of four-vector formulation of angular momentum and space-time and momentum-energy diagrams.

The book is divided into two parts, Part A consisting of theory and techniques, and Part B containing applications to molecules containing very heavy atoms.

Part B is intended for any chemist (experimental or theoretical) interested in the properties of molecules containing heavy atoms. It is not theoretically demanding and does not contain any sophisticated relativistic quantum mechanical formalism. It can be read independently of Part A, but if one needs to apply relativistic quantum mechanical techniques to these species, Part A would be very useful. Part B of the book is organized into four major chapters, all of which contain specific results on the electronic states of molecules containing very heavy atoms and the impact of relativistic effects on the electronic properties of these species accentuating the relative trends. Thus, Part B not only contains discussion of the importance of relativistic effects but specific experimental and theoretical results on the spectroscopic properties and the electronic structure of molecules containing very heavy atoms. Chapter 1 begins with relativistic effects on the electronic structure of transition metal clusters. In particular, clusters of Cu, Ag, and Au triad as well as Pd and Pt are compared. Electronic structure of open-shell transition metal clusters such as Rh_3 and Ir_3 are also contained in this chapter.

Chapter 2 deals with the electronic and spectroscopic properties of heteronuclear diatomics of main group *p*-block elements from Ga to Bi. The diatomic hydrides, halides, and chalcogenides of these species are considered, in particular. Although this chapter evolved from a review article I wrote [*Chem. Rev.* **89**, 1801 (1989)], a substantial amount of new material has been added and molecules for which new data were obtained after the publication of the review were added. Chapter 3 deals with the clusters of very heavy main group *p*-block elements from Ga to Po. While some of the material on dimers of Ga-Tl is derived from another previous review [*Chem. Rev.* **90**, 93 (1990)], a substantial amount of new information and data were added for those dimers, trimers, tetramers, and other clusters for which electronic structure computations were made after 1989. The chapter also contains relative periodic trends and tells how relativistic effects lead to anomalous behavior for the sixth row clusters. The last chapter pertains to relativistic effects on molecules containing lanthanide and actinide atoms. I have written a review on this topic before [*Handbook on the Physics and Chemistry of Rare Earths*, **18**, 29 (1994)], and a substantial portion of the material of this chapter was derived from this review. However, numerous editorial changes both to the text and references were made and some sections of the previous review were modified. The chapters of Part B contain main group, transition metal, and lanthanide and actinide species, thus providing a complete overview of relativistic effects for all these species.

I must express my indebtedness to Professor Kenneth S. Pitzer of University of California, Berkeley who as my post-doctoral mentor introduced me to the area of relativistic quantum chemistry. His continued encouragement for several years has certainly been a valuable inspiration. I would like to thank many of my past post-doctoral associates and PhD students who made valuable contributions to this area; without their contributions this book could not have evolved into this shape. Specifically I would like to thank Drs. Dingguo Dai, D. A. Chapman, K. K. Das, P. Y. Feng, Devashis Majumdar, Lida Latifzadeh-Mosudipour, and M. Al-Zahrani. I wish to express my appreciation to Professors Mu-Zhen Liao, Sheng H. Lin, Szcepan Roszak, Walter S. Koski, Russel M. Pitzer, Walter C. Ermler, and Philip A. Christiansen for their contributions to this area which immensely helped in the evolution of this book. I would also like to thank Professor P. Pyykkö for providing a copy of his book containing a complete list of references, and Professors M. Zerner, N. Rosch, R. M. Pitzer, H. Stoll, R. Buenker, and several others for their contributions to this area which made possible several comparisons. The continuous support by the U.S. Department of Energy for the work on transition metal clusters, and the National Science Foundation for work on main group molecules is greatly appreciated; without this support the results contained in Part B of this book could not have been computed. Finally, I express my deep appreciation to my wife Gomathy and daughters Nila and Sangeetha for their understanding and support during the writing of this book.

KRISHNAN BALASUBRAMANIAN

Arizona State University

RELATIVISTIC EFFECTS IN CHEMISTRY

Part B

1

RELATIVISTIC EFFECTS IN SMALL TRANSITION-METAL CLUSTERS

In this chapter we have chosen a few transition-metal clusters to discuss both experimental and theoretical aspects of such clusters, connect the theoretical results with the experimental findings, and finally, discuss the relative periodic trends and the impact of relativity on the properties of such clusters. In general, several anomalies are found for the third-row transition-metal clusters, due primarily to relativistic effects, although the lanthanide contraction also comes into play. However, as we discuss in Chapter 4, on lanthanides, even a significant portion of the lanthanide contraction itself is attributed to relativity.

Specific clusters that we shall consider in this section are the small clusters of the coinage metal atoms such as Cu, Ag, Au, Pt, and Pd. The electronic structure of small coinage metal clusters, primarily the ones containing Cu, Ag, and Au, has been reviewed by the author before (1). However, a significant amount of new material has been added in discussion of these species, especially in comparing the work done after 1987. We have also included larger clusters and the properties of Pt, Pd, Rh, and Ir clusters.

1.1 RELATIVISTIC EFFECTS IN TRANSITION-METAL ATOMS

Before we discuss the properties of molecules containing transition-metal atoms, it would be interesting to consider some properties of the atoms themselves from the standpoint of periodic trends. We shall also discuss how relativity influences the periodic trends of such atomic properties.

We have already discussed in Part A, Chapter 1 how the ionization energy of the gold atom is impacted by relativity. That is, the first ionization potential of the gold atom is anomalously higher compared with silver and copper.

This high ionization potential of the gold atom is attributed to the relativistic mass–velocity correction, which contracts the $6s$ orbital of the gold atom, thereby causing the $6s$ orbital to be relatively more stable. This enhanced stability of the $6s$ orbital leads to anomalously high ionization energy for the gold atom. The relativistic mass–velocity stabilization of the $6s$ orbital of the gold atom also results in anomalous properties of gold-containing compounds, a phenomenon referred to as *gold anomaly*.

The entire third-row transition-metal atoms from Hf to mercury, and the subsequent atoms such as Tl, Pb, and Bi, are substantially influenced by relativity. Traditionally, the differences between these elements and their lighter analogs have been attributed to lanthanide contraction, which is believed to be due to the incomplete shielding of the nucleus by $4f$ electrons, since the $4f$ orbitals are diffuse. However, as we discuss in Chapter 4 (when chapter number only is given—viz., Chapter 1, 2, 3, or 4—the reference is to a chapter in this volume, Part B; cross-references to chapters in the preceding volume will include the term Part A), the lanthanide contraction itself is due in part to relativity. Let us consider a few anomalies in the atomic properties of the third-row transition-metal atoms.

The ground-state electronic configuration of the platinum atom is [Xe-$4f^{14}$core]$5d^9 6s^1$, in contrast with the ground-state electronic configuration of the palladium atom, which is [Kr-core]$4d^{10}$. Thus the platinum atom exhibits a $^3D_3$ ground state in contrast with a $^1S_0$ ground state for the palladium atom. This anomaly of the platinum atom is due to the fact that the $6s$ orbital of the platinum atom is relativistically stabilized because of the variation of electron mass with electron speed. Thus, although closed-shell $4d^{10}$ is favored for Pd, relativistic stabilization of $6s$ leads to an open-shell $5d^9 6s^1$ configuration for Pt. The $^1S_0$–$^3D_3$ splitting ($^1S_0$ is lower) for Pd is 6564 cm^{-1}, while the $^3D_3$ state is lower for Pt and the $^3D_3$–$^1S_0$ energy separation for Pt is 6140 cm^{-1} (113).

While the ground electronic configuration of the rhodium atom is [Kr-core]$4d^8 5s^1$, the ground state of the iridium atom is [Xe-$4f^{14}$-core]$5d^7 6s^2$. Again, the $6s$ orbital takes up two electrons due to the relativistic mass–velocity stabilization. Although both atoms have the same $^4F_{9/2}$ ground states, they have different electronic configurations due to relativity. This is reflected in the properties of rhodium and iridium compounds in that the $6s$ participation in bonding is enhanced in Ir, compared to the $5s$ participation in the case of Rh. For example, the electronic states of the iridium clusters exhibit enhanced $6s$ Mulliken populations compared to the corresponding states of the rhodium clusters.

Another comparison that reveals the manifestation of relativistic mass–velocity effect is the ground-state electronic configuration of the ruthenium and osmium atoms. While the ground electronic configuration of Ru is [Kr-core]$4d^7 5s^1$, the ground electronic configuration of Os is [Xe-$4f^{14}$-core]$5d^6 6s^2$. The ground state of the Ru^+ is [Kr-core]$4d^7$, while the ground state of Os^+ is [Xe-$4f^{14}$-core]$5d^6 6s^1$.

The feature thus observed is general for all the transition-metal atoms Hf–Pt, in that the 6*s* orbital contains one more electron compared to the corresponding second-row atoms Zr–Pd. However, La and Sc are similar, since the nuclear charge for La is not substantially higher than that for Hf. The presence of the 14 lanthanide elements increases the nuclear charge of Hf by 15 compared to La. Thus relativistic effects make substantial contributions for elements heavier than Hf.

The relativistic mass–velocity effect also explains the unusually lower binding energy of the mercury dimer compared to the cadmium or zinc dimer. The stabilization of the 6*s* orbital of the mercury atom leaves the $6s^2$ shell relatively inert, and thus two mercury atoms form only a van der Waals bond, while the cadmium and zinc atoms form stronger bond. As seen in basic inorganic chemistry books (178), the Zn_2^{2+}, Cd_2^{2+} ions are unstable, and occur only in melts or solids. In contrast, Hg_2^{2+} is unusually stable (178). To generate Zn_2^{2+}, one has to add Zn to fused $ZnCl_2$, which generates a yellow solution, which on cooling yields a yellow glass that contains Zn_2^{2+}. On the other hand, the Hg_2^{2+} mercurous ion is formed on reduction of mercurous salts in aqueous solution. Compounds such as Hg_2Cl_2 exhibit short Hg–Hg bonds indicating the presence of Hg_2^{2+}. Study of Hg^{I}–Hg^{II} equilibrium indicates that Hg readily reduces Hg^{2+} to Hg_2^{2+}. This is attributed to the unusual stability of Hg_2^{2+}, which is made possible by the relativistic mass–velocity effect. However, any reagent that can lower the activity of Hg_2^{2+} will cause disproportionation of Hg_2^{2+}; for example, when OH^- is added to a solution of Hg_2^{2+}, a dark precipitate consisting of Hg and HgO is formed.

1.2 RELATIVISTIC EFFECTS IN COINAGE-METAL CLUSTERS

There have been several theoretical and experimental investigations of coinage-metal clusters and the clusters of Ir and Rh (1–177). The interest in coinage metals has a long history, since these are noble metals and since ancient chemists endeavored to convert other elements into gold (alchemy). Clusters of copper, silver, and gold are interesting from both fundamental and applied standpoints. The silver and gold clusters, in particular, have catalytic properties, and are also models of surfaces and thin films. The silver clusters are of importance in photographic processes.

The coinage-metal clusters are interesting from a fundamental standpoint, since the bonding in gold is different from that of silver, a phenomenon referred to as *gold anomaly*. Relativistic effects are so large for gold-containing compounds that the chemistry of gold is substantially different from that of silver. For example, the characteristic color of gold in contrast with silver is attributed to relativistic effects (117). As seen in Part A, Chapter 1, the relativistic mass–velocity correction, which is basically a relativistic correction to the kinetic energy term, is very large in the core of the gold atom. This large relativistic effect manifests itself in the valence 6*s* orbital of the gold atom by way of introducing a substantial contraction of this orbital. This, in turn, leads to a substantial difference in the gold chemistry and gold anomalies. As discussed in Part A, Chapter 1, the color of gold is due to an electronic transition from the occupied 5*d* bands to nearly Fermi 6*s* bands (117). The stabilization of the 6*s* orbital of the gold atom leads to a decrease in the 5*d* bandgap of gold, resulting in an optical transition in the visible region as opposed to UV; that is, nonrelativistic gold would have a more silverlike color.

Experimental progress in the study of coinage-metal clusters has been enhanced by the advent of powerful methods, such as the matrix-isolation methods, supersonic nozzle expansion methods, sputtering methods and ion trapping methods. The progress in the experimental studies of transition-metal molecules has been reviewed by Weltner and Van Zee (2) as well as others (see other publications listed in Ref. 2), and Morse (3). In summary, a number of spectroscopic methods ranging from electron spin resonance (ESR) spectroscopy of the matrix-isolated clusters, to Raman and visible optical spectroscopy of these clusters have been employed to investigate these clusters. The next subsection of this chapter briefly summarizes the experimental findings, while Section 1.2.2 discusses the theoretical investigations on these clusters. Section 1.2.3 critically compares different coinage-metal dimers, trimers, and tetramers to arrive at periodic trends and the importance of relativistic effects in these clusters. Section 1.3 considers the electronic structure of Pd and Pt clusters, while Section 1.4 compares Rh and Ir clusters.

1.2.1 Experimental Studies on Cu_n, Ag_n, and Au_n

1.2.1.1 Cu_2 The binding energy of Cu_2 was obtained using the Knudsen effusion mass spectrometry as $D_0^\circ(Cu_2) = 2.0$ eV (4, 5). The ionization potential of Cu_2 is also known to be 7.8 ± 0.4 eV (5). The Cu_2 dimer has been investigated spectroscopically both in absorption and emission (6–27). Two band systems, denoted as $A^1\Pi_u$–$X^1\Sigma_g^+$ and $B^1\Sigma_u^+$–$X^1\Sigma_g^+$, have been observed. The vibrational and rotational analyses of these bands have also been performed, which yielded spectroscopic constants of these three states. Table 1.1 shows the spectroscopic constants of Cu_2 obtained using both experimental and theoretical studies.

The assignment of the *B*–*X* system seems to be less ambiguous than the *A*–*X* system, which is assigned to the $^1\Pi_u$–$^1\Sigma_g^+$ transition. However, the rotational constants obtained by Lochet (10) differ substantially from the results of Pesić and Weniger (11). The spectroscopic constants obtained by Lochet, however, seem to suggest that the *A*–*X* system should be attributable to a $A^1\Sigma_u^+$–$X^1\Sigma_g^+$ system. Bondybey's (13) observed bands in LIF seem to support Lochet's assignment. Theoretical calculations of these systems have provided considerable insight into the nature of the *A*–*X* and *B*–*X* systems.

Many matrix-isolation spectroscopic investigations on Cu_2 (20–27) have led to several absorption bands, which are yet to be fully assigned and interpreted.

Table 1.1 Spectroscopic Constants of Cu_2[a]

	R_e, Å		T_e, cm^{-1}	ω_e, cm^{-1}		D_e, eV	
State	Exp.	Theory	Exp.	Exp.	Theory	Exp.	Theory
$X^1\Sigma_g^+$	2.22	2.26	0	266	246	2.0	1.8
$a^3\Sigma_u^+$	2.48	—	15,420	125 ± 25	—	—	—
$A^1\Pi u$?	2.56	—	20,433	192	—	—	—
$A^1\Sigma_u^+$	2.28	—	20,433	192	—	—	—
$B^1\Sigma_u^+$	2.33	—	21,758	246	—	—	—
C	—	—	21,870	221	—	—	—
D	—	—	25,560	160	—	—	—
E	—	—	27,153	232	—	—	—
F	—	—	28,560	116	—	—	—
G	2.73	—	30,753	90	—	—	—
b	—	—	33,120	100	—	—	—
c	—	—	34,870	—	—	—	—
H	—	—	36,782	—	—	—	—
I	—	—	35,000	—	—	—	—
J	—	—	37,440	—	—	—	—

[a]Theoretical results for the ground state are from Reference 93. Experimental values are from Reference 3.

This is further complicated by the fact that the absorption bands are broad and matrix–dimer interactions may be present in the matrix-isolation spectra.

Powers et al. (15) have observed a number of bands using supersonic jet-cooled expansion methods between 3200 and 2590 Å. Complete spectroscopic information (R_e, etc.) does not exist for most of these high-lying states of Cu_2. Further, theoretical calculations are scarce in this regino. Preuss et al. (27) have observed another system using LIF of supersonically expanded copper vapor at 2500 K, which overlaps with the B–X system. These bands have later been identified as due to Cu_2 using resonant two-photon spectroscopic methods (28). The Cu_2 dimer appears to be the most thoroughly studied among the coinage-metal dimers.

1.2.1.2 Experimental Studies on Ag_2 The experimental spectroscopic constants of Ag_2 are summarized in Table 1.2. The D_0° value of Ag_2 has been obtained using the Knudsen mass-spectroscopic methods. A complete list of all the references could be found in Morse (3). The latest value of 1.65 ± 0.03 eV was reported by Hilpert and Gingerich (29).

The spectroscopic studies of Ag_2 include both absorption and emission studies (7, 20, 23, 24, 30–42, 130–137). A number of systems have been observed, which are labeled $A \leftrightarrow X, B \leftrightarrow X, C \leftrightarrow X, D \leftrightarrow X, E \leftrightarrow X$ and $H \leftrightarrow X$ systems. Table 1.2 summarizes all the known experimental R_e, T_e, and ω_e values of Ag_2. There are a large number of matrix spectroscopic and laser-induced fluorescence experiments on Ag_2.

Table 1.2 Spectroscopic Constants of Ag_2 [a]

	R_e, Å		T_e, cm^{-1}		ω_e, cm^{-1}	
State	Exp.	Theory	Exp.	Theory	Exp.	Theory
$X\,^1\Sigma_g^+$	(2.48)	2.58–2.65	0	0	192	178
$A\,^1\Sigma_u^+(0_u^+)$	2.48	—	22,996	—	155	—
$B0_u^+$	2.56	—	35,839	—	153	—
$C^1\Pi_u(1_u)$	2.28	—	37,631	—	171	—
$D0_u^+$	2.33	—	39,014	—	169	—
$E0_u^+$	—	—	40,159	—	146	—
$H0_u^+$	—	—	58,273	—	167	—

[a]MRSDCI $D_e(Ag_2) = 1.31$ eV (120); experimental $D_e(Ag_2) = 1.65$ eV (29). The theoretical R_e and ω_e values are from Reference 93; the experimental summary, from Reference 3. Recent theoretical results in Section 1.2.2.

The ionization potential of Ag_2 has been obtained as 7.56 eV using two-photon ionization of jet-cooled Ag_2 in an unpublished work by Smalley and co-workers (43). The D_0° value of Ag_2^+ has also been estimated as 1.66 eV. Thus it seems that Ag_2^+ has a binding energy comparable to that of Ag_2.

1.2.1.3 Experimental Studies on Au_2 There are fewer experimental studies on Au_2 than on the other coinage-metal dimers. There are many high-temperature mass-spectrometric determinations of D_0° values of Au_2. The most definite D_0° value obtained using both the second- and third-law methods is 2.29 ± 0.02 eV (44).

The gas-phase spectroscopic studies on Au_2 have revealed two systems designated as *A–X* and *B–X* systems analogous to the Ag_2 systems (6, 7, 45, 46). The vibrational and rotational analyses of these systems have provided spectroscopic constants of these states. Klotzbücher and Ozin (46) reported broad absorption bands for Au_2 in an argon matrix in the 31,500-cm^{-1} region and more narrow bands in the 48,000- and 50,500-cm^{-1} regions. The assignments of most of these bands are not definitive. Bishea and Morse (138) studied the jet-cooled spectra of Au_2. Table 1.3 summarizes the spectroscopic constants for Au_2.

1.2.1.4 Experimental Studies on Cu_3 The Cu_3 trimer is the most experimentally studied trimer among the coinage-metal clusters, which was reviewed by Morse (3). Moskovits and Hulse (21) first identified Cu_3 using visible–UV absorption spectroscopy in rare-gas matrices. Both the resonance Raman spectrum (47) and ESR spectrum (48) of Cu_3 were subsequently recorded. Morse et al. (49) investigated Cu_3 using the jet-cooled resonant two-photon ionization method. Crumley et al. (50) studied Cu_3 using LIF, as did Rohlfing and Valentini (51) using the dispersed fluorescence method.

Table 1.3 Experimental Spectroscopic Constants of Au_2[a]

State	R_e, Å		T_e, cm^{-1}	ω_e, cm^{-1}		D_e, eV	
	Exp.	Theory	Exp.	Exp.	Theory	Exp.	Theory
$X\,^1\Sigma_g^+(0_g^+)$	2.47	2.53–2.56	0	191	179	2.29	2.0
$A0_u^+$	2.57	—	19,668	142	—	—	—
$B0_u^+$	2.52	—	25,686	180	—	—	—
C	—	—	31,500	—	—	—	—
D	—	—	48,000	—	—	—	—
E	—	—	50,500	—	—	—	—

[a]Experimental values are from Reference 2. CASSCF/MRSDCI D_e are from Reference 120. The theoretical ω_e is from Reference 93. Recent results in Section 1.2.2.

There was some controversy on the interpretation of the observed optical spectra of Cu_3, which appears to be resolved now. Morse et al. assumed the excited state participating in the observed bands to be of $^2E''$ symmetry, which would undergo Jahn–Teller distortion. The observed spectra were fitted using the theory of Longuet–Higgins. However, as pointed out by Thompson et al. (112), a term ($\propto r\cos 3\phi$) was included erroneously into the Hamiltonian. Revised fits were obtained by Thompson et al. assuming $^2E'' \leftarrow {}^2E'$ transition. The upper state could be described either by the normal mode or Jahn–Teller modes, which need to be distinguished further. The upper state in these bands has not been assigned unambiguously. Theoretical studies on Cu_3 will be discussed later and compared with experiment. Most probably the upper state is a normal mode.

1.2.1.5 Experimental Studies on Ag_3 and Au_3 and Larger Clusters

Most of the experimental studies on the heavier coinage trimers are matrix-isolated ESR spectroscopic studies (52, 53), which were reviewed by Howard et al. (52). These ESR spectroscopic investigations have revealed that the ground state of Ag_3 is probably bent with an isosceles triangular structure. The Raman spectroscopic investigation of Schulze et al. (54) has suggested a linear structure for Ag_3, in accord with a single Raman band at 120.5 cm^{-1}, although the reliability of these results has been questioned (56). Ozin and co-workers (35, 46) have obtained the UV–visible spectra of $Ag_{1/3}Ar$ in rare-gas mixtures generated by condensing silver atoms with the rare-gas mixtures and irradiating the deposits at the frequencies of silver atomic resonance absorption lines. From these spectra, they have speculated that there are "photoisomerization" processes of the type Ag_3 ($D_{\infty h}$) linear to C_{2v} (nonlinear) to D_{3h} (triangular) structure interconversions. This implies existence of nearly degenerate states for Ag_3 with equilateral triangular $(1a_1'^2 1e'^1)\,D_{3h}$ or isosceles forms described by $1a_1^2 1b_2^1$ and $1a_1^2 2a_1^1$ configurations in the D_{3h} group. Hilpert and Gingerich (29) have obtained the atomization energies of Ag_3 to be 2.63 eV for a linear geometry and 2.54 eV for the bent geometry, using the thermodynamic third-law calculations, deduced from high-temperature mass spectrometric measurements. The small difference in the atomization energies of the linear and bent structures suggest the existence of nearly degenerate states for Ag_3. Howard et al. (52, 53) have obtained the first positive ESR identification of a neutral silver cluster and suggested that Ag_3 is probably bent with a $^2B_2$ ground state.

Cheng and Duncan (57) have studied the vibronic spectra and dynamics of jet-cooled Ag_3. The spectrum of Ag_3 with an origin of 26,971 ± 10 cm^{-1} was obtained. The vibronic analyses of the observed bands over 1500 cm^{-1} were considered. The spectrum was consistent with a dynamic Jahn–Teller interaction for a hindered pseudorotor, with an excited-state barrier of 200 cm^{-1}. The vibrational frequencies of the stretching and bending modes were obtained as 179 and 104 cm^{-1}, respectively.

The ionization potential of Ag_3 was estimated to be less than 6.42 eV (58). Cheng and Duncan (57) performed two-color excimer + dye laser experiments. The Ag_3^+ signal was found to be featureless, and an estimate of the vertical ionization potential (IP) of Ag_3 of <6 eV was predicted on the basis of the combined laser energies.

Buckner et al. (59) conducted gas-phase studies on Ag_3^+ and Ag_5^+ in addition to Zn_2^+. Those authors also estimate the IP of Ag_3 and Ag_5 to be less than 7 eV. The Ag_3^+ and Ag_5^+ ions were found to be unreactive with small alkanes, alkenes, and alcohols.

The optical spectra of Au_3 recorded by Klotzbücher and Ozin (46) have revealed two unassigned bands in the regions of 34,246 and 21,231 cm^{-1}. Smalley and co-workers (65) obtained the UV photoelectron spectra of both metal and semiconductor cluster anions. In another study, Smalley and co-workers (66) obtained the UV photoelectron spectra of copper clusters (Cu_x^-) as a function of cluster size. The UV photoelectron spectra of Cu_x^-, Ag_x^-, and Au_x^- obtained by these authors (65) are shown in Figure 1.1. The electron affinities of coinage-metal clusters as a function of cluster size reported by these authors are shown in Figure 1.2. Note the interesting sawtooth-type behavior for the electron affinities of these clusters.

There are a few experimental investigations on other coinage-metal clusters such as Cu_2Ag, Cu_5, Ag_5, $CuAg_4$, and Cu_2Ag_3. Most of these are ESR spectroscopic studies, while Ag_5^+ has been investigated in the gas phase. Most of the ESR spectra studies on coinage-metal clusters have been on trimers and pentamers. Howard et al. (48, 52, 53, 60–64) have reported the ESR spectra of matrix-isolated Cu_5, Ag_5, $CuAg_4$, and Cu_2Ag_3. The homonuclear clusters such as Cu_5 and Ag_5 showed three equivalence classes of nuclei with large hyperfine splitting constants. Thus these spectra appear to be consistent with a distorted trigonal–bypyramidal structure. Smalley and co-workers (146) investigated the negative ion photoelectron spectra of Au_6^-. Based on the UPS spectra which contained a single vibrational progression, they have proposed the ground state structure to be a hexagonal ring structure.

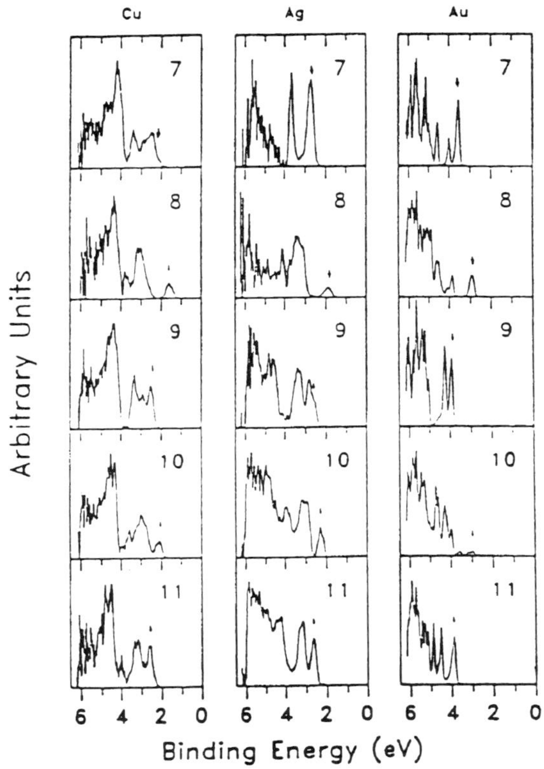

Figure 1.1 Experimental UV photoelectron spectra of Cu_x^-, Ag_x^-, and Au_x^-. (From Ref. 65.)

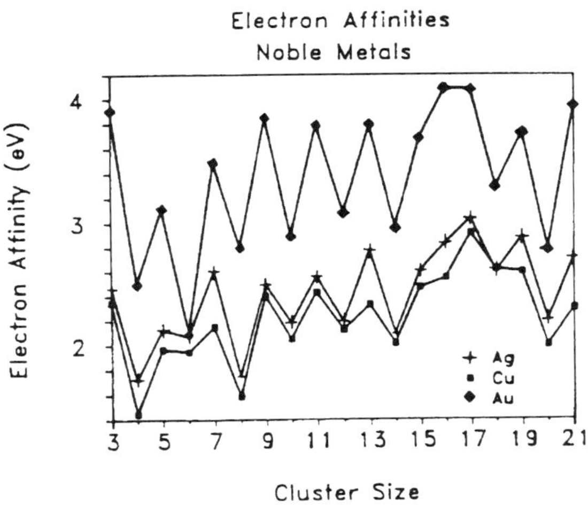

Figure 1.2 Experimental electron affinities of Cu_x, Ag_x, and Au_x, as a function of cluster size (x). (From Ref. 65.)

1.2.2 Theoretical Studies on Cu_n, Ag_n, and Au_n

Semiempirical calculations have been made on the coinage-metal dimers (66–68). Most of the semiempirical calculations have employed the extended Hückel or complete neglect of differential overlap (CNDO) methods. There are also diatomics-in-molecule (DIM) calculations on the coinage-metal clusters (69, 70). There are many density function calculations, which include the $X\alpha$ method on the coinage-metal dimers and trimers (71–76). The density function methods and DIM methods have been especially successful for the ground states of Cu_2, Ag_2, and similar compounds.

A number of *ab initio* calculations employ all-electron methods and effective core potential methods on Cu_2 (77–94). Walch et al. (93) have done SCF/SDCI/CPF computations for the ground state of Cu_2 employing a $[9s7p4d3f1g]$ basis set. They carried out comparable 19*e*-NRECP calculations using the $[4s3p3d]$ valence Gaussian basis set. The ECPs with this basis set were found to give good results, comparable to the large basis-set calculations. The best theoretical results, obtained for the ground state of Cu_2, are also listed in Table 1.1.

There are several theoretical calculations on Ag_2 (93, 95–100, 125–129). A theoretical study by Zhang and the author (125) considered both the ground and excited states of Ag_2, while previous studies focused mainly on the ground state of Ag_2 (97, 100, 126–129). Relativistic effects are significant for both Ag_2 and Au_2. Nonrelativistic calculations are far from adequate, especially for Au_2. The bond lengths obtained by all these calculations for the ground state of Ag_2 are longer than those of the experimental results. The best $TZ + 6f/5f$ SCF/SDCI calculations of Walch et al. (93) yield a bond length of 2.60 Å for Ag_2 compared to an uncertain experimental value of 2.48 Å (see Table 1.2). A D_e of 1.3 eV was obtained by Balasubramanian and Feng (120) using a relativistic CASSCF/MRSDCI method, which correlated all the 22 electrons of Ag_2, compared to an experimental value of 1.7 eV (29).

Zhang and the author (125) investigated 15 electronic states of Ag_2 using the CASSCF/MRSDCI techniques that included both 4*d* correlation effects and 5*p* orbitals of the silver atoms. The computed spectroscopic constants can be compared with the experiments results for some of the excited states that are experimentally characterized. Properties of a few other yet to be observed electronic states were also predicted by the computations of Zhang and the author (125).

There are a few experimental spectroscopic studies on Ag_2, but these have been complicated by the presence of three isotopomers, namely, 26.85% of $^{107}Ag_2$, 49.93% of $^{107}Ag^{109}Ag$, and 23.32% of $^{109}Ag_2$, resulting in the overlapping of spectral bands (130–136). Simard and Hackett (131) studied the $A \leftarrow X$ system of the silver dimer by using a narrow-bandwidth continuous-wave ring dye laser to excite fluorescence in jet-cooled silver dimers. Subsequently, Krämer et al. (132) obtained the sub-Doppler excitation spectra of the A–X system, which led to the determination of accurate molecular constants for the X and A states. Beutel et al. (133) treated the $B \leftarrow X$ system with a narrow-band frequency-doubled pulsed tunable dye laser in a collimated supersonic Ar/Ag_2 beam. They demonstrated that the A and B states of the silver dimer are $^1\Sigma_u^+$ and $^1\Pi_u$, respectively, while the ground state X had already been known to be $^1\Sigma_g^+$, which was supported by the rare-gas matrix experiments (134, 135). Earlier, Pesić and Vujisić (130), incorrectly assigned the B state to $^1\Sigma_u^+$ due to difficulties in the rotational analysis. Ginter and co-workers (136) also studied the spectra of Ag_2, which have led to the characterization of a few excited electronic states.

In the CASSCF calculations made by Zhang and the author (125), all $3d$ shells of the two Ag atoms were kept inactive in that excitations from the $4d$ shells were not allowed, but these orbitals were allowed to relax for both the ground and excited states. But all of the $5p$ orbitals of both the Ag atoms were included uniformly in the active space for all the states, since some of the excited states resulted from promotion of an electron from the $5s$ to the $5p$ orbital. Separate CASSCF calculations were made for each state of the D_{2h} symmetry and different spin multiplicity. These computations were followed by the FOCI (first-order configuration interaction) method, which included all configurations obtained by distributing eight electrons in the internal space in all possible ways, and seven electrons in the internal space with one electron in the external space in all possible ways.

Multireference singles + doubles configuration interaction (MRSDCI) calculations were also made by Zhang and the author (125) on the $X^1\Sigma_g^+$ ground state and the $A^1\Sigma_u^+$ excited state of Ag_2. The MRSDCI included all configurations in the CASSCF with coefficients $\geq$0.07, and single and double excitations of eight electrons.

Figure 1.3 shows the potential-energy curves of the low-lying electronic states of the silver dimer computed by Zhang and the author (125) at the FOCI level of theory. As seen from Figure 1.3, $^1\Sigma_u^+$ is the lowest among these to which an optical transition from the ground state is allowed. This state has been characterized as the A state experimentally. Figure 1.3 also shows the potential-energy curves of the $^1\Sigma_u^+$,$^3\Pi_u$,$^3\Sigma_g^+$, and $^1\Pi_u$ excited states to have deep minima, while the other states have shallow minima.

Table 1.4 shows the spectroscopic constants computed by Zhang and the author (125) for several electronic states of Ag_2 together with experimental data. All the theoretical results come from the FOCI calculations with the exception of the $X^1\Sigma_g^+$ and $A^1\Sigma_u^+$ states for which MRSDCI results are shown.

Simard and Hackett (131) used a narrow-bandwidth, continuous-wave ring dye laser to excite fluorescence in jet-cooled silver dimers to overcome the difficulty of spectral blending. They assigned the 0–0 and 2–0 bands of the $A(0_u^+)$–$X(^1\Sigma_g^+)$ system, and obtained R_e = 2.53096 Å and ω_e = 192.4 cm^{-1} for the ground state and R_e = 2.6549 Å, T_e = 22,977 cm^{-1}, and ω_e = 155 cm^{-1} for the $A^1\Sigma_u^+$ state. Krämer et al. (132) reported sub-Doppler spectra of the $A^1\Sigma_u^+ \leftarrow X^1\Sigma_g^+$ system of Ag_2 and deduced $R_e(^1\Sigma_g^+)$ = 2.53 Å, $R_e(^1\Sigma_u^+)$ = 2.65 Å, and $T_e \approx$ 23 130 cm^{-1}. Pesić and Vujisić (130) obtained too short a bond

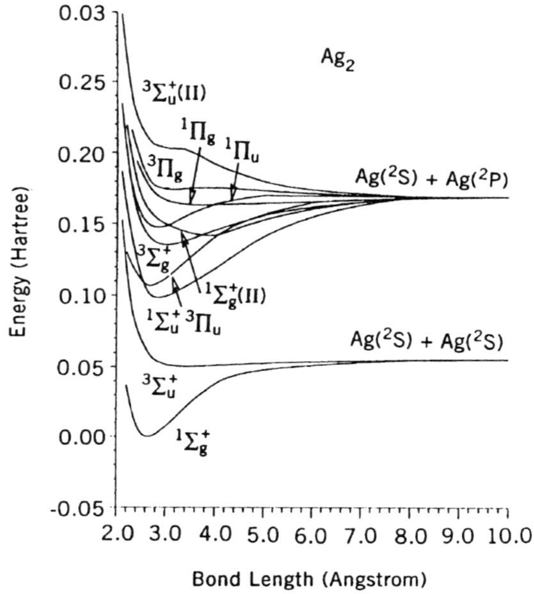

Figure 1.3 Computed potential-energy curves of Ag_2. (From the work of Zhang and Balasubramanian, Ref. 125.)

Table 1.4 Theoretical Spectroscopic Constants for Ag_2 Together with Available Experimental Data[a]

	Theory			Exp.		
State	R_e, Å	T_e, cm^{-1} [a]	ω_e, cm^{-1}	R_e, Å	T_e, cm^{-1}	ω_e, cm^{-1}
$X\,^1\Sigma_g^+$	2.644	0	160	2.53096	0	192
$^3\Sigma_u^+$	3.419	11,055	34	—	—	—
$A\,^1\Sigma_u^+$	2.857	21,597	112	2.6549	22,977	155
$^3\Pi_u$	2.697	23,403	173	—	—	—
$^3\Sigma_g^+$	3.024	29,629	86	—	—	—
$B^1\Pi_u$	2.848	32,297	111	2.6232	35,829	151.4
$^3\Pi_g$	3.150	38,225	54	—	—	—
$^3\Pi_g$(II)	3.613	45,581	—	—	—	—
$^3\Sigma_u^+$(II)	3.524	46,425	94	—	—	—
$^1\Sigma_u^+$(II)	2.959	46,691	76	—	—	—
$^3\Sigma_g^+$(II)	2.481	47,350	193	—	—	—
$^3\Pi_u$(II)	2.579	50,346	165	—	—	—
$^1\Pi_u$(II)	2.580	53,088	180	—	—	—

[a]Computed properties of electronic states with $T_e > 45,000$ cm^{-1} are not expected to be accurate. Theoretical values are from Reference 125. All experimental data for the X and A states from Reference 131 and for the B state, from Reference 133.

length of 2.47 Å, which was disputed by Simard and Hackett (131), who have suggested a longer bond. The results computed by Zhang and the author (125) are r_e = 2.644 Å, ω_e = 160 cm^{-1} for the $^1\Sigma_g^+$ ground state, and r_e = 2.857 Å, T_e = 21 597 cm^{-1}, and ω_e = 112 cm^{-1} for the $^1\Sigma_u^+$ excited state. Although the bond lengths are 10% longer and ω_e values 25% smaller than the experimental values, the differences are expected in view of large electron correlation effects.

The $B \leftarrow X$ system of Ag_2 was reported by Beutel et al. (133), and Pesić and Vujisić (130). Beutel et al. (133) reassigned the B state to $^1\Pi_u$, which was supported by the computations of Zhang and the author (125). The experimental results for the B state (R_e = 2.623 Å, T_e = 35,829 cm^{-1}, and ω_e = 151 cm^{-1}) compared well with the values (R_e = 2.848, T_e = 32,297 cm^{-1}, and ω_e = 111 cm^{-1}) computed by Zhang and the author (125). The FOCI computations of Zhang and the author yield the energy separations of the $^1\Sigma_u^+(\Pi)$ and $^1\Pi_u(\Pi)$ states as 46,691 and 53,088 cm^{-1}, respectively. While these results are likely to be far less accurate for these states, transitions from the ground state to these states should be observable. There are no experimental spectroscopic constants available for the other excited states.

Table 1.5 shows the Mulliken populations for the low-lying electronic states of the silver dimer. It is seen that all the d populations are only a bit smaller than 10.0 in all the states. For the other excited states, the p populations increase to 0.43–0.56, while the s populations drop down to 0.46–0.60, which means that these states are formed by promoting an electron from the 5s orbital to the 5p orbital. The same conclusion can be drawn from analysis of the leading configurations and orbital characteristics.

Figure 1.4 shows the orbital contours of the $3\sigma_g, 3\sigma_u$, and $2\pi_u$ orbitals, obtained from the FOCI natural orbitals of the $^1\Sigma_g^+, ^1\Sigma_u^+$, and $^1\Pi_u$ states, respectively. As seen from Figure 1.4, the $3\sigma_g$ orbital of the $^1\Sigma_g^+$ state is composed of both 5s and 5p. The antibonding combination of 5p minimizes the charge

Table 1.5 Mulliken Populations of the Low-Lying Electronic States of Ag_2 from FOCI Calculations

	Gross Population		
State	s(Ag)	p(Ag)	d(Ag)
$^1\Sigma_g^+$	0.933	0.088	9.978
$^3\Sigma_u^+$	0.909	0.094	9.996
$^1\Sigma_u^+$	0.600	0.434	9.965
$^3\Pi_u$	0.463	0.562	9.975
$^3\Sigma_g^+$	0.551	0.468	9.981
$^1\Pi_u$	0.471	0.555	9.974
$^1\Pi_g$	0.478	0.525	9.996
$^3\Pi_g$	0.455	0.557	9.988

Source: Reproduced from Reference 125.

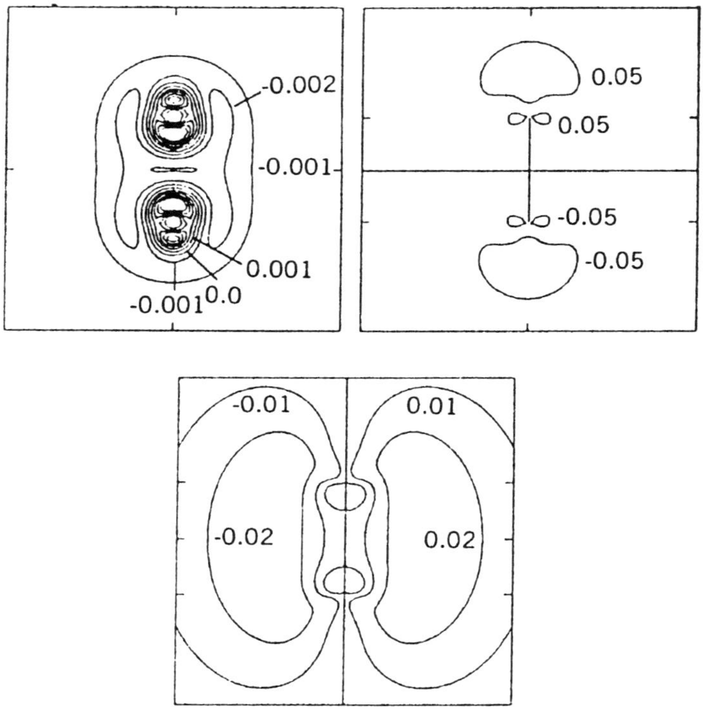

Figure 1.4 Orbital contours for the $3\sigma_g, 3\sigma_u$, and $2\pi_u$ orbitals in the ${}^1\Sigma_g^+, {}^1\Sigma_u^+$, and ${}^1\Pi_u$ states, respectively. The contours are generated from 0 to 1 bohr$^{-3/2}$ in steps of 0.02 bohr$^{-3/2}$. The units of orbital contours are in bohr$^{-3/2}$. (From Ref. 125.)

density near the molecular axis. The $3\sigma_g$ orbital of the $A^1\Sigma_u^+$ state is akin to this qualitatively, but the $5p$ character is enhanced in this orbital for the A state. The $3\sigma_u$ orbital of the $X^1\Sigma_g^+$ state has greater charge density between the two atoms, while the corresponding orbital of the $A^1\Sigma_u^+$ state has enhanced density away from the Ag–Ag bond (see Fig. 1.4). This suggests that the bond length of the $A^1\Sigma_u^+$ state is longer than that of the $X^1\Sigma_g^+$ state, which is reflected in Figure 1.4.

The $2\pi_u$ orbital, which is occupied in the $B^1\Pi_u$ state of Ag_2, is composed primarily of the $5p_x$ or $5p_y$ orbitals of the Ag atoms perpendicular to the Ag–Ag molecular axis (Fig. 1.4). The electronic density is thus centered more perpendicular to the molecular axis in this orbital. The diffuse nature of the $2\pi_u$ orbital of the B state is also noted (Fig. 1.4).

Figure 1.5 shows the contour plot of the differences between the charge densities of the X and A states, while in Figure 1.6, the contour plot of the differences in the total charge densities of the X and B states is shown. The difference contours illustrate how in the X-state density is accumulated between the Ag–Ag atoms, while in the a state the density is accumulated away from

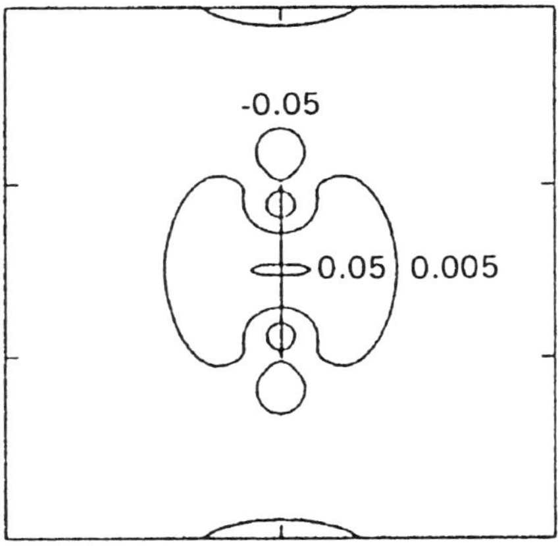

Figure 1.5 Total charge density difference plot for the $A^1\Sigma_u^+$ and the $X^1\Sigma_g^+$ states. The units are in electrons/bohr^3, with maximum value of 0.025 in an increment of 0.0001. (From Ref. 125.)

the Ag–Ag bond. A similar trend is seen in Figure 1.6. The elongation of the bond in both A and B states is seen compared to the ground state is reflected in Figures 1.5 and 1.6.

The $a^1\Sigma_u^+$ state arises from the $3\sigma_g^1 3\sigma_u^1$ configuration, where the $3\sigma_g$ orbital of this state is somewhat different from that of the $X^1\Sigma_g^+$ state, in that the $3\sigma_g$ orbital for the $^1\Sigma_u^+$ state has a significant $5p$ character (see Fig. 1.4). This orbital is composed of both Ag(5s) and Ag(5p) orbitals (bonding combination). The $3\sigma_u$ orbital, on the other hand, is composed of Ag(5s) and Ag(5p), too (antibonding combination), but the $5p$ contribution is dominant in this orbital, while it is slightly less in the $3\sigma_g$ orbital. The $B^1\Pi_u$ state of Ag_2 arises from the

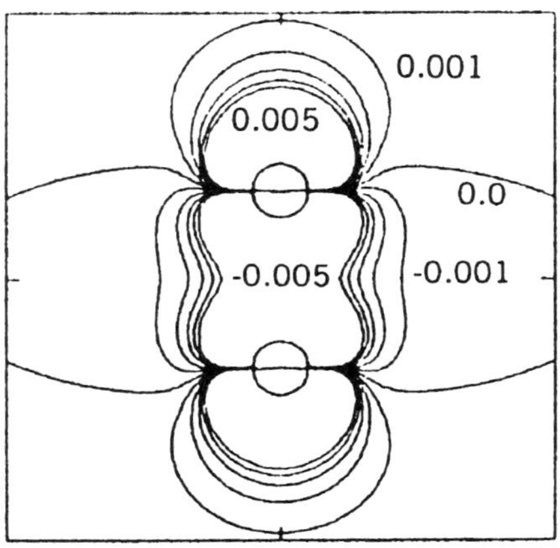

Figure 1.6 Total charge density difference plot for the $B^1\Pi_u$ and $X^1\Sigma_g^+$ states. The units are in electrons/bohr^3 with maximum value of 0.025 in an increment of 0.0001. (From Ref. 125.)

$3\sigma_g 2\pi_u$ open-shell configuration. The $3\sigma_g$ orbital of this state is similar, but the *d* contribution is slightly increased, indicating the reduced core *d* population for this state (see Table 1.5).

The electronic properties of the gold dimer (Au_2) have been investigated by a number of authors (93–107, 138). The interest in Au_2 arises from the large relativistic correction, which contracts the outer 6*s* valence orbital of the gold atom, leading to "gold anomaly," and the substantial difference in the chemistry of gold compared with silver. The theoretical spectroscopic constants of the ground state of Au_2 are included in Table 1.3. Balasubramanian et al. (108) used a relativistic CASSCF/MRSDCI method for Au_2 to obtain a D_e value of 46 kcal/mol, in good agreement with the Hilpert–Gingerich (29) experimental value of 53 kcal/mol. We shall consider the excited electronic states of Au_2, subsequent to our discussion on the ground deexcited states of the trimers.

Consider the properties of trimers of noble metals. There are many theoretical calculations on Cu_3 (78, 81, 93, 109–111). Bauschlicher and co-workers (93, 110) summarized the various theoretical calculations and experimental investigations on Cu_3. Both all-electron and ECP methods have been employed to study Cu_3. Walch and Laskowski (109) used an all-electron nonrelativistic SDCI method to investigate the ground state of Cu_3. Their theoretical symmetric stretch frequency (ω_s) of 224 cm^{-1} supported the assignment of the observed value of 270 cm^{-1} to symmetric stretching. Truhlar et al. (112) reexamined the fluorescence spectrum of Rohlfing and Valentini (51). They obtained a pseudorotation barrier of 95 cm^{-1} and a Jahn–Teller stabilization energy of 221 cm^{-1} by fitting the observed spectrum. There is, however, some uncertainty in the possible values of ω_s (354 and 270 cm^{-1}). Walch et al. (93) obtained a Jahn–Teller stabilization energy of 280 cm^{-1} below the D_{3h} cusp for the $^2B_2$ ground state. The ground state of Cu_3 appears to be the $^2B_2$ state, with an isosceles triangle geometry. There is a very low-lying $^2A_1$ state, which is the other Jahn–Teller component of $^2E'$ (equilateral triangle). Table 1.6 summarizes the experimental and theoretical constants for Cu_3. There is another theoretical study on dimers and trimers of noble-metal atoms (124).

Regarding Ag_3, the calculations by Walch et al. (93) employed the SCF/SDCI methods, while the relativistic CASSCF/MRSDCI/RCI calculations of Balasubramanian and Liao (99) included up to 420,000 configurations and all the 33 electrons in the active space and spin–orbit effects. Table 1.7 summarizes the spectroscopic constants of Ag_3 obtained by these authors.

Balasubramanian and Liao (103) carried out relativistic CASSCF/SOCI calculations on Au_3 without *d* correlation effects, which were estimated from calculations on Au_2. In a more exhaustive study, Balasubramanian and Liao (99) carried out CASSCF/MRSDCI calculations, which included all 33 electrons in the active space, and up to 420,000 configurations. The *d* correlation effects were found to be quite significant. The potential-energy parameters, which

Table 1.6 Spectroscopic Properties of Cu_3

			ω_e, cm^{-1}					
			Stretch			δ J–T[a], cm^{-1}		
State	R_e, Å	θ_e, deg	Exp.	Theory	Bend	Theory	Exp.	D_e, eV[b]
$^2B_2$	2.367	68.5	270?	213	109	555[c]	221[e]	1.12
—	—	—	354?	—	—	280[d]	—	—
$^2A_1$	2.515	55	—	224	151	384[c]	126[e]	1.12
—	—	—	—	—	—	221[d]	—	—

[a]Jahn–Teller stabilization energy.
[b]D_e (eV) for $Cu_3 \rightarrow Cu_2 + Cu$ (93).
[c]SDCI+Q result (109).
[d]CPF + relativistic result (93).
[e]Reference 112; fitting of the spectrum from Reference 51.

included the effects of *d* correlation, are shown in Table 1.8. The atomization energy was calculated using a full MRSDCI method, which included up to 1.5 million configurations (108). As seen from Table 1.8, the best theoretical AE of 76 kcal/mol is in good agreement with an experimental value of 88 kcal/mol (29). Table 1.9a shows the energy separations of the electronic states of Ag_3 and Au_3, including the spin–orbit effects. As seen from Table 1.9a, the spin–orbit contribution to the $^2A_1$–$^2B_2$ splitting of Au_3 is comparable with the energy separation.

Figure 1.7 shows the actual geometries of the bent and linear electronic states of both Ag_3 and Au_3. As seen from this figure, both the trimers are more stable in bent isosceles triangular geometries than linear geometries. It should be pointed out that the *d* correlation effects in Au_3 are so large that the ground state turns out to be linear for the $^2B_2$ state, if *d* correlation effects are not included.

The author and Das (142) have studied the excited electronic states of the

Table 1.7 Spectroscopic Properties of Ag_3

			ω_e, cm^{-1} [b]				AE^c, kcal/mol	
State	R_e, Å[a]	θ_e, deg[a]	Stretch	Bend	δ J–T[b]	E, kcal/mol[a]	Theory	Exp.
$^2B_2$	2.72	63.7	134	68	498	0.0	48	61
$^2A_1$	2.88	54.2	161	99	390	0.74	47	—

[a]From Reference 99; relativistic CASSCF/MRSDCI (33 electrons).
[b]From Reference 93; CPF method.
[c]AE values of Ag_2–Ag_4 calculated using the CASSCF/MRSDCI method (120); experimental AE from Reference 29.

Table 1.8 Potential-Energy Parameters for Au_3[a]

				AE, kcal/mol	
State	R_e, Å	θ_e, deg	E, kcal/mol[a]	Theory	Exp.
$^2B_2$	2.60–2.72	65.7	0.0	76	88
$^2A_1$	2.72	56.4	0.54(0.16)	75.4	—

[a]All the parameters are from Reference 99. Theoretical *AE* calculated using CASSCF/MRSDCI, which included 1.5 million configurations (108). Experimental *AE* from Reference 29. Number in parentheses includes Davidson's correction.

gold trimer (Au_3). Table 1.9b shows the geometries and energy separations of 15 electronic states of Au_3 obtained by the author and Das (142). As seen from this table, there are several excited electronic states for Au_3 with both bent isosceles triangle and linear geometries. We note that there are also C_s stationary points, but these are essentially Jahn–Teller components of the first two ($^2B_2, ^2A_1$) pairs of electronic states, which arise from the $^2E'$ state in the D_{3h} symmetry. The state labeled $^2B_2$ (linear) is the linear local minimum in the potential-energy surface of the $^2B_2$ state, which correlates with $^2\Sigma_u^+$ in $D_{\infty h}$ symmetry, and is only ≈0.5 eV above the bent mimimum. The linear minimum of the $^2A_1$ surface is, however, considerably higher in energy. These four states are bound valence states, in that the predominant electronic configuration contributing to these states arises from a valence electronic configuration with nearly complete 5*d* electronic shells.

All 11 electronic states above 1.47-eV energy separation, starting from $^2B_2(\Pi)$ (linear), arise from the excitation of one of the 5*d* shell electrons of the gold atoms. The existence of such a low-lying excited state arising from 5*d*–6*s* excitation of the Au atom is not surprising, since the atomic $(5d^96s^2)^2D_{5/2}(5d^{10}6s^1)^2S_{1/2}$ energy separation of the Au atom is only 9161 cm^{-1} (1.14 eV) (113). The computed energy separations of the excited states of Au_3 in Table 1.9 (142) that involve 5*d* electron excitations are probably larger than the atomic 1.14-eV excitation energy, since the results in Table 1.9 do not

Table 1.9a Energy Separations of Electronic States of Ag_3 and Au_3, Including Spin–Orbit Effects[a]

	State	R_e, Å	θ_e, deg	E, kcal/mol
Au_3	$^2B_2\alpha$	2.60	65.7	0
Au_3	$^2A_1\alpha$	2.72	56.4	0.16(2.03)
Ag_3	$^2B_2\alpha$	2.72	63.7	0
Ag_3	$^2A_1\alpha$	2.88	54.2	0.69

[a]Number in parentheses includes Davidson's correction.
Source: Reference 99.

Table 1.9b Properties of Excited Electronic States of Au_3

State	R_e, Å	θ_e, deg	E, eV
$^2B_2$	2.60	65.7	0.0
$^2A_1$	2.72	56.4	0.03
$^2B_2$ (linear)	2.60	180.0	0.48
$^2A_1$ (linear)	2.52	180.0	1.46
$^2B_2$(II) (linear)	2.51	180.0	1.47
$^2A_2$ (linear)	2.52	180.0	1.52
$^2A_1$(II) (linear)	2.62	180.0	1.59
$^2B_1$ (linear)	2.60	180.0	2.18
$^2A_2$(II) (linear)	2.61	180.0	2.27
$^2B_2$ (linear)	2.56	66.6	2.38
$^2B_1$(II) (linear)	2.64	180.0	2.64
$^2A_2$	2.62	65.3	2.89
$^2B_1$	2.738	56	2.98
$^2A_2$(II)	2.61	71	3.09
$^2B_1$(II)	2.72	61	3.33

Source: Reproduced from Reference 142.

include the effects of spin-orbit coupling. In fact, the ^2D–^2S energy separation of the gold atom is 14,070 cm^{-1} (1.7 eV) in the absence of spin–orbit coupling, addressed subsequently.

Since there are several nearly degenerate MOs (molecular orbitals) composed of Au(5d), several closely spaced electronic states in the 1.47–1.59-, 2.18–2.38-, and 2.6–3.09-eV regions are found. Analyses of the natural orbitals involved in the excitation of electrons forming these states reveal that all these excited states of Au_3 arise from the 5d core shells to valence space excitations. This is in contrast with small Ag clusters for which excited states are formed by promoting the 5s electron to 5p, since the 5p–5s excitation energy is smaller than the 4d–5s excitation energy for the Ag atom (113). This is due to the fact that the 6s orbital of the Au atom is stabilized significantly by the relativistic mass–velocity effect. Hence, the $5d^9 6s^2$ configuration becomes more favorable than the $5d^{10} 6p^1$ configuration. The excited states of Au_2 and Au_3 arise from the promotion of electrons from 5d to 6s, rather than 6s to 6p.

The RCI calculations, which included spin–orbit coupling, revealed changes in energy separations of up to 0.5 eV (142). The contamination of different electronic states was especially noticeable when there were nearly degenerate electronic states. This is best illustrated by the ground state at $\theta = 60°$ geometry. Both $^2B_2$ and $^2A_1$ electronic states in the D_{3h} group become $^2E'$ state. This means that in the C_{2v}^2 double group, the $^2B_2$ state with open-shell spin β will mix substantially with the $^2A_1$ state with spin α, since both these states correlate with the E state in the C_{2v}^2 double group. The results of the RCI computations reveal that the E(I) state is 60% $^2B_2$ and 29% ^2A, at $\theta = 60°$, while the E(II)

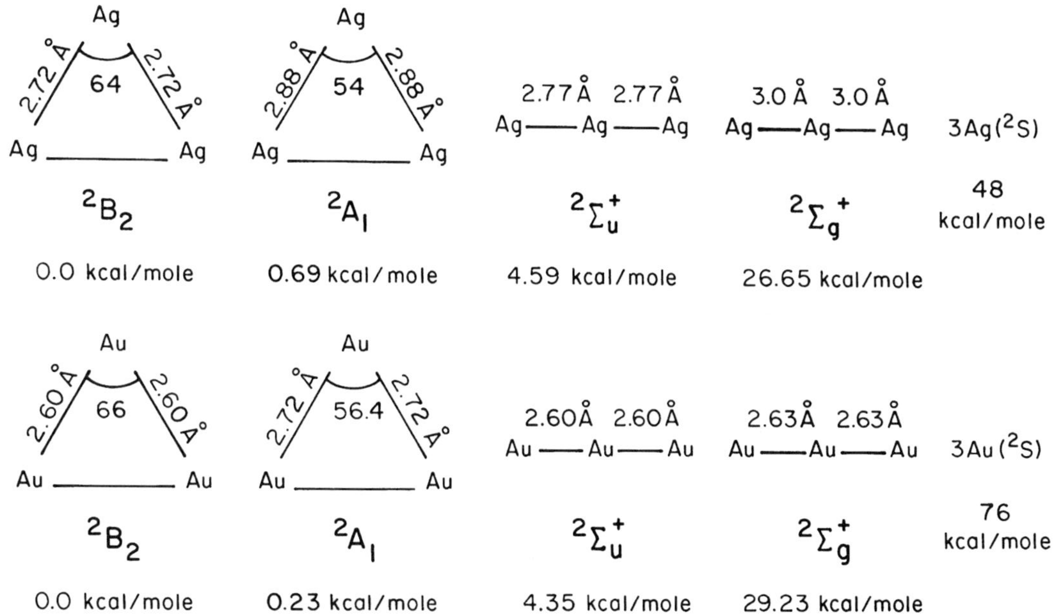

Figure 1.7 The geometries of bent and linear states of Ag_3 and Au_3. (From Ref. 99.)

state is 58% $^2A_1$ and 30% $^2B_2$. The E(I)–E(II) energy splitting of ~0.25 eV is considerably larger than the $^2B_2$–$^2A_1$ Jahn–Teller pseudorotation barrier (~0.07 eV), which could lead to the quenching of the Jahn–Teller distortion for Au_3.

The excited electronic states of Au_3 exhibited interesting behavior when spin–orbit coupling was included. For example, a bent $^2A_2$ state was significantly contaminated with another bent $^2B_1$ state. Hence, the excited states of Au_3 exhibit both spin–orbit and other relativistic effects. The contribution of the leading configuration in the RCI in several cases was found to be ≥60%.

The Au_3 UPS spectra exhibited six well-resolved peaks that correspond to both the ground and excited states of Au_3 (140). Previous theoretical (143, 144) and experimental (145) studies have revealed that the ground state of Au_3^- is a linear $^1\Sigma_g^+$ state with $r_e = 2.58$ Å at the CASSCF/MRSDCI level of theory (143), which correlated all 34 electrons. This suggests that apart from the ground state of neutral Au_3, all excited states with linear geometries would have larger Franck–Condon factors, since the ground-state geometry of Au_3^- is linear. Therefore, peaks corresponding to neutral linear states are likely to be more intense, provided the linear states exhibit substantial wells in the potential-energy surfaces. Furthermore, if the major configurations contributing to the excited electronic states (weight ≥30%) differ by one electron excitation with respect to the $^2B_2$ ground state, then transitions to these excited states are likely to be more intense compared to states, which are composed predominantly of configurations involving two or more excitations relative to the neutral ground state.

Table 1.10 shows the observed peaks labeled A–E in the negative UPS of Au_3^-. Also shown in the same table are the transition energies obtained from the

Table 1.10 Tentative Assignment of the UPS Spectra of Au_3^-[a]

Peak	Transition Energy		Assignment
	Exp.	Theory[a]	
A	3.9	3.8	E(I) 60% $^2B_2$, 29% $^2A_1$
B	4.2	4.05	E(II) 58% $^2A_1$, 30% $^2B_2$
C	4.7	4.5	$^2B_2$ (linear)
D, D'	5.0–5.5	5.1–5.4	$^2A_1$ (linear)
One broad and a	—	—	$^2B_2$(II) (linear)
satellite peak	—	—	$^2A_2$ (linear)
	—	—	$^2A_1$(II) (linear)
E	5.9	6.0	$^2B_1$ (linear)
Several peaks	>6.0	>6.0	$^2A_2$(II), $^2B_1$(II), and third roots of other state

[a]All energies relative to Au_3^- $^1\Sigma_g^+$ linear ground state. The theoretical value for the a state was scaled by a factor of 1.35. The theoretical energy separations of some excited states include spin–orbit effects.

Source: Reproduced from Reference 142.

UPS spectra and the corresponding theoretical values, relative to the ground state of Au_3. Note that the theoretical value of 3.8 eV for the lowest state is obtained using a scaling factor of 1.35, as described in Reference 143. The assignment of peaks in the UPS spectra of Au_3^- indicated in Table 1.10 is tentative in that it is based on computed energy separations consistent with the peak, probability of the electronic transition as determined by the geometry, and the nature of the electronic configurations contributing to the electronic state in question. The experimental transition energies also have uncertainties of ±0.1 eV.

As evidenced from Table 1.10, the first peak corresponds to the ground state of Au_3. The energy separations relative to Au_3^- correspond to the adiabatic electron affinity of Au_3. Smalley and co-workers (140) find a second peak 0.3 eV above the first peak, which is labeled the *B* peak. This peak has a smaller intensity, compared to the *a* peak and other peaks. The spin–orbit splitting of the *E*(I) and *E*(II) states near $\theta = 60°$ is ~0.25 eV. The linear $^2\Sigma_u^+$ state ($^2B_2$ in C_{2v})—which is a probable candidate for the *B* peak—would, however, lead to a more intense peak because of its compatible linear geometry with the ground state of Au_3^-. These arguments suggest that the *B* peak is due to the first excited bent spin–orbit state. Table 1.10 shows the composition of the *A* and *B* states.

The *C* peak at an experimental energy separation of 4.7 eV is attributed to the linear state of neutral Au_3. This corresponds to a linear local potential well of the $^2B_2$ state. The computed energy separation of 4.5 eV is in good agreement with the experimental value of 4.7 eV. Note that the *A* peak is computed 0.1 eV lower than the experimental value relative to Au_3^-, even after scaling. The *C* peak is quite intense and, in fact, appears to be slightly more intense than the *A* peak. This suggests that the most probable geometry of the *C* state is linear. Therefore, the *C* peak is attributed to the $^2\Sigma_u^+$ linear state.

As seen from Table 1.10, there are several excited electronic states for Au_3, which involve excitation of the 5*d* electrons. It is believed that the intense peak at 5.2 eV, in fact, may consist of more than one peak, as evidenced from a shoulder at 5.1 eV. These peaks are labeled *D* and *D′*. The *D* peak may be a result of electronic transitions to more than one excited state of Au_3 in this region. As seen from Table 1.10, several linear excited states (four) are probable candidates for the *D* and *D′* peaks. The fact that all of these states have linear geometries is consistent with the strong intensity of the *D* peak, which is twice as intense as the *A* peak.

The $^2A_1$ (linear), $^2B_2$(II) linear, $^2A_2$ linear, and $^2A_1$(II) linear states of Au_3 differ by a single electronic excitation relative to the $^2B_2$ ground state. Consequently, all states listed in Table 1.10 are probable candidates for the *D* and *D′* peaks. Since more than one state appears to be contributing to this peak, the *D* peak is unusually intense and also exhibits a shoulder.

The last well-resolved peak (*E*) at 5.9 eV is consistent with the linear excited $^2B_1$ state. The linear geometry of the $^2B_1$ state is also consistent with the intensity of the *E* peak, which is at least comparable to the ground-state peak (*A*).

There are several unresolved peaks in the UPS spectra in the 6-eV region. Some of these peaks are attributed to noise, while some may be real and could be a result of several excited states that we compute in the 6-eV region [$^2A_2$(II), $^2B_1$(II), etc.].

The leading configurations of the electronic states reveal that, with the exception of $^2B_2, ^2A_1, ^2B_2$(lin), and $^2A_1$(lin) states, all other states are generated from excitations from the 5*d* shells of the gold atom into the active space. This is consistent with the qualitative description of the excited states of Au_3.

Das and the author (115) have performed extensive studies on the ground state and also several excited electronic states of the gold dimer (Au_2). Subsequent to this theoretical work, Bishea and Morse (138) obtained the jet-cooled spectra of Au_2 together with AgAu. Thus there is some experimental information on a few excited states of Au_2, and experimental evidence to support strong relativistic effects in Au_2, which influence not only the ground state of Au_2 but also several excited electronic states. Spin–orbit coupling makes significant contribution to the excited electronic states of Au_2 (115). The importance of relativistic effects for both the ground and excited electronic states of Au_2 is considered here.

Table 1.11 shows the spectroscopic properties (R_e, T_e, ω_e) for the electronic

Table 1.11 Spectroscopic Properties of the λ–*s* Electronic States of Au_2

State	R_e, Å	T_e (FOCI), cm^{-1}	T_e (MRSDCI), cm^{-1}	ω_e, cm^{-1}
$X\,^1\Sigma_g^+$	2.518	0	0	193
$^3\Sigma_u^+$	2.775	14,980	15,264	99
$^3\Sigma_g^+$	2.496	16,860	23,154	193
$^3\Pi_u$	2.503	18,307	24,684	191
$^1\Sigma_u^+$	2.633	20,366	26,386	160
$^1\Pi_u$	2.535	22,739	28,405	174
$^3\Delta_g$	2.595	23,486	29,311	156
$^1\Delta_g$	2.593	24,293	30,143	153
$^1\Sigma_g^+$(II)	2.590	24,580	33,454	138
$^3\Delta_u$	2.677	25,691	31,176	135
$^1\Delta_u$	2.685	26,348	31,431	129
$^3\Pi_g$	2.932	34,035	33,442	89
$^3\Pi_u$(II)	2.616	36,785	52,019	159
$^3\Pi_g$(II)	2.721	40,768	50,679	121
$^1\Pi_g$(II)	2.685	43,386	52,310	123
$^3\Delta_u$(II)	2.887	43,604	50,093	92
$^3\Delta_g$(II)	2.931	43,858	49,932	88
$^1\Delta_u$(II)	2.858	44,262	50,369	86
$^1\Delta_g$(II)	2.912	44,420	50,270	77
$^1\Pi_u$(II)	2.707	45,151	52,437	121
$^3\Phi_g$	2.519	45,796	—	180

Source: Reproduced from Reference 115.

states of Au_2 obtained by Das and the author (115) without the inclusion of spin–orbit interaction using the CASSCF/FOCI/MRSDCI methods. The potential-energy curves of some of these states are shown in Figures 1.8 and 1.9. Das and the author computed the electronic states, including spin–orbit coupling, and the potential-energy curves of the $X0_g^+, A0_u^+$, and $B0_u^+$, including spin–orbit coupling are shown in Figure 1.10. Table 1.12 shows the spectroscopic properties of 24 ω–ω states of Au_2 computed by Das and the author (115).

The computed properties (ω_e = 193 cm^{-1} and R_e = 2.517 Å) of the ground state ($X0_g^+$) are in excellent agreement with the experimental values of ω_e = 191 cm^{-1} and R_e = 2.47 Å. The dissociation energy of the $X0_g^+$ state is obtained from the CASSCF/MRSDCI/RCI calculations as 2.05 eV. This value is smaller than the experimental value of 2.30 eV obtained by Kordis et al. (44) using both the second- and third-law methods.

The properties of the excited states of Au_2 computed by Das and the author compare well with the experimental results (6, 45, 138). For example, exper-

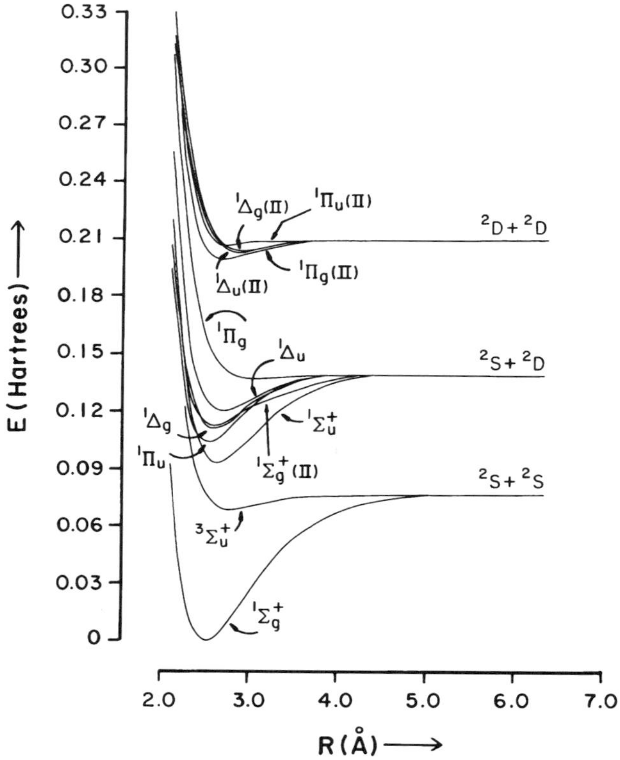

Figure 1.8 Potential-energy curves of the electronic states of Au_2 in the absence of spin–orbit coupling. (From Ref. 115.)

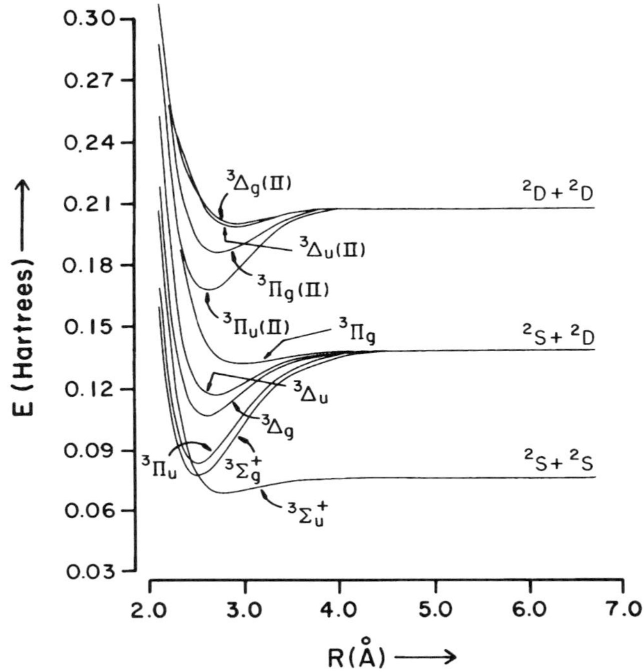

Figure 1.9 Potential-energy curves of the excited electronic states of Au_2 in the absence of spin–orbit coupling. (From Ref. 115.)

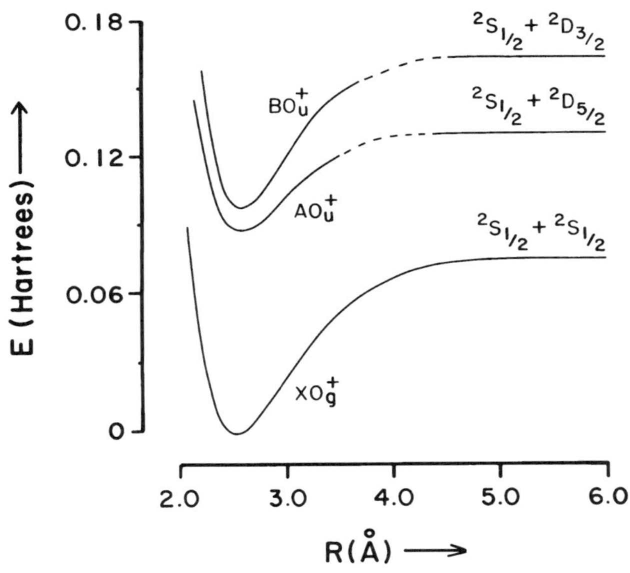

Figure 1.10 Potential-energy curves of the electronic states of Au_2 including spin–orbit coupling. (From Ref. 115.)

Table 1.12 Spectroscopic Properties of Au_2 with Spin–Orbit Coupling

	R_e, Å		T_e, cm^{-1}		ω_e, cm^{-1}	
State	Theory	Exp.	Theory	Exp.	Theory	Exp.
$X0_g^+$	2.517	2.47	0	0	193	191
1_u	2.776	—	15,348	—	98	—
0_u^-	2.774	—	15,380	—	98	—
2_u	2.529	—	20,926	—	180	—
0_g^-	2.580	—	21,281	—	158	—
1_g	2.583	—	21,388	—	156	—
0_u^-(II)	2.618	—	21,692	—	192	—
1_u(II)	2.573	—	22,055	—	170	—
$A0_u^+$	2.574	2.57	24,605	19,668	145	142
3_g	2.590	—	24,833	—	154	—
2_g	2.639	—	24,989	—	151	—
2_g(II)	2.566	—	25,340	—	133	—
1_u(III)	2.536	—	26,063	—	195	—
3_u	2.644	—	26,647	—	149	—
2_u(II)	2.681	—	27,005	—	144	—
2_u(III)	2.634	—	27,520	—	135	—
$B0_u^+$	2.549	2.52	28,828	25,686	214	180
0_g^+(II)	2.650	—	30,831	—	145	—
1_g(II)	2.618	—	34,335	—	144	—
1_g(III)	2.928	—	36,046	—	90	—
0_g^+(III)	2.929	—	36,142	—	89	—
0_g^-(III)	2.929	—	36,142	—	89	—
1_u(IV)	2.748	—	36,252	—	118	—
1_g(IV)	2.619	—	50,468	—	129	—

Source: Reference 115.

iments reveal the existence of an excited $A0_u^+$ state, which extrapolates to the $Au(^2S_{1/2})$ + $Au(^2D_{5/2})$ dissociation limit. The $B0_u^+$ state correlates to the separated atom limit of $Au(^2S_{1/2})$ + $Au(^2D_{3/2})$. The D_e values of the $A0_u^+$ and $B0_u^+$ states were deduced from the spectra as 1.0 and 1.78 eV, respectively, using the dissociation energy of the ground state and the experimental atomic excitation energies to $^2D_{5/2}$ and $^2D_{3/2}$ states (45). The D_e values for the $A0_u^+$ and $B0_u^+$ states obtained by Das and the author (115) (1.17 and 1.82 eV) compared quite favorably with the experimental results. Other properties such as r_e and ω_e for the $A0_u^+$ state were found to be in excellent agreement with the experimental values. For the $B0_u^+$ state, the results calculated by Das and the author (R_e = 2.549 Å and ω_e = 214 cm^{-1}) are somewhat larger compared to the experimental values (R_e = 2.52 Å and ω_e = 180 cm^{-1}). The 1_u and 0_u^- spin–orbit components of $^3\Sigma_u^+$ were found to have T_e values of 15,348 and 15,380 cm^{-1}, respectively, by Das and the author. These states are found to be weakly bound compared to the ground state. Experimental gas-phase spectroscopic studies (45, 138) have revealed the existence of A–X and B–X systems. The a and B states were found to have T_e values of 19,668 and 25,686 cm^{-1}, respectively. Das and the author (115) found that there are $a0_u^+$ and $B0_u^+$ states with the computed T_e values of 24,605 and 28,828 cm^{-1}, respectively. The 0_u^+ components of $^1\Sigma_u^+$ and $^3\Pi_u$ mix strongly in the RCI, thereby altering the spectroscopic properties of these states to a large extent.

Among the dipole-allowed transitions of Au_2, the $A0_u^+ \leftrightarrow X0_g^+$ and $B0_u^+ \leftrightarrow X0_g^+$ transitions are expected to be very intense, and were computed to have T_e values of 24,605 and 28,828 cm^{-1}, respectively by Das and the author (115). The vibrational and rotational analyses of the gas-phase spectroscopic data confirm the existence of these two bands arising from $0_u^+ \leftrightarrow 0_g^{+\,1}\Sigma_g^+$ transitions near 19,668- and 25,686-cm^{-1} regions. The other important dipole-allowed transitions from or to the ground states are $1_u \leftrightarrow X0_g^+$, 1_u(II) $\leftrightarrow X0_g^+$, 1_u(III) $\leftrightarrow X0_g^+$, and 1_u(III) $\leftrightarrow X0_g^+$. These transitions were computed by Das and the author to occur in the 15,348-, 22,055-, 26,063-, and 36,252-cm^{-1} regions, respectively.

Klotzbücher and Ozin (46) reported the absorption spectra of matrix-isolated Au_2. They found broad absorption bands at 27,400 and 31,500 cm^{-1}, respectively, and narrow and intense bands at 48,000 and 50,500 cm^{-1}. They assigned the first of these bands to the $B0_u^+ \leftarrow X0_g^+$ system, but the remaining bands were unassigned. The computations of Das and the author (115) tentatively assigned the second broad absorption band observed by Klotzbücher and Ozin (46) to the $1_u \leftrightarrow X0_g^+$ system. Other dipole-allowed transitions originating from the excited electronic states could be explored using the laser-induced fluorescence (LIF) experiments, if the excited state participating in the transition is long-lived. For example, the 1_g(IV) $\leftrightarrow 1_u$ transition was computed to have a transition energy of 35,120 cm^{-1} by Das and the author. The 1_g(II) $\rightarrow 1_u$ and 1_g(III) $\rightarrow 1_u$ emissions were computed to have transition energies in the 18,987–20,698-cm^{-1} region. The 0_g^+(II) $\leftrightarrow 1_u$ and 0_g^+(III) $\leftrightarrow 1_u$ transition energies were computed by Das and the author as 15,483 and 20,794 cm^{-1}, respectively.

Bishea and Morse (138) obtained the jet-cooled spectra of the gold dimer (Au_2) as well as AgAu. These spectra led to the characterization of X, a, A', A, B', and B states, where X is the ground state. The assignments of all these states are fully consistent with the computed results of Das and the author (115), including spin–orbit coupling. The theoretical T_e values of the A', A, B', and B states computed by Das and the author are 15,438, 22,055, 24,605, and 28,828 cm^{-1}, respectively, compared to 16,679.8, 18,115, 19,668.2, 24,024.6, and 25,685.5 cm^{-1}, respectively reported by Bishea and Morse (138). The theoretical vibrational frequencies were found to be somewhat higher compared to the experimental results. The r_e values for four states are experimentally established for Au_2. They are 2.4715, 2.5669, 2.570, 2.5197 Å, for the X, A, B', and B states, respectively, for Au_2. This compares with the results of 2.517, 2.574, 2.536, and 2.549 Å, computed by Das and the author (115). The mixed nature of the $A0_u^+$ and $B0_u^+$ states has also been established experimentally, thereby confirming that the spin–orbit mixing in the excited electronic states of Au_2 is quite large.

As seen from Figure 1.9, the potential-energy curves of the $^3\Sigma_u^+$ and $^3\Pi_u$ states cross. Thus 0_u^- and 1_u components of the $^3\Sigma_u^+$ state could interact with the corresponding components of the $^3\Pi_u$ state in the presence of spin–orbit coupling. Therefore these states would undergo predissociation through these interactions, while the 0_u^+ and 2_u components would survive predissociation. It is thus expected that the 0_u^+ and 2_u components of the $^3\Pi_u$ state would have longer lifetimes.

Analysis of the coefficients of the leading configurations of the electronic states of Au_2 reveals that although some of the low-lying λ–*s* states are described predominantly by their leading configurations, other electronic states are considerably more complex, as they are represented by many configurations. The description becomes even more complex when spin–orbit coupling is included. The ground state ($X0_g^+$) is quite pure, as it arises mainly from $^1\Sigma_g^+(\sigma_g^2)$ (96%). The spin–orbit contaminations are very significant for the $A0_u^+$ and $B0_u^+$ states. For these states, the $^3\Pi_u(\sigma_u\pi_g^3)$ state mixes almost equally with the $^1\Sigma_u^+(\sigma_g\sigma_u)$ state. The 2_u state is composed predominantly of $^3\Pi_u$ with small contamination from the $^3\Delta_u(2_u)$ and $^1\Delta_u(2_u)$ states. The 2_u(II) state arises from the $^3\Delta_u$ state, while 2_u(III) is described by $^1\Delta_u$. The mixing of $^3\Delta_g$ and $^1\Delta_g$ in the lowest root of 2_g is significantly large. Although the lowest root of 1_u is described predominantly by $^3\Sigma_u^+(\sigma_g\sigma_u)$, the upper roots, namely, 1_u(II) and 1_u(III), have significant $^3\Pi_u$–$^1\Pi_u$ contaminations. The $^3\Delta_u$ state arising from the $\sigma_u\delta_g^3$ configuration contributes up to 8% to the 1_u(II) state, while in the 1_u(III) state there is 15% contribution from $^3\Sigma_u^+(\sigma_u\sigma_g)$. For the 0_g^+(II) state, the contamination of the $^1\Sigma_g^+$(II) state with the $^3\Pi_g$ state is quite significant.

The Mulliken population analysis of the various λ–*s* electronic states reveals that the ground state ($^1\Sigma_g^+$) and $^3\Sigma_u^+$ have gross *s* populations near unity and *d* populations near 10.0. The *p* populations for these two states are quite small. However, the computations of Das and the author (115) reveal that for all other excited states, the *d* populations deviate considerably from 10.0, and the *s* and *p* populations are quite large, since the excited states arise from excitation of the *d* electron of the gold atom. The $^3\Pi_u$(II) and $^1\Pi_u$(II) states have larger *p* populations, while the $^3\Delta_u$(II), $^1\Delta_u$(II), $^3\Delta_g$(II), and $^1\Delta_g$(II) states have larger *s* populations. Therefore, these states exhibit a strong mixing of $5d^{10}6s^1, 5d^96s^2$, and $5d^{10}6p^1$ configurations of the gold atoms.

The differences in the spectroscopic properties of Au_2 and Ag_2 can be attributed primarily to relativistic effects. The excited states of Ag_2 arise from the $5s \rightarrow 5p$ excitation, while the excited states of Au_2 arise from $5d \rightarrow 6s$ promotion (35). While the first excited state of $Au_2(^3\Sigma_u^+)$ is identical to Ag_2, all other excited states appear in a different order. For example, the $^3\Sigma_g^+$ and $^3\Pi_u$ states are below the $^1\Sigma_u^+$ excited state for Au_2, while the opposite trend is seen in Table 1.5 for Ag_2. This is comprehensible, since with the exception of $^1\Sigma_u^+$ all low-lying excited states of Au_2 dissociate into Au(^2S) + Au($5d^96s^1, ^2D$), in contrast with Figure 1.3 for Ag_2.

Balasubramanian et al. (108) have completed 44 electron CASSCF/MRSDCI calculations on Au_4. The CI calculations included excitations of 36 electrons into external orbitals. Fourteen electronic states of Au_4 were considered. The ground state of Au_4 was found to be closed-shell $^1A_1$ state with a rhombus equilibrium geometry. Two excited states of $^3B_2$ symmetry with rhombus and square geometries were found to lie 0.98 and 1.12 eV above the ground state. The spin–orbit effects were included using the relativistic configuration interaction method (RCI). The gold tetramer was found to be much more stable compared with the trimer and dimer.

Table 1.13 shows the calculated equilibrium geometries of 14 electronic states of Au_4 and their energy separations. The computed geometries of some of these states are shown in Figure 1.11. As seen from Table 1.13, the ground state of Au_4 is unambiguously of $^1A_1$ symmetry (^1A_g in D_{2h} symmetry) of rhombus geometry. The four sides have the same length of 2.69 Å at the MRSDCI level.

The $^3B_2$ excited state of Au_4 exhibits two stationary points, one with a rhombus structure ($\Theta_e = 67°$) and the other with a square geometry. They are separated by only 0.09 eV. The square structure in the $^1A_1$ state is a saddle point for the conversion of equivalent acute-angle rhombus into obtuse structures.

Balasubramanian et al. (108) have carried out MRSDCI calculations for the three lowest-lying electronic states of Au_4, namely, $^1A_1$ (rhombus), $^3B_2$ (rhombus), and $^3B_2$ (square). The MRSDCI geometries and energy separations for these three states are compared with POLCI in Table 1.14. As seen from this table, the geometries obtained by the POLCI method are in good agreement with the MRSDCI geometries. For the rhombus structure, only the bond angle changes by about 2%. The energy separations are, however, more sensitive to higher-order *d* correlations not included in the POLCI method. The energy separation of the $^3B_2$ (rhombus) state changes by 10%, while the $^3B_2$ (square) state changes by 15%.

The spin–orbit effects could be important for Au_4. Table 1.15 shows the spin–orbit splittings of various states of Au_4. As seen from this table, the spin–orbit contributions to various states of Au_4 are between 0.01 and 0.022 eV. Thus the spin–orbit effects were found to be nonnegligible.

The CASSCF/MRSDCI method was utilized on the various electronic states of Ag_4 by Balasubramanian and Feng (107). Figure 1.12 shows the geometries of the two lowest states of Ag_4 mainly for comparison with Au_4. In addition, the low-lying electronic states of Cu_4 have been investigated (107).

The hexamers of Cu, Ag, and Au have been studied both experimentally (141, 146, 147) and theoretically (148, 149). Comparison of the properties of these clusters could provide some additional insight into the nature of relativistic effects for these clusters. Taylor et al. (146) considered the negative ion photodetachment spectra of Au_6^-. Jarrold and Creegan (141) studied Cu_6 among other copper clusters. There is another experimental work on these clusters by Ho et al. (147). Liao and the author (148, 149) carried out *ab initio* theoretical studies, including *d*-electron correlation effects on several possible geometries and electronic states of Cu_6, Ag_6, and Au_6 as well as Cu_6^+, Ag_6^+, and Au_6^+.

Table 1.13 Geometries and Energy Separations of Electronic States of Au_4 Obtained Using CASSCF/POLCI Methods[a]

CASSCF					POLCI				
Geometry	State	R_{12}, Å	θ/R_{23}, deg/Å	E, eV	Geometry	State	R_{12}, Å	θ/R_{23}, deg/Å	E, eV
Rhombus	$^1A_1$	2.75	57	0.0	Rhombus	$^1A_1$	2.69	60.4	0.0
Rhombus	$^3B_2$	2.72	70.2	0.78	Rhombus	$^3B_2$	2.69	66.9	0.88
Square	$^3B_2$	2.72	90	0.80	Square	$^3B_2$	2.68	90	0.97
Linear	$^1\Sigma_g^+$	2.43	2.48	0.93	Linear	$^1\Sigma_g^+$	2.40	2.50	1.26
Rhombus	$^1B_2$	2.77	71	0.95	Rhombus	$^1B_2$	2.73	66	1.36
Square	$^1B_2$	2.77	90	0.95	Tetrahedron	$^3B_2$	2.79	109.471	1.42
Square	$^1A_1$	2.65	90	1.24	Square	$^1A_1$	2.70	90	1.43
Tetrahedron	$^3B_2$	2.83	109.471	1.29	Square	$^1B_2$	2.72	90	1.46
Tetrahedron	$^1A_1$	2.86	109.471	1.42	Tetrahedron	$^1A_1$	2.83	109.471	1.79
Rhombus	$^3A_1$	2.99	52	1.50	Rhombus	$^3A_1$	2.84	55.7	1.97
Linear	$^3\Sigma_g^+$	2.57	2.50	1.53	Linear	$^3\Sigma_g^+$	2.57	2.50	2.03
Tetrahedron	$^1B_2$	2.78	109.471	2.18	Tetrahedron	$^1B_2$	2.78	109.471	2.03
Square	$^3A_1$	2.94	90	2.52	Square	$^3A_1$	2.80	90	3.08
Tetrahedron	$^3A_1$	3.02	109.471	5.94	Tetrahedron	$^3A_1$	3.02	109.471	5.09

[a] R_{12} stands for the four equal sides of rhombus; for linear states, R_{12} represents the corner bond lengths, while R_{23} represents the center bond length. q is the shortest Au–Au–Au bond angle. All results in this table are from Reference 108.

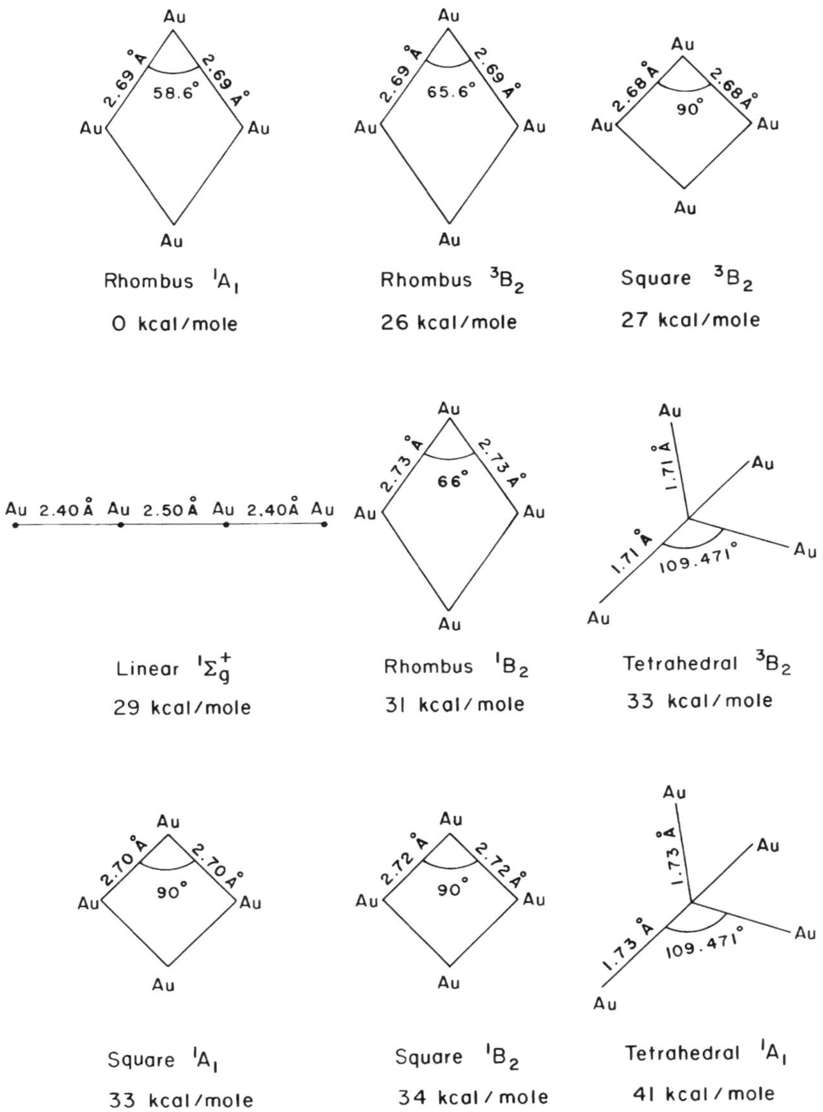

Figure 1.11 Geometries of some of the low-lying states of Au_4. (From Ref. 108.)

We employed CAS-MCSCF followed by the FOCI, SOCI calculations without d-electron correlations and multireference CI (MRCI) calculations including d-electron correlations. We computed the geometries, energy separations, and ionization potentials. The theoretical study of Liao and the author (149) was motivated by the previous studies of Bonačić-Koutecký et al. on alkali metal and related clusters (150, 151).

Table 1.14 Comparison of POLCI and MRSDCI Geometries and Energy Separations of Au_4

		POLCI			MRSDCI		
State	Geometry	r_e, Å	θ_e, deg	E, eV	r_e, Å	θ_e, deg	E, eV
$^1A_1$	Rhombus	2.69	60.4	0.0	2.69	58.6	0.0
$^3B_2$	Rhombus	2.69	66.9	0.88	2.69	65.6	0.98
$^3B_2$	Square	2.68	90	0.97	2.58	90	1.12

[a]All results in this table are from Reference 108.

The author and Liao (148) have suggested that the capped pentagon is the lowest for the neutral cluster, but the hexagonal ring structure is favored for the anion. Hence, Liao and the author considered four possible structures for Cu_6–Ag_6, namely, capped pentagon, tripyramidal, hexagonal ring, and octahedron. In addition, the triangular (D_{3h}) structure used by Bonačić-Koutecký et al. (2) for Na_6 was included in the study of Liao and the author (149). The four structures considered by Liao and the author are shown in Figure 1.13. The first model is a capped pentagon (CP) with an overall C_{5v} symmetry. There are two kinds of bond lengths in the CP structure. In this structure, the cap–origin distance is independent of the metal–metal distance in the pentagon (Fig. 1.13). The second model is tripyramidal (TP) with C_{2v} symmetry, and $r_{1,2} = r_{1,3} = r_{1,6} = r_{5,1} = r_{5,3} = r_{5,6} = r_{5,4} = r_{2,6} = r_{2,3} = r_{6,3} = r_{6,4} = r_{3,4}$. The third model is a ring (R) with D_{6h} symmetry, which looks like benzene with $r_{1,2} = r_{2,3} = r_{3,4} = r_{4,5} = r_{5,6} = r_{6,1}$. The fourth model is an octahedron (O) with O_h symmetry and $r_{1,2} = r_{1,3} = r_{1,4} = r_{1,5} = r_{6,2} = r_{6,3} = r_{6,4} = r_{6,5} = r_{2,3} = r_{3,5} = r_{5,4} = r_{4,2}$. Liao and the author used the relativistic effective core potentials (RECPs) that included the outer $(n-1)d^{10}ns^1$ shells in the valence

Table 1.15 Effect of Spin–Orbit Coupling Term on Energy Separations of Electronic States of Au_4[a]

Geometry	State	E, eV
Rhombus	$^1A_1(A_1)$	0.0
Rhombus	$^1A_1$	0.010
Rhombus	$^3B_2(A_2)$	0.984
Rhombus	$^3B_2(B_1)$	0.991
Rhombus	$^3B_2$	0.993
Rhombus	$^3B_2(A_1)$	1.006

[a]The labels in parentheses designate the symmetry in the double group, including the spin–orbit term. For comparison, we also include states without the spin–orbit terms (without label in parentheses).

Source: From Reference 108.

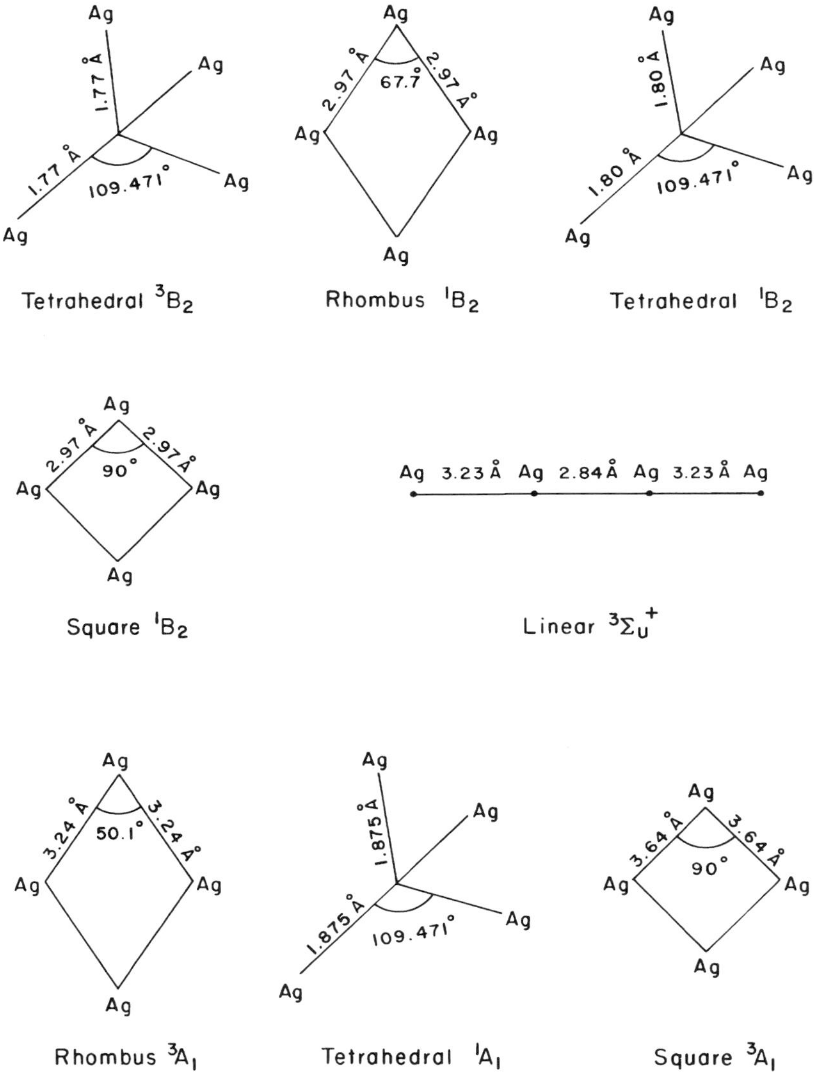

Figure 1.12 Geometries of three low-lying states of Ag_4. (From Ref. 107.)

space uniformly for Cu, Ag, and Au, where *n* is the principal quantum number. Valence Gaussian basis sets of (3*s*3*p*3*d*) quality were used for all Cu, Ag, and Au. The CASSCF followed by three levels of CI calculations for all three clusters and their positive ions. The CASSCF calculations included the six valence *s* electrons in the active space. All valence *s* orbitals (six of them) of the group IB atoms were included in the active space. The core *d* electrons were kept inactive

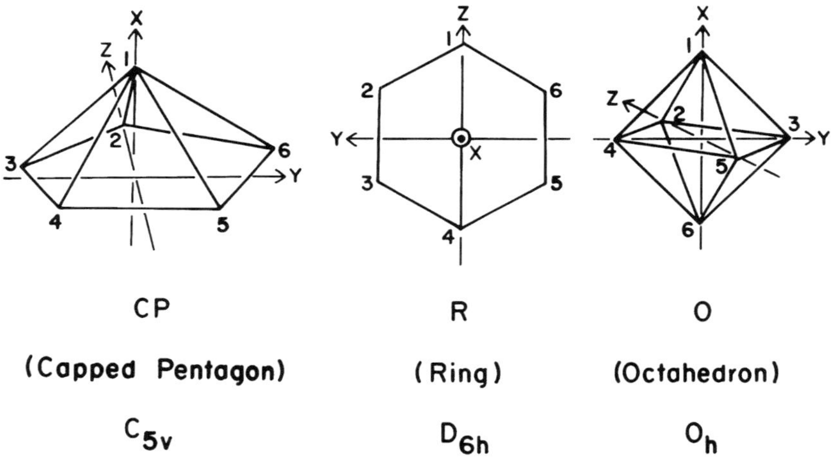

Figure 1.13 Topology of the calculated IR-M_6 clusters. (From Ref. 149.)

at the CASSCF stage, although these d orbitals were allowed to relax. The CI calculations were made following the CASSCF, with and without correlating d electrons. The first two CI procedures, labeled the FOCI and SOCI, correlated only the valence ns electrons. The effects of d-electron correlations were incorporated using the multireference singles + doubles CI (MRSDCI) method. For all structures, the 48 outermost electrons were uniformly correlated. The corresponding 47 electrons of the positive ions were correlated uniformly for all the structures of the ionic species. For Au_6 and Au_6^+, more exhaustive MRSDCI calculations were done.

Tables 1.16–1.18 show the geometries and energy separations for Cu_6, Ag_6, and Au_6 and their positive ions, respectively, as computed by Liao and the author (149). As seen from Table 1.16, at all levels of theory that included d correlation effects, the capped pentagon (C_{5v}) is the lowest for Cu_6. Although the TP structure is favored at the SOCI level, the SOCI does not include electron correlation effects from the $3d$ shells of the Cu atoms. Without inclusion of d correlation effects, Cu_6 is similar to Li_6 and Na_6, which exhibits a more stable tripyramidal structure compared to the capped pentagon.

As seen from Table 1.16, the lowest states of CP, TP, and ring structures are singlets, while the lowest state of the octahedron is a $^3T_{2g}$ state, which would undergo Jahn–Teller distortion. The geometries are somewhat less sensitive to electron correlations for Cu_6, except that the MRCI, which includes $3d$ correlation effects, introduces nonnegligible changes in bond lengths. The energy separations are more sensitive to electron correlations, as seen from Table 1.16. For example, the TP structure is 1 eV higher than CP at the CASSCF level, 0.5 eV higher at the FOCI level, 0.1 eV lower at the SOCI level, and 0.8 eV higher at the MRCI level of theory. It was thus concluded by Liao and the author that

Table 1.16 Geometries and Energy Separations of Electronic States of Cu_6 and Cu_6^+[a,b]

Cluster	Structure	Symmetry	State	CASSCF		FOCI		SOCI		MRCI	
				R_e, Å	E, eV	R_e, Å	E, eV	R_e, Å	E, eV	R_e, Å	E, eV
Cu_6	CP	C_{5v}	$^1A_1$	{2.55, 0.91}	0.0	{2.55, 0.91}	0.0	{2.55, 0.91}	0.0	{2.53, 0.86}	0.0
	TP	C_{2v}	$^1A_1$	2.61	1.05	2.62	0.51	2.63	0.14	2.56	0.83
	R	D_{6h}	$^1A_{1g}$	2.50	1.43	2.50	0.91	2.49	0.28	—	—
	O	O_h	$^3T_{2g}$	2.63	2.23	2.65	1.02	2.64	0.35	2.56	1.10
Cu_6^+	CP	C_{5v}	$^2E_1$	{2.51, 1.20}	5.69	{2.49, 1.15}	5.49	{2.50, 1.16}	5.25	{2.42, 0.98}	5.51
	R	D_{6h}	$^2E_{1u}$	2.57	6.90	2.58	6.61	2.55	6.38	—	—
	R	D_{6h}	$^2A_{1g}$	2.69	7.76	2.69	7.44	2.66	7.32	—	—
	O	O_h	$^4A_{1u}$	3.01	8.79	2.98	8.51	2.92	8.28	—	—

[a]For the CP structure, the first distance is $r_{23} = r_{34} = r_{45} = r_{56} = r_{62}$. The second distance is the origin to the atom distance (see Fig. 1.13).

[b]$R_e = r_{1,2} = r_{1,3} = r_{1,6} = r_{5,1} = r_{5,3} = r_{5,6} = r_{5,4} = r_{2,6} = r_{2,3} = r_{6,3} = r_{6,4} = r_{3,4}$ for the structure TP (Fig. 1.13). $R_e = r_{1,2} = r_{2,3} = r_{3,4} = r_{4,5} = r_{5,6} = r_{6,1}$ for the structure R (Fig. 1.13) $R_e = r_{1,2} = r_{1,3} = r_{1,4} = r_{1,5} = r_{6,2} = r_{6,3} = r_{6,4} = r_{6,5} = r_{2,3} = r_{3,5} = r_{5,4} = r_{4,2}$ for the structure O. (Fig. 1.13).

Source: Reproduced from Reference 149.

Table 1.17 Geometries and Energy Separations of Ag_6 and Ag_6^{+a}

Cluster	Structure	Symmetry	State	CASSCF		FOCI		SOCI		MRCI	
				R_e, Å	E, eV	R_e, Å	E, eV	R_e, Å	E, eV	R_e, Å	E, eV
Ag_6	CP	C_{5v}	$^1A_1$	{2.70, 1.27}	0.0	{2.71, 1.29}	0.0	{2.71, 1.30}	0.0	{2.68, 1.23}	0.0
	TP	C_{2v}	$^1A_1$	2.82	0.55	2.82	0.45	2.82	0.39	2.79	0.29
	R	D_{6h}	$^1A_{1g}$	2.66	1.39	2.65	1.36	2.65	1.24	—	—
	O	O_h	$^3T_{2g}$	2.83	1.22	2.84	1.00	2.84	0.94	2.80	0.86
Ag_6^+	CP	C_{5v}	$^2E_1$	{2.73, 1.33}	4.98	{2.74, 1.35}	5.14	{2.74, 1.35}	5.38	{2.70, 1.28}	4.88

[a]See footnotes *a* and *b* and sourcenote for Table 1.16.

Table 1.18 Geometries and Energy Separations of Au_6 and Au_6^+[a]

Cluster	Structure	Symmetry	State	CASSCF		FOCI		SOCI		MRCI	
				R_e, Å	E, eV	R_e, Å	E, eV	R_e, Å	E, eV	R_e, Å	E, eV
Au_6	CP	C_{5v}	$^1A_1$	{2.72, 1.31}	0.0	{2.76, 1.40}	0.0	{2.78, 1.44}	0.0	{2.71, 1.30}	0.0
	TP	C_{2v}	$^1A_1$	2.83	1.00	2.83	0.87	2.84	0.76	2.80	0.69
	R	D_{6h}	$^1A_{1g}$	2.70	1.01	2.69	0.88	2.72	0.73	2.66	1.39
	O	O_h	$^3T_{2g}$	2.86	1.77	2.87	1.50	2.87	1.40	2.81	1.53
Au_6^+	CP	C_{5v}	$^2E_1$	{2.76, 1.39}	6.75	{2.76, 1.40}	6.77	{2.77, 1.41}	6.92	{2.72, 1.32}	6.48

[a]See footnotes for Table 1.16.

higher-order valence correlation effects and $3d$ correlation effects are important in determining the relative ordering of various geometries.

Three possible structures of Cu_6^+ (CP, ring, octahedron) were considered by Liao and the author. Since the lowest possible configuration of the octahedron contains three electrons in a t_{1u} orbital, the $^4A_{1u}$ state is the lowest for Cu_6^+ with an octahedral geometry. As seen from Table 1.16, the CP structure is the lowest for Cu_6^+. The adiabatic IP of Cu_6 was computed by Liao and the author at the highest (MRCI) level as 5.51 eV. Theoretical studies on smaller clusters (Cu–Cu_4) have suggested that the MRCI IPs have to be scaled by a factor of 1.25 to obtain agreement with experiment. This is a uniform trend in that the theoretical IP of the Cu atom is lower by the same factor compared to experiment. This scaling is thus attributed to static atomic correlation effects of the copper atom. The scaled IP of 6.9 eV obtained by Liao and the author (149) was found to be in good agreement with the value of 7.15 ± 0.75 eV obtained by Jarrold and Creegan (141) experimentally.

Table 1.17 shows the geometries and energy separation of Ag_6 and Ag_6^+ for the various structures obtained by Liao and the author (149). Since for Cu_6^+, the ring and octahedral structures are significantly higher than the CP structure, Liao and the author considered only the CP structure for Ag_6^+. The computed IP can, at most, be 0.5 eV in error due to different geometry. As evidenced from Table 1.17, for Ag_6, too, the CP structure is favored over other structures, except that at the highest MRCI level of theory the TP structure is only 0.3 eV higher than the CP structure. This trend is somewhat different from Cu_6, for which the TP structure is 0.8 eV higher at the MRCI level. The ring structure is at least 1 eV higher than the CP structure at all levels of theory, eliminating this structure as a possible candidate for the ground state of Ag_6. The computed MRCI IP of Ag_6 is 4.88 eV. On the basis of other theoretical works on smaller clusters, a scaling factor of 1.11 was invoked by Liao and the author (149) for IPs of all Ag_n clusters. The scaled IP of 5.43 eV is consistent with the experiment of Cheng and Duncan (57), who have suggested that the vertical IP of Ag_6 should be between 5.6 and 6.4 eV.

Table 1.18 shows the computed geometries and energy separations of four structures of Au_6 and Au_6^+ (CP). As seen from Table 1.18, the CP structure prevails substantially over alternatives for Au_6. The TP structure, which was found to be only 0.3 eV above the CP structure for Ag_6, is 0.7 eV higher for Au_6. In contrast with Cu_6 and Ag_6, higher-order correlation effects do not make substantial contributions for the TP structure. The ring structure is somewhat destabilized relative to the CP structure at the MRCI level, similar to Ag_6 and Cu_6.

Cu_6^+ exhibits $^2E_{1u}$ and $^2A_{1g}$ electronic states with a ring structure; the $^2A_{1g} \leftarrow {}^2E_{1u}$ transition is dipole-allowed, and corresponds to the $^2A_1 \leftarrow {}^2E_1$ transition of the CP structure. Hence, the energy gap between the $^2E_{1u}$ and $^2A_{1g}$ states of Cu_6^+ (ring) provides an estimate for the $^2A_1 \leftrightarrow {}^2E_1$ transition of the CP structure, which is also an optically allowed transition. At the CASSCF and SOCI levels, the T_e values of this system for the ring structure are 0.86

and 0.94 eV. The adiabatic IP of Au_6 varies between 6.5 and 6.9 eV, depending on the level of theory; at the highest level it is 6.48 eV. Liao and the author used a scaling factor of 1.10 for the IPs of the gold clusters based on previous theoretical studies on Au–Au_4 (120). The scaled IP of Au_6, 7.2 eV, is likely to be lower than the experimental value.

The IPs of Cu_6, Ag_6, and Au_6 exhibit an interesting trend. Both the unscaled (5.5, 4.9, 6.5 eV) and scaled IP values (6.9, 5.4, 7.2 eV) follow the order IP (Au_6) > IP(Cu_6) > IP(Ag_6), but the scaled IP(Au_6) is only 0.3 eV higher than the IP(Cu_6), while the unscaled values differ by 1 eV. Thus the scaled IP of Au_6 obtained by Liao and the author was considered as a lower bound, and it was anticipated by these authors that the IP of Au_6 to be near 7.5 eV. Cheng and Duncan (57) suggested that the vertical IP of Au_6 is lower than 7.9 eV.

The leading configuration of the CP structure is $1a_1^2 1e_1^4$ similar to the configuration and orbital energy diagram obtained by Bonačić-Koutecký et al. (2) for Na_6 (CP structure). It is easy to show that the same configuration correlates with $1a'^2 2a'^2 1a''^2$ in the C_s group, $1a_{1g}^2 1e_{1u}^4$ for the ring structure (D_{6h}), $1a_1^2 1b_1^2 2a_1^2$ (or $1a_1^2 1b_2^2 2a_1^1$ depending on the orientation of the TP structure) for the TP structure, and $1a_{1g}^2 1t_{tu}^4$ for the octahedral structure. The positive ion is formed by the removal of an electron from the HOMO (highest occupied MO), which leads to different electronic states of different spin multiplicities for the various geometries. For the CP structure, this leads to the $^2E_1$ state, while the ring structure exhibits the lowest state as $^2E_{1u}$. The octahedral geometry of the positive ion with $1a_{1g}^2 1t_{1u}^3$ results in a $^4A_{1u}$ state. The lowest states of Cu_6^+, Ag_6^+, and Au_6^+ are all $^2E_1$ states, and will thus undergo Jahn–Teller distortions from their ideal CP structures. If the distortion is large, the photoelectron spectrum of the hexamers is likely to show vibrational progression.

1.2.3 Comparison of Coinage-Metal Clusters: Cu_n, Ag_n, and Au_n

The variation in bonding and electronic properties of coinage-metal clusters is best understood by first considering the atomic properties of copper, silver, and gold. Figure 1.14 shows the ^2S–$^2P_{1/2}$ splitting of copper, silver, and gold. The ^2S and $^2P_{1/2}$ states arise from the $(n-1)d^{10}ns^1$ and $(n-1)d^{10}np^1$ configurations, respectively. All the atomic energy separations are from Reference 113. Figure 1.15 shows the $d^{10}s^1(^2S)$–$d^9s^2(^2D)$ splittings of noble-metal atoms. The ionization potentials of the three atoms are compared in Figure 1.16.

All three figures (Figs. 1.14–1.16) demonstrate the anomalous behavior of the gold atom in the coinage-metal group. The ^2S–^2P splitting and the ionization potential of gold dramatically increases, while the 25-$(5d^9 6s^2)^2D$ splitting of the gold atom dramatically decreases, compared with silver. This behavior for gold does not follow the trend from copper to silver; that is, the IP decreases in moving down from copper to silver but shoots up dramatically in moving from silver to gold. The rationalization for this gold anomaly is derived from relativity. As mentioned in the introduction, the large relativistic effects in the inner core of the gold atom manifest themselves in the valence as stabiliza-

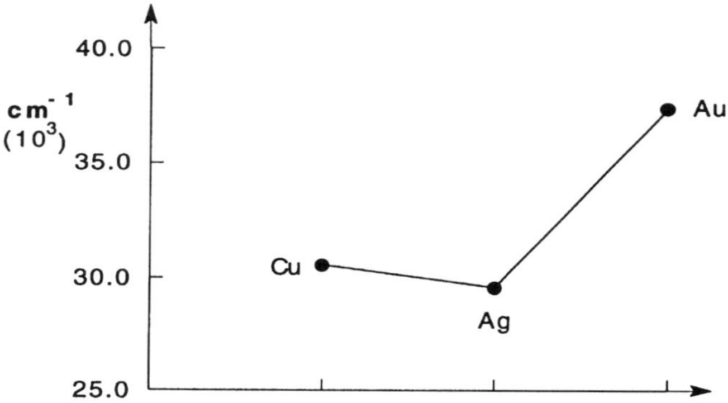

Figure 1.14 Comparison of experimental ^2S–$^2P_{1/2}$ splittings of copper, silver, and gold.

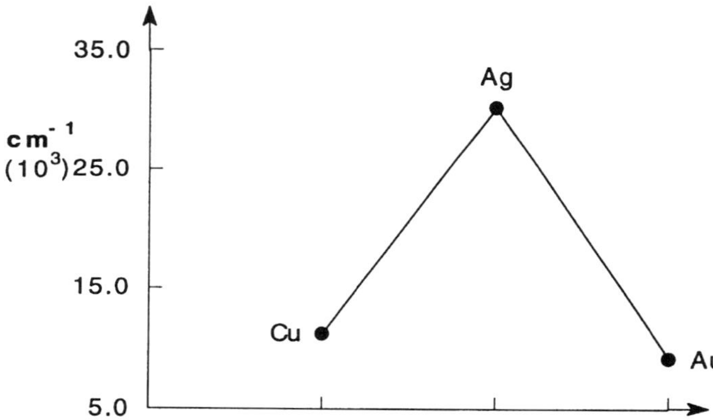

Figure 1.15 Comparison of $d^{10}s(^2S)$–$d^9s^2(^2D)$ splittings of copper, silver, and gold.

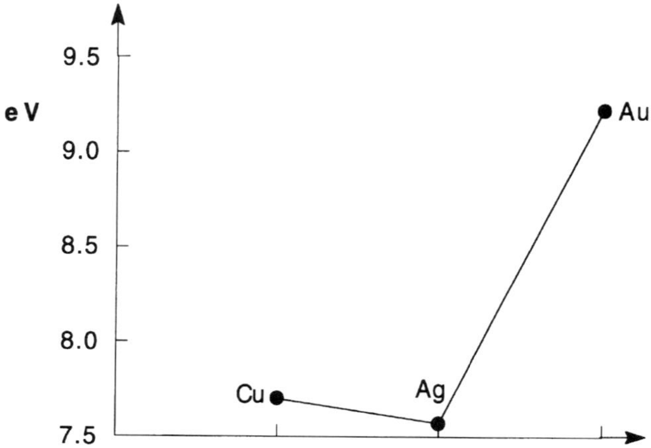

Figure 1.16 Experimental ionization potentials of copper, silver, and gold.

tion and contraction of the outer valence 6*s* orbital of the gold atom owing to the relativistic mass–velocity contraction. The relativistic stabilization of the 6*s* orbital of the gold atom leads to a substantially larger IP and ^2S–^2P splitting for gold compared with silver. On the other hand, promotion of one of the *d* electrons into the *s* orbital becomes more favorable for gold, compared to silver.

Figure 1.17 compares the experimental dissociation energies of the three dimers. As seen from this figure, the D_e decreases in moving down the periodic table from copper to silver but increases in moving from silver to gold. The increase in D_e for Au_2 is purely a consequence of the relativistic mass–velocity contraction (Fig. 1.17).

The equilibrium bond lengths of the Cu_2–Au_2 triad are plotted and compared with the Ge_2–Pb_2 triad in Figure 1.18. The bond lengths of Ge_2–Sn_2 are from References 121–123. As seen from this figure, the bond lengths of the Ge_2–Pb_2 triad increase monotonically with only a slight lowering of the Sn_2–Pb_2 slope, compared with the Ge_2–Sn_2 slope. This dramatically contrasts with the Cu_2–Au_2 triad, for which the Ag_2–Au_2 slope is negative, while the Cu_2–Ag_2 slope is positive. Thus relativistic mass–velocity contraction effects are the largest for the noble-metal triads.

The theoretical atomization energies of Cu_3, Ag_3, and Au_3 clusters are plotted in Figure 1.19. The AE of Cu_3 is from Reference 110. The AE decreases in going from Cu_3 to Ag_3, but increases dramatically in going from Ag_3 to Au_3. This increase is so large that the AE of Au_3 is even larger than Cu_3 (see Fig. 1.19). This, too, is a consequence of the relativistic mass–velocity contraction.

Figure 1.20 compares the geometries of the Cu_3–Au_3 triad in both $^2B_2$ and $^2A_1$ states (the two Jahn–Teller components). The curve labeled "side" corresponds with the variation of the two equal sides of the isosceles triangle in moving from Cu_3 to Au_3. The curve labeled "base" represents the base of the triangle. As seen from this figure, in every case the bond length increases in moving from Cu_3 to Ag_3 but decreases in moving from Ag_3 to Au_3. Note that the base bond lengths are longer for the $^2B_2$ state, while the side bond lengths

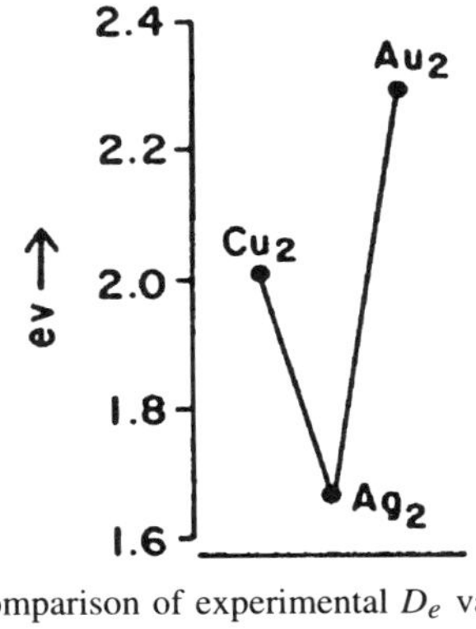

Figure 1.17 Comparison of experimental D_e values of Cu_2–Au_2.

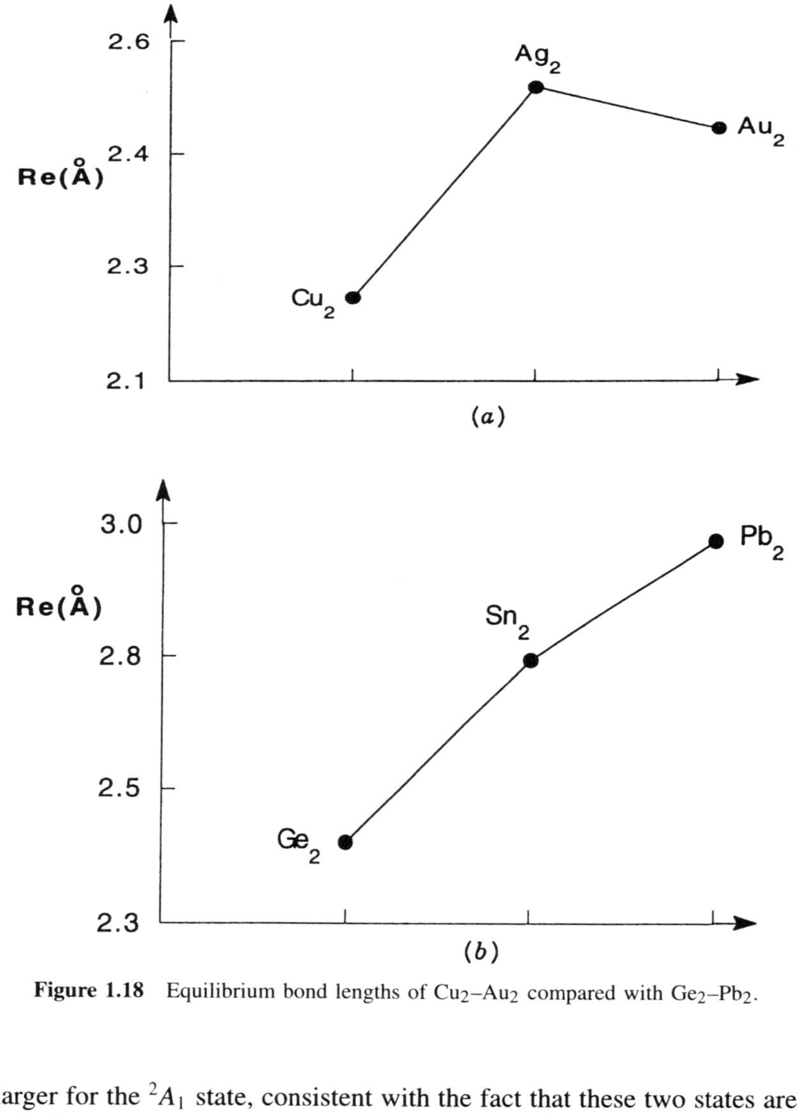

Figure 1.18 Equilibrium bond lengths of Cu_2–Au_2 compared with Ge_2–Pb_2.

are larger for the $^2A_1$ state, consistent with the fact that these two states are the Jahn–Teller components of an equilateral triangular $^2E'$ state.

As seen from Figures 1.11 and 1.12, both Ag_4 and Au_4 have $^1A_1$ ground states with a rhombus geometry. The Au–Au bond lengths of the sides of the rhombus in the ground state of Au_4 are 2.69 Å, while the corresponding bond lengths of Ag_4 are 2.87 Å. Similarly, the total binding energy of Au_4 is expected to be substantially larger for Au_4 compared with Ag_4.

It is also of interest to compare the gross atomic Mulliken populations (Table 1.19) of the coinage-metal trimers, since they provide some insight into the relative importance of the s, d, and p populations. A striking contrast in the

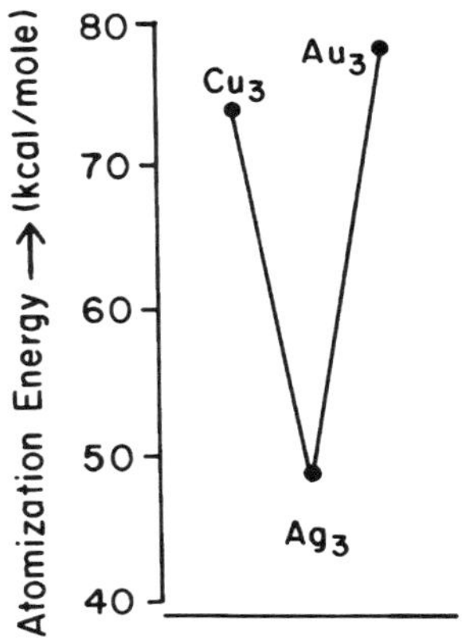

Figure 1.19 Plot of the theoretical atomization energies of Cu_3, Ag_3, and Au_3.

Mulliken populations of Ag_3 and Au_3 is that the gross metal *p* population is 0.29*e* lower for Au_3. This is consistent with Figure 1.14, which shows that the ^2S–^2P splittings are much larger for gold compared with silver. Thus the next *p* shell participation is more important for Ag_3 compared with Au_3. The metal *d* populations are somewhat smaller (0.177–0.183*e*) for Au_3 compared with the

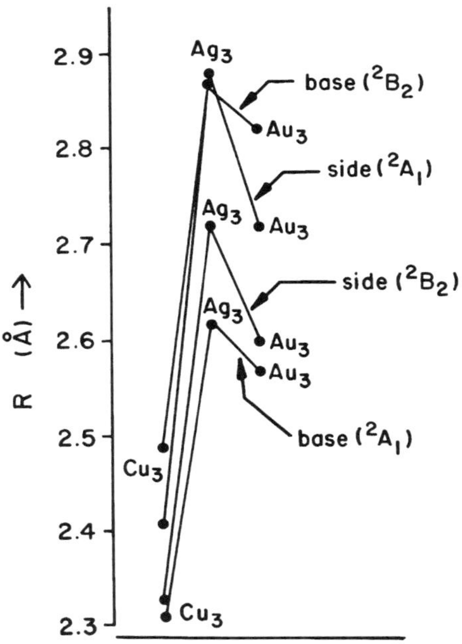

Figure 1.20 Plots of the side and base lengths of the isosceles triangle for Cu_3–Au_3 in both $^2B_2$ and $^2A_1$ states.

Table 1.19 Mulliken Population Analysis for Au_3 and Ag_3

State	Gross Population								Overlap[b]
	M	M_2[a]	M(s)	M_2(s)	M(p)	M_2(p)	M(d)	M_2(d)	
Au_3									
$^2B_2$	10.84	22.16	0.741	2.167	0.147	0.226	9.950	19.769	0.608
$^2\Sigma_u^+$	11.32	21.68	1.178	1.808	0.119	0.010	10.025	19.859	0.632
$^2A_1$	11.14	21.86	1.143	1.802	0.109	0.261	9.885	19.800	0.114
$^2\Sigma_g^+$	11.16	21.84	0.849	1.859	0.313	0.016	9.999	19.964	0.512
Ag_3									
$^2B_2$	10.81	22.19	0.602	1.838	0.227	0.438	9.979	19.917	0.812
$^2\Sigma_u^+$	11.17	21.83	1.004	1.815	0.160	0.060	10.005	19.955	0.867
$^2A_1$	11.09	21.91	0.971	1.607	0.167	0.386	9.960	19.918	0.572
$^2\Sigma_g^+$	11.15	21.85	0.787	1.783	0.359	0.101	10.005	19.964	0.661

[a] M_2 stands for the two equivalent base atoms.
[b] The overlaps are for the side atoms of the isosceles triangle.

Source: Reference 99.

d populations for Ag_3. This is consistent with Figure 1.15, which compares the ^2S–^2D splittings of Cu–Au. Note that this splitting is substantially lowered for gold.

On the basis of these findings, it is predicted that the excited states of Au_3 arise from the promotion of an electron from $5d^{10}$ into $6s$ rather than the promotion of a $6s$ electron into $6p$. Consequently, the excited states of Au_3 would have substantially different character compared with the corresponding states of Ag_3. Walch (114) has carried out calculations on the excited states of Ag_3, while Das and Balasubramanian (115) have carried out CASSCF/MRSDCI calculations on Au_3^- and the excited states of Au_3. All the calculations on the excited states of Au_3 are consistent with this. The electron affinities of Cu, Cu_2, and Cu_3 have been studied both theoretically (118) and experimentally (119).

Balasubramanian and Feng (120) investigated the ionization potentials of Au_x and Ag_x (x = 1,4) using the relativistic CASSCF/MRSDCI methodology, which correlated the *d* shells of the metal atoms to a high order. Figure 1.21 shows the plot of the ionization potentials of the atom to tetramer for both silver and gold. It is interesting to note that the IPs exhibit a sawtooth behavior suggesting that the IPs of odd clusters are lower than those of even clusters, since ionization of the odd clusters leads to stable closed-shell $^1A_1$ ground states for smaller clusters, while ionization of the even clusters destroys the closed-shell

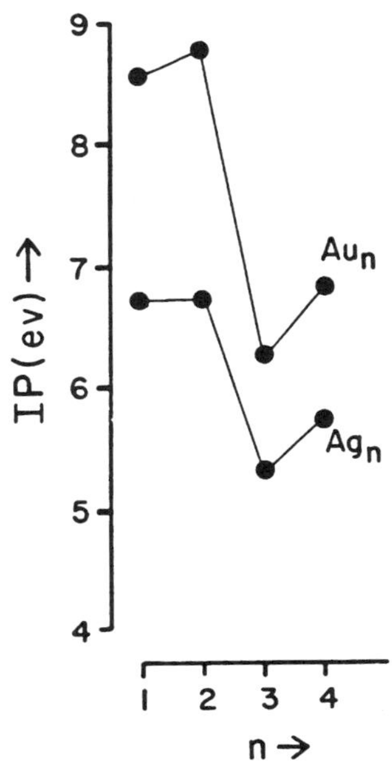

Figure 1.21 Theoretical CASSCF/MRSDCI IPs of Au_x and Ag_x (x = 1,4). (From Ref. 120.)

Table 1.20 Magnitude of Spin–Orbit Mixings in Various States of Ag_3, Au_3, and Au_4

System	State	Percent of Contribution[a]
Au_3	$^2A_1\alpha$	$^2A_1(92)$, $^2B_2(1)$
Au_3	$^2B_2\alpha$	$^2B_2(93)$, $^2A_1(0.6)$
Ag_3	$^2A_1\alpha$	$^2A_1(93)$, $^2B_2(0.02)$
Ag_3	$^2B_2\alpha$	$^2B_2(94)$, $^2A_1(0.00014)$
Au_4	$^1A_1$	$^1A_1(96)$, $^3B_2(0.1)$
Au_4	$^3B_2(A_1)$	$^3B_2(91)$, $^1A_1(5)$, $^1A_1(\text{II})(0.4)$

[a]The weights in percent of the RCI wavefunctions are shown in parentheses.

ground states of the dimer and tetramer. The electron affinities as a function of cluster size (Fig. 1.2) exhibit similar behaviors for the same reason.

In comparing the ionization potentials of silver and gold clusters, it is found that the gold clusters have considerably larger IPs, due mainly to the relativistic mass–velocity stabilization of the outer 6*s* orbital of the gold atom.

The CASSCF/MRSDCI vertical IP of Ag_3 (5.31 eV) calculated by Balasubramanian and Feng (120) is in very good agreement with an experimental prediction of less than 6 eV by Cheng and Duncan (57) using a two-color excimer + dye laser experiment. The theoretical CASSCF/MRSDCI IP of Ag_2 is 6.73 eV (120) compared with an unpublished experimental value of 7.56 eV, obtained by Smalley and co-workers (43). Although IPs and EAs cannot be calculated exactly using *ab initio* methods, the relative trends of the calculated IPs should be more reliable.

The relative importance of the spin–orbit contaminations in Ag_3, Au_3, and Au_4 is also of interest. Table 1.20 compares the percentages of the various configurations in the relativistic CI wavefunctions of the electronic states of Ag_3, Au_3, and Au_4. As seen from Table 1.20, the spin–orbit contamination effects are quite important for Au_4 $[^3B_2(A_1)]$ and Au_3. The closed-shell $^1A_1$ ground state of Au_4 is slightly contaminated with the open-shell $^3B_2$ state.

1.3 ELECTRONIC STRUCTURE OF PALLADIUM AND PLATINUM CLUSTERS

1.3.1 Pt_2

Table 1.21 shows the spectroscopic properties (R_e, T_e) for the electronic states of Pt_2 in the absence of spin–orbit interaction as obtained using the CASSCF/FOCI method by the author (152). As seen from Table 1.21, the CASSCF/FOCI level of theory predicts that the $^3\Pi_u$ state arising from the $\sigma_u\pi_g^3$ open-shell configuration to be the lowest and a $^3\Sigma_g^-$ state almost nearly degenerate ($T_e \sim 371\ \text{cm}^{-1}$). The higher-order correlations can change not only the

Table 1.21 FOCI Spectroscopic Properties of Pt_2 in the Absence of Spin–Orbit Interaction

State	R_e, Å	T_e, cm^{-1}	ω_e, cm^{-1}
$^3\Pi_u$	2.398	0.0	305
$^3\Sigma_g^-$	2.428	371	219
$^1\Sigma_g^+$	2.459	1,792	175
$^3\Delta_g$	2.486	1,852	206
$^3\Phi_u$	2.583	2,200	239
$^3\Sigma_g^-$(II)	2.459	2,718	175
$^1\Pi_u$	2.454	2,718	193
$^1\Gamma_g$	2.558	2,863	197
$^3\Sigma_u^+$	2.578	3,186	189
$^3\Gamma_u$	2.570	3,189	193
$^1\Delta_g$	2.544	3,339	155

Source: Reproduced from Reference 152.

spectroscopic properties listed in Table 1.21 but also the ordering of the electronic states, since there are many states below 3000 cm^{-1}. The calculated r_e and ω_e values are quite reliable, since higher-order correlations are unlikely to change these properties by more than 5%.

Table 1.22 shows the spectroscopic properties of Pt_2, including spin–orbit interaction. As seen from this table, the spin–orbit splitting of the electronic states of Pt_2 is so large that the ordering of the electronic states listed in Table 1.21 without spin–orbit effects change dramatically. For example, the 5_u–$^3\Gamma_u$ and 0_g^+–$^3\Sigma_g^-$ (δ_u^2) splittings are 8139 and 8282 cm^{-1}, respectively, with the corresponding Ω component being lower. The ^5u–^3u splitting, arising from the $^3\Gamma_u$ state, was found to be 15,192 cm^{-1}. Thus the spin–orbit contribution to the energies of electronic states is quite significant. However, the effect of spin–orbit interaction on r_e (0.002–0.008 Å) and ω_e (1–15 cm^{-1}) values are somewhat smaller.

As seen from Table 1.22, there are many electronic states for Pt_2 below 10,000 cm^{-1}. The lowest state, which includes spin–orbit interaction, is the $^3\Sigma_g^-(0_g^+)$ state. The $^3\Gamma_u(5_u)$ state is only 614 cm^{-1} above the 0_g^+ state. This table is, however, not complete in that many more electronic states are feasible for Pt_2, although some of these states may be repulsive.

Taylor et al. (153) investigated the spectra of Pt_2 using resonance two-photon ionization and dissociation spectroscopy. Their spectra exhibited overall qualitative agreement in that the large density of electronic states computed by the author (152) was confirmed by the work of Taylor et al. (153). The theoretically computed vibrational frequencies were also found to be in good accord with the experimentally deduced results for the states, for which vibrational analysis could be carried out. Although the experimental r_e values could not be deduced from these spectra, since no rotational analysis was feasible, Muller et al. (154)

Table 1.22 Spectroscopic Properties of Pt_2

State	R_e, Å	T_e, cm^{-1}	ω_e, cm^{-1}
$\delta_u^2\ ^3\Sigma_g^-(0_g^+)$	2.456	0.0	189
$\delta_u^3\delta_g^3\ ^3\Gamma_u(5_u)$	2.570	614	193
$\delta_u^3\pi_g^3\ ^3\Phi_u(4_u)$	2.580	1,074	240
$\pi_g^2\ ^3\Sigma_g^-(0_g^+)$	2.423	3,064	215
$\sigma_u\pi_g^3\ ^3\Pi_u(2_u)$	2.396	3,214	306
$\sigma_u\pi_g^3\ ^3\Pi_u(1_u)$	2.398	3,556	305
$\sigma_u\delta_u^3\ 3\Delta_g(2_g)$	2.493	5,838	202
$\pi_g^2\ ^3\Sigma_g^-(1_g)$	2.428	5,935	219
$\sigma_u\pi_g^3\ ^3\Pi_u(0_u^-)$	2.400	7,500	304
$\sigma_u\pi_g^3\ ^3\Pi_u(0_u^+)$	2.400	7,501	304
$\delta_u^2\ ^1\Gamma_g(4_g)$	2.558	7,886	196
$\sigma_u\delta_u^3\ ^3\Delta_g(3_g)$	2.485	8,164	208
$\delta_u^3\delta_g^3\ ^3\Sigma_u^+(1_u)$	2.578	8,180	191
$\delta_u^3\delta_g^3\ ^3\Gamma_u(4_u)$	2.570	8,259	192
$\delta_u^2\ ^3\Sigma_g^-(1_g)$	2.459	8,282	175
$\delta_u^3\delta_g^3\ ^3\Sigma_u^+(0_u^-)$	2.578	8,755	188
$\sigma_u\pi_g^3\ ^1\Pi_u(1_u)$	2.455	9,556	197
$\pi_g^2\ ^1\Sigma_g^+(0_g^+)$	2.453	9,765	180
$\sigma_u\delta_u^3\ ^3\Delta_g(1_g)$	2.487	10,810	206
$\sigma_u\delta_u^3\ ^1\Delta_g(2_g)$	2.539	11,286	170
$\delta_u^3\pi_g^3\ ^3\Phi_u(3_u)$	2.553	15,234	277
$\delta_u^3\delta_g^3\ ^3\Gamma_u(3_u)$	2.571	15,806	191

Source: Reproduced from Reference 152.

confirmed the author's (152) computed bond lengths for Pt_2 and Pt_3 through a scanning electron microscopy study on Pt_2 and Pt_3 on pyrollitic graphite.

Basch et al. (158), using a small bond-pair SCF treatment without spin–orbit interaction, found two degenerate states: $^1\Gamma_g$ and $^3\Gamma_u$. As seen from Table 1.21, although these are not the lowest states in the absence of spin–orbit interaction, the $^1\Gamma_g$ and $^3\Gamma_u$ are nearly degenerate.

The dissociation energy of Pt_2 in the absence of spin–orbit interaction based on the difference in energies of the $r_{\min}$ of $^3\Pi_u$ and 8.00 Å obtained using the FOCI method is 2.31 eV. The corresponding value for the $^3\Gamma_g^-$ state is 2.27 eV. Both the values are orders of magnitude better than the D_e values obtained by Basch et al. (158) for the lowest states (0.92–0.93 eV), indicating that correlation corrections are significant for Pt_2. The spin–orbit interaction would influence the D_e values to a considerable extent, since the molecular spin–orbit splitting and atomic spin–orbit splittings do not cancel. The $^3D_3$–^3D splitting for the platinum atom was calculated by the current author to be 0.01974 hartree a.u. (atomic unit) or 4376 cm^{-1}. Thus the spin–orbit splitting lowers to the two platinum atoms relative to ^3D by 8752 cm^{-1}. The 0_g^+ state is lowered with respect to $^3\Sigma_g^+$ by 8282 cm^{-1}. Thus the dissociation energy of $^3\Sigma_g^-(0_g^+)$, which includes spin–orbit correction, is about 1.97 eV. The D_e of the $^3\Gamma_u(5u)$ state, including spin–orbit correction, is 1.84 eV. The D_e value of the $^3\Pi_u(2u)$ state, however, is found to be only 1.54 eV, indicating considerable weakening of the bond in the $^3\Pi_u$ state as a result of spin–orbit interaction. Thus, for all the lowest-lying states, the spin–orbit interaction destabilizes the bond by 0.07–0.77 eV, where the lowest destabilization was observed for the 5_u component of the $^3\Gamma_u$ state.

The experimental D_0° values for Pt_2 range from 2.88 to 3.77 eV (155, 156). The 3.71 eV was obtained based on a number of assumptions on the ground state and excited states of Pt_2, which do not appear to be consistent with the author's calculations. Taylor et al. (153) recomputed the thermodynamic partition function of Pt_2 using the spectroscopic constants computed by the author. The corrected experimental D_0° value reported by those authors (153) for Pt_2 is 3.15 eV. This value is in very good agreement with a D_0 value of 2.88 eV deduced by Miedema and Gingerich (156). Since theoretical calculations of the CASSCF/FOCI level typically yield 80–90% of the experimental D_0° values for the main-group dimers, and correlation corrections are expected to be larger for Pt_2, a difference of about 25% due to higher-order correlation and basis-set corrections was anticipated by the author. Thus a lower limit for the D_0° value of Pt_2 was suggested by the author as 2.88 eV.

Analysis of the coefficients of the leading configurations for the electronic states of Pt_2 reveals (152) that only for the $^3\Gamma_u$ and $^3\Sigma_u^+$ states the leading configuration has a weight of 85–87%. For all other electronic states, the leading configurations make smaller contributions, indicating that the electron correlation corrections are large for these states. The correlation effects are especially large for the $^3\Sigma_g^+$ and $^1\Delta_g$ states.

Table 1.23 shows the weights of various λ–*s* states contributing to the relativistic ω–ω states of Pt_2. As seen from that table, the low-Ω states are heavily contaminated, while the high-Ω states are relatively pure. In particular, the spin–orbit contaminations are the largest for the 0_g^+ states, as seen from the mixing of $^3\Sigma_g^-$ and $^3\Sigma_g^+$ states. Likewise, the 1_u states arising from $\sigma\pi^3$ configurations have significant $^3\Pi$–$^1\Pi$ contamination. The mixing of $^3\Phi_u$ and $^1\Phi_u$ in the ^3u state is a bit smaller. The r_e and ω_e values of all the electronic states arising from $\delta_g^3\delta_u^3$ configurations are not altered much by spin–orbit interaction. However, their energies are altered substantially by spin–orbit contribution. This is consistent with the usual trend that the changes in r_e and ω_e values in the electronic states are brought about primarily through spin–orbit contamination with other λ–*s* states. Since the $\delta_u^3\delta_g^3$ electronic states are pure, the shifts in r_e and ω_e are somewhat small. As evidenced from Table 1.22, the changes in r_e and ω_e values are appreciable for the 0_g^+ and ^1u states ($\sigma\pi^3$) as a result of off-diagonal spin–orbit contamination.

Table 1.23 Weights (Shown at Equilibrium Geometries) of the Various λ–s States Contributing to Low-Lying Electronic States of Pt_2

State	Contribution
$\delta_u^2\ {}^3\Sigma_g^-(0_g^+)$	$32\%\ {}^3\Sigma_g^-(\delta_u^2), 30\%\ {}^1\Sigma_g^+(\delta_u^2), 17\%\ {}^1\Sigma_g^+(\delta_g^2), 15\%\ {}^3\Sigma_g^-(\delta_g^2)$
$\delta_u^3\delta_g^3\ {}^3\Gamma_u(5_u)$	$93\%\ {}^3\Gamma_u(\delta_u^3\delta_g^3)$
$\delta_u^3\pi_g^3\ {}^3\Phi_u(4_u)$	$78\%\ {}^3\Phi_u(\delta_u^3\pi_g^3), 17\%\ {}^3\Phi_u(\delta_g^3\pi_u^3)$
$\pi_g^2\ {}^3\Sigma_g^-(0_g^+)$	$56\%\ {}^3\Sigma_g^-(\pi_g^2), 21\%\ {}^1\Sigma_g^+(\pi_g^2), 12\%\ {}^3\Sigma_g^-(\pi_u^2), 7\%\ {}^1\Sigma_g^+(\pi_u^2)$
$\sigma_u\pi_g^3\ {}^3\Pi_u(2_u)$	$81\%\ {}^3\Pi_u(\sigma_u\pi_g^3), 10\%\ {}^3\Pi_u(1\sigma_g\pi_u^3), 2\%\ {}^3\Pi_u(2\sigma_g\pi_u^3)$
$\sigma_u\pi_g^3\ {}^3\Pi_u(1_u)$	$55\%\ {}^3\Pi_u(\sigma_u\pi_g^3), 26\%\ {}^1\Pi_u(\sigma_u\pi_g^3), 7\%\ {}^3\Pi_u(1\sigma_g\pi_u^3),$ $4\%\ {}^1\Pi_u(1\sigma_g\pi_u^3)$
$\sigma_u\delta_u^3\ {}^3\Delta_g(2_g)$	$76\%\ {}^3\Delta_g(\sigma_u\delta_u^3), 12\%\ {}^1\Delta_g(\sigma_u\delta_u^3), 3\%\ {}^3\Delta_g(2\sigma_g\delta_g^3),$ $3\%\ {}^1\Delta_g(2\sigma_g\delta_g^3)$
$\pi_g^2\ {}^3\Sigma_g^-(1_g)$	$79\%\ {}^3\Sigma_g^-(\pi_g^2), 17\%\ {}^3\Sigma_g^-(\pi_u^2)$
$\sigma_u\pi_g^3\ {}^3\Pi_u(0_u^-)$	$80\%\ {}^3\Pi_u(\sigma_u\pi_g^3), 11\%\ {}^3\Pi_u(1\sigma_g\pi_u^3)$
$\sigma_u\pi_g^3\ {}^3\Pi_u(0_u^+)$	$81\%\ {}^3\Pi_u(\sigma_u\pi_g^3), 11\%\ {}^3\Pi_u(1\sigma_g\pi_u^3)$
$\delta_u^2\ {}^1\Gamma_g(4_g)$	$51\%\ {}^1\Gamma_g(\delta_u^2), 42\%\ {}^1\Gamma_g(\delta_g^2)$
$\sigma_u\delta_u^3\ {}^3\Delta_g(3_g)$	$78\%\ {}^3\Delta_g(\sigma_u\delta_u^3), 13\%\ {}^3\Delta_g(1\sigma_g\delta_g^3)$
$\delta_u^3\delta_g^3\ {}^3\Sigma_u^+(1_u)$	$93\%\ {}^3\Sigma_u^+$
$\delta_u^3\delta_g^3\ {}^3\Gamma_u(4_u)$	$93\%\ {}^3\Gamma_u$
$\delta_u^2\ {}^3\Sigma_g^-(1_g)$	$65\%\ {}^3\Sigma_g^-(\delta_u^2), 30\%\ {}^3\Sigma_g^-(\delta_g^2)$
$\delta_u^3\delta_g^3\ {}^3\Sigma_u^+(0_u^-)$	$93\%\ {}^3\Sigma_g^+$
$\sigma_u\pi_g^3\ {}^1\Pi_u(1_u)$	$53\%\ {}^1\Pi_u(\sigma_u\pi_g^3), 26\%\ {}^3\Pi_u(\sigma_u\pi_g^3), 9\%\ {}^1\Pi_u(1\sigma_g\pi_u^3),$ $3\%\ {}^3\Pi_u(1\sigma_g\pi_u^3)$
$\pi_g^2\ {}^1\Sigma_g^+(0_g^+)$	$51\%\ {}^1\Sigma_g^+(\pi_g^2), 23\%\ {}^3\Sigma_g^-(\pi_g^2), 18\%\ {}^1\Sigma_g^+(\pi_u^2), 5\%\ {}^3\Sigma_g^-(\pi_u^2)$
$\sigma_u\delta_u^3\ {}^3\Delta_g(1_g)$	$78\%\ {}^3\Delta_g(\sigma_u\delta_u^3), 13\%\ {}^3\Delta_g(1\sigma_g\delta_g^3), 4\%\ {}^3\Delta_g(2\sigma_g\delta_g^3)$
$\sigma_u\delta_u^3\ {}^1\Delta_g(2_g)$	$37\%\ {}^1\Delta_g(\sigma_u\delta_u^3), 32\%\ {}^3\Delta_g(\sigma_u\delta_u^3), 12\%\ {}^1\Delta_g(1\sigma_g\delta_g^3),$ $7\%\ {}^3\Delta_g(1\sigma_g\delta_g^3), 5.5\%\ {}^1\Delta_g(2\sigma_g\delta_g^3),\ {}^3\Delta_g(1\sigma_g\delta_g^3)$
$\delta_u^3\pi_g^3\ {}^3\Phi_u(3_u)$	$85\%\ {}^3\Phi_u(\delta_u^3\pi_g^3), 4\%\ {}^1\Phi_u(\delta_u^3\pi_g^3)$
$\delta_u^3\delta_g^3\ {}^3\Gamma_u(3_u)$	$93\%\ {}^3\Gamma_u$

Source: Reproduced from Reference 152.

1.3.2 Pd_2

The author (159) has carried out CASSCF/FOCI/MRSDCI/RCI computations on the electronic states of the palladium dimer (Pd_2). The related Pd_3 cluster has also been studied by the author (160) using a comparable level of theory. First, the results on Pd_2 are discussed. Subsequently, the electronic properties of Pd_2 are compared with the properties of Pt_2 from the standpoint of relativistic effects. Table 1.24 shows the calculated MRSDCI/RCI properties of 41 electronic states of Pd_2 computed earlier by the author (159). At the MRSDCI/RCI

Table 1.24 Spectroscopic (MRSDCI/RCI) Properties of Pd_2

State	R_e, Å	T_e, cm^{-1}	ω_e, cm^{-1}
$^3\Sigma_u^+(1_u)$	2.48	0	159
$^3\Sigma_u^+(0_u^-)$	2.48	9	161
$^3\Sigma_u^+$	2.48	214	160
$^1\Sigma_g^+(0_g^+)$	2.87	1,257	121
$^1\Sigma_g^+$	2.87	1,328	121
$^3\Pi_g(1_g)$	2.55	1,649	160
$^3\Pi_g(0_g^+)$	2.58	1,770	159
$^3\Pi_g$	2.55	1,888	160
$^3\Pi_g(0_g^-)$	2.55	2,531	160
$^3\Delta_g(2_g)$	2.53	2,575	173
$^3\Delta_g(3_g)$	2.55	2,629	169
$^3\Delta_u(3_u)$	2.62	2,679	154
$^3\Delta_u(2_u)$	2.62	2,825	159
$^3\Pi_u(2_u)$	2.51	3,075	175
$^3\Pi_u(1_u)$	2.50	3,356	176
$^1\Pi_g$	2.50	3,536	162
$^3\Delta_g$	2.55	3,718	169
$^1\Pi_g(1_g)$	2.50	3,787	161
$^3\Delta_u$	2.62	3,905	154
$^3\Pi_u$	2.51	3,930	175
$^1\Delta_g$	2.64	3,948	151
$^1\Delta_u$	2.61	3,991	188
$^3\Pi_u(0_u^-)$	2.53	4,052	179
$^3\Gamma_u(4_u)$	2.63	4,212	166
$^3\Gamma_u(5_u)$	2.63	4,346	166
$^3\Pi_u(0_u^+)$	2.51	4,354	175
$^1\Sigma_u^+(0_u^+)$	2.51	4,443	155
$^1\Sigma_u^+$	2.52	4,724	154
$^1\Pi_u$	2.57	4,800	168
$^1\Delta_u(2_u)$	2.61	4,968	188
$^1\Delta_g(2_g)$	2.62	5,075	147
$^3\Delta_u(1_u)$	2.62	5,105	145
$^1\Pi_u(1_u)$	2.56	5,183	169
$^3\Delta_g(1_g)$	2.55	5,286	169
$^1\Delta_g$(II)	2.68	6,037	184
$^3\Pi_u$(II)	2.61	6,127	147
$^3\Sigma_u^+$(II)	2.69	6,202	152
$^3\Gamma_u$	2.63	6,728	166
$^3\Pi_g$(II)	2.69	7,147	175
$^3\Delta_g$(II)	2.69	8,078	178
$^3\Gamma_u(3_u)$	2.64	9,560	166

Source: Reproduced from Reference 159.

levels of calculation, the ground state of Pd_2 is the $^3\Sigma_u^+$ state arising from the $1\sigma_g^2 1\sigma_u 2\sigma_g \delta_g^4 \delta_u^4 \pi_g^4 \pi_u^4$ electronic configuration. The spin–orbit splitting to first order is zero for the $^3\Sigma_u^+$ state arising from the $1\sigma_u 2\sigma_g$ configuration, and thus only a small splitting of 9 cm^{-1} arises from second-order spin–orbit contamination.

The properties of the $^1\Sigma_g^+$ state are of interest, since this state arises predominantly from the $d^{10}(^1S_0)$ state of the Pd atom. A pure single-configuration treatment would produce a repulsive curve for the $^1\Sigma_g^+$ state. At the CASSCF level of theory, the $^1\Sigma_g^+$ state was found to have a bond length of 2.75 Å and a ω_e value of 95 cm^{-1}, when the CASSCF was done starting from $^3\Sigma_u^+$ orbitals. A CASSCF with atomic guess leads to a longer bond and higher energy. The FOCI calculations improve the $^1\Sigma_g^+$ bond a little. The *d*–*d* correlation effects are important for the $^1\Sigma_g^+$ state. As can be seen from Table 1.24, the MRSDCI calculations lead to a longer bond but a slightly deeper well for the $^1\Sigma_g^+$ state.

Although the spin–oribit effects for Pd_2 are significant, they are substantially smaller compared to Pt_2. Most of the electronic states follow Hund's rule with the exception of $^3\Delta_g(2_g)$, $^3\Delta_u(2_u)$, and a few other states. For these two states the $^3\Delta_g(2_g)$–$^1\Delta_g(2_g)$ contamination is so large that it lowers the 2_g component more than the 3_g component relative to $^3\Delta_g$. Thus the $^3\Delta_g(2_g)$ and $^3\Delta_u(2_u)$ states are lower than the $^3\Delta_g(3_g)$ and $^3\Delta_u(3_u)$ states.

The spin–orbit splitting is large for the high-λ states arising from $\delta_g^3\delta_u^3$ configuration. The spin–orbit separation of the $^3\Gamma_u$ state [$^3\Gamma_u(5_u)$–$^3\Gamma_u(3_u)$ splitting] is 5210 cm^{-1}. The $^3\Gamma_u(4_u)$ state is lower than $^3\Gamma_u(5_u)$ mainly because of the $^3\Gamma_u(4_u)$–$^1\Gamma_u(4_u)$ mixing in the 4_u component of $^3\Gamma_u$.

The ordering of electronic states predicted by the CASSCF/MRSDCI/RCI method differs from the work of Shim and Gingerich (161), who used a nonrelativistic SCF/valence CI treatment. The bond lengths predicted by the SCF/valence CI calculations are longer compared to the author's results, while the ω_e and D_e values are lower than the values obtained from the author's calculations (159). Most of the difference arises from the electron correlation effects.

The calculated MRSDCI D_e value, including Davidson's correction for higher-order unlinked cluster configurations, is 2.05 eV for the $^3\Sigma_u^+$ state with respect to the ^3D atoms. The spin–orbit effects lower the D_e by 0.19 eV, leading to a dissociation energy of 1.86 eV for the $^3\Sigma_u^+(1_u, 0_u^-)$ state, with respect to the separated $^3D_3$ atoms. In another theoretical study on PdH (162), it was found that the CASSCF/FOCI/RCI calculations yield a $^1S_0$–$^3D_3$ separation of 4826 cm^{-1} compared to an experimental value of 6564 cm^{-1} (113). The $^1S_0$ state is the lowest state at the CASSCF/FOCI level of theory. The dissociation energy of the $^1\Sigma_g^+$ state with respect to the separated d^{10} atoms is only 0.5 eV, at the MRSDCI/RCI level, since the calculated ^1S–^3D splitting is 4826 cm^{-1}. The accuracy of the computed D_e should be about 70% of the experimental value, and thus a D_0° value of 2.42 eV was recommended by the author (159) for the $^3\Sigma_u^+$ ground state with respect to the separated ^3D atoms.

The $^1\Sigma_g^+$–$^1\Sigma_u^+$ transition of Pd_2 should be especially of interest, and it is observable in the 3186-cm^{-1} region. Most of the allowed transitions of Pd_2 have adiabatic transition energies less than 8000 cm^{-1}, primarily because the author's investigation (159) was restricted to these states. There are a large number of electronic transitions for Pd_2. Thus accurate resolution of the spectral bands may not be feasible.

Subsequent to the author's theoretical work on Pd_2 (159), Lineberger and co-workers (163) observed the negative photodetachment spectra of Pd_2^-, which led to the characterization of the excited electronic states of the neutral Pd_2. Overall, the density of the electronic states, computed by the author, was confirmed by the work of Lineberger and co-workers. The computed $^3\Sigma_u^+(1_u)$–$^3\Sigma_u^+(0_u^-)$ energy splitting of 4443 cm^{-1} compared quite favorably with a value of 4008 cm^{-1} reported by Lineberger and co-workers (163) for the same splitting. The computed trends for the equilibrium bond lengths were also found to be in good agreement with the results deduced from the spectra.

As shown by the author (159), the electronic states of Pd_2 are quite mixed in character in that leading configurations in the MRSDCI calculations of the electronic states of Pd_2 without the spin–orbit term have weights below 70%. The exceptions to this are the $^3\Sigma_u^+$, $^1\Sigma_g^+$, $^3\Gamma_u$, and $^3\Sigma_u^+$(II) states. All other states are quite mixed in character. The $^1\Delta_g$, $^1\Delta_u$, and $^1\Sigma_u^+$ states are most mixed, since the weights of the leading configurations are about 47–53% for these states.

The spin–orbit contaminations in the RCI wavefunctions seem to be significant, especially in some of the low-Ω states. The high-Ω states, such as $^3\Delta_u(3_u)$, $^3\Delta_g(3_g)$, and $^3\Gamma_u(5_u)$, were found to be relatively pure. The low-Ω states, such as $^3\Pi_g(1_g)$, $^3\Pi_g(0_g^+)$, $^3\Delta_u(2_u)$, $^3\Pi_u(1_u)$, $^1\Pi_g(1_g)$, $^1\Pi_u(1_u)$, and $^3\Pi_u(0_u^+)$, are spin–orbit-contaminated. For example, the $^3\Pi_g(1_g)$ state is actually made up of 63% $^3\Pi_g(2\sigma_g\pi_g)$, 10% $^1\Pi_g(2\sigma_g\pi_g)$, 14% $^3\Pi_g(1\sigma_u\pi_u)$, and 4% $^1\Pi_g(1\sigma_u\pi_u)$. The $^3\Pi_g(0_g^+)$ state is 53% $^3\Pi_g(2\sigma_g\pi_g)$, 18% $^3\Pi_g(1\sigma_u 1\pi_u)$, 10% $^1\Sigma_g^+(1\sigma_g^2 2\sigma_g^2)$, 4% $^1\Sigma_g^+(1\sigma_g 2\sigma_g)$, and 3% $^1\Sigma_g^+(1\sigma_g^2\sigma_g^2)$. Thus the mixing of $^1\Sigma_g^+$ with $^3\Pi_g(0_g^+)$ is quite significant. The singlet–triplet mixing in the $^3\Delta(2_u)$ state was also significant. The $^3\Delta_u(2_u)$ state is 37% $^3\Delta_u(2\sigma_g\delta_u)$, 23% $^1\Delta_u(2\sigma_g\delta_g)$, 23% $^3\Delta_u(1\sigma_u\delta_g)$, and 20% $^1\Delta_u(1\sigma_u\delta_g)$. The $^3\Pi_u(1_u)$ state is made up of 44% $^3\Pi_u(1\sigma_u\pi_g)$, 16% $^3\Pi_u(2\sigma_g\pi_u)$, 13% $^1\Pi_u(1\sigma_u\pi_g)$, 9% $^3\Pi_u(1\sigma_g\pi_u)$, 4% $^1\Pi_u(1\sigma_g\pi_u)$, and 4% $^1\Pi_u(2\sigma_g\pi_u)$. Thus many of the electronic states of Pd_2 are quite contaminated.

The Mulliken population analysis of the electronic states of Pd_2 made by the author reveals that the $^3\Sigma_u^+$, $^3\Pi_g$, $^3\Gamma_u$, $^3\Delta_u$, $^1\Pi_g$, $^1\Delta_u$, and $^1\Sigma_u^+$ states have d populations between 9.1 and 9.36. The mixing of d^9s^1 and d^{10} configurations seems to be evident in these states, although these states arise predominantly from the d^9s^1 configuration of the atom. The $^3\Pi_u$, $^1\Pi_u$, and $^1\Delta_g$ states of Pd_2 were found to be a bit purer, although the gross d populations in the $^3\Pi_u$ and $^1\Pi_u$ states are smaller than 9.0. For all the states of Pd_2, the p population is somewhat low (0.04–0.09).

It would be interesting to compare Pd_2 and Pt_2 from the standpoint of relativistic effects. In the previous section, the properties of Pt_2 were discussed.

There are many differences between the electronic states of Pt_2 and Pd_2. The Pt–Pt bond lengths are much shorter compared with the Pd–Pd bond lengths in various states. For example, the $^3\Pi_u$ state has a bond length of 2.4 Å for Pt_2, while the corresponding value for Pd_2 is 2.51 Å. Similarly, the $^3\Gamma_u$ state has a bond length of 2.57 Å for Pt_2, while it is 2.64 Å for Pd_2. The shorter bond lengths in Pt_2 compared to Pd_2 are attributed to relativistic mass–velocity contraction of the 6s orbital of Pt, which leads to a shorter Pt–Pt bond. The $^1\Sigma_g^+$ state is much more strongly bound for Pt_2 (R_e = 2.45 Å), and arises from the π_g^2 configuration, while for Pd_2 the R_e value of the $^1\Sigma_g^+$ state is 2.87 Å and arises predominantly from the $1\sigma_g^2 1\sigma_u^2$ configuration. This difference is due primarily to the fact that the ground state of Pt is $(d^9s^1)^3D_3$, while the ground state of Pd is $(d^{10})^1S_0$.

The differences in the ground states of the two metal atoms combined with the large spin–orbit splittings in Pt lead to substantial differences in the electronic states for Pd_2 and Pt_2. Thus the electronic and spectroscopic properties of Pt_2 have little similarity to the corresponding states of Pd_2. For Pt_2, the open-shell configurations arising from high λ values are stabilized, while for Pd_2, the low-λ configurations are stabilized. The ground state of Pt_2 is $^3\Sigma_g^-(0_g^+)$ arising from δ_u^2; a very low-lying $^3\Gamma_u(5_u)$ state, arising from $\delta_g^3\delta_u^3$, was found only 614 cm^{-1} above the ground state. Similarly, the $\delta_u^3\pi_g^3$ states are lower in energy for Pt_2. The electronic states of Pt_2 are, in general, more strongly bound with respect to the separated ground-state atoms.

The spin–orbit splittings of the electronic states of Pt_2 are much larger. For example, the $^3\Gamma_u(5_u)$–$^3\Gamma_u(3_u)$ splitting in Pt_2 is 15,192 cm^{-1}, while the corresponding splitting in Pd_2 is only 5214 cm^{-1}. Thus the spin–orbit contaminations and splittings are an order of magnitude larger in Pt_2 compared to Pd_2.

1.3.3 Pt_4 and Pd_4

Experimental spectroscopic studies on the platinum and palladium clusters are restricted mainly to the dimers. However, Lineberger and co-workers (163) have studied the trimers of these species. There are a few studies on small platinum and palladium clusters (152, 159, 160, 164–168). Salahub and co-workers (164) considered small clusters of Pd and Rh in the context of modeling CO chemisorption on the Pd and Rh surfaces using the density function theory. Rubio et al. (165) considered Pt_3 and Pt_4 using a one-electron pseudoatom model for Pt. Dai and the author (168) studied the electronic states of these clusters with several geometries, using high-level *ab initio* CASSCF/MRSDCI/RCI schemes, including relativistic effects.

Four possible geometries for Pd_4 and Pt_4 were considered by Dai and the author (168). The tetrahedral and linear structures were considered in the C_{2v} group, while the rhombus and square structures were studied in the D_{2h} group. Relativistic effective core potentials, which retained the outer $4d^{10}$ shell of the palladium atoms, and $5d^96s^1$ shells of the platinum atoms in the valence space were employed. Dai and the author (168) started with the $(3s3p4d)$ basis sets for the two atoms, which were contracted to the ($3s2p3d$) basis sets. The first two (large exponent) p and d functions were contracted.

The multiconfiguration self-consistent-field (MCSCF) calculations were carried out using the complete active space MCSCF (CASMCSCF) method. Dai and the author (168) considered a restricted active space MCSCF method based on several preliminary sets of calculations with different choices of active orbitals. The active space comprised eight a_1, five b_2, five b_1, and four a_2 orbitals for all structures, except the linear structure (in C_{2v}), for which the active space consisted of seven a_1, seven b_2, four b_1, and four a_2 orbitals. All 40 electrons were distributed in all possible ways among these 22 orbitals. Dai and the author computed the properties of 44 electronic states of Pd_4 and 51 electronic states of Pt_4 at the CASSCF level. Among these electronic states, 11 low-lying states of Pd_4 and 10 low-lying states of Pt_4 were fully optimized and considered at higher levels of theory.

The MRSDCI calculations were carried out for some low-lying states to include the effects of higher-order electron correlation. All configurations in the CASSCF with coefficients $\geq$0.15 were included as reference configurations in the MRSDCI. All possible single and double excitations were allowed from these reference configurations. The effect of unlinked quadruple clusters for the MRSDCI was estimated through the multireference Davidson correction, referred to as the MRSDCI+Q results. The CASSCF calculations included up to 6678 configuration spin functions (CSFs), while the MRSDCI calculations included up to 4.1 million CSFs.

1.3.3.1 Geometries and Energies of Electronic States of Pd_4 Table 1.25 shows the geometries and energy separations of 44 electronic states of the palladium tetramer, which exhibit tetrahedral, rhombus, square, or linear geometries (168). Among these states, as seen from Table 1.25, nine electronic states are within 0.07 eV of the $^1A_1$ tetrahedral state of Pd_4, at the CASSCF level. The Pd–Pd side bond lengths are near 2.83–2.86 Å for most of these electronic states. The square geometries are generally higher compared to the rhombus structures. For example, the $^1A_{1g}$ state, with a square geometry and a Pd–Pd side bond length of 2.768 Å, is 0.43 eV higher in energy, compared to the rhombus structure.

As seen from Table 1.25, the ^1A_g, $^3B_{2u}$, $^1B_{2g}$, ^1A_u, $^3B_{2g}$ and ^3A_u states of the rhombus geometry are quite close in energy. The high-spin states such as $^5B_{3g}$, $^5B_{3u}$, and so on are considerably higher in energy. Note that the ground state of the palladium atom is the ^1S state arising from the $4d^{10}$ configuration. Since this is a complete shell, significant bonding is not feasible from a pure $4d^{10}$ state. The promotion of an electron to $5s$ results in a ^3D state arising from $4d^95s^1$. The low-lying electronic states for Pd_4 thus arise from a combination of the palladium $4d^95s^1$ and $4d^{10}$ configurations. The $5p$ orbitals also make appreciable contribution to the bonding near equilibrium geometries.

Table 1.26 shows the MRSDCI results, while Figure 1.22 shows the optimized MRSDCI geometries. The Pd–Pd bond lengths change considerably in

Table 1.25 Geometries and Energy Separations of 44 Electronic States of Pd_4 at the CASSCF Level

Geometry	State	R_1, Å[a]	θ, deg	E, eV	Geometry	State	R_1, Å[a]	R_2, Å[a]	θ, deg	E, eV
Tetrahedron	$^1A_1$	2.832	109.47	0.000	Square	^3E_u	2.772		90.0	0.46
	$^3T_1$	2.836	109.47	0.008		$^1B_{2g}$	2.761		90.0	0.47
Rhombus	^1A_g	2.839	60.5	0.02		^1E_u	2.773		90.0	0.49
	$^3B_{2u}$	2.855	60.4	0.04		$^1A_{2u}$	2.782		90.0	0.49
Tetrahedron	$^1A_2$	2.854	109.47	0.05		$^3B_{2g}$	2.754		90.0	0.50
Rhombus	$^1B_{2g}$	2.864	60.2	0.06		$^1B_{1u}$	2.778		90.0	0.50
	^1A_u	2.865	60.2	0.06		^1E_g	2.778		90.0	0.50
	$^3B_{2g}$	2.861	60.0	0.07		$^3A_{2u}$	2.778		90.0	0.51
	^3A_u	2.862	60.1	0.07		$^3B_{1u}$	2.773		90.0	0.52
	^3A_g	2.856	59.7	0.13		^3E_g	2.773		90.0	0.52
	$^1B_{3u}$	2.870	61.2	0.14		$^3A_{1g}$	2.770		90.0	0.55
	$^1B_{2u}$	2.871	61.2	0.15	Rhombus	$^5B_{3g}$	2.963		55.8	0.62
	$^1B_{1g}$	2.877	61.6	0.15		$^5B_{3u}$	2.963		55.5	0.68
	$^3B_{3u}$	2.868	60.5	0.15		$^5B_{2g}$	2.964		55.5	0.68
	$^1B_{1u}$	2.861	61.1	0.15		$^5B_{2u}$	2.943		56.6	0 68
	$^3B_{1u}$	2.853	60.8	0.15		^5A_g	2.949		56.3	0.69
Tetrahedron	$^3A_1$	2.854	109.47	0.16		$^5B_{1g}$	2.971		55.5	0.70
Rhombus	$^1B_{3g}$	2.860	61.5	0.17		^5A_u	2.971		55.8	0.76
Tetrahedron	$^1T_1$	2.863	109.47	0.19		$^5B_{1u}$	2.976		55.2	0.77
Rhombus	$^3B_{3g}$	2.855	61.7	0.20	Linear	$^1\Sigma_g^+$	3.038	2.737	180.0	1.11
	$^3B_{1g}$	2.868	61.2	0.21	Tetrahedron	$^5T_1$	2.913		109.47	2.10
Square	$^1A_{1g}$	2.768	90.0	0.45		$^5A_1$	2.946		109.47	2.24

[a] R_1 stands for four equal sides of rhombus, square, and tetrahedron; for the linear geometry, R_1 represents the central bond length, while R_2 represents two equal terminal bond lengths.

Source: Reproduced from Reference 168.

Table 1.26 Geometries and Energy Separations of 11 Electronic States of Pd_4 at the MRSDCI and MRSDCI+Q Levels

Geometry	State	R_e, Å[a]	θ, deg	E, eV[b]
Tetrahedron	$^3T_1$	2.686	109.47	0.000 (0.00)
	$^1A_1$	2.696	109.47	0.017 (0.05)
	$^1A_2$	2.722	109.47	0.16 (0.25)
Rhombus	^1A_g	2.665	60.9	0.65 (0.77)
	$^3B_{2u}$	2.712	62.6	0.68 (0.88)
	^3A_u	2.700	62.4	0.73 (0.92)
	$^1B_{2g}$	2.697	62.4	0.74 (0.93)
	^1A_u	2.700	62.4	0.74 (0.94)
	$^3B_{2g}$	2.647	62.4	0.75 (0.91)
Square	$^1A_{1g}$	2.625	90.0	1.05 (1.23)
	^3E_u	2.639	90.0	1.13 (1.30)

[a]For the definition of R_1, see Table 1.25.

[b]The values in parentheses are the Davidson corrected energies.

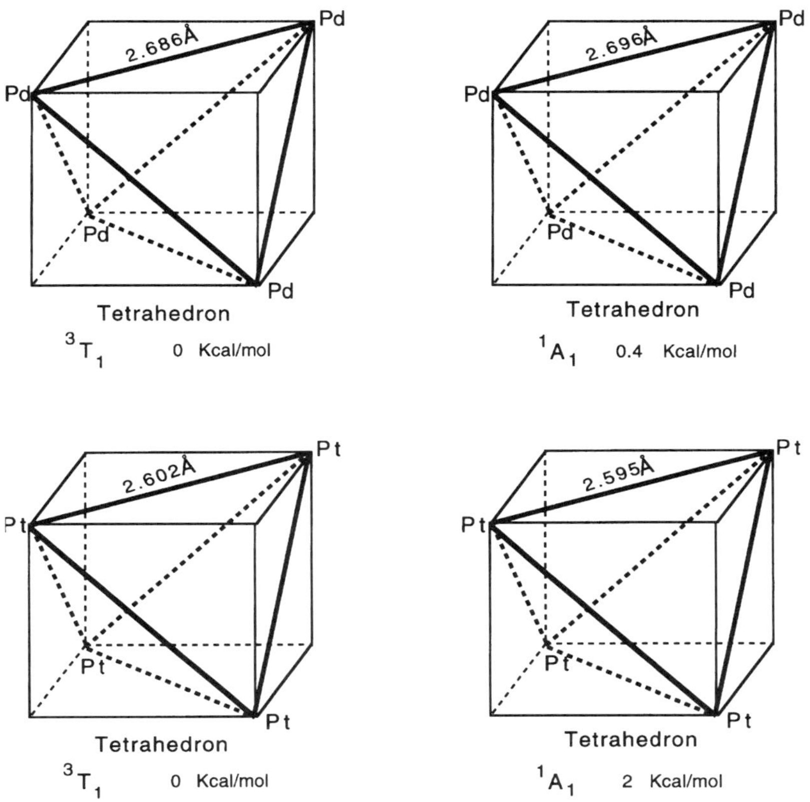

Figure 1.22 MRSDCI geometries of Pd_4 and Pt_4. (From Ref. 168.)

going from the CASSCF to a more accurate MRSDCI. For example, the Pd–Pd bond length of the $^1A_1$ tetrahedral state changes from 2.83 to 2.69 Å, at the MRSDCI level. The ordering of the electronic states changes, too, at the MRSDCI level. Notably, the $^1A_1$ state is slightly lower at the CASSCF level, while the $^3T_1$ state with the tetrahedral geometry becomes lower at the MRSDCI level. The energy separation between the $^1A_1$ tetrahedral structure and the $^1A_1$ rhombus structure changes from 0.02 to 0.63 eV. Higher-order electron correlation effects stabilize the tetrahedral geometry of Pd_4 compared to the rhombus structure. The Pd–Pd side bond lengths of the rhombus structure change from 2.839 to 2.665 Å. The Pd–Pd–Pd acute bond angle does not change much between the CASSCF and MRSDCI levels. The unlinked quadruple corrections tend to stabilize the tetrahedral structure compared to other geometries of Pd_4, as seen from Table 1.26. This is reflected in higher MRSDCI+Q energy separations of the excited states compared to the MRSDCI level.

The dissociation energy to dissociate Pd_4 into two Pd_2 dimers at the MRSDCI and MRSDCI+Q levels was computed by Dai and the author (168) as 77 and 89.7 kcal/mol, respectively. The D_e of the $^3\Sigma_u^+$ state of Pd_2 relative to two Pd(^3D) atoms was computed as 48 kcal/mol at the MRSDCI level (159) using the theoretical ^3D–^1S energy separation of 14 kcal/mol. The total binding energy of Pd_4 relative to four Pd($^1S_0$) atoms was thus computed as 117 and 129.7 kcal/mol at the MRSDCI and MRSDCI+Q levels, respectively, with respect to four Pd($^1S_0$) atoms.

Goursot et al. (164) considered the ground state of Pd_4 with the DFT method. Their DFT results on Pd_2 agreed with the previously reported MRSDCI results on Pd_2 (159). Goursot et al. obtained the $^3\Sigma_u^+$ as the ground state with r_e of 2.46 Å, which compares quite well with Balasubramanian's (159) $r_e = 2.48$ Å for Pd_2 obtained using the MRSDCI technique. Goursot et al. (164) obtained the ground state of Pd_4 to be the open-shell $^3T_1$ state with a Pd–Pd bond length of 2.75 Å. This agrees very well with the $^3T_1$ ground state computed by Dai and the author (168) with a Pd–Pd bond length of 2.69 Å. Goursot et al. have computed binding energy per atom for Pd_4 at the DFT level as 1.58 eV or a total binding energy of 6.3 eV for Pd_4 compared to the MRSDCI and MRSDCI+Q results of 5.07 and 5.63 eV obtained by Dai and the author (168).

1.3.3.2 Geometries and Energies of Electronic States of Pt_4 Table 1.27 shows the results of Pt_4 with rhombus geometries at the CASSCF level as obtained by Dai and the author (168). At the CASSCF level, the $^5B_{3g}$ state with a rhombus geometry is lower, but there are five different states of rhombus geometry within 0.09 eV of each other. Table 1.28 shows the geometries and energy separations of the electronic states of Pt_4 with tetrahedral and linear geometries at the CASSCF level. At the CASSCF level, the tetrahedral states are higher than the $^5B_{3g}$ state with the rhombus equilibrium geometry. The $^1\Sigma_g^+$ state with a linear geometry is also higher in energy.

The MRSDCI results for Pt_4 in Table 1.29 reveal that the Pt–Pt side bond lengths of the $^3T_1$ (tetrahedral geometry) state change from 2.64 Å at the

Table 1.27 Energy Separations of 42 Electronic States of Pt_4 with the Rhombus and Square Geometries at the CASSCF Level

Rhombus State	R_1, Å[a]	θ, deg	E, eV	Square State	R_1, Å[a]	E, eV
$^5B_{3g}$	2.690	61.2	0.00	$^1A_{1g}$	2.529	0.25
$^3B_{2u}$	2.686	61.3	0.04	^3E_u	2.623	0.91
$^3B_{3g}$	2.690	61.2	0.09	$^5B_{2g}$	2.618	0.93
^1A_g	2.696	56.1	0.09	$^3B_{2g}$	2.550	0.95
$^3B_{1u}$	2.701	56.1	0.09	$^1B_{2g}$	2.555	0.98
$^5B_{2g}$	2.682	61.2	0.12	$^3A_{2u}$	2.540	1.00
$^1B_{3g}$	2.715	61.1	0.13	^5E_g	2.613	1.03
$^5B_{3u}$	2.685	61.1	0.13	$^5B_{1u}$	2.614	1.04
$^3B_{2g}$	2.703	58.9	0.15	^3E_g	2.618	1.07
$^3B_{3u}$	2.704	58.9	0.16	$^5A_{2u}$	2.609	1.09
^5A_u	2.672	61.8	0.18	$^3B_{1u}$	2.618	1.11
$^5B_{1g}$	2.674	61.8	0.18	$^5A_{1g}$	2.600	1.14
^3A_u	2.677	61.8	0.22	^5E_u	2.600	1.17
$^1B_{2g}$	2.706	59.0	0.23	^1E_g	2.630	1.17
$^1B_{3u}$	2.707	59.0	0.23	$^1A_{2u}$	2.630	1.18
^1A_u	2.700	56.7	0.24	$^3A_{1g}$	2.578	1.21
$^1B_{1g}$	2.702	56.6	0.25	$^1B_{1u}$	2.625	1.21
$^3B_{1g}$	2.679	61.8	0.25	^1E_u	2.600	1.24
$^5B_{2u}$	2.680	60.2	0.29	—	—	—
^5A_g	2.663	63.9	0.31	—	—	—
$^5B_{1u}$	2.698	60.4	0.32	—	—	—
^3A_g	2.691	59.9	0.34	—	—	—
$^1B_{2u}$	2.718	60.0	0.36	—	—	—
$^1B_{1u}$	2.719	59.6	0.37	—	—	—

[a]See Table 1.25.

CASSCF level to 2.60 Å at the MRSDCI level. Likewise, the $^1A_1(T_d)$ state changes from 2.642 to 2.595 Å. The rhombus Pt–Pt side bond lengths change from 2.696 to 2.632 Å at the MRSDCI level. While electron correlation effects have less influence on the geometries of the electronic states of Pt_4, the energy separations of the electronic states change substantially at the MRSDCI level, in that the low-spin electronic states are more stabilized at the higher MRSDCI level of theory. This suggests that higher-order correlation effects are more important for energy separations. The dissociation energy for the process

$$Pt_4(^3T_1) \rightarrow 2\,Pt_2(^3\Sigma_g^-)$$

was computed by Dai and the author as 4.49 and 5.5 eV at the MRSDCI and MRSDCI+Q levels of theory, respectively. Since both Pt_4 and Pt_2 are in triplet electronic states, the spin–orbit contributions would be comparable for both clusters. The D_e for the platinum dimer was computed as 2.31 eV at the FOCI level, which lowered to 1.84 eV, when spin–orbit effects were included. As

Table 1.28 Energy Separations of 9 Electronic States of Pt_3 with the Tetrahedral and Linear Geometries at the CASSCF Level

Geometry	State	R_1, Å[a]	R_2, Å[a]	E, eV
Tetrahedron	$^3T_1$	2.640	—	0.05
	$^1A_1$	2.642	—	0.11
	$^1A_2$	2.648	—	0.19
	$^1T_1$	2.655	—	0.48
	$^5A_1$	2.620	—	0.54
	$^3A_1$	2.677	—	0.62
	$^5A_2$	2.687	—	0.94
	$^5T_2$	2.680	—	1.22
Linear	$^1\Sigma_g^+$	2.832	2.600	1.07

[a]See Table 1.25.

discussed before, Taylor et al. (153) corrected the thermodynamic D_e of Pt_2 using the spectroscopic properties of low-lying electronic states of Pt_2 computed by Balasubramanian (152) before. The corrected experimental D_0 value for Pt_2 (3.15 eV (152)) was used by Dai and the author (168) to compute the atomization energy of Pt_4 as 11.8 eV. This value is much higher than the corresponding value for Pd_4, suggesting that relativistic effects stabilize the Pt–Pt bonds compared to Pd–Pd bonds.

Rubio et al. (165) studied Pt_4 in a model that treats the Pt atom as a one-electron pseudoatom. In this approach, $1s$–$5d$ shells are incorporated into the core. Rubio et al. (165) obtained the rhombus structure with a ^1A_g state to be the lowest with a Pt–Pt side bond length of 2.839 Å, and the shorter diagonal bond length of 2.73 Å, compared with the values of 2.63 and 2.44 Å, respectively, obtained by Dai and the author (168). At the MRSDCI level, the rhombus structure is higher than the tetrahedral structure, while Rubio et al. (165) find the opposite trend and a dif-

Table 1.29 Geometries and Energy Separations of 10 Electronic States of Pt_4 at the MRSDCI and MRSDCI+Q Levels

Geometry	State	R_e, Å[a]	θ, deg	E, eV
Tetrahedron	$^3T_1$	2.602	109.47	0.00(0.00)
	$^1A_1$	2.595	109.47	0.08(0.13)
	$^1A_2$	2.602	109.47	0.22(0.38)
Rhombus	^1A_g	2.632	55.1	0.47
	$^3B_{1u}$	2.632	55.1	0.47
	$^5B_{3g}$	2.603	62.3	0.47
	$^3B_{2u}$	2.623	59.0	0.60
	$^3B_{3g}$	2.633	62.4	0.62
Square	$^1A_{1g}$	2.472	90.0	0.64
Rhombus	$^1B_{3g}$	2.638	61.4	0.67

[a]See Table 1.25.

ferent ($^3A_2$) state for the tetrahedral structure. The binding energy obtained by Rubio et al. is significantly lower (~96.9 kcal/mol for the rhombus structure). These differences could be attributed to the treatment of Pt as a one-electron pseudoatom (165) and the fact that the participation of $5d$ orbitals in both bonding and electron correlation effects are quite significant for Pt_4.

1.3.3.3 Comparison of Pt_4 and Pd_4

Both $^3T_1$ and $^1A_1$ states of Pt_4 and Pd_4 have $1a_1^2 1t_2^6 1e^4 2t_2^6 2a_1^2 3t_2^6 2e^4 1t_1^6 2t_1^4$ as the leading configuration, in accord with the DFT results of Goursot et al. (164), who find that the ground state of Pd_4 to have a $1a_1^2 1t_2^6 1e^4 2t_2^6 2a_1^2 3t_2^6 2e^4 1t_1^6 2t_1^4$ configuration. However, the $^1A_1$ state exhibits multiconfigurational character. The rhombus structures of both Pt_4 and Pd_4 also exhibit multiconfigurational character. The other electronic states are considerably more complex, requiring at least multiconfigurational representations for proper description.

The Mulliken populations of the low-lying electronic states of Pd_4 and Pt_4 obtained by Dai and the author (168) are listed in Tables 1.30 and 1.31, respectively. In the $^3T_1$ states, Pd_4 and Pt_4 exhibit the populations of $4d^{9.38}5s^{0.52}5p^{0.11}$ and $5d^{9.25}6s^{0.63}6p^{0.12}$, respectively. Goursot et al. (164) computed the Pd population to be $4d^{9.34}5s^{0.54}5p^{0.12}$ for the $^3T_1$ ground state of Pd_4, in agreement with the computed results of Dai and the author. While the Mulliken populations of the tetrahedral electronic states of Pt_4 and Pd_4 are similar, large differences in the rhombus, linear, and square structures are found. In general, the Pt($5d$) population is considerably smaller than Pd($4d$) in these electronic states (see Tables 1.30 and 1.31). For example, the ^1A_g state with a rhombus geometry exhibits Pt ($5d^{8.81}6s^{1.18}6p^{0.09}$) compared to Pd ($4d^{9.82}5s^{0.10}5p^{0.04}$) for one of the atoms. The other (M_2) atom exhibits Pt ($5d^{8.88}6s^{0.81}6p^{0.23}$) and Pd ($4d^{9.07}5s^{0.84}5p^{0.11}$) populations. The sum of the d populations of the two atoms (M_1 and M_2) for Pd_4 and Pt_4 are 18.89 and 17.69, respectively. The corresponding s populations are 0.94 and 1.99, respectively. The difference is consistent with Pt($5d^9 6s^1$) and Pd($4d^{10}$) ground states. That is, the Pt($6s$) population is enhanced in the rhombus structures of Pt_4, while the Pd($4d$) population is enhanced in the case of Pd_4, due to the relativistic mass–velocity effect, which stabilizes the $6s$ orbital of Pt, leading to enhanced total metal s population in the case of Pt_4 compared to Pd_4.

The electronic states of both Pt_4 and Pd_4 exhibit enhanced metal (p) populations. For Pt_4 electronic states, the Pt($6p$) populations are near 0.2 for rhombus structures, and 0.1 for the tetrahedral structures. Consequently, participation of the $6p$ orbitals is important, especially for the rhombus structures.

1.4 ELECTRONIC STRUCTURE OF Rh_3 AND Ir_3

1.4.1 Rh_3

There are very few theoretical and experimental studies on Rh_3 (169–171). The related Rh_2 has also been studied (172–176). Van Zee et al. (171) obtained the

Table 1.30 Mulliken Population Analysis of Some Low-Lying Electronic States of Pd_4[a]

		Gross Population							
Geometry	State	M_1	$M_1(s)$	$M_1(p)$	$M_1(d)$	M_2	$M_2(s)$	$M_2(p)$	$M_2(d)$
Tetrahedron	$^3T_1$	10.000	0.515	0.107	9.378	—	—	—	—
	$^1A_1$	10.000	0.513	0.107	9.380	—	—	—	—
	$^1A_2$	10.000	0.514	0.107	9.379	—	—	—	—
Rhombus	^1A_g	9.969	0.102	0.044	9.823	10.031	0.842	0.117	9.072
	$^3B_{2u}$	10.021	0.103	0.048	9.870	9.979	0.906	0.130	8.943
	^3A_u	10.006	0.096	0.045	9.865	9.994	0.902	0.125	8.967
	$^1B_{2g}$	10.004	0.095	0.044	9.865	9.996	1.049	0.124	8.833
	^1A_u	10.005	0.093	0.044	9.868	9.995	0.909	0.125	8.961
	$^3B_{2g}$	9.998	0.105	0.043	9.850	10.002	0.891	0.127	8.984
Square	$^1A_{1g}$	10.000	0.448	0.086	9.466	—	—	—	—
	^3E_u	10.000	0.510	0.090	9.400	—	—	—	—
Linear	$^1\Sigma_g^+$	10.274	1.134	0.165	8.975	9.726	0.637	0.045	9.044

[a]See Table 1.25. M_1 stands for one of the four atoms for the tetrahedron and square geometries. For the rhombus structure, M_1 stands for one of the two atoms along the z axis, while M_2 stands for one of the two atoms along the y axis. For the linear geometries, M_1 is one of the nonterminal atoms and M_2 is one of the terminal atoms.

Table 1.31 Mulliken Population Analysis of Some Low-Lying Electronic States of Pd_4[a]

		Gross Population							
Geometry	State	M_1	$M_1(s)$	$M_1(p)$	$M_1(d)$	M_2	$M_2(s)$	$M_2(p)$	$M_2(d)$
Tetrahedron	$^3T_1$	10.000	0.627	0.121	9.252	—	—	—	—
	$^1A_1$	10.000	0.626	0.086	9.288	—	—	—	—
	$^1A_2$	10.000	0.615	0.089	9.296	—	—	—	—
Rhombus	^1A_g	10.075	1.177	0.086	8.812	9.925	0.810	0.233	8.882
	$^3B_{1u}$	10.075	1.167	0.086	8.822	9.925	0.799	0.240	8.886
	$^5B_{3g}$	10.064	1.075	0.087	8.902	9.936	0.688	0.192	9.056
	$^3B_{2u}$	10.043	1.042	0.085	8.916	9.957	0.726	0.205	9.026
	$^3B_{3g}$	10.078	1.085	0.088	8.905	9.922	0.676	0.206	9.040
	$^1B_{3g}$	10.074	1.078	0.088	8.908	9.926	0.686	0.210	9.030
Square	$^1A_{1g}$	10.000	0.773	0.105	9.122	—	—	—	—
Linear	$^1\Sigma_g^+$	10.603	1.462	0.125	9.016	9.397	0.500	0.028	8.869

[a]See Table 1.25. For M_1 and M_2, see Table 1.30.

ESR spectra of Co, Rh, and Ir trimers in Ar/Kr matrices together with their diatomic positive ions. The ESR spectra of Rh_3 reveal that Rh_3 probably has $S = \frac{5}{2}$ spin. Analysis of the hyperfine intensity patterns appears to suggest that the structures in the rare-gas matrices are equilateral triangles with the unpaired electrons, in essentially nonbonding atomic $4d$ orbitals of Rh. Das and Balasubramanian (169) carried out *ab initio* theoretical studies on low-spin electronic states of Rh_3, especially the quartet and doublet states. Among the eight electronic states studied by these authors, $^2A_2$ and $^2A_1$ Jahn–Teller components of $^2E''$ and $^2E'$, respectively, were found to be the lowest. Subsequently, Dai and the author (170) considered a more detailed theoretical investigation of the electronic states of the rhodium trimer, which included high-spin electronic states using a high level of theory.

Dai and the author (170) have employed a complete active space multiconfiguration self-consistent field (CAS-MCSCF) method to generate the orbitals for higher-order MRSDCI calculations. Both CAS-MCSCF and CI calculations were made using the relativistic effective core potentials (RECPs), which retained the outer $4d^8 5s^1$ shells of the three Rh atoms in the valence space. The CASSCF calculations included 27 valence electrons, which were distributed in the active space consisting of six a_1, four b_2, and three each of b_1 and a_2 orbitals in the active space, which were labeled the 6433-CASSCF; for all octet states 7433 orbitals were used in the active space, while for the decet states the complete active space of 7533 orbitals was employed. The CI calculations, which included 27 electrons, were carried out at the multireference SDCI (MRSDCI) level. The MRSDCI calculations included single and double excitations from all configurations in the CASSCF with coefficients ≥0.10, and 3.1 million CSFs.

Table 1.32 shows the equilibrium geometries and energy separations (T_e) of 16 electronic states of Rh_3 (170). As seen from Table 1.32, at the lowest CASSCF level of theory, the four octet states are virtually degenerate lying within 0.5 eV below the sextet electronic states. At the MRSDCI and MRSDCI+Q levels of theory, clearly the sextet electronic states of Rh_3 prevail as seen from Table 1.32.

The final complete geometries of both octet and sextet states of Rh_3 are shown in Figure 1.23, which shows that the $^6A_1$ and $^6A_2$ states have similar geometries with very acute Rh–Rh–Rh apex angles of <51°. The r_e values for the two equal sides of the isosceles triangle are near 2.6 Å, while the base bond lengths are near 2.2 Å for all four sextet electronic states, which suggests enhanced bonding along the base of the isosceles triangle compared to the two equal sides.

The octet electronic states, with very acute apex angles of 48.6–48.7°, have significantly longer side bond lengths, which are consistent with the very angle of 49°. The $^8A_1$ and $^8A_2$ states, on the other hand, have apex angles close to 60°, especially for the $^8A_1$ state ($\theta_e \sim 59.7°$).

The energy separations of the dectet states ($S = \frac{9}{2}$), computed by Dai and the author (170) at the CASSCF level, are 0.5–0.6 eV above the octet electronic

Table 1.32 Geometries and Energy Separations for Electronic States of Rh_3[a]

	CASSCF			CI			
State	R_e, Å	θ_e, deg[b]	E, eV[c]	R_e, Å[a]	θ_e, deg[b]	E_{SDCI}, eV[d]	E_{MRSDCI}, eV[d]
$^6A_1$	2.755	49.5	0.000	2.615	50.8	0.00 (0.00)	0.00 (0.00)
$^6A_2$	2.754	49.5	0.015	2.601	50.7	0.01	0.02
$^6B_2$	2.762	49.2	0.024	2.561	52.0	0.05	0.08
$^6B_1$	2.760	49.2	0.012	2.620	49.3	0.15	0.38
$^8B_1$	2.921	52.8	−0.47	2.822	48.7	−0.31 (−0.19)	0.57 (0.84)
$^8B_2$	2.922	52.7	−0.47	2.851	48.6	−0.13 (−0.03)	—
$^8A_2$	2.926	52.6	−0.47	2.393	57.4	0.92 (0.47)	—
$^8A_1$	2.915	52.9	−0.47	2.317	59.7	1.50 (0.67)	—
$^2A_2$	—	—	—	2.535	61.6	—	2.64
$^2A_1$	—	—	—	2.558	60.7	—	2.67
$^4A_2$[e]	—	—	—	2.646	56.0	—	2.74
$^4B_2$[e]	—	—	—	2.637	56.5	—	2.75
$^4A_1$[e]	—	—	—	2.652	26.7	—	2.77
$^4B_1$[e]	—	—	—	2.644	56.7	—	2.78
$^2B_1$[e]	—	—	—	2.637	55.9	—	2.81
$^2B_2$[e]	—	—	—	2.647	56.2	—	2.84

[a] R_e represents the bond length between the apex atom and one of the two equivalent base atoms.
[b] θ_e stands for the Rh–Rh–Rh apex angle.
[c] The CASSCF of sextet, octet, and dectet states included 6433, 7433, and 7533 orbitals in the active space, respectively, as described in the text.
[d] The values in parentheses are the Davidson corrected energies.
[e] Geometries and energy separating from Reference 169.

Source: Reproduced from Reference 170.

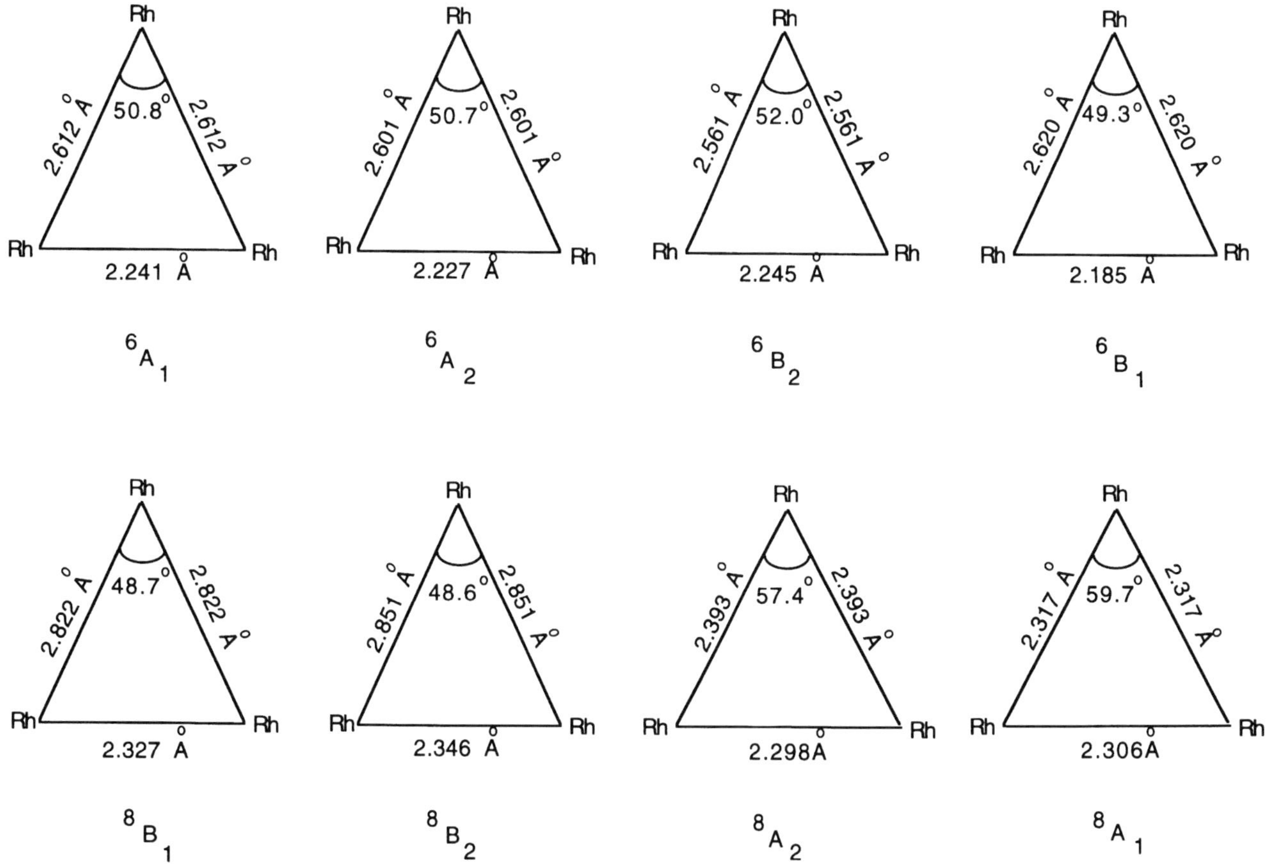

Figure 1.23 The computed geometries of Rh_3. (From Ref. 170.)

states. The general trend is that electron correlation effects tend to stabilize the low-spin states, while the spin exchange energy tends to stabilize the high-spin states. Hence, at the lower level of theory the high-spin states are lower in energy, while at the higher level of theory the low-spin states are stabilized; yet the spin exchange energy is so large that the sextet states prevail at the highest level of theory. Since the dectet states are 0.6 eV higher than the octet states at the CASSCF level, they are eliminated as possible candidates for the ground state of Rh_3.

Comparison of the CASSCF and MRSDCI geometries of the electronic states in Table 1.32 reveals that the apex angles of the sextet states increase slightly, accompanied by decrease in the two equal sides of the isoceles triangle. The $^8B_1$ and $^8B_2$ states exhibit an opposite trend, as seen from Table 1.32. The $^8A_2$ and $^8A_1$ electronic states exhibit geometry changes approaching equilateral triangles at the MRSDCI level. Consequently, the Rh–Rh side bond lengths decrease significantly in going from the CASSCF to the MRSDCI level of theory.

The computed results of Dai and the author (170) at the highest level of theory (MRSDCI+Q) support the ESR prediction by Van Zee et al. (171) of a spin $S = \frac{5}{2}$ state for Rh_3. Their hyperfine intensity pattern seems to support an equilateral triangle for Rh_3, since departure from D_{3h} to C_{2v} would result in a 1 : 1 : 2 : 2 : 1 : 1 or a 1 : 2 : 1 : 1 : 2 : 1 pattern. At the highest level of theory, the Rh–Rh–Rh bond angle computed by Dai and the author (170) at the equilibrium geometry is 50.8°. However, the small $^6A_1$–$^6B_2$ energy separation of 0.02–0.08 eV suggests that if these two states are Jahn–Teller components of the $^6E'(D_{3h})$ state, then the barrier to pseudorotation is rather small. The $^6A_2$–$^6B_1$ energy is also quite small. The error bar in the computations of Dai and the author is too large to resolve the barrier to pseudorotation. If the actual barrier to pseudorotation is insignificant, then Rh_3 would be floppy, in which case the computations would not contradict the observed hyperfine pattern, since one would see an averaged D_{3h} structure. The atomization energy of the lowest $^6A_1$ state for the Rh_3 (169) was computed as 249 kcal/mol.

Although the atomic Rh spin–orbit splitting of 0.15 eV is quite small, it is sufficient to quench the Jahn–Teller distortion. Note that the Jahn–Teller stabilization energy is rather small for Rh_3. If the spin–orbit splitting is larger than the Jahn–Teller stabilization energy, then the system would not undergo Jahn–Teller distortion. Thus it is suggested that spin–orbit coupling can contribute significantly to quenching Jahn–Teller distortion, thereby leading to a D_{3h} structure. This could explain the observed ESR hyperfine pattern, which is consistent with the D_{3h} structure.

The $4d$ and $5s$ valence orbitals of the three rhodium atoms correlate with seven a_1, five b_2, three b_1, and three a_2 orbitals. If one starts with the D_{3h} equilateral triangular structure for Rh_3, 27 valence electrons would lead to $^6E'$, $^6E''$, $^4E'$, $^4E''$, $^2E'$, and $^2E''$ among other possible states. Other electronic states of $A_1'(A_1'')$, $A_2'(A_2'')$ symmetries in D_{3h} are also possible, but these states would not undergo Jahn–Teller distortion, and hence, their equilibrium geometries should be near-equilateral-triangular.

The sextet and two octet states deviate substantially from the ideal D_{3h} structures. This appears to suggest that they are Jahn–Teller components with very small barriers for pseudorotations. The $^6E'$ state correlates with $^6A_1$ + $^6B_2$ in C_{2v}, while the $^6E''$ state becomes $^6A_2$ + $^6B_1$.

Analysis of the composition of the open-shell orbitals supports Van Zee et al.'s (171) observed small *hf* splitting of 2.3 G, which is consistent with the five unpaired electrons being predominantly in the 4*d* orbitals of Rh. Table 1.33 shows the dipole moments of the various electronic states of Rh_3, while Table 1.34 shows the corresponding Mulliken populations. All the sextet states have small positive dipole moments consistent with the excess electronic charge densities on the base atoms for the sextet (Table 1.34), while the octet states have negative dipole moments consistent with the excess electronic densities (Table 1.34) on the apex atoms. From Table 1.34, it can be seen that all the sextet states have smaller 5*s* populations at the apex and base atoms compared to the octet states, which exhibit an opposite trend. The 5*p* populations of the base atoms of the octet states are noticeably large (>0.3), while the 5*p* populations of the sextet states are near 0.13–0.14. The apex atoms exhibit smaller 4*d* populations compared to the base atoms in the sextet states, while in the octet states the populations are nearly the same.

In the $^6A_1$ ground state of Rh_3, the Rh atom has ($4d^{7.92}5s^{0.86}5p^{0.12}$) population at the apex site and ($4d^{8.24}5s^{0.67}5p^{0.14}$) population at the base sites. The enhanced 4*d* populations at the base sites are consistent with the geometry of the $^6A_1$ state in Figure 1.23. All sextet states exhibit multiple bonding at the base of the isosceles triangle.

1.4.2 Ir_3

Dai, Liao, and the author (177) have used techniques similar to the ones employed for Rh_3 to study Ir_3. In addition, the spin–orbit effects of Ir_3 were introduced through the relativistic configuration interaction (RCI) scheme. The

Table 1.33 Dipole Moments for Bent Minimum of Rh_3

State	μ, D
$^6A_1$	0.28
$^6A_2$	0.29
$^6B_2$	0.17
$^6B_1$	0.23
$^8B_1$	−0.43
$^8B_2$	−0.36
$^8A_2$	−0.67
$^8A_1$	−0.60

Source: Reproduced from Reference 170.

Table 1.34 Mulliken Population Analysis for the Sextet and Octet Electronic States of Rh_3

State	Rh^a	Rh^b	$Rh^a(s)$	$Rh^b(s)$	$Rh^a(p)$	$Rh^b(p)$	$Rh^a(d)$	$Rh^b(d)$	Overlap
$^6A_1$	8.911	9.044	0.863	0.670	0.124	0.136	7.924	8.238	0.274
$^6A_2$	8.905	9.048	0.860	0.672	0.125	0.137	7.919	8.239	0.270
$^6B_2$	8.929	9.036	0.928	0.664	0.120	0.133	7.881	8.239	0.224
$^6B_1$	8.944	9.028	0.925	0.648	0.123	0.136	7.895	8.244	0.263
$^8B_1$	9.070	8.965	0.930	0.772	0.195	0.292	7.945	7.901	0.310
$^8B_2$	9.066	8.967	0.927	0.791	0.191	0.300	7.949	7.876	0.321
$^8A_2$	9.061	8.929	1.045	0.765	0.290	0.320	7.726	7.889	0.378
$^8A_1$	9.076	8.962	1.019	0.763	0.302	0.331	7.755	7.868	0.355

[a] Rh^a represents the apex atom.
[b] Rh^b represents one of the two equivalent base atoms.

Source: Reproduced from Reference 170.

RCI calculations were made near the potential well of the $^8A_2$ ground state. These calculations were carried out in the C_{2v}^2 double group. For example, the $^2A_1$ and $^2A_2$ states, which contain leading configurations with one open shell, were occupied by an electron with spin α and mixed with the $^2B_1$ and $^2B_2$ states with an open-shell spin of β. For the $^4A_1$ and $^4A_2$ states there are the spin function combinations in the C_{2v}^2 double group with $\alpha\alpha\beta$, $\alpha\beta\alpha$, $\beta\alpha\alpha$, and $\beta\beta\beta$, while for the $^4B_1$ and $^4B_2$ states the flipped spin combinations of $\beta\beta\alpha$, $\beta\alpha\beta$, $\alpha\beta\beta$, and $\alpha\alpha\alpha$ were included. For each of the sextet states there are 16 combinations of spin functions, while for each of the octet states, the number of combinations of spin functions increases to 64. For example, the $^8A_2$ state exhibits the following open-shell spin function combinations: $1\,\beta\beta\beta\beta\beta\beta\beta, 21\,\alpha\alpha\beta\beta\beta\beta\beta$ + permutations, $35\,\alpha\alpha\alpha\alpha\beta\beta\beta$ + permutation, and $7\,\alpha\alpha\alpha\alpha\alpha\alpha\beta$ + permutations. Thus the RCI calculations had to include all of these open-shell spin combinations that arise from different electronic states, which lead to a total of 343 reference configurations and 28,267 determinants.

Table 1.35a shows the equilibrium geometries and energy separations of the electronic states of Ir_3 obtained by Dai et al. (177), while Figure 1.24 shows the complete equilibrium geometries, including the Ir–Ir base bond lengths for Ir_3. The effects of higher-order electron correlation are significant for all the electronic states of Ir_3, and thus the CASSCF method was used only as a zeroth-

Table 1.35a Geometries and Energy Separations for the Electronic States of Ir_3 at the MRSDCI and MRSDCI+Q Levels

State	r_e, Å[a]	θ_e, deg[b]	MRSDCI	MRSDCI+Q
$^8A_2$	2.632	51.6	0.00	0.00
$^8B_1$	2.608	52.1	0.12	0.10
$^8A_1$	2.665	50.7	0.18	0.40
$^8B_2$	2.566	57.9	0.53	—
$^6B_1$	2.541	54.7	0.28	0.52
$^6A_2$	2.586	56.7	0.41	0.51
$^6B_2$	2.391	66.5	0.44	0.47
$^6A_1$	2.539	54.8	0.53	0.67
$^4B_1$	2.511	55.0	0.36	0.51
$^4A_2$	2.424	64.6	0.48	0.68
$^4A_1$	2.485	56.3	0.60	0.72
$^4B_2$	2.522	60.1	0.66	1.00
$^2B_1$	2.510	55.5	0.37	0.53
$^2A_1$	2.534	57.0	0.50	0.81
$^2B_2$	2.431	62.1	0.51	0.80
$^2A_2$	2.492	57.1	0.54	0.77

[a] r_e stands for the distance between the apex Ir atom and on the two equivalent base atoms.
[b] θ_e represents the Ir–Ir–Ir apex bond angle.

Source: Reproduced from Reference 177.

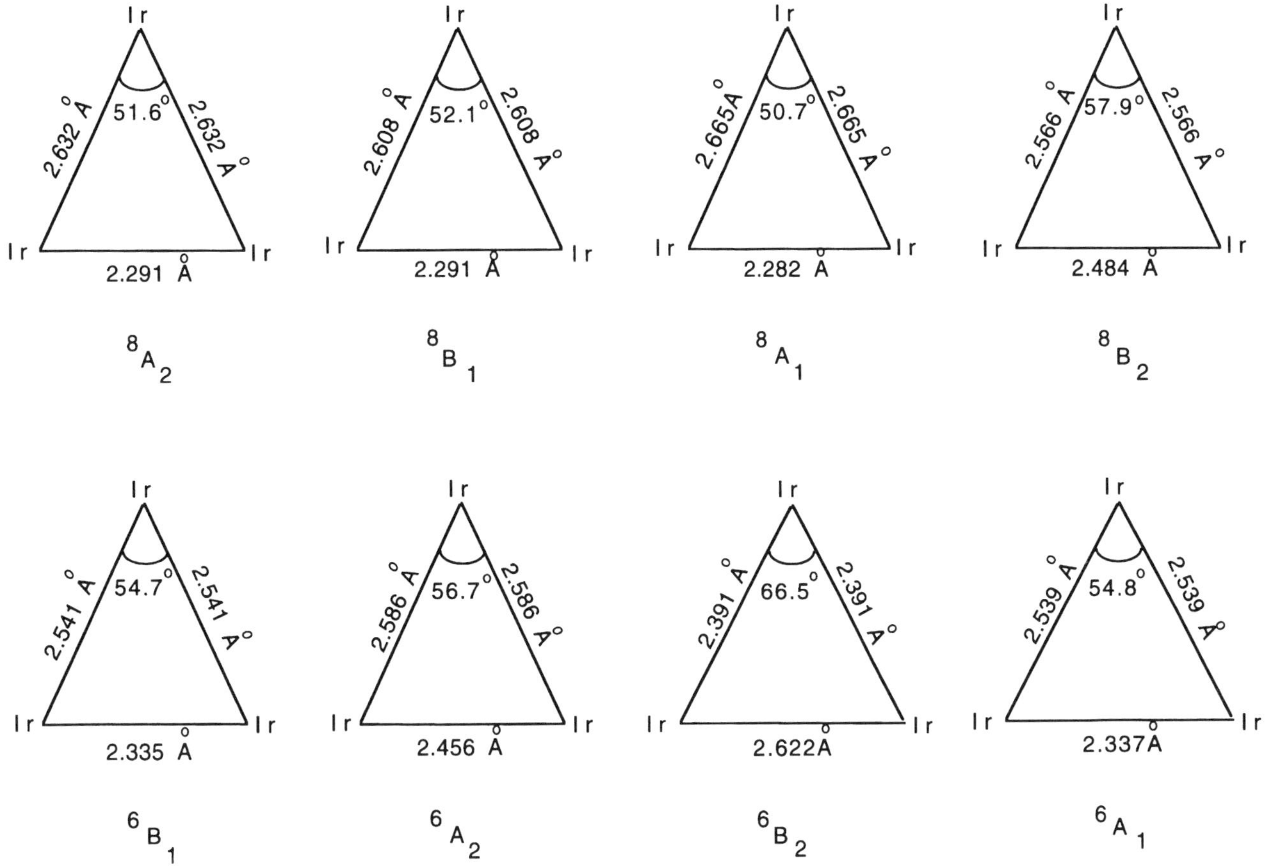

Figure 1.24 The computed geometries of Ir_3. (From Ref. 177.)

order approximation for the more accurate MRSDCI computations. The energy separations were found to be especially sensitive to higher-order electron order approximation for the more accurate MRSDCI computations. The energy separations were found to be especially sensitive to higher-order electron correlation effects. As seen from Table 1.35a, the $^8A_2$ state is found to be the ground state at both MRSDCI and MRSDCI+Q levels, with an isosceles triangular equilibrium geometry, an apex angle of 51.6°, and equal side bond lengths of 2.632 Å. The first excited $^8B_1$ state is, however, only 0.12 and 0.10 eV higher than $^8A_2$ at the MRSDCI and MRSDCI+Q levels, respectively. All octet states are found to have lower energy compared to the low-spin electronic states.

As seen from Figure 1.24, all the electronic states of Ir_3 exhibit very acute apex angles. Hence, the two equal sides of the isosceles triangle are longer than the Ir–Ir base bond lengths in these states. Evidently, there is multiple Ir–Ir bonding at the base of the isosceles triangle. The electronic states such as $^8B_2$ and $^6A_2$ deviate only a little from the ideal D_{3h} structures. This deviation could also be attributed to the truncated MRSDCI expansion, which could lead to slight symmetry breaking.

Table 1.35b shows the effect of spin–orbit (SO) coupling of the low-lying electronic states of Ir_3 near the equilibrium geometries. As seen from this table, the spin–orbit mixing of different states and different spin functions is substantial.

The ground state of Ir_3, including the spin–orbit effect, is an E state in the C_{2v}^2 double group with an apex bond angle of 55.0° and the two equal side bond lengths of 2.715 Å. The energy separation decreases by −0.41 eV, as a consequence of spin–orbit coupling for the ground state.

As seen from Table 1.35b, the spin–orbit mixing of different electronic states is very significant. Another feature is that the spin–orbit effects tend to favor the $^8B_2$ and $^8A_1$ electronic states compared to the $^8A_2$ state. Consequently, the E(I) state, which is the lowest electronic state in the presence of spin–orbit coupling, is a mixture of $^8B_2$, $^6B_2$, $^8A_1$, $^6A_1$, and so on (see Table 1.36b). Four different kinds of spin functions, which can yield the $^8B_2$ or $^6B_2$ states in the C_{2v}^2 group—namely, $3\alpha 4\beta$, $5\alpha 2\beta$, $1\alpha 6\beta$, and 7α—play important roles in the low-lying relativistic states. The constitution of the spin–orbit states vary as a function of the bond angles. For example, at $\theta = 52°$, the E(I) state is predominantly $^8A_1$, $^6A_1$ (79.3%) with only 0.9% contribution from the $^8A_2$ state, while at $\theta = 55°$, the E(I) state is predominantly $^8B_2$, $^6B_2$ (77.5%). The two states undergo avoided crossing, as a consequence of strong spin–orbit coupling.

It is evident from Table 1.35a that some of the low-lying electronic states of Ir_3 are Jahn–Teller components of ideal D_{3h} electronic states. For example, in the D_{3h} symmetry, an $^8E''$ electronic state would yield $^8B_1$ and $^8A_2$ states of C_{2v} symmetry. Likewise, an $^8E'$ electronic state in D_{3h} symmetry would yield $^8B_2$ and $^8A_1$ electronic states in the C_{2v} symmetry. A similar correlation can be obtained for the quartet and doublet electronic states. If the low-lying pairs of $^8A_2$ and $^8B_1$ electronic states are, indeed, Jahn–Teller components, then the energy separation between them (0.1 eV) is the barrier to pseudorotation, which is much smaller than the spin–orbit stabilization energies, reported in Table 1.35b. For the

Table 1.35b Properties of the Low-Lying Electronic States of Ir_3 Including Spin–Orbit Effect

				Weight in Percentage									
				$^8B_2$				$^8A_1$				$^8A_2$	$^6B_2$
Relativistic States	θ_e, deg	R_e, Å	E, eV[a]	7α	$5\alpha 2\beta$	$3\alpha 2\beta$	$1\alpha 6\beta$	7β	$2\alpha 5\beta$	$4\alpha 3\beta$	$6\alpha 1\beta$	$4\alpha 3\beta$	$2\alpha 3\beta$
E(I)	55.0	2.715	−0.41	—	21.8	51.2	4.5	—	4.6	4.6	1.4	—	—
—	—	—	—	—	3.4[b]	7.4[b]	—	—	17.0[b]	59.7[b]	2.5[b]	0.9[b]	—
E(II)	55.6	2.709	−0.39	3.2	49.1	—	13.7	0.7	5.3	—	5.7	—	—
E(III)	54.9	2.721	−0.34	1.9	—	17.2	63.3	3.9	2.9	—	—	—	—
E(IV)	55.5	2.726	−0.30	88.4	5.3	—	—	—	—	—	2.0	—	—
E(V)	54.5	2.735	−0.20	—	—	63.9	—	—	0.8	8.0	—	—	14.9

[a]The energies are relative to the $^8A_2$ around state without spin–orbit coupling.
[b]*E*(I) state at the apex bond angle of 52°.

Source: Reproduced from Reference 177.

$^8B_2$ and $^8A_1$ components, the energy separation in the absence of spin–orbit coupling is 0.16 eV, while the spin–orbit stabilization is 0.41 eV as seen from Table 1.35b. It is thus evident that the spin–orbit stabilization energy would overcome the Jahn–Teller stabilization energy for these structures. Thus it is anticipated that the low-lying electronic states of Ir_3 would not undergo pseudorotation.

Spin–orbit mixing of different electronic states could alter the Ir–Ir–Ir apex bond angles as seen from Table 1.35b. For example, in the E(I) state significant spin–orbit contamination is noted among the $^8B_2$, $^6B_2$, $^8A_1$, and $^6A_1$ states. Hence, the apex angle of the $^8B_2$ state becomes smaller, due to the spin–orbit contamination.

As was noted in Section 1.4.1, Van Zee et al. (171) observed the ESR spectra of Co_3, Rh_3, and Ir_3 in Ar/Kr matrices, although there is uncertainty in the case of Ir_3 regarding its identity. This is because two Ir isotopes occur naturally (^{191}Ir, ^{193}Ir). Thus there are more than one possible isotopomers for the trimers. Moreover, rather small nuclear moments of the two isotopes results in considerable overlapping and broadening of the lines. Hence Van Zee et al. (171) tentatively attributed the observed broad and temperature-sensitive band at 930 G to Ir_3. These authors suggest that if, indeed, the observed bands are due to Ir_3, then the effective $g_\perp^e \cong 7.3$. For larger D, the $g_\perp \cong 1.83$, while for smaller D, the $g_\perp$ increased toward 2. Thus these authors suggested that the most probable ground state of Ir_3 exhibits $S = \frac{7}{2}$, although for a heavy molecule such as Ir_3, the possibility of $S = \frac{5}{2}$ cannot be ruled out. As seen from Table 1.35, the octet electronic states ($S = \frac{7}{2}$) are favored over the sextet states, in the absence of spin–orbit coupling. However, when spin–orbit coupling is included, Ir_3 is in an intermediately coupled electronic state, which is a mixture of both octet and sextet states consistent with the observed $g_\perp$ values.

Table 1.36 shows the dipole moments of the various electronic states of Ir_3, while Table 1.37 shows the corresponding Mulliken populations. All octet and

Table 1.36 Dipole Moments for Bent Minima of Ir_3 at the Equivalent Structure

State	μ, D	State	μ, D
$^8A_2$	−1.00	$^4B_1$	0.32
$^8B_1$	−1.12	$^4A_2$	−0.16
$^8A_1$	−1.02	$^4A_1$	−0.09
$^8B_2$	−0.96	$^4B_2$	0.01
$^6B_1$	−0.42	$^2B_1$	0.29
$^6A_2$	−0.03	$^2A_1$	−0.03
$^6B_2$	−0.38	$^2B_2$	0.01
$^6A_1$	−0.14	$^2A_2$	−0.41

Source: Reproduced from Reference 177.

Table 1.37 Mulliken Population Analysis for Electronic States of Ir_3[a]

State	Ir_1[b]	Ir_2[b]	$Ir_1(s)$	$Ir_1(p)$	$Ir_1(d)$	$Ir_2(s)$	$Ir_2(p)$	$Ir_2(d)$
$^8A_2$	9.112	8.944	1.291	0.163	7.658	0.870	0.186	7.888
$^8B_1$	9.146	8.927	1.239	0.168	7.739	0.945	0.199	7.783
$^8A_1$	9.110	8.945	1.273	0.157	7.680	0.871	0.182	7.892
$^8B_2$	9.134	8.933	1.160	0.178	7.795	0.876	0.177	7.880
$^6B_1$	9.032	8.984	0.626	0.136	8.270	0.848	0.174	7.965
$^6A_2$	9.066	8.967	0.605	0.125	8.336	0.876	0.155	7.936
$^6B_2$	9.103	8.948	0.643	0.162	8.298	0.868	0.160	7.930
$^6A_1$	9.036	8.982	0.608	0.141	8.287	0.884	0.168	7.930
$^4B_1$	8.986	9.007	0.722	0.148	8.116	0.787	0.165	8.055
$^4A_2$	9.010	8.995	0.772	0.162	8.076	0.770	0.149	8.076
$^4A_1$	9.006	8.997	0.752	0.142	8.112	0.796	0.162	8.039
$^4B_2$	8.998	9.001	0.757	0.157	8.085	0.784	0.157	8.060
$^2B_1$	8.986	9.007	0.740	0.150	8.096	0.788	0.163	8.056
$^2A_1$	9.005	8.997	0.748	0.151	8.106	0.757	0.161	8.079
$^2B_2$	8.994	9.003	0.728	0.165	8.101	0.763	0.158	8.082
$^2A_2$	9.038	8.981	0.777	0.152	8.109	0.751	0.165	8.065

[a] $5s^2 5p^2$ shells of Ir atom are omitted in this table.

[b] Ir_1 represents the apex atom, while Ir_2 represents one of the two equivalent base atoms.

Source: Reproduced from Reference 177.

sextet states have negative dipole moments consistent with the excess electronic densities (Table 1.37) on the apex atom. We noticed that in Rh_3 the octet states have negative dipole moments from −0.36 to −0.67, while all sextet states have small positive dipole moments (0.17–0.29). This proves that more electronic densities are transferred from the base atoms to the apex atom in Ir_3 compared to Rh_3, which results in more negative dipole moments (from −0.96 to −1.12) for all octet states of Ir_3, while all the sextet states of Ir_3 exhibit opposite polarities of dipole moments.

As one may see from Table 1.37, all sextet states have smaller $6s$ populations, especially at the apex atoms compared to all octet states. Since the relativistic mass–velocity effect stabilizes the $6s$ orbitals, the octet electronic states with enhanced $6s$ populations are favored for Ir_3 over the sextet states. The $6p$ populations of all electronic states in this table are between 0.13 and 0.20. The apex atoms of the octet states exhibit smaller $5d$ population compared to the base atoms, while for all sextet, quartet, and doublet states, the apex atoms exhibit more $5d$ populations than do the base atoms.

An interesting contrast between the Mulliken populations of the electronic states of Ir_3 and Rh_3 is the comparison between $6s$ Mulliken populations of all the electronic states of Ir_3 and the corresponding $5s$ Mulliken populations in the case of Rh_3. For example, the $^8A_2$ electronic state of Rh_3 has an overall Mulliken population of Rh ($4d^{7.73}5s^{1.045}5p^{0.29}$) at the apex site and Rh ($4d^{7.87}5s^{0.763}5p^{0.32}$) at the base (see Table 1.34). This is compensated by slightly smaller $5d$ and $6p$ populations in the case of Ir_3. The enhanced $6s$ populations for all the electronic states of Ir_3 can be explained in terms of the relativistic mass–velocity contraction of the $6s$ orbital of the Ir atom, which stabilizes this orbital, thus leading to enhanced $6s$ populations.

REFERENCES

1. K. Balasubramanian, *J. Molec. Struct. THEOCHEM* **202**, 291 (1989).
2. W. Weltner, Jr. and R. J. Van Zee, *Annu. Rev. Phys. Chem.* **35**, 291 (1984); M. Kappes, *Chem. Rev.* **88**, 369 (1988); V. Bonačić-Koutecký, P. Fantucci, and J. Koutecký, *Chem. Rev.* **91**, 1035 (1991); M. W. Ruckman and M. Strongin, *Acc. Chem. Res.* **27**, 250 (1994); K. M. Ervin and W. C. Lineberger, in *Advances in Gas Phase Ion Chemistry*, Vol. 1, N. G. Adams and L. M. Babcock, eds., JAI Press, Greenwich, CT, 1991; R. R. Corderman and W. C. Lineberger, *Annu. Rev. Phys. Chem.* **30**, 347 (1979).
3. M. D. Morse, *Chem. Rev.* **86**, 1049 (1986); in *Advances in Metal and Semiconductor Clusters*, Vol. I, *Spectroscopy and Dynamics*, M. Duncan, ed. JAI Press Inc., Greenwich, CT 1993.
4. E. A. Rohlfing and J. J. Valentini, *J. Chem. Phys.* **84**, 6560 (1986).
5. K. Hilpert, *Ber. Bunsenges Phys. Chem.* **83**, 161 (1979).
6. B. Kleman and S. Lindqvist, *Ark. Fys.* **8**, 333 (1954).
7. J. C. R. Ruamps, *J. Ann. Phys.* (Paris) **4**, 1111 (1959).

8. N. Aslund, R. F. Barrow, W. G. Richards, and D. N. Travis, *Ark. Fys.* **30**, 171 (1965).
9. T. V. R. Rao and R. V. J. Lakshman, *J. Quant. Spectrosc. Radiat. Transfer* **11**, 1157 (1971).
10. J. Lochet, *J. Phys. B* **11**, L55 (1978).
11. D. S. Pesić and S. C. R. Weniger, *Hebd. Senaces. Acad. Sci.* **237**, 602 (1971).
12. R. E. Steele, *J. Molec. Spectrosc.* **61**, 477 (1976).
13. V. E. Bondybey, *J. Chem. Phys.* **77**, 3771 (1982).
14. V. E. Bondybey and J. H. English, *J. Phys. Chem.* **87**, 4647 (1983).
15. D. E. Powers, S. G. Hansen, M. E. Geusic, D. L. Michalopoulos, and R. E. Smalley, *J. Chem. Phys.* **78**, 2866 (1983).
16. A. D. Smirnov, N. E. Kuzmenko, and Y. Y. Kuzyakov, *Opt. Spectrosc.* **48**, 111 (1980); **47**, 141 (1979).
17. V. E. Bondybey, G. P. Schwartz, and J. H. English, *J. Chem. Phys.* **78**, 11 (1983).
18. S. S. S. Rao and S. V. K. Rao, *Indian J. Pure Appl. Phys.* **16**, 923 (1978).
19. N. L. Singh, *Phys. Rev.* **99**, 1624 (1955).
20. G. A. Ozin, *Appl. Spectrosc.* **30**, 573 (1976).
21. M. Moskovits and J. E. Hulse, *J. Chem. Phys.* **67**, 4271 (1977).
22. G. A. Ozin, H. Huber, D. McIntosh, S. Mitchell, J. G. Norman, and C. Noddleman, *J. Am. Chem. Soc.* **101**, 3504 (1979).
23. G. A. Ozin, S. A. Mitchell, S. M. Matton, and J. Garcia-Prieto, *J. Phys. Chem.* **87**, 4666 (1983).
24. R. Grinter, S. Armstrong, U. A. Jayasooriya, J. McCombie, D. Norris, and J. P. Springall, *Faraday Symp. Chem. Soc.* **14**, 94 (1980).
25. V. E. Bondybey, *J. Chem. Phys.* **77**, 3771 (1982).
26. V. E. Bondybey and J. H. English, *J. Phys. Chem.* **87**, 4647 (1983).
27. D. R. Preuss, S. A. Pace, and J. L. Gole, *J. Chem. Phys.* **71**, 3553 (1979).
28. D. E. Powers, S. G. Hansen, M. E. Geusic, A. C. Puiu, J. B. Hopkins, J. G. Dietz, M. A. Duncan, P. R. R. Langridge-Smith, and R. E. Smalley, *J. Phys. Chem.* **86**, 2556 (1982).
29. K. Hilpert and K. A. Gingerich, *Ber. Bunsenges Phys. Chem.* **84**, 739 (1980).
30. R. C. Maheswari, *Indian J. Phys.* **37**, 368 (1963).
31. S. P. Choong, L. S. Wang, and Y. S. Lim, *Nature* (London) **209**, 1300 (1966).
32. C. M. Brown and M. L. Ginter, *J. Molec. Spectrosc.* **69**, 25 (1978).
33. V. I. Srdanvov and D. S. Pesić, *J. Molec. Spectrosc.* **90**, 27 (1981).
34. S. A. Mitchell and G. A. Ozin, *J. Am. Chem. Soc.* **100**, 6776 (1978).
35. G. A. Ozin and H. Huber, *Ingor. Chem.* **17**, 155 (1978).
36. S. A. Mitchell, G. A. Kenney-Wallace, and G. A. Ozin, *J. Am. Chem. Soc.* **103**, 6030 (1981).
37. S. A. Mitchell and G. A. Ozin, *J. Phys. Chem.* **88**, 1425 (1984).
38. T. Welker and T. P. Margin, *J. Chem. Phys.* **70**, 5683 (1979).
39. W. Schulze, H. O. Becker, and H. Abe, *Chem. Phys.* **35**, 777 (1978).
40. W. Schulze and H. Abe, *Faraday Symp. Chem. Soc.* **14**, 87 (1980).

41. W. Schrittenlacher, H. H. Rotenmunal, W. Schroeder, and D. M. Kolb, *Surf. Sci.* **156**, 777 (1985).
42. D. M. Gruen and J. K. Bates, *Inorg. Chem.* **16**, 2450 (1977).
43. J. B. Hopkins, P. R. R. Langridge-Smith, M. D. Morse, and R. E. Smalley, unpublished results.
44. J. Kordis, K. A. Gingerich, and R. Sayse, *J. Chem. Phys.* **61**, 5114 (1974).
45. L. L. Ames and R. F. Barrow, *Trans. Faraday Soc.* **63**, 39 (1974).
46. W. E. Klotzbücher and G. A. Ozin, *Inorg. Chem.* **19**, 3767 (1980).
47. D. P. DiLella, K. V. Taylor, and M. Moskovits, *J. Phys. Chem.* **87**, 524 (1983).
48. J. A. Howard, K. F. Preston, R. Sutcliffe, and B. Mile, *J. Chem. Soc., Chem. Commun.* 1449 (1983).
49. M. D. Morse, J. B. Hopkins, P. R. R. Langridge-Smith, and R. E. Smalley, *J. Chem. Phys.* **79**, 5316 (1983).
50. W. H. Crumley, J. S. Haden, and J. L. Gole, *J. Chem. Phys.* **84**, 5250 (1986).
51. E. A. Rohlfing and J. J. Valentini, *Chem. Phys. Lett.* **126**, 113 (1986).
52. J. A. Howard, R. Sutcliffe, and B. Mile, *Surf. Sci.* **156**, 214 (1985).
53. J. A. Howard, K. F. Preston, and B. Mile, *J. Am. Chem. Soc.* **103**, 6226 (1981).
54. W. Schulze, H. U. Becker, R. Minkwitz, and K. Manzel, *Chem. Phys. Lett.* **55**, 59 (1978).
55. G. A. Ozin, H. Huber, and S. A. Mitchell, *Inorg. Chem.* **18**, 2932 (1979).
56. M. Moskovits and G. A. Ozin, *J. Phys. Chem.* **72**, 2267 (1980).
57. P. Y. Cheng and M. A. Duncan, *Chem. Phys. Lett.* **152**, 341 (1988).
58. K. LaiHing, P. Y. Cheng, and M. A. Duncan, *J. Phys. Chem.* **91**, 6521 (1987).
59. S. W. Buckner, J. R. Gord, and B. S. Freiser, *J. Chem. Phys.* **88**, 3678 (1988).
60. J. A. Howard, K. F. Preston, R. Sutcliffe, and B. Mile, *J. Phys. Chem.* **87**, 536 (1983).
61. J. A. Howard, R. Sutcliffe, and B. Mile, *J. Am. Chem. Soc.* **105**, 1364 (1983).
62. J. A. Howard, R. Sutcliffe, J. S. Tse, and B. Mile, *Chem. Phys. Lett.* **94**, 561 (1983).
63. J. A. Howard, R. Sutcliffe, and B. Mile, *J. Phys. Chem.* **87**, 2268 (1983).
64. J. A. Howard, R. Sutcliffe, and B. Mile, *J. Phys. Chem.* **88**, 2183 (1984).
65. O. Cheshnovsky, C. L. Pettiette, and R. E. Smalley, in *UPS of Metal and Semiconductor Clusters, Ion and Cluster Ion Spectroscopy*, J. P. Maier, ed., Elsevier, Amsterdam, 1988.
66. C. L. Pettiette, S. H. Yang, M. J. Craycraft, J. Concejaco, R. T. Laaksonen, O. Cheshnovsky, and R. E. Smalley, *J. Chem. Phys.* **88**, 5377 (1988); L. S. Zheng, C. M. Karner, P. J. Brucat, C. L. Pettiette, M. J. Craycraft, and R. E. Smalley, *J. Chem. Phys.* **85**, 1681 (1986).
67. A. B. Anderson, *J. Chem. Phys.* **68**, 1744 (1978).
68. R. C. Baetzold, *J. Chem. Phys.* **55**, 4355 (1971).
69. R. C. Baetzold, *J. Chem. Phys.* **68**, 555 (1978).
70. S. C. Richtsmeier, J. L. Gole, and D. A. Dixon, *Proc. Natl. Acad. Sci.* (USA) **77**, 5611 (1980).

71. D. Post and E. J. Baerends, *Chem. Phys. Lett.* **86**, 176 (1982).
72. D. Guenzburger, *Chem. Phys. Lett.* **86**, 316 (1982).
73. F. W. Averill and G. S. Painter, *Phys. Rev. B* **32**, 2141 (1985).
74. G. S. Painter and F. W. Averill, *Phys. Rev. B* **28**, 5536 (1983).
75. B. Delley, D. E. Ellis, A. J. Freeman, E. J. Baerends, and D. Post, *Phys. Rev. B* **27**, 2132 (1983).
76. J. Flad, G. Igle-Mann, H. Preuss, and H. Stoll, *Chem. Phys.* **90**, 257 (1984).
77. P. Joyes and M. Leleyter, *J. Phys. B* **6**, 150 (1973).
78. C. Bachmann, J. Demuynck, and A. Veillard, *Faraday Symp. Chem. Soc.* **14**, 170 (1980).
79. H. Tatewaki and S. Huzinaga, *J. Chem. Phys.* **72**, 399 (1980).
80. H. Tatewaki, Y. Sakai, and S. Huzinaga, *J. Comput. Chem.* **2**, 96 (1981).
81. H. Tatewaki, E. Miyoshi, and T. Nakamura, *J. Chem. Phys.* **76**, 5073 (1982).
82. E. Miyoshi, H. Tatewaki, and T. Nakumura, *J. Chem. Phys.* **78**, 815 (1983).
83. M. Witko and H. O. Beckman, *Molec. Phys.* **47**, 945 (1982).
84. H. Tatewaki, Y. Sakai, and S. Huzinaga, *J. Comput. Chem.* **2**, 96 (1981).
85. C. W. Bauschlicher, S. P. Walch, and P. E. M. Siegbahn, *J. Chem. Phys.* **76**, 6015 (1982).
86. C. W. Bauschlicher, Jr., *Chem. Phys. Lett.* **97**, 204 (1983).
87. K. Raghavachari, K. K. Sunil, and K. D. Jordan, *J. Chem. Phys.* **83**, 4633, (1985).
88. R. L. Martin, *J. Chem. Phys.* **78**, 5840 (1983).
89. T. Ziegler, J. G. Snijders, and E. J. Baerends, *J. Chem. Phys.* **74**, 1271 (1981).
90. M. Pelissier, *J. Chem. Phys.* **79**, 2099 (1983).
91. H. J. Werner and R. L. Martin, *Chem. Phys. Lett.* **113**, 451 (1985).
92. C. W. Bauschlicher, S. P. Walch, and P. E. M. Siegbahn, *J. Chem. Phys.* **78**, 3347 (1983).
93. S. P. Walch, C. W. Bauschlicher, and S. R. Langhoff, *J. Chem. Phys.*, **85**, 5980 (1986).
94. P. Scharf, S. Brode, and R. Ahlrichs, *Chem. Phys. Lett.* **113**, 447 (1985).
95. H. Basch, *Faraday Symp. Chem. Soc.* **14**, 149 (1980).
96. H. Basch, *J. Am. Chem. Soc.* **103**, 4657 (1981).
97. I. Shim and K. A. Gingerich, *J. Chem. Phys.* **79**, 2903 (1983).
98. A. D. McLean, *J. Chem. Phys.* **79**, 3392 (1983).
99. K. Balasubramanian and M. Z. Liao, *Chem. Phys.* **127**, 313 (1988).
100. R. B. Ross and W. C. Ermler, *J. Chem. Phys.* **89**, 5202 (1985).
101. Y. S. Lee, W. C. Ermler, and K. S. Pitzer, *J. Chem. Phys.* **73**, 360 (1980).
102. Y. S. Lee, W. C. Ermler, and K. S. Pitzer, *J. Chem. Phys.* **70**, 288 (1979).
103. K. Balasubramanian and M. Z. Laio, *J. Chem. Phys.* **86**, 5587 (1987).
104. Y. S. Lee and A. D. McLean, *J. Chem. Phys.* **76**, 2903 (1982).
105. A. D. McLean and Y. S. Lee, *Stud. Phys. Theor. Chem.* **21**, 219 (1982).
106. G. L. Malli, *Topics in Molecular Organization and Engineering*, Reidel, Dordrecht, the Netherlands, 1990.

107. K. Balasubramanian and P. Y. Feng, *J. Phys. Chem.* **94**, 1536 (1990); K. K. Das and K. Balasubramanian, *J. Phys. Chem.* **94**, 8379 (1990).

108. K. Balasubramanian, P. Y. Feng, and M. Z. Liao, *J. Chem. Phys.*, **91**, 3561 (1989).

109. S. P. Walch and B. C. Laskowski, *J. Chem. Phys.* **84**, 2734 (1986).

110. S. R. Langhoff, C. W. Bauschlicher, Jr., S. P. Walch, and B. C. Laskowski, *J. Chem. Phys.* **85**, 7211 (1986).

111. G. H. Jeung, M. Pelissier, and J. C. Barthelat, *Chem. Phys. Lett.* **97**, 369 (1983).

112. D. G. Truhlar, T. C. Thompson, and C. A. Mead, *Chem. Phys. Lett.* **127**, 287 (1986); T. C. Thompson, D. G. Truhlar, and C. A. Mead, *J. Chem. Phys.* **82**, 2392 (1985).

113. C. E. Moore, *Atomic Energy Levels*, U.S. National Bureau of Standards, 1971.

114. S. P. Walch, *J. Chem. Phys.* **87**, 6776 (1987).

115. K. K. Das and K. Balasubramanian, *J. Molec Spectrosc.* **140**, 280 (1990).

116. K. Balasubramanian and P. Feng, *J. Phys. Chem.* **94**, 1536 (1990).

117. J. P. Desclaux and P. Pyykkö, *Acc. Chem. Res.* **12**, 276 (1979).

118. C. W. Bauschlicher, S. R. Langhoff, and P. R. Taylor, *J. Chem. Phys.* **88**, 1041 (1988).

119. D. G. Leopold, J. Ho, and W. C. Lineberger, *J. Chem. Phys.* **86**, 1715 (1987).

120. K. Balasubramanian and P. Y. Feng, *Chem. Phys. Lett.* **159**, 452 (1989).

121. K. S. Pitzer and K. Balasubramanian, *J. Chem. Phys.* **86**, 3068 (1982).

122. K. Balasubramanian and K. S. Pitzer, *J. Chem. Phys.* **78**, 321 (1983).

123. K. Balasubramanian, *J. Molec. Spectrosc.* **123**, 228 (1987).

124. C. W. Baschlicher, Jr., S. R. Langhoff, and H. Partridge, *J. Chem. Phys.* **91**, 2412 (1989).

125. H. Zhang and K. Balasubramanian, *J. Chem. Phys.*, **98**, 7092 (1983).

126. H. Stoll, P. Funetealba, M. Dolg, J. Flad, L. von Szentpaly, and H. Preuss, *J. Chem. Phys.* **79**, 5532 (1983); P. Schwerdtfeger, M. Dolg, W. H. E. Schwarz, G. A. Bowmaker, and P. W. D. Boyd, *J. Chem. Phys.* **91**, 1762 (1989).

127. R. L. Martin, *J. Chem. Phys.* **86**, 5027 (1987).

128. J. L. Martin and W. Andreoni, *Phys. Rev.* **A82**, 3637 (1983).

129. J. Andzelm, E. Radzio, and D. R. Salahub, *J. Chem. Phys.* **83**, 4573 (1985).

130. D. S. Pesić and B. R. Vujisić, *J. Mol. Spectrosc.* **146**, 561 (1991); V. I. Srdanov and D. S. Pesić, *J. Mol. Spectrosc.* **90**, 27 (1991).

131. B. Simard and P. A. Hackett, *Chem. Phys. Lett.* **186**, 415 (1991).

132. H.-G. Krämer, V. Beutel, A. M. James, P. R. R. Langridge-Smith, K. Weyers, and W. Demtröder, *Chem. Phys. Lett.* **193**, 331 (1992).

133. V. Beutel, M. Kuhn, and W. Demtröder, *J. Molec. Spectrosc.* **155**, 343 (1992).

134. D. L. Robbins, K. F. Willey, C. S. Yeh, and M. A. Duncan, *J. Phys. Chem.* **96**, 4824 (1992).

135. D. Roser, R. Pellow, M. Eyring, M. Vala, J. Lignieres, and J. C. Rivoal, *Chem. Phys.* **166**, 393 (1992).

136. C. M. Brown and M. L. Ginter, *J. Molec. Spectrosc.* **69**, 25 (1978).

137. D. Roser, R. Pellow, M. Eyring, M. Vala, J. Liginieres, and J. C. Rivoal, *Chem. Phys.* **166**, 393 (1992).
138. G. A. Bishea and M. D. Morse, *J. Chem. Phys.* **95**, 5646 (1991).
139. S. Taylor, E. M. Spain, and M. D. Morse, *J. Chem. Phys.* **92**, 2698, 2710 (1990).
140. K. Taylor, O. Cheshovsky, and R. E. Smalley, *UPS Spectra of Au_3 and Au_2*, unpublished manuscript.
141. M. F. Jarrold and K. M. Creegan, *Chem. Phys. Lett.* **166**, 116 (1990).
142. K. Balasubramanian and K. K. Das, *Chem. Phys. Lett.* **186**, 577 (1991).
143. K. K. Das and K. Balasubramanian, *Chem. Phys. Lett.* **176**, 571 (1991).
144. C. W. Bauschlicher, S. R. Langhoff, and H. Partridge, *J. Chem. Phys.* **91**, 2412 (1989).
145. M. F. Jarrold and K. M. Creegan, *Internatl. J. Mass. Spectrosc. Ion Proc.* (1991).
146. K. J. Taylor, C. Jia, J. Conceicao, L. S. Wang, B. R. Johnson, P. J. Norlander, and R. E. Smalley, *J. Chem. Phys.* **93**, 7515 (1990).
147. J. Ho, K. M. Ervin, and W. C. Lineberger, *J. Chem. Phys.* **93**, 6987 (1990).
148. K. Balasubramanian and D. W. Liao, *J. Chem. Phys.* **94**, 5233 (1991).
149. D. W. Liao and K. Balasubramanian, *J. Chem. Phys.* **94**, 5233 (1991).
150. V. Bonačic-Koutecký, P. Fantucci, and J. Koutecký, *J. Chem. Phys.* **93**, 802 (1990).
151. K. Przybylski, J. Koutecký, V. Bonačic-Koutecký, P. V. Ragué-Schleyer, and M. F. Guest, *J. Chem. Phys.* **94**, 5533 (1991).
152. K. Balasubramanian, *J. Chem. Phys.* **87**, 6573 (1988).
153. S. Taylor, G. W. Lemire, Y. M. Hamrick, A. Fu, and M. D. Morse, *J. Chem. Phys.* **89**, 5517 (1988).
154. U. Muller, K. Satter, J. Xhie, N. Venkateswaran, and G. Raina, *Z. Phys. D.* **19**, 319 (1991).
155. S. K. Gupta, B. M. Nappi, and K. A. Gingerich, *Inorg. Chem.* **20**, 966 (1981).
156. A. R. Miedema and K. A. Gingerich, *J. Phys. B* **12**, 2081 (1979).
157. K. Jansson and R. Scullman, *J. Molec. Spectrosc.* **61**, 299 (1976).
158. H. Basch, D. Cohen, and S. Topiol, *Israel J. Chem.* **19**, 233 (1980).
159. K. Balasubramanian, *J. Chem. Phys.* **90**, 6310 (1988).
160. K. Balasubramanian, *J. Chem. Phys.* **91**, 307 (1989).
161. I. Shim and K. A. Gingerich, *J. Chem. Phys.* **80**, 5107 (1984).
162. K. Balasubramanian, P. Y. Feng, and M. Z. Liao, *J. Chem. Phys.* **87**, 3981 (1987).
163. K. M. Ervin, J. Ho, and W. C. Lineberger, *J. Chem. Phys.* **89**, 4514 (1988).
164. A. Goursot, I. Papai, and D. R. Salahub, *J. Am. Chem. Soc.* **114**, 7452 (1992).
165. J. Rubio, S. Zurita, J. C. Barthelat, and F. Illas, *Chem. Phys. Lett.* **217**, 283 (1994).
166. I. Papai, J. Mink, R. Fournier, and D. R. Salahub, *J. Phys. Chem.*, **97**, 9986 (1993).
167. H. Wang and E. A. Carter, *J. Phys. Chem.* **96**, 1197 (1992).
168. D. Dai and K. Balasubramanian, *J. Chem. Phys. Lett.* **103**, 648 (1995).
169. K. K. Das and K. Balasubramanian, *J. Chem. Phys.* **93**, 625 (1990).
170. D. Dai and K. Balasubramanian, *Chem. Phys. Lett.* **195**, 207 (1992).

171. R. J. Van Zee, S. Li, and W. Weltner, *Chem. Phys. Lett.* **195**, 214 (1992).
172. K. Balasubramanian and D. S. Liao, *J. Phys. Chem.* **93**, 3989 (1989).
173. G. A. Ozin and A. L. J. Hanlan, *Inorg. Chem.* **18**, 1781 (1979).
174. D. L. Cocke and K. A. Gingerich, *J. Chem. Phys.* **60**, 1958 (1974).
175. I. Shim and K. A. Gingerich, *J. Chem. Phys.* **78**, 563 (1983).
176. F. Illas, J. Rubio, J. Cañellas, and J. M. Ricart, *J. Chem. Phys.* **93**, 2603 (1990).
177. D. Dai, M. Z. Laio, and K. Balasubramanian, *Chem. Phys. Lett.* **249**, 141 (1996).
178. F. A. Cotton and G. W. Wilkinson, *Basic Inorganic Chemistry*, Wiley, New York, 1976.

2

RELATIVISTIC EFFECTS IN HETERONUCLEAR DIATOMICS OF MAIN-GROUP *p*-BLOCK ELEMENTS

In this chapter we consider the impact of relativistic effects on the main-group heteronuclear diatomics such as hydrides, halides, and chalconides constituting *p*-block elements. Numerous papers have appeared in the literature (1–386) concerning both theoretical computations and techniques pertinent to such molecules, and experimental spectra of such species, which have led to a wealth of information on the spectroscopic constants of these molecules. A part of the current chapter is derived from a review article by the author (2), but some sections have been shortened, and some new sections were added, especially pertaining to molecules, such as TlCl, PbBr, PbI, SbF, SbO, and BiO, which were not theoretically studied prior to 1989. We have also included some more recent developments that were not contained in the original review paper. We shall focus on the spectroscopic properties, potential-energy curves, dipole moments, and transition moments of heavy main-group molecules containing *p*-block elements. Although comprehension of relativistic effects would be the ultimate objective, the first few sections of this chapter are reviews of known experimental and theoretical information on these species.

Although the early spectroscopic investigations of heavy hydrides, halides, and chalconides date back to the 1920s, a complete and definitive understanding of the electronic states and the observed electronic transitions was accomplished recently. Much of the problem centered around not having any theoretical calculations or insight into the electronic energy levels, especially of very heavy molecules. This has been overcome, in recent years, with the advent of *ab initio* relativistic quantum-mechanical methods that have provided reliable and accurate spectroscopic constants, and potential-energy curves of almost diatomic molecule in the periodic table (1–31). These theoretical calculations, together with the experimental spectroscopic and high-temperature thermodynamic studies, have provided a wealth of information, which could form a sound basis for obtaining relative periodic trends, and gaining insight into the impact of relativity on the properties of such species. We find several anomalies, especially with sixth-row *p*-block compounds that are attributed to be consequences of relativistic corrections. We discuss these trends in the last section of this chapter. First, the individual spectroscopic properties of diatomic hydrides, halides, and chalconides are considered.

2.1 SPECTROSCOPIC PROPERTIES AND POTENTIAL-ENERGY CURVES OF HEAVY HYDRIDES

2.1.1 GaH

Early experimental studies on GaH by various authors (e.g., 33–39) provided the electronic spectrum of GaH. Reference 32 contains a summary of the references for the work done before 1977 for the molecules considered here. Poynor et al. (38) and Kronekvist et al. (39) studied the $3\Pi_{0^-} \rightarrow X^1\Sigma^+$ emission and $A^1\Pi$–$X^1\Sigma^+$ systems, respectively; the latter workers suggested that there should be a barrier due to some avoided crossing. The line widths in the *A*–*X* system were found to increase with increasing *J*, suggesting a tunneling effect and a small barrier prior to dissociation. On the basis of the predissociation of the *A*–*X* bands, the D_e of GaH was estimated as <22,900 cm^{-1}. Urban et al. (31) obtained the infrared spectrum of GaH in its $^1\Sigma^+$ ground state using a laser diode spectrometer, which resulted in the measurement of 113 transitions, and accurate ground-state Dunham parameters and an improved R_e = 1.6621 Å. Lakshminarayana and Shetty (332) carried out the rotational analysis of the $^3\Pi$–$X^1\Sigma^+$ bands of GaD.

An early theoretical calculation on GaH is due to Hurley et al. (22), who obtained the R_e and ω_e values of the ground state of GaH in an attempt to gauge the RECPs. Kim and the author (42) made relativistic CASSCF/SOCI calculations on several valence and Rydberg electronic states of GaH. The spectroscopic properties, dipole moments, and potential-energy curves of several electronic states were reported, among which 17 new electronic states were found (42).

Table 2.1 shows the calculated spectroscopic properties, and Figure 2.1 shows the actual potential-energy curves (PECs) obtained by Kim and the author (42). The experimental spectroscopic constants were known for the $X^1\Sigma^+$ state and the spin–orbit components of the $^3\Pi$ state for some time. As seen from Table 2.1, the theoretical constants for the ground state are almost in exact agreement with the experimental values. The spin–orbit interaction was not introduced in these theoretical calculations (42), since its effect was found to be rather small.

An experimentally observed system, designated as the *A*–*X* system, was

Table 2.1 Theoretical Spectroscopic Constants for GaH[a]

State	R_e, Å	T_e, cm^{-1}	ω_e, cm^{-1}
$X\,^1\Sigma^+$(I)	1.662 (1.663)	0	1,612 (1,605)
$a^3\Pi$(I)	1.603 (1.633)	16,836 (17,300)	1,559 (1,631–1,640)
A^1(II)	1.780	24,206	—
$^3\Sigma^+$	1.935	39,271	2,245
$^1\Sigma^+$(II)	3.582	40,933	447
$^3\Sigma^-$	1.564	45,649	1,772
$^1\Sigma^{+\prime}$(II)	1.747	46,632	931
$^3\Pi$(II)	2.091	46,737	1,218
$^1\Sigma^+$(III)	4.609	50,573	219
$^1\Delta$(I)	1.541	51,319	1,907
$^1\Delta$(II)	1.553	51,982	1,835
$^1\Sigma^{+\prime}$(III)	1.612	52,076	2,192
$^3\Pi$(III)	1.737	54,243	1,660
$^3\Delta$(I)	3.440	64,415	—
$^3\Pi$(IV)	1.813	65,420	831
$^1\Delta$(III)	1.615	72,554	1,567
$^3\Delta$(II)	2.039	79,284	917
$^3\Delta$(III)	2.033	82,295	956
$^3\Sigma^-$	1.606	83,271	1,497
$^1\Delta$(IV)	2.321	85,703	1,744

[a]Numbers in parentheses are experimental values. D_0° (GaH) = 2.81 (theory), 2.8 eV (experiment).

Source: Reference 42.

found to be predissociated. The A state was tentatively assigned to a $^1\Pi$ state with an experimental T_0 value of 22,745 cm^{-1} obtained by Kronekvist et al. (39). The experimental R_e value of the A state was 1.82 Å compared to a theoretical value of 1.78 Å. The $^1\Pi$ curve has a very shallow minimum with a theoretical T_e value of 24,206 cm^{-1}, compared to an experimental value of 22,745 cm^{-1}. The theoretical value was a bit high, as anticipated for the level of electron correlation and basis sets included in those calculations.

The predissociation of the experimental A state was found to be consistent with the potential-energy curve in Figure 2.1, which has a shallow short-range minimum and a long-range minimum separated by a very small barrier. The D_e calculated by using the predissociation of the $^1\Pi$ state should, however, be quite accurate, since the predissociation energy and the energy of ^2P + ^2S atoms are very close. Kim and the author (42) estimated that the D_e of 2.84 eV obtained by using extrapolation of the predissociation of the A–X bands should be only 0.04 eV lower, yielding a corrected value of 2.80 eV. A direct theoretical SOCI D_e value of the $X^1\Sigma^+$ state of 2.81 eV is almost in exact agreement with this experimental result.

Kronekvist et al. (39) noted that the observed line width in the A–X bands

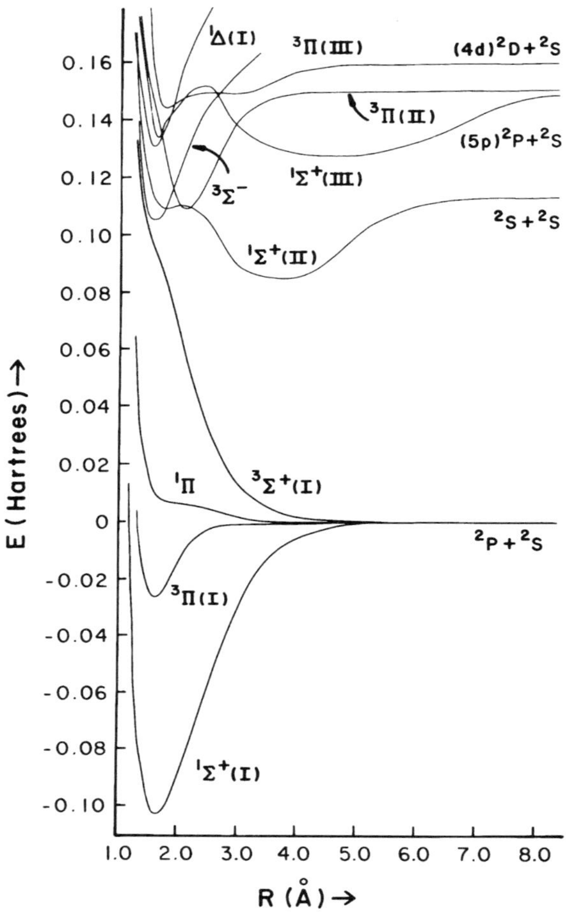

Figure 2.1 Potential-energy curves for the electronic states of GaH. (Reprinted from Ref. 42; copyright 1989, Academic Press.)

increased with increasing rotational quantum number J, indicating a tunneling effect. These authors predicted that there should be an avoided crossing between a deep $^1\Pi$ PEC dissociating into an excited configuration and another repulsive $^1\Pi$ dissociating into the ground-state configuration. It was found that the $^1\Pi$ state arising from $1\sigma 2\sigma^2 1\pi$ mixes more with $^1\Pi(1\sigma^2 2\sigma 1\pi)$ at short distance than at long distance. The origin of the short-range minimum appears to be due to this mixing but it is not very large at the minimum.

The theoretical dipole moments of the electronic states of GaH are shown in Table 2.2. The dipole moment of the ground state of GaH in Table 2.2 (0.46 D) was found to be in good agreement with a SCF/SDCI/CPF value of 0.37 D obtained by Pattersson and Langhoff (43). It is interesting to note that the dipole moment of the A state has opposite sign compared to the ground state.

Table 2.2 Dipole Moments Near Equilibrium Geometries of the Various Electronic States of GaH[a]

State	μ_e, D	State	μ_e, D	State	μ_e, D
$^1\Sigma^+$(I)	−0.460	$^3\Pi$(II)	0.239	$^3\Pi$(IV)	0.684
$^3\Pi$(I)	−0.221	$^1\Sigma^+$(III)	0.010	$^1\Delta$(III)	0.117
$^1\Pi$	0.107	$^1\Delta$(I)	0.572	$^3\Delta$(II)	0.424
$^3\Sigma^+$	2.799	$^1\Delta$(II)	0.745	$^3\Delta$(III)	0.745
$^1\Sigma^+$(II)	−1.151	$^3\Sigma^{+\prime}$(III)	0.417	$^1\Sigma^-$	−0.424
$^3\Sigma^-$	−0.285	$^3\Pi$(III)	0.935	$^1\Delta$(IV)	0.117
$^1\Sigma^{+\prime}$(II)	−1.395	$^3\Delta$(I)	0.221	—	—

[a]Positive polarity means positive charge on the hydrogen atom.

Source: Reference 42.

The ground state of GaH arises from the $1\sigma^2 2\sigma^2$ configuration, while the excited $^3\Pi$ and $^1\Pi$ states arise predominantly from the $1\sigma^2 2\sigma 1\pi$ configuration. The nature of the other electronic states in Table 2.1 is more complex. The complex nature of the excited states of GaH such as $X^1\Sigma^+$(II) and $X^1\Sigma^+$(III) led to avoided crossings, resulting in multiple minima in the potential-energy curves of these states (Fig. 2.1), but these are not due to relativistic effects. Further details on this molecule can be found in the original paper (42).

2.1.2 GeH

Kleman and Werhagen (44, 45) observed the first emission spectrum of GeH in a King furnace led to the characterization of two systems designated as *a–X* and *A–X* systems. Barrow et al. (46) and Howell (47) studied the *B–X* system, from which the rotational constants of the *B* state were obtained. Klynning and Lindgren (48) carried out the rotational analysis of the *A–X* system. The *A* state was found to be predissociated, probably due to the crossing of a repulsive state dissociating into the ground-state atoms. The *A–X* system was also studied by an electron impact method (51). GeH is also interesting from the standpoint of growth of germanium films by UV laser-assisted chemical vapor deposition (CVD) (49, 50). Multiphoton dissociation of the GeH_4 molecule also revealed the *B–X* emission spectrum of GeH in the UV.

The dipole moment of the ground state of GeH has been a topic of considerable discussion (52–55). An earlier experimental value of 1.24 ± 0.1 D obtained by Brown et al. (52) using laser magnetic resonance (LMR) spectroscopy of GeH was questioned by four theoretical investigations (53–55, 57), all of which, suggested that the dipole moment of GeH to be 0.09–0.16 D (Ge^+H^-). However, a subsequent study by Brown et al. has yielded a value in closer agreement with the experiment.

The author and Li (57) used the CASSCF/SOCI/RCI methods to obtain the properties of 12 electronic states of GeH, 10 of which were found to be bound.

The three experimentally observed bands (*a–X*, *A–X*, *B–X*) were assigned, and the uncertainties in the experimental T_e and ω_e values of these states were resolved. The D_e of GeH (2.81 eV) obtained with a CASSCF/SOCI calculation employing a large Gaussian basis set did not agree with an experimental D_e of ~3.3 eV obtained from the predissociation in the $A^2\Delta$ state (48).

Table 2.3 shows the theoretical spectroscopic constants for ten electronic states of GeH obtained by the author and Li (57). The results in Table 2.3 were obtained with the CASSCF/FOCI method employing a 4*s*4*p*4*d* valence Gaussian basis set. Figure 2.2 shows the potential-energy surfaces of 12 electronic states of GeH in the absence of the spin–orbit term. Among the states reported by the author and Li, only four states have been observed experimentally through the *a–X*, *A–X*, and *B–X* systems. The *a*, *A*, and *B* states were assigned to $^4\Sigma^+$, $^2\Delta$, and $^2\Sigma^+$, respectively.

Uncertain experimental T_e value of 16,747 cm^{-1} for the a state is a bit high compared with the theoretical calculations, which yielded a T_e of 14,000 cm^{-1}. The R_e value of 1.60 Å for the a state compared well with an approximate experimental value of 1.58 Å. As seen from Table 2.3, the theoretical T_e and R_e values for the $A^2\Delta$ state (26,663 cm^{-1} and 1.66 Å, respectively) are in good agreement with these values. The experimental ω_e value of the *A* state is uncertain because of the predissociation of the observed bands. The ω_e value of 1185 cm^{-1} listed in Huber and Herzberg (32) is somewhat low compared to the theoretical ω_e (1302 cm^{-1}) (57). Since the theoretical ω_e of the ground state (1807 cm^{-1}) was found to be within 1% of the experimental result, it was believed that the theoretical ω_e of the $^2\Delta$ state should be reasonable. Klynning and Lindgren (48) reported a somewhat more accurate ω_e value of 1437 ± 24 cm^{-1} from the rotational analysis.

The experimentally observed system designated as $B \leftarrow X$ was tentatively assigned to the $^2\Sigma^+$–$^2\Pi$ system (48). The bands in the $B \leftarrow X$ system could not be resolved completely, presumably since these bands were predissociated. An

Table 2.3 Theoretical Spectroscopic Properties of GeH[a]

State	R_e, Å	T_e, cm^{-1}	ω_e, cm^{-1}	D_e, eV
$X^2\Pi_{1/2}$	1.62	0	1,806	2.34
$^2\Pi_{3/2}$	1.62	869	1,810	2.30
$a^4\Sigma^-$	1.60	13,914	1,637	0.74
$A^2\Delta$	1.66	26,663	1,302	0.26
$B^2\Sigma^+$(II)	1.83	38,528	1,362	0.56
$^2\Sigma^+$(III)	1.67	46,267	2,107	3.06
$^2\Pi$(II)	1.59	49,239	1,999	—
$^4\Sigma^-$(II)	1.95	49,461	1,567	1.84
$^2\Pi$(III)	1.81	50,958	—	1.37
$^4\Delta$	1.64	67,455	1,399	0.44

Source: Reference 57.

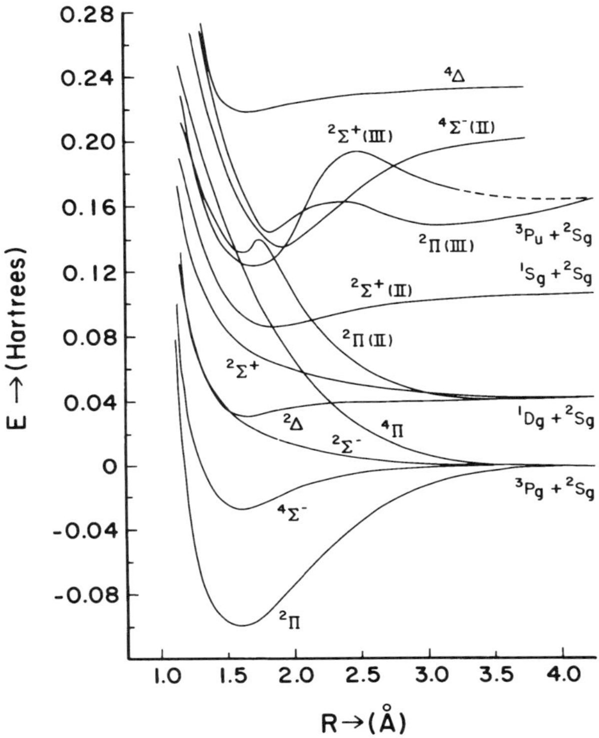

Figure 2.2 Potential-energy curves for the low-lying electronic states of GeH. (Reprinted from Ref. 57; copyright 1988, Academic Press.)

approximate experimental T_e value of 41,074 cm^{-1} for this state was found to be, however, in reasonable agreement with the theoretical T_e value of 38,528 cm^{-1} for the $^2\Sigma^+$(II) state. As seen from Figure 2.2, the first $^2\Sigma^+$ state is repulsive, and dissociates into Ge(^1D) + H(^2S). The B state was assigned to the second $^2\Sigma^+$ state dissociating adiabatically into Ge(^1S) + H(^2S). The theoretical R_e value of this state (1.83 Å) was noted to be considerably longer than the R_e of the $X^2\Pi$ ground state. The recommended R_e and ω_e values for the R state are 1.79 Å and 1430 cm^{-1}, respectively (57).

Klynning and Lindgren (48) noted that the $A^2\Delta$–$X^2\Pi$ bands were predissociated probably as a result of the crossing of a repulsive curve dissociating into the ground-state atoms. These authors hypothesized that the $^4\Pi$ state is responsible for predissociation. As seen from Figure 2.2, the $^4\Pi$ repulsive curve crosses with the $^2\Delta$ curve, confirming the Klynning–Lindgren hypothesis (48).

The D_e value of GeH calculated from predissociation of the A–X system was found to be between 22,900 and 26,600 cm^{-1}. The theoretical calculations (57) disagreed with the Klynning–Lindgren (48) assumption that the $^2\Sigma^+$ state dissociating into Ge(^1S) + H(^2S) would almost be repulsive. The $^2\Sigma^+$ state correlating with Ge(^1S) + H(^2S) was found to be bound; it is the B state participating in the B–X system. The $^2\Sigma^+$ state dissociating adiabatically into Ge(^1D) + H(^2S) was found to be repulsive. The best theoretical calculations gave a D_e of 2.81 eV compared to the lower and upper bounds of the experimental values of 2.84 and 3.30 eV.

Figure 2.3 shows the RCI dipole moment curves for the electronic states of GeH obtained by Chapman et al. (55). It is interesting to note that the curves including the spin–orbit effects are parallel to the $X^2\Pi$ dipole moment curve obtained without the spin–orbit term. Note that μ_e reaches a maximum ($^2\Pi$), and changes sign at long distance. The transition moments of the various electronic transitions are shown in Figure 2.4. The transition labeled $\frac{1}{2} \rightarrow \frac{1}{2}$(II) corresponds to the a $^4\Sigma^-_{1/2} \leftarrow X^2\Pi_{1/2}$ transition. In the absence of the spin–orbit term this transition is forbidden. Even for GeH, the transition moment for this transition becomes large at distances >4.5 bohr, indicating the importance of relativistic effects in this region. The intensity ratio of the $\frac{3}{2}$(I) $\rightarrow$ $\frac{3}{2}$(II) : $\frac{1}{2}$(I) $\rightarrow$ $\frac{1}{2}$(II) : $\frac{1}{2}$(I) $\rightarrow$ $\frac{3}{2}$(II) transitions was theoretically found to be 1 : 0.4 : 0.01

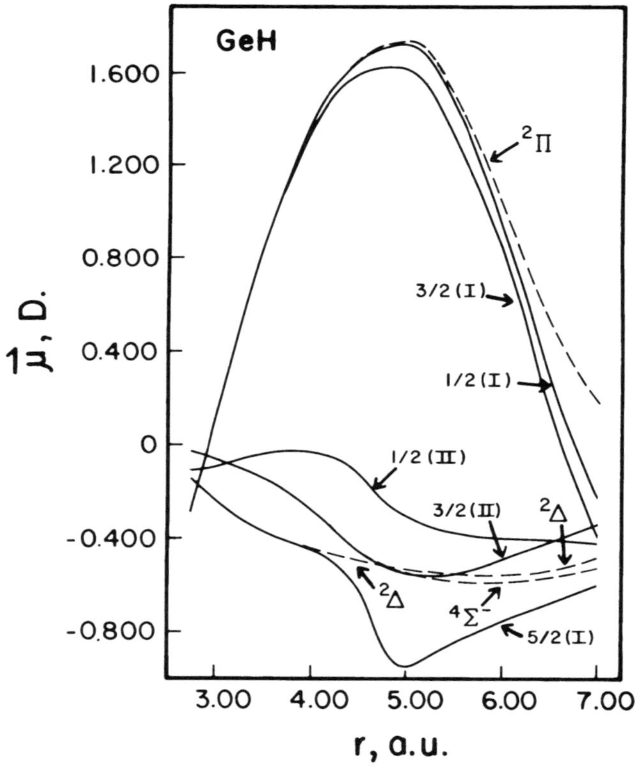

Figure 2.3 RCI dipole moment curves for the low-lying electronic states of GeH. (Reprinted from Ref. 55; copyright 1988, American Institute of Physics.)

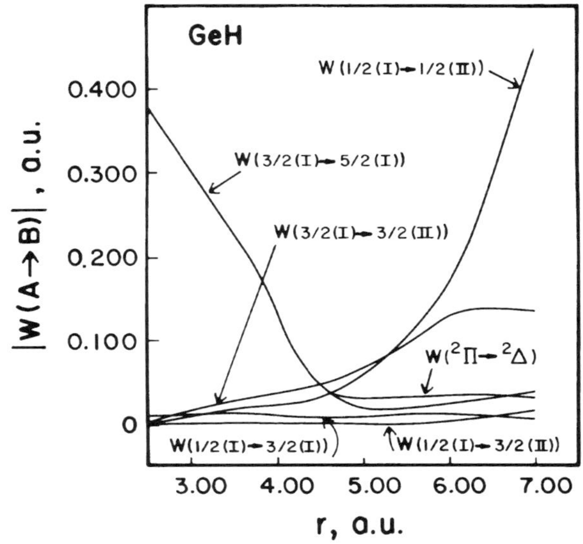

Figure 2.4 Electronic transition moment curves for the low-lying electronic states of GeH. (Reprinted from Ref. 55; copyright 1988, American Institute of Physics.)

for GeH, in good agreement with the experimental intensity ratios of the (0–0) bands without the rotational line strength factors.

2.1.3 AsH

The known experimental spectroscopic data, and tentative assignments (as suggested in References 32, 58–60, and 312) are shown in Table 2.4. Pettersson and Langhoff (43) performed SCF/SDCI/CPF calculations on the spectroscopic constants of the ground state of AsH. The author and Nannegari (61) have completed CASSCF/SOCI calculations on AsH. Results of these calculations are summarized in Table 2.4.

Most of the information on the $^3\Pi$ excited state of AsH was obtained from the $A \leftarrow X$ transition with band origins near 29,282, 29,822, and 30,518 cm^{-1}. These bands have been tentatively assigned to the $^3\Pi_2$–$X^3\Sigma^-$, $^3\Pi_1$–$X^3\Sigma^-$, and $^3\Pi_{0+}$–$X^3\Sigma^-$ systems. On the basis of the theoretical calculations on the isoelectronic SbH (158) and BiH (180, 181), it was predicted that there should be $^1\Delta$, $^1\Sigma^+$, $^5\Sigma^-$ (repulsive), $^3\Pi$, and $^1\Pi$ low-lying electronic states for AsH. The author and Nannegari (61) have carried out CASSCF/SOCI calculations on these states of AsH.

Berkowitz (312) deduced an accurate D_0(AsH) to be 64.6 kcal/mol using high-temperature photoionization mass spectrometric methods. In this investigation, and in a more recent investigation, Berkowitz and Cho (313) studied the related species AsH_2, AsH_3, and similar compounds. The singlet–triplet

Table 2.4 Spectroscopic Constants of AsH[a]

State	R_e, Å		ω_e, cm^{-1}		T_e, cm^{-1}		D_e, eV		μ_e,[c] D
	Theory	Exp.	Theory	Exp.	Theory	Exp.	Theory	Exp.	
$X^3\Sigma^-$	1.528	[1.534]	2,194	2,130	0	0	2.60[b]	2.80	0.178
$^1\Delta$	1.525	—	2,190	—	9,968	—	3.03	—	0.205
$^1\Sigma^+$	1.522	—	2,197	—	17,114	—	3.02	—	0.185
$A^3\Pi$	1.577	[1.58]	1,398	1,208	32,244	[30,000]	0.26	—	0.779
$^1\Pi$	2.467	—	441	—	36,608	—	—	—	0.211
$^3\Pi$(II)	2.187	—	1,386	—	41,461	—	—	—	−1.315
$^1\Pi$(II)	2.561	—	710	—	44,319	—	—	—	−0.225
$^1\Pi$(III)	1.603	—	3,606	—	63,288	—	—	—	−1.872
$^3\Pi$(III)	1.875	—	3,264	—	67,848	—	—	—	0.717
$^1\Delta$(II)	1.553	—	2,042	—	72,881	—	—	—	0.525
$^1\Sigma^+$(II)	1.552	—	2,059	—	75,312	—	—	—	0.523

[a] All theoretical values are from Reference 61. All experimental values are from Reference 58 except the D_0° (AsH), which is from Reference 312.
[b] A superior value of 2.71 eV is obtained including 4*f* functions (314).
[c] Positive polarity means As^-H^+ polarity.

energy separation of AsH_2^+ was also crudely estimated. The present author (314) carried out CASSCF/SOCI calculations employing large Gaussian basis sets that included $4f$-type functions on the ground state of AsH, as well as electronic states of AsH_2, AsH_2^+, SbH_2^+, and BiH_2^+. The CASSCF/SOCI calculations yielded R_e = 1.528 Å and D_e = 62.4 kcal/mol. The theoretical D_e(HAs–H) = 69.1 kcal/mol, in excellent agreement with an experimental D_0(HaS–H) = 66.5 kcal/mol deduced by Berkowitz (312).

The author and Nannegari (61) used the CASSCF/SOCI calculations employing a ($4s4p4d$) valence Gaussian basis set for 16 low-lying electronic states of AsH. Their computed potential-energy curves of AsH are shown in Figure 2.5. Among the 16 states computed, 11 bound states were found, for which spectroscopic constants were calculated. The experimentally observed predissociation and Λ doubling (58) in the A–X system were explained on the basis of computed potential-energy curves. The dipole moment curves for the low-lying states of AsH were also obtained.

2.1.4 SeH and SeH⁺

Radford (62) obtained the first EPR spectrum of SeH, and determined the rotational and spin–orbit constants for the ground state, which were further refined by Carrington and co-workers (63). Lindgren (64) observed diffuse spectra of SeH and SeD in the 300–3250-Å region by employing flash photolysis of SeH_2. The observed bands in this region were tentatively assigned to an $A^2\Sigma$–$X^2\Pi_i$ transition. Donovan et al. (65) observed bands at ~55,866, 66,800, 69,589, and 71,174 cm^{-1} attributed to Rydberg transitions. These authors calculated the ionization potential of SeH to be 9.8 eV. Smyth et al. (66) and Bollmark et al. (67) obtained the spin–orbit constant of the ground state of SeH.

The vacuum UV spectra of SeH (70), the far-infrared laser magnetic spectrum of SeD ($X^2\Pi_{3/2}$) (71), and the fine transitions in the LMR spectrum of SeH (70–72) have been studied. These studies have yielded refined R_e and spin–orbit and spin–rotation constants for SeH. Gibson et al. (73) reported the photoionization yield curves of SeH, the dissociation energy D_0 for the SeH–H system, the dissociation energy for SeH, and the ionization potential of SeH.

A few theoretical investigations on SeH are the CNDO/2-FPP calculations on the polarizability (74), the SCF/SDCI calculations of Pettersson and Langhoff (43) on the ground state, and the relativistic calculations of the author and co-workers (75) on SeH and SeH^+.

In another study on SeH (75), CASSCF/SOCI/RCI calculations were made on the $X_1^2\Pi_{3/2}, X_2^2\Pi_{1/2}, A^2\Sigma_{1/2}^+$, $\frac{3}{2}$(II), and $\frac{5}{2}$ states of SeH. Spectroscopic properties of the bound states (R_e, T_e, ω_e) were computed for the electronic states of SeH. The calculated dissociation energy for the ground state, including spin–orbit effects, was 3.18 eV. The spin–orbit splitting constant for the ground state was calculated to be -1934 cm^{-1}. The $\frac{5}{2}$ and $\frac{3}{2}$(II) excited states of SeH, which are yet to be observed, exhibit interesting avoided crossings. The RCI

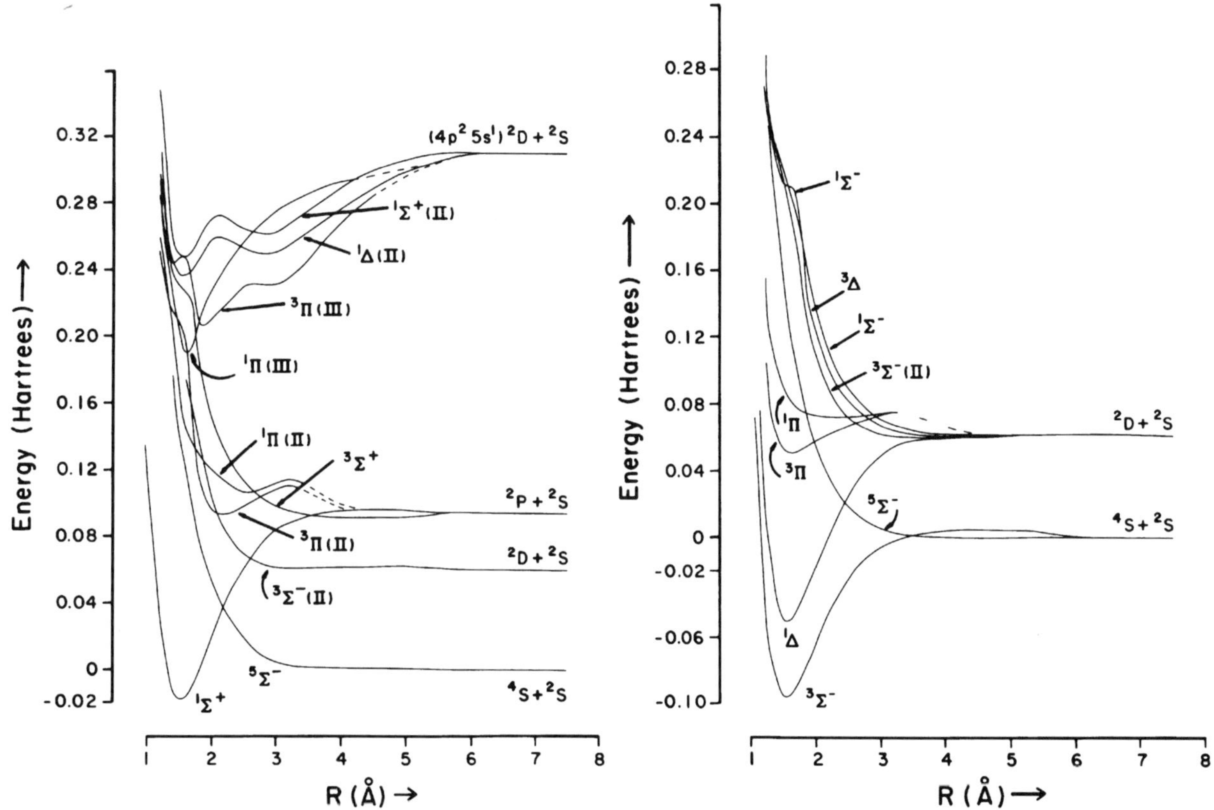

Figure 2.5 Potential-energy curves for the low-lying singlet electronic states of AsH. (Reprinted from Ref. 61; copyright 1989, Academic Press, Inc.)

calculations were also carried out on five low-lying electronic states of SeH^+ (75).

Table 2.5 shows the theoretical and available experimental spectroscopic properties of the low-lying bound electronic states of SeH and SeH^+. Figure 2.6 shows the potential-energy curves of the low-lying states of SeH, obtained by using the SOCI and RCI methods, respectively. As seen from Table 2.5, the calculated R_e, ω_e, and D_e values of the ground state are in very good agreement with the experimental values, although the experimental ω_e is somewhat uncertain, since the $A^2\Sigma^+$–$X^2\Pi_i$ spectral bands were found to be diffuse. The $^2\Pi_{3/2}$–$^2\Pi_{1/2}$ spin–orbit splitting was calculated (75) to be 1934 cm^{-1}, compared to the experimental value (71) of 1764 cm^{-1}. The diffuse spectra observed by Lindgren (64) led to an experimental estimate of the T_e value of the upper state (31,500 cm^{-1}). The theoretical T_e value (75) for the $^2\Sigma^+_{1/2}$ state was found to be 32,872 cm^{-1}, in good agreement with this estimate. The earlier tentative assignment of observed bands in this region to the $^2\Sigma^+_{1/2}$–$^2\Pi_{3/2}$ transition was thus confirmed by the author and co-workers (75).

Lindgren (64) obtained an approximate $D_0 = 3.2 \pm 0.2$ eV for SeH by extrapolating the predissociated bands to Se(^1D) + H(^2S). A more definitive D_e of 3.22 eV was obtained by Gibson et al. (73) using photoionization techniques. The theoretical D_e of 3.18 eV was found to be in excellent agreement with these values (75).

It appears that the $\frac{1}{2}$(II)–$X_2(\frac{1}{2})$ transition has not been observed, which was predicted to be in the region of 30,938 cm^{-1} (75). In order to observe this transition, the $X_2(\frac{1}{2})$ state should be populated appreciably. This transition cannot probably be observed in emission, since the $\frac{1}{2}$(II) state was found to be predissociated.

The theoretical adiabatic ionization energy of SeH was found to be 9.05 eV, while the corresponding SeH–SeH^+ separation at long distance was calculated as 8.86 eV (75). The long-distance separation corresponds to the ionization potential of the Se atom, which is experimentally known to be 9.75 eV (306). Thus the theoretical atomic IP was found to be in 9% error compared to the experimental result. The corrected theoretical IP of 9.88 eV for SeH was found to be in excellent agreement with an adiabatic IP of 9.845 ± 0.003 eV reported by Gibson et al. (73).

The ground state of SeH was found to be predominantly $^2\Pi(1\sigma^2 2\sigma^2 1\pi^3)$ at near-equilibrium distances. At larger internuclear separations (R = 7.00 bohr), mixing with other configurations such as $1\sigma^2 2\sigma 3\sigma 1\pi^3$ and $1\sigma^2 3\sigma 2 1\pi^3$ increases. The spin–orbit mixings with other λ–s states such as $^4\Sigma^-_{3/2}$ became significant at larger internuclear distances. The $\frac{1}{2}$(II) and $\frac{5}{2}$ states exhibited interesting avoided crossings. At shorter distances the $\frac{1}{2}$(II) state was found to be predominantly $^2\Sigma^+(1\sigma^2 2\sigma 1\pi^4)$, while at 5.00 bohr this state became predominantly $^4\Sigma^-_{1/2}(1\sigma^2 2\sigma^2 3\sigma 1\pi^2)$. This led to the predissociation shoulder in the $\frac{1}{2}$(II) curve (see Fig. 2.6). This explained why the experimental $A^2\Sigma^+$–$X^2\Pi_i$ absorption bands were found to be diffuse (64). The $\frac{5}{2}$ state was found to be predominantly $^2\Delta_{5/2}$ at short distances but was mainly $^4\Pi_{5/2}$ at intermediate and long

Table 2.5 Spectroscopic Properties of Low-Lying States of SeH and SeH^{+a}

		R_e, Å		T_e, cm^{-1}		ω_e, cm^{-1}		D_e, eV	
System	State	Theory	Exp.	Theory	Exp.	Theory	Exp.	Theory	Exp.
SeH	$X_1\,{}^2\Pi_{3/2}$	1.467	1.464	0.0	0.0	2,371	$(2400)^b$	3.18	3.22
SeH	$X_2\,{}^2\Pi_{1/2}$	1.467	—	1,934	1,764	2,374	—	2.94	—
SeH	${}^2\Pi$	1.466	—	1,063	—	2,376	—	3.19	—
SeH	A(1/2(II))	1.65	—	32,872	$(31{,}500)^b$	2,285	—	—	—
SeH	${}^2\Sigma^+$	1.641	—	32,965	—	2,407	—	—	—
SeH^+	$X\,{}^3\Sigma^-_{0+}$	1.58	—	0.0	—	2,084	—	2.33	—
SeH^+	${}^1\Delta_2$	1.57	—	11,990	—	2,143	—	0.84	—
SeH^+	$X\,{}^1\Sigma^+_{0+}$	1.57	—	23	—	2,135	—	1.63	—
SeH^+	${}^3\Pi_{0+}$	1.73	—	31,978	—	1,548	—	0.64	—

[a] μ_e(SeH) = −0.587 D (Se^-H^+).[43] All other theoretical spectroscopic constants in this table are from Reference 75.
[b] Experimental values are uncertain.

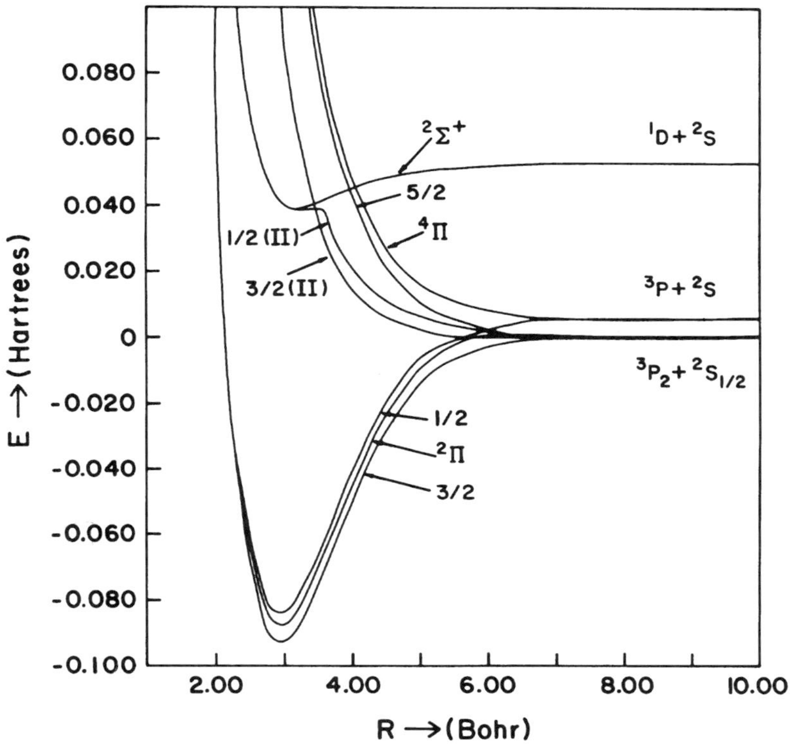

Figure 2.6 Potential-energy curves for the low-lying states of SeH. (Reprinted from Ref. 75; copyright 1987, Elsevier Science Publishers B.V.)

distances. However, this avoided crossing did not lead to a sudden change in the shape of the $\frac{5}{2}$ curve: The $\frac{3}{2}$(II) state was found to be predominantly $^4\Sigma^-_{3/2}$ at short distance. At long distance, however, the $^4\Pi$ and $^2\Pi$ states arising from the $1\sigma^2 2\sigma 3\sigma 1\pi^3$ configuration dominated (75).

The $^3\Sigma^-_{0+}$ ground state of SeH^+ was found to be almost a pure $^3\Sigma^-$ state at R_e. The 2 state of SeH^+ exhibited an interesting avoided crossing. At near R_e the 2 state was found to be $^1\Delta$, while at 5.00 bohr the $^5\Sigma^-$ state crosses with $^1\Delta$. This led to the avoided crossing of $^1\Delta_2$ with $^5\Sigma^-_2$. Thus at long distances the 2 state became $^5\Sigma^-_2$. This leads to a barrier in the potential-energy surface of the 2 state. The 0^+(III) and 1(II) states of SeH^+ exhibited interesting trends as a function of distance. The inflections in the 1(II) surface were attributed to the changes in the contributions of various λ–*s* states as a function of distance (75).

2.1.5 HBr and HBr

The spectroscopic investigations of low-lying electronic states of the HBr molecule have been the subjects of many experimental investigations (75–94).

The vacuum UV absorption spectra of HBr were studied by Price (76) and by Barrow and Stamper (77, 78). Ginter and Tilford (79, 80) reexamined the absorption spectra of HBr in the 66,000–79,500-cm^{-1} region, and assigned the observed bands to transitions from the ground state to the upper states that originated from the $(5s\sigma)\pi^3$, $(5p\sigma)\pi^3$, and $(5p\pi)\pi^3$ excited configurations.

The absorption continua near 56,500 cm^{-1} were attributed mainly to transitions between the ground state and the repulsive $A^1\Pi$ state, arising from the $(\sigma * \pi^3)$ configuration. Mulliken (82, 83) analyzed the electronic spectra of heteronuclear polar diatomic molecules.

The ground state of HBr has been studied experimentally and theoretically (89–94). *Ab initio* spectroscopic constants (91), dissociation energies, and electric dipole moments (90, 91) of the ground state have been obtained, and were found to be in excellent agreement with known experimental results.

The HBr^- anion and other halide ions have been the subjects of a large number of studies (95–121, 123–129), since they can be formed by electronic attachment. The mechanisms that govern the HBr^- production and electron attachment processes are not entirely clear. In the fixed finite dipole model (110), an infinite number of bound states of an electron are predicted for attachment of an electron into a dipole field that exceeds 1.625 D, while no bound states of the electron are predicted for attachment into dipole fields below this value. In the absence of nonadiabatic effects, HBr^- ($\mu = 0.82$ D) was predicted to be unstable with respect to autodetachment of the electron, since its dipole moment is known to be below the critical value of 1.625 D.

The formation of H^- ions by dissociative attachment between 3 and 9 eV was observed in HBr and HI. The ^2P ground states of the halogen atoms are split by spin–orbit coupling into the $^2P_{3/2}$ and $^2P_{1/2}$ atomic states, providing two exit channels for H^- production. The upper $^2\Sigma^+$ and $^2\Pi$ states of HBr^- and HI^- are similarly split up by the spin–orbit coupling into their components ($^2\Sigma^+_{1/2}$, $^2\Pi_{3/2}$, and $^2\Pi_{1/2}$). The $^2\Sigma^+_{1/2}$ and $^2\Pi_{3/2}$ components should dissociate into the lower $H^-(^1S_0) + X(^2P_{3/2})$ limit, while the $^2\Pi^{1/2}$ state should dissociate into the upper $H^-(^1S_0) + X(^2P_{1/2})$ asymptote. The excited $^2\Sigma^+_{1/2}$ state of HBr^- was suggested to dissociate through both the $H^-(^1S_0) + Br(^2P_{3/2})$ and $H^-(^1S_0) + Br(^2P_{1/2})$ channels.

Chapman et al. (129) carried out accurate CASSCF/SOCI calculations employing large diffuse Gaussian basis sets on HBr^- and HI^-. The theoretical SCF/RCI spectroscopic constants (94) for the electronic states of HBr and the corresponding experimental values are given in Table 2.6. The actual potential-energy curves of 12 electronic states of HBr obtained by Chapman et al. (94) are shown in Figure 2.7. Figure 2.8 shows the dipole moment curves for the $X^1\Sigma^+$ ground state of HBr obtained by using three different theoretical methods and the plot of known experimental values. Figure 2.9 shows the dipole moment curves for the excited electronic states of HBr.

As seen from Figure 2.8, the ECP dipole moment curve exhibits the correct behavior between 2 and 3.8 bohr. The experimental μ becomes divergent from theoretical results for $R > 3.7$ bohr, while there are no CEPA μ values for long

Table 2.6 Spectroscopic Constants for HBr[e]

State	R_e, Å		ω_e, cm^{-1}		$\vec{\mu}$,[a] D		D_e, eV		T_{vert},[b] cm^{-1}	T_e, cm^{-1}	
	Theory	Exp.[c]	Theory	Exp.	Theory	Exp.[c]	Theory	Exp.[c]	Theory	Theory	Exp.[c]
$X\,^1\Sigma^+$	1.455	1.414	2,645	2,648	0.870	0.819[d]	3.72	3.76	0	0	0
$b^3\Pi_2$	1.488	[1.473]	2,445	—	−1.057	—	3.04	—	68,300	67,862	[67,663.0]
$b^3\Pi_1$	1.489	1.442	2,496	[2,444]	−1.752	—	2.74	—	71,154	70,276	67,180
$b^3\Pi_{0^+}$	1.470	[1.455]	2,518	2,452	−1.036	—	2.44	—	73,042	72,646	68,998
$C\,^1\Pi_1$	1.51	1.46	2,512	2,552	−1.951	—	—	—	74,889	73,963	70,578

[a] Dipole moments are tabulated at R_e for the respective states. Positive value of $\vec{\mu}$ indicates polarity of H^+Br^-.
[b] T_{vert} vertical transition energy measured from r_e of the ground state.
[c] From Reference 32; numbers in square brackets are not certain.
[d] Reference 85.
[e] All the theoretical values are from Reference 94.

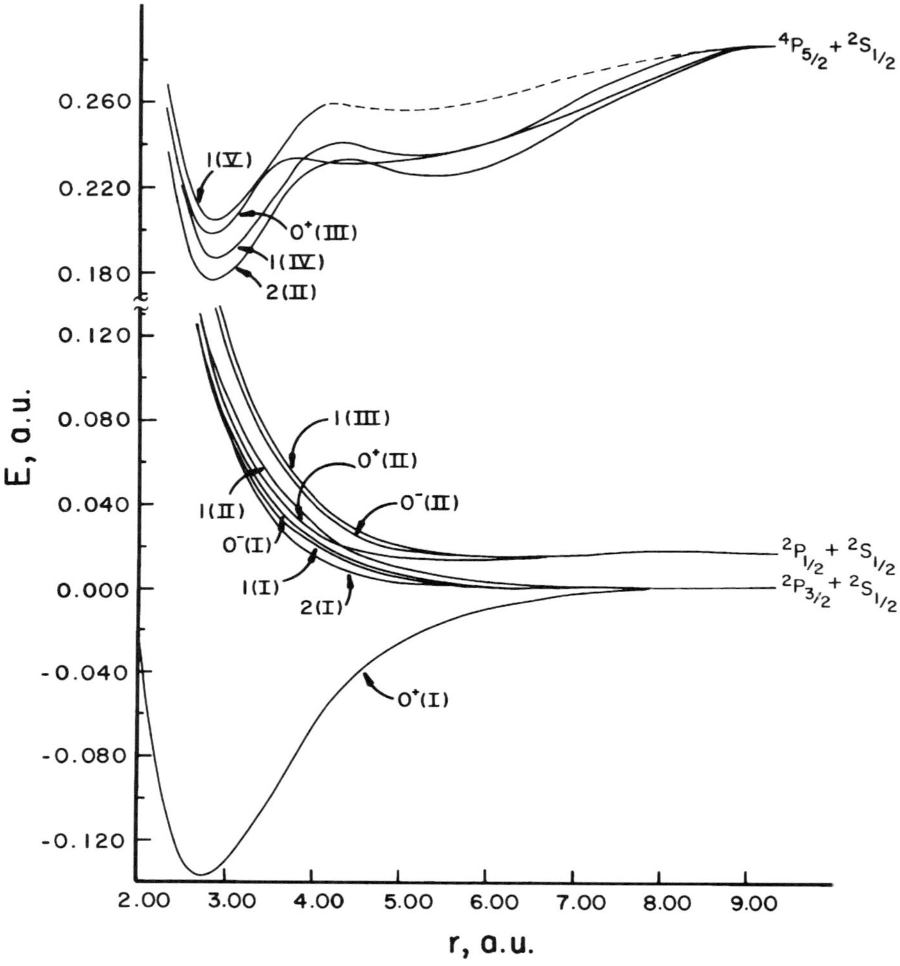

Figure 2.7 Potential-energy curves for the low-lying states of HBr. (Reprinted from Ref. 94; copyright 1987, Elsevier Science Publishers B.V.)

distances. The RCI curve and the MCSCF curve of Ogilvie et al. (92) exhibit a maximum. The ECP dipole moments of HBr are within 6% of the all-electron CEPA results (91).

The theoretical spectroscopic constants in Table 2.6 are in excellent agreement with available experimental results. It is clear from Table 2.6 and Figure 2.7 that the ground state is well separated from the other bound states of HBr. Most of the other states that can possibly be seen near-visible–UV regions were found to be repulsive (Fig. 2.7).

As is well known, the ground state of HBr arises from the $1\sigma^2 2\sigma^2 1\pi^4$ configuration. The bromine–hydrogen bond results primarily from interactions

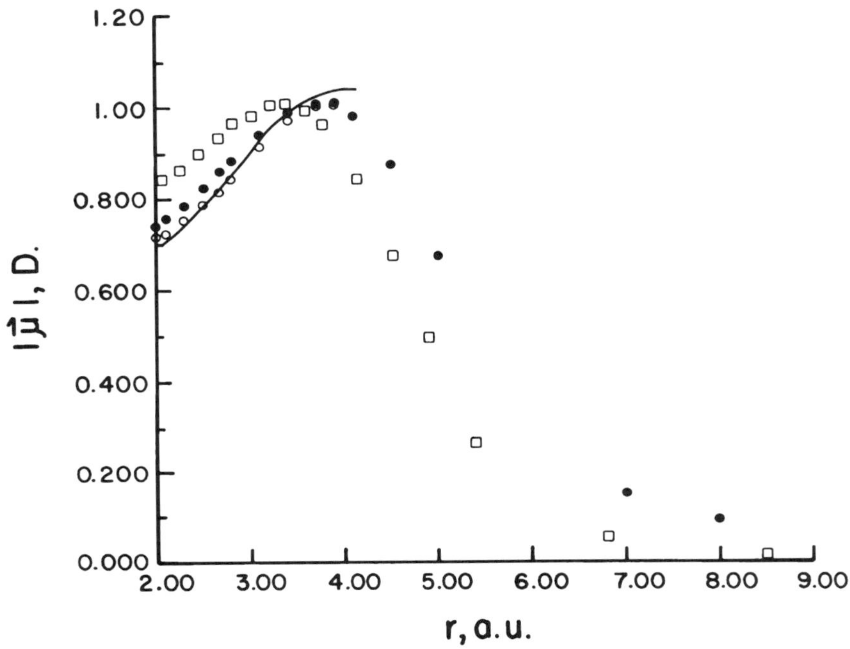

Figure 2.8 Dipole moment curves for the ground state of HBr using CEPA (0), RCI (●), and MCSCF (□) methods. (Reprinted from Ref. 94; copyright 1987, Elsevier Science Publishers B.V.)

between the Br $4p_z$ and H $1s$ orbitals to form the 2σ orbital. The 1σ orbital is predominantly of Br $4s$ character. The π electrons are localized on Br and are essentially nonbonding; promotion of a π electron into the lowest virtual antibonding 3σ orbital gives rise to $^3\Pi_i$ and $^1\Pi$ λ–s states, which are split by spin–orbit interaction into $2,1,0^+$, 0^-, $1(^1\Pi)$ ω–ω states. These states form the repulsive manifold of the curves shown in Figure 2.7. The excitation of a 2σ electron from the ground-state configuration into the virtual 3σ orbital results in the $^1\Sigma^+$ and $^3\Sigma^+$ states. The $^1\Sigma^+$ state dissociates into the ionic species (Br^- + H^+), while the $^3\Sigma^+$ state dissociates into the ground-state limit. [See the 0^-(II) and 1(III) curves of Fig. 2.7.] Between the ground state and the terms that arise from the $1\sigma^2 2\sigma^2 3\sigma 1\pi^3$ and $1\sigma^2 2\sigma 3\sigma 1\pi^4$ ($^3\Sigma^+$) configurations, one spectral band was found experimentally (32). The absorption continuum with a maximum near 56,500 cm^{-1} was assigned to the $A^1\Pi \leftarrow X$ absorption (94). The A state corresponds to the 1(II) curve in Figure 2.7. The calculated vertical transition energy to the 1(II) state was 57,100 cm^{-1} (94) in good agreement with experiment.

The 0^+(II), 1(I), 1(II), 1(III), 0^-(I), 0^-(II), and 2(I) ω–ω states were found to be opposite in polarity (Br^+H^-) to the ground state (Br^-H^+). Thus, these states may be characterized as charge-transfer states that arise from excitation into the antibonding 3σ orbital. Figure 2.10 shows electronic transition moment curves

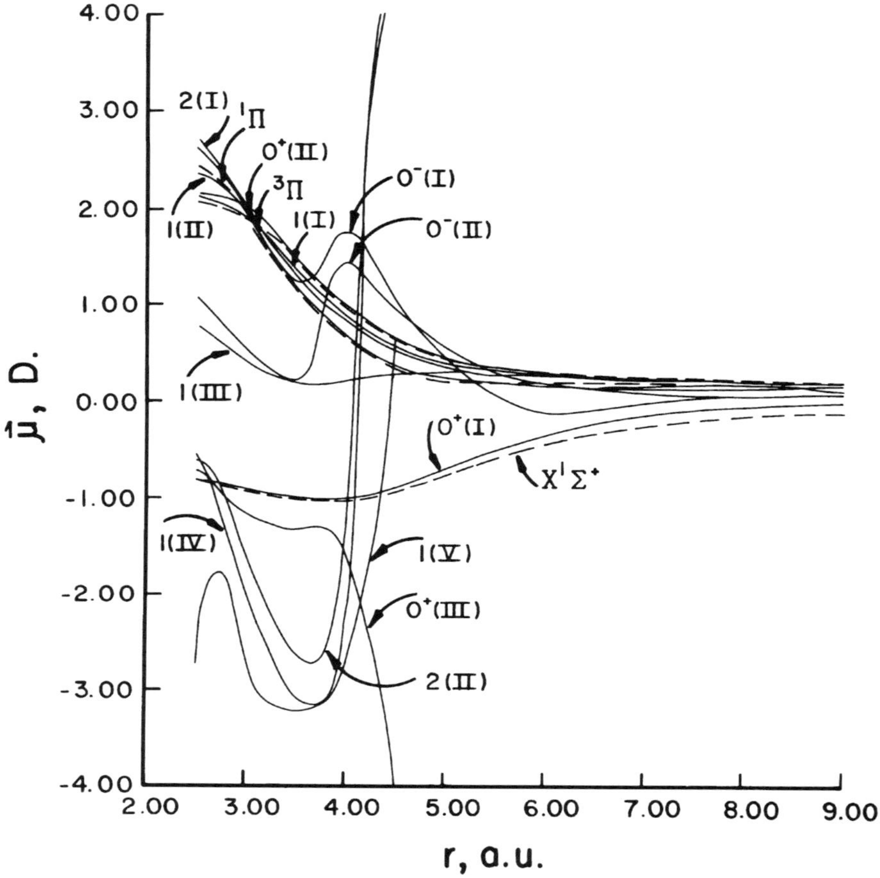

Figure 2.9 Dipole moment curves for the excited states of HBr. (Reprinted from Ref. 94; copyright 1987, Elsevier Science Publishers B.V.)

for the parallel $0^+(\mathrm{I}) \rightarrow 0^+(\mathrm{II})$ and perpendicular $0^+(\mathrm{I}) \rightarrow 1(I), 0^+(\mathrm{I}) \rightarrow 1(\mathrm{II})$, and $0^+(\mathrm{I}) \rightarrow 1(\mathrm{III})$ transitions. The transition moments without the spin–orbit terms for the $A^1\Pi$–$X^1\Sigma^+$ transitions are also shown in Figure 2.10. The transition moment $W[0^+(\mathrm{I}) \rightarrow 1(\mathrm{II})]$ is much larger than the other moments $W[0^+(\mathrm{I}) \rightarrow 1(\mathrm{I})]$, $W[0^+(\mathrm{I}) \rightarrow 0^+(\mathrm{II})]$, and $W[0^+(\mathrm{I}) \rightarrow 1(\mathrm{III})]$ at the ground state R_e. At this distance, comparison of the transition moments with and without the spin–orbit coupling term $[W(0^+(\mathrm{I}) \rightarrow 1(\mathrm{II})], W[X^1\Sigma^+ \rightarrow A^1\Pi)]$ gives the magnitude of spin–orbit contamination in the 1(II) state (Fig. 2.9). At 2.73 bohr, the 1(II) wavefunction was found to be 64% $^1\Pi_1$ from the $1\sigma^2 2\sigma^2 3\sigma 1\pi^3$ configuration and 11% $^3\Pi_1$. Consequently, the transition moment $W[0^+(\mathrm{I}) \rightarrow 1(\mathrm{II})]$ is lowered by about 5% relative to $W(X^1\Sigma^+ \rightarrow A^1\Pi)$, due to singlet–triplet mixing in the 1(II) state. This state exhibited interesting behavior at long distances.

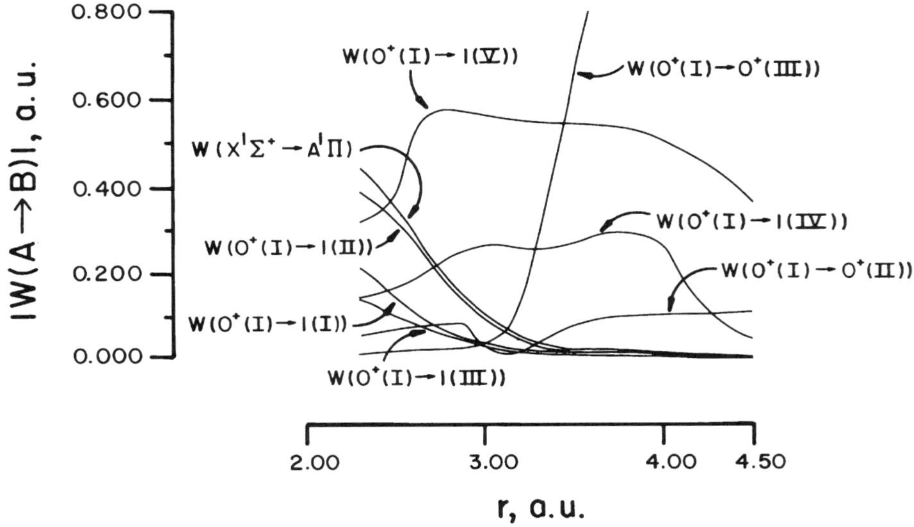

Figure 2.10 Electronic transition moment curves for the excited states of HBr. (Reprinted from Ref. 94; copyright 1987, Elsevier Science Publishers B.V.)

Up to 5.00 bohr, this state was found to be predominantly $^1\Pi_1(1\sigma^2 2\sigma^2 3\sigma 1\pi^3)$; near 7.00 bohr, however, the $^1\Pi$ curve underwent an avoided crossing with the $^3\Sigma_1^+$ state from the $1\sigma^2 2\sigma 3\sigma 1\pi^4$ configuration. Thus, for distances larger than 7.00 bohr, the 1(II) state became predominantly $^3\Sigma_1^+$, while at these distances the 1(III) state became predominantly $^1\Pi$, which is reflected in the transition moment curves.

The excitation of a π electron from the ground-state configuration of HBr into the 5σ Rydberg orbital generates $^3\Pi_i$ and $^1\Pi\lambda$–s states. These curves correspond to the 2(II), 1(IV), 0^+(III), and 1(V) curves (Fig. 2.7). Ginter and Tilford (79) discussed transitions from the ground state of HBr into the $^3\Pi_i$ and $^1\Pi$ states derived from a $(5s\sigma)\pi^3$ configuration, which were designated as the $b^3\Pi_i \leftarrow X^1\Sigma^+$ and $C^1\Pi \leftarrow X^1\Sigma^+$ bands. The ω–ω states arising from the $1\sigma^2 2\sigma^2 5\sigma 1\pi^3$ configuration dissociate into the $(^4P_{5/2} + {}^2S_{1/2})$ asymptote at long distances. The calculated $^2P_{3/2}$–$^4P_{5/2}$ splitting (94) of 63,164 cm^{-1} was found to be in excellent agreement with the experimental value of 63,529 cm^{-1} (306). Since the $J = \frac{3}{2}$ and $\frac{1}{2}$ states of ^4P are within only 3447 cm^{-1} of $^4P_{5/2}$, states dissociating into these limits could mix strongly.

The theoretical spectroscopic constants for the 2(II), 1(IV), 1(V), and 0^+(III) states are given in Table 2.6. The theoretical constants are in generally good agreement with the experimental values, although the 1(V) wavefunction was not considered to be very reliable past 4.50 bohr by Chapman et al. (94).

In Figure 2.10, the electronic transitions labeled $0^+(\mathrm{I}) \rightarrow 1(\mathrm{IV})$, $0^+(\mathrm{I}) \rightarrow 0^+$-(III), and $0^+(\mathrm{I}) \rightarrow 1(\mathrm{V})$ correspond to the $b^3\Pi_1 \leftarrow X^1\Sigma^+, b^3\Pi_{0^+} \leftarrow X^1\Sigma^+$, and $C^1\Pi_1 \leftarrow X^1\Sigma^+$ transitions assigned by Ginter and Tilford (79).

The intensity estimates for the (0–0) bands of the $b^3\Pi_2 \leftarrow X^1\Sigma^+, b^3\Pi_1 \leftarrow X^1\Sigma^+, b^3\Pi_{0^+} \leftarrow X^1\Sigma^+$, and $C^1\Pi_1 \leftarrow X^1\Sigma^+$ transitions of HBr were presented by Ginter and Tilford (79). The ratio of the intensities of $C^1\Pi_1 \leftarrow {}^1\Sigma^+$ (0–0) to $b^3\Pi_1 \leftarrow X^1\Sigma^+$ (0–0) bands was found to be 5 : 1, while the relative intensity of the (0–0) bands of $b^3\Pi_1 \leftarrow X^1\Sigma^+$ to $b^3\Pi_{0^+} \leftarrow X^1\Sigma^+$ was found to be 1 : 0.02 without rotational line strengths. The ratio of theoretical intensities for transitions $0^+(\mathrm{I}) \rightarrow 1(\mathrm{V})$ to $0^+(\mathrm{I}) \rightarrow 1(\mathrm{IV})$ is 4.6 : 1, in good agreement with experiment. The intensity ratio for the (0–0) bands of $0^+(\mathrm{I}) \rightarrow 1(\mathrm{IV})$ and $0^+(\mathrm{I}) \rightarrow 0^+(\mathrm{III})$ near R_e for these states was found to be approximately 1 : 0.01 (94), again a reasonable agreement with experiment.

Figure 2.11 shows the RCI potential-energy curves for the $\frac{1}{2}$(I), $\frac{1}{2}$(II), $\frac{1}{2}$(III), and $\frac{3}{2}$(I) states of HBr^- and the potential-energy curves for their probable parent HBr^- states. The electron affinity of HBr $\{EA_{AD} = E[HX(R_e^{HX})] - E[HX^-(R_e^{HX^-}]\}$ was calculated as −0.326 eV (RCI), while the SOCI results gave a value of −0.485 eV. Both results indicated that the HBr^- anion should be unstable with respect to electron autodetachment. The formation of H^- ions by dissociative electron attachment between 5 and 10 eV in HBr has been the topic of considerable discussion.

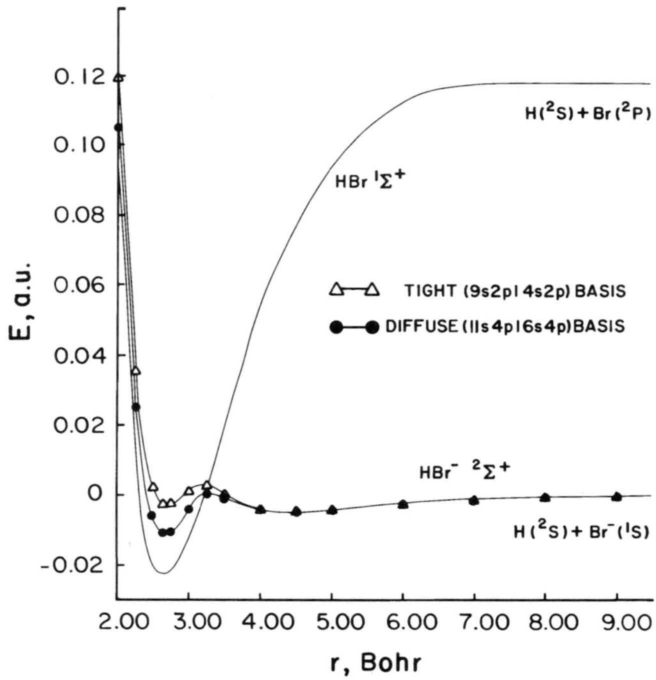

Figure 2.11 Potential-energy curves for the low-lying states of HBr and HBr^-. (Reprinted from Ref. 129; copyright 1988, American Institute of Physics.)

2.1.6 InH

The author (141) carried out CASSCF/SOCI calculations on several electronic states of InH. Table 2.7 shows the experimental spectroscopic constants for InH collected by Huber and Herzberg (32) for the ground state, together with the theoretical constants (141) for six electronic states. The computed potential-energy curves are shown in Figure 2.12. In summary, many experimental works (130–140) on InH have led to the observation of the *a*–*X* and *A*–*X* transitions. The *a* and *A* states were both found to be predissociated. The D_0° value of 2.48 eV was estimated from the predissociation of the *A*–*X* and *a*–*X* bands. The theoretical SOCI D_e value of InH was found to be 2.60 eV (141). The Λ-type doublings were found in some of the electronic states of InH. As seen from Table 2.7, overall, the computed results agree quite well for the two of the experimentally characterized electronic states.

Veseth and Lofthus (139) suggested the existence of a $^1\Sigma^+$ state that was presumed to be repulsive and responsible for the anomalies seen in the Λ doubling of the $a\ ^3\Pi_1$ component. The $^1\Sigma^+$ state was also suggested as being responsible for the predissociation of all the $a^3\Pi$ components. On the basis of the selection rules for predissociation, the $^3\Sigma_{0^-}^+$ and $^3\Sigma_1^+$ states can interact with the $^3\Pi_{0^-}$ and $^3\Pi_1$ components through spin–orbit coupling. Thus the $^3\Sigma^+$ state can predissociate these components. The predissociation of $^3\Pi_2$ and $^3\Pi_{0^+}$ through $^3\Sigma^+$ cannot occur unless the molecule is rotating. Consequently, if $^3\Pi_2$ and $^3\Pi_{0^+}$ states are predissociated by $^3\Sigma^+$, they should show dependence of predissociation rate on the rotational *J* quantum number. The author's calculations (141) reveal that there is a $^3\Sigma^+$ curve for InH that dissociates into the ground-state atoms. For GaH, the $^3\Sigma^+$ curve was found to be repulsive, and lies above the $a^3\Pi$ and $A^1\Pi$ curves.

Bahnmaier et al. (315) obtained the infrared spectrum of InH in its $^1\Sigma^+$ ground state using a diode laser spectrometer. A total of 83 transitions were measured, yielding the most accurate Dunham parameters for the ground state and an improved R_e = 1.83776 Å.

Table 2.7 Spectroscopic Constants for InH[a]

State	R_e, Å	T_e, cm^{-1}	ω_e, cm^{-1}	μ_e, D	D_e, eV
$X\,^1\Sigma^+$	1.838(1.820)	0	1,476(1,524)	(0.460)	2.48(2.60)
$a^3\Pi_{0^-}$	1.776	16,230	[1,303]	—	—
$a^3\Pi_{0^+}$	1.779	16,278	1,459	—	—
$a^3\Pi_1$	1.768	16,942	1,415	—	—
$a^3\Pi_2$	1.753	17,800	[1,301]	—	—
$A^1\Pi$	[2.09]	22,655	[142]	—	—

[a]Experimental values are from Reference 32. Theoretical values in parentheses are from Reference 141. The dipole moment was obtained for a theoretical CASSCF/SOCI calculation of Balasubramanian (141). The numbers in square brackets are uncertain.

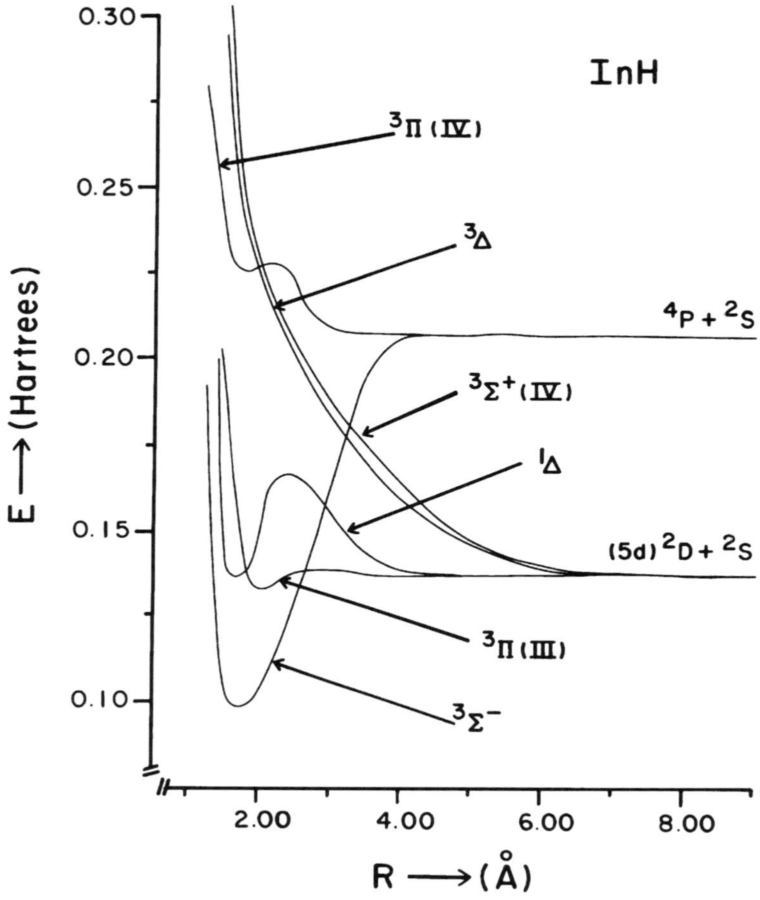

Figure 2.12 Potential-energy curves of InH obtained without the spin–orbit term. (Reprinted from Ref. 194; copyright 1988, American Institute of Physics.)

2.1.7 SnH

The early electronic spectra of SnH were recorded by Watson and Simon (142, 143) in 1939–1940. The spectra contained bands in the red and two red-degrading bands in the violet. The red bands and the violet bands were assigned to $^2\Sigma^+$–$^2\Pi$ and $^2\Delta$–$^2\Pi$ transitions, respectively ($^2\Pi$ is the ground state). Kleman (131) reinterpreted the red bands as due to a $^4\Sigma^-$–X^{Π} transitions by analogy to lighter group IV hydrides. Klynning and co-workers (145, 146) also recorded the absorption spectra of SnH and SnD. Energy expressions for these bands were fitted. Centrifugal distortion of the spin–orbit and spin–rotation interactions for several states of SnH was considered. Theoretical analysis of the spin–orbit splitting of the $A^2\Delta$ and $a^4\Sigma^-$ states was considered by Veseth (151).

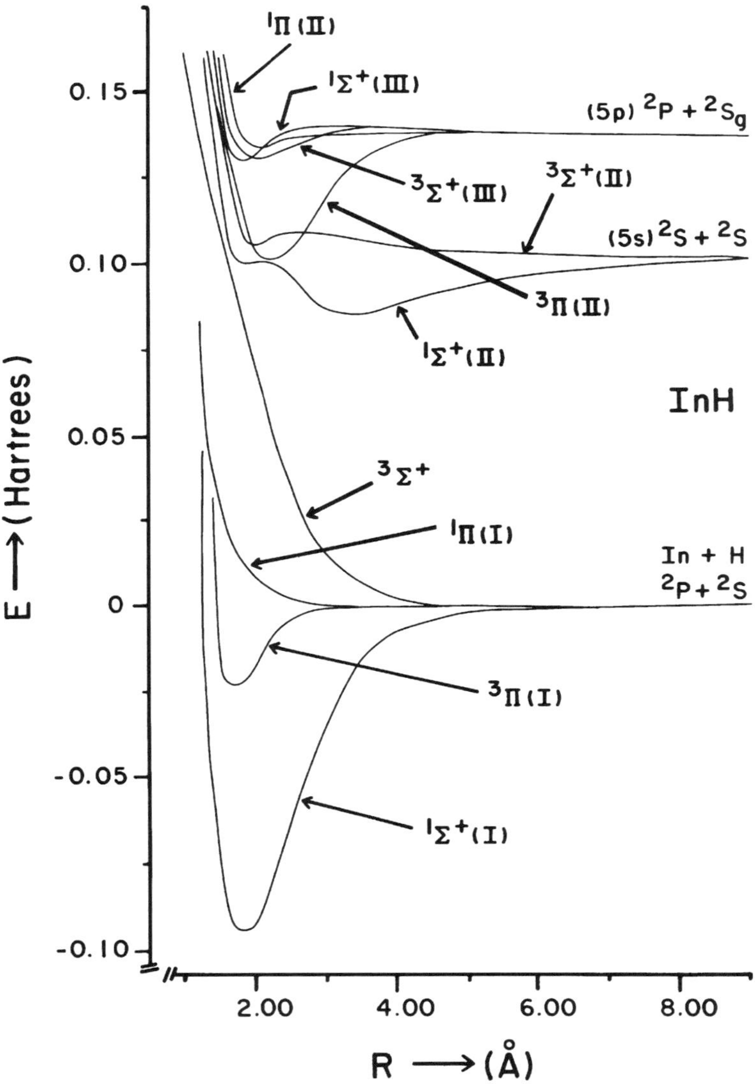

Figure 2.12 (*Continued*)

The experimental works and their analysis of SnH are contained in References 142–152.

The author and Pitzer (152) made relativistic SCF/RCI calculations including the spin–orbit term for eight low-lying electronic states of SnH. The spectroscopic constants and the potential-energy curves for these states were calculated by using this method employing a double-ζ STO basis set. Chapman et al. (55), in a latter investigation employed a more accurate triple-ζ STO basis as well as CASSCF/SOCI method in conjunction with large valence Gaussian basis sets. The dipole moment curves for the electronic states of SnH in addition to the transition moment curves as a function of internuclear distance have now been obtained (55). Table 2.8 shows the spectroscopic constants of SnH and comparison with known experimental results. Figure 2.13 shows the SCF/RCI potential-energy curves of the low-lying electronic states of SnH. As seen from Table 2.8, the agreement between theory and experiment is very good at the level of calculations employed. The calculated total excitation energy to the $^4\Sigma^-$ term is within 15% of the observed value.

The $A^2\Delta$–$X^2\Pi$ bands of SnH exhibited (145) predissociation with line broadening beginning at the energy of 25,177 cm^{-1}. The SCF/RCI calculations of the author and Pitzer (152) indicated that the $\frac{5}{2}$ state (shown as the uppermost curve in Fig. 2.13) is predominantly $^2\Delta_{5/2}$ at a bond distance of 3.4 bohr but has a broad maximum (Fig. 2.13) arising from an avoided crossing with the $^4\Pi_{5/2}$ state. Thus some quantum tunneling was anticipated, which should result in line broadening to a lower energy for SnH than for SnD. Huber and Herzberg (32) reported an upper estimate for D_0° of SnH as 2.73 eV obtained when dissociation to the $^3P_2$ + $^2S_{1/2}$ atoms was assumed. The calculations of the author and Pitzer (152) revealed a potential maximum substantially above that energy. Hence the dissociation energy of SnH was estimated as 2.3 eV (152).

A more recent and sophisticated CASSCF/SOCI investigation of the electronic states of SnH was carried out by Chapman et al. (55). The RCI ω_e value obtained for the ground state of SnH using a triple-ζ STO basis was found to be 1689 cm^{-1}, agreeing well with the experimental value in Table 2.8 (1715 cm^{-1}). The RCI ω_e values for the excited states of SnH also did not differ significantly from the earlier results (152). The dipole moment (μ_e) of the SnH molecule in the $^2\Pi_{1/2}$ ground state was calculated to be 0.398 D, while the corresponding

Table 2.8 Spectroscopic Properties of SnH[a]

	T_e, cm^{-1}	A or 6λ, cm^{-1}	R_e, Å	ω_e, cm^{-1}
$X^2\Pi$				
Theory	0	2,376	1.83	1,600
Experimental	0	2,179	1.78	1,715
$a^4\Sigma^-$				
Theory	13,265	328	1.78	1,560
Experimental	15,580	270	1.77	1,521
$A^2\Delta$				
Theory ($\frac{5}{2}$)	26,460	—	1.96	700
Experimental	23,790	(20)	1.85	1,080

[a]Theoretical results are from Reference 152. Experimental constants are from Reference 32.

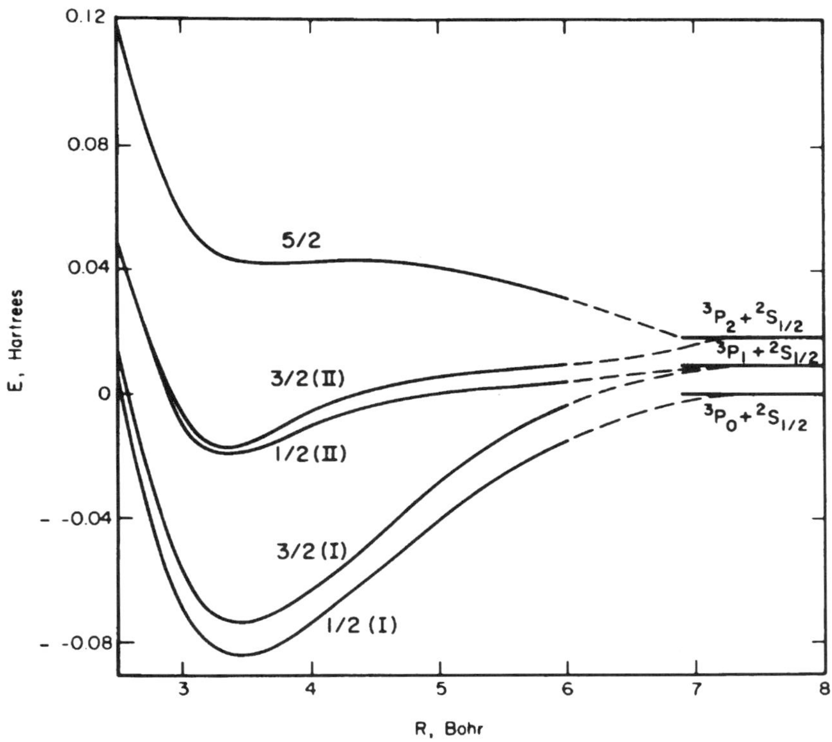

Figure 2.13 Potential-energy curves for the electronic states of SnH. (Reprinted from Ref. 152; copyright 1984, Academic Press, Inc.)

μ_e for the $^2\Pi_{3/2}$ state was calculated as 0.393 D. The dipole moment of the ground state of SnH was calculated by Pettersson and Langhoff (53) using the SCF/SDCI method as 0.357 D with Sn^+H^- polarity.

The calculated dipole moment curves of the various electronic states of SnH and PbH are shown together in Figure 2.14. Figure 2.15 shows the electronic transition moment curves as a function of the internuclear distance for SnH and PbH as obtained by Chapman et al. (55). Note that the dipole moment curves of the $^2\Pi_{1/2}$ and $^2\Pi_{3/2}$ states of both SnH and PbH reach a maximum, eventually reaching zero at long distance. As seen from Figure 2.14, the spin–orbit effects on the dipole moments of the $^2\Pi_{3/2,\,1/2}$ components of PbH are substantially larger.

Comparison of the transition moment curves in Figure 2.15 with the corresponding curves in Figure 2.4 for GeH clearly demonstrates the large relativistic effects in PbH. In particular, note the dramatically large transition moment for the $\frac{1}{2}(\mathrm{I}) \rightarrow \frac{1}{2}(\mathrm{II})$ transition of PbH. The intensity ratios of the $\frac{3}{2}(\mathrm{I}) \rightarrow \frac{3}{2}(\mathrm{II}) : \frac{1}{2}(\mathrm{I}) \rightarrow \frac{1}{2}(\mathrm{II}) : \frac{1}{2}(\mathrm{I}) \rightarrow \frac{3}{2}(\mathrm{II})$ transitions were calculated as

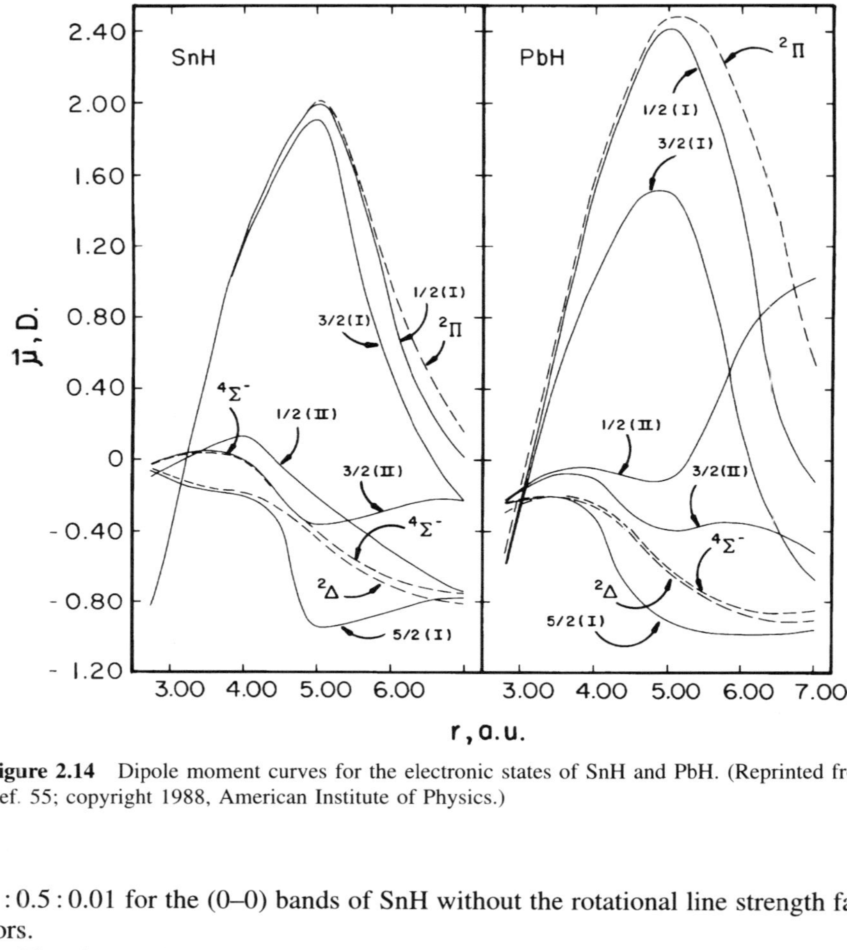

Figure 2.14 Dipole moment curves for the electronic states of SnH and PbH. (Reprinted from Ref. 55; copyright 1988, American Institute of Physics.)

1 : 0.5 : 0.01 for the (0–0) bands of SnH without the rotational line strength factors.

The observed red bands of SnH were resolved into two components by Klynning et al. (145), the first of these was assigned to $^{4}\Sigma^{-}_{3/2} \leftarrow {}^{2}\Pi_{3/2}$, while the second was assigned to the $^{4}\Sigma^{-}_{1/2} \leftarrow {}^{2}\Pi_{1/2}$ transition. The theoretical transition energies for the SnH bands (14, 122, and 12,479 cm^{-1}) (55) agreed with the interpretation of Klynning et al. (145).

2.1.8 SbH

The SbH molecule was studied in the flash photolysis of SbH (153–155) Bollmark and Lindgren (153) obtained the absorption bands of SbH in the 3300–3450-Å region, which were assigned to the $A^{3}\Pi$–$X^{3}\Sigma^{-}$ systems. The rotational constant and R_e of the ground state were obtained from rotational analysis. Basco and Yee (154) studied the $A^{3}\Pi_{0^+}$–$X^{3}\Sigma^{-}$ system and noted that only

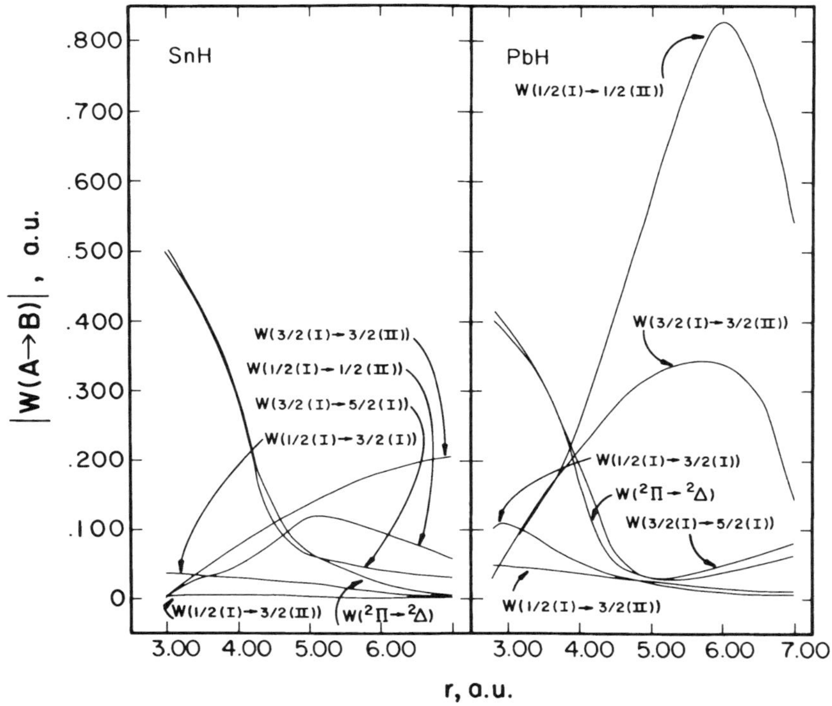

Figure 2.15 Electronic transition moment curves for the electronic states of SnH and PbH. (Reprinted from Ref. 55; copyright 1988, American Institute of Physics.)

the $A^3\Pi_{0^+}$–$X^3\Sigma^-$ system has rotational structure. Bollmark and Lindgren (155) carried out the rotational analysis of the absorption spectra of SbH and calculated the spin–orbit splitting of the ground state (0^+–1 splitting) as 660 cm^{-1}. These authors proposed that there should be BO^+ and CO^+ states that cause perturbation to the $^3\Pi_{0^+}$ bands, leading to predissociation of the $^3\Pi$ state. The other experimental works on SbH can be found in References 156 and 157.

The author, Tanpipat, and Bloor (158) have performed relativistic SCF/RCI calculations on nine low-lying ω–ω and four λ–s states (without spin–orbit coupling) of SbH. A double-ζ STO basis with the 14-electron RECPs was employed for the antimony atom. These calculations were useful in the interpretation of the experimentally observed bands.

Table 2.9 shows the spectroscopic constants of the low-lying electronic states of SbH, while Figures 2.16 and 2.17 show the potential-energy curves of the electronic states of SbH with and without the spin–orbit term, respectively.

The A–X system observed in absorption was assigned to $0^+(\text{III})(^3\Pi_{0^+})$–$X(0^+)$ transition. The theoretical and experimental T_e values are in very good agreement (Table 2.9). The spin–orbit splitting (0^+–1 splitting) of the ground state

Table 2.9 Spectroscopic Properties of SbH[a]

	R_e, Å		T_e, cm^{-1}		ω_e, cm^{-1}	
State	Theory	Exp.	Theory	Exp.	Theory	Exp.
$X_1 0^+$(I)	1.81	1.72	0	0	1,763	—
$X_2 1$	1.81	—	696	655	1,762	—
2	1.81	—	9,362	—	1,784	—
0^+(II)	1.80	—	15,772	—	2,061	—
$A_3 0^+$(III)	1.87	[2.03]	30,788	30,116	1,085	—
0^+(IV)	—	—	34,891	—	—	—
$^3\Sigma^-$	1.81	—	919	—	1,768	—
$^1\Delta$	1.81	—	9,607	—	1,786	—
$^1\Sigma^+$	1.79	—	17,876	—	1,804	—

[a] All theoretical results are from Reference 158.

was calculated as 697 cm^{-1} (158), compared to a value of 660 cm^{-1} obtained by Bollmark and Lindgren (155) from the rotational analysis.

Theoretical calculations of the author et al. (158) predicted the spectroscopic properties of a number of states [1, 2, 0^+(II), 0^+(IV)] of SbH that are yet to be observed. The 0^+(II)–0^+(I) system corresponds to a $^1\Sigma^+_{0^+}$–$^3\Sigma^-_{0^+}$ transition and should be observable, since the spin–orbit effects were found to be nonnegligible. Theoretical calculations predicted this transition in the region of 15,770 cm^{-1} (158). Similarly, the 0^+(IV)–$X0^+$(I) transition should be observable in the 35,000-cm^{-1} region. Bollmark and Lindgren (155) observed a C state that perturbed the bands in the $A^3\Pi_{0^+}$–$X0^+$ system with a T_0 value of 30,566 cm^{-1}. Thus, the 0^+(IV) state could possibly be the C state.

The theoretical dissociation energy of the SbH radical was calculated as 2.26 eV (158). There is no experimental D_e value for the SbH radical at present. Since SCF/RCI calculations yield 85% of the experimental D_e value due to neglect of higher-order correlation corrections, it was predicted that the D_e of SbH should be 2.7 ± 0.2 eV (158).

The electronic states of SbH exhibited avoided crossings analogous to those of BiH, which resulted in exotic shapes of potential curves. For example, the barrier in the 2 curve is due to the avoided crossing of $^1\Delta_2$ with $^5\Sigma^-_2$. Similarly, the 0^- state became predominantly $^5\Sigma^-_{0^-}$ at long distances.

The fact that only the $A^3\Pi_{0^+}$–$^3\Sigma^-_{0^+}$ system shows rotational structure can be explained on the basis of theoretical calculations of the author and colleagues. As seen from Figures 2.16 and 2.17, the $^5\Sigma^-$ curve, which dissociates into the ground-state atom, is repulsive. This state contains $\Omega = 2, 1, 0^-$ components. Consequently, only the $^3\Pi_{0^+}$ state cannot interact with $^5\Sigma^-$ repulsive state. This should explain the experimental findings that only the A–X system shows rotational structure among the $^3\Pi\ \Omega$ components.

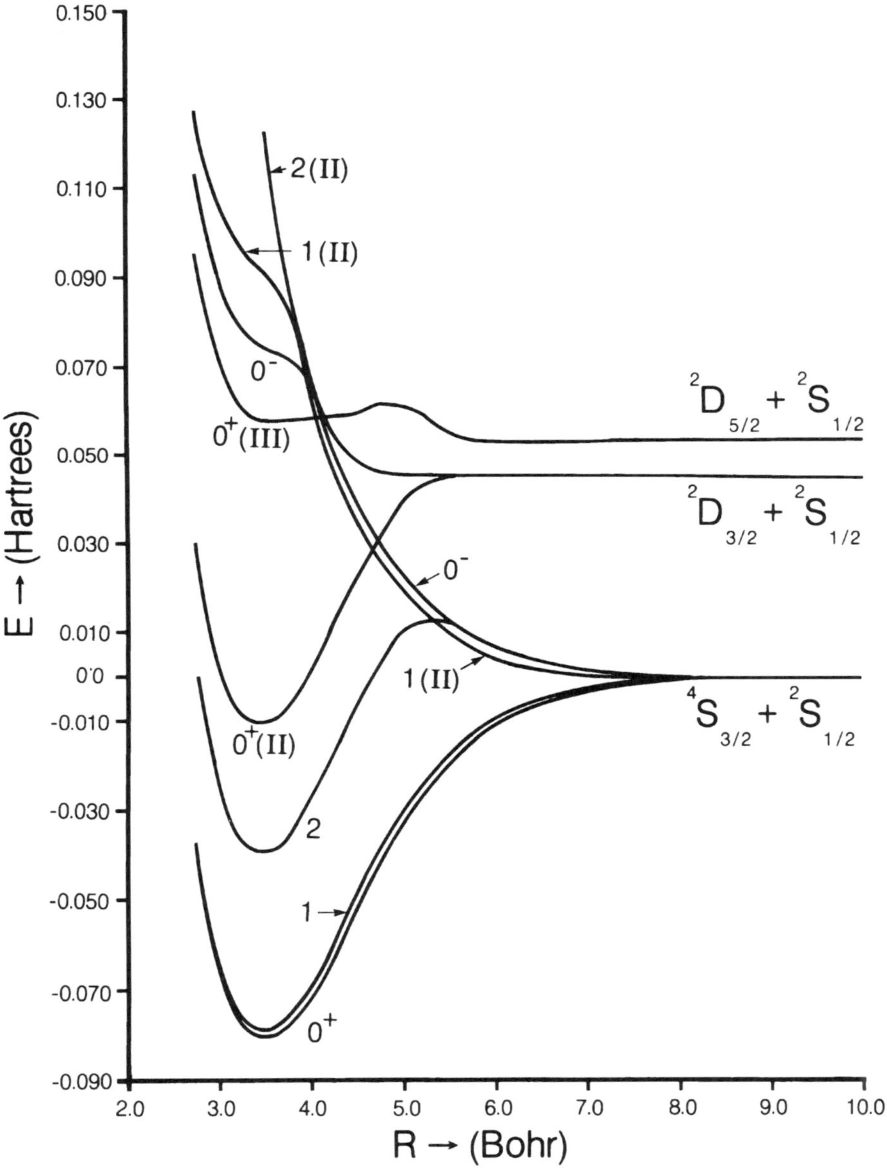

Figure 2.16 Potential-energy curves for the electronic states of SbH with spin–orbit coupling. (Reprinted from Ref. 158; copyright 1987, Academic Press, Inc.)

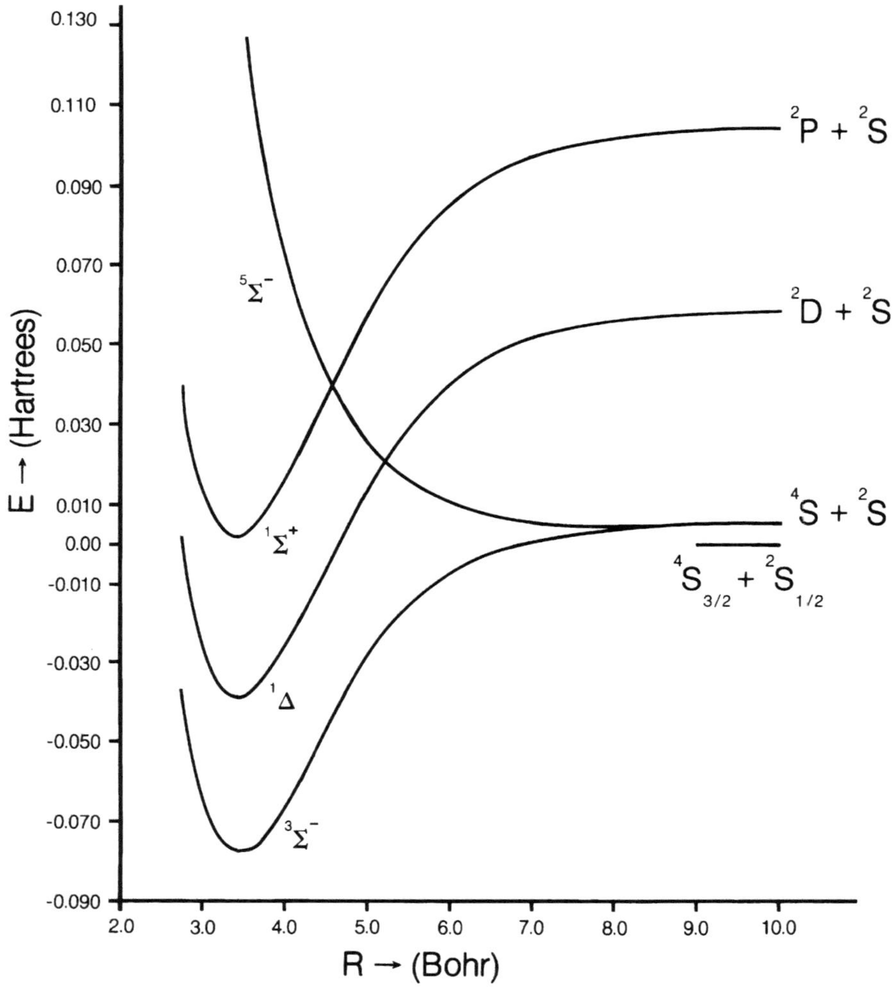

Figure 2.17 Potential-energy curves for the electronic states of SbH without spin–orbit coupling. (Reprinted from Ref. 158; copyright 1987, Academic Press Inc.)

2.1.9 TeH

Little et al. (159) studied the vacuum UV spectra of TeH, although the resolution was insufficient to resolve the rotational structure. The spin–orbit constant for the ground state of TeH was obtained from the ESR spectra. Freidhoff et al. (160) studied the negative-ion photodetachment spectra of TeH^-, which revealed the electron affinity of TeH.

The author, Han, and Liao (161) carried out relativistic SCF/RCI calculations on five low-lying electronic states of TeH. The spectroscopic constants and the potential-energy curves of the low-lying states of TeH were obtained.

Table 2.10 Spectroscopic Constants of Low-Lying States of TeH[a]

	T_e, cm^{-1}		R_e, Å		ω_e, cm^{-1}		D_e, eV	
State	Theory	Exp.	Theory	Exp.	Theory	Exp.	Theory	Exp.
$X_1\frac{3}{2}$	0.0	0.0	1.767	1.74	1,839	—	2.31	—
$X_2\frac{1}{2}$	3,989	3,830	1.77	—	1,845	—	1.81	—
	2,449	—	1.77	—	1,855	—	2.35	—

[a]All theoretical values are from Reference 161.

Table 2.10 shows the theoretical spectroscopic constants together with available experimental data on the low-lying states of TeH. Figure 2.18 shows the calculated potential-energy curves of the $\frac{3}{2}$, $\frac{1}{2}$, $\frac{3}{2}$(II), $\frac{1}{2}$(II), $\frac{5}{2}$, $^2\Pi$, and $^4\Pi$ states. There are no experimental ω_e values for the $\frac{3}{2}$ and $\frac{1}{2}$ states. The calculated dissociation energy of the $X_1(\frac{3}{2})$ ground state is (2.31 eV) 0.04 eV lower than the corresponding $^2\Pi\lambda$–s state. The D_e value of the $\frac{1}{2}$ state is, however, much lower (1.81 eV), due to the significant spin–orbit splitting and contamination of this state. We predicted that the D_e value of the ground state should be 2.75–3.00 eV, although the experimental D_e is not known at present.

Since the $\frac{1}{2}$(II), $\frac{3}{2}$(II), and $\frac{5}{2}$ curves are repulsive (Fig. 2.18), transitions from the two spin–orbit components (X_1, X_2) of the ground state to these states can be seen only in absorption. This is consistent with the known experimental spectra of TeH. There are four absorption bands below 55,000 cm^{-1}, which are labeled $B \leftarrow X_1, C \leftarrow X_1, C \leftarrow X_2$, and $D \leftarrow X_1$. The author, Han, and Liao (161) calculated the vertical separations of the $\frac{3}{2}$(II), $\frac{1}{2}$(II), and $\frac{5}{2}$ states at 3.25 bohr (near R_e of $\frac{3}{2}(\mathrm{X}_1)$). These splittings are 28,568, 39,000, and 52,010 cm^{-1}, respectively. The $\frac{1}{2}$(II)–$\frac{1}{2}$ splitting at this distance is 24,563 cm^{-1}. On the basis of this splitting, the $B \leftarrow X_1$ transition was tentatively assigned to the $\frac{3}{2}$(II) $\leftarrow \frac{3}{2}$ transition.

The $\frac{1}{2}$(II) and $\frac{5}{2}$ states of TeH exhibited interesting avoided crossings. At short distances, the $\frac{1}{2}$(II) state was found to be predominantly $^2\Sigma^+(1\sigma^2 2\sigma 1\pi^4)$. At about 4.00 bohr, it became predominantly $^4\Sigma^-(1\sigma^2 2\sigma^2 3\sigma 1\pi^2)$, although other states such as $^2\Sigma^-$ and $^2\Sigma^+$ also made appreciable contributions. This avoided crossing led to the shoulder in the $\frac{1}{2}$(II) curve (Fig. 2.18). The $\frac{5}{2}$ state was found to be predominantly $^2\Delta_{5/2}$ at short distances but became mainly $^4\Pi_{5/2}$ at intermediate and long distances. The $\frac{3}{2}$(II) state was found to be predominantly $^4\Sigma^-_{3/2}$ at short distances. At long distances, however, the $^4\Pi$ and $^2\Pi$ states arising from the $1\sigma 2^2\sigma 3\sigma 1\pi^3$ configuration dominated.

2.1.10 HI and HI

Among the heavy hydrogen-containing molecules, HI and HBr have been the most studied species for a number of years (76–129, 162–166). The electronic spectra of HI basically consisted of continuous absorption bands near 23,000,

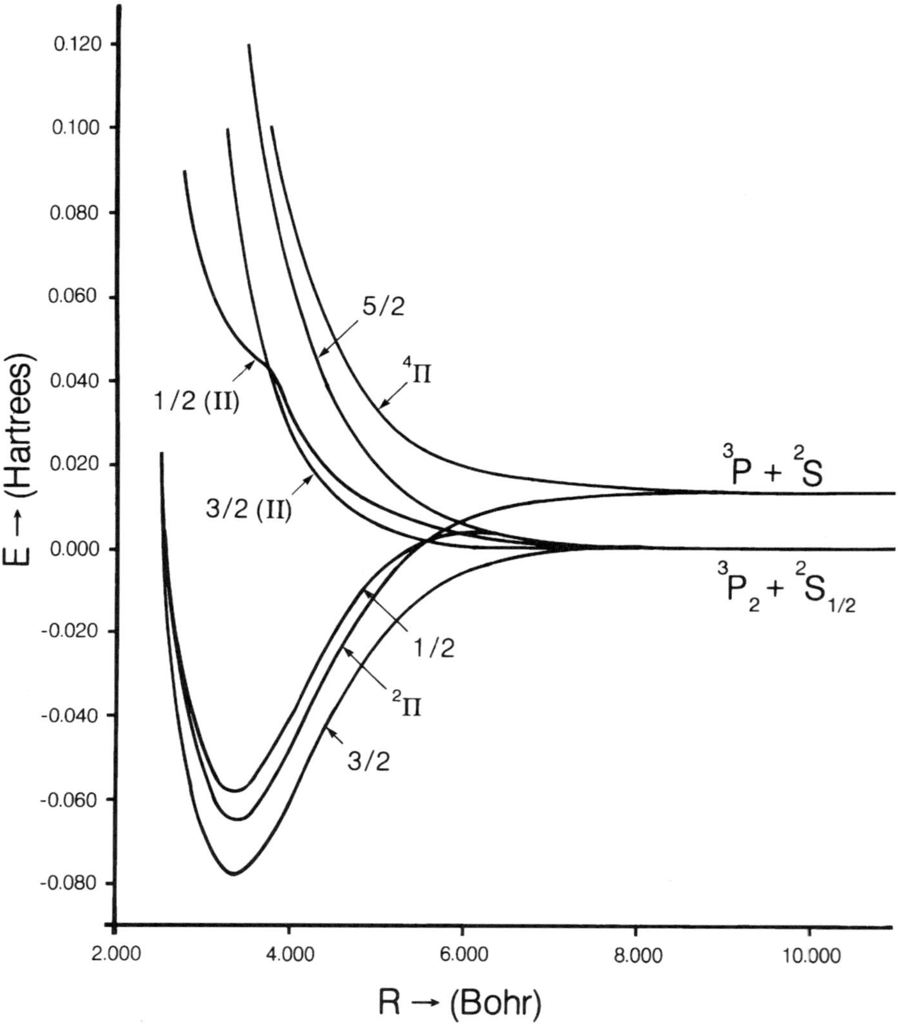

Figure 2.18 Potential-energy curves for the *u* electronic states of TeH. (Reprinted from Ref. 161; copyright 1987, American Institute of Physics.)

46,000, and 55,555 cm^{-1}. The UV continuum spectrum of HI contained a maximum at 45,000 cm^{-1} and was found to be broad and featureless. Mulliken (82, 83) interpreted the observed spectra of HI as due to $^3\Pi_{0^+} \leftarrow {}^1\Sigma^+_{0^+}, {}^3\Pi_1 \leftarrow {}^1\Sigma^+_{0^+}$, and $^1\Pi_1 \leftarrow {}^1\Sigma^+_{0^+}$ transitions. Clear et al. (164) carried out a photofragmentation investigation of HI.

The first *ab initio* relativistic SCF/RCI calculation on the electronic states of HI was made by Chapman, the author, and Lin (122). In this investigation a double-ζ + polarization STO basis set was used for the iodine atom in conjunction with the RECPs. The potential-energy curves of seven low-lying electronic states dissociating into $I(^2P_{1/2})$ + H and $I(^2P_{3/2})$ + H were obtained.

The spectroscopic constants for the electronic states of HI are summarized in Table 2.11. Figure 2.19 shows the potential-energy curves for six low-lying states of HI, including the spin–orbit term as obtained by Chapman, the author, and Lin (122). The atomic $^2P_{3/2}$–$^2P_{1/2}$ splitting of the iodine atom calculated by these authors (6971 cm^{-1}) was found to be in very good agreement with the experimental splitting of 7603 cm^{-1}. As seen from Table 2.11, the agreement between the theoretical constants for the ground state and the experimental values is also very good. The theoretical D_e value for the 0^+(I) ground state of HI (3.03 eV) was found to be almost in exact agreement with an experimental value of 3.05 eV.

The experimental vertical transition energies listed in Table 2.11 are from Ref 164. Clear et al. (164) demonstrated the composite nature of the observed absorption continuum of HI, which was attributed to three overlapping transitions $[1 \leftarrow X, 0^+(\mathrm{II}) \leftarrow X, 1(\mathrm{II}) \leftarrow X]$ whose maxima were found to be located at approximately 39,500, 43,000, and 47,000 cm^{-1}, respectively. Table 2.11 shows these values compared with the theoretical results. In general, theoretical results are expected to be a bit higher because of limitations in basis sets and electron correlation effects.

As mentioned in reviewing HBr^- calculations, Chapman, the author, and Lin (129) carried out both SCF/RCI and CASSCF/SOCI calculations employing a large Gaussian basis set on HI^- and comparable calculations on the ground state of HI. The HI molecule was predicted to have a slightly positive adiabatic electron affinity compared to −0.485 eV for HBr.

Figure 2.20 shows the potential-energy curves for the ground states of HI and HI^-. The diffuse Gaussian basis sets yielded the first minimum in the HI^- ground state, which was found to be absent when a tight basis set was used

Table 2.11 Spectroscopic Properties of Low-Lying States of HI[a]

	T^b_{vert}, cm^{-1}		R_e, Å		ω_e, cm^{-1}	
State	Theory	Exp.[c]	Theory	Exp.	Theory	Exp.
$X0^+$(I)	0.0	0.0	1.66	1.61	2,939	2,309
$a0^+$(II)	49,605	43,000	—	—	—	—
$a1$(I)	46,532	39,500	—	—	—	—
$A1$(II)	54,763	47,000	—	—	—	—

[a]All theoretical results are from Reference 122.
[b]T_{vert} is the vertical separation of the excited states from the ground state at the minimum R value.
[c]Approximate values obtained from measurements taken near the probable absorption curve maxima.

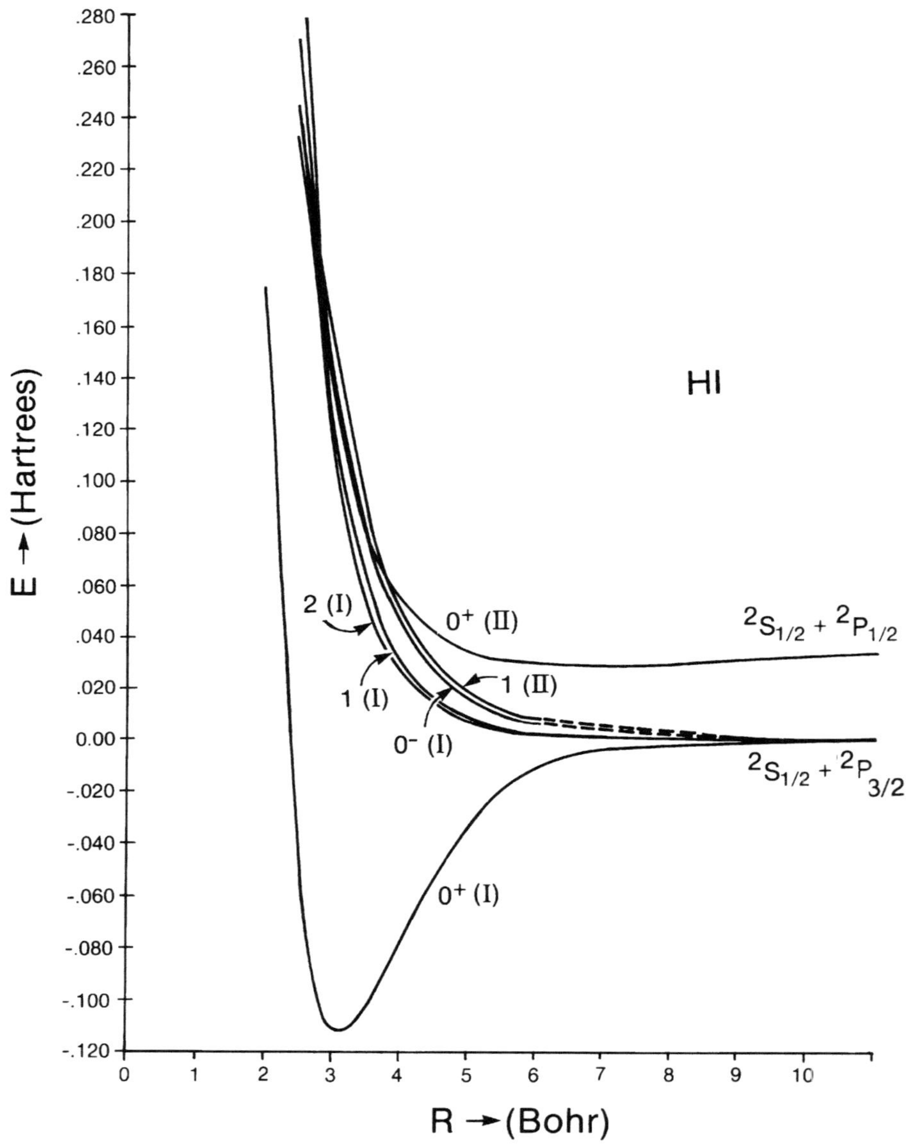

Figure 2.19 Potential-energy curves for the electronic states of TeH. (Reprinted from Ref. 122; copyright 1985, Elsevier Science Publishers B.V.)

(Fig. 2.20). The HI/HI$^-$ curves cross at 3.45 bohr (CASSCF/SOCI). The $^2\Sigma^+_{1/2}$ ground state of HI$^-$ was found to be 0.1–0.2 eV bound at 5 bohr. The theoretical RCI electron affinity of the halogen atom was found to be 2.83 eV compared to an experimental value of 3.06 eV. The CASSCF/SOCI calculations yielded a slightly better EA of 2.99 eV for the iodine atom.

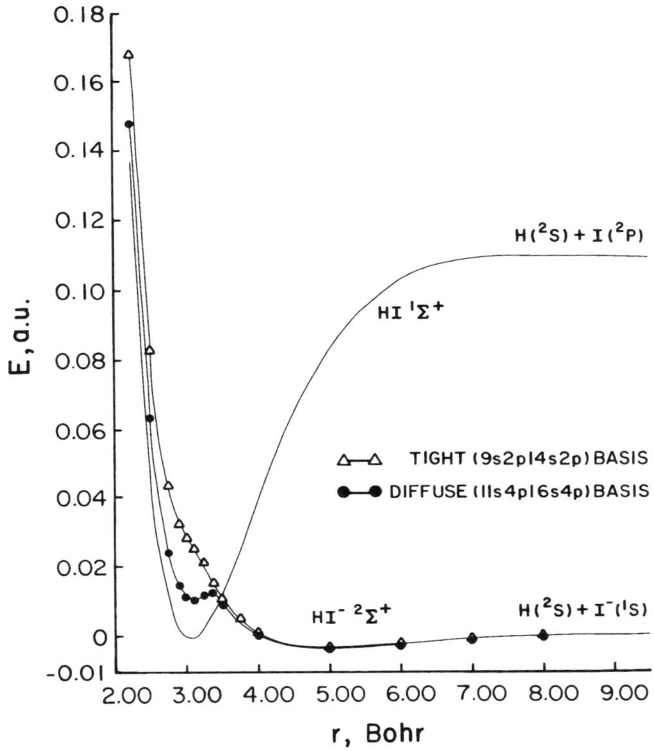

Figure 2.20 Potential-energy curves for the ground states of HI and HI^-. (Reprinted from Ref. 129; copyright 1988, American Institute of Physics.)

The theoretical positive electron affinity for HI was found to be consistent with the mass-spectrometric investigations of Spence et al. (128) in which evidence for the stability of HI^- was found. Spence et al. (128) suggested that the $^2\Sigma^+$ curve should be deep and broad enough to support at least one vibrational level below the HI $^1\Sigma^+$ curve. On the basis of this, these authors concluded that the electron affinity of HI is positive and should exceed 0.007 eV.

The exceptionally large rate constant for thermal electron attachment of HI observed by Adams et al. (96) was explained by Chapman et al. (129) using the Langevin model. The attachment rate constant data were found to exhibit a temperature dependence [$\beta = (3.3 \pm 1) \times 10^{-12}$ cm^3/s at 300 K; $\beta = (2.8 \pm 1) \times 10^{-10}$ cm^3/s] at 510 K for HBr, while the rate coefficients for HI remained constant [$\beta = (3.5 \pm 1) \times 10^{-7}$ cm^3/s] at 300 and 510 K. These data suggested that the attachment process was governed by two different mechanisms in HBr and HI. The thermal-averaged Langevin rate constants, $k_L(\varepsilon)$, were calculated for HBr and HI, and the results were found to be of the same order (10^{-7} cm^3/s) as the experimental β values for HI.

The potential-energy curves of the excited states of HI^- and HI are shown

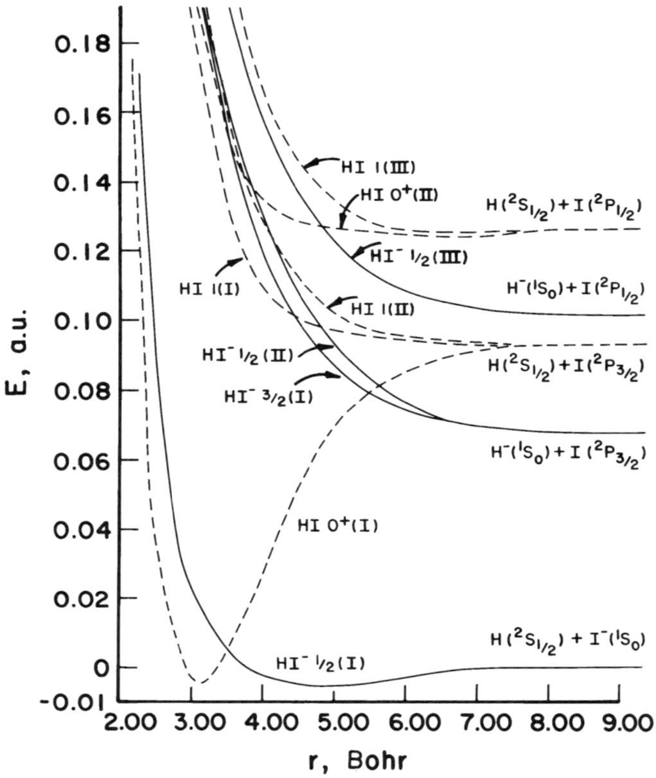

Figure 2.21 Potential-energy curves for the low-lying states of HI and HI^-. (Reprinted from Ref. 129; copyright 1988, American Institute of Physics.)

in Figure 2.21. Le Coat et al. (97, 98) deduced the potential-energy curves for some of the excited states of HI^- [$\frac{1}{2}$(III), $\frac{3}{2}$, $\frac{1}{2}$(III)] from dissociative electron attachment experiments. The spin–orbit effects are more important for HI^-, since the $^2P_{3/2}$–$^2P_{1/2}$ spin–orbit splitting of I (7603 cm^{-1}) (306) exceeds the electron affinity of the hydrogen atom. Consequently, the crossings of ionic and neutral curves differ for HI^- substantially compared to HBr^-. The H^- + I($^2P_{3/2}$) limit is below the H + I($^2P_{3/2}$) neutral asymptote. The HI^- $\frac{1}{2}$(II) and $\frac{3}{2}$(I) curves are the only curves that crossed with the ground-state curve of HI.

The nature of the HI^- curves can be understood if one compares the parent (HI) electronic states. The 1(III) state of HI was found to be predominantly of $^3\Sigma_1^+$ ($1\sigma^2 2\sigma 3\sigma 1\pi^4$) character, while the HI^- $\frac{1}{2}$(II) state was found to be predominantly $^2\Sigma_{1/2}^+$($1\sigma^2 2\sigma 3\sigma^2 1\pi^4$). Hence electron attachment to the 3σ orbital of the neutral $^3\Sigma_1^+$ state could generate the negative ion in the $\frac{1}{2}$(II) state. The $\frac{1}{2}$(II) state of HI^- was found to lie below the neutral 1(II) curve (Fig. 2.21), leading to the $\frac{1}{2}$(II) state as a Feshbach resonance. On the contrary, the $\frac{3}{2}$(I) and $\frac{1}{2}$(III) states are components of the $^2\Pi(1\sigma^2 2\sigma^2 3\sigma^2 \pi^3)$ states, which were found to lie above their parent $^3\Pi,^1\Pi(1\sigma^2 2\sigma^2 3\sigma 1\pi^3)$ states, and, consequently, the $\frac{3}{2}$(I) and $\frac{1}{2}$(III) states cannot be considered as Feshbach resonances.

2.1.11 TlH

The thallium hydride molecule has been the subject of many relativistic theoretical calculations (28, 167, 168), since it is the simplest sixth-row p-block hydride and has a $^1\Sigma^+$ closed-shell ground state. Lee et al. (167) carried out relativistic SCF calculations in ω–ω coupling for TlH. In a subsequent study Christiansen and Pitzer (168) made MCSCF LCAS-MS spinor calculations on TlH. Christiansen, the author, and Pitzer (28) carried out the first RCI calculations on TlH that included spin–orbit effects and electron correlation corrections simultaneously in a RCI scheme. In this investigation Christiansen, the author, and Pitzer considered the spectroscopic properties and potential-energy curves for six electronic states of TlH.

The experimental spectroscopic constants on five electronic states of TlH were obtained through the electronic spectrum obtained in both absorption and emission (169–172). Grundström and Valberg (169) as well as Neuhaus and Muld (171) studied the $A \leftrightarrow X$ systems. Larsson and Neuhaus (172) obtained several bands with origins at 23, 556 and 23,655 cm^{-1}, which were attributed to $D \leftarrow X$ and $C \leftarrow X$ systems, respectively. Ginter and Battino (37) obtained two potential-energy curves for TlH from the experimental data.

Table 2.12 shows the spectroscopic constants for TlH obtained from the theoretical calculations on Christiansen, the author, and Pitzer (28) and the experimental values for comparison. Figure 2.22 shows a comparison of the RCI potential-energy curves and the curves deduced from the Rydberg–Kein–Rees (RKR) method of fitting experimental data. As seen from Figure 2.22, the overall shapes of the potential curves are very similar. The experimental curve is deeper, with a shorter R_e. The behavior of the 0^+(II) state is especially interesting and compares remarkably well with experiment. Figure 2.23 shows the potential-energy curves of six electronic states of TlH. The 0^-, 1, 2, and 0^+(II) states were found to be predominantly components of a $^3\Pi$ state near the inner well, while the 1(II) state was found to be predominantly composed of a $^1\Pi$ state.

The 0^+ and 0^+(II) states of TlH were found to exhibit avoided crossings. At short distances the 0^+ state was found to be predominantly $^1\Sigma^+$, while it became $^3\Pi_{0^+}$ at longer distances. The reverse behavior was found to be true for the $A0^+$(II) state. That is, at short distances $A0^+$(II) was found to be $^3\Pi_{0^+}$ and became $^1\Sigma^+_{0^+}$ at long distances. The inner wells in the 0^-, 1, and 2 states are results of an avoided crossing. In the outer region the metal atom was described by the $6s^2 6p$ configuration, while in the inner region there was considerable $6s6p^2$ character, allowing a bond between $6s$ of Tl and H $1s$. A number of spectral lines observed by Larsson and Neuhaus (172) were tentatively assigned

Table 2.12 Spectroscopic Constants for TlH[a]

State	R_e, Å		T_e, cm^{-1}		ω_e, cm^{-1}		D_e, eV	
	Theory	Exp.	Theory	Exp.	Theory	Exp.	Theory	Exp.
$X0^+$(I)	1.99	1.87	0	0	1,300	1,391	1.81	1.97
0^-(I)	1.95	—	16,600	—	795	—	0.24	—
$A0^+$(II)	1.91	1.91	17,100	17,723	1,000	760	0.61	0.74
2	1.90	—	21,800	—	740	—	0.04	—
$C1$(II)	3.1	2.9	23,400	24,180	200	140	—	—

[a] All theoretical values are from Reference 28.

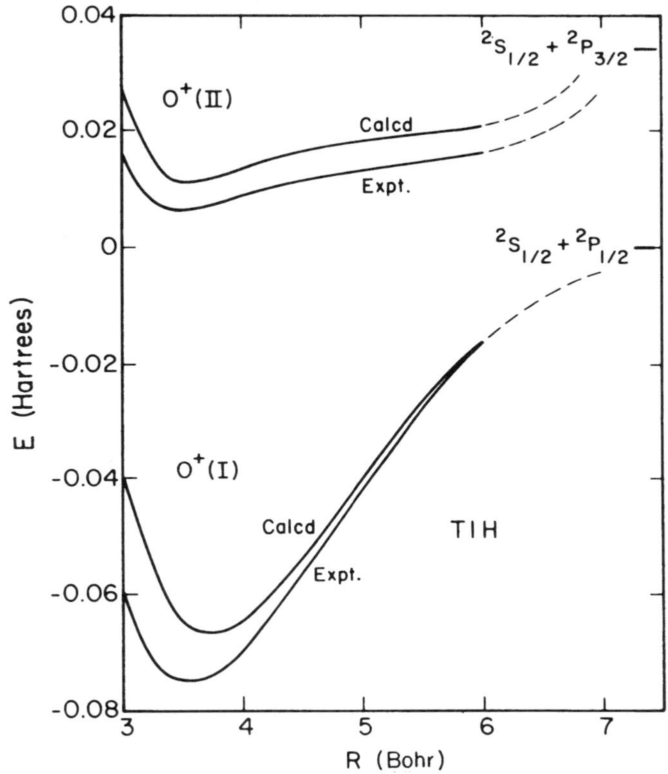

Figure 2.22 Experimental and theoretical potential-energy curves for the low-lying states of TlH. (Reprinted from Ref. 28; copyright 1982, American Institute of Physics.)

to the 2–X and $C1$(II)–X transitions. The approximate R_e and ω_e values of the 1(II) state were deduced as 2.9 Å and 140 cm^{-1}, respectively (see Table 2.12).

The merging (crossing) of the 1(II) and 2 curves (Fig. 2.23) supported the hypothesis of Larsson and Neuhaus (172), who assigned some of the spectral lines to the $2 \leftarrow 0^+$ transition as a result of perturbation with the 1(II) state. The low-potential maxima were found in the 2 and 1(II) states.

2.1.12 PbH

Watson (173) was the first to observe complex red and near-infrared bands for PbH. Watson and Simon (142, 143) assigned these bands to A–X and $B \leftarrow X$ systems, although the ground state of PbH was then regarded as $^2\Sigma^+_{1/2}$. However, now it is clearly established that the ground state of PbH is $X^2\Pi_{1/2}$. Gerö (174) reassigned the A and B states to two $^2\Sigma^+_{1/2}$ states, although later Kleman (131) argued that these should be components of a $^4\Sigma^-$ state. Howell (47) reported

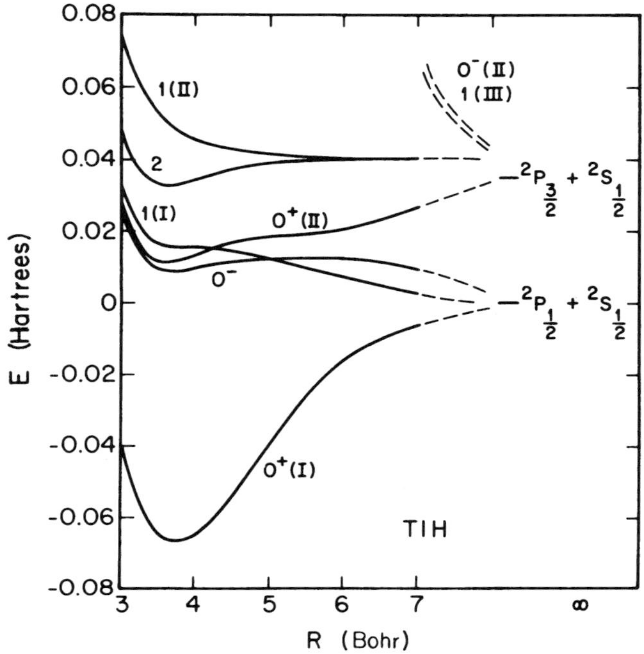

Figure 2.23 Theoretical potential-energy curves for the low-lying states of TlH. (Reprinted from Ref. 28; copyright 1982, American Institute of Physics.)

the $C \leftarrow X$ system in the region of 26,200 cm^{-1}, although these bands have not yet been fully analyzed.

The author and Pitzer (175) made relativistic SCF/RCI calculations on five electronic states of PbH employing a double ζ STO basis with 14e RECPs for the lead atom. The calculated RCI spectroscopic constants are summarized in Table 2.13 and compared with experimental values, whenever available. Figure 2.24 shows the RCI potential-energy curves of PbH. The ground state of

Table 2.13 Spectroscopic Properties of a Few Low-Lying Bound States of PbH

	R_e, Å		T_e, cm^{-1}		ω_e, cm^{-1}	
State	Theory	Exp.	Theory	Exp.	Theory	Exp.
$X\frac{1}{2}$(I)	1.95	1.84	0.0	0.0	1,418	1,564
$\frac{3}{2}$(I)	1.92	—	6,846	—	1,457	—
$A\frac{1}{2}$(II)	2.68	(2.36)	17,213	17,590	391	500
$B\frac{3}{2}$(II)	2.44	(2.60)	20,341	18,030	442	479

[a] All theoretical results are from Reference 175.

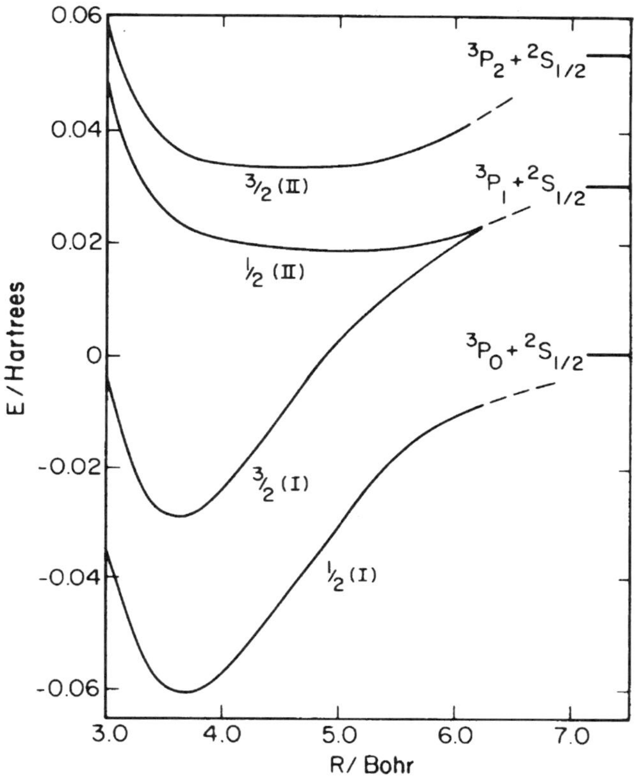

Figure 2.24 Potential-energy curves of the *g* states of PbH. (Reprinted from Ref. 175; copyright 1984, American Chemical Society.)

PbH was found to be clearly $^2\Pi_{1/2}$, confirming Howell's (47) reinterpretation of Watson's spectra (173).

The excited $A\frac{1}{2}$(II) and $B\frac{3}{2}$(II) states were found to be predominantly $^4\Sigma^-_{1/2}$ and $^4\Sigma^-_{3/2}$ at near R_e distances of these states. The dissociation energy of PbH was calculated as 1.64 eV (175), in very good agreement with an experimental value of 1.59 eV (32) obtained from the predissociation in the B state assuming dissociation into $^3P_1$ + ^2S.

Chapman et al. (55) calculated the dipole moment of PbH in the $X^2\Pi_{1/2}$ ground state as 0.937 D employing a double-ζ STO basis set. The dipole moment of the $^2\Pi_{3/2}$ state was found to be 0.673 D (Pb^+H^-). The transition moment curves and the electronic dipole moment curves of PbH were also obtained and compared with those of SnH (see Figs. 2.14 and 2.15).

Schwerdtfeger et al. (316) studied the ground states of PbH^+, PbH, PbH_2, and PbH_4 using semiempirical relativistic and nonrelativistic pseudopotentials and the CISD method for inclusion of electron correlation. The calculated R_e(PbH) = 1.839 Å was found to be almost in exact agreement with the experimental value. The ground state of PbH^+ was found to be the $^1\Sigma^+$ state, analogous to the ground state of TlH.

The author studied the $^1A_1$–$^3B_1$ energy separations of GeH_2, SnH_2, and PbH_2 in a recent investigation (317). He found that the geometry of the $^3B_1(A_1)$ state changed substantially as a result of the spin–orbit effects via mixing of $^3B_1(A_1)$ with the $^1A_1(A_1)$ ground state.

2.1.13 BiH and BiH⁺

The BiH molecule is one of the most complex and thoroughly investigated (theoretically) species among the heavy hydrides (180, 181). The author (181), in a preliminary theoretical investigation, studied four low-lying electronic states of BiH. In a subsequent detailed study (180) the spectroscopic constants and potential-energy curves of 10 low-lying ω–ω states and 5 λ–s states of BiH were investigated. BiH is the most complex among the main-group hydrides because of both the $6s^26p^3$ open-shell configuration of the Bi atom and the large spin–orbit effects. Consequently, there are many low-lying electronic states for BiH.

The experimental spectroscopic investigations of BiH date back to the early 1930s (176–179). In particular, Heimer (178, 179) Khan and Khan (157), Neuhaus (177), and Lindgren and Nilsson (156) investigated the electronic spectra of BiH. These investigations yielded information on the $B \rightarrow X, B \rightarrow A, D \rightarrow C$, and $E \leftarrow X$ systems of BiH. The rotational analyses of the $E \leftarrow X$ system were made by Khan and Khan (157) and Lindgren and Nilsson (156). The abrupt disrupture of the bands in the $E \leftarrow X$ system clearly indicated that the E state must be predissociated; it was hypothesized that a repulsive L state dissociating into $Bi(^2D_{3/2}) + H(^2S)$ should be responsible for this predissociation. An upper bound for the D_e of BiH ground state was estimated by assuming the predissociation of $E \leftarrow X$ into $Bi(^2P_{3/2})$ + H. From the hyperfine structure of the A state, the spin–orbit constant of the $^3\Sigma^-$ ground state of BiH was deduced to be 4917 cm^{-1}. Bopegedera et al. (333) recorded diode laser spectra of BiH and BiD. From the rotational line positions these authors have determined the R_e value of BiH accurately as 1.80867 Å.

The spin–orbit effects were found to be substantial for BiH, leading to many avoided crossings. As shown by the author (180), these avoided crossings resulted in shoulders, barriers, multiple minima, and other exotic features in the potential-energy curves of the excited states of BiH.

The ground state of BiH was found to be a $^3\Sigma^-_{0^+}$ state arising from the $1\sigma^22\sigma^21\pi^2$ configuration, where the 2σ orbital should be the bonding $Bi(6p_z)$ + $H(1s)$ orbital, while the 1π orbital should be a nonbonding Bi $6p$ orbital. Thus the spin–orbit splitting of the $^3\Sigma^-$ state arising from this configuration should be large. The author carried out SCF/RCI calculations on BiH employing a STO basis set (180, 181). Ramos et al. (334) have also carried out relativistic calculations on a few sixth-row diatomic hydrides, including BiH. Comparison

Table 2.14 Spectroscopic Properties of Low-Lying States of BiH[a]

	R_e, Å		T_e, cm^{-1}		ω_e, cm^{-1}	
State	Theory	Exp.	Theory	Exp.	Theory	Exp.
$X0^+$(I)	1.90	1.81	0.0	0.0	1,619	1,636
$A1$	1.89	1.79	5,737	4,917	1,630	1,669
2	1.89	—	13,469	—	1,630	—
$B0^+$(II)	1.88	1.78	26,286	21,263	1,585	1,643
2(II)	2.58	—	23,450	—	1,529	—
$E0^+$(IV)	2.51	2.18	38,780	32,940	1,000	1,106
$^3\Sigma^-$	1.88	—	8,013	—	1,687	—
$^1\Delta$	1.92	—	16,494	—	1,296	—
$^1\Sigma^+$	1.86	—	25,670	—	1,705	—
$^1\Sigma^+(\pi^4)$	1.74	—	—	—	1,890	—

[a] All theoretical values are from Reference 180.

of the results and techniques used by Ramos et al. and the RCI technique has been done elsewhere.

Table 2.14 shows the spectroscopic constants of BiH obtained from the RCI calculations, including the spin–orbit term. Figure 2.25 shows the potential-energy curves of BiH, including the spin–orbit effects. Figure 2.26 shows comparable potential-energy curves without the spin–orbit term. In Figure 2.26, although the ^4S state of the Bi atom (dissociation limit) is not split by the spin–orbit term, it can interact with $^2D_{3/2}$ when spin–orbit integrals are included. Thus the limit labeled $^4S_{3/2}$ + $^2S_{1/2}$ (Fig. 2.26) includes spin–orbit integrals, while ^4S + ^2S does not include these integrals.

The RCI calculations of the author (180) enabled the assignment of the experimentally observed X, A, B, and E states to the 0^+, 1, 0^+(II), and 0^+(IV) states, respectively. Among these the 0^+ and 1 states are type c analogs of $^3\Sigma^-_{0^+}$ and $^3\Sigma^-_1$ at near-equilibrium geometries. The 0^+(IV) state (E) exhibited avoided crossing of $^3\Pi_{0^+}(\sigma^2\sigma^*\pi)$ with $^1\Sigma^+_{0^+}(\sigma^2\pi^2)$ among other avoided crossings. The $^1\Sigma^+(\sigma^2\pi^2)$ state would dissociate into ^2P(Bi) + ^2S(H) atoms in the absence of spin–orbit interactions. The $^3\Pi$ state, however, dissociates into ^2D(Bi) + ^2S(H) and is a repulsive state. The ^2P(Bi) state is much higher than the ^2D(Bi) state. Thus these two states cross in the absence of spin–orbit interactions, and therefore their 0^+ components exhibited avoided crossings.

The repulsive 1(II) state, which dissociates into ground-state atoms, crosses the 2(II) state close to the equilibrium bond distance (see Fig. 2.25). The 1(II) state also crosses with the 0^+(II) state, but at a considerably longer distance than the R_e value of the 0^+(II) state. The 2(II) state is thus predissociated and should be short-lived. The 0^+(II) and 2 states have barriers as a result of avoided crossings. The $^3\Sigma^-$ and $^3\Pi$ states, which dissociate to the Bi(^{2}D) state, cross $^1\Sigma^+$, which dissociates to the ^2P state. In the presence of spin–orbit interactions

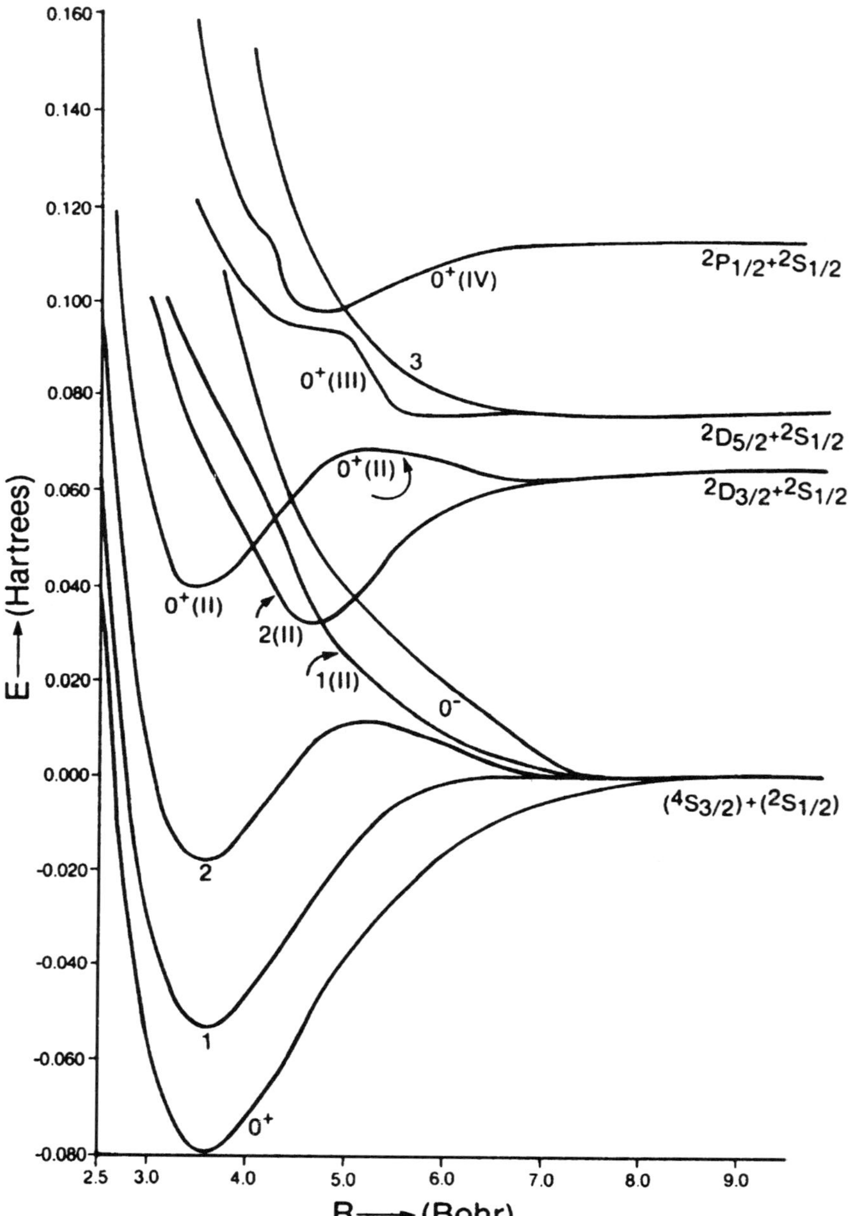

Figure 2.25 Potential-energy curves for the electronic states of BiH with spin–orbit coupling. (Reprinted from Ref. 180; copyright 1986, Academic Press, Inc.)

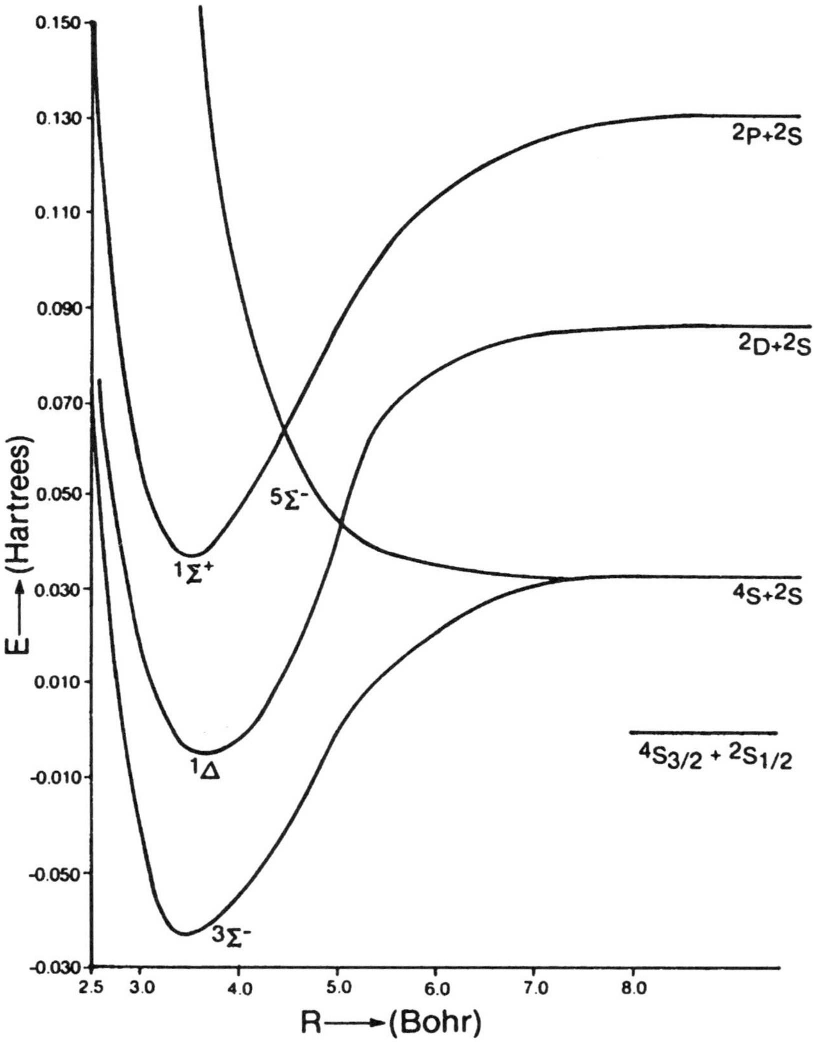

Figure 2.26 Potential-energy curves for the electronic states of BiH without spin–orbit coupling. (Reprinted from Ref. 180; copyright 1986, Academic Press, Inc.)

this led to avoided crossings. Thus 0^+(II), which is predominantly $^1\Sigma^+(\sigma^2\pi^2)$ at near-equilibrium geometries, is forced to dissociate into $^2D_{3/2}$(Bi) + $^2S_{1/2}$(H) as a result of avoided crossings. The barrier in the 2 state arises from the avoided crossing of the bound $^1\Delta_2$ with the repulsive $^5\Sigma_2^-$ dissociating into the ground-state atoms.

The 0^+(III) state, which has not been experimentally characterized yet, is repulsive but at long distance (6.0 bohr) has a shallow minimum. At 3.5 bohr the 0^+(III) state is 73% $^3\Pi_{0^+}(\sigma\pi^3)$, 13% $^1\Sigma^+_{0^+}(\sigma^2\pi^2)$, and 4% $^3\Sigma^-_{0^+}(\sigma^2\pi^2)$. Thus, it is predominantly $^3\Pi_{0^+}$ in type c coupling. The $X(0^+)$ state was found to be 76% $^3\Sigma^-_{0^+}(\sigma^2\pi^2)$ and 16% $^1\Sigma^+_{0^+}(\sigma^2\pi^2)$. Thus the 0^+(III) $\leftarrow X(\mathrm{O}^+)$ transition is a type c analog of the $^3\Pi \leftarrow {}^3\Sigma^-$ transition at the same distance. This transition is an allowed case a transition in the perpendicular direction. Also, mixing of $^3\Sigma^-_{0^+}$ and $^1\Sigma^+_{0^+}$ in the 0^+(III) state should contribute to the z component of the transition moment indirectly, since $^3\Pi_{0^+}$ mixes with 1 $^1\Sigma^+_{0^+}$. For the isoelectronic NH molecule, the $^3\Pi(\sigma\pi^3)$ state is bound, and is observed both in emission and in absorption ($A \leftrightarrow X$) at 29820 cm^{-1} (32). The theoretical separation of the 0^+(III) state with respect to the $X(0^+)$ state is about 43,895 cm^{-1} at 3.5 bohr.

The experimental $E0^+ \leftarrow X0^+$ absorption bands break off abruptly, which was attributed by Lindgren and Nilsson (156) to the predissociation of the $E0^+$ state caused by the crossing of a repulsive 1 state with the $E0^+$ state at a distance longer than the equilibrium geometry to cause predissociation in the v = 2 level. These authors suggested that the most probable candidate for this predissociation is the repulsive $^5\Sigma^-_1$ state. The 1(II) state (Fig. 2.25) is predominantly $^5\Sigma^-_1$ at distances greater than 4.5 bohr. Since the predissociation of the $E0^+$ state occurs at about 5 bohr, it is unlikely that this is caused by the $^5\Sigma^-_1$ state. This is because the 1(II) state is much too low in energy at this distance to cross the 0^+(IV) state. The author (180) suggested that the predissociation of the $E0^+$ state should be attributed to crossing of the 1(III) or 1(IV) curves, which dissociate into $^2D_{3/2}$ + $^2S_{1/2}$ atoms.

The experimental dissociation energy of the BiH molecule was calculated by using the predissociation of the $v = 2$ vibrational band (in the $E \leftarrow X$ system) of the $E0^+$ state. It was assumed that this dissociated into $^2D_{3/2}$ + $^2S_{1/2}$ atoms. The 0^+(III) state actually dissociates into $^2D_{5/2}$ + $^2S_{1/2}$ atoms and hence will keep the $E0^+$(IV) state higher than the $^2D_{5/2}$ + $^2S_{1/2}$(Bi + H) atoms at all distances (see Fig. 2.25). Thus the assumption of the predissociation limit of $E0^+$ as Bi($^2D_{3/2}$) + H($^2S_{1/2}$) atoms should yield a $D_0{}^{306}$ 4019 cm^{-1} [Bi($^2D_{5/2}$)-Bi($^2D_{3/2}$) splitting] higher than the true value. Consequently, the D_0 value calculated by extrapolating the $v = 2$ band of $E0^+$ to $^2D_{3/2}$ + $^2S_{1/2}$ (2.89 eV) should be corrected to 2.39 eV. The corrected value was found to be in very good agreement with the author's (180) calculated separation of the minimum of $X(0^+)$ with respect to the energy of $X(0^+)$ at 9.0 bohr (2.17 eV).

The author (180) calculated a high-lying bound $^1\Sigma^+$ state arising from the π^4 electronic configuration. This state was found to lie about 83 160 cm^{-1} above the ground state for the iso-valence-electronic NH (32). The calculated R_e and ω_e values are shown in Table 2.14. This state should be in the far-ultraviolet region by comparing with isovalence-electronic NH. Electronic spectra of the BiH molecule in the ultraviolet and far-ultraviolet regions have not been studied at all.

The BiH molecule is one of the most fascinating candidates to exemplify the important impacts of the spin–orbit coupling on the electronic states. The nature of the electronic states of BiH was investigated by analyzing the RCI wavefunctions. The squares of the coefficients of the various λ–s states in the

RCI wavefunctions of the electronic states of BiH were collected and plotted against R. The plots of two very interesting states—0^+(II) ($B0^+$) and 0^+(IV) ($E0^+$)—are shown in Figures 2.27 and 2.28, respectively. The nature of various electronic states of BiH is described in the ensuing paragraphs.

The ground state $X(0^+)$ was found to be a mixture of ${}^3\Sigma^-_{0+}$ (76%) and ${}^1\Sigma^+_{0+}$ (16%) near the equilibrium bond distance. At an intermediate distance (6.0 bohr) the 0^+(I) state became 54% ${}^3\Sigma^-_{0+}$, 12% ${}^1\Sigma^+_{0+}$, 14% ${}^3\Sigma^-_{0+}$ ($\sigma^{*2}\pi^2$), 3% ${}^1\Sigma^+_{0+}$ ($\sigma^{*2}\pi^2$), 6% ${}^3\Pi_{0+}(\sigma^2\sigma^*\pi)$, and 5% ${}^3\Sigma^-_{0+}$ ($\sigma\alpha\sigma^*\alpha\pi_x\beta\pi_y\beta$–$\sigma\beta\alpha^*\beta\pi_x\alpha\pi_y\alpha$).

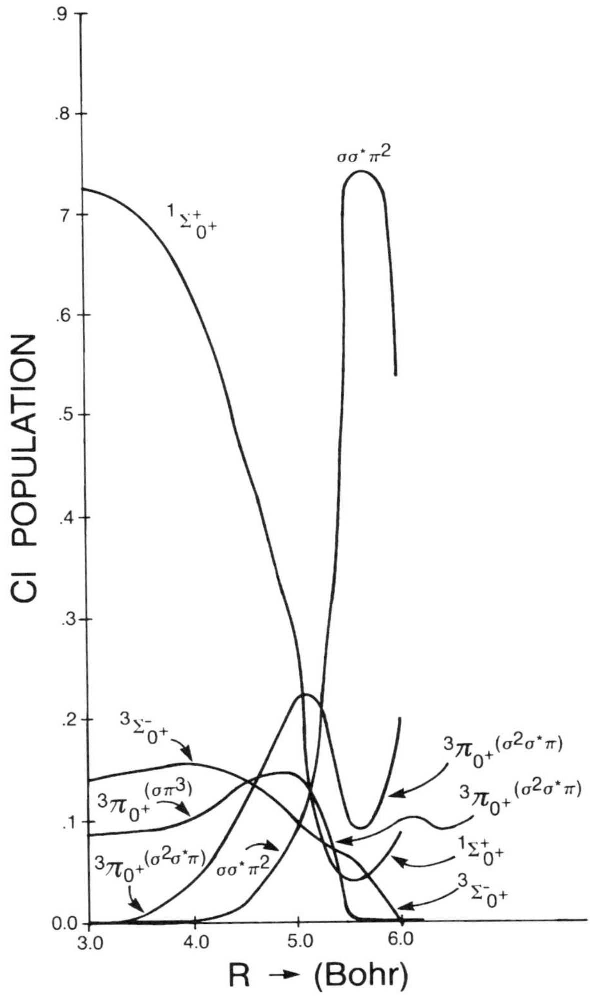

Figure 2.27 Fractional CI populations for the 0^+(II) electronic states of BiH. (Reprinted from Ref. 180; copyright 1986, Academic Press, Inc.)

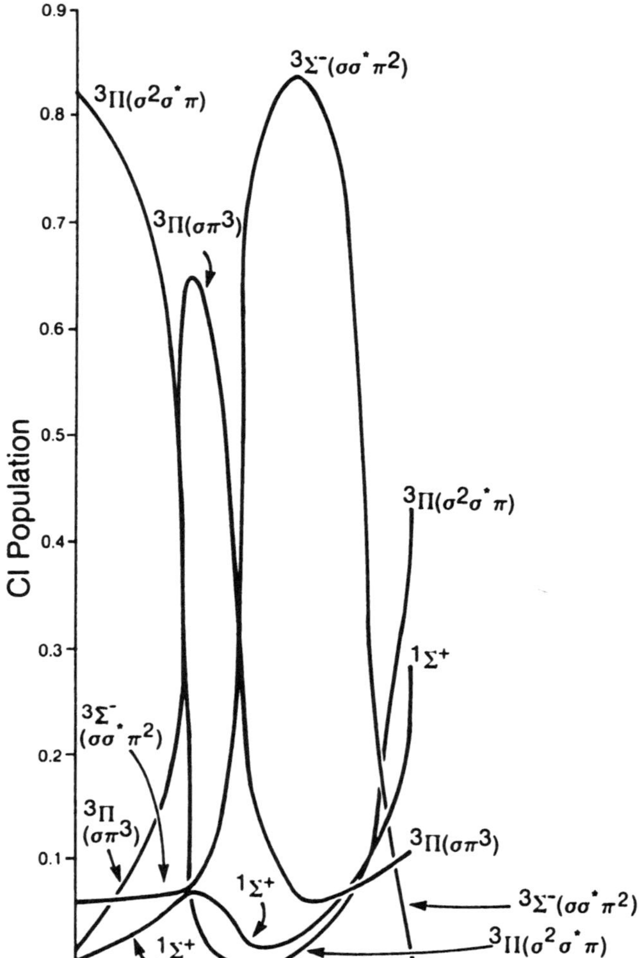

Figure 2.28 Fractional CI populations for the 0^+(IV) electronic states of BiH. (Reprinted from Ref. 180; copyright 1986, Academic Press, Inc.)

The 0^+(II) state ($B0^+$) exhibited an interesting behavior (see Fig. 2.27) (180). As seen from Figure 2.27, at near-equilibrium geometries this state is predominantly $^1\Sigma^+_{0^+}$ (70%). The $^3\Sigma^-_{0^+}$ and $^3\Pi_{0^+}(\sigma\pi^3)$ configurations account for 15% and 8% of the weight, respectively. At 6.0 bohr, this state became 39% $^3\Sigma^-_{0^+}$ (arising from $\sigma\alpha\sigma^*\alpha\pi_x\beta\pi_y\beta$ and $\sigma\beta\alpha^*\beta\pi_x\alpha\pi_y\alpha$), 20% $^3\Pi_{0^+}(\sigma^2\sigma^*\pi)$, 10% $^1\Sigma^+_{0^+}(\sigma^2\pi^2)$ 8% $^3\Pi_{0^+}$), 9% $^3\Pi_{0^+}(\sigma\sigma^{*2}\pi)$, and 2% $^3\Sigma^-_{0^+}(\sigma^2\pi^2)$. At 9.0 bohr it is 67% $^3\Sigma^-_{0^+}(\sigma\alpha\sigma^*\alpha\pi_x\beta\pi_y\beta\text{–}\sigma\beta\alpha^*\beta\pi_x\alpha\pi_y\alpha)$, 5% $^1\Sigma^+_{0^+}(\sigma^2\pi^2)$, 8% $^3\Pi_{0^+}(\sigma^2\sigma^*\pi)$, and 10% $^3\Pi_{0^+}(\sigma\sigma^{*2}\pi)$.

The 1(I) ($A1$) state of BiH was found to be 88% $^3\Sigma_1^-$ and 2% $^3\Pi_1$. At long distances the $^5\Sigma_1^-(\sigma\sigma^*\pi^2)$ made a substantial contribution. At 6.0 bohr the 1 state was found to be 46% $^3\Sigma_1^-$, 18% $^5\Sigma_1^-$, 8% $^3\Pi^1(\sigma\pi^3)$, and 13% $^3\Sigma_1^-(\sigma^{*2}\pi^2)$. The 2 state was found to be 90% $^1\Delta_2$ near the equilibrium distance, while at 6.0 bohr it became 54% $^5\Sigma_2^-$, 7% $^1\Delta_2$, 4% $^3\Pi_2(\sigma^2\sigma^*\pi)$, and 12% $^3\Pi_2(\sigma\pi^3)$ Thus $^5\Sigma_2^-$ dominated at long distances, which explains the barrier in this state. The 2(II) state exhibited an opposite behavior. It was found to be 40% $^5\Sigma_2^-$ and 44% $^3\Pi_2(\sigma\pi^3)$ at 4.25 bohr. At 4.5 bohr it was found to be 50% $^5\Sigma_2^-$ and 35% $^3\Pi_2$. The second avoided crossing arises from the crossing of $^1\Delta$ with $^5\Sigma^-$. At 6.0 bohr the 2(II) state was found to be 55% $^1\Delta_2$, 23% $^5\Sigma_2^-$ and 0.7% $^3\Pi_2$.

The 0^- state was found to be 54% $^3\Pi_{0^-}(\sigma^2\sigma^*\pi)$, 28% $^5\Sigma_2^-(\sigma\sigma^*\pi^2)$, 9% $^3\Sigma/a/al(+, 0^-)$ $(\sigma\sigma^*\pi^2)$, and 4% $^3\Pi_{0^-}(\sigma\sigma^{*2}\pi)$ at 4.0 bohr. At 6.0 bohr it became 62% $^5\Sigma_{0^-}^-(\sigma\sigma^*\pi^2)$, 12% $^3\Pi_{0^-}(\sigma^2\sigma^*\pi)$, 15% $^3\Sigma_{0^-}^+(\sigma\sigma^*\pi^2)$, and 9% $^3\Pi_{0^-}(\sigma\sigma^{*2}\pi)$. Thus, the 0^- state exhibits not only an avoided crossing but also a large spin–orbit contamination at long distances.

The 0^+(III) and 0^+(IV)(E0^+) states exhibited several interesting relativistic avoided crossings. The CI weight of the 0^+(IV) state is shown in Figure 2.28. At 3.5 bohr the 0^+(III) state was found to be 73% $^3\Pi_{0^+}(\sigma\pi^3)$, 13% $^1\Sigma_{0^+}^+(\sigma^2\pi^2)$, and 4% $^3\Sigma_{0^+}^+(\sigma^2\pi^2)$. However, at 4.0 bohr it became 73% $^3\Pi_{0^+}(\sigma^2\sigma^*\pi)$, 16% $^1\Sigma_{0^+}^+(\sigma^2\pi^2)$, and 1% $^3\Pi_{0^+}(\sigma\pi^3)$. Thus at this distance the 0^+(III) state exhibited an avoided crossing of $^3\Pi_{0^+}(\sigma^2\sigma^*\pi)$ with $^3\Pi_{0^+}(\sigma\pi^3)$. This results in the first shoulder in the potential-energy curve of 0^+(III) in Figure 2.25. At 4.0 bohr 0^+(IV) was predominantly $^3\Pi_{0^+}(\sigma^2\sigma^*\pi)$, and at 4.25 bohr it became predominantly $^3\Pi_{0^+}(\sigma\pi^3)$. At 5.0 bohr the 0^+(III) state was found to be 84% $^3\Pi_{0^+}(\sigma\alpha\sigma^*\alpha\pi_x\beta\pi_y\beta-\sigma\beta\alpha^*\beta\pi_x\alpha\pi_y\alpha)$, 6.4% $^3\Pi_{0^+}(\sigma\pi^3)$, and 3% $^3\Sigma_{0^+}^-(\sigma^2\pi^2)$.

Thus at 5.0 bohr the 0^+(III) curve has a second shoulder as a result of the avoided crossing of repulsive $^3\Sigma_{0^+}^-(\sigma\sigma^*\pi^2)$ with $^3\Pi_{0^+}(\sigma^2\sigma^*\pi)$. At 6.0 bohr the 0^+(III) has a shallow minimum, and at this distance it is 42% $^3\Sigma/a/al(-,0^+)(\sigma\sigma^*\pi^2)$ 12% $^3\Sigma_{0^+}^-(\sigma^2\pi^2)$, 15% $^3\Pi_{0^+}(\sigma\pi^3)$, and 11% $^3\Pi_{0^+}(\sigma^2\sigma^*\pi)$. The experimentally observed $E0^+$ exhibits several avoided crossings (see Fig. 2.25).

Subsequent to the author's theoretical study on BiH, Fink et al. (342) made a nice experimental study of BiH with a Fourier transform spectrometer. This study facilitated a detailed investigation of the $A1$–$X0^+$ emission system of BiH. Fink et al. (342) also studied the $a2$–$X21$ transition, but this is of considerably smaller intensity. Accurate spectroscopic constants of the ground state of the BiH molecule have been obtained by Urban et al. (343) as well as Bernath and co-workers (333, 344) with the aid of laser diode spectroscopy and Fourier transform spectroscopy. These techniques have yielded very accurate Dunham coefficients for the ground state of BiH.

After the appearance of the theoretical study on BiH by the author, Alekseyev et al. (345) carried out a spin–orbit configuration interaction study on the potential-energy curves and radiative lifetimes of the low-lying states of the BiH molecule. These authors carried out SCF/MRDCI computational followed by the spin–orbit CI. The MRDCI computations were done using the configuration selection scheme and the subsequent energy extrapolation procedure. Typically the selected configuration space in the MRDCI included between 1100 and 2030 configurations, although the total number of generated configurations varied between 17,000 and 34,000 configurations. The basis set employed is of $(3s3p1d)$ quality for the bismuth atom in conjunction with RECPs that included only the outermost $6s^2 6p^3$ shells of the bismuth atom. The hydrogen atom's basis set was $(5s1p/3s1p)$ quality. A threshold of 5.0 μhartree was used in the configuration selection procedure. The spin–orbit CI was done in the double group of the C_{2v} group, in which selected spatial configurations are multiplied by the spin functions, so that the total representation would transform according to the desired irreducible representation in the double group. The CI expansions in the spin–orbit CI are typically 4–6 times larger than the corresponding MRDCI made without spin–orbit coupling.

First we compare the techniques employed by Alekseyev et al. (345) and the author (180). Both groups used a single-configuration SCF technique to generate the orbitals for the CI computations. However, the basis sets used by the author are Slater-type orbitals, which are ideally suited for diatomics, compared to the Gaussian basis sets used in the work of Alekseyev et al. (345). The main differences are in the CI expansion. No configuration selection was invoked in the author's method either in the normal CI or in RCI, whereas Alekseyev et al. used the configuration selection procedure in the MRDCI. The RCI method employed by the author (180) was primitive since it was done in 1985, when the RCI method was still at its infancy. Thus the RCI expansion was limited to a few thousand configurations. Thus, the author was restricted in terms of the level of excitations that can be allowed for the various references included in the RCI. Generally the lower roots of a given symmetry are better represented than the upper roots. This, we consider as one source of difficulty in the RCI employed at that time. Alekseyev et al. used the selected configurations in the MRDCI of the various states in C_{2v} symmetry with appropriate spin functions so as to yield the overall state of desired symmetry in the double group. Thus, it appears that the upper roots in the spin–orbit CI may have been treated somewhat better in the Alekseyev et al. method. These authors have obtained several more electronic states of BiH that could not be obtained by the author before. Furthermore, the lifetimes of the electronic states have also been computed by these authors.

Table 2.15 shows the spectroscopic constants of several electronic states of BiH computed by Alekseyev et al. (345) together with the experimental results for some of the electronic states, and the results of the previous computational study by the author in 1985. This table is reproduced from Alekseyev's work without modification, and, thus, the assignments of electronic states in that table are according to Alekseyev et al. As seen from Table 2.15, overall, there is a very good agreement between the results computed in 1985 by the author and the results of Alekseyev et al. in 1994. It should also be noted that there are two other computations on the ground state of BiH by the author's group and Dolg et al. (347). The ground-state computation of BiH in 1990 by Dai and the

Table 2.15 Spectroscopic Properties of BiH from Various Calculations and Experiment (Bond Lengths r_e, Transition Energies T_e, and Vibrational Frequencies ω_e)

	R_e, Å			T_e, cm^{-1}			ω_e, cm^{-1}		
	Calc.			Calc.			Calc.		
State	Ref. 180[b]	Ref. 345	Exp.[a]	Ref. 180[b]	Ref. 345	Exp.[a]	Ref. 180[b]	Ref. 345	Exp.[a]
$X_1 0^+$	1.90	1.867	1.805	0	0	0	1,619	1,632	1,699.5[g]
	1.847[e]	—	1.80867[d]	—	—	—	1,780[e]	—	—
	1.858[e]	—	—	—	—	—	1,756[e]	—	—
	1.869[f]	—	—	—	—	—	—	—	—
$X_2 1$	1.89	1.854	1.791	5,737	4,303	4,917	1,630	1,618	1,669[c]
	1.840[e]	—	—	—	—	—	—	—	—
$a2$	1.89	1.855	—	13,469	11,906	—	1,630	1,719	—
$B0^+$	1.88	1.851	1.7795	26,286	22,496	21,263	1,585	1,620	1,643[c]
$E0^+$	2.51	2.283	2.1772	38,780	33,364	32,940	1,000	1,215	1,106[c]
0^+(IV)	—	2.669	—	—	35,575	—	—	2,169	—
0^+(V)	—	2.271	—	—	43,395	—	—	1,528	—
2(II)	2.58	2.521	—	23,450	24,100	—	1,529	1,013	—
1(III)	—	2.823	—	—	26,735	—	—	559	—

[a]Reference 32 unless otherwise indicated.
[b]Reference 180 unless otherwise indicated.
[c]Value quoted is $\Delta G(\frac{1}{2})$.
[d]Reference 333.
[e]Reference 347.
[f]Reference 346.
[g]References 343–345.

author was made in conjunction with other molecules such as BiH_3, SbH_3, and AsH_3. The work of Dolg et al. was designed primarily to calibrate the effective core potentials for the various atoms generated by these authors and was thus restricted to the ground state.

A notable contrast between the results for Alekseyev et al. and those of the author is in the state labeled $E0^+$ by Alekseyev et al. This state was considered as the fourth root of the 0^+ state by the author based on the computed energy separation compared to the T_e value of the observed $E0^+$ state. However, the computed bond length of the 0^+(IV) state is much longer (2.51 Å) compared to the experimental value of 2.1772 Å. Furthermore, the energy separation computed by Alekseyev et al. came out to be closer to the experimental value for the third root of the 0^+ state, although, as seen from Table 2.15, the fourth root is also quite close in energy. However, the equilibrium distance computed by Alekseyev et al. for this state (2.283 Å) is much closer to the experimental value. Consequently, Alekseyev et al. have favored the assignment of the observed $E0^+$ state of BiH to the third root of 0^+ symmetry. As mentioned before, a somewhat smaller RCI expansion had to be used in view of the code restrictions in 1985 for BiH. This restricted excitations from the reference configurations. Thus less excitations had to be allowed in some of the reference configurations, which were less important for the lower roots, but these reference configurations may become more important in the upper roots. We believe this to be the primary reason for the discrepancy of the computed properties in the upper roots. This was also documented by Alekseyev et al. through a critical comparison of the composition of the RCI wavefunctions of the author and Alekseyev et al.

Alekseyev et al. have computed the dipole moments, transition moments, and radiative lifetimes of the electronic states of BiH. Since this information is quite useful for a global understanding of the electronic states of BiH, we reproduce in Tables 2.16–2.18, Alekseyev et al.'s computed dipole moments, transition

Table 2.16 Electric Dipole Moment Values μ (in EA_0) of Various States ($v = 0$) of the BiH Molecules[a]

State	μ
$X_1 0^+$	0.3760
$X_2 1$	0.2710
$a2$	0.2252
$B0^+$	0.1269
2(II)	0.2551
$E0^+$	0.7924
0^+(IV)	0.3653

[a]All values correspond to Bi^+H^- polarity.

Source: Reference 345.

Table 2.17 Computed Transition Dipole Moments μ (in ea_0) Between X_10^+, $B0^+$, $E0^+$, 0^+(IV), and X_21 States of the BiH Molecule at Various Bond Lengths r (in a_0)

	μ				
r	B–X_1	E–X_1	0^+(IV)–X_1	B–X_2	E–X_2
2.10	0.002211	0.025385	0.568352	0.140559	0.265966
2.25	0.009629	0.014769	0.560785	0.126591	0.264906
2.50	0.020839	0.004477	0.523467	0.106066	0.256665
2.60	0.024686	0.011663	0.510182	0.098166	0.252268
2.75	0.031356	0.024520	0.494939	0.087462	0.241040
2.80	0.034120	0.029334	0.490895	0.084009	0.236704
3.00	0.044473	0.067947	0.471172	0.070870	0.214385
3.20	0.056855	0.429285	0.164177	0.058847	0.019177
3.25	0.060333	0.448679	0.089680	0.056253	−0.019504
3.40	0.072195	0.447743	0.034552	0.048373	−0.051396
3.50	0.080949	0.449388	0.028208	0.042883	−0.060307
3.60	0.090804	0.444523	0.027996	0.037283	−0.064798
3.75	0.110120	0.442531	0.038669	0.028617	−0.070733
3.90	0.134080	0.437353	0.055958	0.019074	−0.074189
4.00	0.151173	0.430749	0.079300	0.012273	−0.075510
4.10	0.173723	0.427678	—	0.004423	−0.077509
4.25	0.208810	0.411526	0.209590	−0.006564	−0.077255
4.40	0.248199	0.395571	0.239521	−0.018533	−0.075572
4.50	0.271506	0.383820	0.252021	−0.025385	−0.073086
4.60	0.295802	0.367566	0.268520	−0.032217	−0.070567
4.75	0.324873	0.341733	0.296840	−0.040047	−0.066191
4.90	0.347345	0.285739	0.336757	−0.044854	−0.060014
5.00	0.360893	0.158160	0.412387	−0.046879	−0.051566
5.10	0.371075	−0.105764	0.428020	−0.048116	−0.025570
5.25	0.379667	−0.211846	0.383862	−0.048311	−0.012349
5.50	0.385371	−0.223271	0.355266	−0.046197	−0.010042
6.00	0.346954	−0.197617	0.308324	−0.036977	−0.007528
6.50	0.264784	−0.153847	0.213900	−0.025931	−0.003475
7.00	—	−0.102315	0.122735	—	—
8.00	—	−0.029338	0.028318	—	−0.001836
9.00	0.020507	−0.007505	0.006482	−0.002658	−0.000916

Source: Reproduced from Reference 345.

moments as a function of internuclear distance, and the radiative lifetimes of the electronic states, respectively. Among the low-lying electronic states the $E0^+$ state has the highest dipole moment (0.7924 D). As seen from Table 2.18, the $E0^+$ and 0^+(IV) states have very short lifetimes primarily because of transitions to the $X_1(0^+)$ state. The $X_2(1)$ and $a2$ states have longer lifetimes, as expected. Note that the transition from the $a2$ state to the $X_1(0^+)$ state is forbidden. It would have been interesting to compare the radiative lifetimes of the excited

Table 2.18 Calculated Radiative Lifetimes (in μs) of Excited States ($v' = 0$) of BiH; Partial Lifetimes τ_p for Transitions to $X_1 0^+$, $X_2 1$, $B0^+$, and $a2$, Respectively, and Total Lifetime τ

State	$\tau_p(X_1 0^+)$	$\tau_p(X_2 1)$	$\tau_p(B0^+)$	$\tau_p(a2)$	τ
$X_2 1$	16,100	—	—	—	16,100
$a2$	—	545	—	—	545
$B0^+$	5.2	24.3	—	—	4.3
$E0^+$	0.12	5.3	28.5	—	0.12
0^+(IV)	0.17	>40	78	—	0.17
2(II)	—	660	—	12	12
1(III)	141	1.6	—	—	1.6

Source: Reproduced from Reference 345.

states with and without spin–orbit coupling. Unfortunately, Alekseyev et al. do not report the radiative lifetimes of the excited states in the absence of spin–orbit coupling. However, since the $E0^+$ is composed predominantly of $^3\Pi$ states, and some $^1\Sigma^+$ state, we believe that the transition to this state from the ground $^3\Sigma^-$ state is allowed even in then absence of spin–orbit coupling, and thus this state would be short-lived even in the absence of spin–orbit coupling. However, the $B0^+$ state would be a pure $^1\Sigma^+$ state nonrelativistically, while it becomes an admixture of $^1\Sigma^+$ and $^3\Pi$ relativistically. Therefore, the lifetime of this state would be influenced to a considerable extent by relativistic effects. We expect the lifetime of this state to become shorter by relativistic effects.

Although the BiH^+ ion has not been investigated experimentally at all, the spectroscopic constants and the potential-energy curves of many electronic states of the ion were obtained by the author (182). The SCF/RCI calculations of the author (182) revealed the existence of eight low-lying ω–ω states and five λ–s states of BiH^+. The ionization potential of Bi was calculated by the RCI method to be 6.98 eV.

Table 2.19 shows the RCI spectroscopic constants of the BiH^+ ion. Figure 2.29 shows the RCI potential-energy curves of BiH^+. Since the ground state

Table 2.19 Spectroscopic Properties of BiH^{+a}

State	R_e, Å	T_e, cm^{-1}	ω_e, cm^{-1}	D_e, eV
$X\frac{1}{2}$(I)	1.89	0.0	1,608	1.05
$\frac{3}{2}$(I)	1.87	11,384	1,653	1.37
$\frac{1}{2}$(II)	2.24	17,388	801	0.48
$\frac{3}{2}$(II)	2.37	24,537	504	0.25
$^2\Pi$	1.87	9,846	1,751	2.17

[a] All results are from Reference 182.

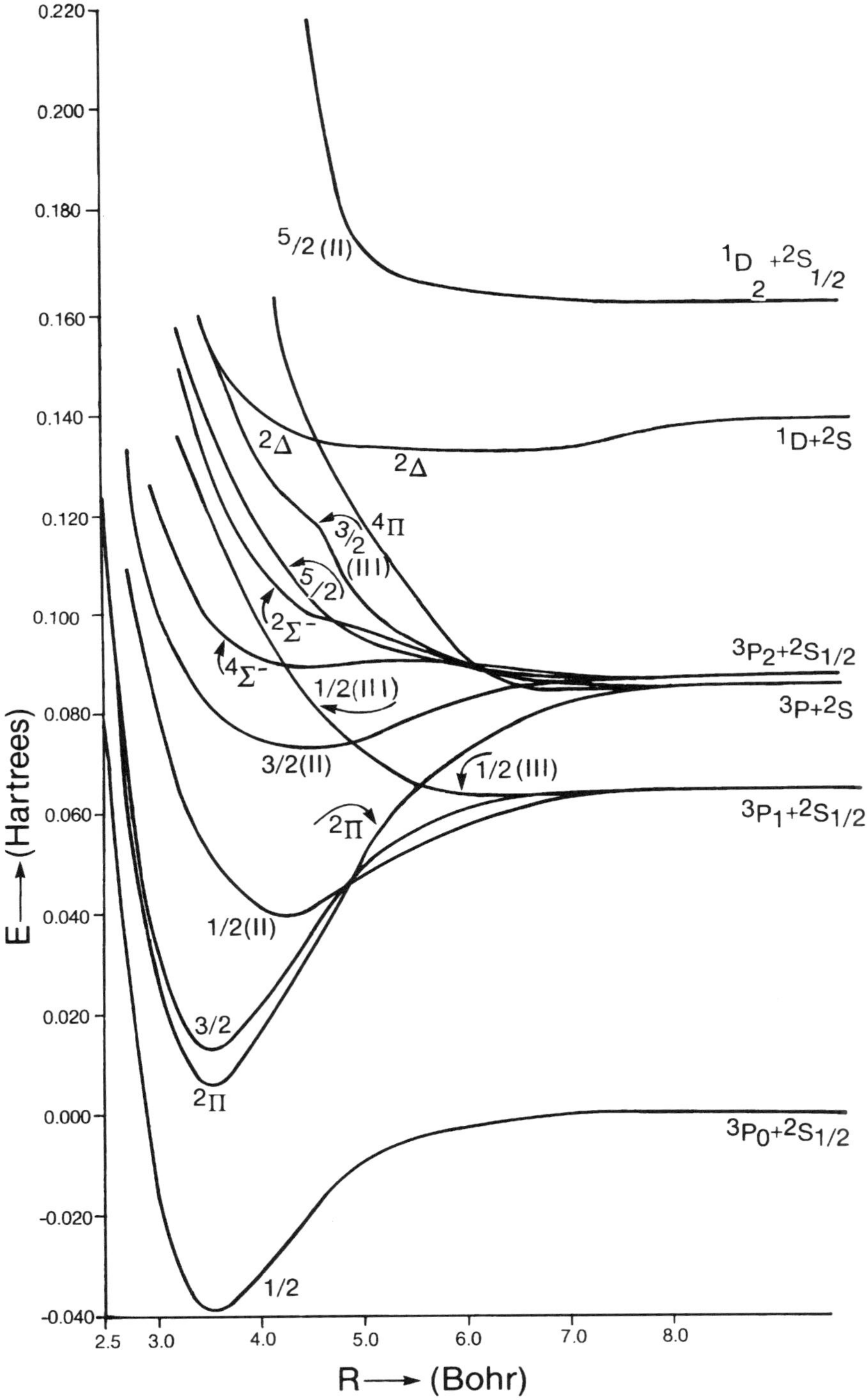

Figure 2.29 Potential-energy curves for the electronic states of BiH^+. (Reproduced from Ref. 182; copyright 1990, American Chemical Society.)

of the neutral BiH molecule is $^3\Sigma^-(0^+)$ arising from $1\sigma^2 2\sigma^2 1\pi^2$, the ionization of the highest occupied nonbonding 1π orbital of BiH should lead to a $^2\Pi_{1/2}$ ground state for BiH^+ arising from the $1\sigma^2 2\sigma^2 1\pi$ configuration. The removal of one of the 2σ electrons should result in a large array of states arising from the $1\sigma^2 2\sigma 1\pi^2$ configuration ($^4\Sigma^-, ^2\Delta, ^2\Sigma^-, ^2\Sigma^+$, etc.). All these states were investigated by the author (182).

The ionization potential of the Bi atom was also calculated by the author (182) as 6.98 eV, in very good agreement with an experimental value of 7.29 eV (306). Further using the asymptotic separations of $X(\frac{1}{2})$–$\frac{3}{2}$, $\frac{1}{2}$–$\frac{5}{2}$, and $\frac{1}{2}$–$\frac{5}{2}$(II) states, the author (182) calculated the $^3P_0$–$^3P_1$, $^3P_0$–$^3P_2$, and $^3P_0$–$^1D_2$ splittings of the Bi^+ ion as 13,937, 19,138, and 36,038 cm^{-1}, respectively. These values were found to be in very good agreement with the corresponding experimental values (306) of 13,324, 17,030, and 33,939 cm^{-1}, respectively.

The D_e of BiH^+ was calculated as 1.05 eV (182). This is much smaller than the D_e value of the 0^+ state of BiH, which was calculated as 2.2 eV (180). Since the π orbital of BiH is essentially nonbonding, it was expected that the ionization of the π shell should not alter the D_e value too much. Thus the difference in the D_e values of BiH and BiH^+ was puzzling. However, at dissociation limits the spin–orbit stabilizations of the electronic states of the neutral atom and ion are substantially different. The Bi neutral atom has a ^4S ground state that cannot be split by spin–orbit interaction but is lowered by 0.03266 hartree. The Bi^+ ion has a $^3P1$–s state that is split apart into $^3P_0$, $^3P_1$, and $^3P_2$. The author's calculations (182) indicated that the $^3P_0$ state is 0.0859 hartree lower than ^3P. Thus the $^3P_0$ state of Bi^+ is lowered considerably compared to Bi as well as BiH^+. This appears to be the reason for the lower D_e for BiH^+ (since Bi^+ is more stabilized by spin–orbit interaction taken in BiH^+). This is further justified if one compares the D_e value of the $^3\Sigma^-$ state of BiH with that of the $^2\Pi$ state of BiH^+ calculated without spin–orbit integrals. The D_e value of $^3\Sigma^-$ is 2.06 eV compared to the $^2\Pi$ state of BiH^+, which is 2.17 eV. Thus it is clear that the spin–orbit interaction reduces the D_e of BiH^+.

2.2 SPECTROSCOPIC PROPERTIES AND POTENTIAL-ENERGY CURVES OF HEAVY HALIDES

The interest (183–232) in heavy metal and other halides arises from a number of reasons, one of which is that many of the halides such as MF and MCl, are generated in chemiluminescent reactions of the following type:

$$M + X_2 \rightarrow MX^* + X$$

Some of the heavier halides have been investigated by Parson and co-workers (214). In the chemiluminescent reactions of M + X_2, the MX_2 triatomic (C_{2v}) is generated, which eventually forms MX* in the excited state. The excited MX chemiluminesces, emitting photons. Thus, many of the heavy main-block halides and chalconides have been investigated as candidates for chemical lasers.

The photoionization of nonvolatile materials such as thallous halides has been the topic of many investigations (185–191). Berkowitz and Chupka (185) examined the ion-pair processes in thallous halides and the high-energy processes that ionize the molecule. The electronic spectra of some of these species are not completely understood because of the complexities of the electronic states. The potential-energy curves of these species exhibit peculiar shapes such as barriers. There are many recent experimental investigations (318–323) on group IIIA halides, and in particular InCl. In the most recent investigation on InCl, Hoeft and Nair (318) obtained the rotational spectra of InCl. From these data, the vibrational constants and Dunham potential constants are derived for InCl. Glenewinkel-Meyer et al. (324) studied the emission spectra of 10 group IIIA monohalide ions, MX^+ (M = B,Al,Ga,In; X = F,Cl,Br) in the visible and near-UV regions. The emission spectra of all these species were obtained through the following chemiluminescent reaction:

$$M^+ + X_2 \rightarrow MX^{+*} + X$$

On the basis of the observed spectra, these authors concluded that the excited states of these ions are considerably displaced compared to the ground states. For 6 of the 10 species studied, these authors observed narrow-band systems, which they assigned to $C^2\Pi$–$X^2\Sigma^+$. The narrow features that accompanied this system for $GaCl^+$, $GaBr^+$, and $InBr^+$ were attributed to $D^2\Sigma^+$–$X^2\Sigma^+$ transitions. All these results were compared with existing *ab initio* calculations on AlF^+ (325). The potential-energy curves of the $X0^+$, $A0^+$, and $B1$ states of the InI diatomic have been constructed from experimental data and rotational analyses (327–329).

2.2.1 TlF

The spectroscopic investigations on TlF include those of Boizova and Butkow (193), Howell (194), and Barrow and co-workers (192). These investigations have revealed the existence of three excited states of TlF, ${}^3\Pi_{0^+}$, ${}^3\Pi_1$, and ${}^1\Pi_1$, in addition to the closed-shell ${}^1\Sigma^+_{0+}$ ground state. The rotational analyses of the A–X and B–X systems have also been carried out. The microwave spectra, the Stark–Zeeman spectra, and the rotational spectra of TlF have also been obtained.

The photoionization studies of the vapors of thallous halides have been considered by Trenin and co-workers (183, 184) and Berkowitz and Chupka (185). The vaporization reactions of thallous halides have been investigated by Cubicciotti (186). The dimeric and tetrameric species of thallous halides have been detected in the vapor phase. The infrared spectra and the structure of matrix-isolated Tl_2F_2 and Tl_2Cl_2 species have been investigated by Brom and Fransen (188), who proposed a linear structure for Tl_2F_2(F–Tl–Tl–F).

Murad et al. (190) calculated the dissociation energies of group IIIA halides using high-temperature mass-spectroscopic methods. Because the thermodynamic D_e values are lower than the spectroscopically derived D_e values for TlF, these authors suggested that there should be barriers in the excited potential-energy curves of TlF. This was also supported by the photoionization mass-spectroscopic investigation of Berkowitz and Walter (191), who examined TlF, TlCl, and TlBr. The threshold for the formation of Tl^+ was used to calculate the D_e values of these species in this method.

The author (198) carried out SCF/RCI calculations on nine low-lying ω–ω states and the corresponding λ–s states of TlF. The spectroscopic properties and the potential-energy curves of these states were calculated. Table 2.20 shows the spectroscopic constants of TlF obtained by the author (198) and the experimental data, whenever available. Figures 2.30 and 2.31 show the theoretical potential-energy curves (198) of the low-lying electronic states of TlF.

The ground state of TlF arises from the $1\sigma^2 2\sigma^2 3\sigma^2 1\pi^4$ configuration, where the 1σ and 2σ orbitals arise from Tl(6s) and F(2s), while the 3σ orbital is a mixture of Tl($6p_z$) with F($2p_z$). The 1π orbital is predominantly F(2p). Thus, the 2π orbital is predominantly the Tl(6p) nonbonding orbital. The low-lying excited electronic states of TlF arise from the $1\sigma^2 2\sigma^2 3\sigma^1 1\pi^4 2\pi^1$ configuration, which leads to ω–ω states of the $^3\Pi$ and $^1\Pi$ states.

As seen in Table 2.20, theoretical calculations confirmed the earlier assignment of the X, A, and B states of TlF to 0^+, 0^+(II), and $1(^3\Pi_1)$ states, respectively. The 1(II) state in Table 2.20 is the C state in the $C \rightarrow X(0^+)$ emission system.

Absorption continua in the region of 45,400 cm^{-1} and above 50,000 cm^{-1} were assigned to the 0^+(IV) state, which was found to be predominantly $^1\Sigma^+(II)_{0^+}$ arising from $\sigma^2\pi^3\pi^*$, which was mixed significantly with $\sigma^2\pi^3\pi^*(^3\Sigma_{0^+})$. There are also several 1 states arising from $\sigma^2\pi^3\pi^*$ configura-

Table 2.20 Calculated Spectroscopic Properties of Low-Lying Bound States of TlF and Available Experimental Values[a]

	R_e, Å		T_e, cm^{-1}		ω_e, cm^{-1}	
State	Theory	Exp.	Theory	Exp.	Theory	Exp.
$X0^+$(I)	2.04	2.08	0	0	592	477
$A0^+$(II)	2.12	2.05	40,462	35,186	462	436
0^-(I)	2.10	—	40,891	—	345	—
$B1$(I)	2.12	2.08	42,612	36,863	326	366
$C1$(II)	2.69	—	45,164	45,546	396	346
2(I)	2.10	—	46,245	—	340	—
2(II)	2.92	—	49,700	—	264	—
0^-(II)	2.90	—	50,053	—	279	—

[a]All results are from Reference 182.

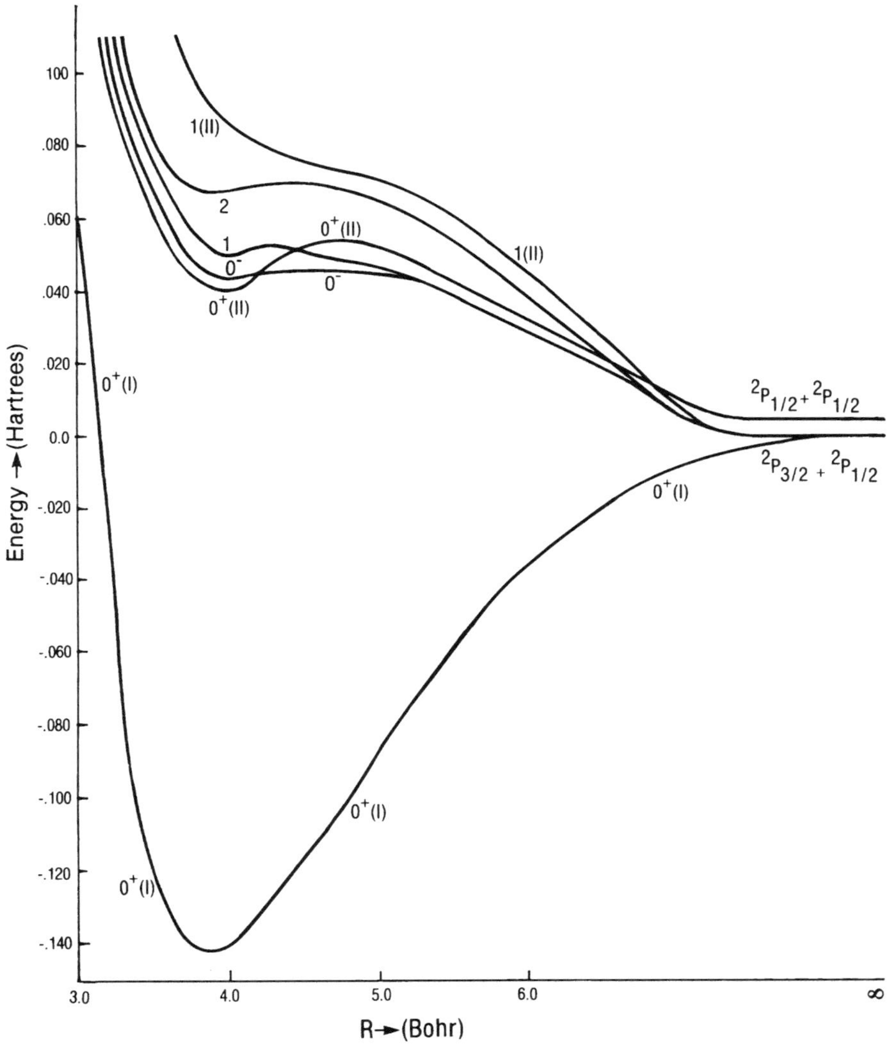

Figure 2.30 Potential-energy curves for some low-lying states of TlF. (Reprinted from Ref. 198; copyright 1985, American Institute of Physics.)

tions. These states could also be attributed to those continua of bands in the region above 50,000 cm^{-1}.

In the $^3\Pi_{0^+}$–X system no bands higher than $v' = 3$ for $^3\Pi_1$ [0^+(II)] were observed. Howell (194) postulated the existence of a repulsive state that crosses with $^3\Pi_{0^+}$ and $^3\Pi_1$ so as to produce predissociation in both $^3\Pi_{0^+}$ and $^3\Pi_1$. As seen from Figure 2.30, the $^3\Pi_{0^-}$ [i.e., 0^-(I)] crosses with 0^+(II) and 1 [the crossing with 0^+(II) is at a shorter distance compared to the crossing with the

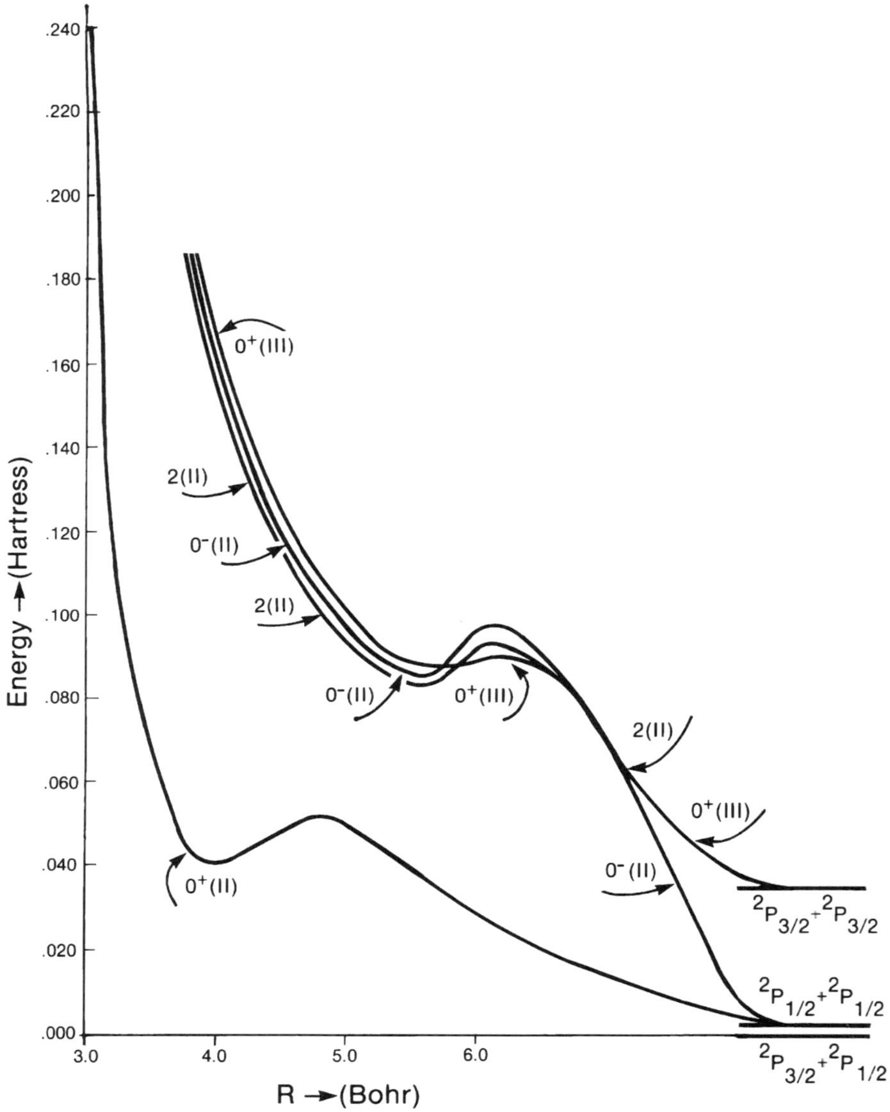

Figure 2.31 Potential-energy curves for some excited states of TlF. (Reprinted from Ref. 198; copyright 1985, American Institute of Physics.)

1 state]. Thus the 0^-(I) state may be responsible for the predissociation in $^3\Pi_1$. However, predissociation via a tunneling mechanism through the small barrier or via rotationally induced coupling to $B^3\Pi_1$ should not be ruled out. Howell's (194) observed spectra showed the continuum to be of lower intensity compared to the bands in these systems.

Rao and Rao (199) reported band spectra of TlF in the 28,500–29,500-cm^{-1} region that were abruptly cut off. The vibrational frequency of the upper state calculated by these authors was about 194 cm^{-1}, suggesting that this is a shallow minimum. This was attributed to a possible long-distance shallow minimum by the author (198).

The existence of maxima in the potential-energy curves of the excited II states of TlF was predicted by many authors (190, 191) This prediction was based on the higher spectroscopic dissociation energies obtained from Birge–Sponer extrapolation of the excited Π states. The spectroscopic dissociation energies for these molecules were thus higher than the thermochemical values. As seen in Figures 2.30 and 2.31, the excited Π states, 0^+(II), 0^-, 1, 2, and 1(II), and the 0^-(II), $0^+(m)$, and 2(II) states have maxima in their potential-energy curves. These states are bound but their dissociation limits are way down, and, thus, they have to go through maxima.

Howell (194) misassigned the A and B states of TlF to $^3\Pi_{0+}$ and $^3\Pi_1$ arising from the $\sigma^2\pi^3\sigma^*$ configuration. The author's (198) SCF calculations indicated that the σ^* orbital is very highly antibonding, while the $\pi*$ orbital is the nonbonding thallium $6p$ orbital. The CI calculations confirmed that the $^3\Pi$ state arising from $\sigma^2\pi^3\sigma^*$ is much higher in energy than $^3\Pi$ arising from the $\sigma^2\pi^4\sigma^*$ configuration. Thus the $A0^+$ and $B1$ states are $^3\Pi_{0+}$ and $^3\Pi_1$, arising from the $\sigma\pi^4\pi^*$ configuration. On the basis of the electronic properties of isoelectronic BF and AlF molecules. Murad et al. (190) correctly concluded that the lowest-lying $^3\Pi$ and $^1\Pi$ states should arise from the $\sigma\pi^4\sigma^*$ configuration.

The theoretical dissociation energy for the 0^+(I) state of 3.86 eV (198) was found to be in reasonable agreement with the photoionization mass-spectrometric value of Berkowitz and Walter (4.57 eV) (191).

2.2.2 PbF

The early (1930s) investigations (200, 201) of the band spectrum of PbF showed conflicting evidence on the ground state of PbF. The electronic spectra of PbF did not show the typical doublet-type structures of the lighter analogs of that group such as SnF. Thus the early assignments of the observed spectra were very difficult. The spin–orbit splitting of the lead atom is so large that the ground state of PbF was found to be split apart by 8000 cm^{-1}. The absorption spectra of PbF were studied by Morgan (200) and Rochester (201). A part of the earlier analysis of the spectra of PbF was incorrect as shown by Barrow and co-workers (202) in a later investigation. The electronic spectra of PbF exhibited Λ doubling due to the large spin–orbit coupling. Lumley and Barrow (205) carried out the rotational analyses of the observed B–X_2, B–X_1, and A–X_1 systems. The assignments of the A and B states remained uncertain. The other experimental works on PbF are contained in References 203, 204, 206, and 207. Mélen and Dubois (331) assigned 39 unattributed electronic states of GeX, SnX, and PbX (X = F,Cl,Br,I) to Rydberg configurations. Ionization potentials and quantum defects have been derived.

The author (208) carried out SCF/RCI calculations on six ω–ω states and five λ–s states of PbF. These calculations enabled the assignment of the observed X_1, X_2, A, and B states. In addition, the spectroscopic properties of some of states of PbF that have not been observed to date were calculated.

Table 2.21 shows the theoretical and experimental spectroscopic constants of PbF. Figure 2.32 shows the SCF/RCI potential-energy curves of PbF with and without the spin–orbit coupling term. As seen from Table 2.21, the spin–orbit splitting of the $^2\Pi_{1/2}$ and $^2\Pi_{3/2}$ states was calculated as 7895 cm^{-1} (208), compared to an experimental value of 8264 cm^{-1} (32). The calculated properties of the ground state are also in excellent agreement with experiment.

The A state observed in the A–X_1 system was assigned to the $^2\Sigma^+_{1/2}$ state arising from the $\sigma^2\sigma^*\pi^4$ configuration. The calculated T_e and R_e values of this state confirmed this assignment. The properties of the $^4\Sigma^-_{3/2}$ and $^4\Sigma^-_{1/2}$ states are shown in Table 2.21. It seems that these states have not yet been observed experimentally.

Lumley and Barrow (205) discussed the possible assignments of the B state observed in the B–X_2 and B–X_1 systems. These possibilities included B to be the lowest Rydberg $^2\Sigma^+$ state, $^4\Sigma^-$, or $^2\Pi_{1/2}$. The R_e of $^4\Sigma^-_{1/2}$ is larger than the R_e of the ground state (Table 2.21). The experimentally observed B state has an R_e smaller than the R_e of the ground state. The calculated ω_e of the $^4\Sigma^-_{1/2}$ state (239 cm^{-1}) is quite different from the ω_e of the B state (606 cm^{-1}). The separation of $^4\Pi(\sigma^2\pi^3\pi^{*2})$ from $X^2\Pi$ at 4.0 bohr was calculated as 42,227 cm^{-1} (208) and $^2\Pi(\sigma^2\pi^3\pi^{*2})$ would thus be even higher. The author (208) calculated the properties of the first Rydberg state ($^2\Sigma^+_{1/2}$) arising from the $\sigma^2\sigma_R\pi^4$ configuration with a triple-ζ Pb s and p basis set. The T_e value calculated for this Rydberg state agreed with the experimental value. Thus the B state was assigned to a $^2\Sigma^+_{1/2}$ Rydberg state.

2.2.3 BiF

The BiF molecule appears to be the most studied among the p-block sixth-row fluorides (215–232). More recent experimental and theoretical works are discussed subsequently. The first spectrum of BiF was recorded in emission by Howell (215), while Morgan (216) studied BiF in absorption. The A–X emission system observed by Howell (215) was studied by many authors subsequently (217–228). Kuijpers and Dynamus (229) obtained the rotational constants of the ground state of BiF through the millimeter wave spectrum of BiF.

The author (232) carried out SCF/RCI calculations employing a double-ζ basis set on the low-lying electronic states of BiF. The theoretical calculations were not consistent with Jones and McLean's (230) suggestion of a 0^+ excited state in the 9216-cm^{-1} region. Further, theoretical calculations on BiF indicated that the bands observed by Avasthi (226), Chakko and Patel (228), and Murthy et al. (227) in the 580–6600-Å region were misassigned by these authors. The correct assignments of these bands were made by the author (232).

Table 2.21 Calculated and Available Experimental Spectroscopic Properties of PbF[a]

	R_e, Å		T_e, cm^{-1}		ω_e, cm^{-1}		D_e, eV	
State	Theory	Exp.	Theory	Exp.	Theory	Exp.	Theory	Exp.
$X_1\,^2\Pi_{1/2}\frac{1}{2}$(I)	2.08	2.06	0	0	540	503	3.75	3.69 ± 0.09
$X_2\,^2\Pi_{3/2}\frac{3}{2}$(I)	2.06	2.03	7,895	8,264	547	520	2.78	—
$A^2\Sigma^+_{1/2}\frac{1}{2}$(II)	2.13	2.16	24,510	22,557	502	[395][c]	0.78	—
$^4\Sigma^-_{3/2}\frac{3}{2}$(II)	2.53	—	34,440	—	267	—	0.38	—
$B^2\Sigma^+_{1/2}$ [b]	1.99	1.98	35,941	35,664	724	606	—	—
$^4\Sigma^-_{1/2}$	2.37	—	35,998	—	239	—	0.28	—
$^2\Pi$	2.07	—	5,168	—	553	—	4.43	—
$^2\Sigma^+$	2.16	—	24,218	—	487	—	2.09	—

[a]The theoretical results are from Reference 208.
[b]The spectroscopic properties of the $B^2\Sigma^+_{1/2}$ (Rydberg) state were calculated with an extended triple-ζ s and p basis set for the lead atom and a comparison calculation for the ground state. See text for discussion.
[c]Experimental value is uncertain.

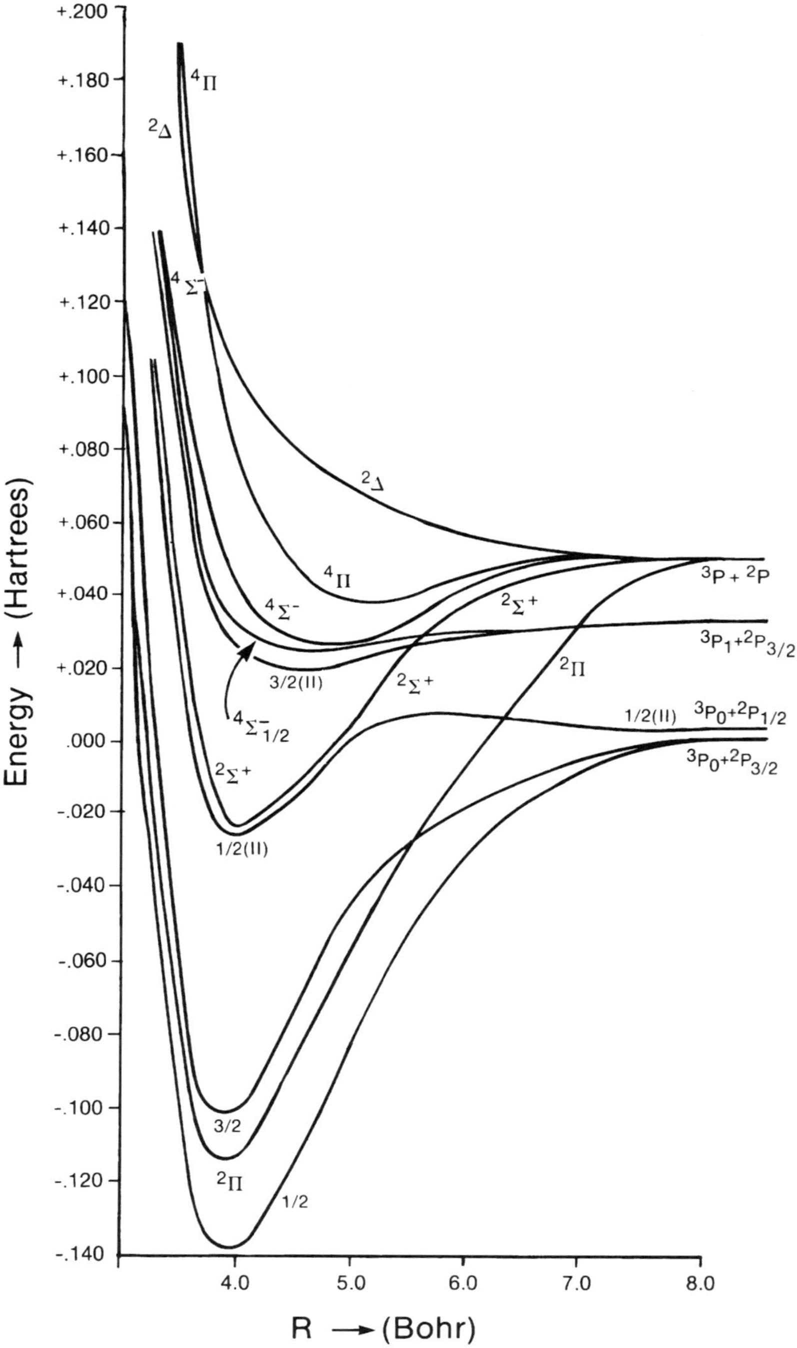

Figure 2.32 Potential-energy curves for some low-lying states of PbF. (Reprinted from Ref. 208; copyright 1985, American Institute of Physics.)

The rotational analyses of the A–X and B–X systems were done by Jones and McLean (230), who pointed out the difficulties in the rotational analyses as a result of predissociation and perturbation of the levels arising from the interaction of two 0^+ states in this region. These authors attempted to assign the bands in the 5800–6000-Å region by comparing the BiF molecule with the isoelectronic well-characterized SbF molecule. They suggested the existence of a $b0^+$ state about 9216 cm^{-1} above the ground state, which was designated as $X0^+$.

Table 2.22 shows the theoretical (232) and known experimental spectroscopic properties of the low-lying states of BiF. Figure 2.33 shows the theoretical potential-energy curves of BiF (232). The properties of the $^3\Sigma^{-1}\Delta$, and $^3\Sigma^+\lambda$–s states calculated without the spin–orbit term are also reported in Table 2.22. The $X0^+$ ground state was found to be a mixture of $^3\Sigma^-_{0^+}$ and $^1\Sigma^+_{0^+}$ states, whereas the 1(I) state was found to be the 1 component of the $^3\Sigma^-$ state. The 2 state was predominantly $^1\Delta_2$ at near-equilibrium geometries, while the 0^+(II) state was found to be a mixture of $^1\Sigma^+$, $^3\Sigma^-$, and $^3\Pi(\sigma^2\sigma^*\pi^4\pi^*)$, with the $^1\Sigma^+$ state making a predominant contribution. The vibrational frequency of the 0^+(II) state was found to be considerably influenced by avoided crossings.

The second 0^+ state [0^+(II)] was found to be about 25,900 cm^{-1} above the ground state, and no excited state of 0^+ symmetry was found below this state. Thus the assignment of the b state to 0^+ by Jones and McLean (230) as well as Chakko and Patel (228), who designate the b state by the X_3 state, was shown to be incorrect (232). The experimental b or X_3 state was found to be 9216 cm^{-1} above the ground state. The only state with a T_e value in this region is the 1(I) state (Table 2.22). The electronic bands around 6600 Å observed by Avasthi (226) and the electronic bands around 6200 Å observed by Murthy et al. (227) were assigned to the A'–$X_2(2)$ and A_1–$X_2(2)$ transitions, where the X_2 state was assumed to be a 2 state above the ground state. Jones and McLean (230) called this the $a(2)$ state with a T_e value of about 6000 cm^{-1}. The theoretical calculations revealed that there is no such state (Table 2.22). The author (232) suggested, on the basis of calculated properties of low-lying states of BiF, that

Table 2.22 Spectroscopic Properties of BiF[a]

	R_e, Å		T_e, cm^{-1}		ω_e, cm^{-1}	
State	Theory	Exp.	Theory	Exp.	Theory	Exp.
$X0^+$	2.087	2.052	0	0	502	511
1	2.075	—	7,280	9,216	501	535
2	2.075	—	14,595	—	511	—
$A0^+$(II)	2.076	2.15	25,931	22,960	467	381
$^3\Sigma^-$	2.068	—	9,441	—	514	—
$^1\Delta$	2.06	—	18,247	—	496	—
$^1\Sigma^+$	2.06	—	23,544	—	537	—

[a] All theoretical results are from Reference 232.

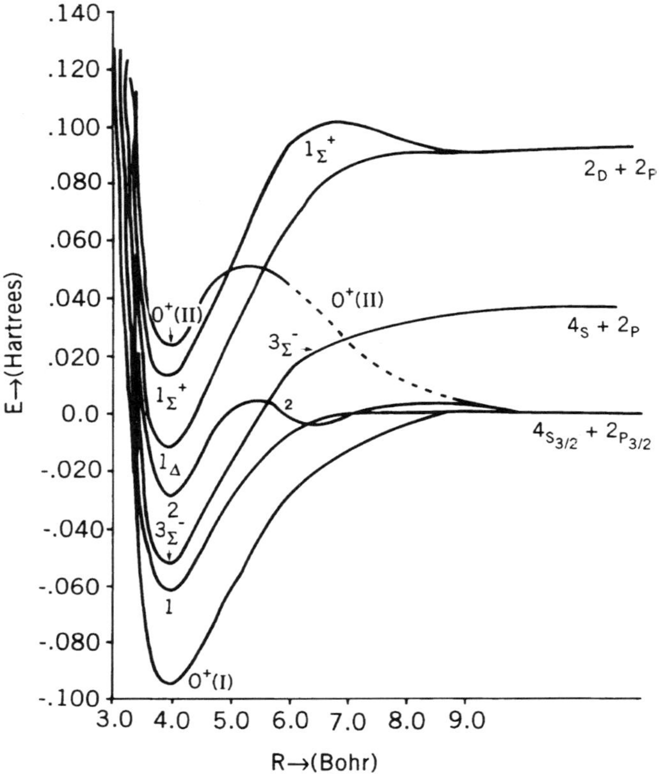

Figure 2.33 Potential-energy curves for the electronic states of BiF. (Reprinted from Ref. 232; copyright 1985, Elsevier Science Publishers B.V.)

the electronic bands observed by Avasthi (226), Murthy et al. (227), and Chakko and Patel (228) may all be due to the same electronic transition.

Jones and McLean (230) designated the 0^+ and 1 states arising from $^3\Sigma^-(\sigma\pi^4\pi^{*2})$ as the X_1 and X_2 states, while the $^1\Delta_2$ state was designated *a*; the *b* state was assigned to $^1\Sigma^+(b0^+)$ arising from the electronic configuration $\sigma^2\pi^4\pi^{*2}$. The experimentally observed *b* state is 9216 cm^{-1} above the ground state. The author's (232) calculations indicated that the $^1\Sigma^+(0^+)[0^+(\text{II})]$ state was 25,900 cm^{-1} above the ground state. The bands observed by Chakko and Patel (228) as well as Murthy et al. (227) in this region were assigned to the $0^+(\text{II}) \rightarrow 1$ transition by the author (232). The difference between the calculated T_e values of these two states (18,650 cm^{-1}) was found to be in good agreement with an experimental value of 16,570 cm^{-1} (223). Thus, the lowest excited electronic state above the ground state is a 1 state and is 7280 cm^{-1} above the ground state. Consequently, the author (232) showed that the hypothesis of an X_2 state (*a*2) below this state is false. The experimentally observed A state in the A–X system was assigned to the 0^+(II)–0^+(I) transition. The A state was found to be a mixture of $^1\Sigma^+(\sigma^2\pi^4\pi^{*2})$ and $^3\Sigma^-(\sigma^2\pi^4\pi^{*2})$.

Theoretical calculations supported the existence of a B state very close to the $A0^+$(II) state that perturbed the bands in the A–X system, making rotational analysis of this system difficult. The 0^+(II) state was found to exhibit an avoided crossing within 1 bohr from its equilibrium geometry. This avoided crossing resulted from the crossing of a $^3\Pi(\sigma^2\sigma^*\pi^3\pi^{*2})$ state with $^1\Sigma^+(\sigma^2\pi^4\pi^{*2})$. Thus the theoretical and experimental R_e and ω_e values of the A state are not in very good agreement. It seems that a more accurate rotational analysis of the A state should be pursued.

The theoretical D_e value of BiF was found to be about 2.63 eV (232). An upper limit for the D_e of 3.63 eV for the ground state was obtained by Jones and McLean (230) by extrapolating the $v' = 6$ vibrational level to the predissociation limit of the A state. Gaydon (231) reported a D_0 value of 2.65 ± 0.3 eV for BiF. This value was found to be in more reasonable agreement with the theoretically calculated value of 2.63 eV (232).

The excited states of BiF exhibited avoided crossings (232). As seen from Figure 2.33, the 2 and 0^+(II) curves contain barriers that are due to these avoided crossings. The potential-energy curve of the 2 state also contains a second shallow minimum.

These features in the potential-energy curves of BiF are best understood by considering the weights of various λ–s states in the RCI wavefunction of BiF as a function of distance.

The highest occupied orbital of BiF was found to be a bonding orbital of Bi $6s$, $6p_z$ and F $2p_z$ making the dominant contribution. The occupied π orbital was found to be mostly on the fluorine atom. The π^* orbital was predominantly on the Bi atom, but the fluorine p orbitals made a nonnegligible contribution. The contribution of the spin–orbit interaction to the electronic states arising from the lowest $\sigma^2\pi^4\pi^{*2}$ electronic configuration should thus be quite high, since the π^* orbital is predominantly on Bi.

At 4.25 bohr, the RCI wavefunction of the 0^+(I) state consisted of 66% $^3\Sigma^-$, 20% $^1\Sigma^+$, and 5.5% $^3\Pi$. At longer distances, contributions from $^3\Sigma^-(\sigma\sigma^*\pi^4\pi^{*2})$ and $^3\Sigma^-(\sigma^{*2}\pi^4\pi^{*2})$ became important. The 0^+(II) (A) state consisted of 23% $^3\Sigma^-$, 58% $^1\Sigma^+$, 5% $^3\Pi(\sigma^2\sigma^*\pi^4\pi^*)$, and 2% $^3\Pi(\sigma\pi^4\pi^{*3})$ at 4.0 bohr. On the contrary, at 5.0 bohr the 0^+(II) state was found to be 18% $^1\Sigma^+$, 12% $^3\Sigma^-$, 14% $(\sigma^2\sigma^*\pi^3\pi^{*2})$, 18% $^3\Pi(\sigma^2\sigma^*\pi^4\pi^*)$, and 9% $^3\Sigma^-(\sigma^2\sigma^*\pi^4\pi^{*2})$. At 5.5 bohr, this state became predominantly $\sigma^2\sigma^*\pi^3\pi^{*2}$. The 2 state was 79% $^1\Delta_2$ and 5% $^3\Pi_2(\sigma^2\sigma^*\pi^4\pi^*)$ at 4.0 bohr. However, at 6.0 bohr it became 8% $^1\Delta_2$, 3% $^3\Pi(\sigma^2\sigma^*\pi^4\pi^*)$ 17%, $^3\Pi(\sigma\pi^4\pi^{*2})$, 52% $^5\Sigma_2^-(\sigma^2\sigma^*\pi^4\pi^{*2})$, and 8% $^1\Delta(\sigma\sigma^*\pi^4\pi^{*2})$, and at longer distances, $^5\Sigma_2^-$ dominated. The barrier and the second minimum in the 2 state were attributed to this avoided crossing.

After the author's paper (232) appeared in 1986, several experimental and theoretical papers have appeared on the BiF molecule. The dissociation energy of the BiF molecule has been a topic of several studies. Gole and co-workers (350) as well as Ross et al. (351) discussed the dissociation energy of the BiF molecule. As noted before, Jones and McLean (230) obtained an upper limit of 3.63 eV, while Gaydon estimated an experimental value of 2.65 eV. The author's computation yielded a D_e of 2.63 eV, but it was argued by the author that this is a lower bound, since higher-order electron correlation effects would increase the dissociation energy. Gole and co-workers (350) obtained a surprisingly high D_e of 5 eV. This value was also in contrast with a value of 3.76 eV reported by Berkowitz and co-workers (352). Alekseyev et al. (348) report a D_e value of 3.1 eV. A relativistic study that includes electron correlation effects to a very high order would be warranted to establish the D_e of BiF unequivocally.

Gole and co-workers (350) studied the spectra of BiF in the 565–740-nm regions, from which they inferred that there are two different transitions, both of which terminated to the $X_2 1$ state. These were assigned to the $A'1$–$X_2 1$ and $A''0^+$–$X_2 1$ systems. From analysis of their spectra, these authors confirmed the author's prediction that the first excited 0^+ state lies near 25,000-cm^{-1} region. The measured T_e value of this state by Gole and co-workers is 22,552 cm^{-1}.

Fink et al. (349a) in 1991 studied the $X_2 1$–$X_1 0^+$ electronic band systems of the bismuth monohalide molecules (BiF–BiI). In this work the $X_2 1$ states of these species were excited by near-resonant electronic-to-electronic energy exchange from metastable O_2 molecule in a fast-flow system. The T_e value of the $X_2 1$ state was measured by Fink et al. (349) as 6753 cm^{-1}, in very good agreement with the author's 1986 computed result of 7280 cm^{-1}. The vibrational frequencies of the $X_1 0^+$ ground state and the $X_2 1$ excited states obtained by Fink et al. are 510.49 and 540.98 cm^{-1}, respectively, in very good agreement with the author's (232) 1986 computed result of 502 and 501 cm^{-1}, respectively. One interesting outcome of this experimental work is experimental confirmation of the theoretical finding by the author in 1986 that there is no 0^+ state for BiF in the 7000 cm^{-1} region, and the second 0^+ state is 25,000 cm^{-1} above the ground state, while the experimental studies prior to the author's work in 1986 had misassigned the first excited state of BiF to an 0^+ state.

Alekseyev et al. (348) carried out a theoretical study of the electronic states of the BiF molecule using the SCF/MRDCI/spin–orbit CI techniques employing relativistic effective core potentials. They used a $(3s3p1d)$ valence Gaussian basis set for the Bi atom together with the RECPs that included the $6s^2 6p^3$ shells of the Bi atom in the valence space replacing the remaining electrons in the core RECPs. For the fluorine atom, a valence $(4s4p1d)$ Gaussian basis set was employed. A single configuration SCF approach was used for the generation of orbitals for the MRDCI computations. The configuration selection and energy extrapolation schemes were used in the MRDCI, which reduced the number of configurations to be included in the CI expansion drastically. The same configuration selection scheme was also used in the spin–orbit CI.

Table 2.23 shows the spectroscopic constants computed by Alekseyev et al. (348) together with the author's (232) previous computed results in 1986 and the available experimental results on BiF as of 1993. As seen from this table, the overall agreement between different theoretical approaches to study

Table 2.23 Spectroscopic Properties of BiF from Two Computations and Experiment (Bond Lengths r_e, Transition Energies T_0, Vibrational Frequencies ω_e, and Radiative Lifetimes τ for $v' = 0$)

	r_e, Å			T_0, cm^{-1}			ω_e, cm^{-1}			τ, μs	
	Calc.		Exp.	Calc.		Exp.	Calc.			Calc.	Exp.
State	Ref. 232	Ref. 348	Ref. 230	Ref. 232	Ref. 348	Ref. 349	Ref. 232	Ref. 348	Exp.	Ref. 348	Ref. 349
$X_1 0^+$	2.087	2.238	2.052	0	0	0	502	476	508	—	—
$X_2 1$	2.075	2.217	—	7,280	6,937	6,768	501	503	538	1050	1400
$a2$	2.075	2.212	—	14,600	13,930	11,555	511	481	527	80400	—
$A_2 0^+$ (A)	2.076	2.383	2.151	25,913	23,029	22,956	467	342	383	1	2
$A_1 0^-$ (A″)	—	2.414	—	—	23,225	22,544	—	355	412	95	50
$A_3 1$ (A′)	—	2.425	—	—	26,155	24,382	—	321	~400	6	16
$B0^+$	—	2.247	2.05	—	27,405	26,056	—	549	620	10	3

Source: Reproduced from Reference 348.

the electronic states of BiF is quite satisfactory. Alekseyev et al. (348) computed the properties of more electronic states compared to the author's work in 1986. Some quantitative differences could be found, however. The equilibrium bond lengths computed by Alekseyev et al. (348) are uniformly longer for all of the electronic states compared to the author's results and the experiment. Therefore, the vibration frequencies computed by Alekseyev et al. (348) are also uniformly smaller compared to the experiment and the author's previous work. For example, the ground-state bond distance computed by Alekseyev et al. is 0.15 Å longer than the value reported by the author and 0.188 Å longer than the experimental value. Of course, the author used Slater-type orbital basis sets together with numerical RECPs for the Bi atom that retained the outer $5d^{10}6s^26p^3$ shells in the valence space. Alekseyev et al. also made computations using the 15-electron RECPs that retained the outer $5d^{10}6s^26p^3$ shells in the valence space. They obtained improved results with the 15-electron RECPs. For example, the equilibrium bond distance for the ground state was 2.1128 Å with the 15-electron RECPs compared to the author's result of 2.087 Å and the experimental value of 2.052 Å. Thus part of the errors in the computed results of Alekseyev et al. are attributed to the use of five-electron RECPs, while the other part has to do with the basis set.

A notable difference between the computations of the author in 1986 and those of Alekseyev et al. (348) is in the composition of the excited 0^+ electronic state. They find this state to be contaminated significantly by the $^3\Pi$ state at distances longer than the equilibrium bond distance of this state. For example, at near $r = 5.0$ atomic units, Alekseyev et al. find this state to be 72.5% $^3\Pi$ and 19% $^1\Sigma^+$, while the author's (232) reported value is only 18% for the $^3\Pi$ state. One source of this discrepancy could be that somewhat less excitation might have been allowed for the $^3\Pi$ state in the RCI reported by the author, since the primary focus of his work is the low-lying electronic states. However, the basis set and RECPs employed by Alekseyev et al. produce uniformly longer bond lengths. Hence, it is not fully understood whether this contrast is due to the errors caused in the bond lengths in the Alekseyev et al. work or due to the limited excitation in the RCI employed by the author.

Alekseyev et al. (348) also computed the transition moments as a function of internuclear distances for the $A(1)$–$X(1)$, $A(0^+)$–$X(1)$, $A(0^+)$–$X(0^+)$, and $B(0^+)$–$X(0^+)$ transitions, among several other transitions. The lifetimes of some of the electronic states of BiF have also been computed. Among the low-lying electronic states, the $X_2(1)$ state has a fairly long lifetime of 1.4 ms as measured by the experimental study of Fink et al. This is also consistent with the theoretical value of a millisecond computed by Alekseyev et al. The lifetime of the $A(0^+)$ electronic state was computed as 1.0 μs by Alekseyev et al. compared with an experimental value of 2.0 μs measured by Fink et al. The total lifetime of the $B(0^+)$ excited state was computed by Alekseyev et al. as 10 μs compared to an experimental value of 2.97 μs. Evidently, computations of the lifetimes of the excited states of these species seems to be quite challenging.

2.2.4 BiI

Alekseyev et al. (366) computed the potential-energy curves and spectroscopic constants of the bismuth iodide molecule (BiI). They have employed techniques very similar to those that they employed for the BiF molecule. Analogous to the BiF molecule, BiI has a $X_1(0^+)$ ground state and a $X_2 1$ excited state, both of which arise from the same $^3\Sigma^-$ nonrelativistic electronic state. Their competed spectroscopic constants are shown in Table 2.24. As seen from Table 2.24, at the present time there is very little experimental information on the BiI molecule primarily pertaining to the ground state, $X_2 1$ state, and the $B0^+$ state. The computed properties agree well, but analogous to their computations on BiF, the bond lengths of the electronic states are longer and the vibrational frequencies are smaller. The lifetimes of the electronic states computed by Alekseyev et al. (366) are shown in Table 2.25. Das et al. (367) studied the analogous SbI molecule.

2.2.5 TlCl

Recently, Li et al. (368) studied the ground and excited electronic states of the diatomic thallium chloride molecule using the single-configuration SCF/MRDCI methods including the effects of spin–orbit coupling. They have used the RECPs for the thallium atom included the $5d^{10}6s^26p^1$ shells of the thal-

Table 2.24 Computed Spectroscopic Properties of BiI (Excitation Energy T_e, Equilibrium Bond Length r_e, and Vibrational Frequency ω_e) Together with Corresponding Experimental Data

	T_e, cm^{-1}		r_e, Å		ω_e, cm^{-1}	
State	Calc.[c]	Exp.	Calc.[c]	Exp.	Calc.[c]	Exp.
$X_1\,^3\Sigma^-_{0+}$	0	0	2.886	2.8005[a]	156	164[a]
$X_2\,^3\Sigma^-_1$	5,096	(6,190)[a]	2.878	—	152	171[a]
	—	6,182[b]	—	—	—	169[b]
$a^1\Delta_2$	12,336	—	2.861	—	155	—
2(III)	21,827	—	3.444	—	127	—
0^-(III)	23,175	—	3.587	—	100	—
$B0^+$	24,148	23,389[a]	2.898	—	170	198[a]
2(IV)	24,226	—	3.724	—	116	—
0^-(IV)	24,262	—	3.904	—	102	—
0^+(IV)	25,691	—	3.095	—	217	—
2(V)	28,664	—	3.231	—	159	—

[a]Reference 32.
[b]Reference 346.
[c]Reference 363.

Source: Reproduced from Reference 366.

Table 2.25 Calculated Radiative Lifetimes of Excited States of BiI ($v' = 0$); Partial Lifetimes (in seconds) τ_0, τ_1, and τ_2 for Transitions to $X_1\,{}^3\Sigma^-_{0+}$, $X_2\,{}^3\Sigma^-_1$, and $a^1\Delta_2$, Respectively; and Total Lifetime τ[a]

State	τ_0	τ_1	τ_2	τ
$X_2\,{}^3\Sigma^-_1$	20.7(−3)	—	—	20.7(−3)[b]
$a^1\Delta_2$	—	6.1(−3)	—	6.1(−3)
2(III)	—	169(−6)	≤4.2(−6)	≤4.1(−6)
0^-(III)	—	75.4(−6)	—	75.4(−6)
$B0^+$	61.5(−6)	12.1(−6)	—	10.1(−6)
2(IV)	—	7.6(−3)	—	7.6(−3)
0^+(IV)	0.7(−6)	50.9(−6)	—	0.7(−6)
2(V)	—	17.2(−6)	1.1(−6)	1.0(−6)

[a]Values in parentheses denote powers of 10. Theoretical values from Reference 366.
[b]Experimental value is 20 ± 4 ms from E. H. Fink and O. D. Shestakov, unpublished.

lium atom in the valence space. These authors have also noted that the RECPs that included only three electrons in the valence space ($6s^26p^1$) did not produce a potential minimum at all for the TlCl molecule. However, these RECPs were found to be satisfactory for the computation of the properties of small thallium clusters by the author's group.

Li et al. (368) obtained the potential-energy curves of the low-lying electronic states, their spectroscopic properties, and the radiative lifetimes of some of these states. Analogous to the ground state of TlF, the TlCl molecule has a closed-shell ${}^1\Sigma^+$ ground state and thus yields only a 0^+ ground state, when spin–orbit effects were included. The first excited ${}^3\Pi$ state splits into 0^+, 0^-, 1, and 2 spin–orbit states similar to TlF, which was discussed before. In contrast to TlF, the 1 spin–orbit state of the ${}^3\Pi$ state does not form a strong minimum in the Li et al. computation, while the author finds a minimum for the 1 spin–orbit component of the ${}^3\Pi$ state. In fact this state has been observed experimentally for TlF and is denoted as the B1 state (see Table 2.20).

Table 2.26 shows the spectroscopic constants computed by Li et al. together with the available experimental data as of 1994. As seen from this table, there is a good agreement between the computed results of Li et al. and the experimental results. The equilibrium bond distances computed by Li et al. are somewhat longer (0.04–0.09 Å) for the low-lying electronic states. The computed vibrational frequencies are surprisingly higher than the experimental values, considering that the computed bond distances are longer. The T_e values of most of the electronic states are in very good agreement with the experimental results. The first excited state of TlCl is a ${}^3\Pi$ state, which yields 0^+, 0^-, 1, and 2 spin–orbit components. Li et al. have reported the computed potential energy curves of the 0^+, 1, and 2 spin–orbit states. From these curves, it can be inferred that the 0^+ spin–orbit state arising from ${}^3\Pi$ has a somewhat stronger potential minimum compared to the 1 and 2 spin–orbit states. Li et al. have not reported the spectroscopic constants of the A 1 and 2 states arising from ${}^3\Pi$.

Table 2.26 Spectroscopic Constants of TlCl

	T_e, cm^{-1}		ω_e, cm^{-1}		r_e, Å	
State	Exp.[a]	Calc.[b]	Exp.[a]	Calc.	Exp.[a]	Calc.[b]
$X0^+$	0	0	285	287	2.485	2.535
$A0^+$	31,054	30,890	210	248	2.437	2.476
E_10^+	55,491	54,602	221	275	2.77[c]	2.863
	55,473[b]	—	245[b]	—	—	—
E_20^+	56,444[b]	55,546	95[b]	101	3.10[c]	3.052
$F1$	57,886	56,787	285	290	2.75[c]	2.97

[a]Experimental values from References 32, 369–371.
[b]Calculated values from Reference 366.

Li et al. have observed several avoided crossings in the excited 0^+ states of TlCl. These authors have attributed these avoided crossings to the predissociation effects observed by Tiemann and co-workers. The potential-energy curves of these states computed by Li et al. show very irregular shapes due to avoided crossings.

The lifetimes of a few vibrational levels have also been computed for the electronic states of TlCl. It is interesting to note that although the $A0^+$–$X0^+$ transition is forbidden in the absence of relativistic effects, as this corresponds to a ^3P–$^1S^+$ transition, it becomes allowed through spin–orbit coupling. Table 2.27 shows the radiative lifetimes of the low-lying electronic states of TlCl. As seen from this table, the lifetimes of the $A0^+$ state for a few vibrational levels are in the microsecond region, and the lifetimes generally decreases for the higher vibrational quantum numbers. The lifetimes of the E_10^+ state are considerably shorter, due to the ^3P–^3P dipole matrix elements. The total lifetime is thus shorter as a consequence of the E_1–A transition. Thus, there is a high probability of predissociation for relatively high vibrational and rotational quantum numbers. The E_20^+ state is of predominantly $^1\Sigma^+$ character, and thus couples strongly to the ground state. The contribution to the E_2–A transition process was found to be negligible by Li et al. Since E_20^+ and the $X0^+$ states have sub-

Table 2.27 Radiative Lifetimes (in sec.) Calculated for the Vibrational Levels (v') of the $A0^+$, E_10^+, and E_20^+ States of TlCl (Numbers in Parentheses Refer to Two Powers of Ten)

State	v' =	0	1	2	3	4
$A0^+$		0.808(−6)	0.708(−6)	0.602(−6)	0.520(−6)	0.452(−6)
E_10^+		0.146(−7)	0.188(−7)	0.369(−7)	0.327(−7)	0.278(−7)
E_20^+		0.655(−6)	0.621(−7)	0.159(−7)	0.820(−8)	0.633(−8)

Source: Reproduced from Reference 368.

stantially different bond lengths, the Franck–Condon factor for this transition is lower, leading to a relatively long lifetime of 0.655 μs in the lowest vibrational level. The lifetimes of the excited state, however, decrease rapidly as expected.

2.2.6 PbCl

The electronic spectroscopy of lead halides has been the topic of a number of investigations for many years (200, 201, 205, 206, 233–237). Analogous to the spectra of PbH, the band spectra of lead halides showed conflicting evidence for the nature of the ground state, since these spectra did not exhibit the doublet character typical of the lighter analogs.

The absorption spectrum was studied by Morgan (200), who assigned the observed spectra to the $A \leftrightarrow X_1$ system. The vibrational analysis of this system yielded the ω_e and T_e values for the A state. Rochester (201) studied the band spectrum of PbCl, which facilitated the calculation of the vibrational frequencies of both the X_1 and A states.

The B–X_1 system in the ultraviolet region was first observed by Wieland and Newburgh (233), although the bands were diffuse and predissociated. These authors extrapolated these bands to predict the D_e of PbCl as 3.05 and 3.40 eV, where the latter value was considered somewhat improbable by these authors (233). Rao and Rao (235) carried out the rotational analysis of four bands designated by these authors as (4,0), (6,0), (1,1), and (3,1). The rotational constants obtained from this analysis, however, were inconsistent with the theoretical calculations of the author (238), which suggested that a revised rotational analysis was necessary for PbCl. The electron impact study of $PbCl_2$ provided the ionization potential and electron affinity for PbCl (236). Cordes and Gehrke (234) carried out the revised vibrational analysis for the B–X_1 system.

The author (238) carried out the first SCF/RCI calculations on six low-lying ω–ω states and five λ–s states ($^2\pi$, $^4\Sigma^-$, $^2\Delta$, $^4\Sigma^-$, $^4\Pi$) of PbCl. The spectroscopic constants and the potential-energy curves of these states were calculated.

Table 2.28 shows the theoretical spectroscopic properties of bound low-lying states of PbCl and the corresponding available experimental values. Figure 2.34

Table 2.28 Spectroscopic Properties of PbCl[a]

	R_e, Å		T_e, cm^{-1}		ω_e, cm^{-1}	
State	Theory	Exp.	Theory	Exp.	Theory	Exp.
$X_1\,^2\Pi_{1/2}$	2.61	—	0	0	304	304
$X_2\,^2\Pi_{3/2}$	2.58	—	8,473	8,272	313	322
$A^2\Sigma^+_{1/2}$	2.84	—	22,916	21,865	220	229
$\frac{3}{2}$(II)	2.97	—	29,915	—	199	—
$^2\Pi$	2.59	—	5,640	—	313	—

[a]All theoretical results are from Reference 238.

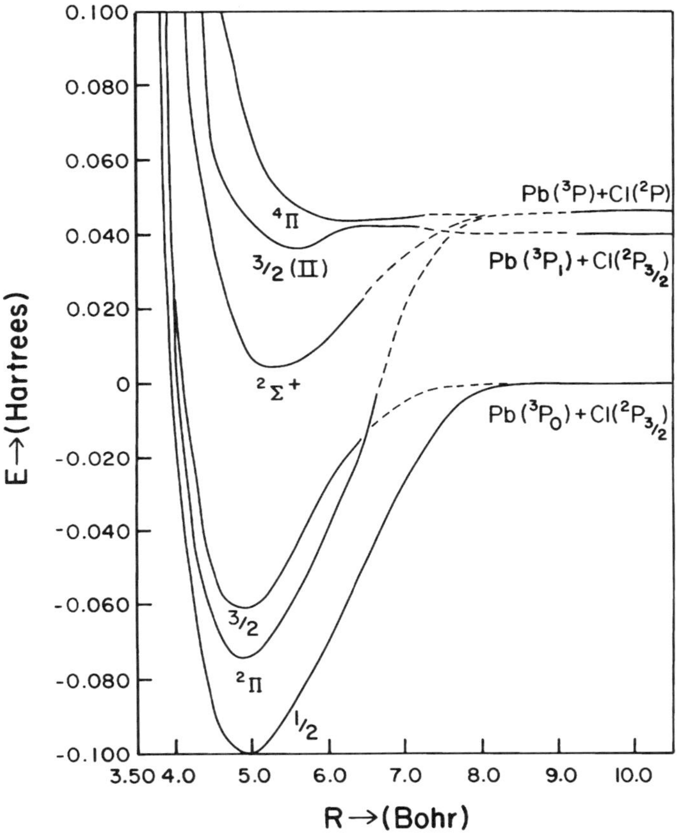

Figure 2.34 Potential-energy curves for the electronic states of PbCl. (Reprinted from Ref. 238; copyright 1988, Academic Press, Inc.)

shows the potential-energy curves of the low-lying electronic states of PbCl. As seen from Table 2.28, the vibrational frequencies (ω_e) of the low-lying bound electronic states are in excellent agreement with the known experimental values for the X_1, X_2, and A states. The ground-state spin–orbit splitting of 8473 cm^{-1} was found to be in excellent agreement with an experimental value of 8272 cm^{-1} (32).

An unambiguous assignment of the A state observed in the A–X_1 system was not possible, since many electronic states could be candidates for the A state. The A state could be ${}^2\Sigma^+_{1/2}(1\sigma^2 2\sigma^2 3\sigma^2 4\sigma 1\pi^4)$, ${}^4\Sigma^-_{1/2}(1\sigma^2 2\sigma^2 3\sigma 1\pi^4 2\pi^2)$, or ${}^2\Sigma^\pm_{1/2}(1\sigma^2 2\sigma^2 3\sigma 1\pi^4 2\pi^2)$. The author's theoretical calculations (238) revealed that the ${}^4\Sigma^-$ and ${}^2\Sigma^\pm$ electronic states arising from $1\sigma^2 2\sigma^2 3\sigma 1\pi^4 2\pi^2$ are not very strongly bound. The only probable electronic state in the A–X_1 region was found to be the ${}^2\Sigma^+_{1/2}$ state arising from the $1\sigma^2 2\sigma^2 3\sigma^2 4\sigma 1\pi^4$ configuration. The A–X_1 system was thus assigned by the author (238) to the ${}^2\Sigma^+_{1/2}$–${}^2\Pi_{1/2}$ transition.

Although Table 2.28 does not report the properties of the $\frac{1}{2}$(III) state, the author's theoretical calculations (233) indicated the existence of a $\frac{1}{2}$(III) state [predominantly $^4\Sigma^-_{1/2}(1\sigma^2 2\sigma^2 3\sigma 1\pi^4 2\pi^2)$] 200 cm^{-1} below the $^4\Sigma^-_{3/2}$ state. The B–X system observed in the ultraviolet region was found to be analogous to the B–X system of PbF (see Section 2.2.4). The T_e value of the B state of PbF was calculated to be about 35,941 cm^{-1} (238), which agreed well with an experimental value of 35,664 cm^{-1} (32). The B–X system of PbCl was observed in the same (35,199-cm^{-1}) region as PbF. On the basis of the earlier calculations on PbF (208), the author (238) concluded that the B state of PbCl also should be a Rydberg state arising from the $1\sigma^2 2\sigma^2 3\sigma^2 5\sigma 1\pi^4$ configuration.

The tentative rotational analysis of the bands in the A–X_1 system made by Rao and Rao (235) was found to be incorrect by the author (238). Rao and Rao (235) reported R_e values obtained from the rotational analysis for the X_1 and A states as 2.195 and 2.338 Å, respectively. However, as seen from Table 2.28, the theoretical equilibrium bond lengths of the X_1 and A states are much longer; for the X_1 state there is a difference of about 0.411 Å between the value reported by Rao and Rao (235) and the theoretical results. Huber and Herzberg (32) considered the results of Rao and Rao (235) as only tentative and did not report their results for R_e. Accurate experimental results are available for the isoelectronic (SnF, SnCl) as well as (SiF, SiCl) pairs (32). The author (238) compared the bond lengths of the ground states of SnF (1.944 Å), SnCl (2.361 Å), SiF (1.60 Å), and SiCl (2.06 Å), which suggested a difference of about 0.42–0.46 Å between the bond lengths of fluorides and chlorides. Since the bond length of PbF was established by both theoretical (208) and experimental investigations (32) to be about 2.06 Å, the R_e value of the ground state of PbCl should be at least 2.52 Å, eliminating the 2.191 Å value reported by Rao and Rao (235). Further, the bond length of PbCl should be longer than the corresponding value for SnCl (2.36 Å). Thus, it was concluded by the author that the correct R_e value of the ground state of PbCl should be at least 2.45 Å, suggesting that a revised rotational analysis of the A–X bands should be carried out.

The D_e of PbCl was calculated by the author (238) as 2.72 eV. The experimental D_e value of 3.0 eV was calculated by extrapolating the B–X bands, since these bands were predissociated (32). On the basis of the theoretical results of the author (230), it was predicted that the D_e of PbCl should be <3.0 eV.

The $X_1(\frac{1}{2})$ ground state of PbCl at its equilibrium geometry was found to be 91% $^2\Pi_{1/2}(1\sigma^2 2\sigma^2 3\sigma^2 1\pi^4 2\pi)$ and 2% $^2\Sigma^+_{1/2}(1\sigma^2 2\sigma^2 3\sigma^2 4\sigma^1 \pi^4)$. The $X_2(\frac{3}{2})$ state was found to be predominantly $^2\Pi_{3/2}$ at the equilibrium geometry, while the $\frac{3}{2}$(II) state was found to be 78% $^4\Sigma^-_{3/2}$, 4% $^2\Pi_{3/2}$, 2.5% $^4\Sigma^-_{3/2}(1\sigma^2 2\sigma^2 4\sigma 1\pi^4 2\pi^4)$, and 4% $^2\Pi_{3/2}(1\sigma^2 2\sigma^2 3\sigma 4\sigma 1\pi^4 2\pi)$. At long distances, the $\frac{3}{2}$(II) state was found to exhibit an avoided crossing; at 6.0 bohr the $\frac{3}{2}$(II) state was 76% $(1\sigma^2 2\sigma^2 3\sigma^2 4\sigma 1\pi^4 2\pi^2$: $^4\Pi_{3/2}$, $^1\Pi_{3/2}$, $^2\Delta_{3/2})$, 9% $(1\sigma^2 2\sigma^2 3\sigma 1\pi^4 2\pi^2$, and 3% $(1\sigma^2 2\sigma^2 3\sigma^2 1\pi^4 2\pi)$. The A state [$\frac{1}{2}$(II)] was found to be 80% $^2\Sigma^+_{1/2}(1\sigma^2 2\sigma^2 3\sigma^2 4\sigma 1\pi^4)$, 5% $^2\Pi^-_{1/2}(1\sigma^2 2\sigma^2 3\sigma^2 1\pi^4 2\pi)$, and 2% $^2\Sigma^+_{1/2}$ (Rydberg; $1\sigma^2 2\sigma^2 3\sigma^2 5\sigma 1\pi^4$) at its equilibrium geometry. At 6.0 bohr, this state was 66% ${}^2\Sigma^+_{1/2}(1\sigma^2 2\sigma^2 3\sigma^2 4\sigma 1\pi^4)$, 6% ${}^2\Pi_{1/2}(1\sigma^2 2\sigma^2 3\sigma^2 1\pi^4 2\pi)$, and 7% ${}^4\Sigma^-_{1/2}(1\sigma^2 2\sigma^2 3\sigma 1\pi^4 2\pi^2)$.

The $\frac{1}{2}$(III) state exhibited an avoided crossing of Rydberg ${}^2\Sigma^+_{1/2}$ with ${}^4\Sigma^-_{1/2}$. At short distances (3.50–4.00 bohr) it was found to be the Rydberg state arising from ${}^2\Sigma^+_{1/2}(1\sigma^2 2\sigma^2 3\sigma^2 5\sigma 1\pi^4)$, while at 5.00 bohr the $\frac{1}{2}$(III) state consisted of 63% $1\sigma^2 2\sigma^2 1\pi^4 3\sigma\alpha 2\pi^2({}^3\Sigma^-_0)$, 16% $1\sigma^2 2\sigma^2 1\pi^4 3\sigma\beta 2\pi^2({}^3\Sigma^-_0)$, 3% ${}^2\Sigma^+_{1/2}(1\sigma^2 2\sigma^2 3\sigma^2 4\sigma 1\pi^4)$, and 8% ${}^2\Sigma^+_{1/2}(1\sigma^2 2\sigma^2 3\sigma^1 1\pi^4 2\pi^2)$.

2.2.7 PbBr

The PbBr radical has been the focus of several experimental studies that have led to both the absorption and emission spectra since the 1930s (32, 380, 381). Earlier studies on PbBr revealed the existence of the $X_1^2\Pi_{1/2}$, $A_{1/2}$, and $B^2\Sigma^+$ states of PbBr, although only the T_e and ω_3 values of those states were known before (32). Ziebarth et al. (380) obtained the near-infrared chemiluminescence spectra of PbBr that yielded accurate ω_e, $\omega_e x_e$, and T_e values of the X_1 and X_2 states. The equilibrium bond lengths of PbBr are unestablished at the present time. The spectroscopic constants of the X_1 and X_2 electronic states of lead halides were measured from the X_2–X_1 emission spectra near 1.2 μm. Those measurements yielded some of the spectroscopic constants of the $X_2(\frac{3}{2})$ state of PbBr.

The spin–orbit effects and relativistic effects are known to be quite significant for molecules containing very heavy atoms. As discussed in the earlier sections of this chapter, the analogous PbF and PbCl molecules exhibit large relativistic effects. The experimentally observed small Λ doubling in the ground state of PbBr requires some explanation through quantum-mechanical studies. Furthermore, the nature of the low-lying electronic states and their equilibrium bond distances (r_e) need to be determined.

Benavides-Garcia and the author (382) have employed the complete active space multiconfiguration followed by first-order, second-order, and relativistic configuration interaction (CASSCF/FOCI/SOCI/RCI) methods to compute the spectroscopic constants (r_e, T_e, ω_e, D_e, μ_e) and the potential-energy curves. The inner core electrons are replaced by relativistic effective core potentials (RECPs), and the valence electrons are treated explicitly. Valence Gaussian sets of the type ($3s3p1d$) were employed for Pb and Br atoms. The outermost $6s^2 6p^2$ shells of the Pb atom and the outermost $4s^2 4p^5$ shells of the Br atom were explicitly retained in the calculations. The FOCI was employed to obtain not only the spectroscopic constants of low-lying electronic states of PbBr but also to obtain the potential-energy curves of these states. The SOCI calculations included up to 607,068 configurations employed to compute the spectroscopic constants of the ground state and first excited state near the minimum. The relativistic configuration interaction (RCI) method was employed to study the effects of spin–orbit coupling. In this method, spin–orbit electronic states that have the same Ω values were allowed to mix.

Table 2.29 shows the spectroscopic constants of the low-lying electronic states of the PbBr radical obtained using the FOCI and SOCI methods, while Figure 2.35*a* shows the FOCI potential-energy curves. The SOCI calculations were performed only for the ground state and the first excited state near their equilibrium bond distances, in the ranges of 2.5–2.8 Å for the $^2\Pi$ state and 2.8–3.1 Å for the $^2\Sigma^+$ state. From this table we can establish that the ground state corresponds to a $^2\Pi$ state, which is followed by the excited $^2\Sigma^+$ state analogous to the PbF and PbCl radicals discussed before. It is seen from Table 2.29 that the equilibrium bond distances are longer in the SOCI than in the FOCI level, and the ω_e values are smaller in the SOCI than in the FOCI level for both the $^2\Pi$ and $^2\Sigma^+$ states. The electronic states marked with an asterisk do not exhibit pronounced wells and, thus the energy separations were computed at 2.6 Å.

The increases in the dipole moments of the two lowest states of PbBr in Table 2.29 are consistent with the increases in the corresponding r_e values. The overall total populations of Pb and Br are consistent with the Pb^+Br^- polarity of bond deduced from the dipole moments. The 6*s* population of Pb is close to 2 because of the relativistic stabilization of the 6*s* orbital of the lead atom. A notable contrast between the $^2\Pi$ and $^2\Sigma^+$ states is in the Pb–Br overlap population. The overlap of the $^2\Sigma^+$ state is much smaller compared to $^2\Pi$ because the leading configuration of these two states are $3\sigma^2 2\pi^1$ and $3\sigma^2 4\sigma^1$, respectively. The 2π orbital is primarily a nonbonding Pb(6*s*) orbital, while the 4σ orbital is a somewhat antibonding orbital composed of Pb(6*p*) and Br(5*p*). Therefore, the Pb–Br overlap is reduced for $^2\Sigma^+$.

Figure 2.35*b* shows the potential-energy curves of the electronic states of the PbBr radical plotted as a function of the bond distance. Among the states considered here, only the $^2\Pi$ and $^2\Sigma^+$ states exhibit pronounced minima. The crossing of the $^4\Pi$ state with the $^4\Sigma^-$ state at 2.7 Å is quite interesting in that the corresponding $\frac{1}{2}$ and $\frac{3}{2}$ spin–orbit components would undergo avoided crossing. In Table 2.29 the vertical energy separation calculated for the $^4\Sigma^-$ state is lower than the value calculated for the $^4\Pi$ state because the bond distance used was 2.6 Å.

Table 2.30 presents the spectroscopic properties of the PbBr radical, when the spin–orbit coupling is included. It can be inferred from Table 2.29 that numerous electronic states arise for PbBr, when spin–orbit coupling is considered. However, many of these states would be repulsive in nature. The $X_1^2\Pi_{1/2}$–$X_2^2\Pi_{3/2}$ interval is quite large as Ziebarth and co-workers (380) experimentally determined this splitting for the lead monohalide radicals, PbX (X = F, Cl,Br,I). The spectroscopic constants reported by Ziebarth for the $^2\Pi_{1/2}$ and $^2\Pi_{3/2}$ of the PbBr molecule are compared with the ones theoretically computed in this study (see Table 2.30) and found to be in good agreement. The $\frac{1}{2}$(II), an upper root of $\frac{1}{2}$, was assigned as the $^2\Sigma^+_{1/2}$ state, the $\frac{3}{2}$(II) was assigned to the $^4\Sigma^-_{3/2}$ state, but we did not determine the spectroscopic constants of all these states. The $\frac{5}{2}$ state was assigned to $^4\Pi_{5/2}$, and its upper root $\frac{5}{2}$(II) was assigned to $^4\Delta_{5/2}$, but again contamination of those states was found.

Table 2.29 Spectroscopic Properties of PbBr Radical Using FOCI and SOCI Methods

	FOCI					SOCI			
State	R_e, Å	ω_e, cm^{-1}	T_e, cm^{-1}	D_e, eV	μ_e, D[b]	R_e, Å	ω_e, cm^{-1}	T_e, cm^{-1}	μ_e, D[a]
$^2\Pi$	2.635	225	0	2.911	3.061	2.679	200	0	4.107
$^2\Sigma^+$	2.944	156	16,017	1.024	3.318	2.962	148	14,834	5.098
$^4\Sigma^-$ [b]	—	—	27,110	—	—	—	—	—	—
$^4\Pi$[b]	—	—	30,247	—	—	—	—	—	—
$^2\Pi$(II)[b]	—	—	37,897	—	—	—	—	—	—
$^2\Delta$[b]	—	—	41,959	—	—	—	—	—	—
$^4\Delta$[b]	—	—	44,628	—	—	—	—	—	—

[a]Positive polarity means Pb^+Br^-.

[b]Bond distance used to compute the energy separation is 2.6 Å.

Source: Reproduced from Reference 379.

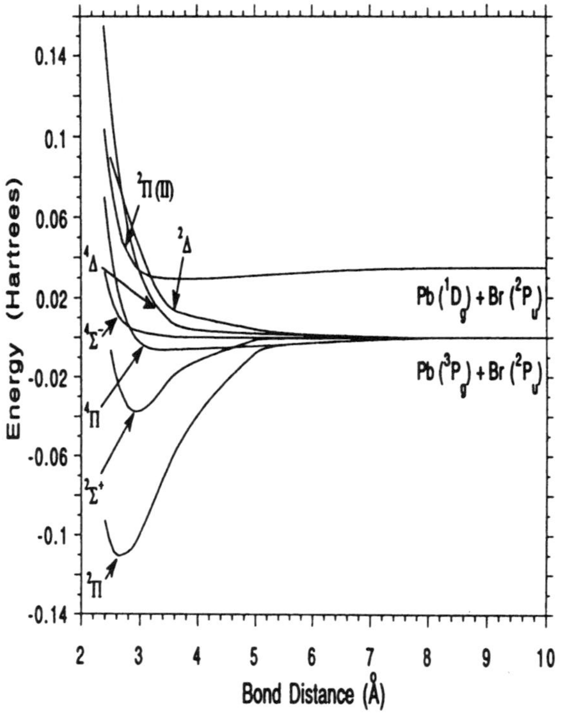

Figure 2.35a Potential-energy curves for the electronic states of (*a*) PbBr (reprinted from Ref. 382; copyright 1993, Elsevier Science Publishers B.V.); (*b*) PbI (reprinted from Ref. 383; copyright 1993, Academic Press, Inc.)

Table 2.30 Spectroscopic Properties of PbBr Including Spin–Orbit Coupling[a,b]

Electronic State	R_e, Å	ω_e, cm^{-1} Theory	ω_e, cm^{-1} Exp.	T_e, cm^{-1} Theory	T_e, cm^{-1} Exp.
$^2\Pi_{1/2}$	2.712	195	206.77	0	0
$^2\Pi_{3/2}$	2.674	205	218.35	7,220	8,198
$^2\Sigma^+_{1/2}$	2.904	154	153	21,600[b]	20,884
$^4\Pi_{5/2}$	3.054	152	—	49,095	—

[a]The properties of $^2\Pi_{1/2}$ and $^2\Pi_{3/2}$ states were computed using the $^3\Sigma^-$ FOCI vector of $PbBr^-$, while the properties of $^2\Sigma^+_{1/2}$ and $^4\Pi_{5/2}$ were computed using the $^2\Sigma^+$ FOCI vector of PbBr.

[b]An approximate T_e of 21,600 cm^{-1} is deduced by correcting the SOCI T_e of $^2\Sigma^+$ for spin–orbit correction. This value includes the effect of higher-order electron correlations to a greater extent compared to the direct RCI value of 25,950 cm^{-1}

Source: Reproduced from Reference 379.

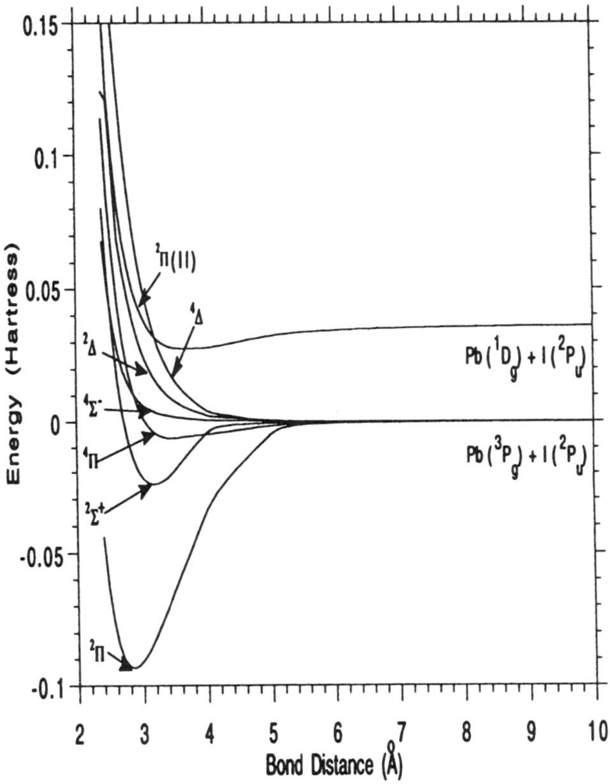

Figure 2.35b *(Continued)*

Analysis of the RCI wavefunctions of the low-lying electronic states of PbBr reveals the following details. The $X_1\frac{1}{2}$ ground state of PbBr is composed of 90% $^2\Pi_{1/2}$ and 3% $^2\Sigma^+(1\sigma^2 2\sigma^2 3\sigma^2 1\pi^4 4\sigma)$. Consequently, contamination of $^2\Sigma^+$ with $^2\Pi$ in the $X_1\frac{1}{2}$ ground state is not negligible. The $\frac{3}{2}$ state of PbBr, on the other hand, was found to be relatively pure (94% $^2\Pi$).

Ziebarth et al. (380) have noted that the Λ doubling is somewhat smaller for heavier halides compared to the lighter analogs. They suggest that this could be attributed to partial cancellation of the effect of $A^2\Sigma^+$ by a nearby state, possibly $^2\Sigma^-$. The FOCI and RCI computation reveal that the $^2\Sigma^-$ state is much higher than the $A^2\Sigma^+$ state. In particular, Figure 2.35*a* shows the $^4\Sigma^-$ and $^2\Delta$ states arising from $1\sigma^2 2\sigma^2 3\sigma^2 1\pi^4 2\pi^2$ electronic configuration. The $^2\Sigma^-$ is certainly higher than $^4\Sigma^-$ and thus is not a very attractive candidate to quench the effect of $A^2\Sigma^+$ in the Λ doubling of the ground state. For heavier halides we expect mixing of $^2\Sigma^+$ with other states due to spin–orbit coupling.

The RCI computations reveal that the $A_{1/2}$ state of PbBr is actually 83% $^2\Sigma^+_{1/2}$ and 10% $^2\Pi_{1/2}(X_1)$; consequently, the A state is heavily contaminated by the $X_1(^2\Pi_{1/2})$ state. We did not find any contribution from $^4\Sigma^-_{1/2}$, $^2\Sigma^-_{1/2}$, or other higher-lying $\frac{1}{2}$ states in the $A_{1/2}$ state. This may be due to the use of $^2\Sigma^+$ FOCI vector.

The X_2 state of PbBr was observed by Ziebarth et al. (380) through the chemiluminescent reaction. Pb + $Br_2 \rightarrow PbBr^*(X_2)$. Using the ΔH_0° for this reaction and the established $D_0^\circ(Br_2)$, these authors deducted the D_0° (PbBr) as 2.987 eV. The CASSCF/FOCI method yielded a D_e of 2.911 eV. The higher-order correlation effects are expected to increase the D_e, but spin–orbit coupling would destabilize the radical by the same magnitude. Consequently, the two effects appear to cancel out, resulting in a FOCI D_e, which is very close to the experimental D_0° of 2.987 eV deduced by Ziebarth et al.

The $\Omega = \frac{5}{2}$ lowest root is predominantly $^4\Pi_{5/2}$ at short internuclear distances. However, as the distance increases, the contribution from $^4\Delta_{5/2}$ increases. This is consistent with Figure 2.35*a*, which shows that $^4\Delta$ and $^4\Pi$ become closer to each other as in the internuclear separation increases.

2.2.8 PbI

Spectroscopic studies of PbI have already been summarized in the previous section on PbBr. The most recent experimental work of Ziebarth and co-workers (380) deals with the $X_2^2\Pi_{3/2}$–$X_1^2\Pi_{1/2}$ emission spectra of the lead monohalide molecules near 1.2 μm and determined, for the first time, the electronic states and vibrational constants of the X_2 state of PbI. Previous studies on PbI are summarized in References 380 and 32.

Benvides-Garcia and the author (383) have studied the low-lying electronic states of PbI. Relativistic effective core potentials (RECPs) with complete active space self-consistent field (CASSCF) methods were used to generate a starting set of orbitals for the inclusion of high-order electron correlation effects. Valence Gaussian sets of the type $(3s3p1d)$ were used for the Pb and I atoms. In the CASSCF method, the outermost $6s^26p^2$ shells of the Pb atom and the outermost $5s^25p^5$ shells of the I atom were explicitly retained in the CASSCF calculations. The FOCI calculations were done after CASSCF to compute the potential-energy curves. The SOCI method was applied to the ground state and the first excited state near the minimum after the FOCI. The RCI method was used to study the effects of spin–orbit coupling. The spin–orbit electronic states that have the same Ω values were allowed to mix.

Table 2.31 shows the spectroscopic constants of the low-lying electronic states of PbI computed by the FOCI and SOCI methods in the absence of spin–orbit coupling. The SOCI calculations were employed only for the ground state and the first excited state, namely, $^2\Pi$ and $^2\Sigma^+$, respectively, near their minima, in the ranges of 2.7–3.0 Å for the $^2\Pi$ state and 3.0–3.3 Å for the $^2\Sigma^+$ state.

The bond distance used for the states marked with asterisks in Table 2.31 corresponded to 2.8 Å, suggesting that the energy separations reported for them are vertical energy separations at the ground-state equilibrium bond distance of

Table 2.31 Spectroscopic Properties of PbI Molecule Using FOCI and SOCI Methods

	FOCI					SOCI			
State	R_e, Å	ω_e, cm^{-1}	T_e, cm^{-1}	D_e, eV	μ_e, D[a]	R_e, Å	ω_e, cm^{-1}	T_e, cm^{-1}	μ_e, D[a]
$^2\Pi$	2.849	165	0	2.541	2.515	2.871	161	0	3.721
$^2\Sigma^+$	3.180	115	15,103	0.668	2.946	3.180	116	14,306	4.739
$^4\Sigma^{-b}$	—	—	23,392	—	—	—	—	—	—
$^4\Pi^b$	—	—	24,322	—	—	—	—	—	—
$^2\Delta^b$	—	—	30,167	—	—	—	—	—	—
$^2\Pi(II)^b$	—	—	32,632	—	—	—	—	—	—
$^4\Delta^b$	—	—	38,841	—	—	—	—	—	—

[a]Positive polarity means Pb^+I^-.
[b]Bond distance corresponds to 2.8 Å.

2.8 Å. Figure 2.35*b* shows the potential-energy curves for the electronic states of PbI plotted as a function of the internuclear distance. The $^2\Pi$ state exhibits a shorter equilibrium bond distance and a smaller dipole moment compared to the $^2\Sigma^+$ state. The smaller dipole moment for $^2\Pi$ is consistent with shorter R_e. As seen from this figure, only the $^2\Pi$ and $^2\Sigma^+$ states exhibit deeper potential minima. Analogous to PbBr, the $^4\Pi$ and $^4\Sigma^-$ states cross.

The T_e value of the $^2\Sigma^+_{1/2}$ state can be deduced from the SOCI result of Benavides-Garcia and the author (383) reported in Table 2.31. Note that the $^2\Sigma^+$ state arises from the $1\sigma^2 2\sigma^2 3\sigma^2 1\pi^4 4\sigma 1$ configuration and dissociates into Pb($^3P_0$) + I($^2P_{1/2}$). The $\frac{1}{2}$ ground state dissociates into Pb($^3P_0$) + I($^2P_{3/2}$). The spin–orbit contribution for the $^2\Sigma^+$ state should arise primarily from the interaction with $^2\Pi_{1/2}$. The $^2\Pi_{1/2}$ will thus be lowered and $^2\Sigma^+_{1/2}$ will be raised. This means that the T_e value of the $\frac{1}{2}$(II) state should increase in relation to the $^2\Sigma^+$–$^2\Pi$ splitting by 7600 cm^{-1}. The contamination of $^2\Sigma^+$ with $^2\Pi$ should bring about a decrease in the R_e of the $\frac{1}{2}$(II) state and an increase in the ω_e compared to the $^2\Sigma^+$ state.

Table 2.32 shows the spectroscopic properties for the $^2\Pi$ when spin–orbit coupling is included in the calculation, where the data are compared to the spectroscopic constants reported by Ziebarth and co-workers (380). When spin–orbit interaction is included at the RCI step, the $^2\Pi$ state splits into $^2\Pi_{1/2}$ and $^2\Pi_{3/2}$ components. A very good agreement between experiment and theory is found, as seen from this table.

The $X_1\frac{1}{2}$ ground state of PbI is 88% $^2\Pi$ and 3% $(4\sigma 1\pi^4)^2\Sigma^+$. The $X_2\frac{3}{2}$ state is 87% $^2\Pi$ and only 2% $^4\Sigma^-$. The excited states of PbI exhibit interesting trends and contrast with lighter analogs. Because of the large spin–orbit contributions of both Pb and I atoms, the excited states are contaminated. It should, however, be noted that we used the $^3\Sigma^-$ FOCI orbitals for the RCI computations, and thus the π orbitals are improved compared to the $^2\Sigma^+$ vector, in which, the 2π orbital is not occupied. This could favor $^4\Sigma^-$ compared to $^2\Sigma^+$. The *A* state of PbI, which is relatively purer for PbF, becomes quite contaminated for PbI. In fact, we found the $^2\Sigma^+_{1/2}$ to undergo avoided crossing with the $^4\Sigma^-_{1/2}$ and $^2\Sigma^-_{1/2}$ manifold for states arising from the $3\sigma 2\pi^2$ electronic configuration.

Table 2.32 Spectroscopic Properties of PbI Including Spin–Orbit Coupling

Electronic State	R_e, Å Theory	ω_e, cm^{-1} Theory	ω_e, cm^{-1} Exp.[a]	T_e, cm^{-1} Theory	T_e, cm^{-1} Exp.[a]
$^2\Pi_{1/2}$	2.885	153	161	0	0
$^2\Pi_{3/2}$	2.859	162	169	8,255	7,969.15
$\frac{1}{2}$(II)($^4\Sigma^-_{1/2}$)	3.43	103	—	21,423	—
$\frac{1}{2}$(III)($^2\Sigma^+$)	—	—	142	23,073	22,529

[a]Theoretical constants from Reference 380.
[b]Experimental constants from Reference 377.

The computed T_e value for the $\frac{1}{2}$(III) state of PbI, which is 23,073 cm^{-1}, is in remarkable agreement with the experimental value of 22,529 cm^{-1}. Note that the corresponding SOCI $^2\Sigma^+$–$^2\Pi$ T_e of 14,306 cm^{-1} differs sharply from the $\frac{1}{2}$(III)–$\frac{1}{2}$ separation of 23,073 cm^{-1} mainly because of the dramatic mixing of the $^2\Sigma^+$ state with other electronic states due to spin–orbit coupling as discussed above.

At 3.1 Å the $\frac{1}{2}$(II) state is composed of 41% $^2\Sigma^+$ ($4\sigma 1\pi^4$), 32% $^4\Sigma^-$, $^2\Sigma^-$($3\sigma 1\pi^4 2\pi^2$), and 6% $^2\Sigma^+$($3\sigma 1\pi^4 2\pi^2$). However, at 3.4 Å the $\frac{1}{2}$(II) state is composed of 19% $^2\Sigma^+$($4\sigma 1\pi^4$), 44% $^4\Sigma^-$, $^2\Sigma^-$($3\sigma 1\pi^4 2\pi^2$), and 8% $^2\Sigma^+$($3\sigma 1\pi^4 2\pi^2$). As a result the two states undergo avoided crossing. The $\frac{1}{2}$(III) state, on the other hand, is composed of 68% $^4\Sigma^-$, $^2\Sigma^-$($3\sigma 1\pi^4 2\pi^2$), and 14% $^2\Sigma^+$($3\sigma 1\pi^4 2\pi^2$). However, at 3.4 Å, this state is composed of 47% ($4\sigma 1\pi^4$), 13% ($3\sigma 1\pi^4 2\pi^2$), and 8% $^2\Pi_{1/2}$($1\pi^4 2\pi$). This is the state that couples to the ground state, as evidenced by the large mixing of the X_1 state. Consequently, this state is assigned to the experimental A state. The $\frac{1}{2}$(II) root, which is below this, as seen from Table 2.32, with a long RCI bond length, is probably not observable, due to much smaller transition moment for the $X_1(\frac{1}{2}) \rightarrow \frac{1}{2}$(II) transition.

Ziebarth and co-workers (380) noted that the Λ doubling is substantially quenched for the heavier halides compared to the lighter halides in the $X_1\frac{1}{2}$ ground state. They suggested that this was probably due to the cancellation of the effect of $A^2\Sigma^+$ with a nearby $^2\Sigma^-$ state arising from the $3\sigma 2\pi^2$ configuration. Of course, the lowest of the states arising from $3\sigma 2\pi^2$ is the $^4\Sigma^-$ state, which yields $\Omega = \frac{3}{2}$ and $\frac{1}{2}$ states. As seen from Table 2.32, the $\Omega = \frac{1}{2}$ component of $^4\Sigma^-$ and $^2\Sigma^-$ (combined) is in fact lower than the $A\frac{1}{2}$ state. The $^4\Sigma^-_{1/2}$ as well as $^4\Sigma^-_{3/2}$ states of PbI are yet to be observed. However, the observed $A\frac{1}{2}$ state is a substantial mixture of $^2\Sigma^+$($1\pi^4 4\sigma$), ($^4\Sigma^-$,$^2\Sigma^-$)($3\sigma 2\pi^2$), and other states as suggested above. Consequently, the quenching of Λ doubling in PbI was attributed to the $\frac{1}{2}$ state arising from the $3\sigma^2\pi^2$ electronic configuration of PbI. It is a mixture of both $^4\Sigma^-_{1/2}$ and $^2\Sigma^-_{1/2}$. The upper root $\frac{3}{2}$(II) was assigned to a $^4\Sigma^-_{3/2}$ state, while the $\frac{5}{2}$ state was assigned as a $^4\Pi_{5/2}$. Some of these states were found to be repulsive.

Ziebarth et al. (380) have deduced the D_0° value of PbI from the heat of formation of the chemiluminescent reaction:

$$\text{Pb} + \text{I}_2 \rightarrow \text{PbI} + \text{I}$$

Using the enthalpy of this reaction and $D_0^\circ(\text{I}_2)$, Ziebarth et al. deduced the D_0°(PbI) as 2.53 eV. The CASSCF/FOCI method yielded a D_e of 2.54 eV (382). Higher-order electron correlations would increase the D_e, while the spin–orbit effect would decrease the D_e. The two effects tend to cancel out, so that an FOCI value is close to experiment.

2.2.9 SnCl

The electronic spectra of SnCl were observed by Jevons (239) and Ferguson (240) in the late 1920s and later by Fowler (241). Sarma and Venkateswarlu

(242) identified a band system in the 6440–4880-Å region and assigned the bands observed by Ferguson (240) and by Jevons (239) to the $^2\Delta$–$^2\Pi$ and $^2\Sigma$–$^2\Pi$ systems. Hastie et al. (243) Oldershaw and Robinson (244), and Richter (245) studied the absorption spectrum of SnCl in the UV region. Vibrational analyses of the *C*–*X* systems have been carried out (245).

Chatalic et al. (246) carried out the rotational analysis of the *B*–*X* and *A*–*X* bands and questioned the earlier assignment of the systems in the 28000-cm^{-1} region to the $^2\Delta$–$^2\Pi$ transition. They reassigned these bands to the $\frac{3}{2}$ and $\frac{1}{2}\Omega$ components of the $^4\Sigma^-$–$^2\Pi$ system. The other experimental work on SnCl is described in Reference 247. Mulliken (248), in his general theoretical paper on electronic states and band spectra of diatomics, interpreted the bands observed by Jevons (239) as due to the ^2S–$^2P_{1/2}$ atomic transition of Sn^+.

The author (249) carried out SCF/RCI calculations employing a double-ζ STO basis set for the Sn and Cl atoms. Spectroscopic properties and potential-energy curves of seven electronic states of SnCl were obtained.

Table 2.33 shows the theoretical spectroscopic constants of SnCl compared with available experimental results. Figure 2.36 shows the actual potential-energy curves of the low-lying electronic states of SnCl. The experimental T_e value of the A' state (19,418 cm^{-1}) was found to be in very good agreement with the theoretical T_e value of the $^2\Sigma^+_{1/2}$ state (20,956 cm^{-1}). This combined with the good agreement between the calculated and experimental R_e and ω_e values enabled the assignment of the A' state to a $^2\Sigma^+_{1/2}$ state.

The electronic state (*A*) in the $A \leftrightarrow X$ system was first assigned to a $^2\Delta$ state. Later Chatalic et al. (246) reinterpreted the observed bands in the *A*–*X* system as due to the Ω components of a $^4\Sigma^-$ state ($\frac{3}{2}$, $\frac{1}{2}$). The theoretical results in Table 2.33 support Chatalic et al.'s reinterpretation of the *A* state. The *B* state in the *B*–*X* system was again assigned to a $^2\Sigma^+_{1/2}$ Rydberg state, based on comparison with PbCl.

The theoretical dissociation energy (D_e) of SnCl was calculated as 3.34 eV. There are no experimental D_e values available for SnCl. The author (249) predicted that the experimental D_e of SnCl should be 3.7–3.9 eV.

Table 2.33 Spectroscopic Properties of SnCl[a]

	R_e, Å		T_e, cm^{-1}		ω_e, cm^{-1}		D_e, eV
State	Theory	Exp.	Theory	Exp.	Theory	Exp.	Theory
$X_1\,^2\Pi_{1/2}$	2.479	2.361	0	0	361	357	3.34
$X_2\,^2\Pi_{1/2}$	2.477	2.356	2,769	2,357	360	364	2.99
$^2\Pi$	2.479	—	1,607	—	359	—	3.14
$A'^2\Sigma^+_{1/2}$	2.62	2.619	20,956	19,418	259	232	—
$^2\Sigma^+$	2.62	—	20,773	—	259	—	—
$A^4\Sigma^-_{1/2}$	2.50	—	25,864	28,963	308	301	—
$^4\Sigma^-$	2.49	—	25,092	—	308	—	—

[a]All theoretical values are from Reference 28.

At the equilibrium geometry, the ground state X_1–($^2\Pi_{1/2}$) was found to be 95% $^2\Pi_{1/2}$ and 0.4% $^2\Sigma^+_{1/2}(1\sigma^2 2\sigma^2 3\sigma 1\pi^4)$. The A' $^2\Sigma^+$ state was found to be 90% $^2\Sigma^+_{1/2}(1\sigma^2 2\sigma^2 3\sigma 1\pi^4)$ and 0.52% $^2\Pi_{1/2}$. The $^4\Sigma^-_{1/2}(A)$ state was found to be 87% $^4\Sigma^-_{1/2}$ and 0.4% $^2\Sigma^+_{1/2}$-$(1\sigma^2 2\sigma^2 3\sigma 1\pi^4)$.

2.2.10 SnBr

Jevons and Bashford (384) made the first spectroscopic study on SnBr, which led to the observation of the $A \rightarrow X$ and $B \leftrightarrow X_1$ systems. Sarma and Venkateswarlu (242) as well as Oldershaw and Robinson (385) observed several bands attributed to the $C \leftarrow X_1, D \leftarrow X_1$, and $E \leftarrow X_1$ systems, although definitive assignments could not be made. The bond distances and dissociation energy of the electronic states are not available at present, although the vibrational frequencies have been deduced from the experimental spectra. The lowest X_1, X_2 states and the A' excited state of SnBr have been assigned at present.

The author (386) carried out relativistic *ab initio* SCF/RCI computations on the low-lying electronic states of SnBr. Relativistic effective core potentials (RECPs) for both tin and bromine atoms that retained the outer $4d^{10}5s^2 5p^2$ shells of the tin atom and the outer $4s^2 4p^5$ shells of Br were employed. A double-ζ Slater-type orbital (STO) basis sets were employed for both tin and bromine atoms. The employed state-averaged self-consistent field method (SCF) was employed to generate the orbitals for the $^2\Pi$ state. In this manner the degeneracy of the singly occupied π orbital was retained. Following SCF, RCI computations that included both electron correlation effects and spin–orbit coupling were carried out.

Table 2.34 shows the computed spectroscopic constants together with the experimental values, whenever available. The computed $^2\Pi_{1/2}$–$^2\Pi_{3/2}$ spin–orbit splitting of 2936 cm^{-1} is in good agreement with the experimental value of 2463 cm^{-1}. Note that similar level of theory yielded a theoretical value of 2769 cm^{-1} compared to the experimental value of 2357 cm^{-1} for SnCl (see previous section). The computed vibrational frequency for SnBr (202 cm^{-1}) is somewhat lower than the experimental value of 247 cm^{-1}, but the experimental value is an averaged value. As seen from Table 2.34, at present there are no r_e values for the ground state of SnBr. Analogous computation on SnCl yielded theoretical r_e values 5% larger than experimental values. Consequently, the computed r_e values should be at least 5% larger than true values.

Chatalic et al. (246) as well as Jevons and Bashford (384) observed a number of weak emission bands attributed to the $A' \rightarrow X$ system with ν_{00} near 18,676 cm^{-1}. Chatalic et al. suggest X_1 as the lower-state component and assign several weak bands in the 6300-Å region. On this the basis of, a T_e value of 18,717 cm^{-1} is deduced for the A' state. The computed spectroscopic constants of the $^2\Sigma^+_{1/2}$ state with a T_e of 16,272 cm^{-1} supports the previous assignment of the A state to $^2\Sigma^+_{1/2}$.

Figure 2.36*a* shows the computed potential-energy curves for SnBr. There is an excited $^4\Sigma^-$ state lying above the $A'^2\Sigma^+$ state, which form the $\sigma\pi^2$ con-

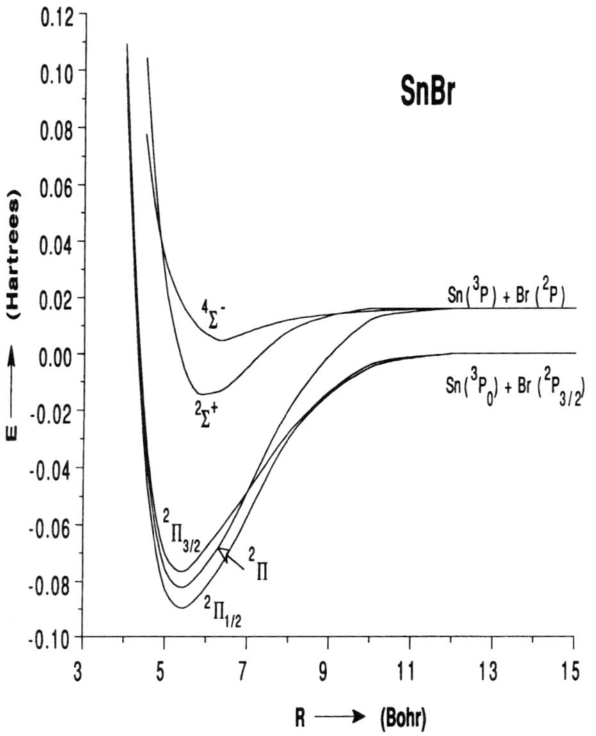

Figure 2.36a Potential-energy curves for the electronic states of (*a*) SnCl (reprinted from Ref. 249; copyright 1988, Academic Press, Inc.); (*b*) SnBr (reprinted from Ref. 386; copyright 1993, Academic Press, Inc.)

figuration. The same configuration also yields $^2\Delta$, $^2\Sigma^+$, and $^2\Sigma^-$ states. Table 2.34 shows the corrected vertical energy separation for the $^2\Delta$ state. A series of bands attributed to $A \rightarrow X$, $B \leftrightarrow X_1$, and $B \rightarrow X_2$ systems have been observed. The A state is deduced to have a tentative T_e of 27,063 cm^{-1}. This was tentatively assigned before to $^2\Delta$. The computations support the existence of $^2\Delta_{3/2}$, $^4\Sigma^-_{3/2}$ and $^4\Sigma^-_{1/2}$ states in this region. Since the A–X system was assigned to the $^4\Sigma^-_{1/2}$–$^2\Pi_{1/2}$ transition for SnCl, the possibility of the A state of SnBr being $^4\Sigma^-_{1/2}$ cannot be ruled out. The B $^2\Sigma^+_{1/2}$ state has been observed in both absorption and emission in the $B \leftrightarrow X_1$ system. A T_e value of 33,062 cm^{-1} has been tentatively deduced for this system. On the basis of PbF, this state should be a $^2\Sigma^+$ Rydberg state.

The D_e value for the ground state of SnBr is theoretically computed as 2.5 eV using the SCF/RCI method. Previous studies on similar molecules have shown that such levels of theory yields only <90% of the experimental value. Therefore, a lower bound for the D_e (SnBr) is predicted as 2.75 eV. At present, there are no experimental D_e values for SnBr.

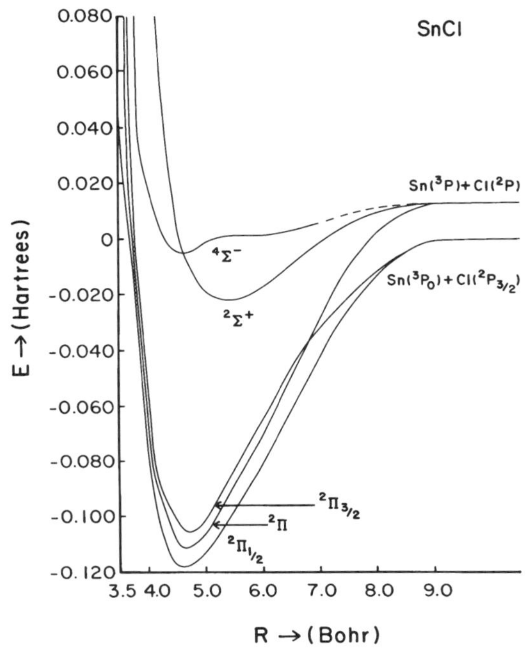

Figure 2.36b *(Continued)*

The $^2\Pi_{1/2}$ ground state of SnBr was found to be 93% $^2\Pi(1\sigma^2 2\sigma^2 1\pi^4 2\pi)$ and 1% $^2\Sigma^+(1\sigma^2 2\sigma^2 3\sigma 1\pi^4)$. The $A\ ^2\Sigma^+_{1/2}$ state was found to be 87% $^2\Sigma^+(1\sigma^2 2\sigma^2 3\sigma 1\pi^4)$ and 2% $^2\Pi_{1/2}(1\sigma^2 2\sigma^2 1\pi^4 2\pi)$. Thus spin–orbit mixing between different λ–*s* states is small but nonnegligible. However, spin–orbit splitting is quite large.

Table 2.34 Spectroscopic Constants of SnBr[a]

	r_e, Å		T_e, cm^{-1}		ω_e, cm^{-1}	
State	Theory	Exp.	Theory	Exp.	Theory	Exp.
$X_1\,^2\Pi_{1/2}$	2.87	—	0.0	0.0	202	(247)
$X_2\,^2\Pi_{3/2}$	2.86	—	2,936	(2,463)	200	(242)
$A'^2\Sigma^+_{1/2}$	3.16	—	16,272	(18,717)	182	164
$A^2\Delta$	—	—	28,000	(27,063)	—	—

[a]Theoretical r_e values are expected to be 5% larger. This listed theoretical energy separation for the $A\ ^2\Delta$ state is the corrected vertical energy separation.

2.2.11 SbF

Latifzadeh and the author (353) have computed the potential-energy curves and spectroscopic constants of the diatomic antimony fluoride molecule. Prior to this work, Boustani et al. (354) performed SCF/MRDCI computations on the same molecule. In this section we shall discuss and compare the results of these two investigations together with the experimental data on this species.

Latifzadeh and the author (353) used the CASSCF followed by the first-order CI(FOCI) method for the computation of the potential-energy curves. The spectroscopic constants were computed using the full second-order CI method. Relativistic effective potentials were used for both the antimony and F atoms. A valence Gaussian basis sets for Sb and F with two sets of six-component $3d$ functions were employed on each atom in the work.

Boustani et al. (354) also used the same RECPs and a $(3s3p)$ basis set but augmented with only one set of $3d$ functions. Their basis set for the fluorine atom also contains a single set of $3d$ functions. These authors used the MRDCI method in conjunction with the configuration selection scheme for computing the spectroscopic constants. The spin–orbit interaction was also included but was found to be quite small. They have also computed the transition moments and the radiative lifetimes of these electronic states. The techniques used by Latifzadeh and the author are superior in the sense that (a) better basis sets were used for both Sb and F with two sets of $3d$ functions, (b) a full CASSCF method was used for the generation of the orbitals followed by a full second-order CI (SOCI) that included more than a million configurations is used, and (c) no configuration selection was invoked in the work of Latifzadeh and the author (353). More electronic states were considered by Latifzadeh and the author compared to Boustani et al. (354). We shall first consider the results obtained by Latifzadeh and the author (353) for SbF and then compare the results of the two groups.

The potential-energy curves of the electronic states of SbF, dissociating into the Sb(^4S_u) + F(^2P_u) and Sb(^2D_u) + F(^2P_u) limits computed by Latifzadeh and the author (353) using the CASSCF/FOCI methods, are presented in Figure 2.37. As seen from Figure 2.37, only the $X^3\Sigma^-$, $a^1\Delta$, $b^1\Sigma^+$, $A^3\Pi$, $A^1_3\Pi$, $C^3_3\Pi$(II), $d^3\Sigma^-$(II), and $e^1\Pi$(II) curves exhibit potential minima. In some contrast to PF and AsF, there are two potential minima for the $C^3_3\Pi$(II) state and a shallow potential minimum for the $e^1\Pi$(II) state with an R_e value of 1.850 Å. The two potential minima of the $C^3_3\Pi$(II) state are computed at R_e = 1.855 Å with higher energy and R_e = 2.752 Å with lower energy.

The ground electronic state of SbF, analogous to other group VA halides, is the $X^3\Sigma^-$ state with the $1\sigma^2 2\sigma^2 3\sigma^2 1\pi^4 2\pi^2$ electronic configuration. The 1π orbital is predominantly F($2p$), while the 2π orbital is expected to be mainly Sb($5p$). The $A^3\Pi$ and $A^1_3\Pi$ excited states arise by transferring an electron from the 2π orbital into the 4σ orbital, yielding the $1\sigma^2 2\sigma^2 3\sigma^2 4\sigma^1 1\pi^4 2\pi^1$ configuration. As seen from Figure 2.37, the crossing of the $a^1\Delta$ and the $A^3\Pi$ curves takes place at $R \cong 2.60$ Å. Likewise, the $b^1\Sigma^+$ and $A^3\Pi$ curves cross at $R \cong$

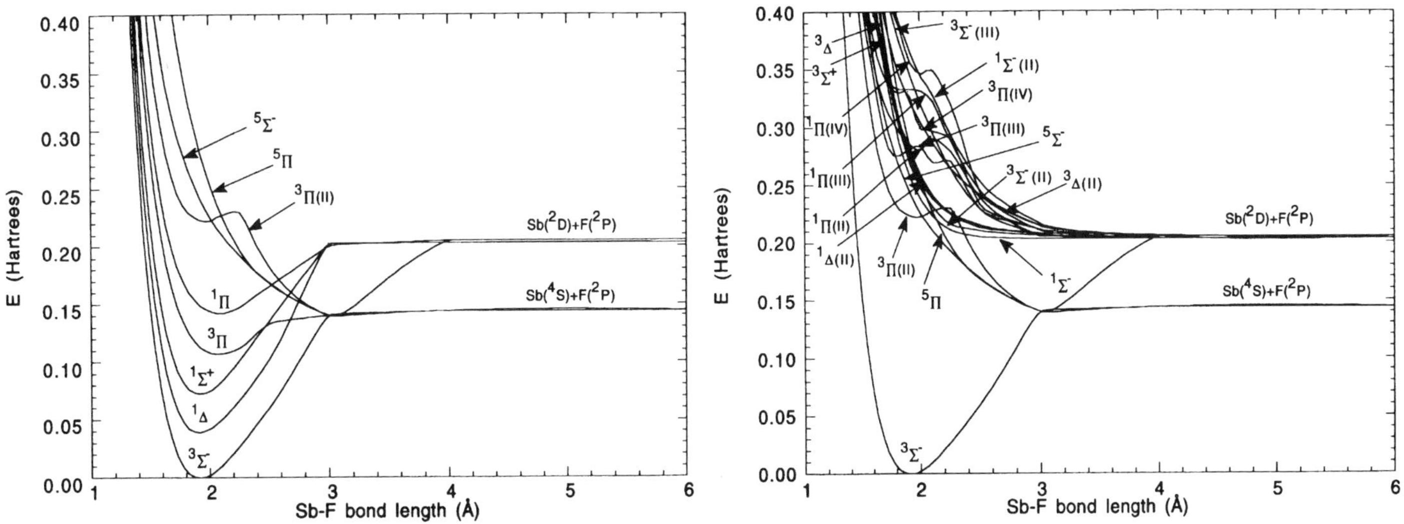

Figure 2.37 Potential-energy curves for the electronic states of SbF. (Reprinted from Ref. 353; copyright 1995, Elsevier Science Publishers B.V.)

2.40 Å. Since spin–orbit coupling of the antimony atom is nonnegligible, these curve crossings would lead to the spin–orbit mixing of the corresponding components. For example, the $\Omega = 2$ state of the $^3\Pi$ state would couple with the $^1\Delta_2$ through the spin–orbit coupling. Likewise, the $\Omega = 0^+$ spin–orbit component of the $^3\Pi$ state will mix strongly with the $^1\Sigma^+$ state in the region of curve crossing. Consequently, these curves would undergo avoided crossing in this region.

The calculated FOCI spectroscopic constants of the bound electronic states of SbF are presented in Table 2.35. It was noted that the $e^1\Pi$(II) state exhibits a potential minimum at the FOCI level. However, the $C_3^3\Pi$(II) state exhibits two potential minima at 1.855 and 2.752 Å, respectively, and thus, in anticipation that further improvement in the basis set can yield another shallow potential well near 2.752 Å, the energy separation of the $e^1\Pi$(II) state is reported in Table 2.35. As seen from Table 2.35, the energy separation of the excited $C_3^3\Pi$(II) state at $R_e = 1.855$ Å is in reasonable agreement with the experimental value.

Table 2.36 shows the spectroscopic constants of the bound electronic states of SbF determined at the SOCI level. The most notable difference in comparing the properties at the FOCI and SOCI levels is in the T_e values of the excited electronic states, which decrease at the SOCI level of theory.

The electronic spectra of SbF have been obtained and analyzed by several researchers (355–365). Although several of the observed spectral bands have been assigned, some of the excited electronic states are still unassigned. The spectrum of SbF consists of four groups of bands in the visible region (*A* system) and four groups of bands in the ultraviolet region (*C* system). Although most of the experimental studies are concerned with the analysis of these spectra, it should be mentioned that the work of Wang et al. (365) deals with the

Table 2.35 Spectroscopic Constants of SbF at the CASCF/FOCI Level[a]

State	R_e, Å	T_e, eV	Experimental T_e, eV[c]
$X\,^3\Sigma^-$	1.918	0	0
$a\,^1\Delta$	1.916	1.09	0.845
$b\,^1\Sigma^+$	1.917	1.99	1.69
$A\,^3\Pi$	2.083	2.92	3.07–2.65
$A_3\,^1\Pi$	2.092	3.89	3.56
$C_3\,^3\Pi$(II)	2.752	4.85	—
$d\,^3\Sigma^-$(II)	3.005	5.21	—
$e\,^1\Pi$(II)	(2.752)[b]	5.98	—
$C_3\,^3\Pi$(II)	1.855	6.11	5.56
$e\,^1\Pi$(II)	1.850	8.09	—

[a]At the minimum of $C_3\,^3\Pi$(II) state.
[b]Reference 21.
Source: Reproduced from the work of Latifzadeh and Balasubramanian (353).

Table 2.36 Spectroscopic Constants of SbF at the SOCI Level

					Experimental[c]		
State	R_e, Å	T_e, eV	μ_e, D[c]	ω_e, cm^{-1}	R_e, Å	T_e, eV	ω_e, cm^{-1}
$X^3\Sigma^-$	1.924	0	−2.30	606.0	1.914	0	605–612.6
$a^1\Delta$	1.921	1.00 (0.982)[a]	−2.18	611.7	1.912	0.845	616.3
$b^1\Sigma^+$	1.920	1.70 (1.69)[a]	−2.13	611.3	1.909	1.69	615.5
$A^3\Pi$	2.073	2.79 (2.71)[a]	−2.69	417.2	2.074	3.07–2.65[b,d,e]	416–420.3
$A_3\,{}^1\Pi$	2.075	3.70 (3.56)[a]	−2.27[f]	407.8	2.062	3.56[c]	411.1

[a]The value in the parenthesis includes Davidson correction.
[b]Reference 30.
[c]Reference 14.
[d]Reference 16.
[e]Reference 32.
[f]Negative polarity means Sb^+F^- polarity.
Source: Reproduced from the work of Latifzadeh and Balasubramanian (353).

emission spectra of SbF in the near-infrared, which lead to the characterization of two new emission systems.

Howell and Rochester (355) obtained the spectrum of the SbF molecule using an active-nitrogen method of excitation. The spectrum consisted of an intense band system in the visible region and a wide-band system in the ultraviolet region. Subsequently, Howell and Rochester (355) observed two groups of emission band systems of SbF in the UV and visible regions. The ultraviolet bands consisted of three systems, namely, the C_1, C_2, and C_3 systems, all of which extended from 2200 to 2450 Å (5.64–5.06 eV). The visible bands were composed of triplet band systems, designated as the A_1, A_2, and A_3 systems, observed between 3200 and 5000 Å (3.88–2.48 eV).

Rao and Rao (363) obtained a set of bands in the 3140–3340-Å region, which was assigned to a system designated by these authors as the B system. An analysis of this system yielded two low-lying electronic states designated by these authors as X_1 and X_2 with vibrational frequencies of 612.6 and 616.9 cm^{-1}, respectively. The X_1 and X_2 states were identified as $^3\Sigma^-$ and $^1\Sigma^+$, states, respectively, by these authors. Patel and Abraham (360) reported the ultraviolet bands of the SbF molecule using a *hf* discharge method. A vibrational analysis of the observed bands was carried out that resulted in the vibrational constants of the states. They proposed that the excited electronic state in the C_1 system belongs to $^1\Pi$. They (359) assigned the (0,0) emission band at 44,798.89 cm^{-1} to the $C_3^1\Pi \rightarrow b^1\Sigma^+$ transition for the C_3 system, through rotational analysis. The same authors (358) obtained a group of bands in the $\lambda \cong$ 4050–5450-Å

(3.06–2.28-eV) region, overlapping with the A_1 and A_2 systems. They identified the (0,0) band at 22,493.68 cm^{-1}, but no definitive assignments were made by these authors, although the upper state of the system was considered to be similar to that of the A_2 system reported a new emission spectrum in the $\lambda \cong$ 2600–2700-Å (4.77–4.59-eV) region. A vibrational analysis of the observed bands yielded the vibrational frequencies. On the basis of a fine-structure analysis, they suggested that the emission band is attributed to the C_1 system that corresponds to the $^3\Pi_1 \rightarrow {}^1\Delta_2$ electronic transition. The $^3\Pi_1$ excited state was assigned to either the C_1 system or the B system observed in $\lambda \cong$ 3140–3340-Å (3.95–3.17-eV) region by Rao and Rao (363).

Chakravorty et al. (364) studied the emission spectra of the SbF molecule in the 4730–5770-Å region. The vibrational constants of different electronic states were determined through vibrational analysis. They tentatively assigned the bands in the visible region (the A-band systems) to the $^1\Sigma^+$, $^1\Sigma^-$, $^1\Delta$, $^3\Delta$, $^3\Sigma^-$, and $^3\Sigma^+$ states having $\cdots(x\sigma)^2(\omega\pi)^3(\upsilon\pi)^3$ electronic configuration, which arises by transferring one electron from a bonding orbital ($\omega\pi$) to the corresponding antibonding ($\upsilon\pi$) orbital. However, as seen from the computations, this leading configuration yields the $^1\Delta$(II), $^1\Sigma^+$(II), $^3\Sigma^+$, $^1\Sigma^-$, $^3\Delta$, and a higher root of the $^3\Sigma^-$ state, none of which exhibit any potential minima, analogous to other group VA halides, ruling out this possibility. These researchers assigned the upper state of the C system (C_3, C_2) to the $^3\Pi$ states by a rotational analysis and using the absorption spectrum of the analogous BiF molecule.

Vasudev and Jones (356) obtained the emission spectra of the SbF molecule in the near-ultraviolet and visible regions through rotational analysis. They classified the various states into three groups. The first group with the electronic configuration of $\cdots(\omega\pi)^4(x\sigma)^2(\upsilon\pi)^2$(I) corresponds to the $X^3\Sigma^-$, $a^1\Delta$, and $b^1\Sigma^+$ states. The second group consisted of $A_3 1$, $A_2 2$, and $A0^+$, which arise from the $(\omega\pi)^4(x\sigma)^2(\upsilon\pi)^1(u\sigma)^1$ configuration (II) with a larger R_e value compared to the ground state. The third group involves two electron jumps giving less intensity band. As a consequence of these investigations, the A-band system can be attributed to the $^3\Pi$ or $^1\Pi$ states according to the electronic configuration (II), and the C_3 system corresponds to the $^3\Pi$(II) state (R_e = 1.855 Å) having a different leading configuration.

As seen from Table 2.36, the $A(A_1, A_2)^3\Pi$, and $A_3^1\Pi$ states have R_e values in the 2.073–2.075-Å range, which are longer than the R_e of the ground electronic state. The energy separations of the $A^3\Pi$ state relative to the $X^3\Sigma^-$ state are 2.79 (SOCI) and 2.71 eV (SOCI + Q), both of which are within the experimental range of 2.65–3.074 eV (32, 353, 355, 361). The T_e value for the $A_3^1\Pi$ state is 3.70 and 3.58 eV at the SOCI and SOCI + Q levels, respectively, in good agreement with the experimental value of 3.56 eV (32, 356). The good agreement between the computed results and the experiment also confirms the previous assignments of the spin–orbit states arising from the $A^3\Pi$ state.

Prevot et al. (361) performed the rotational analysis of the absorption and emission spectra of the C_2 and C_3 systems in the UV region. In the emission spectra, the 1–0 band of the $C_3 \rightarrow b$ transition was found to overlap with the strongest bands of the $C^3\Pi_u$–$B^3\Pi_g$ system of N_2 present in the air. In the absorption spectra, the $C_2 \rightarrow X$ bands were studied without satisfactory rotational analysis because of mixing of the vibrational branches. The strongest bands in the UV region were attributed to the energy separations of the C_3–$b^1\Sigma^+$, C_3–$X^3\Sigma^-$, and C_3–$a^1\Delta$ systems at 31,148.9 ± 0.1 (3.86 eV), 44,802.6 ± 0.1 (5.56 eV), and 37,983.53 ± 0.05 cm^{-1} (4.71 eV), respectively. These values are consistent with the computed values of 4.12, 6.11, and 5.03 eV at the FOCI level predicted for the energy separation of the $C_3^3\Pi$(II)–$b^1\Sigma^+$, $C_3^3\Pi$(II)–$X^3\Sigma^-$, and $C_3^3\Pi$(II)–$a^1\Delta$ systems, respectively. It should be noted that the $C_3^3\Pi$(II) state at its shorter R_e value (1.855 Å) has been considered for the comparison with experiment.

Wang et al. (365) observed a new band in the emission spectrum of the SbF molecule in the near-infrared region. The band origin was located at 1.59 eV, and was assigned to the (0,0) band of the $b^1\Sigma^+ \rightarrow X^3\Sigma^-$ transition. According to Table 2.36, the SOCI energy separations of the $a^1\Delta$ and $b^1\Sigma^+$ states relative to the $X^3\Sigma^-$ ground state are somewhat higher than the spectral results. However, the SOCI + Q method is in much better agreement, indicating this to be primarily a consequence of the electron correlation problem.

The dipole moments of the electronic states of the SbF molecule are reported in Table 2.36 at the SOCI level at their respective equilibrium geometries. As seen from this table, all the low-lying bound electronic states exhibit Sb^+F^- polarity. At the present time, there are no experimental measurements on the dipole moments of the electronic states of SbF.

The dissociation energy of SbF is computed by setting the Sb–F bond length to 8 Å, using the CASSCF/FOCI/SOCI methods. The computed D_e values are 3.87 (SOCI) and 3.98 eV (SOCI + Q), which are consistent with the experimentally deduced values of 4.17 and 4.4 eV reported by other researchers (32). While the level of theory employed here generally underestimates the D_e values, it is believed that the higher value of 4.4 eV is somewhat improbable.

The Boustani et al. (354) computed MRDCI results are shown in Table 2.37 for the spectroscopic constants. It is noted from Table 2.37 that the spin–orbit splitting of the ground state is quite small, although the excited $^3\Pi$ state experiences slightly larger splitting (3600 cm^{-1}). Overall, the spectroscopic properties computed by Boustani et al. (354) are found to be in reasonable agreement with both experiment and the CASSCF/SOCI constants. However, there are quantitative differences. Their computed vibrational frequencies of all the electronic states are uniformly smaller than the experimental results with the exception of the $c1$ state, for which an opposite trend is found in Table 2.37. The energy separations of the excited electronic states computed by Boustani et al. (354) are generally larger than the experiment and the values reported by Latifzadeh and the author (353) in Table 2.36. This is especially striking for the $b^1\Sigma^+$ and the $A^3\Pi$ and the $^1\Pi$ electronic states. Boustani et al. have not reported the dipole moments or the dissociation energy of SbF, and thus no comparison can be made with these properties.

Table 2.38 shows the lifetimes of the electronic states of SbF computed by Boustani et al. (354), together with the partial lifetimes for the transitions of the

Table 2.37 The MRDCI and Experimental Spectroscopic Properties of SbF (Bond Lengths r_e, Transition Energies T_e, and Vibrational Frequencies ω_e

	r_e, Å		T_e, cm^{-1}		ω_e, cm^{-1}	
State	MRDCI[a]	Exp.[b]	MRDCI[a]	Exp.[b]	MRDCI[a]	Exp.[b]
$X_1 0^+$	1.9286	1.9177	0	0.0	575	605.0
$X_2 1$	1.9279	1.9132	744	796.3	576	612.6
$a2$	1.9090	1.9129	8,534	6,816.0	572	616.3
$b0^+$	1.9054	1.9098	15,162	13,651.1	574	615.5
$A_1 0^-$	2.0853	2.0747	22,590	21,407.5	364	416
$A_2 0^+$	2.0814	2.0747	22,685	21,407.5	367	416
$A_3 1$	2.0819		23,834	22,589.4	373	418.9
$A_4 2$	2.0791	2.0636	25,811	24,788.4	389	420.3
$c1$	2.0693	2.0624	31,580	28,706.6	435	411.1

[a]From Reference 354.
[b]Reference 32.

$X_1 0^+$ and $X_2 1$ states of SbF. As seen from Table 2.38, the lifetimes of the $X_2 1$ and $a2$ states are considerably longer than the other electronic states. The larger lifetime of the $X_2 1$ state (4 times) compared to BiF is primarily a consequence of relativistic effects. That is, since spin–orbit coupling is relatively small for SbF compared to BiF, the spin–orbit contamination with other electronic states is also considerably smaller. The $a2$ state can radiate only to the $X_2 1$ state, but the two states have $^1\Delta$ and $^3\Sigma^-$ symmetries in the absence of spin–orbit coupling. Consequently, this transition is spin- and spatially forbidden in the absence of spin–orbit coupling. Although the transition between the $X_2 1$ and $a2$

Table 2.38 MRDCI Lifetimes of Excited States of SbF ($p' = 0$): Partial Lifetimes (in seconds) τ_0 and τ_1 for Transitions to $X_1 0^+$ and $X_2 1$, Respectively, and Total Lifetimes τ

State	τ_0	τ_1	τ
$X_2 1$	16.38	—	16.38
$a2$	—	0.49	0.49
$b0^+$	2.23×10^{-3}	58.5×10^{-6}	57.00×10^{-6}
$A_1 0^-$	—	3.04×10^{-6}	3.04×10^{-6}
$A_2 0^+$	32.2×10^{-6}	5.76×10^{-6}	4.89×10^{-6}
$A_3 1$	4.57×10^{-6}	0.114×10^{-3}	4.39×10^{-6}
$A_4 2$	—	3.31×10^{-6}	3.31×10^{-6}
$c1$	22.2×10^{-6}	22.6×10^{-6}	1.55×10^{-6b}

[a]Includes transition probabilities to the $a2$ and $b0^+$ lower states.

Source: From Reference 354.

states becomes relativistically allowed, due to relatively small spin–orbit effects for SbF, the lifetime of the *a*2 state (0.49 s) is considerably longer compared to other states.

2.3 SPECTROSCOPIC PROPERTIES AND POTENTIAL-ENERGY CURVES OF HEAVY GROUP IV CHALCOGENIDES AND THEIR IONS

The interest in heavy group IV oxides and sulfides arises from the fact that these species are generated in chemiluminescent reactions of atoms such as Pb and Sn with molecules such as N_2O, O_3, and OCS (250–253). Examples of many chemiluminescent reactions of the heavy group IV atoms are shown in the following reactions:

$$\mathrm{Pb} + \mathrm{O_3} \rightarrow \mathrm{PbO}^* + \mathrm{O_2}$$

$$\mathrm{Pb} + \mathrm{N_2O} \rightarrow \mathrm{PbO}^* + \mathrm{N_2}$$

$$\mathrm{Sn} + \mathrm{N_2O} \rightarrow \mathrm{SnO}^* + \mathrm{N_2}$$

$$\mathrm{Pb} + \mathrm{OCS} \rightarrow \mathrm{PbS}^* + \mathrm{CO}$$

$$\mathrm{Sn} + \mathrm{OCS} \rightarrow \mathrm{SnS}^* + \mathrm{CO}$$

These five reactions were investigated for their suitabilities as chemical laser pumping reactions. Consequently, a wealth of information has accumulated on the spectroscopic properties of both the ground state and the excited electronic states of molecules such as SnO, SnS, PbO, and PbS (254–270) the author and Pitzer (271, 285) studied the spectroscopic properties and potential-energy curves of SnO and PbO. Later the author pursued this line of study further to complete calculations on SnS, PbS, and PbSe and their positive ions (273, 285, 304). The experimental photoelectron spectra of heavy group IV chalcogenides were also interpreted by using the results of these calculations.

2.3.1 SnO and SnO^+

The spectroscopic properties, electronic dipole moments, and photoelectron spectra of SnO have been the subjects of many investigations for several years (254–270). The chemiluminescent spectra of SnO have also demonstrated the existence of several emission systems in the visible and ultraviolet regions. The tin oxide molecule was also investigated as a candidate for chemical lasers because of the high photon yields (~50%) of the Sn + N_2O reaction (262). The observed bands in the visible and near-UV regions for SnO, however, were not interpreted satisfactorily before the theoretical calculations of the author and Pitzer (271).

The author and Pitzer (271) performed SCF/CI calculations on seven λ–s states of SnO employing a double-ζ STO basis set. The theoretical spectroscopic

Table 2.39 Spectroscopic Properties of SnO[a]

State	T_e, cm^{-1}	R_e, Å	ω_e, cm^{-1}
$X\,^1\Sigma^+(0^+)$	0	1.94	800
		(1.83)	(823)
$^3\Sigma^+(0^-, 1)$	18,606	2.13	542
	(20,622?)	—	—
$^3\Delta(1, 2, 3)$	20,561	2.16	493
$^3\Sigma^-(0^+, 1)$	22,750	2.15	530
	(20,622?)	—	(554?)
$^1\Sigma^-(0^-)$	22,890	2.14	540
$A^3\Pi(0^+, 0^-, 1, 2)$	24,100	2.06	770
B	(24,333, 0^+)	(2.01, 0^+)	(555, 0^+)
	(24,890, 1)	(1.99, 1)	(560, 1)
$D^1\Pi(1)$	26,700	2.06	710
	(29,624)	(1.95)	(574)

[a]Numbers in parentheses are experimental values from Reference 32. The theoretical values are from Reference 271.

constants for SnO together with available experimental data are shown in Table 2.39. The potential-energy curves of these states are shown in Figure 2.38.

The SnO molecule is expected to have a closed-shell $^1\Sigma^+$ ground state arising from the $1\sigma^2 2\sigma^2 3\sigma^2 1\pi^4$ configuration, where the highest occupied 3σ orbital should be a mixture of metal p_z with O p_z, while the highest occupied 1π orbital should be predominantly the oxygen $2p$ nonbonding orbital. Thus, the 2π orbital should be predominantly on the tin atom. The theoretical investigation by the author and Pitzer (271) considered the $^1\Sigma^+$ ground state; the $^3\Sigma^+$, $^3\Sigma^-$, $^3\Delta$, and $^1\Sigma^-$ states arising from the $1\sigma^2 2\sigma^2 3\sigma^2 1\pi^3 2\pi$ configuration; and the $^3\Pi$ and $^1\Pi$ states arising from the $1\sigma^2 2\sigma^2 3\sigma 1\pi^4 2\pi$ configuration.

Deutsch and Barrow (260) observed and analyzed the emission bands from the $^3\Pi(0^+)$ and $^3\Pi(1)$ states to the $^1\Sigma^+$ state designated as the *A–X* and *B–X* systems. As seen from Table 2.39, the theoretical T_e values of these states are in reasonable agreement with the experimental results. Capelle and Linton (262) obtained additional bands for the system near 20,000 cm^{-1} reported earlier by Joshi and Yamdagni (261) at 19,000 cm^{-1}. Capelle and Linton (262) assigned this system to the $^3\Sigma^+(1) \rightarrow {}^1\Sigma^+$ transition. However, the experimental T_e of 20,622 cm^{-1} was found to be higher than the theoretical value of 17,650 cm^{-1} obtained by the author and Pitzer (271). Thus the author and Pitzer (271) argued for the possibility of this system state as either $^3\Sigma^-(0^+)$ or $^3\Sigma^-(1)$.

The calculations of the author and Pitzer (271) supported the earlier assignment of the *D–X* system [$D(T_e)$ = 29,624 cm^{-1}] to $^1\Pi(1)$ [1(V)], although the calculated T_e of 26,700 cm^{-1} was considerably lower. The author and Pitzer (271) attributed the difference to the fact that it is the fifth excited 1 state, and thus theoretical results may not be very accurate.

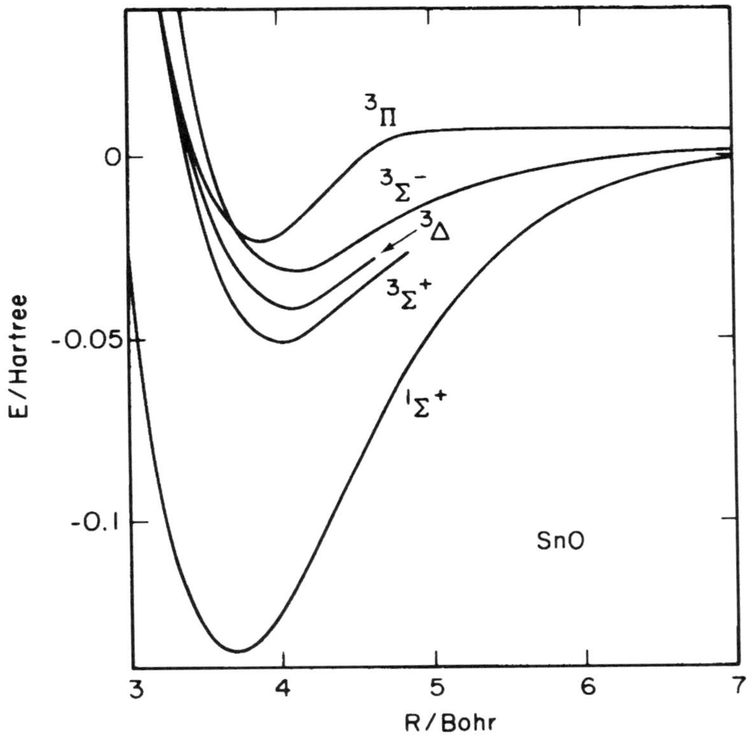

Figure 2.38 Potential-energy curves for the electronic states of SnO. (Reprinted from Ref. 271; copyright 1983, Elsevier Science Publishers B.V.)

The theoretical D_e of 3.70 eV was found to be considerably smaller than the experimental value (32) of 5.49 eV. This was mainly because of a single-configuration SCF followed by CI, which does not take into account all the electron correlation effects.

The He(I) photoelectron spectra of the SnO molecule was recorded by Dyke et al. (272) in the $X(^1\Sigma^+)$ ground state. This photoelectron spectrum revealed the existence of low-lying states of SnO^+ of $^2\Pi$ and $^2\Sigma^+$ character. Dyke et al. (272) also performed Dirac–Fock SCF-$X\alpha$ calculations to determine the ordering of the two observed states of SnO^+.

The author (273) carried out SCF/RCI calculations on the low-lying electronic states of SnO^+ as well as the heavier PbO^+, PbS^+, and $PbSe^+$ ions. Three low-lying states, including the spin–orbit coupling, were found for all these species. Table 2.40 shows the spectroscopic constants for SnO^+ and the ionization potentials of SnO as obtained by different methods. Figure 2.39 shows the potential-energy curves of the two low-lying electronic states of SnO^+.

As seen from Table 2.40, the agreement between SCF/RCI results of the author (273) and Dyke et al.'s (272) photoelectron spectrum is excellent. The

Table 2.40 Spectroscopic Constants of SnO^+ and Ionization Potentials of SnO[a]

		R_e, Å		T_e, cm^{-1}		ω_e, cm^{-1}	
Method[b]	State	Theory	Exp.	Theory	Exp.	Theory	Exp.
RCI	$^2\Pi$	2.05	2.00	0	0	695	700
RCI	$^2\Sigma^+$	1.96	—	3616	—	708	—

Method[b]	Splitting	IP, eV	Method[b]	Splitting	IP, eV
SCF	$^1\Sigma^+$–$^2\Pi$	6.73	KT	$^1\Sigma^+$–$^2\Pi$	10.15
RCI	$^1\Sigma^+$–$^2\Pi$	8.44	KT	$^1\Sigma^+$–$^2\Sigma^+$	9.79
RCI	$^1\Sigma^+$–$^2\Sigma^+$	8.76	—		—

[a] All theoretical results are from Reference 273. The experimental values are from Reference 272.
[b] RCI = relativistic CI; KT = Koopman's theorem.

D_e values of SnO^+ in the $^2\Pi$ and $^2\Sigma^+$ states are 2.95 and 2.50 eV, respectively, compared to the corresponding experimental values of 3.23 and 2.71 eV, respectively.

The vertical ionization potentials of SnO to generate SnO^+ in the $^2\Pi$ and $^2\Sigma^+$ states were calculated to be 8.44 and 8.76 eV, respectively. Agreement with the corresponding experimental values of 9.98 and 10.12 eV is not exact, but it is well known that the IPs and EAs cannot be calculated exactly by *ab initio* methods. However, the SCF/RCI method predicts the correct ordering of the electronic states of SnO^+ in contrast with Koopman's theorem, which predicts the ground state of SnO^+ to be $^2\Sigma^+$ incorrectly. Koopman's theorem is also off in the magnitudes of the IPs of the two states. The author (273) explained the breakdown of Koopman's theorem based on the large orbital relaxation effects.

The potential-energy curves of the $^2\Sigma^+$ and $^2\Pi$ states cross near the repulsive wall of the $^2\Pi$ state. Thus the corresponding $\frac{1}{2}$ components undergo avoided crossings. The adiabatic splitting of the $^2\Sigma^+$-$^2\Pi$ states (0.34 eV) was found to be a bit larger than the vertical separation at the R_e of SnO molecule (0.32 eV). The bond length of the $^1\Sigma^+$ state of the SnO molecule was calculated as 1.94 Å, while the R_e of the $^2\Pi$ state of SnO^+ was calculated as 2.05 Å. Thus the bond length is increased by 0.11 Å when SnO is ionized.

2.3.2 PbO and PbO+

The PbO molecule has been studied experimentally in both the visible and UV regions (274–284). An intense system labeled $A0^+$–$X0^+$ in the emission has been characterized. A few other less intense systems labeled $a \rightarrow X$ and $b \rightarrow X$ have also been observed. In addition to these, the B, C, C', D, E, and G states of PbO have been characterized in the $B \leftrightarrow X$, $C \leftarrow X$, $C' \leftarrow X$, $D \leftrightarrow X$, $E \leftrightarrow X$, $F \leftarrow X$, and $G \leftarrow X$ systems. The dissociation energy of PbO was obtained by using the mass-spectrometric method (279) as well as from the chemilumines-

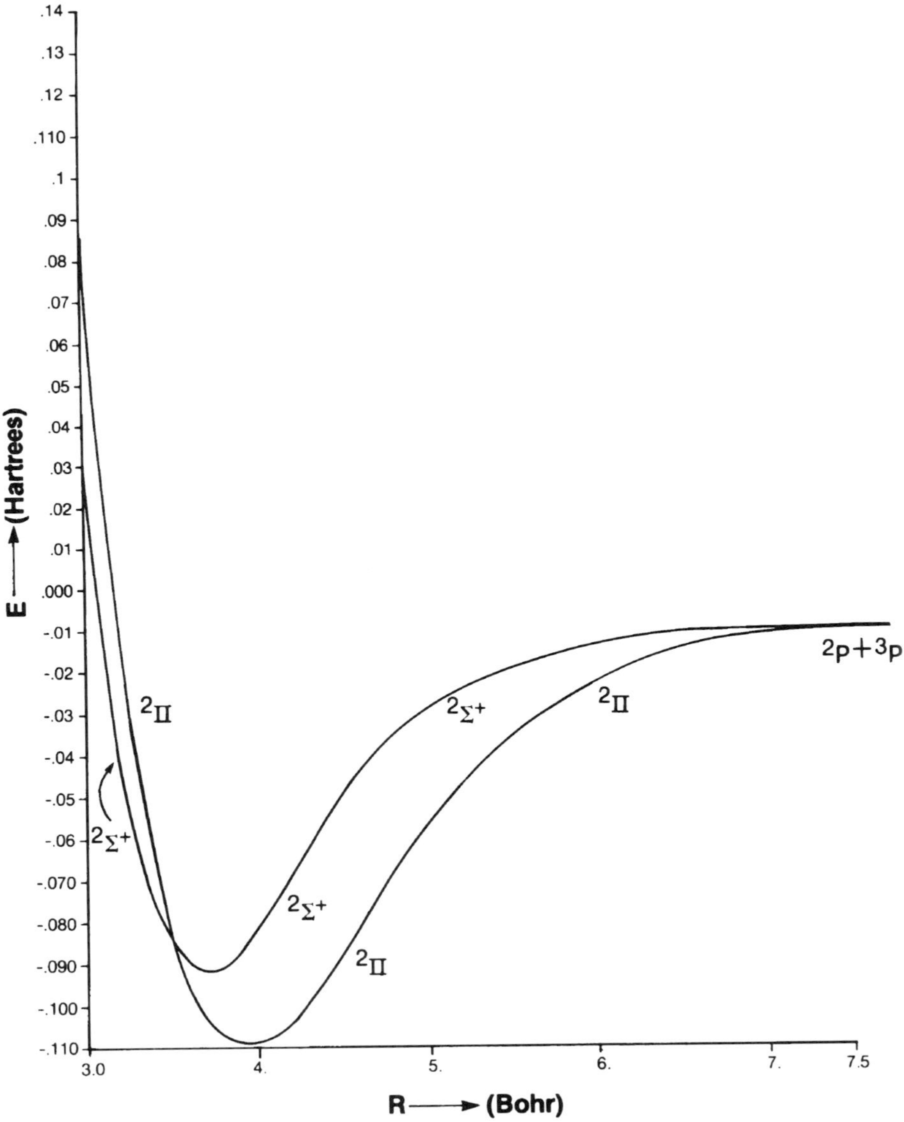

Figure 2.39 Potential-energy curves for the electronic states of SnO^+. (Reprinted from Ref. 273; copyright 1984, American Chemical Society.)

cence spectra of Pb + O_3 under single-collision conditions by Oldenborg et al. (281). The experimental D_e of PbO obtained by these methods was found to be between 3.74 and 3.83 eV.

The dipole moment of the PbO molecule was measured as $\mu_{el}(v = 0) = 4.64$ D using the Stark spectra (267). The relative intensities (309) and transition probabilities (310) of the B–X bands have also been deduced from the experimental information. The RKR potential functions have been derived for the A and X states (280).

The theoretical calculations on PbO were made by the author and Pitzer (285) using the SCF/RCI method, which employed a double-ζ STO basis set. These authors found 11 low-lying electronic states of PbO. The spectroscopic constants of eight λ–s states were also obtained without the spin–orbit term for comparison.

Table 2.41 shows the spectroscopic constants of PbO together with available experimental results. Figure 2.40 shows the theoretical potential-energy curves for PbO calculated by the author and Pitzer (285).

In analogy with SnO, the ground state of PbO was found to be a $^1\Sigma^+_{0^+}$ state arising from the $1\sigma^2 2\sigma^2 3\sigma^2 1\pi^4$ configuration. The excited states of PbO arise from the $1\sigma^2 2\sigma^2 3\sigma 1^1 \pi^4 2\pi$ and $1\sigma^2 2\sigma^2 3\sigma^2 1\pi^3 2\pi$ configurations.

The $A0^+$–$X0^+$ bands observed in emission with a T_e of 19,863 cm^{-1} were assigned by the author and Pitzer (285) to a $^3\Pi_{0^+}$–$X^1\Sigma^+$ system, where the the-

Table 2.41 Spectroscopic Properties of PbO[a]

	R_e, Å		T_e, cm^{-1}		ω_e, cm^{-1}	
State	Theory	Exp.	Theory	Exp.	Theory	Exp.
$X\,^1\Sigma^+$(I), 0^+(I)	2.02	1.92	0	0	715	721
$b?^3\Sigma^+0^-$	2.23	—	14,461	16,454?	—	441?
$a1$	2.23	2.12	14,551	16,025	487	482
$b?^3\Delta_2$	2.23	—	15,205	16,454?	540	441?
1(II)	2.24	—	15,360	—	472	—
$^1\Delta$[2(II)]	2.24	—	16,035	—	451	—
$^3\Pi0^-$(III)?	2.14	—	18,758	—	576	—
$A0^+$(II)	2.14	2.09	18,890	19,863	528	444
$B1$(III)	—	2.07	—	22,285	—	498
$C\,^3\Sigma^-0^+$(III)	2.23	—	20,747	23,820	612	532
$C'1$(IV)	—	—	—	24,947	—	494
$D^1\Pi$[1(V)]	2.14	2.05	27,215	30,199	521	530
E	—	2.18	—	34,454	—	454
$^1\Sigma^+$(II)[0^+(IV)]	—	—	—	—	—	—
$^1\Sigma^+$	2.02	—	682	—	706	—
$^3\Sigma^+$	2.21	—	16,610	—	503	—
$^3\Delta$	2.24	—	18,267	—	485	—
$^3\Sigma^-$	2.23	—	20,292	—	600	—
$^1\Sigma^-$	2.22	—	20,477	—	594	—
$^3\Pi$	2.13	—	22,469	—	514	—
$^1\Pi$	2.15	—	24,771	—	505	—
$^1\Sigma^+$(II)	2.22	—	39,202	—	703	—

[a]All theoretical results are from Reference 285.

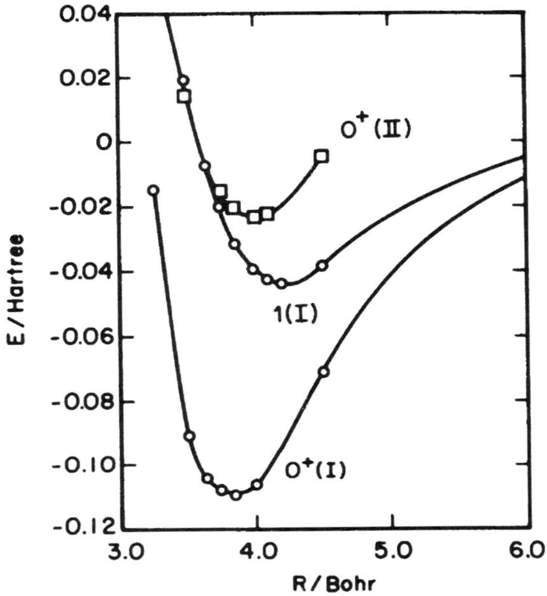

Figure 2.40 Potential-energy curves for the electronic states of PbO. (Reprinted from Ref. 285; copyright 1983, American Chemical Society.)

oretical T_e value of the A state was found to be 18,890 cm^{-1}. The A state was found to be actually a mixture of the $^3\Pi_{0^+}$ and $^1\Sigma^+_{0^+}$ states. The agreement between the theoretical and experimental R_e and ω_e values of the A state also appeared to be good.

The earlier assignment of the $a1$ state to $^3\Sigma^+_1$ was confirmed by the calculations of the author and Pitzer (285). However, the spin–orbit components (0^-, 1) of the $^3\Sigma^+$ state were found to be contaminated substantially with other λ–s states. The 0^- state was found to be 75% $^3\Sigma^+_{0^-}$ and 25% $^3\Sigma^-_{0^-}$, while the 1 state was found to be a $\frac{3}{4}\ ^3\Sigma^+_1 - \frac{1}{4}\ ^3\Sigma^-_1$ mixture. In Table 2.41, two possible assignments for the b state ($^3\Sigma^+_{0^-}\ ^3\Sigma^+_{0^-}$ or $^3\Delta_2$) are suggested.

An emission system labeled $B \rightarrow X0^+$ with a T_e value 22,885 cm^{-1} was assigned to $^3\Pi_1$ by several investigators. The assignment of the B state to $^3\Pi_1$ was supported by the theoretical calculations of the author, although this state was found to be certainly a mixture of $^3\Pi_1$, $^1\Pi_1$, and $3\Sigma^-_1$. The C state was assigned to $^3\Sigma^-_{0^+}$. The theoretical ω_e value of C (613 cm^{-1}) was in good agreement with the experimental value. The $D \rightarrow X0^+$ and $E \rightarrow X0^+$ emissions have also been observed experimentally. The experimental T_e values of these states are 30,199 and 34,454 cm^{-1}. The D and E states correspond theoretically to the 1(V) and 0^+(IV) states. These states were found to be predominantly $^1\Pi_1$ and $^1\Sigma^+$(II) states.

The calculated dissociation energy of 3.0 eV (285) for the PbO molecule was found to be in reasonable agreement with the experimental value of 3.83 eV (32). The calculated (285) R_e and ω_e values for the ground state were in good agreement with the experimental values and the values obtained by Basch et al. using a MCSCF calculation (286). The calculated ω_e and R_e values for the excited states were found to be in good agreement with the available experimental results.

The three lowest-lying electronic states of PbO^+ were investigated by the author (273) using a SCF/RCI scheme including the spin–orbit term. The ionization potentials of PbO obtained by various methods together with the theoretical spectroscopic constants are shown in Table 2.42. Figure 2.41 shows the potential–energy curves for the low-lying electronic states of PbO^+.

As seen from Table 2.42, the $\frac{3}{2}$ and $\frac{1}{2}$ components of the $^2\Pi$ state of PbO^+ are nearly degenerate, since the 1π orbital of PbO is predominantly the nonbonding 0(2p) orbital. Since the spin–orbit splitting of the oxygen atom is small, the $^2\Pi_{3/2,1/2}$ components are close in energy.

The RCI $^1\Sigma^+$-$^2\Pi$ and $^1\Sigma^+$-$^2\Sigma^+$ neutral–ionic splittings should be about 15% lower than the experimental values, due to the errors introduced by electron correlation effects. The IPs of PbO are thus predicted as 9.32 ($^2\Pi$) and 9.8 eV ($^2\Sigma^+$), respectively. There are no experimental IPs available for PbO. Koopman's theorem, however, breaks down for PbO in that it predicted the wrong ordering for the electronic states of PbO^+.

As seen from Figure 2.41, the $\frac{1}{2}$ state of PbO^+ contains a shoulder. This shoulder resulted from an avoided crossing. In the absence of the spin–orbit coupling term, the $^2\Sigma^+$ and $^2\Pi$ states cross. Since both states yield one state of the same symmetry ($\frac{1}{2}$), when spin–orbit coupling is included, the $\frac{1}{2}$ components of these states undergo an avoided crossing. Thus, the $\frac{1}{2}$ state was found to be predominantly $^2\Sigma^+_{1/2}$ at short distances but became $^2\Pi_{1/2}$ near the well and long distances. At 3.8 bohr, the $\frac{1}{2}$ state of PbO^+ was found to be 75% $^2\Pi_{1/2}$ and 16% $^2\Sigma^+_{1/2}$.

Table 2.42 Spectroscopic Constants for PbO^+ and Ionization Potentials of PbO[a]

Method	State	R_e, Å	T_e, cm^{-1}	ω_e, cm^{-1}
RCI	$\frac{3}{2}$	2.18	0	590
RCI	$\frac{1}{2}$	2.16	46	610
RCI	$\frac{1}{2}$(II)	2.09	4979	712

Method	Splitting	IP, eV	Method[b]	Splitting	IP, eV
SCF	$^1\Sigma^+$–$^2\Pi$	6.23	KT	$^1\Sigma^+$–$^2\Pi$	9.82
RCI	$^2\Sigma^+$–$^2\Pi$	8.10	KT	$^1\Sigma^+$–$^2\Sigma^+$	9.31
RCI	$^1\Sigma^+$–$^2\Sigma^+$	8.52	—	—	—

[a] All results are from Reference 273.
[b] KT = Koopman's theorem.

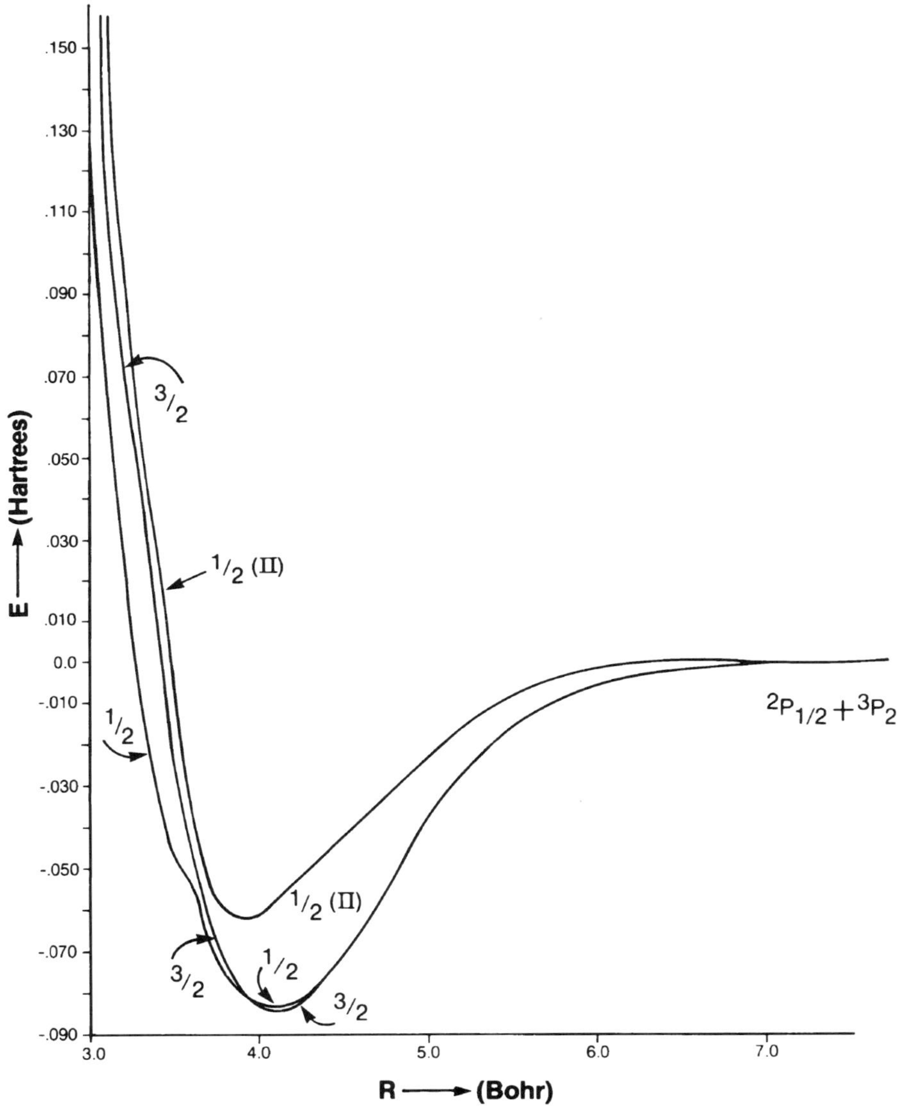

Figure 2.41 Potential-energy curves for the electronic states of PbO^+. (Reprinted from Ref. 273; copyright 1984, American Chemical Society.)

2.3.3 SnS

The absorption and emission spectra of SnS observed below 35,000 cm^{-1} yielded information on the $a \rightarrow X$, $A \leftrightarrow X$, $C' \leftrightarrow X$, $D \leftrightarrow X$, and $E \leftrightarrow X$ systems (287–294). Greenwood et al. (292) studied the electronic spectrum of SnS using the Sn + OCS chemiluminescence reaction. Two electronic states excited in chemiluminescence were characterized (a, A) and assigned to a $^3\Sigma_1^+$ and $A^3\Pi_{0^+}$. These authors made rotational analyses of the bands seen in absorption, which enabled tentative assignment of these to the Ω components of $^3\Sigma^-$ and $^3\Pi_1$.

The author (295) carried out SCF/CI calculations that employed a double-ζ STO basis on eight λ–s states of SnS. The spectroscopic constants and the potential-energy curves of these states were computed. Since the SnS molecule is isovalence-electronic with SnO, its ground state should be a $^1\Sigma^+$ state arising from the $1\sigma^2 2\sigma^2 3\sigma^2 1\pi^4$ electronic configuration. The excited states of SnS should arise from the $1\sigma^2 2\sigma^2 3\sigma^2 1\pi^3 2\pi$ and $1\sigma^2 2\sigma^2 3\sigma 1\pi^4 2\pi$ configurations.

Table 2.43 shows the theoretical spectroscopic constants for SnS together with experimental results. The theoretical potential-energy curves of SnS are shown in Figure 2.42 in the absence of the spin–orbit coupling term.

The first excited $a^3\Sigma^+$ electronic state of SnS was observed by Greenwood et al. (292) in chemiluminescence. The theoretical T_e values of all the excited states are about 15% higher than the experimental values, since the ground state was correlated more than the excited states. If this correction is applied to all the excited states, the assignments of the observed states are possible.

The $^3\Delta_{3,2,1}$ electronic states have not yet been observed; the transitions to the 3 and 2Ω components of this state are forbidden from the ground state. The T_e value of this state (24,000 cm^{-1}), when corrected for the differential correlation errors, yielded a T_e of 20,400 cm^{-1} in the absence of spin–orbit interaction.

Greenwood et al. (292) observed five systems of SnS, for which rotational analyses were made. Since the transitions of systems 5 and 4 had three-branch bands, these authors assigned the upper states of systems 4 and 5 to $\Omega = 1$ components. Systems 2 and 3 were assigned to $\Omega = 0^+$. The first system with an experimental T_e value of 18,144 cm^{-1} was assigned to $a^3\Sigma_1^+$. Greenwood

Table 2.43 Spectroscopic Properties of SnS[a]

	R_e, Å		T_e, cm^{-1}		ω_e, cm^{-1}	
State	Theory	Exp.	Theory	Exp.	Theory	Exp.
$X^1\Sigma^+$	2.32	2.21	0.0	0.0	449	487
$a^3\Sigma^+(0^-, 1)$	2.55	—	21,864	18,144	356	338
$C^3\Sigma^- 0^+, 1)$	2.62	—	25,632	(22,380, 0^+)	308	—
C'	—	—	—	(22,480, 1)	—	—
$^3\Delta(3,2,1)$	—	—	~24,000	—	—	—
$^1\Sigma^-(0^-)$	2.62	—	25,685	—	307	—
$A^3\Pi$	2.42	—	28,658	22,021, 0^+	377	—
B	—	—	—	23,320, 1	—	—
$D^1\Pi$	2.48	2.36	32,707	28,337	371	331
$E^1\Sigma^+$(II)	2.63	—	41,238	33,037	336	294

[a]All theoretical results are from Reference 295.

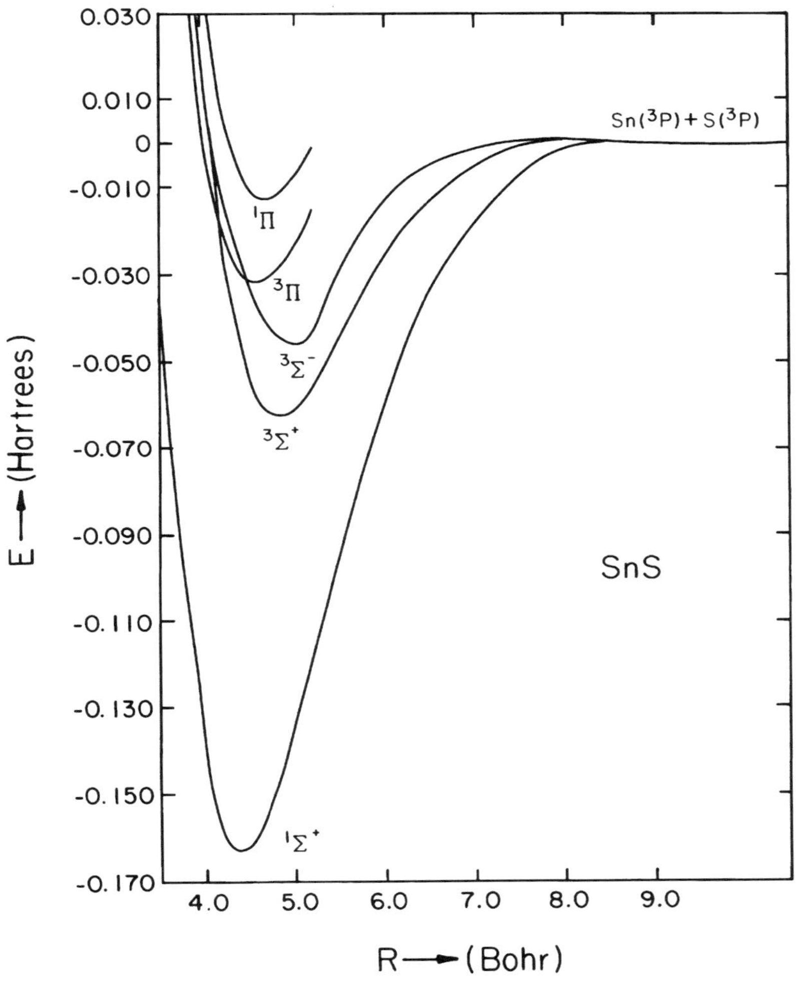

Figure 2.42 Potential-energy curves for the electronic states of SnS. (Reprinted from Ref. 295; copyright 1987, Elsevier Science Publishers B.V.)

et al. (292) assigned the bands in systems 2–5 on the basis of comparison to the ordering of lead compounds. System 5 (T_e = 23,320 cm^{-1}) was assigned to the $B^3\Pi_1$-$X^1\Sigma^+$ system, while system 4 (T_e = 22,480 cm^{-1}) was assigned to the $C'^3\Sigma_1^-$–$X^1\Sigma^+$ system; system 3 (T_e = 22,390 cm^{-1}) was assigned to $C^3\Sigma_{0^+}^-$–$^1\Sigma^+$ and system 2 (T_e = 22,021 cm^{-1}) to $A^3\Pi_{0^+}$–$^1\Sigma^+$. The theoretical calculations of the author (295), when corrected for the differential correlation errors in T_e values, supported all these assignments.

The vibrational frequencies of the A, B, C, and C' states could not be calculated accurately from experimental results, since the observed bands were perturbed. Further, the $A0^+$ and $C0^+$ states were found to be only 380 cm^{-1} apart, implying that mixing of $^3\Sigma^-_{0+}$ and $^3\Pi_{0+}$ could be large.

The calculated dissociation energy of 4.48 eV (295) for the $^1\Sigma^+$ state in the absence of spin–orbit interaction was found to be in very good agreement with the thermochemical value of 4.77 eV (32). The approximate D_0° (4.48 eV) (32) derived from the spectroscopic continuous absorption (assuming dissociation into ^1D + ^1D) was also not unreasonable.

2.3.4 PbS

The first electronic spectrum of PbS was obtained in absorption by Rochester and Howell (296) in the regions of 2600–4500 and 3100–8000 Å. Vago and Barrow (276) later studied the UV absorption systems of PbS, while Barrow et al. (297) studied four absorption systems of PbS, which they assigned to the $a \leftarrow X$, $A \leftarrow X$, $B \leftarrow X$, and $D \leftarrow X$ systems. Nixon and co-workers (298, 299) studied the emission and vibronic spectra of PbS in inert gas and SF_6 matrices, while Burkin et al. (300) studied the laser-induced fluorescence spectra of PbS. Greenwood et al. (294) carried out a revised vibrational analysis of the a–X system as well as rotational analyses of bands of the C–X and C'–X systems. Knöckel et al. (301) recorded the microwave optical double resonance (MODR) spectra in a molecular beam of the X and A states of PbS. The dipole moment of PbS was also measured (302, 308). Maki and Lovas (330) recently recorded the infrared spectrum of PbS ($\Delta v = 2$ transitions). These spectra have provided improved Dunham constants for PbS.

Thermodynamic studies of PbS using mass-spectrometric techniques revealed the D_0° values for these systems (303). Through analysis of experi-

Table 2.44 Spectroscopic Properties of PbS[a]

	R_e, Å		T_e, cm^{-1}		ω_e, cm^{-1}	
State	Theory	Exp.	Theory	Exp.	Theory	Exp.
$X0^+$(I)	2.40	2.29	0.0	0.0	432	429
$a1$(I)	2.63	2.56	17,533	14,893	321	286
0^-(I)	2.64	—	17,581	—	312	—
$A0^+$(I)	2.61	2.51	22,256	18,853	282	261
0^-(II)	2.55	—	24,486	—	324	—
$C0^+$(III)	2.60	—	24,698	23,213	395	304
$^1\Sigma^+$	2.40	—	1,645	—	446	—
$^3\Sigma^+$	2.62	—	20,691	—	332	—
$^3\Sigma^-$	2.68	—	24,144	—	295	—
$^3\Pi$	2.52	—	28,377	—	340	—
$^1\Pi$	2.57	—	31,611	—	327	—

[a] All theoretical results are from Reference 304.

mental data, potential-energy curves of some low-lying states of PbS have been constructed (264).

The author (304) made relativistic SCF/RCI calculations on six ω–ω states [0^+, 1, 0^-, 0^-(II), 0^+(II), 0^+(III)] and six λ–s states without the spin–orbit term for PbS. The spectroscopic constants of PbS calculated this way are shown in Table 2.44 and compared with experiments. Figures 2.43 and 2.44 show the

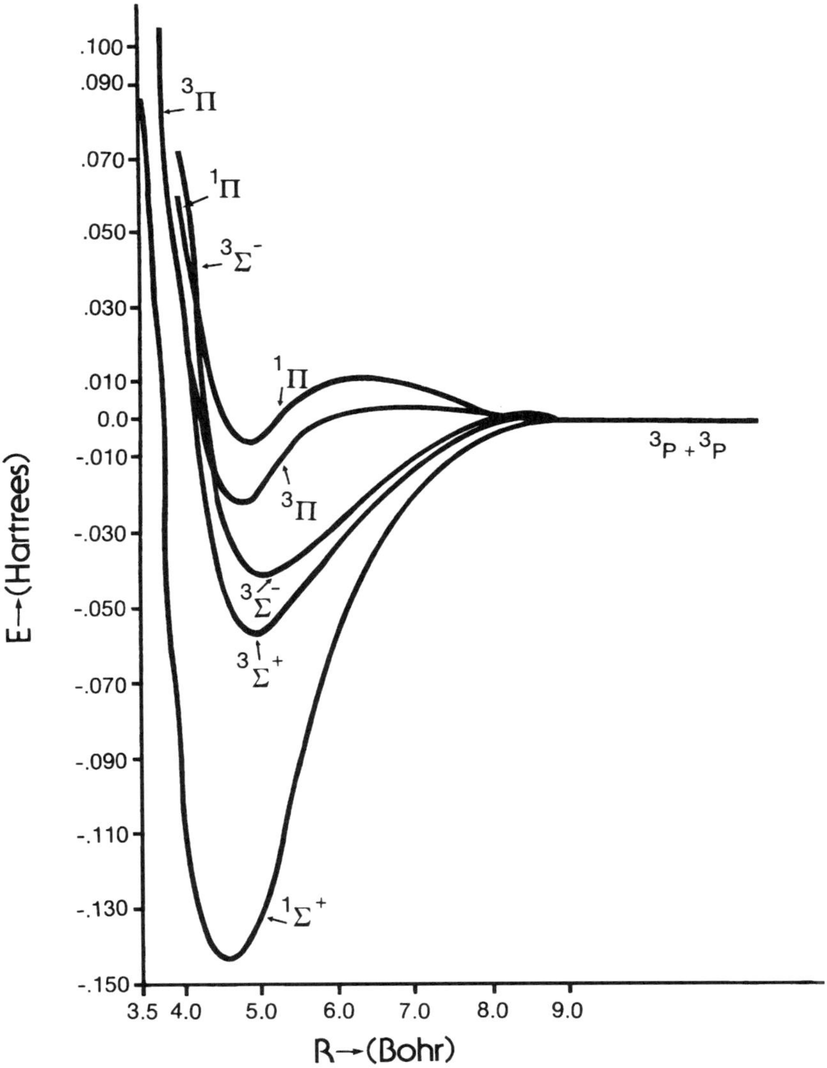

Figure 2.43 Potential-energy curves for the electronic states of PbS without spin–orbit term. (Reprinted from Ref. 304; copyright 1986, American Institute of Physics.)

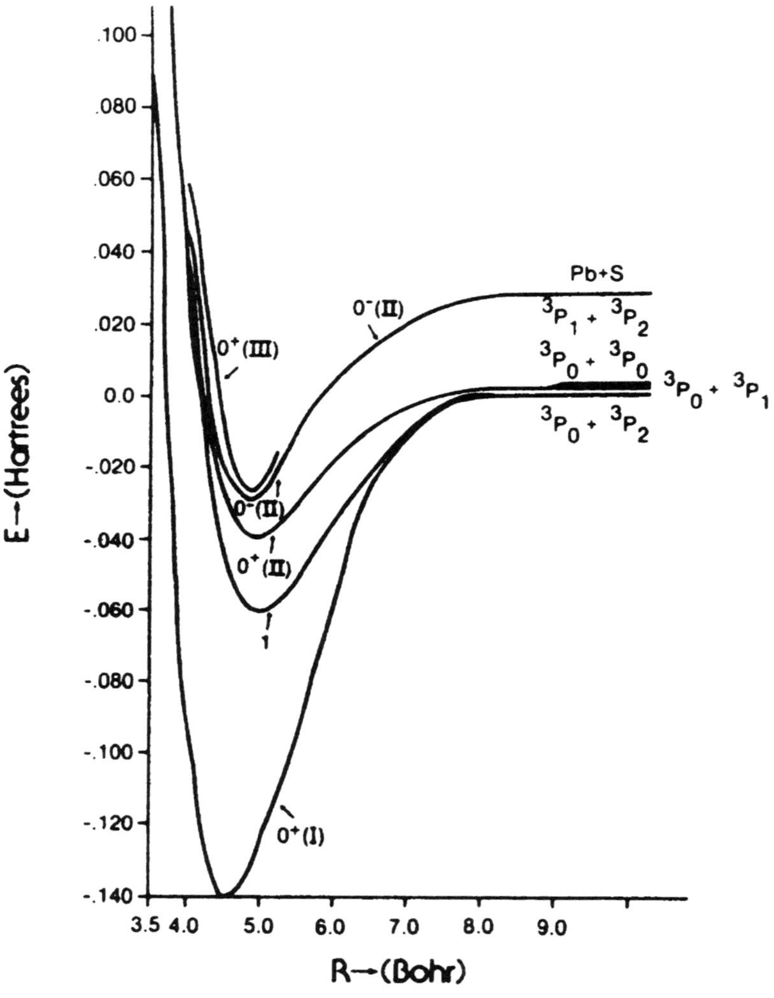

Figure 2.44 Potential-energy curves for the electronic states of PbS with spin–orbit term. (Reprinted from Ref. 304; copyright 1986, American Institute of Physics.)

theoretical RCI potential-energy curves for the electronic states of PbS without and with spin–orbit effects, respectively.

The X ground state is a $^1\Sigma^+_{0+}$ state arising from the $1\sigma^2 2\sigma^2 3\sigma^A 21\pi^4$ electronic configuration. The a state is the 1 component of $^3\Sigma^-$, while the A state is the second root of the 0^+ calculations. The A state was found to be a mixture of $^3\Pi_{0^+}$ and $^3\Sigma^-_{0+}$ with $^3\Pi_{0^+}$ making a predominant contribution. The C state was also found to be of 0^+ symmetry and corresponded to the third root of the 0^+ theoretical calculations (304).

The theoretical (304) D_e value for the $X0^+$ state of PbS was 3.83 eV, com-

Table 2.45 Spectroscopic Constants of PbS^+ and Ionization Potentials of PbS[a]

Method	State	R_e, Å	T_e, cm^{-1}	ω_e, cm^{-1}
RCI	$\frac{3}{2}$	2.56	0	363
RCI	$\frac{1}{2}$	2.56	506	335

Method	Splitting	IP, eV
RCI	$^1\Sigma^+ – ^2\Pi_{3/2}$	7.87
KT	$^1\Sigma^+ – ^2\Pi$	8.77
KT	$^1\Sigma^+ – ^2\Sigma^+$	8.98

[a]All results are from Reference 273. KT—Koopman's theorem.

pared to an experimental thermochemical value of 3.49 eV (32). Thus the theoretical D_e value is in very good agreement with the experimental value.

The theoretical SCF/RCI spectroscopic properties and ionization potentials of PbS^+ are shown in Table 2.45, while the potential-energy curves of the two lowest states of PbS^+ are shown in Figure 2.45. Again, because of electron correlation errors, the theoretical IP should be 15% less than the experimental value, and thus the IP of PbS should be about 9 eV.

Wang et al. (326), in a joint experimental–theoretical study, obtained the photoelectron spectra of PbTe, PbSe, SnTe, and SnSe. CASSCF/MRSDCI/RCI calculations including up to 200,000 configurations were also carried out on the $^1\Sigma^+$ ground states of all four species and the $^2\Pi_{3/2,1/2}$ and $^2\Sigma^+_{1/2}$ states of the positive ions. The theoretical calculations on these species facilitated interpretation of the experimental photoelectron spectra and construction of the potential-energy curves. All four positive ions exhibited avoided crossings of the $^2\Pi_{1/2}$ and $^2\Sigma^+_{1/2}$ components. The theoretical RCI potential-energy curves of PbTe and $PbTe^+$ are shown in Figure 2.46. The theoretical R_e and ω_e values were within 2–6% of the experimental values in all cases. The theoretical IPs of all these species were also found to be within 0.6–6% of the experimental results. The rather broad and diffuse nature of the observed spectral bands for the first two envelopes ($^2\Pi_{3/2} \leftarrow {}^1\Sigma^+$, $^2\Pi_{1/2} \leftarrow {}^1\Sigma^+$) compared to the third peak ($^2\Sigma^+_{1/2} \leftarrow {}^1\Sigma^+$) was explained by using theoretical potential-energy curves. The experimental and theoretical spectroscopic constants for PbSe and $PbSe^+$ are shown in Table 2.46 as an example illustrating the good agreement between theory and experiment.

2.3.5 SbO

The antimony oxide diatomic molecule has a long history of over 60 years. Since the reporting of the first spectrum of the antimony oxide molecule, numerous spectroscopic studies have appeared. Reference 32 summarizes all of the

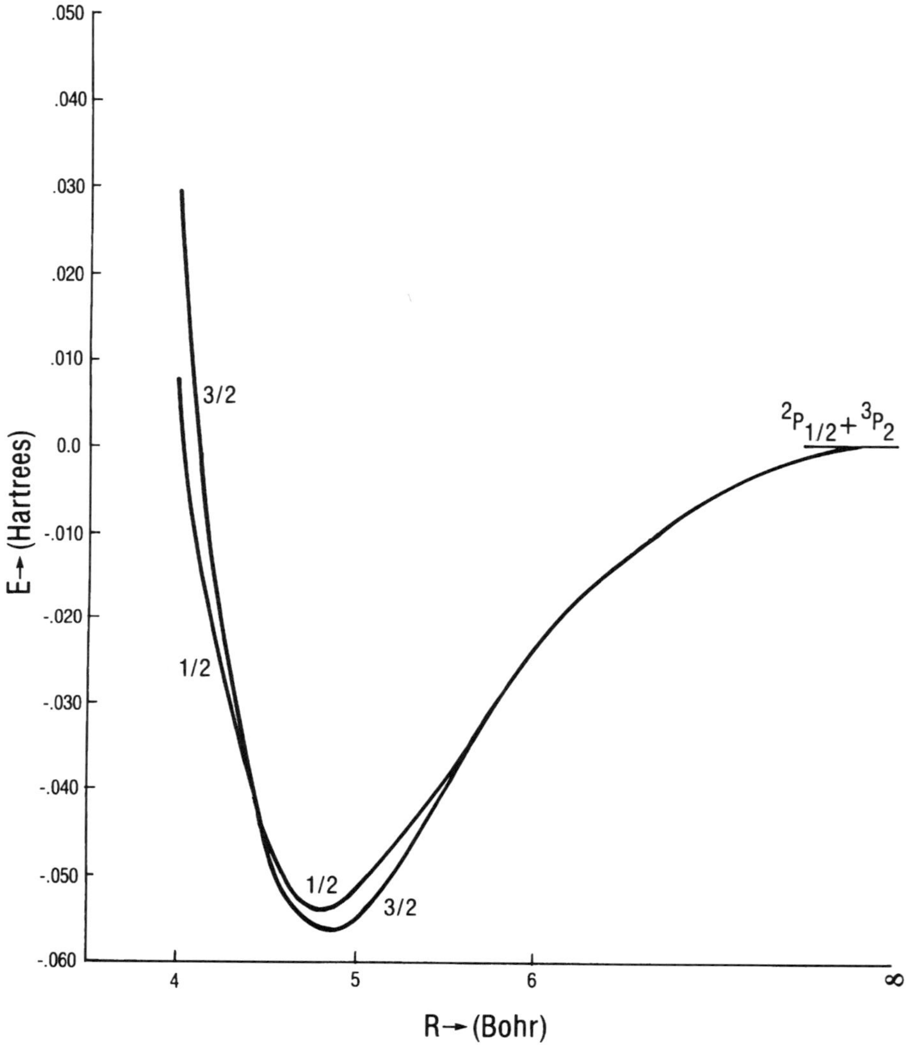

Figure 2.45 Potential-energy curves for the electronic states of PbS^+. (Reprinted from Ref. 273; copyright 1984, American Institute of Physics.)

works done on this molecule till 1977. More recent spectroscopic and theoretical studies can be found in References 372–375. Balfour and Ram (373, 374) reinvestigated the *A*–*X* and *C*–*X* systems and observed a new system in the emission denoted as *H*–*X* by these authors. The ground electronic state of SbO is unambiguously established as the $X^2\Pi$ state. Eleven excited electronic states lying below the 35,000-cm^{-1} region have been identified for SbO. Fink and Shestakov (375) reported a recent high-resolution spectroscopic study on SbO.

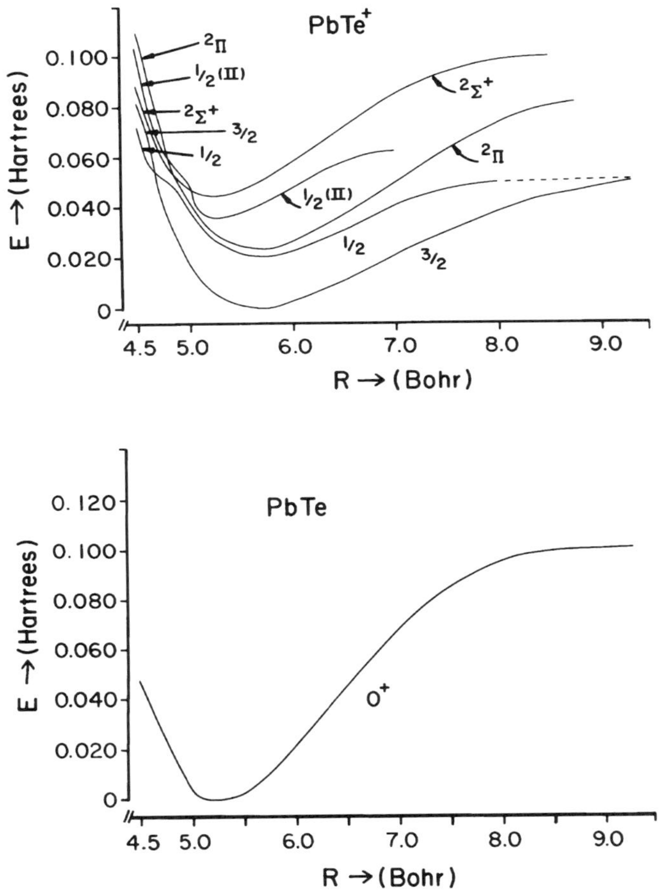

Figure 2.46 Potential-energy curves for the electronic states of PbTe and $PbTe^+$ with spin–orbit term. (Reprinted from Ref. 326; copyright 1989, American Institute of Physics.)

Quite recently, Alekseyev et al. (372) reported theoretical studies on the low-lying electronic states of SbO. In this section we shall discuss their computed results together with comparison of experimental constants for SbO. Alekseyev et al. (372) used SCF followed by MRDCI techniques including spin–orbit effects. They used relativistic RECPs for the antimony and oxygen atoms together with valence Gaussian basis sets.

The computed spectroscopic constants of Alekseyev et al. (372) together with the known experimental spectroscopic constants are shown in Table 2.47. As seen from this table, the agreement between the computed spectroscopic constants and the experimental values is very good. The bond lengths are slightly longer (0.01 Å), although only three of the electronic states in Table 2.47 have experimental bond lengths. The calculated vibrational frequencies are uniformly

Table 2.46 Spectroscopic Constants of PbSe and $PbSe^{+}$[a]

		R_e, Å		ω_e, cm^{-1}		IP(ad), eV		D_e, eV		IP(vert), eV	
Species	State	Theory	Exp.	Theory	Exp.	Theory	Exp.	Theory	Exp.	Theory	Exp.
PbSe	0^+	2.447	2.40	268	278	—	—	3.5	3.1	—	—
$PbSe^+$	$^2\Pi_{3/2}$	2.66	2.58	207	210	8.72	8.67	1.38	1.7	8.82	8.8
$PbSe^+$	$^2\Pi_{1/2}$	2.67	2.55	205	210	9.04	9.00	1.07	1.44	9.14	9.1
$PbSe^+$	$^2\Sigma^+_{1/2}$	2.53	2.45	222	242	9.64	9.54	—	—	—	—

[a] All values from Reference 326. The theoretical constants are from CASSCF/MRSDCI/RCI calculations; experimental results of the positive ion are from photoelectron spectra, while those of the neutral species are from Reference 32.

Table 2.47 Calculated and Experimental Spectroscopic Properties of SbO (Transition Energies T_e, Bond Lengths r_e, and Vibrational Frequencies ω_e)

State	T_e, cm^{-1} Calc.[c]	T_e, cm^{-1} Exp.	r_e, Å Calc.[c]	r_e, Å Exp.	ω_e, cm^{-1} Calc.[c]	ω_e, cm^{-1} Exp.
$X_1\,^2\Pi_{1/2}$	0	0	1.8364	1.8258[a]	754	816,[a] 819[b]
$X_2\,^2\Pi_{3/2}$	2,080	2,272[a,b]	1.8358	—	751	814[b]
$a_1\,^4\Pi_{3/2}$	14,207	16,862[d]	2.0666	—	542	558[d]
$a_2\,^4\Pi_{1/2}$	14,390	17,152[d]	2.0673	—	541	559[d]
$a_3\,^4\Pi_{5/2}$	14,535	—	2.0678	—	540	—
$a_4\,^4\Pi_{1/2}$	14,802	17,698[d]	2.0652	—	545	556[d]
$A_1\,^2\Pi_{3/2}$	19,476	20,668,[a] 20,794[b]	2.0627	—	566	569,[a] 570[b]
$A_2\,^2\Pi_{1/2}$	20,091	20,801,[a] 21,467[b]	2.0702	—	546	566[b]
$b_1\,^4\Sigma^-_{1/2}$	22,455	24,174[d]	1.9581	—	632	610[d]
$b_2\,^4\Sigma^-_{3/2}$	22,836	—	1.9473	—	611	—
$l_1\,^2\Phi_{7/2}$	24,695	—	2.0750	—	507	—
$l_2\,^2\Phi_{5/2}$	24,863	—	2.0790	—	500	—
$B\,^2\Sigma^{1+}_{1/2}$	26,616	26,494[a]	1.9418	—	557	582[a]
$C\,^2\Sigma^-_{1/2}$	29,683	29,747,[a] ~ 29,750[c]	2.0078	1.997[a]	591	571,[a] 569[c]
$H_1\,^2\Pi_{3/2}$	30,425	30,315,[a] ~ 30,495[c]	2.0865	—	495	~546[c]
$H_2\,^2\Pi_{1/2}$	31,371	—	2.0948	—	501	—
$J_1\,^2\Delta_{3/2}$	32,348	—	2.0023	—	681	—
$J_2\,^2\Delta_{5/2}$	32,399	—	1.9944	—	610	—
$c_1\,^4\Sigma^{1-}_{1/2}$	33,192	—	2.2492	—	533	—
$c_2\,^4\Sigma^{1-}_{3/2}$	34,041	—	2.1449	—	856	—
$D\,^2\Pi_{1/2}$	...	34,544[a]	...	2.073[a]	...	506[a]

[a]Reference 32.
[b]Reference 373.
[c]Reference 374.
[d]Reference 375.
[e]Calculated values from Reference 372.

smaller than the experimental values, a trend that is expected at the level of theory employed by Alekseyev et al. (372) and the use of full-core RECPs, which retained only the $5s^2 5p^3$ shells of the antimony atom in the valence space.

As seen from Table 2.47, the splitting between the $X_1(0^+)$ and $X_2(1)$ states is computed as 2080 cm^{-1} compared to the experimental value of 2272 cm^{-1}. Most of this splitting should come from the spin–orbit splitting of the antimony atom as the singly occupied p orbital of SbO is mostly on the antimony atom. The energy separations of the excited electronic states are computed with less accuracy as seen from Table 2.47. The T_e values differ by as much as 3000 cm^{-1}. This can be attributed primarily to the level of electron correlation treatment in the SCF/MRDCI techniques used by Alekseyev et al. (372). It is believed that the spin–orbit effects are accounted for more accurately compared to the electron correlation effects in this treatment. However, Alekseyev et al. (372) reported an impressive number of electronic states.

The spin–orbit splitting of the excited $A^2\Pi$ state is considerably smaller (615 cm^{-1}) primarily because the open-shell p orbital participating in this state is primarily on the oxygen atom. Since the spin–orbit splitting of the oxygen atom is considerably smaller, the splitting of the A spin–orbit states is also small.

Alekseyev et al. (372) computed the D_e of SbF to be in the range of 28,000–29,000 cm^{-1} or 3.5–3.6 eV. This value should be an underestimate because of the basis sets and the limited electron correlation effects in the MRDCI, which seems to be reasonable for the computation of the equilibrium properties compared to the dissociation energy. A spectroscopic value of 4.29 eV has been estimated from the predissociation of the $D^2\Pi$ state. An upper limit of 4.39 eV has also been estimated (32). Alekseyev et al. (372), however, consider their estimate of 3.5–3.6 eV accurate to within a few tenths of an electronvolt.

Alekseyev et al. (372) computed the radiative lifetimes of several electronic states of SbO, which are shown in Table 2.48. As seen from this table, the $X_2(^2P_{3/2)}$ and $b_2[^4S^-(\frac{3}{2})]$ states have relatively longer lifetimes compared to other excited electronic states. Fink et al. (375) measured the lifetimes of some of the electronic states of SbO experimentally. The $A^2\Pi$ excited state is found to have large Einstein coefficients due to parallel transitions to the ground state. The computed $v' = 0$ partial lifetimes of 17.7 and 11.5 μs compare quite favorably with the experimental values of 20 and 13 μs reported by Fink et al. Since the $^4\Pi$–$^2\Pi$ transition is spin-forbidden, the spin–orbit states of the $^4\Pi$ state have longer lifetimes. The $B^2\Sigma^+(\frac{1}{2})$ state has the shortest lifetime, since the transition from this state to the ground state is strongly allowed. However, the computed result is in large discrepancy with the experiment (a factor of 4). Alekseyev et al. (372) attribute this discrepancy to nonadiabatic effects arising from the overlap of the B state with the $b_1\ ^4\Sigma^-(\frac{1}{2})$ state. The $C^2\Sigma^-(\frac{1}{2})$ state possesses even shorter lifetime, as seen from Table 2.48. The computed value by Alekseyev et al. (372) agrees quite well with the experimental value of Fink et al. (377).

Table 2.48 Radiative Lifetimes (in seconds) of Excited States of SbO ($v' = 0$); Calculated Partial Lifetimes τ_1 and τ_2 for Transitions to $X_1\,^2\Pi_{1/2}$ and $X_2\,^2\Pi_{3/2}$, Respectively, as Well as Calculated and Experimental Total Lifetime τ

	Calculated[a]			Experiment[b]
State	τ_1	τ_2	τ	τ
$X_2\,^2\Pi_{3/2}$	0.29	—	0.29	—
$a_1\,^4\Pi_{3/2}$	6.4×10^{-3}	4.7×10^{-3}	2.7×10^{-3}	—
$a_2\,^4\Pi_{1/2}$	356×10^{-6}	37.6×10^{-3}	353×10^{-6}	190×10^{-6}
$a_3\,^4\Pi_{5/2}$	—	5.4×10^{-3}	5.4×10^{-3}	—
$a_4\,^4\Pi_{1/2}$	5.0×10^{-3}	10.0×10^{-3}	3.3×10^{-3}	—
$A_1\,^3\Pi_{3/2}$	1.8×10^{-3}	17.9×10^{-6}	17.7×10^{-6}	20×10^{-6}
$A_2\,^2\Pi_{1/2}$	11.6×10^{-6}	1.4×10^{-3}	11.5×10^{-6}	13×10^{-6}
$b_1\,^4\Sigma^-_{1/2}$	63.9×10^{-6}	352×10^{-6}	54×10^{-6}	$>20 \times 10^{-6}$
$b_2\,^4\Sigma^-_{3/2}$	0.27	144×10^{-6}	144×10^{-6}	—
$B^2\Sigma^+_{1/2}$	1.25×10^{-6}	1.9×10^{-6}	0.75×10^{-6}	3.0×10^{-6}
$C^2\Sigma^-_{1/2}$	0.60×10^{-6}	1.9×10^{-6}	0.45×10^{-6}	0.45×10^{-6}
$H_1\,^2\Pi_{3/2}$	12.7×10^{-6}	3.0×10^{-6}	2.4×10^{-6}	—
$H_2\,^2\Pi_{1/2}$	31.3×10^{-6}	80.7×10^{-6}	22.6×10^{-6}	—

[a]Reference 372.
[b]Reference 375.

2.3.6 BiO

Analogous to antimony oxide, bismuth oxide has a rich spectroscopic history dating back to 1927. The most recent experimental work due to Fink et al. (377) and the theoretical work due to Alekseyev et al. (376) have summarized the pertinent information on BiO. A review by Rai and Rai (379) summarizes the spectroscopic information on BiO up to 1984. Fink et al. (377) also investigated the lifetimes of some electronic states of BiO.

Bismuth oxide is quite interesting from a relativistic standpoint, due to the large ground state and excited state spin–orbit splitting. Analogous to BiF, the ground-state relativistic splitting of BiO is so large (7089 cm^{-1}) that one may think of the X_2 state as an entirely new electronic state, although both X_1 and X_2 states become degenerate in the absence of relativistic effects. Only those excited states of BiO that have open-shell orbitals, which are composed predominantly of the oxygen atom exhibit less spin–orbit splitting. The lifetimes of the excited electronic states of BiO are also in contrast with those of SbO, due primarily to relativistic effects.

Table 2.49 summarizes the theoretical and experimental information available up to now on the electronic states of BiO. The theoretical results are from the SCF/MRDCI computations of Alekseyev et al. (376), including spin–orbit coupling. The experimental constants are from different sources, as seen from Table 2.49. The calculated equilibrium bond distances are generally longer, and

Table 2.49 Calculated and Experimental Spectroscopic Properties of BiO (Transition Energies T_e, Bond Lengths r_e, and Vibrational Frequencies ω_e)

	T_e, cm^{-1}		r_e, Å		ω_e, cm^{-1}	
State	Calc.[c]	Exp.[a]	Calc.[c]	Exp.[a]	Calc.[c]	Exp.[b]
$X_1\,{}^2\Pi_{1/2}$	0	0	2.0464	1.934	648	692 (692)
$X_2\,{}^2\Pi_{3/2}$	6,809	7,089	2.0425	—	634	688
$A_1\,{}^4\Pi_{3/2}$	10,653	11,830	2.1864	—	599	550
$A_2\,{}^4\Pi_{1/2}$	12,258	14,195	2.2081	2.1426	498	505 (509)
$A_3\,{}^4\Pi_{5/2}$	15,445	—	2.1931	—	528	—
$A_4\,{}^4\Pi_{1/2}$	15,760	—	2.2088	—	497	—
$G^2\Pi_{3/2}$	18,688	20,178	2.1912	—	562	502
$H^2\Pi_{1/2}$	18,874	20,470	2.2005	—	481	472 (472)
$I^4\Sigma^-_{1/2}$	20,848	21,983	2.1810	—	536	507
$J^4\Sigma^-_{3/2}$	23,983	25,599	2.1588	—	507	490
$L_1\,{}^2\Phi_{5/2}$	24,045	—	2.2261	—	541	—
$L_2\,{}^2\Phi_{7/2}$	24,045	—	2.2413	—	477	—
$K^2\Sigma^+_{1/2}$	25,205	26,753	2.1772	—	449	420
$B^2\Sigma^-_{1/2}$	27,784	28,740	2.1892	2.09	479	483 (483)
$C^2\Delta_{3/2}$	29,465	30,210	2.1932	2.1102	451	492 (465)

[a]Reference 32.
[b]Reference 378 (values in parentheses are from Reference 32).
[c]Reference 376.

the vibrational frequencies are smaller compared to the experiment. Although Alekseyev et al. (376) attribute this to the use of full-core RECPs for the bismuth atom that retained only the outer $6s^2 6p^3$ shells, it is believed that the longer bond distances (almost 0.1 Å) are due to both RECPs and electron correlation effects, including the electron correlation corrections from the semicore $5d$ orbitals of the bismuth atom. More accurate computations that include correlation effects from the $5d$ are warranted. Indeed, such computations were done on TlF by the author, and it was shown that the correlation effects from the $5d$ shells are significant.

As seen from Table 2.49, the energy separations of 10 excited electronic states of BiO appear to be known. Overall, the agreement in the computed and experimental energy separations is quite good, as seen from this table. One of the most interesting outcomes of the theoretical computations is that the compositions of the A_1 and A_2 states are predominantly ${}^4\Pi$ rather than the previously assumed ${}^2\Pi$ state, although significant contributions from two excited ${}^2\Pi$ states are also present in this state. The G and H states are inverted $\frac{3}{2}$ and $\frac{1}{2}$ spin–orbit components arising from the $\sigma^2\pi^3\pi^{*2}$ configuration. The splitting between these states is quite small, due to the small spin–orbit splitting of the oxygen atom, as noted before. The same configuration also produces an $A^4\Pi$ state. Some of the excited states undergo avoided crossing as discussed by Alekseyev et al. (372).

Table 2.50 shows the radiative lifetimes of the excited electronic states of BiO($v' = 0$). The computed partial lifetimes for the transitions to the $X_1(\frac{1}{2})$ and $X_2(\frac{3}{2})$ states are also shown in that table. As seen from this table, the computed lifetimes are in large disagreement with the measured values. Alekseyev et al. (376) attribute this difference in part to the longer bond lengths (0.1 Å) at the level of theory employed by them. For the $X_2(\frac{3}{2})$ state a smaller computation that used only 37 × 37 secular equations yields a value of 8.2 ms, in contrast to the measured value of 0.48 ms reported by Fink and Shestakov (375). Alekseyev et al. (376) noted that improvement of their secular equations to 68 × 68 size yielded an improved value of 2.7 ms reported in Table 2.50, which is still in disagreement with the experiment.

Although the lifetimes computed by Alekseyev et al. (376) disagree with the experimental values in an absolute sense, they yield meaningful relative trends for the electronic states. The lifetime of the $A_3(\frac{5}{2})$ state is strikingly very long (6200 μs), while the lifetime of the other spin–orbit states arising from the same ^4P state are relatively short. This is primarily a consequence of the fact that the transition from the $\Omega = \frac{5}{2}$ state to the $X_1(\frac{1}{2})$ state is forbidden even

Table 2.50 Radiative Lifetimes of Excited States of BiO ($v' = 0$) Calculated Partial Lifetimes τ_1 and τ_2 for Transitions to $X_1\,^2\Pi_{1/2}$ and $X_2\,^2\Pi_{3/2}$, Respectively, as Well as Calculated and Experimental Total Lifetimes τ

	Calculated[a]			Experimental[b]
State	τ_1, μs	τ_2, μs	τ, μs	τ, μs
$X_2\,^2\Pi_{3/2}$	2,700	—	2,700	480
$A_1\,^4\Pi_{3/2}$	700	1,600	500	—
$A_2\,^4\Pi_{1/2}$	10.8	—	10.8	9.3
$A_3\,^4\Pi_{5/2}$	—	6,200	6,200	—
$A_4\,^4\Pi_{1/2}$	165	—	165	—
$G\,^2\Pi_{3/2}$	725	14.6	14.3	~15
$H\,^2\Pi_{1/2}$	8.3	—	8.3	15
$I\,^4\Sigma^-_{1/2}$	5.9	—	5.9	16
$J\,^4\Sigma^-_{3/2}$	250	25.3	12.6[c]	4.9
$K\,^2\Sigma^+_{1/2}$	1.0	76	0.96[c]	~2
$B\,^2\Sigma^-_{1/2}$	1.2	24.1	1.1[c]	0.55
$C\,^2\Delta_{3/2}$	1.6	1.2	0.68	0.81

[a]Reference 376.
[b]Reference 378.
[c]Other transitions have been included in the calculation of the total lifetime τ for this state.

relativistically, while the corresponding transitions from other spin–orbit states to the ground state are allowed. Thus, the dramatic variation of the lifetime of these states is due to relativity. Nonrelativistically, all the spin–orbit components of the ^4P state would have the same lifetime, while relativistically the lifetime of the $\omega = \frac{5}{2}$ state is dilated, while the lifetimes of the other spin–orbit states are relativistically contracted.

The $A_2^4\Pi(\frac{1}{2})$–$X_1^2\Pi(\frac{1}{2})$ transition moment was shown to be quite strong by Alekseyev et al. (376). Note that nonrelativistically this transition is forbidden by spin. Good agreement is found between theory and experiment, but this agreement is fortuitous considering large discrepancies in the bond lengths and the lifetime of the $X_2(\frac{3}{2})$ state. The G and H states have somewhat short lifetimes, which is confirmed by the experimental study of Fink et al. (377).

We note some contrasts in the lifetimes of the electronic states of BiO with those of SbO. Such a comparison also sheds some light on the nature of relativistic effects insofar as lifetimes are concerned. The $X_2(\frac{3}{2})$ state of the SbO has a computed lifetime of 0.29 s, compared to a value of 0.27 ms. Unfortunately, experimental lifetime for the $X_2(\frac{3}{2})$ state of SbO does not exist, since it is hard to measure this value. Even if one assumes the same level of disagreement between experiment and theory as noted for BiO (2.7 ms vs. 0.48 ms), the lifetime of the X_2 state of SbO is considerably longer than that of the corresponding state of BiO, which is primarily a consequence of relativity. While the $^4P_{5/2}$ states of SbO and BiO have similar lifetimes of 5400 and 6200 μs, this is not the case with the other spin–orbit states arising from ^4P. For example, the $^4P_{3/2}$ states of SbO and BiO have lifetimes of 2700 and 500 μs, respectively, indicating that the corresponding state of BiO is very short-lived, due to significant transition probability to the ground state. Likewise the two $^4P_{1/2}$ spin–orbit components have substantially different lifetimes for SbO and BiO, with the BiO states having considerably shorter lifetimes primarily due to relativity.

2.4 COMPARISON OF THE SPECTROSCOPIC PROPERTIES OF HEAVY HYDRIDES

In this section we compare the spectroscopic constants of heavy hydrides, obtain periodic trends, and explain the deviations observed, especially in the sixth-row hydrides. We focus principally on comparing the ground-state spectroscopic properties (R_e, ω_e, D_e, μ_e) and the energy separations (T_e) of select excited states with respect to the ground state.

Table 2.51 contains the ground-state spectroscopic constants of all the heavy hydrides in a periodic tabular form. There are no experimental or theoretical data available on PoH and HAt mainly because there are not many stable isotopes of these elements, although $^{210}Po_2$ has been studied (32). As seen from Table 2.51, interesting trends emerge in comparing the spectroscopic constants of even the ground states of these species. For example, the dissociation ener-

Table 2.51 Periodic Table of the Ground-State Properties of Hydrides[a]

GaH	GeH	AsH	SeH	HBr
$^1\Sigma^+$, 1.662 Å 1,612 cm^{-1}, 2.81 eV $\mu_e = +0.46$	$^2\Pi_{1/2}$, 1.62 Å 1,806 cm^{-1}, 2.34 eV $\mu_e = 0.09$	$^3\Sigma^-$, 1.534 Å 2,130 cm^{-1}, 2.8 eV $\mu_e = -0.2$	$^2\Pi_{3/2}$, 1.62 Å 2,371 cm^{-1}, 3.2 eV $\mu_e = -0.587$	$^2\Sigma^+$, 1.455 Å 2,645 cm^{-1}, 3.72 eV $\mu_e = -0.82$
InH	**SnH**	**SbH**	**TeH**	**HI**
$^1\Sigma^+$, 1.837 Å 1,048 cm^{-1}, 2.48 eV $\mu_e = +0.675$	$^2\Pi_{1/2}$, 1.83 Å 1,600 cm^{-1}, 2.3 eV $\mu_e = 0.398$	$^3\Sigma^-_{0+}$, 1.72 Å 1,763 cm^{-1}, 2.7 eV	$^2\Pi_{3/2}$, 1.76 Å 1,839 cm^{-1}, 2.31 eV	$^1\Sigma^+_{0+}$, 1.66 Å 2,939 cm^{-1}, 3.03 eV $\mu_e = -0.45$
TlH	**PbH**	**BiH**	—	—
$^1\Sigma^+_{0+}$, 1.95 Å 1,300 cm^{-1}, 1.81 eV $\mu_e = 1.231$	$^2\Pi_{1/2}$, 1.95 Å 1,418 cm^{-1}, 1.64 eV $\mu_e = 0.937$	$^3\Sigma^-_{0+}$, 1.81 Å 1,619 cm^{-1}, 2.2 eV	— — —	— — —

[a]The value in Å is R_e; the values in cm^{-1} and eV are ω_e and D_e, respectively. μ_e is the dipole moment in Debye units, with the convention of positive value, meaning that the positive charge is on the heavy atom.

Source: Reference 2.

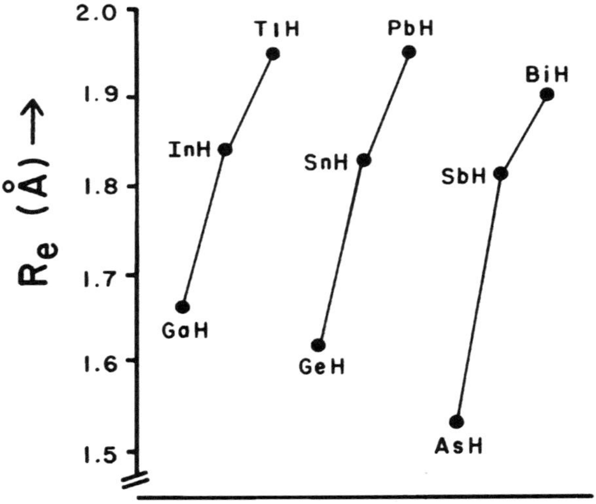

Figure 2.47 Plots of R_e values of the GaH–TlH, GeH–PbH, and AsH–BiH triads. (Reprinted from Ref. 2; copyright 1989, American Chemical Society.)

gies, D_e, decrease dramatically in going from the fifth row to the sixth row. The trends in R_e, D_e, μ_e, and so on are best understood by plotting these for the various groups.

Figure 2.47 shows the plot of R_e for each of the three groups for which a complete set of data is available. Note the upward tilt in the slope of the InH–TlH curve compared to the GaH–InH curve. Similar behaviors are observed in the GeH–PbH and AsH–BiH triads.

Figure 2.48 shows the plot of the D_e values for all three triads. The slopes tilt more downward for the D_e values in moving from the fifth row to the sixth row. This trend is approximately opposite of the trend in R_e values, since increases in R_e values imply weakening of the bond and, thus, D_e values should decrease correspondingly.

Figure 2.49 shows plots of the dipole moments for the GaH–TlH, GeH–PbH, and AsH–BiH triads. The dipole moments of InH, TlH, and HAt were obtained by Kim and the author (305) in an investigation that considered the periodicities in the dipole moments of these species. As seen from Figure 2.49 and Table 2.51, there is a marked increase in the dipole moment for TlH and PbH compared to their lighter analogs. This clearly demonstrates the increased electropositive character of the elements in the sixth row. All the above trends can be explained on the basis of relativistic effects and a comparison of the atomic properties of the elements in this group.

Figure 2.50 shows the $(ns^2np^1)^2P$–$(ns^1np^2)|^4P$ energy separations for B–Tl and C^+–Pb^+. All the atomic energy separations were taken from Reference 306. As seen from Figure 2.50, this separation increases up to Ga (Ge^+), then decreases, and again increases sharply in moving from In (Sn^+) to Tl (Pb^+).

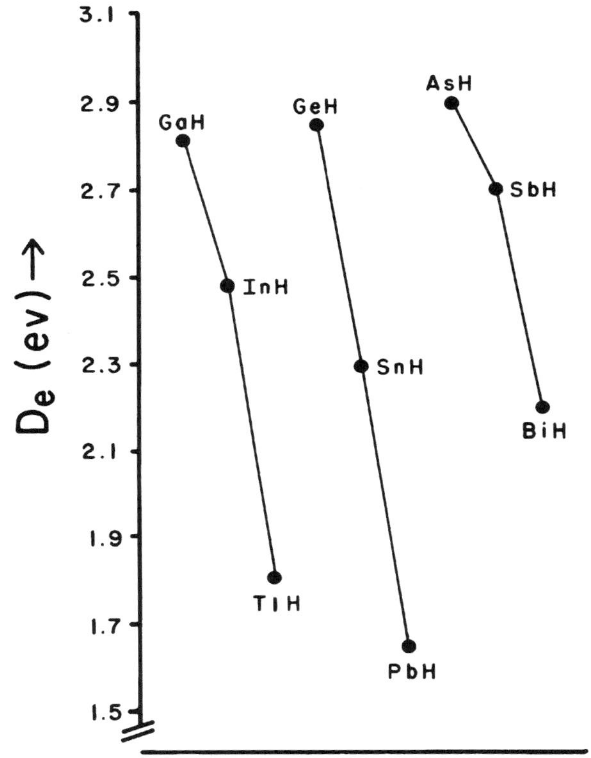

Figure 2.48 Plots of D_e values of the GaH–TlH, GeH–PbH, and AsH–BiH triads. (Reprinted from Ref. 2; copyright 1989, American Chemical Society.)

This is mainly a consequence of the "inert-pair effect," which is really due to the relativistic mass–velocity effect, considered in previous chapters. The relativistic contraction of the outer valence 6*s* orbital of these elements leads to an enhanced stability for the $6s^2$ shell, thereby making this shell somewhat inert compared to the other valence ns^2 shells of the lighter *p*-block elements.

The enhanced stability of the $6s^2$ shell (the inert-pair effect) manifests itself in many ways in the atomic and molecular properties. For example, the dramatic increase in the $(ns^2np^1)2P$–$(ns^1np^2)^4P$ energy separation of Tl compared to In (Fig. 2.50) is due to the relativistic stabilization of the $6s^2$ shell in Tl. Most of the observed trends in Figures 2.47–2.49 are attributed to this. The bonding in TlH and PbH is considerably weakened compared to that in their lighter analogs, since stronger bonds are formed through $ns^2np^1 - ns^1np^2$ hybridization.

The spin–orbit coupling is also very relevant to energy separations and bonding of very heavy molecules. Figure 2.51 shows the $^3P_0$–$^3P_2$ energy separations of elements C through Pb. As seen from this figure, the spin–orbit splitting goes up dramatically as one moves down the periodic table from Sn to Pb.

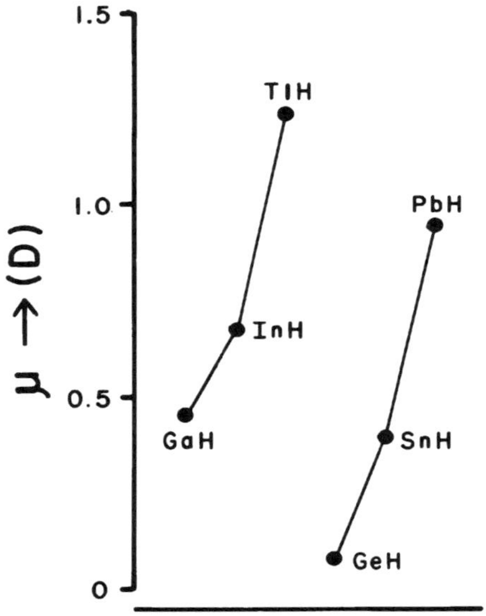

Figure 2.49 Plots of dipole moments of the GaH–TlH, GeH–PbH, and AsH–BiH triads. (Reprinted from Ref. 2; copyright 1989, American Chemical Society.)

The changes in the spectroscopic constants in Table 2.51 are due in part to the spin–orbit splitting.

Table 2.52 shows the spin–orbit splittings for the ground states of the GeH–PbH to SeH–TH species. As seen from Table 2.52, the spin–orbit splittings increase dramatically in moving from SnH to PbH and SbH to BiH. The spin–orbit effects are so large in BiH and PbH that they cause many avoided

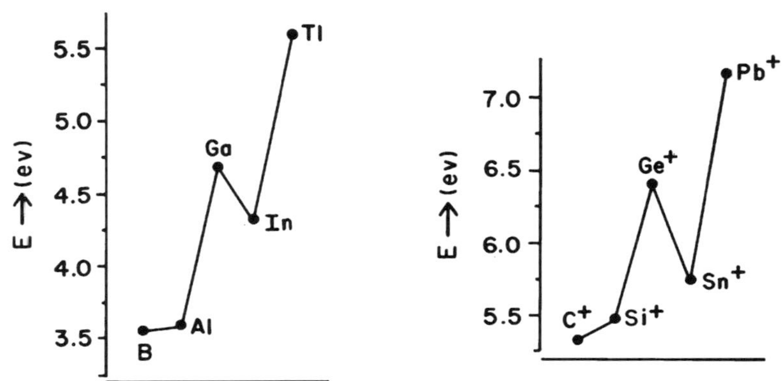

Figure 2.50 $(ns^2np^1)^2P$–$(ns^1np^2)^4P$ energy separations of B–Tl and C^+–Pb^+. (Reprinted from Ref. 13; copyright 1989, American Chemical Society.)

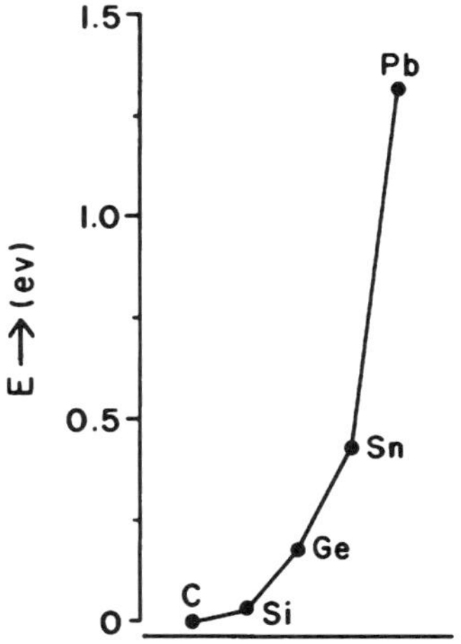

Figure 2.51 $^3P_0$–$^3P_2$ spin–orbit energy separations of C–Pb. (Reprinted from Ref. 13; copyright 1989, American Chemical Society.)

crossings in the potential-energy curves of these species. These features were discussed in section 2.1.13.

Next we compare the energy separations of a select low-lying excited electronic state and the ground state for the hydrides. Figure 2.52 shows the $^3\Pi_{0^+}$–$^1\Sigma^+_{0^+}$ energy separations of GaH–TlH. As seen from this figure, there is a pronounced increase in this energy splitting in moving from InH to TlH, whereas it decreases in moving from GaH to InH. In general, energy separations of the corresponding states should decrease with increase in principal quantum number in the absence of spin–orbit and other relativistic effects. This trend is obeyed by GaH and InH pairs. The anomaly in TlH is mainly a consequence of the relativistic effects discussed earlier.

Table 2.52 Periodic Table of Spin–Orbit Splittings for Heavy Hydrides

GeH $^2\Pi_{1/2}$–$^2\Pi_{3/2}$ 869 cm^{-1}	AsH $^3\Sigma^-_{0+}$–$^3\Sigma^-_1$ 59 cm^{-1}	SeH $^2\Pi_{3/2}$–$^2\Pi_{1/2}$ 1,934 cm^{-1}
SnH $^2\Pi_{1/2}$–$^2\Pi_{3/2}$ 2,180 cm^{-1}	SbH $^3\Sigma^-_{0+}$–$^3\Sigma^-_1$ 655 cm^{-1}	TeH $^2\Pi_{3/2}$–$^2\Pi_{3/2}$ 3,900 cm^{-1}
PbH $^2\Pi_{1/2}$–$^2\Pi_{3/2}$ 6,946 cm^{-1}	BiH $^3\Sigma^-_{0+}$–$^3\Sigma^-_1$ 5,000 cm^{-1}	— — —

Source: Reference 2.

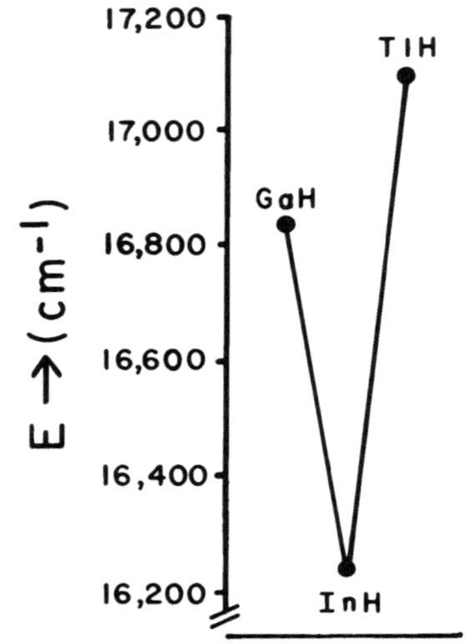

Figure 2.52 $^3\Pi_{0^+}-^1\Sigma^+$ energy separations for GaH–TlH. (Reprinted from Ref. 2; copyright 1989, American Chemical Society.)

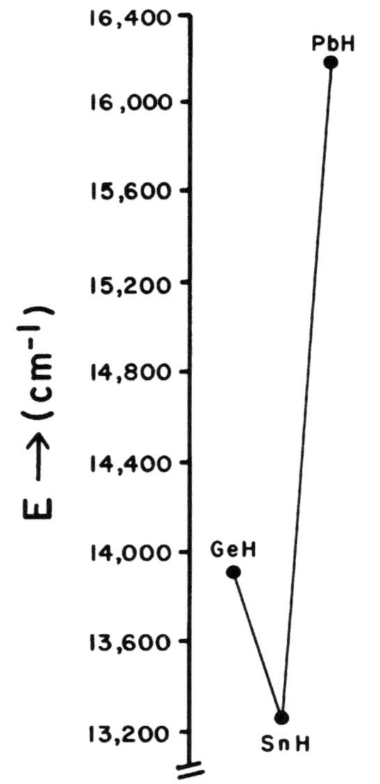

Figure 2.53 $^4\Sigma^--^2\Pi_{1/2}$ energy separations for GeH–PbH. (Reprinted from Ref. 2; copyright 1989, American Chemical Society.)

Figure 2.53 shows the $^4\Sigma^-_{1/2}$–$^2\Pi_{1/2}$ energy separations of the GeH–PbH triad. Again this separation decreases in moving from GeH to SnH but increases dramatically in moving from SnH to PbH. The spin–orbit effects are also considerably larger for PbH. Figure 2.54 shows the $^3\Pi_{0^+}$–$^3\Sigma^-_{0^+}$ separations for the AsH–BiH triad. For BiH, this state is so contaminated by the spin–orbit coupling that the designation $^3\Pi_{0^+}$ is inappropriate. In the original manuscript (180) this state was designated as 0^+(IV) (fourth root of the O^+ RCI calculations). This energy separation once again dramatically increases in moving from SbH to BiH, while it is not substantially different for SbH compared to AsH. The

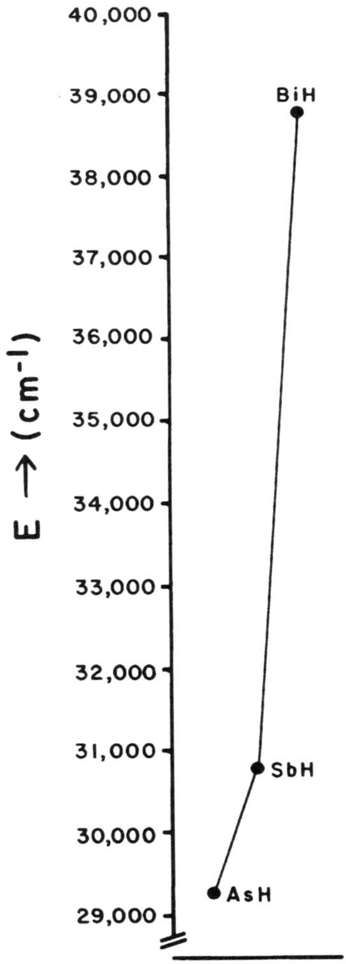

Figure 2.54 $^3\Pi_{0^+}$–$^3\Sigma^-_{0^+}$ energy separations for AsH–BiH. (Reprinted from Ref. 2; copyright 1989, American Chemical Society.)

dramatic increase in BiH is due to both mass–velocity stabilization effects and the large spin–orbit effects. In any event, relativistic effects make a dramatic impact in the sixth-row compounds.

The comparison of the properties of HBr and HI is also of interest. The $^3\Pi_{0^+}$–$X^1\Sigma^+$ vertical energy separations of the two species are 73,042 and 49,605 cm^{-1}. The $^3\Pi_{0^+}$ states of both species are repulsive, and thus vertical separations are compared. The much smaller energy separation for HI is consistent with the experimentally observed electronic spectra of these two species.

2.5 COMPARISON OF THE SPECTROSCOPIC PROPERTIES OF HEAVY HALIDES

Table 2.53 comprises the ground-state spectroscopic constants of heavy fluorides in periodic tabular form. The spectroscopic properties of many low-lying electronic states of TlF, PbF, and BiF were calculated by the author (198, 208, 232). O'Hare et al. (307) carried out theoretical calculations on the ground state of AsF. The dissociation energy (D_0°) in Table 2.53 for AsF is from this reference. These authors also calculated the electronic dipole moment of AsF as 1.75 D and the electron affinity as 1.1 eV. The level of electron correlation treatment available then was not adequate enough to calculate the dipole moments and electron affinities, in general, accurately.

As seen from Table 2.53, the D_e values uniformly drop in moving from GaF to TlF. The experimental electronic dipole moments of GaF–TlF are available (see Table 2.53). Comparison of the dipole moments of these three species clearly reveals the increased electropositivity of the thallium atom in this group.

Table 2.53 Periodic Table of Known Spectroscopic Constants of Heavy Fluorides[a]

GaF $^1\Sigma^+$, 1.774 Å 622 cm^{-1}, 5.98 eV μ_e = +2.45 D	GeF $^2\Pi_{1/2}$, 1.745 Å 666 cm^{-1}, 5.0 eV	AsF $^3\Sigma^-_{0+}$, 1.736 Å 686 cm^{-1}, 4.2 eV
InF $^1\Sigma^+$, 1.985 Å 535 cm^{-1}, 5.25 eV μ_e = +3.40 D	SnF $^2\Pi_{1/2}$, 1.944 Å 578 cm^{-1}, 4.90 eV	SbF $^3\Sigma^-_{0+}$, 1.92 Å 706 cm^{-1}, 4.4 eV
TlF $^1\Sigma^+$, 2.04 Å 592 cm^{-1}, 4.57 eV μ_e = 4.20 D	PbF $^2\Pi_{1/2}$, 2.08 Å 540 cm^{-1}, 3.75 eV	BiF $^3\Sigma^-_{0+}$, 2.09 Å 502 cm^{-1}, 2.63 eV

[a]The value in Å is Re; the values in cm^{-1} and eV are ω_e and D_e, respectively. μ_e is the dipole moment in Debye with positive value meaning positive change at the heavy atom.

Source: Reference 2.

Comparison of the spin–orbit coupling constants is also of considerable interest. Table 2.54 shows the spin–orbit splittings of the GeF–PbF and AsF–BiF triads. The $^2\Pi_{1/2}$–$^2\Pi_{3/2}$ spin–orbit splitting dramatically increases from 2317 to 8264 cm^{-1} in moving from SnF to PbF, while the corresponding $^3\Sigma^-_{0+}$–$^3\Sigma^-_1$ splitting increases from 796 to 9216 cm^{-1} in moving from SbF to BiF. Thus the spin–orbit effects are substantially larger for the sixth-row fluorides compared to the fifth-row fluorides.

The energy separations of low-lying excited electronic states with respect to the ground state are also of considerable interest. Figure 2.55 compares the $^3\Pi_{0^+}$–$^1\Sigma^+_{0+}(X)$ energy separations of the GaF–TlF triad. All three energy separations are from experimental spectroscopic data. As seen from Figure 2.55, this energy separation decreases in moving from GaF to InF but increases in moving from InF to TlF. The $^3\Pi_{0^+}$ state arises from a regular $^3\Pi$ and, thus the 0^+ component should be lower than $^3\Pi_1$. The main difference between TlF and InF should be in the magnitude of spin–orbit contamination. For TlF, the author (198) found that both the ground state ($^1\Sigma^+_{0+}$) and the $^3\Pi_{0^+}$ state were actually mixed. That is, the $^3\Pi_{0^+}$ state was found to be a mixture of $^3\Pi_{0^+}$ and $^1\Sigma^+_{0+}$. This mixing lowers the $^1\Sigma^+_{0+}$ state but increases the energy of the $^3\Pi_{0^+}$ state. The larger $^3\Pi_{0^+}$–$^1\Sigma^+_{0+}$ splitting for TlF should thus be due to spin–orbit contamination.

Figure 2.56 shows the $^4\Sigma^-_{1/2}$–$X^2\Pi_{1/2}$ energy splittings for GeF–PbF. Again, this splitting decreases in moving from GeF to SnF but increases in moving from SnF to PbF. Figure 2.57 shows the $^1\Sigma^+_{0+}$–$X^3\Sigma^-_{0+}$ energy separations of the AsF–BiF triad. Note the dramatic increase in this energy separation for BiF. It was very clear that the large change in BiF is a consequence of spin–orbit contamination. For BiF, the $X^3\Sigma^-_{0+}$ ground state was found (232) to be 66% $^3\Sigma^-$ and 20% $^1\Sigma^+_{0+}$. The state labeled $A0^+$(II) was found to be 58% $^1\Sigma^+_{0+}$, 23% $^3\Sigma^-_{0+}$, and 7% $^3\Pi_{0^+}$. This substantial mixing of different states lowers the energy of

Table 2.54 Spin–Orbit Coupling Constants for the GeF–PbF and AsF–BiF Triads

GeF	AsF
$^2\Pi_{1/2}$–$^2\Pi_{3/2}$	$^3\Sigma^-_{0+}$–$^3\Sigma^-_1$
934 cm^{-1}	139 cm^{-1}
SnF	SbF
$^2\Pi_{1/2}$–$^2\Pi_{3/2}$	$^3\Sigma^-_{0+}$–$^3\Sigma^-_1$
2,317 cm^{-1}	796 cm^{-1}
PbF	BiF
$^2\Pi_{1/2}$–$^2\Pi_{3/2}$	$^3\Sigma^-_{0+}$–$^3\Sigma^-_1$
8,264 cm^{-1}	9,216 cm^{-1}

Source: Reference 2.

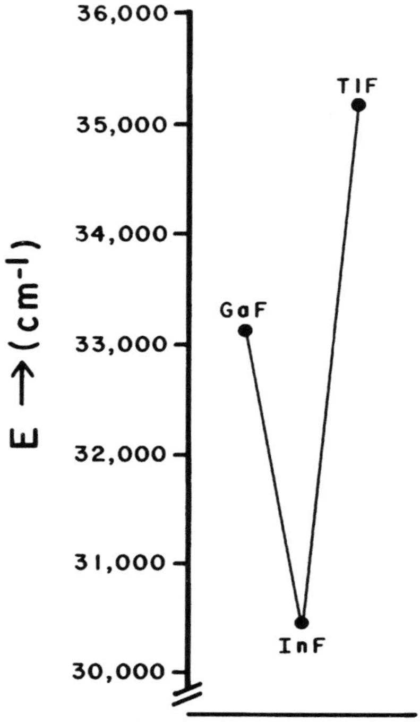

Figure 2.55 $^3\Pi_{0^+}-^1\Sigma^+$ energy separations for GaF–TlF. (Reprinted from Ref. 2; copyright 1989, American Chemical Society.)

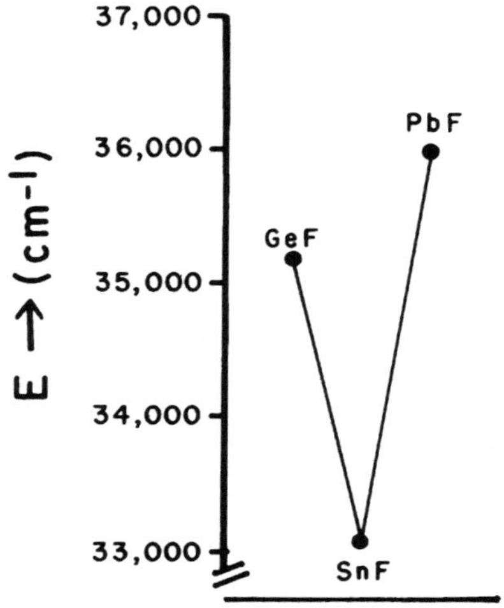

Figure 2.56 $^4\Sigma^- - ^2\Pi_{1/2}$ energy separations for GeF–PbF. (Reprinted from Ref. 2; copyright 1989, American Chemical Society.)

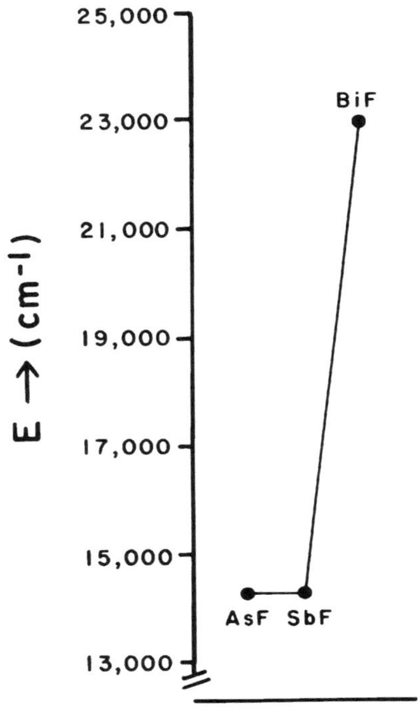

Figure 2.57 $^1\Sigma^+_{0^+}$–$^3\Sigma^-_{0^+}$ energy separations for AsF–BiF. (Reprinted from Ref. 2; copyright 1989, American Chemical Society.)

the $^3\Sigma^-_{0^+}(X)$ ground state and raises the energy of the $^1\Sigma^+_{0^+}$ state substantially, leading to a much larger energy separation for BiF.

Although all the heavy chlorides have not been studied theoretically, the spectroscopic properties of PbCl and SnCl have been obtained (238–249). A comparison of these two species seems to provide some interesting information. For both SnCl and PbCl the ground state is of $^2\Pi_{1/2}$ symmetry. The spin–orbit splitting for the $^2\Pi$ state is 2357 cm^{-1} for SnCl and 8272 cm^{-1} for PbCl (experimental). The bond length of the ground state of PbCl (238) is 2.61 Å, compared to a value of 2.48 Å for SnCl (249). The $^2\Sigma^+_{1/2}$–$^2\Pi_{1/2}$ energy separations for SnCl and PbCl are 19,415 and 21,865 cm^{-1}, respectively. For GeCl, this energy separation, although not known precisely, since the observed bands are diffuse, should be about 22,600 cm^{-1} (32). Again, the trend observed for energy separations in fluorides is observed for chlorides.

2.6 COMPARISON OF HEAVY CHALCOGENIDES

Since SCF/RCI calculations have been completed only on SnO, SnS, PbO, and PbS species, the comparison of chalcogenides here is restricted to oxides and

Table 2.55 Comparison of Ground-State Properties of Heavy Group IV Oxides and Sulfides[a,b]

GeO	GeS
$^1\Sigma^+$, 1.625 Å	$^1\Sigma^+$, 2.01 Å
986 cm^{-1}, 6.78 eV	576 cm^{-1}, 5.67 eV
μ = 3.27 D	μ = 2.00 D
SnO	SnS
$^1\Sigma^+$, 1.83 Å	$^1\Sigma^+$, 2.21 Å
815 cm^{-1}, 5.49 eV	487 cm^{-1}, 4.77 eV
μ = 4.32 D	μ = 3.17 D
PbO	PbS
$^1\Sigma^+$, 1.922 Å	$^1\Sigma^+$, 2.29 Å
721 cm^{-1}, 3.83 eV	429 cm^{-1}, 3.49 eV
μ = 4.64 D	μ = 3.59 D?, 4.02 D?

[a]The value in Å is R_e; the values in cm^{-1} and eV are ω_e and D_e, respectively. μ_e is the dipole moment in Debye units, with the convention of positive value meaning that positive charge is on the metal atom.

[b]Two values are listed for the dipole moment of PbS. The 3.59 value is from Reference 308, while the 4.02 value is from Reference 302.

Source: Reference 2.

sulfides. For comparison, the author took the available experimental information (32) on GeO and GeS. Table 2.55 lists the spectroscopic constants on the ground states of the GeO–PbO and GeS–PbS triads in a periodic tabular form. As seen from this table, the dissociation energies D_e decrease uniformly in going down the periodic table, although the decrease is larger in comparing PbO with SnO and PbS with SnS.

The experimental dipole moments of the heavy group IV oxides and sulfides are available for comparison. The dipole moments of these species in general increase as one goes down the periodic table. This is consistent with the theoretical trend for the hydrides (GeH–PbH). For both oxides and sulfides the polarity of the bond is M^+O^-. The experimental dipole moment of PbS appears to be somewhat uncertain (see Table 2.55). The value of 3.59 D was obtained from the Stark effect of the rotation spectrum by Hoeft et al. (308), while the value of 4.02 D was obtained by Murty and Curl (302) around the same time. In any case, the dipole moments of sulfides are lower than those of oxides, as expected. In general, the sulfides have about 1.2–1.3 D lower electronic dipole moments than oxides. Since the dipole moment of PbO is 4.64 D, we predict that the dipole moment of PbS should be about 3.4–3.5 D. Thus the value obtained by Hoeft et al. (308) through the Stark effect of the rotational spectra (3.59 D) appears to be more reliable. This is further supported by the fact that the difference in μ_e values of SnO and PbO should be approximately the same as that of the SnS and PbS pairs. Thus the μ_e value of PbS should be 0.32 D

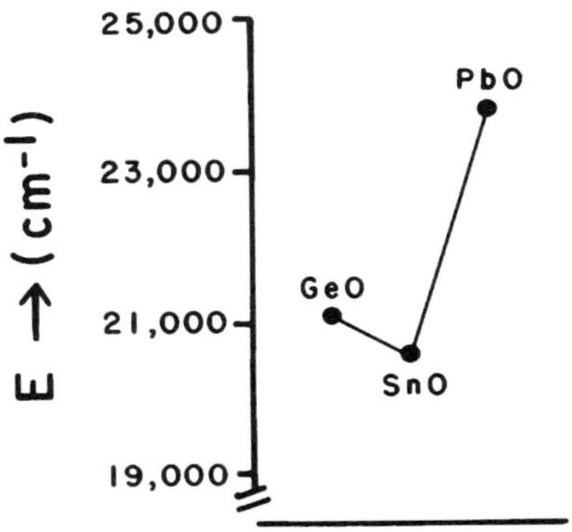

Figure 2.58 $^3\Sigma^-_{0^+}-{}^1\Sigma^+_{0^+}$ energy separations for GeO–SnO. (Reprinted from Ref. 2; copyright 1989, American Chemical Society.)

larger than the corresponding value of SnS or 3.49 D. All these arguments seem to support the μ_e obtained by Hoeft et al. (308).

Next we compare the energy separations of some low-lying electronic states of oxides and sulfides. The comparisons of the $^3\Pi_{0^+}$–$X^1\Sigma^+_{0^+}$ and $^3\Sigma^-_{0^+}$–$X^1\Sigma^+_{0^+}$ energy separations seem to provide interesting trends.

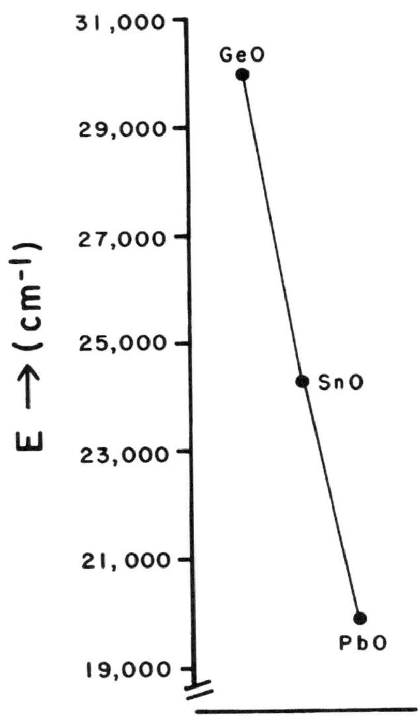

Figure 2.59 $^3\Pi_{0^+}-{}^1\Sigma^+$ energy separations for GeO–PbO. (Reprinted from Ref. 2; copyright 1989, American Chemical Society.)

Figure 2.58 shows a comparison of the ${}^3\Sigma^-_{0^+}$–${}^1\Sigma^+_{0^+}$ energy separations for the GeO–PbO triad. The ${}^3\Sigma^-_{0^+}$ state arises from the $\pi^3\pi^*$ configuration, where the π^* orbital consists predominantly of the heavy atom, while the π orbital is composed predominantly of the oxygen atom. Note the increase in the energy separation in moving from SnO to PbO. Figure 2.59 shows the ${}^3\Pi_{0^+}$–${}^1\Sigma^+_{0^+}$ adiabatic energy separations (T_e) for the GeO–PbO triad. The T_e value for GeO is approximate (32). The trend is somewhat different for the ${}^3\Pi_{0^+}$–${}^1\Sigma^+_{0^+}$ energy separation. This separation monotonically decreases in moving from GeO to PbO. The ${}^3\Pi$ state arises from the $\sigma\pi^4\pi^*$ configuration, where the a orbital is the bonding orbital consisting of the metal p_z orbital and the oxygen p_z orbital. The main difference between the ${}^3\Sigma^-$ and ${}^3\Pi$ states is that ${}^3\Sigma^-$ arises from promotion of a nonbonding oxygen electron into a nonbonding metal orbital. Thus the trend is dominated by relativistic effects. The ${}^3\Pi$ state, on the contrary, arises from the promotion of a bonding σ electron into a nonbonding metal orbital. Since the σ orbital itself consists of the heavy-atom orbital, the relative trend is not so much determined by relativistic effects. It is rather determined by electron correlation effects and the general trend of decrease in energy separations as one goes down the periodic table.

REFERENCES

1. K. S. Pitzer, *Acc. Chem. Res.* **12,** 271 (1979); *Internatl. J. Quantum Chem.* **25,** 131 (1984).
2. K. Balasubramanian, *Chem. Rev.* **89,** 1801 (1989).
3. P. Pyykkö and J. P. Desclaus, *Acc. Chem. Res.* **12,** 276 (1979).
4. M. Krauss and W. J. Stevens, *Annu. Rev. Phys. Chem.* **35,** 357 (1984).
5. P. Pyykkö, *Adv. Quantum Chem.* **11,** 353 (1978).
6. P. Pyykkö, ed., "Proceedings of the Symposium on Relativistic Effects in Quantum Chemistry," *Internatl. J. Quantum Chem.* **25** (1984).
7. P. A. Christiansen, W. C. Ermler, and K. S. Pitzer, *Annu. Rev. Phys. Chem.* **36,** 407 (1985).
8. K. Balasubramanian and K. S. Pitzer, *Adv. Chem. Phys.* **67,** 287 (1987).
9. P. Pyykkö, *Chem. Rev.* **88,** 563 (1988).
10. W. H. E. Schwarz, in *Theoretical Models of Chemical Bonding*, Z. Maksic, ed., Springer-Verlag, Berlin, 1990.
11. G. Malli, *Stud. Phys. Theor. Chem.* **21,** 199 (1982).
12. P. Pyykkö, *Relativistic Theory of Atoms and Molecules*, Springer-Verlag, Berlin, New York, 1986.
13. K. Balasubramanian, "Feature Article," *J. Phys. Chem.* **93,** 6585 (1989).
14. K. Balasubramanian, *J. Molec. Struct.* (*Theochem.*) (special issue) **102,** 291 (1989).
15. Y. S. Lee, W. C. Ermler, and K. S. Pitzer, *J. Chem. Phys.* **15,** 5861 (1977).

16. Y. S. Lee, W. C. Ermler, and K. S. Pitzer, *J. Chem. Phys.* **73,** 360 (1980).
17. P. A. Christiansen, Y. S. Lee, and K. S. Pitzer, *J. Chem. Phys.* **71,** 4445 (1979).
18. Y. S. Lee, W. C. Ermler, and K. S. Pitzer, and A. D. McLean, *J. Chem. Phys.* **70,** 288 (1979).
19. L. Kahn, P. Baybutt, and D. G. Truhlar, *J. Chem. Phys.* **65,** 3826 (1976).
20. W. C. Ermler, Y. S. Lee, P. A. Christiansen, and K. S. Pitzer, *Chem. Phys. Lett.* **71,** 70 (1981).
21. L. F. Pacios, and P. A. Christiansen, *J. Chem. Phys.* **82,** 2664 (1985).
22. M. M. Hurley, L. F. Pacios, P. A. Christiansen, R. B. Ross, and W. C. Ermler, *J. Chem. Phys.* **84,** 6840 (1986).
23. L. A. LaJohn, P. A. Christiansen, R. B. Ross, T. Atashroo, and W. C. Ermler, *J. Chem. Phys.* **87,** 2812 (1987).
24. J. J. Cohen, W. R. Wadt, and P. J. Hay, *J. Chem. Phys.* **71,** 2955 (1979).
25. P. J. Hay and W. R. Wadt, *J. Chem. Phys.* **82,** 270 (1985).
26. W. R. Wadt and P. J. Hay, *J. Chem. Phys.* **82,** 284 (1985).
27. P. J. Hay and W. R. Wadt, *J. Chem. Phys.* **82,** 299 (1985).
28. P. A. Christiansen, K. Balasubramanian, and K. S. Pitzer, *J. Chem. Phys.* **76,** 5087 (1982).
29. K. Balasubramanian and K. S. Pitzer, *J. Chem. Phys.* 78, 321 (1983).
30. K. S. Pitzer and K. Balasubramanian, *J. Phys. Chem.* **86,** 3068 (1982).
31. W. J. Stevens, H. Basch, and M. Krauss, *J. Chem. Phys.* **81,** 6026 (1984).
32. K. P. Huber, and G. Herzberg, *Molecular Spectra and Molecular Structue.* IV. *Constants of Diatomic Molecules*, Van Nostrand Reinhold, New York, 1979.
33. W. R. S. Garton, *Proc. Phys. Soc.* (London) **A64,** 509 (1951).
34. H. Neuhaus, *Nature* **180,** 433 (1957).
35. H. Neuhaus, *Ark. Fys.* **14,** 551 (1959).
36. M. L. Ginter and K. K. Innes, *J. Molec. Spectrosc.* **7,** 64 (1967).
37. M. L. Ginter and R. J. Battino, *Chem. Phys.* **42,** 3222 (1965).
38. P. C. Poynor, K. K. Innes, and M. L. Ginter, *J. Molec. Spectrosc.* **23,** 237 (1967).
39. M. Kronekvist, A. Langerquist, H. Neuhaus, *J. Molec. Spectrosc.* **39,** 516 (1971).
40. T. Larsson and H. Neuhaus, *Ark. Fys.* **23,** 461 (1963).
41. T. Larsson and H. Neuhaus, *Ark. Fys.* **31,** 299 (1966).
42. G. B. Kim and K. Balasubramanian, *J. Molec. Spectrosc.* **134,** 412 (1989).
43. L. G. M. Pettersson, and S. R. Langhoff, *J. Chem. Phys.* **85,** 3130 (1986).
44. B. Kleman, and E. Werhagen, *Ark. Fys.* **6,** 359 (1953).
45. B. Kleman, and E. Werhagen, *Ark. Fys.* **6,** 399 (1953).
46. R. F. Barrow, G. Drummond, and W. R. S. Garton, *Proc. Phys. Soc.* (London) **66A,** 191 (1953).
47. H. G. Howell, *Proc. Phys. Soc.* (London) **57,** 37 (1945).
48. L. Klynning and B. Lindgren, *Ark. Fys.* **32,** 575 (1966).
49. J. F. Osmudsen, C. C. Abele, and J. G. Eden, *J. Appl. Phys.* **57,** 2921 (1985).
50. J. F. Osmudsen, C. C. Abele, and J. G. Eden, *J. Chem. Phys.* **83,** 2139 (1985).

51. J. Perrin and J. F. M. Aarts, *Chem. Phys.* **80,** 351 (1983).
52. J. M. Brown, K. M. Evenson, and T. J. Sears, *J. Chem. Phys.* **83,** 3275 (1985).
53. L. G. M. Pettersson and S. R. Langhoff, *Chem. Phys. Lett.* **125,** 429 (1986).
54. H. J. Werner and A. D. Buckingham, *Chem. Phys. Lett.* **125,** 433 (1986).
55. D. A. Chapman, J. Li, K. Balasubramanian, and S. H. Lin, *J. Chem. Phys.* **88,** 3826 (1988).
56. L. Veseth, *J. Molec. Spectrosc.* **48,** 283 (1973).
57. K. Balasubramanian and J. Li, *J. Molec. Spectrosc.* **128,** 413 (1988).
58. R. N. Dixon and H. M. Lamberton, *J. Molec. Spectrosc.* **25,** 12 (1968).
59. B. Lindgren, *Phys. Scripta* **12,** 164 (1975).
60. J. M. Brown and A. D. Fackerell, *Phys. Scripta* **25,** 351 (1982).
61. K. Balasubramanian, and V. Nannegari, *J. Molec. Spectrosc.*, **138,** 482 (1989).
62. R. E. Radford, *J. Chem. Phys.* **40,** 2732 (1964).
63. A. Carrington, G. N. Currie, and N. J. D. Lucas, *Proc. Roy. Soc., Ser. A* **A315,** 355 (1970).
64. B. Lindgren, *J. Molec. Spectrosc.* **28,** 536 (1968).
65. B. J. Donovan, D. J. Little, and J. Konstantatos, *J. Chem. Soc., Faraday Trans. 2* **68,** 1812 (1972).
66. C. K. Smyth, J. I. Brauman, and I. John, *J. Chem. Phys.* **56,** 5993 (1972).
67. P. Bollmark, B. Lindgren, B. Rydh, and V. Sassenberg, *Phys. Scripta* **17,** 561 (1978).
68. P. Bollmark, B. Lindgren, and V. Sassenberg, *Phys. Scripta* **21,** 811 (1980).
69. D. I. Cliff, P. B. Davies, B. J. Handy, and B. A. Thrush, *Chem. Phys. Lett.* **75,** 9 (1980).
70. J. M. Brown, A. Carrington, and T. Sears, *Molec. Phys.* **37,** 1837 (1979).
71. J. M. Brown, A. Carrington, and A. D. Fackerell, *Chem. Phys. Lett.* **75,** 13 (1980).
72. J. M. Brown and A. D. Fackerell, *Phys. Scripta* **25,** 351 (1982).
73. S. T. Gibson, J. P. Freene, and J. Berkowitz, *J. Chem. Phys.* **85,** 4815 (1986).
74. M. R. Metzger and C. H. Rhee, *Cryst. Liq. Cryst.* **85,** 1471 (1982).
75. K. Balasubramanian, M. Han, and M. Z. Liao, *Chem. Phys. Lett.* **139,** 551 (1987).
76. W. C. Price, *Proc. Roy. Soc., Ser. A* **167,** 216 (1938).
77. R. F. Barrow and J. G. Stamper, *Proc. Roy. Soc., Ser. A* **263,** 259 (1961).
78. R. F. Barrow and J. G. Stamper, *Proc. Roy. Soc., Ser. A* **263,** 277 (1961).
79. M. L. Ginter and S. G. Tilford, *J. Mol. Roy. Spectrosc.* **34,** 206 (1970).
80. M. L. Ginter and S. G. Tilford, *J. Molec. Roy. Spectrosc.* **37,** 159 (1971).
81. J. G. Stamper and R. F. Barrow, *J. Phys. Chem.* **65,** 250 (1961).
82. R. S. Mulliken, *Phys. Rev.* **50,** 1017 (1936).
83. R. S. Mulliken, *Phys. Rev.* **51,** 310 (1937).
84. F. A. van Dijk and A. Dyrnanus, *Chem. Phys.* **6,** 474 (1974).
85. P. Bernage and P. J. Niay, *Quant. Spectrosc. Radiat. Transfer* **18,** 315 (1977).
86. P. Niay, P. Bernage, C. Coquant, and C. Fayt, *Can. J. Phys.* **55,** 1829 (1977).

87. O. B. Dabbousi, W. L. Meerts, F. H. de Leeuw, and A. Dymanus, *Chem. Phys.* **2,** 473 (1973).
88. H. J. Babrov, A. L. Shabot, and B. S. Rao, *J. Chem. Phys.* **42,** 4124 (1965).
89. P. A. Straub and A. D. McLean, *Theor. Chem. Acta* **32,** 227 (1974).
90. P. Botschwina and W. Meyer, *J. Chem. Phys.* **67,** 2390 (1977).
91. H. J. Werner and P. Rosmus, *J. Chem. Phys.* **73,** 2319 (1980).
92. J. F. Ogilvie, W. R. Rodwell, and R. H. Tipping, *J. Chem. Phys.* **73,** 5221 (1980).
93. J. Andzelm, M. Klobukowski, and E. Radzio-Andzelm, *J. Comput. Chem.* **5,** 146 (1984).
94. D. A. Chapman, K. Balasubramanian, and S. H. Lin, *Chem. Phys.* **118,** 333 (1987).
95. S. H. Alajajian and A. Chutjian, *Phys. Rev.* **A37,** 3680 (1988).
96. N. G. Adams, D. Smith, A. A. Viggiano, J. F. Paulson, and M. J. Henchman, *J. Chem. Phys.* **84,** 6728 (1986).
97. Y. Le Coat, R. Azria, and M. Tronc, *J. Phys.* **B18,** 809 (1985).
98. Y. Le Coat, R. Azria, and M. Tronc, *J. Phys.* **B15,** 1569 (1982).
99. R. Abouaf and D. Teillet-Billy, *Chem. Phys. Lett.* **73,** 106 (1980).
100. R. Azria, Y. Le Coat, D. Simon, and M. Tronc, *J. Phys.* **B13,** 1909 (1980).
101. R. Abouaf, and D. Teillet-Billy, *J. Phys.* **B10,** 2261 (1977).
102. K. Rohr, K. and F. Linder, *J. Phys.* **B9,** 2521 (1976).
103. J. P. Ziesel, I. Nenner, and G. J. Schulz, *J. Chem. Phys.*, **63,** 1943 (1975).
104. D. Tillet-Billy, and J. P. Gauyacq, *J. Phys.* **B17,** 3329 (1984).
105. A. U. Hazi, *J. Phys.* **B16,** 29 (1983).
106. M. Krauss, and W. J. Stevens, *J. Chem. Phys.* **74,** 570 (1981).
107. E. Goldstein, G. A. Segal, and R. W. Wetmore, *J. Chem. Phys.* **68,** 271 (1978).
108. K. D. Jordan, and J. Wendolski, *J. Chem. Phys.* **21,** 145 (1977).
109. H. S. Taylor, E. Goldstein, and G. A. Segal, *J. Phys.* **B10,** 2253 (1977).
110. O. H. Crawford and B. J. D. Koch, *J. Chem. Phys.* **60,** 4512 (1974).
111. F. Fiquet-Fayard, *J. Phys.* **B7,** 810 (1974).
112. K. D. Jordon, A. Herzenberg, and W. Luken, *J. Chem. Phys.* **64,** 2760 (1976).
113. F. Fiquet-Fayard, *Vacuum* **24,** 533 (1974).
114. T. F. O'Malley, and H. S. Taylor, *Phys. Rev.* **A176,** 207 (1968).
115. W. Bondybey, P. K. Pearson, and H. F. Schaefer, III, *J. Chem. Phys.* **57,** 1123 (1972).
116. W. M. Hartman, T. L. Gilbert, K. A. Kaiser, and A. C. Wahl, *Phys. Rev. B* 1140 (1970).
117. A. W. Weiss and M. Krauss, *J. Chem. Phys.* **52,** 4363 (1970).
118. H. H. Michels, R. E. Harris, and J. C. Browne, *J. Chem. Phys.* **48,** 2821 (1968).
119. H. S. Taylor, F. W. Bobrowicz, P. J. Hay, and T. H. Dunning, *J. Chem. Phys.* **65,** 1182 (1976).
120. D. A. Chapman, K. Balasubramanian, and S. H. Lin, *J. Chem. Phys.* **87,** 5325 (1987).
121. W. J. Stevens, A. M. Karo, and J. R. Hiskes, *J. Chem. Phys.* **74,** 3989 (1981).

122. D. A. Chapman, K. Balasubramanian, and S. H. Lin, *Chem. Phys. Lett.* **118,** 192 (1985).
123. P. Bernage and P. Niay, P. *J. Quant. Spectrosc. Radiat. Transfer* **18,** 315 (1977).
124. C. A. Burrus, *J. Chem. Phys.* **60,** 2991 (1974).
125. H. Hotop and W. C. Lineberger, *J. Phys. Chem. Ref. Data* **4,** 539 (1975).
126. B. Liu and P. E. Siegbahn, *J. Chem. Phys.* **68,** 2457 (1968).
127. H. S. Taylor and F. E. Harris, *J. Chem. Phys.* **39,** 1012 (1963).
128. D. Spence, W. A. Chupka, and C. M. Stevens, *J. Chem. Phys.* **76,** 2759 (1982).
129. D. A. Chapman, K. Balasubramanian, and S. H. Lin, *Phys. Rev.* **A38,** 6098 (1988).
130. Grundstruöm, *Z. Phys.* **113,** 721 (1939).
131. D. Kleman, dissertation, Stockholm, (1953).
132. H. Neuhaus, *Z. Phys.* **150,** 4 (1958).
133. H. Neuhaus, *Z. Phys.* **152,** 402 (1958).
134. M. L. Ginter, *J. Molec. Spectrosc.* **11,** 301 (1963).
135. T. Larsson and H. Neuhaus, *Ark. Fys.* **27,** 275 (1964).
136. K. F. Freed, *J. Chem. Phys.* **45,** 1714 (1966).
137. M. L. Ginter, *J. Molec. Spectrosc.* **20,** 240 (1966).
138. T. Larsson, H. Neuhaus, and N. Aslund, *Ark. Fys.* **37,** 114 (1968).
139. L. Veseth and A. Lofthus, *J. Molec. Spectrosc.* **49,** 414 (1974).
140. L. Veseth, *J. Molec. Spectrosc.* **59,** 51 (1976).
141. K. Balasubramanian, *J. Phys. Chem.* **94,** 6582 (1990).
142. W. W. Watson and R. Simon, *Phys. Rev.* **55,** 358 (1939).
143. W. W. Watson and R. Simon, *Phys. Rev.* **57,** 708 (1940).
144. J. T. Hougen, *Can. J. Phys.* **40,** 598 (1962).
145. L. Klynning, B. Lindgren, and N. Aslund, *Ark. Fys.* **30,** 141 (1965).
146. L. Klynning, *Ark. Fys.* **31,** 281 (1966).
147. I. Kopp and J. T. Hougen, *Can. J. Phys.* **45,** 2581 (1967).
148. I. Kovacs and V. M. Korwar, *J. Phys.* **B4,** 759 (1971).
149. I. Kovacs and B. Vujisic, *J. Phys.* **B4,** 1123 (1971).
150. I. Kovacs and P. Pacher, *J. Phys.* **B4,** 1633 (1971).
151. L. Veseth, *Physica* **56,** 286 (1971).
152. K. Balasubramanian and K. S. Pitzer, *J. Molec. Spectrosc.* **103,** 105 (1984).
153. P. Bollmark and B. Lindgren, *Chem. Phys. Lett.* **1,** 480 (1967).
154. N. Basco and K. K. Yee, *Spectrosc. Lett.* **1,** 13 (1968).
155. P. Bollmark and B. Lindgren, *Phys. Scripta* **10,** 325 (1974).
156. B. Lindgren and C. H. Nilsson, *J. Molec. Spectrosc.* **55,** 407 (1975).
157. M. A. Khan and Z. M. Khan, *Proc. Roy. Soc., Ser. A* **88,** 211 (1966).
158. K. Balasubramanian, N. Tanpipat, and J. Bloor, *J. Molec. Spectrosc.* **124,** 458 (1987).
159. D. J. Little, R. J. Donovan, and R. J. Butcher, *J. Photochem.* **2,** 451 (1973).
160. C. B. Freidhoff, J. T. Snodgrass, V. J. Coe, K. M. McHugh, and K. H. Bowen, *J. Chem. Phys.* ***84,*** 1051 (1986).

161. K. Balasubramanian, M. Han, and M. Z. Liao, *J. Chem. Phys.* **86,** 4979 (1987).
162. M. L. Ginter, S. G. Tilford, and A. M. Bass, *J. Molec. Spectrosc.* **57,** 271 (1975).
163. S. G. Tilford, M. L. Ginter, and A. M. Bass, *J. Molec. Spectrosc.* **34,** 327 (1970).
164. R. D. Clear, S. J. Riley, and K. R. Wilson, *J. Chem. Phys.* **63,** 1340 (1975).
165. J. Romand, *Ann. Phys.* (Paris) **4,** 527 (1949).
166. S. Datta and D. N. Kundu, *Proc. Natl. Acad. Sci.* (India) **7,** 311 (1941).
167. Y. S. Lee, W. C. Ermler, and K. S. Pitzer, *J. Chem. Phys.* **73,** 360 (1980).
168. P. A. Christiansen and K. S. Pitzer, *J. Chem. Phys.* **73,** 5160 (1980).
169. B. Grundström and P. Valberg, *Z. Phys.* **108,** 326 (1938).
170. E. M. Bulewicz and T. M. Sugden, *Trans. Faraday Soc.* **54,** 830 (1958).
171. H. Neuhaus and V. Muld, V. *Z. Phys.* **153,** 412 (1959).
172. T. Larsson and H. Neuhaus, *Ark. Fys.* **23,** 461 (1963), **31,** 299 (1966).
173. W. W. Watson, *Phys. Rev.* **54,** 1068 (1938).
174. L. Gerö, *Z. Phys.* **116,** 379 (1940).
175. K. Balasubramanian and K. S. Pitzer, *J. Phys. Chem.* **88,** 1146 (1984).
176. E. Hulthen and H. Nauhaus, *Phys. Rev.* **102,** 1415 (1956).
177. H. Neuhaus, *Z. Naturforsch.* **21a,** 2113 (1966).
178. A. Heimer, *Z. Phys.* **85,** 328 (1935).
179. A. Heimer, *Z. Phys.* **103,** 621 (1936).
180. K. Balasubramanian, *J. Molec. Spectrosc.* **115,** 258 (1986).
181. K. Balasubramanian, *Chem. Phys. Lett.* **114,** 201 (1985).
182. K. Balasubramanian, *J. Phys. Chem.* **90,** 1043 (1986).
183. A. Trenin, *Phys. Rev.* **36,** 147 (1930).
184. A. Trenin and B. Popov, *Z. Phys.* **75,** 338 (1932).
185. J. Berkowitz and W. A. Chupka, *J. Chem. Phys.* **45,** 1287 (1966).
186. D. Cubicciotti, *J. Phys. Chem.* **68,** 1528, 3834, (1964); **69,** 1410 (1965).
187. F. J. Keneshea and D. Cubicciotti, *J. Phys. Chem.* **71,** 1958 (1967).
188. J. M. Brom, Jr., and H. F. Fransen, *J. Chem. Phys.* **54,** 2874 (1971).
189. D. Cubicciotti, *High Temp. Sci.* **2,** 65 (1970).
190. E. Murad, D. L. Hildebrand, and R. P. Main, *J. Chem. Phys.* **45,** 263 (1966).
191. J. Berkowitz, and T. A. Walter, *J. Chem. Phys.* **49,** 1184 (1968).
192. R. F. Barrow, H. F. K. Cheall, P. M. Thomas, and P. B. Zeeman, *Proc. Phys. Soc.* (London) **71,** 128 (1958).
193. Z. V. Boizova and K. V. Butkow, *Phys. Z. Sowjetunion* **5,** 705 (1936).
194. H. G. Howell, *Proc. Roy. Soc.* (London), *Ser. A* **160** 242 (1937).
195. M. Mandel and A. H. Barrett, *Phys. Rev.* **98,** 1159 (1955); A. H. Barrett and M. Mandel, *Phys. Rev.* **109** 1572 (1958).
196. W. Drechsler and G. Graff, *Z. Phys.* **163,** 165 (1961).
197. R. K. Ritchie and H. Lew, *Can. J. Phys.* **43,** 1701 (1965).
198. K. Balasubramanian, *J. Chem. Phys.* **82,** 3741 (1985).
199. J. V. R. Rao and P. T. Rao, *Indian J. Phys.* **29,** 20 (1955).

200. F. Morgan, *Phys. Rev.* **49,** 47 (1936).
201. G. D. Rochester, *Proc. Roy. Soc.* (London), *Ser. A* **153,** 407 (1936), **167,** 567 (1938).
202. R. F. Barrow, D. Butler, J. W. C. Johns, and J. L. Powell, *Proc. Phys. Soc.* (London) **73,** 317 (1959).
203. S. P. Singh, *Indian J. Pure Appl. Phys.* **5,** 292 (1967).
204. O. N. Singh, I. S. Singh, and P. N. Singh, *Can. J. Phys.* **50,** 2206 (1972).
205. D. J. W. Lumley, and R. F. Barrow, *J. Phys.* **B10,** 1537 (1977).
206. K. Zmbov, J. W. Hastie, and J. L. Margrave, *Trans. Faraday Soc.* **64,** 861 (1968).
207. K. M. Rao and P. T. Rao, *Can. J. Phys.* **42,** 690 (1964).
208. K. Balasubramanian, *J. Chem. Phys.* **83,** 2311 (1985).
209. W. J. Rosano and J. M. Parson, *J. Chem. Phys.* **82,** 4401 (1985).
210. M. Mezinger, *Adv. Chem. Phys.* **42,** 1 (1980).
211. M. R. Levy, *Prog. React. Kinet.* **10,** 1 (1979).
212. T. Ishikawa and J. M. Parson, *J. Chem. Phys.* **79,** 4261 (1983).
213. J. M. Parson, J. H. Wang, C. C. Fang, and B. S. Cheong, *Chem. Phys. Lett.* **152,** 330 (1988).
214. B. W. Sohwenz, L. C. Geiger, and J. M. Parson, *J. Chem. Phys.* **74,** 1736 (1981).
215. W. G. Howell, *Proc. Roy. Soc.* (London) *Ser. A* **155,** 141 (1936).
216. F. Morgan, *Phys. Rev.* **49,** 41 (1936).
217. G. D. Rochester, *Phys. Rev.* **51,** 489 (1937).
218. K. C. Joshi, *Proc. Phys. Soc.* (London) **78,** 610 (1961).
219. T. A. Rao and P. T. Rao, *Indian J. Phys.* **36,** 85 (1962).
220. K. M. Rao and P. T. Rao, *Indian J. Phys.* **39,** 572 (1965).
221. T. A. Rao and P. T. Rao, *Can. J. Phys.* **40,** 1077 (1962).
222. M. M. Patel and P. S. Narayanan, *Indian J. Pure Appl. Phys.* **5,** 223 (1967).
223. B. S. Mohanty, D. K. Rai, K. N. Upadhya, and N. L. Singh, *J. Phys.* **B1,** 523 (1968).
224. A. K. Chaudry, K. N. Upadhya, D. K. Rai, and B. S. Mohanty, *J. Phys.* **B1,** 1223 (1968).
225. A. K. Chaudry, K. N. Upadhya, and D. K. Rai, *J. Phys.* **B2,** 628 (1969).
226. M. N. Avasthi, *Spectrosc. Lett.* **3,** 157 (1970).
227. P. S. Murthy, D. V. Rao, Y. P. Reddy, and P. T. Rao, *Spectrosc. Lett.* **8,** 217 (1975).
228. K. J. Chakko and M. M. Patel, *Indian J. Pure Appl. Phys.*, **17,** 189 (1979).
229. P. Kuijpers and A. Dynamus, *Chem. Phys.* **24,** 97 (1977).
230. W. E. Jones and T. D. McLean, *J. Molec. Spectrosc.* **83,** 317 (1980); **90,** 481 (1981).
231. A. G. Gaydon, *Dissociation Energy of Diatomic Molecules*, 3rd ed., Chapman & Hall, London, 1968.
232. K. Balasubramanian, *Chem. Phys. Lett.* **127,** 324 (1986).
233. K. Wieland and R. Newburgh, *Helv. Phys. Acta* **25,** 87 (1952).
234. H. Cordes and F. Gehrke, *Z. Phys. Chem.* **51,** 281 (1966).

235. V. S. Rao and P. T. Rao, *Z. Phys.* **181,** 58 (1966).
236. J. W. Hastie, H. Bollm, and J. D. Morrison, *J. Chem. Phys.* **47,** 1580 (1967).
237. S. P. Singh, *Indian J. Pure Appl. Phys.* **8,** 114 (1970).
238. K. Balasubramanian, *J. Molec. Spectrosc.* **127,** 97 (1988).
239. W. Jevons, *Proc. Roy. Soc.* (London) *Ser. A* **110,** 365 (1926).
240. W. F. C. Ferguson, *Phys. Rev.* **32,** 607 (1928).
241. C. A. Fowler, *Phys. Rev.* **62,** 141 (1942).
242. P. R. K. Sarma and P. Venkateswarlu, *J. Molec. Spectrosc.* **17,** 252 (1965).
243. J. W. Hastie, R. H. Hauge, and J. L. Margrave, *J. Molec. Spectrosc.* **29,** 152 (1969).
244. G. A. Oldershaw and K. Robinson, *J. Molec. Spectrosc.* **32,** 469 (1969).
245. W. Richter, *Z. Phys. Chem.* **71,** 303 (1970).
246. A. Chatalic, D. Iacocca, and G. Pannetier, *J. Chim. Phys. Phys.-Chim. Biol.* **69,** 82 (1972).
247. A. Pannetier and P. Deschamps, *J. Chim. Phys. Phys.-Chim. Biol.* **65,** 1164 (1968).
248. R. S. Mulliken, *Phys. Rev.* **28,** 481 (1925).
249. K. Balasubramanian, *J. Molec. Spectrosc.* **132,** 280 (1988).
250. M. C. Lin, *Adv. Chem. Phys.* **42,** 113 (1980).
251. S. Toby, *Chem. Rev.* **84,** 277 (1984).
252. M. M. Rauhut, *Acc. Chem. Res.* **2,** 80 (1969).
253. F. Q. McCapara, *Rev. Chem. Soc.* **20,** 485 (1966).
254. P. G. Mahanti, *Z. Phys.* **68,** 114 (1931).
255. W. Loomis and T. F. Watson, *Phys. Rev.* **45,** 805 (1934).
256. F. W. Jevons, *Proc. Phys. Soc.* (London) **50,** 910 (1938).
257. B. Eisler and R. F. Barrow, *Proc. Phys. Soc.* (London), *Ser. A* **62,** 740 (1949).
258. R. F. Barrow and J. Rowlinson, *Proc. R. Soc.* (London), *Ser. A* **224,** 374 (1954).
259. A. Lagerqvist, N. E. L. Nilsson, and K. Wigartz, *Ark. Fys.* **15,** 521 (1959).
260. E. M. Deutsch and R. F. Barrow, *Nature* **201,** 815 (1964).
261. M. M. Joshi and R. Yamdagni, *Indian J. Phys.* **33,** 275 (1967).
262. G. A. Capelle and C. Linton, *J. Chem. Phys.* **65,** 5361 (1976).
263. R. Colin, J. Drowart, and G. Verhagen, *Trans. Faraday Soc.* **61,** 1364 (1965).
264. K. P. R. Nair, R. B. Singh, and D. K. Rai, *J. Chem. Phys.* **43,** 3570 (1965).
265. T. Torring, *Z. Naturforsch.* **22a,** 1234 (1967).
266. J. J. Smith and B. Meyer, *J. Molec. Spectrosc.* **27,** 304 (1968).
267. J. Hoeft, F. J. Lovas, E. Tiemann, R. Tischer, and T. Torring, *Z. Naturforsch.* **24a,** 1222 (1969).
268. J. S. Ogen and M. Ricks, *J. Chem. Phys.* **53,** 896 (1970).
269. R. S. Dube and D. K. Rai, *J. Phys.* **B4,** 579 (1971).
270. R. Honerjäger, and R. Tischer, *Z. Naturforsch.* **28a,** 1372 (1973).
271. K. Balasubramanian and K. S. Pitzer, *Chem. Phys. Lett.* **100,** 273 (1983).
272. J. M. Dyke, A. Morris, A. M. Ridha, and J. G. Snijders, *Chem. Phys.* **67,** 245 (1982).

273. K. Balasubramanian, *J. Phys. Chem.* **88,** 5759 (1984).
274. S. Bloomenthal, *Phys. Rev.* **35,** 34 (1930).
275. H. Howell, *Proc. Roy. Soc.* (London), *Ser. A* **153,** 683 (1936).
276. E. G. Vago and R. F. Barrow, *Proc. Phys. Soc.* London **59,** 449 (1947).
277. R. F. Barrow, E. M. Deutsch, and D. N. Travis, *Nature* **191,** 374 (1984).
278. T. Torring, *Z. Naturforsch.* **19a,** 1426 (1964).
279. J. Drowart, R. Colin, and G. Exsteem, *Trans. Faraday Soc.* **61,** 1376 (1965).
280. K. P. R. Nair, R. B. Singh, and D. K. Rai, *J. Chem. Phys.* **43,** 3570 (1965).
281. R. C. Oldenborg, C. R. Dickson, and R. N. Zare, *J. Molec. Spectrosc.* **58,** 283 (1975).
282. M. J. Kurylo, W. Brdun, S. Abramowitz, and M. Krauss, *J. Res. Natl. Bur. Stand., Sect. A* **80,** 167 (1976).
283. C. Linton and H. P. Broida, *J. Molec. Spectrosc.* **62,** 396 (1976).
284. J. M. Brom and W. H. Beatie, *J. Molec. Spectrosc.* **81,** 445 (1980).
285. K. Balasubramanian and K. S. Pitzer, *J. Phys. Chem.* **87,** 4857 (1983).
286. H. Basch, W. J. Stevens, and M. Krauss, *Chem. Phys. Lett.* **74,** 2416 (1981).
287. G. D. Rochester, *Proc. R. Soc.* (London), *Ser. A* **150,** 668 (1935).
288. E. W. Shawhan, *Phys. Rev.* **48,** 421 (1935).
289. A. E. Douglas, L. L. Howe, and R. J. Morton, *J. Molec. Spectrosc.* **7,** 161 (1961).
290. R. Yamdagni and M. M. Joshi, *Indian J. Phys.* **40,** 495 (1966).
291. J. J. Smith and B. Meyer, *J. Molec. Spectrosc.* **27,** 304 (1968).
292. D. J. Greenwood, C. Linton, and R. F. Barrow, *J. Molec. Spectrosc.* **89,** 134 (1981).
293. R. A. Teichmann, III and E. R. Nixon, *J. Molec. Spectrosc.* **65,** 258 (1977).
294. D. J. Greenwood, R. F. Barrow, and C. Linton, *J. Molec. Spectrosc.* **86,** 480 (1981).
295. K. Balasubramanian, *Chem. Phys. Lett.* **139,** 262 (1987).
296. G. D. Rochester and H. G. Howell, *Proc. Roy. Soc.* (London) *Ser. A* **148,** 157 (1935).
297. R. F. Barrow, P. W. Fry, and R. C. LeBargy, *Proc. Phys. Soc.* (London) **81,** 697 (1963).
298. C. P. Marino, J. D. Guerin and E. R. Nixon, *J. Molec. Spectrosc.* **51,** 160 (1974).
299. R. A. Teichman, III and E. R. Nixon, *J. Molec. Spectrosc.* **54,** 78 (1975).
300. B. Burkin, M. Carleer, R. Colin, C. Dreze, and T. Ndikumana, *J. Phys.* **B13,** 3783 (1980).
301. M. Knöckel, T. Kröckerskothen, and E. Tiemann, *Chem. Phys.* **93,** 349 (1985).
302. A. N. Murty, and R. F. Curl, Jr., *J. Molec. Spectrosc.* **30,** 102 (1969).
303. R. Colin, and J. Drowart, *J. Chem. Phys.* **37,** 1120 (1962).
304. K. Balasubramanian, *J. Chem. Phys.* **85,** 1443 (1986).
305. G. B. Kim and K. Balasubramanian, unpublished results.
306. C. E. Moore, *Tables of Atomic Energy Levels*, Vols. I–III, U.S. National Bureau of Standards, Washington, DC, 1971.
307. P. A. G. O'Hare, A. Batana, and A. C. Wahl, *J. Chem. Phys.* **59,** 6495 (1973).

308. J. Hoeft, F. J. Lovas, E. Tiemann, and T. Torring, *Z. Naturforsch.* **25a,** 539 (1970).

309. R. S. Dube, K. N. Upadhya, and D. K. Rai, *J. Quant. Spectrosc. Radiat, Transfer* **10,** 1191 (1970).

310. R. S. Dube, *Curr. Sci.* **40,** 32 (1971).

311. R.-D. Urban, U. Magg, and H. Jones, *Chem. Phys. Lett.* **154,** 135 (1989).

312. J. Berkowitz, *J. Chem. Phys.* **89,** 7065 (1988).

313. J. Berkowitz and H. Cho, *J. Chem. Phys.* **90,** 1 (1989).

314. K. Balasubramanian, *J. Chem. Phys.* **91,** 2443 (1989).

315. A. H. Bahnmaier, R.-D. Urban, and H. Jones, *Chem. Phys. Lett.* **155,** 269 (1989).

316. P. Schwerdtfeger, H. Silberbach, and B. Miehlich, *J. Chem. Phys.* **90,** 762 (1989).

317. K. Balasubramanian, *J. Chem. Phys.* **89,** 5731 (1988).

318. J. Hoeft and K. P. R. Nair, *Chem. Phys. Lett.* **155,** 273 (1989).

319. U. Wolf and E. Tiemann, *Chem. Phys. Lett.* **139,** 191 (1987).

320. K. Perumalswamy, S. B. Rai, and K. N. Upadhya, *Physica* **C141,** 315 (1989).

321. K. Perumalswamy, S. B. Rai, D. K. Rai, and K. N. Upadhya, *Physica* **C132,** 122 (1985).

322. J. Borkowska-Burnecka and W. Zyrnicki, *Physica* **C115,** 415 (1983).

323. V. P. N. Nampoori, M. N. Kamalasanan, and M. M. Patel, *Indian J. Phys.* **B51,** 401 (1977).

324. Th. Glenewinkel-Meyer, A. Kowalski, B. Müller, Ch. Ottinger, and W. H. Beckenridge, *J. Chem. Phys.* **89,** 7112 (1988).

325. J. M. Dyke, C. Kirby, A. Morris, B. W. J. Gravenor, R. Klein, and P. Rosmus, *Chem. Phys.* **88,** 289 (1984).

326. L.-S. Wang, B. Niu, D. A. Shirley, and K. Balasubramanian, *J. Chem. Phys.* **92,** 899 (1990).

327. S. N. Vempati and W. E. Jones, *J. Molec. Spectrosc.* **127,** 232 (1988).

328. S. N. Vempati and W. E. Jones, *J. Molec. Spectrosc.* **122,** 190 (1987).

329. S. N. Vempati and W. E. Jones, *J. Molec. Spectrosc.* **120,** 441 (1986).

330. A. G. Maki and C. J. Lovas, *J. Molec. Spectrosc.* **125,** 188 (1987).

331. F. Mélen and I. Dubois, *J. Molec. Spectrosc.* **124,** 476 (1987).

332. G. Lakshminarayana and J. Shetty, *J. Molec. Spectrosc.* **122,** 417 (1987).

333. A. M. R. P. Bopegedera, C. R. Brazier, and P. F. Bernath, *Chem. Phys. Lett.*, **162,** 309 (1989).

334. A. F. Ramos, N. C. Pyper, and G. L. Malli, *Phys. Rev.* **A38,** 2729 (1988).

335. G. Igelmann, C. Feller, H. J. Flad, A. Savin, H. Stoll, and R. H. Preuss, *Molec. Phys.* **68,** 209 (1989).

336. C. Teichteil and F. Spiegelmann, *Chem. Phys.* **81,** 283 (1981).

337. P. Schwerdtfeger, L. V. Szenpaly, H. Stoll, and H. Preuss, *J. Chem. Phys.* **87,** 510 (1987).

338. P. Schwerdtfeger, *Phys. Scripta* **36,** 453 (1987).

339. M. G. Kozlov, V. I. Fomichev, Y. Y. Donitriev, L. N. Labzovsky, and A. V. Titov, *J. Phys.* **B20,** 4939 (1987).

340. V. Wolf and E. Tiemann, *Chem. Phys. Lett.* **133,** 116 (1987).

341. F. Ito, T. Nakanaga, H. Takeo and H. Jones, *J. Molec. Spectra* **164,** 379 (1994).
342. E. H. Fink, K. D. Setzer, D. A. Ramsay, and M. Vervlote, *J. Molec. Spectrosc.* **142,** 108 (1990).
343. R. D. Urban, P. Polomsky, and J. Jones, *Chem. Phys. Lett.* **181,** 485 (1991).
344. H. G. Hedderich and P. F. Bernath, *J. Molec. Spectrosc.* **158,** 170 (1993).
345. A. B. Alekseyev, R. J. Buenker, H.-P. Libermann, and G. Hirsch, *J. Chem. Phys.* **100,** 2989 (1994).
346. D. Dai and K. Balasubramanian, *J. Chem. Phys.* **93,** 1837 (1990).
347. M. Dolg, W. Kuechle, H. Stoll, H. Preuss, and P. Schwerdtfeger, *Molec. Phys.* **74,** 1265 (1991).
348. A. B. Alekseyev, H.-P. Libermann, I. Boustani, G. Hirsch, and R. J. Buenker, *Chem. Phys.* **173,** 333 (1993).
349. (a) E. H. Fink, K. D. Setzer, D. A. Ramsay, and M. Vervloet, *Chem. Phys., Lett.* **179,** 95 (1991); (b) M. D. Obserlander and J. M. Parson, *J. Chem. Phys.* **97,** 250 (1992).
350. T. C. Devore, L. Brock, R. Kahlscheuer, K. Dulaney, and J. L. Gole, *Chem. Phys.* **155,** 423 (1991).
351. A. J. Ross, R. Bacis, J. D'Incan, C. Effantin, B. Koffend, A. Topouzkhanian, and J. Verges, *Chem. Phys. Lett.* **166,** 539 (1990)¿
352. R. K. Yoo, B. Rucic, and J. Berkowitz, *Chem. Phys.*, **166,** 215 (1992).
353. L. Latifzadeh and K. Balasubramanian, *Chem. Phys. Lett.* (in press).
354. I. Boustani, S. N. Rai, H.-P. Libermann, A. Alekseyev, G. Hirsch, and R. J. Buenker, *Chem. Phys. Lett.* **177,** 45 (1993).
355. H. G. Howell and G. D. Rochester, *Proc. Phys. Soc.* **51,** 329 (1939): G. D. Rochester, *Phys. Rev.* **51,** 486 (1937).
356. R. Vasudev and W. E. Jones, *J. Molec. Spectrosc.* **59,** 442 (1976).
357. K. C. Abraham and M. M. Patel, *J. Phys. B, Atom. Molec. Phys.* **4** 1398 (1971).
358. K. C. Abraham and M. M. Patel, *J. Phys. B, Atom. Molec. Phys.* **3** 882 (1970).
359. K. C. Abraham and M. M. Patel, *J. Phys. B, Atom. Molec. Phys.* **3,** 1183 (1970).
360. M. M. Patel and K. C. Abraham, *Indian J. Pure Appl. Phys.*, **7** 641 (1969).
361. F. Prevot, R. Colin, and W. E. Jones, *J. Molec. Spectrosc.* **56,** 432 (1975).
362. D. M. Haaland, M. R. Robinson, G. W. Koep, E. V. Thomas, and P. R. Eaton, *J. Appl. Spectrosc.* **46,** 1575 (1992).
363. T. A. P. Rao and P. T. Rao, *Indian J. Phys.* **36,** 85 (1962).
364. M. Chakravorty, K. C. Abraham and M. M. Patel, *J. Phys., B, Atom. Molec. Phys.* **6,** 757 (1973).
365. D. W. K. Wang, W. E. Jones, F. Prevot, and R. Colin, *J. Mol. Spectrosc.* **49,** 377 (1974).
366. A. Alekseyev, K. K. Das, H.-P. Libermann, R. J. Buenker, and G. Hirsch, *Chem. Phys.* **198,** 333 (1995).
367. K. K. Das, A. B. Alekseyev, H.-P. Libermann, G. Hirsch, and R. J. Buenker, *Chem. Phys.* **196,** 395 (1995).
368. Y. Li, H.-P. Libermann, G. Hirsch, and R. J. Buenker, *J. Molec. Spectrosc.* **165,** 219 (1994).

369. U. Wolf, F. Bender, S. H. Schaefer, and E. Tiemann, *Chem. Phys.* **102,** 175 (1986).
370. J. Linder, D. Bender, and E. Tiemann, *Chem. Phys.* **105,** 435 (1986).
371. J. Linder and E. Tiemann, *J. Molec. Spectrosc.* **155,** 234 (1992).
372. A. B. Alekseyev, H.-P. Libermann, R. J. Buenker, and G. Hirsch, *J. Chem. Phys.* **102,** 2539 (1995).
373. W. J. Balfour and R. S. Ram, *J. Molec. Spectrosc.* **105,** 246 (1984).
374. W. J. Balfour and R. S. Ram, *J. Molec. Spectrosc.* **130,** 382 (1984).
375. E. H. Fink and O. D. Shestakov, unpublished results.
376. A. B. Alekseyev, H.-P. Libermann, R. J. Buenker, G. Hirsch, and Y. Li, *J. Chem. Phys.* **100,** 8956 (1994).
377. E. H. Fink, K. D. Setzer, D. A. Ramsay, and M. Vervloet, *Chem. Phys. Lett.* **179,** 103 (1991).
378. O. Shestakov, R. Breidohr, H. Demes, K. D. Setzer, E. H. Fink, and W. Zyrnicki, unpublished results.
379. S. B. Rai and D. K. Rai, *Chem. Rev.* **84,** 73 (1984).
380. K. Ziebarth, R. Breidohr, O. Shestakov, and E. H. Fink, *Chem. Phys. Lett.* **190,** 271 (1992).
381. O. Shestakov, A. M. Pravilov, H. Demes, and E. H. Fink, *Chem. Phys.* **165,** 415 (1992).
382. M. Benavides-Garcia and K. Balasubramanian, *Chem. Phys. Lett.* **211,** 631 (1993).
383. M. Benavides-Garcia and K. Balasubramanian, *J. Molec. Spectrosc.* **161,** 552 (1993).
384. W. Jevons and L. A. Bashford, *Proc. Phys. Soc.* **49,** 554 (1937).
385. G. A. Oldershaw and K. Robinson, *Trans. Faraday Society* **64,** 616 (1968).
386. K. Balasubramanian, *J. Molec. Spectrosc.* **157,** 268 (1993).

3

RELATIVISTIC EFFECTS IN MAIN-GROUP CLUSTERS

3.1 INTRODUCTION

This chapter deals with the electronic and spectroscopic properties of small clusters of heavy atoms and metal atoms that have been the topics of many experimental (1–40, 71–114) and theoretical studies (41–70). Subsequently, the properties of the clusters within a group of the periodic table are compared, and it is shown that the relativistic effects play a significant role for the heavier clusters. The author (2) has reviewed the properties of dimers and trimers containing very heavy *p*-block elements in 1990. However, since that time significant progress has been made in the study of these species and larger clusters. This chapter comprises material from both the earlier review on the dimers and trimers, and new material on other clusters and new developments, especially after 1990, that were not included in the earlier review. Thus, there is significant new material, in addition to the material derived from the review article.

Theoretical and experimental works on the clusters of main-group elements have resulted in a wealth of information on the spectroscopic properties and potential-energy curves of many heavy dimers and trimers, as well as the electronic states of larger clusters. Experimental progress in this area has been made possible (1–40, 71–113) by laser vaporization and supersonic jet expansion methods (78–86). Typically, a sample of a foil or sheet of material containing these elements such as a GaAs crystal (97, 100), a Si crystal (98), or Pt foil (114) is vaporized with a high-energy Nd : YAG (neodymium : yttrium aluminum garnet) laser and passed through a supersonic nozzle. The jet-cooled beam of clusters with varied compositions is then photoionized and investigated by laser spectroscopic methods. A number of other techniques such as sputtering methods (87–89), rare-gas matrix isolation methods (32), particle bombardment methods (20, 115, 116) ligand-stripping methods (90–92), and aggregation within zeolites and molecular sieves (93–96) have been used for cluster generation.

There are several *ab initio* calculations (41–54) on such clusters, which have yielded vital data on the spectroscopic constants and have enhanced our knowledge of bonding in these systems. Relativistic effective core potentials (RECP) (44, 55–67) have become quite useful for clusters containing very heavy atoms such as Pb, Sn, Bi, and Sb.

Experimental and theoretical investigations have demonstrated that the geometries and electronic properties of small clusters containing typically 3–10 atoms have no resemblance to the more familiar properties of the bulk. It is this aspect that has made the cluster area, perhaps one of the most exciting and challenging areas for both experimental and theoretical activities. Electron affinities and ionization potentials of these species exhibit dramatic variations as a function of cluster size. The stabilities of these clusters as a function of size exhibit certain numbers (magic numbers) for which clusters are unusually more stable.

The cluster area is relatively young compared to the more established traditional branches of main-group inorganic chemistry or organometallic chemistry. Although a significant amount of work has been done on larger clusters, this chapter is restricted to mainly homonuclear dimers, trimers, and some heteronuclear dimers and trimers of such heavy species. A recent review by Mandich et al. (1) has placed special emphasis on larger clusters. For most of the heavy main-group dimers, theoretical calculations are now available.

Section 3.2 discusses the experimental and theoretical techniques for the investigation of the main-group clusters. Section 3.3 considers the spectroscopic properties and potential-energy curves of the main-group dimers, while Section 3.5 discusses the main-group trimers. Section 3.4 is devoted to the literature review of some mixed clusters, but these species are not discussed in detail. Section 3.6 discusses the relative periodic trends in the various properties of these clusters, and shows how relativity impacts such trends in the properties.

3.2 METHODS OF INVESTIGATION

3.2.1 Experimental Techniques

The generation of heavy main-group clusters can be accomplished through a number of methods, including evaporation methods (71–77), jet expansion methods (78–86), sputtering techniques (87–89), aggregation methods (93–96), rare-gas matrix isolation methods (32), and ligand-stripping methods (91, 92). In recent years, supersonic jet expansion methods have been employed with much success for cluster generation (78–86, 97–101).

Simple evaporation of a source material in an oven can generate small cluster particles of heavy group V elements such as As_2, Sb_2, Bi_2, Bi_4, and Sb_4 (71–77). The oven evaporation method followed by mass-spectrometric analysis can unambiguously determine the cluster size. The use of a Knudsen cell containing an effusion orifice is especially suitable for measuring the binding energies and D_0° values of small clusters using the second- or third-law thermodynamic methods (75–77). Many group III–V clusters have been generated by these methods. The main limitation of a simple evaporation method seems to be size; clusters containing fewer than 10 atoms can be generated by this method. Also, this method is not suitable for heteronuclear clusters and alloy clusters.

A popular method for generating clusters of both main-group and transition-metal atoms appears to be the supersonic jet expansion method (78–86, 97–101). In this method, typically a high-energy laser evaporation method is used to form the vapor of the material. This is then expanded into a small nozzle, resulting in the formation of clusters of varied composition. This technique can be used to generate simple clusters and mixed clusters of both refractory and nonrefractory materials. Laser vaporization followed by gas aggregation can yield cold beams of clusters that can be probed further by laser spectroscopic methods. Typically, a pulsed Nd : YAG laser is employed in this method. Smalley and co-workers (78–86) have pioneered this technique for many main-group and transition-metal clusters. Clusters containing up to 55 atoms of the main-group elements and carbon clusters in the range C_1–C_{190} have been generated by these techniques. Mixed main-group clusters such as Ga_xAs_y, (97–99), In_xP_y, (104, 106) Sn_xBi_y, Pb_xSb_y (84, 110–112), and Ga_xP_y (117) have also been generated by the evaporation of group III–V semiconductor and alloy targets.

Bondybey and co-workers (87–89, 140, 197) have isolated clusters of heavy main-group elements such as Pb_2 and Sn_2, using a sputtering method as well as laser vaporization methods. Clusters of various sizes are sputtered and trapped in rare-gas matrices. The matrix-isolated clusters are then probed by laser-induced fluorescence (LIF) methods. For example, Bondybey and English (87) investigated Pb_2 using the sputtering method. A disadvantage of this method is that the size of the cluster cannot be determined unambiguously.

Cluster formation can be induced via aggregation of atoms on a support such as molecular sieves or a zeolite (93–96). This technique involves simple deposition of elemental vapor onto a target in which the atoms are coalesced into small clusters. Other techniques include the ligand-stripping method (90–92). In this technique, naked clusters are formed by stripping ligands such as CO and H from a more stable organometallic compound. The stripping is normally induced by high-energy collision with species such as rare-gas atoms, a process that can be called *collision-induced dissociation* (CID). For example, As_2^+ can be produced from As_2H_4 by this method (90). The electron-impact method can also be used to produce charge clusters.

Clusters generated by the above-mentioned methods can be probed by a variety of techniques, including laser-induced fluorescence, multiphoton ionization, and dissociation spectroscopy, especially two-photon and two-color methods, resonance multiphoton methods, time-of-flight and quadrupole mass-spectroscopic methods, ion cyclotron resonance mass-spectroscopic methods, Raman spectroscopy, visible–UV electronic spectroscopy, and ESR spectroscopy. The mass-spectral methods are mainly for the detection of clusters, calibration of cluster sizes, and measurement of relative abundance as a function of size. The other spectroscopic techniques provide valuable information on the low-lying electronic states, geometries, stabilities, ionization potentials, and electron affinities as a function of size and the spin multiplicities of the low-lying electronic states.

The negative-ion photoelectron spectroscopic technique has also provided valuable information on the electronic states of neutral species via ionization of an electron of the anion (34, 36, 118–121). The photofragmentation patterns of the cationic (100) and anionic clusters have also provided some fascinating information that is far from understood. For example, the photofragmentation patterns of $Ga_xAs_y^+$ clusters (100) exhibit anomalous patterns that are more reminiscent of metal clusters than of isoelectronic Si_x and Ge_x clusters (98). Similarly, the photofragmentation patterns of Sb_x and Bi_x clusters are dramatically different. The Si_x^+ and Ge_x^+ clusters fragment into daughters containing 6–11 atoms, while metal clusters lose one atom at a time. The photoionization of Ga_xAs_y semiconductor clusters exhibits dramatic even–odd alternations.

3.2.2 Theoretical Techniques of Calculation

All the calculations of dimers and trimers considered here were done with relativistic effective core potentials (44, 55–62). In the earlier investigations of Pb_2, Sn_2, Tl_2, and Bi_2, these potentials were used in conjunction with Slater-type orbital (STO) basis sets of double-ζ + polarization quality. The calculations that employed STO basis sets were done mostly by using the single-configuration self-consistent field (SCF) method or a generalized valence bond (GVB) method in the case of the ground state of Bi_2 followed by relativistic configuration interaction (RCI) calculations. The RCI calculations included the spin–orbit term derived from RECPs as suggested by Ermler et al. (60) and Hafner and Schwarz (63). The calculations including the spin–orbit term are called relativistic CI calculations; they included single and double excitations from a multireference list of configurations. In general, the RCI included all low-lying λ–s states of the same Ω symmetry as reference configurations. For example, the RCI calculations of a state of 0_g^+ symmetry could include as reference configurations those configurations that describe ${}^1\Sigma_g^+(0_g^+)$, ${}^3\Pi_g(0_g^+)$, ${}^3\Sigma_g^-(0_g^+)$, and so on and other states of 0_g^+ symmetry. The RCI thus differs from a normal CI in that in the ordinary CI, ${}^3\Pi_g$ cannot mix with the ${}^1\Sigma_g^+$ or ${}^3\Sigma_g^-$ states.

Most of the theoretical calculations made in the author's laboratory after 1985 were done by using the complete active space MCSCF followed by multireference singles + doubles (CASSCF/MRSDCI/RCI) methods or the first-order CI method employing valence Gaussian basis sets of higher than double-ζ + polarization quality. Recent calculations on In_2, Sb_2, I_2, and so forth, were done with relatively larger basis sets compared to calculations on Ga_2, Ge_2, As_2, and Se_2.

In the CASSCF method, a set of the most important electrons for chemical bonding (active electrons) are distributed in all possible ways among orbitals referred to as the *internal* or *active orbitals*. The active orbitals are normally chosen as the set of orbitals that correlate into valence atomic orbitals at infinite separation of the various atoms in the molecule. The CASSCF method thus provides a zeroth-order starting set of orbitals for inclusion of higher-order correlation effects.

The higher-order electron correlation effects not included in the CASSCF are taken into account by using the configuration interaction method. The CI calculations are carried out by using the second-order CI method (SOCI), the multireference singles + doubles CI method (MRSDCI), and the first-order CI method (FOCI). The first two methods are more accurate compared to the last method. The SOCI calculations included (a) all configurations in the CASSCF, (b) configurations generated by distributing $n-1$ electrons in the internal space and 1 electron in the external space (where n = number of active electrons), and (c) configurations generated by distributing $n-2$ electrons in the internal space and 2 electrons in the external space in all possible ways. The MRSDCI calculations included a subset of configurations determined by the important configurations in the CASSCF (coefficient ≥ 0.07) as reference configurations. Then single + double excitations were allowed in the MRSDCI. A less accurate method compared to MRSDCI labeled POLCI includes all configurations in the MRSDCI except those generated by distributing 2 electrons in the external space.

3.3 SPECTROSCOPIC PROPERTIES AND POTENTIAL-ENERGY CURVES OF HEAVY HOMONUCLEAR DIMERS (Ga_2 TO Bi_2)

3.3.1 Ga_2

Ginter et al. (122) obtained the first absorption spectra of Ga_2 in a furnace in the 19,000–21,000-cm^{-1} region. Douglas et al. (123) reported electronic absorption spectra of A_2, Ga_2, and In_2. Two absorption bands, one in the range 13,000–15,600 cm^{-1} and the other in the range 24,000–30,000 cm^{-1}, were observed. These authors (123) concluded that the ground state of these dimers was $^1\Sigma_g^+$. *Ab initio* calculations on Al_2 (124) have revealed, however, that the $^3\Pi_u$ state is the lowest state, and $^1\Sigma_g^+$ is much higher in energy. Basch et al. (124) reassigned the two electronic bands of Al_2 observed by Douglas et al. (123) to this $^3\Pi_g \rightarrow {}^3\Pi_u$ and $^3\Pi_g(2) \rightarrow {}^3\Pi_u$ transitions. The vibrational frequency of the ground state of Ga_2 was reported by Froben et al. (192). Related B_2 (127, 129) and Al_2 (124, 126, 130) molecules have also been studied.

Thermodynamic studies of Ga_2 and related molecules (131, 132) revealed that the dissociation energy of Ga_2 is about 1.4 eV. Since the electronic states of Ga_2 have not been well characterized, earlier thermodynamic calculations of D_e assumed an incorrect ground state and partition function (132). The Ga_2

dimer has also been observed by laser irradiation of a GaAs crystal (125) and in a supersonic jet beam (97).

The author (133) carried out CASSCF/FOCI calculations on Ga_2 that employed RECPs. The RECPs included the outermost $4s^2 4p^1$ shells in the valence space replacing the rest of the core electrons by relativistic effective core potentials. A ($3s3p1d$) valence Gaussian basis set was employed for the Ga atom. The d exponent was optimized for the ground state of Ga_2 at the CASSCF level. The author (402) also studied many electronic states of Ga_2^+ and Ga_2^-. There was an error in one of the ECP parameter used in the earlier study (133), which led to longer bond lengths; however, this was subsequently corrected.

Table 3.1 shows the possible low-lying states of Ga_2, while Table 3.2 depicts their dissociation limits and atomic energy separations (134) at the dissociation limits. As seen from these tables, there are at least 20 low-lying states of Ga_2 in the absence of the spin–orbit term.

Table 3.3 shows the corrected spectroscopic properties of bound states of Ga_2 obtained by the author (133). Among the 18 electronic states investigated by the author, the $^3\Delta_u$, $^3\Sigma_u^+$, $^3\Sigma_u^+$(II), $^3\Pi_g$, and $^1\Pi_g$ curves were found to be repulsive. The calculated potential-energy curves of some of many states are shown in Figure 3.1. Note that although the earlier work on Ga_2 was performed with a defective ECP, this was subsequently corrected, and Figure 3.1 reports the correct curves. As seen from Table 3.3, the calculated ground state of Ga_2 is $^3\Pi_u$. However, the T_e value of $^3\Sigma_g^-$ is only 410 cm^{-1}. The $^1\Sigma_g^-$ state is 4734 cm^{-1} above the $^3\Pi_u$ state. Evidently, this state does not appear to be the ground state of Ga_2. Consequently, Douglas et al.'s (123) assignment of $^1\Sigma_g^+$ to the ground state of $Ga_2(Al_2)$ is inconsistent with the author's electronic structure calcula-

Table 3.1 A Few Low-Lying Configurations and the Electronic States Arising from Them for Ga_2

MO Configuration	λ–s States
$1\sigma_g^2 1\sigma_u^2 2\sigma_g 1\pi_u$	$^3\Pi_u$, $^1\Pi_u$
$1\sigma_g^2 1\sigma_u^2 1\pi_u^2$	$^3\Sigma_g^-$, $^1\Delta_g$, $^1\Sigma_g^+$
$1\sigma_g^2 1\sigma_u^2 1\pi_u 1\pi_g$	$^3\Delta_u$, $^3\Sigma_u^+$, $^3\Sigma_u^-$, $^1\Delta_u$, $^1\Sigma_u^-$, $^1\Sigma_u^+$
$1\sigma_g^2 1\sigma_u 2\sigma_g 1\pi_u^2$	$^5\Sigma_u^-$, $^3\Sigma_u^-(2)$, $^1\Sigma_u^-$, $^3\Delta_u$, $^3\Sigma_u^+$, $^1\Delta_u$, $^1\Sigma_u^+$
$1\sigma_g^2 1\sigma_u^2 2\sigma_g^2$	$^1\Sigma_g^+$
$1\sigma_g^2 1\sigma_u^2 2\sigma_g 1\pi_g$	$^3\Pi_g$, $^1\Pi_g$
$1\sigma_g^2 1\pi_u^4$	$^1\Sigma_g^+$
$1\sigma_g^2 1\sigma_u^2 1\pi_g^2$	$^3\Sigma_g^-$, $^1\Delta_g$, $^1\Sigma_g^+$
$1\sigma_g^2 1\sigma_u^2 2\sigma_u^2$	$^1\Sigma_g^+$
$1\sigma_g^2 1\sigma_u^2 2\sigma_g 2\sigma_u$	$^3\Sigma_u^+$, $^1\Sigma_u^+$

Table 3.2 Dissociation Relationships for a Few Low-Lying States of Ga_2

λ–s States	Dissociation Limit	
	Atomic Levels	Energy, Exp.[a]
$^3\Sigma_g^-$, $^3\Pi_u$, $^3\Pi_g$, $^3\Delta_u$ $^3\Sigma_u^+(2)$, $^1\Delta_g$, $^1\Pi_u$ $^1\Pi_g$, $^1\Sigma_u^-$, $^1\Sigma_g^+(2)$	$^2P + {}^2P$	0.0
$^3\Pi_g$, $^3\Pi_u$, $^3\Sigma_g^+$, $^3\Sigma_u^+$ $^1\Pi_g$, $^1\Pi_u$, $^1\Sigma_g^+$, $^1\Sigma_u^+$	$^2P + {}^2S(4s^25s)$	24,789

[a]From Reference 134.

tions. Table 3.4 shows the adiabatic and vertical transition energies of allowed electronic dipole transitions for Ga_2. For Ga_2, Douglas et al. (123) reported two absorption bands, one in the 15,590-cm^{-1} region and the other in the 29,934-cm^{-1} region. These bands were assigned by those authors to $A^1\Pi_u \rightarrow X^1\Sigma_g^+$ and $B^1\Sigma_u^+ \rightarrow X^1\Sigma_g^+$, respectively. However, as seen from Table 3.4, the $^1\Pi_u$ state is below the $^1\Sigma_g^+$ state. Further, the most probable candidate for the ground state is $^3\Pi_u$. The author (133) argued that the most consistent allowed electric dipole transition for the 15,590 cm^{-1} is the $A^3\Pi_g \rightarrow X^3\Pi_u$ transition. The absorption bands in the 29,000-cm^{-1} region could be assigned to the $^3\Pi_g(\text{II}) \rightarrow X^3\Pi_u$ transition. The theoretical T_{vert} (29,102 cm^{-1}) is most consistent with the observed bands in this region.

Ginter et al. (122) have reported emission bands in the 19,500–22,000-cm^{-1} region for Ga_2. It can be inferred from Table 3.3 and 3.4 that the most prob-

Table 3.3 FOCI Spectroscopic Properties of Ga_2[a]

State	R_e, Å	T_e, cm^{-1}	ω_e, cm^{-1}	D_e, eV
$X^3\Pi_u$	2.762	0	158	1.18
$^3\Sigma_g^-$	2.506	446	197	1.13
$^1\Pi_u$	2.802	4,037	145	0.68
$^1\Sigma_g^+$	2.954	4,129	129	0.67
$^1\Delta_g$	2.584	4,274	168	0.65
$^1\Sigma_g^+(\text{II})$	2.912	8,267	125	—
$^3\Sigma_u^-$	2.681	20,978	166	—
$^1\Delta_u$	2.702	28,772	210	—
$^3\Pi_g(\text{II})$	3.495	29,877	86	—
$^3\Delta_u(\text{II})$	2.407	30,325	292	—
$^1\Sigma_u^+$	2.757	35,706	210	—

[a]From Reference 133. The A state observed in the $A \leftarrow X$ broad band was a system assigned to the $^3\Pi_g \leftarrow {}^3\Pi_u$ transition. (133)

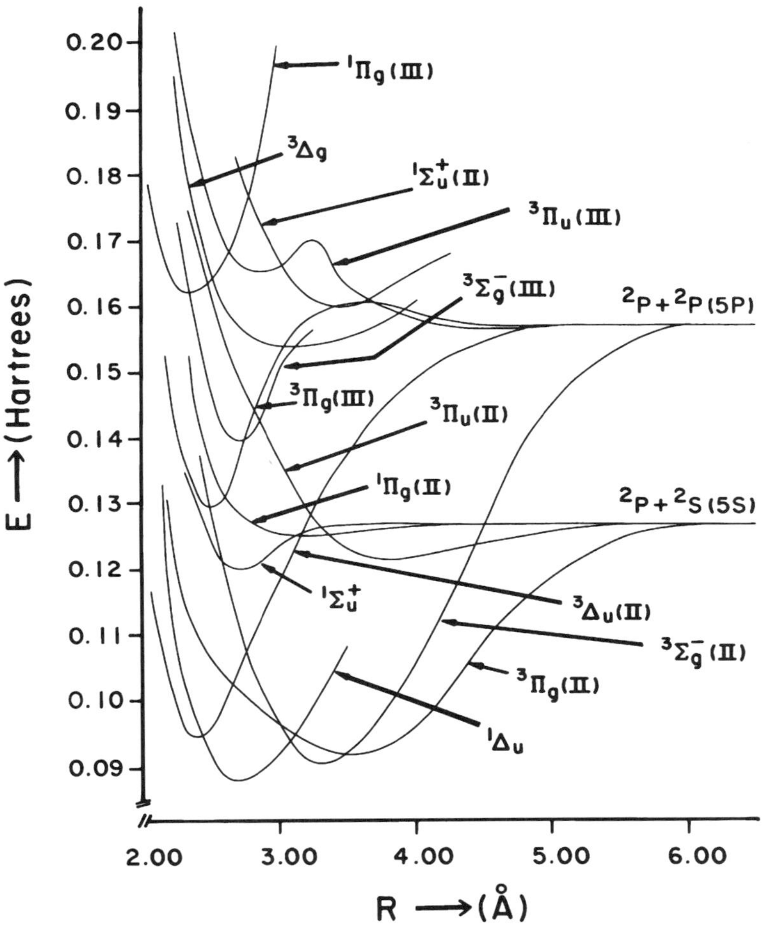

Figure 3.1 Potential-energy curves for the electronic states of Ga_2. (Reprinted from Ref. 133; copyright 1986, American Chemical Society.)

able transition for these bands is the ${}^3\Sigma_u^-(\mathrm{I}) \rightarrow {}^3\Sigma_g^-(\mathrm{I})$ transition. The theoretical transition energy for this transition (21,191 cm^{-1}) was found to be in very good agreement with the observed bands in this region (19,000–21,000 cm^{-1}).

The theoretical dissociation energy of the ground state of Ga_2 obtained by the author (133) was found to be 1.2 eV. This was in reasonable agreement with the thermodynamic value of 1.4 eV (131, 132). However, the thermodynamic value needs to be recalculated by using the correct partition function for Ga_2. The author's calculations on Ga_2 should be useful in calculating the partition function.

A number of low-lying allowed electric dipole transitions exist for Ga_2, many of which have not yet been observed (Table 3.4). Specifically, ${}^3\Sigma_g^- \rightarrow {}^3\Sigma_u^-(\mathrm{II})$

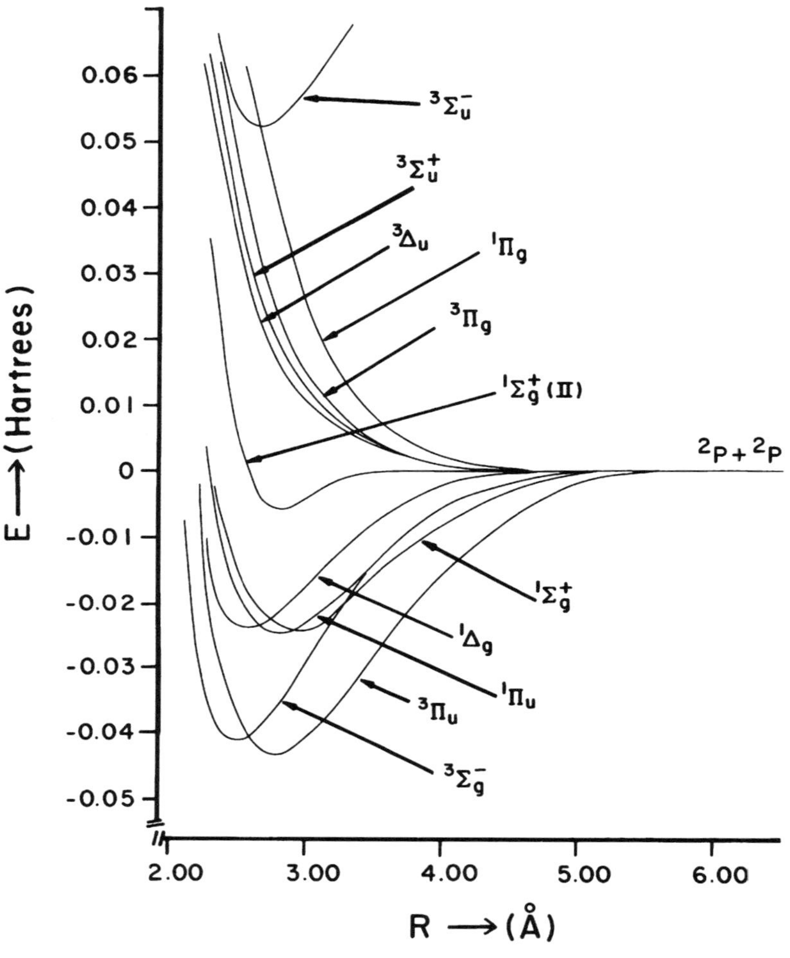

Figure 3.1 *(Continued)*

is a dipole-allowed transition, which falls in the region of 41,000 cm^{-1}. The theoretical transition energy should be slightly higher than the true energy. The $^3\Sigma_g^-(II) \leftarrow X^3\Pi_u$ and $^3\Sigma_g^-(I) \rightarrow X^3\Pi_u$ transitions are allowed in the perpendicular direction. The former transition energy is in the 28,000-cm^{-1} region. Transitions originating from the $^1\Pi_u$ electronic state to the $^1\Pi_g$ as well as $^1\Pi_g$(II) states should also be investigated.

Table 3.5 contains the contributions of various electronic configurations to the electronic states of Ga_2. As seen from Table 3.5, the ground state ($^3\Pi_u$) and $^1\Pi_u$ are predominantly $1\sigma_g^2 1\sigma_u^2 2\sigma_g 1\pi_u$. The $^3\Sigma_g^-$ and $^1\Delta g$ states arise from the $1\sigma_g^2 1\sigma_u^2 1\pi_u^2$ configuration. The $^1\Sigma_g^+$ state exhibits an interesting behavior as a function of internuclear distance. At short and near-equilibrium distances this

Table 3.4 Vertical and Adiabatic Excitation Energies for Dipole-Allowed Transitions

Electronic Transitions	T_{vert}, cm^{-1}	T_e, cm^{-1}
$A^3\Pi_g \leftarrow X^3\Pi_u$	14,147	—[a]
$^3\Sigma_g^- \rightarrow {}^3\Sigma_u^-$	22,052	21,191
$^3\Sigma_g^- \rightarrow {}^3\Sigma_u^-(II)$	41,570	41,307
$^3\Sigma_u^- \rightarrow {}^3\Sigma_g^-$	1,288	410
$^3\Sigma_u^- \rightarrow {}^3\Sigma_g^-(II)$	32,419	27,954
$B^3\Sigma_g^-(II) \rightarrow X^3\Pi_u$	29,102	27,565
$^1\Pi_u \rightarrow {}^1\Sigma_g^+$	1,029	1,026
$^1\Pi_u \rightarrow {}^1\Pi_g$	12,248	—[a]
$^1\Pi_u \rightarrow {}^1\Pi_g(II)$	31,172	29,493

[a]Final electronic state is not bound.
Source: Reference 133.

state is a mixture of $1\sigma_g^2 1\sigma_u^2 1\pi_u^2$ and $1\sigma_g^2 1\sigma_u^2 2\sigma_g^2$ configurations. Consequently, the R_e value of the $^1\Sigma_g^+$ state is quite different from the R_e values of the $^3\Sigma_g^-$ and $^1\Delta_g$ states, which are predominantly $1\sigma_g^2 1\sigma_u^2 1\pi_u^2$. At longer distances, $^1\Sigma_g^+(II)$ exhibits approximately the opposite behavior in that $1\sigma_g^2 1\sigma_u^2 1\pi_u^2$ makes a significant contribution at long distances. The Rydberg configurations contribute substantially to the $^3\Sigma_u^+(III)$, $^3\Sigma_u^+(III)$, and $^3\Delta_u$ states at short distances.

3.3.2 Ge_2

Froben and Schulze (135) carried out a Raman–fluorescence spectroscopic study of Ge_2. Gingerich and co-workers (136) obtained the dissociation energy of Ge_2. Shim et al. (137) and Pacchioni (138) studied the lowest-lying states of Ge_2 using the Hartree–Fock (HF)/CI method. The molecular orbitals for CI calculations were obtained by a single-configuration SCF treatment. However, as pointed out by Shim et al. (137), starting with different MOs for different states led to different results. Thus a single-configuration SCF treatment does not seem to be very reliable for obtaining the spectroscopic properties of excited states. Also, the splitting between the $^3\Sigma_g^-$ and $^3\Pi_u$ states is small for Ge_2, analogous to Ga_2.

The author (139) used the CASSCF/FOCI techniques for Ge_2 employing RECPs that retained the $4s^2 4p^2$ outer shells in the valence space. A ($3s3p1d$) valence Gaussian basis set of the same quality employed for Ga_2 was used for Ge_2. Table 3.6 shows a few low-lying MO configurations and a list of possible λ–s states for Ge_2. Table 3.7 shows the dissociation limits for the low-lying states of Ge_2. As seen from these tables, there are many low-lying states for Ge_2.

Table 3.8 shows the theoretical spectroscopic properties (R_e, T_e, ω_e) of 14 low-lying electronic states of Ge_2 obtained by the author (139). Figure 3.2

Table 3.5 Contributions (%) of Various Configurations to the Electronic States of Ga_2

State	Distance	$1\sigma_g^2 1\sigma_u^2 2\sigma_g\pi_u$	$1\sigma_g^2 1\sigma_u^2 \pi_u^2$	$1\sigma_g^2 1\sigma_u^2 \pi_u\pi_g$	$1\sigma_g^2 1\sigma_u^2 2\sigma_u\pi_u$	$1\sigma_g^2 1\sigma_u^2 2\sigma_g\pi_g$	$1\sigma_g^2 1\sigma_u^2 2\sigma_g^2$	$1\sigma_g^2 1\sigma_u^2 \pi_g^2$	$1\sigma_g^2 1\sigma_u^2 2\sigma_u\pi_g$
${}^3\Pi_u$	3.00	89	—	—	—	—	—	—	1.57
	5.00	70	—	—	—	—	—	—	22
${}^3\Sigma_g^-$	2.70	—	88	—	—	—	—	2	—
	4.00	—	72	—	—	—	—	19.3	—
${}^1\Pi_u$	3.00	88	—	—	—	—	—	—	2.5
	5.00	64	—	—	—	—	—	—	27
${}^1\Delta_g$	2.85	—	85	—	—	—	—	5.6	—
	3.00	—	54	—	—	—	—	37	—
${}^1\Sigma_g^+$	3.00	—	48	—	—	—	32	4.2	—
	3.25	—	38	—	—	—	38	4.2	—
	3.75	—	52	—	—	—	18	15	—
	5.00	—	53	—	—	—	—	38	—
${}^1\Sigma_g^+$(II)	3.00	—	31	—	—	—	40	8.3	—
	3.25	—	37	—	—	—	32.5	12	—
	3.75	—	15	—	—	—	48	10	—
${}^3\Delta_u$	2.80	—	—	—	—	—	—	—	—
${}^3\Pi_g$	3.00	—	—	91	8	73.3	—	—	—
	5.00	—	—	—	31	60	—	—	—

$^3\Sigma_u^-$	2.70	—	68	—	—	—	—	8.6	1	—
	3.00	—	66	—	—	—	—	12	2	—
	4.00	—	84	—	—	—	—	—	—	—
$^1\Pi_g$	3.00	—	—	18	64	—	—	—	—	—
	5.00	—	—	34	58	—	—	—	—	—
$^3\Pi_g$(II)	3.00	—	—	54	5	—	13.7	—	—	—
	5.00	—	—	56	29	—	—	—	—	—
$^3\Sigma_g^-$(II)	4.00	17	—	—	—	66	—	—	—	—
$^3\Sigma_u^-$(II)	2.50	—	4	—	—	—	—	76	—	—
	2.70	—	3	—	—	—	—	74	—	—
	3.00	—	2	—	—	—	—	70	—	—
$^1\Pi_g$(II)	3.00	—	—	47.5	16	—	16	—	—	—
	5.00	—	—	50	29	—	—	—	—	—
$^3\Sigma_u^+$	2.55	—	90	—	—	—	—	—	—	—
	2.80	—	91	—	—	—	—	—	—	—
$^3\Sigma_u^+$(II)	2.25	—	—	—	—	—	—	79	3.3	2
	2.55	—	—	—	—	—	—	43	—	21
	2.80	—	—	—	—	—	—	3	—	49
$^3\Sigma_u^+$(III)	2.55	—	—	—	—	—	—	33	—	24
	2.80	—	—	—	—	—	—	69	7.4	1
$^3\Delta_u$(II)	2.80	—	—	—	—	—	—	76	4	—

Table 3.6 A Few Low-Lying Electronic Configurations of Ge_2 and the λ–s States Arising from Them[a]

MO Configuration	λ–s States
$2\sigma_g^2 1\pi_u^2$	${}^3\Sigma_g^-$, ${}^1\Delta_g$, ${}^1\Sigma_g^+$
$2\sigma_g 1\pi_u^3$	${}^3\Pi_u$, ${}^1\Pi_u$
$1\pi_u^4$	${}^1\Sigma_g^+$
$2\sigma_g 2\sigma_u 1\pi_u^2$	${}^5\Sigma_u^-$, ${}^3\Sigma_u^-(2)$, ${}^1\Sigma_u^-$, ${}^3\Delta_u$, ${}^3\Sigma_u^+$, ${}^1\Delta_u$, ${}^1\Sigma_u^+$
$2\sigma_g^2 2\sigma_u 1\pi_u$	${}^3\Pi_g$, ${}^1\Pi_g$
$2\sigma_g^2 1\pi_u 1\pi_g$	${}^3\Delta_u$, ${}^3\Sigma_u^-$, ${}^3\Sigma_u^+$, ${}^1\Delta_u$, ${}^1\Sigma_u^+$, ${}^1\Sigma_u^-$

[a]The $1\sigma_g^2 1\sigma_u^2$ shell is not shown.

shows the potential-energy curves of some of these states. As seen from Table 3.8, the ground state of Ge_2 is $X^3\Sigma_g^-$; however, the ${}^3\Pi_u$ state is only 767 cm^{-1} above the ${}^3\Sigma_g^-$ state. Since this splitting is small, the possibility of ${}^3\Pi_u$ being the ground state cannot be ruled out, since higher-order correlation corrections not included in the author's calculations (139) may reverse the ordering. Table 3.9 shows the adiabatic transition energies of some allowed electric dipole transitions. The most important transition originating from the ground state (${}^3\Sigma_g^-$) is the ${}^3\Sigma_u^- \leftrightarrow X^3\Sigma_g^-$ transition. The theoretical energy separation for this transition is 20,979 cm^{-1}. The corresponding transition was observed for Sn_2 at 18,222 cm^{-1} (140). It seems that, to date, this transition has not been observed for Ge_2. The spectroscopic bands in this region should be intense and well resolved. The ${}^3\Pi_u \leftrightarrow \Sigma_g^-$ transition is also important. The author's calculations predicted this transition to be in the 13,000-cm^{-1} region. The $3\Pi_u \leftrightarrow^3 \Sigma_g^-$ transition energy for Ge_2 is much lower than the corresponding $0_u^+ \leftrightarrow 0_g^+$ transition for Sn_2 or Pb_2.

The ${}^1\Sigma_g^+$ state may be observable in the ${}^3\Sigma_u^-(0_u^+) \leftrightarrow^1 \Sigma_g^+(0_g^+)$ emission, since Ge_2 is a reasonably heavy molecule. The theoretical energy for this transition is

Table 3.7 Dissociation Relationships for a Few Molecular States of Ge_2

Molecular States	Atomic States	Energy of the Separated Atoms[a]
${}^1\Sigma_g^+(2)$, ${}^1\Sigma_u^-$, ${}^1\Pi_g$, ${}^1\Pi_u$, ${}^1\Delta_g$, ${}^3\Sigma_u^+(2)$, ${}^3\Sigma_g^-$, ${}^3\Pi_g$, ${}^3\Pi_u$, ${}^3\Delta_u$, ${}^5\Sigma_g^+(2)$, ${}^5\Sigma_u^-$, ${}^5\Pi_g$, ${}^5\Pi_u$, ${}^5\Delta_g$	${}^3P + {}^3P$	0.0
${}^3\Sigma_g^+$, ${}^3\Sigma_u^+$, ${}^3\Sigma_g^-(2)$, ${}^3\Sigma_u^-(2)$, ${}^3\Pi_g(3)$, ${}^3\Pi_u(3)$, ${}^3\Delta_g(2)$, ${}^3\Delta_u(2)$, ${}^3\Phi_g$, ${}^3\Phi_u$	${}^3P + {}^1D$	7126

[a]From Reference 134.

TABLE 3.8 Theoretical Spectroscopic Constants of Ge_2[a]

State	R_e, Å	T_e, cm^{-1}	ω_e, cm^{-1}
$^3\Sigma_g^-$	2.44	0	259
$^3\Pi_u$	2.34	767	251
$^1\Delta_g$	2.50	3,982	255
$^1\Sigma_g^+$	2.44	5,521	192
$^1\Pi_u$	2.35	5,794	269
$^1\Sigma_g^+$(II)	2.41	10,840	258
$^3\Delta_u$	2.88	13,534	162
$^3\Pi_g$	2.67	13,995	167
$^3\Pi_g$(II)	2.69	17,373	165
$^1\Pi_g$	2.95	17,465	99
$^3\Sigma_u^-$	2.93	20,979	152
$^3\Sigma_g^-$(II)	3.68	24,634	100
$^1\Delta_u$	2.86	29,078	156
$^1\Pi_u$(II)	3.25	32,663	125

[a]The experimental D_0° = 2.65 eV [see Kingcade et al. (136)] compared to CASSCF/FOCI $D_e(Ge_2)$ = 2.3 eV. All spectroscopic constants are from Reference 139.

15,460 cm^{-1} (139). If the $^3\Sigma_u^-$ state can be populated, then this transition should be observable in emission. In the same way, the $^3\Sigma_u^-(0_u^+) \rightarrow {}^1\Sigma_g^+$(II) transition should be observable in the 10,140-cm^{-1} region. The $^3\Pi_g(\text{II}) \leftrightarrow {}^3\Pi_u$ transition is also an important allowed electric dipole transition. As seen from Table 3.9, this transition occurs at 16,503 cm^{-1}. The electronic spectra that correspond to the $^1\Pi_g \leftrightarrow {}^1\Pi_u$ and $^1\Pi_u(\text{II}) \leftrightarrow {}^1\Pi_g$ transitions should be in the regions of 12,204 and 15,198 cm^{-1}, respectively.

The theoretical CASSCF/FOCI (139) D_e value for the ground state $X^3\Sigma_g^-$ of Ge_2 is 2.29 eV, in reasonable agreement with the experimental value of 2.65 eV, reported by Gingerich and co-workers (136). The small discrepancy between the calculated and observed values could be attributed to higher-order correlation corrections, as well as the use of effective core potentials. The author's D_e (139) was found to be close to the HF/CI calculation of Pacchioni (138), which yielded a D_e value of 2.34 eV. The author's vibrational frequencies of the $^3\Sigma_g^-$ and $^3\Pi_u$ states were in reasonable agreement with those of Shim et al. (137). The value reported by Pacchioni (138) for the $^3\Pi_u$ state (195 cm^{-1}), however, is different from the author's and that of Shim et al. (137). For both Sn_2 and Pb_2 the vibrational frequencies of $^3\Sigma_g^-$ and $^3\Pi_u$ differ by at most 10%. However, Pacchioni's ω_e for 3Π is 17% smaller.

Table 3.10 shows the contributions of various MO configurations to the FOCI wavefunction of the electronic states of Ge_2. As seen from Table 3.10, the $^3\Sigma_g^-$ and $^1\Delta_g$ states are predominantly $2\sigma_g^2 1\pi_u^2$. The $^3\Pi_u$ and $^1\Pi_u$ states are composed predominantly of the $2\sigma_g 1\pi_u^3$ configuration. Like Ga_2, the $^1\Sigma_g^+$ state

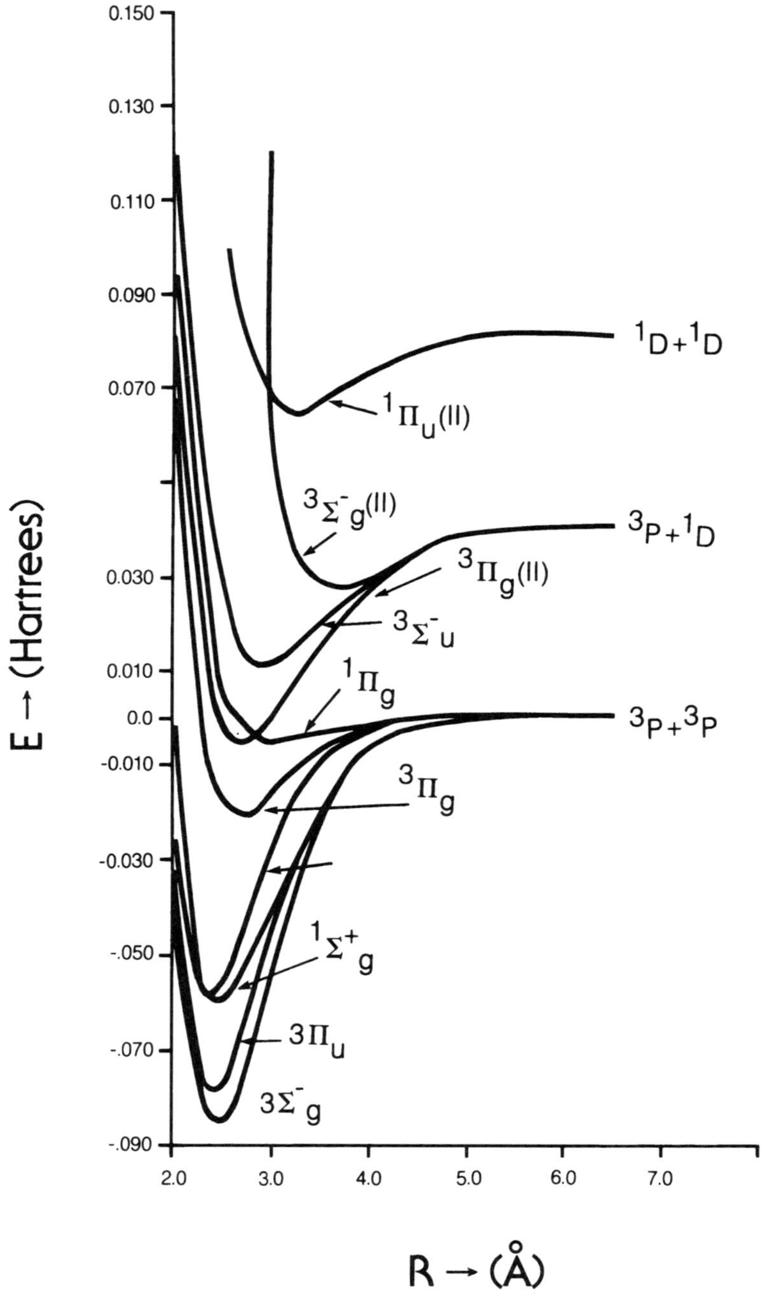

Figure 3.2 Potential-energy curves for the low-lying electronic states of Ge_2. (Reprinted from Ref. 139; copyright 1987, Academic Press, Inc.)

Table 3.9 Adiabatic Transition Energies of Allowed Electric Dipole Transitions of Ge_2

Transition	Energy, cm^{-1}	Transition	Energy, cm^{-1}
$^3\Pi_u \leftrightarrow {}^3\Sigma_g^-$	767	$^3\Pi_g$(II) $\leftrightarrow {}^3\Pi_u$	16,503
$^3\Sigma_g^-$(II) $\leftrightarrow {}^3\Sigma_u^-$	3,665	$^3\Sigma_u^- \leftrightarrow {}^3\Sigma_g^-$	20,979
$^3\Pi_g$(II) $\leftrightarrow {}^3\Delta_u$	3,839	$^1\Pi_u$(II) $\leftrightarrow {}^1\Sigma_g^+$(II)	21,823
$^1\Sigma_g^+$(II) $\leftrightarrow {}^1\Pi_u$	5,579	$^3\Pi_u$(II) $\leftrightarrow {}^3\Sigma_g^-$	24,357
$^1\Pi_g \leftrightarrow {}^1\Pi_u$	12,204	$^1\Delta_u \leftrightarrow {}^1\Delta_g$	25,096
$^3\Pi_g \leftrightarrow {}^3\Pi_u$	13,125	$^1\Pi_u$(II) $\leftrightarrow {}^1\Sigma_g^+$	27,142
$^1\Pi_u$(II) $\leftrightarrow {}^1\Pi_g$	15,198	$^1\Pi_u$(II) $\leftrightarrow {}^1\Delta_g$	28,681

Source: Reference 139.

of Ge_2 exhibits an interesting behavior as a function of internuclear distance. At short distances, it was found to be composed predominantly of the $1\pi_u^4$ configuration. At near-equilibrium geometries this state was found to be a mixture of $2\sigma_g^2 1\pi_u^2$ and $1\pi_u^4$ configurations. The $^1\Sigma_g^+$(II) state is also a mixture of these two configurations, with the latter configuration making a greater contribution at its near-equilibrium geometry, resulting in avoided crossings.

Table 3.10 Contributions of Various Configurations to the Electronic States of Ge_2 at Their Equilibrium Geometries

State	Configurations
$^3\Sigma_g^-$	$2\sigma_g^2 1\pi_u^2$ (85%), $2\sigma_g^2 1\pi_g^2$ (2%)
$^3\Pi_u$	$2\sigma_g 1\pi_u^3$ (80%), $2\sigma_g 1\pi_u 1\pi_g^2$ (7%)
$^1\Delta_g$	$2\sigma_g^2 1\pi_u^2$ (83%), $2\sigma_g^2 1\pi_g^2$ (5%)
$^1\Pi_u$	$2\sigma_g 1\pi_u^3$ (80%), $2\sigma_g 1\pi_u 1\pi_g^2$ (8%)
$^1\Sigma_g^+$	$2\sigma_g^2 1\pi_u^2$ (63%), $1\pi_u^4$ (17%), $2\sigma_g^2 1\pi_g^2$ (6%)
$^1\Sigma_g^+$(II)	$1\pi_u^4$ (52%), $2\sigma_g^2 1\pi_u^2$ (17%), $1\pi_u^2 1\pi_g^2$ (14%)
$^3\Delta_u$	$2\sigma_g^2 1\pi_u 1\pi_g$ (82%), $1\pi_u^3 1\pi_g$ (3%)
$^3\Pi_g$	$2\sigma_g 1\pi_u^2 1\pi_g$ (83%), $2\sigma_g 1\pi_g^3$ (2%)
$^3\Pi_g$(II)	$2\sigma_g 1\pi_u^2 1\pi_g$ (83%), $2\sigma_u 1\pi_u 1\pi_g^2$ (2%)
$^3\Sigma_u^-$	$2\sigma_g^2 1\pi_u 1\pi_g$ (57%), $2\sigma_g 2\sigma_u 1\pi_u^2$ (17%)
	$1\pi_u^3 1\pi_g$ (8%), $2\sigma_g 2\sigma_u 1\pi_g^2$ (4%)
$^3\Sigma_g^-$(II)	$2\sigma_g 2\sigma_u 1\pi_u 1\pi_g$ (47%), $2\sigma_g^2 1\pi_g^2$ (16%), $2\sigma_g^2 1\pi_u^2$ (15%), $1\pi_u^2 1\pi_g^2$ (10%)
$^1\Delta_u$	$2\sigma_g^2 1\pi_u 1\pi_g$ (54%), $2\sigma_g 2\sigma_u 1\pi_u^2$ (17%), $1\pi_u^3 1\pi_g$ (14%), $2\sigma_g 2\sigma_u 1\pi_g^2$ (3%)
$^1\Pi_u$(II)	$2\sigma_g 1\pi_u 1\pi_g^2$ (57%), $2\sigma_g 1\pi_u^3$ (15%), $2\sigma_u 1\pi_u^2 1\pi_g$ (11%)

Source: Reference 139.

3.3.3 As_2

Gibson and MacFarlane (71) obtained the first absorption spectrum of As_2 in the 40,500-cm^{-1} region. The vibrational levels of the excited state participating in these bands were highly perturbed and predissociated, which led to the prediction of crossing of another potential-energy curve with this curve. Subsequently, a number of authors studied the As_2 molecule experimentally (141–154, 158). These studies included both absorption and emission spectra of As_2 in the visible and vacuum UV regions. Particular systems that have been observed include $A \leftrightarrow X$, $C \rightarrow X$, $C \leftarrow A$, $a \rightarrow X$, $B \rightarrow X$, $d \rightarrow X$, and $D \rightarrow c$. Among these, the $A \leftrightarrow X$ system has been extensively studied. Vibrational and rotational analyses of a number of observed systems have been carried out (146, 150). Heimbrook et al. (156) studied the As_2 dimer generated by evaporating gallium arsenide, which was then trapped in solid neon. These authors (156) studied vibronic bands of three systems of As_2 designated as the $c \leftrightarrow X$, $e \rightarrow X$, and $a \leftrightarrow X$ systems. Both the ionization potential (147) and the electron affinity (148) of As_2 have been measured.

Theoretical studies on As_2 include those of Kok and Hall (155), Watanabe et al. (157), and the author (159), although the first two works were restricted only to the ground state of As_2 (155, 157). The author (159) carried out CASSCF/FOCI calculations on 18 electronic states of As_2 using RECPs ($4s^24p^3$ shell of As included explicitly in valence) and a valence ($3s3p1d$) Gaussian basis set. The CASSCF active space included valence $4s$ and $4p$ orbitals of the two As atoms.

Table 3.11 shows the possible low-lying electronic states of As_2 and their dissociation limits. As seen in Table 3.11, because of the $4s^24p^3$ open-shell ground state of the As atom, As_2 has the largest number of possible valence electronic states of all the dimers in that row. The ground state of As_2 is a $^1\Sigma_g^+$

Table 3.11 Dissociation Relationships for a Few Low-Lying States of As_2

Molecular States	Atomic States	Energy (cm^{-1})[a]
$^1\Sigma_g^+$, $^3\Sigma_u^+$, $^5\Sigma_g^+$, $^7\Sigma_u^+$	$^4S + {}^4S$	0.0
$^3\Sigma_g^+$, $^3\Sigma_u^+$, $^3\Pi_g$, $^3\Pi_u$, $^3\Delta_g$, $^3\Delta_u$, $^5\Sigma_g^+$, $^5\Sigma_u^+$, $^5\Pi_g$, $^5\Pi_u$, $^5\Delta_g$, $^5\Delta_u$	$^4S + {}^2D$	10,790
$^3\Sigma_g^-$, $^3\Sigma_u^-$, $^3\Pi_g$, $^3\Pi_u$, $^5\Sigma_g^-$, $^5\Sigma_u^-$, $^5\Pi_g$, $^5\Pi_u$	$^4S + {}^2P$	18,530
$^1\Sigma_g^+(3)$, $^1\Sigma_u^-(2)$, $^1\Pi_g(2)$, $^1\Pi_u(2)$, $^1\Delta_g(2)$, $^1\Delta_u$, $^1\Phi_g$, $^1\Phi_u$, $^1\Gamma_g$, $^3\Sigma_u^+(3)$, $^3\Sigma_g^-(2)$, $^3\Pi_u(2)$, $^3\Pi_g(2)$, $^3\Delta_u(2)$, $^3\Delta_g$, $^3\Phi_u$, $^3\Phi_g$, $^3\Gamma_u$	$^2D + {}^2D$	21,580

[a]Experimental atomic energy separations from Reference 134.

state arising from $1\sigma_g^2 1\sigma_u^2 2\sigma_g^2 1\pi_u^4$, which represents a triple bond between As atoms.

Table 3.12 shows the spectroscopic properties of a number of low-lying states of As_2. Also indicated in that table are the assignments of the states that have been observed experimentally. Figure 3.3 shows the potential-energy curves of low-lying singlet states, and Figure 3.4 shows the potential-energy curves of the triplet states. As seen from Table 3.12, the theoretical bond length of the ground state is ~0.06 Å longer than the experimental value. The slightly longer bond was attributed by the author (159) to higher-order correlation effects that were not included in the FOCI calculations and the use of RECPs.

The strongest electric dipole transition from the ground state is the $A^1\Sigma_u^+ \leftrightarrow X^1\Sigma_g^+$ transition. The first $^1\Sigma_u^+$ state is calculated 43,207 cm^{-1} above the ground state (Table 3.12). This has been characterized experimentally as the $A \leftrightarrow X$ system in the 40,350-cm^{-1} region. The calculated ω_e and R_e values of the A state are in very good agreement with the experimental values. The $^1\Sigma_u^+$ state is a mixture of the $1\sigma_g^2 1\sigma_u^2 2\sigma_g^2 1\pi_u^3 1\pi_g$, $1\sigma_g^2 1\sigma_u^2 2\sigma_g 2\sigma_u 1\pi_u^4$, and $1\sigma_g^2 1\sigma_u^2 1\pi_u 1\pi_g^3$ configurations.

The predissociated $A \leftrightarrow X$ bands have been studied by many authors. The author's (159) theoretical calculations supported the existence of a B state near the A state, which had the same symmetry, since the A state is a mixture of several configurations. The exact nature of the state predissociating the $A \leftrightarrow X$ bands has not yet been established.

Table 3.12 Spectroscopic Properties of As_2[a]

State	Theory	Exp.	Theory	Exp.	Theory	Exp.
$X^1\Sigma_g^+$	2.164	2.103	0	0	394	430
$c^3\Sigma_u^+$	2.418	2.304	11,860	14,500	235	314
$e^3\Delta_u$	2.357	—	19,976	19,915	324	330
$^3\Pi_g$	2.58	—	26,053	—	316	—
$a^3\Sigma_u^- 0_u^+$	2.345	2.279	26,406	24,641	341	337
$^3\Delta_g$	2.756	—	30,676	—	179	—
$^1\Delta_u$	2.313	—	30,743	—	341	—
$^1\Sigma_g^+$(II)	2.561	—	32,722	—	274	—
$^1\Pi_g$	2.298	—	33,262	—	317	—
$^1\Sigma_g^+$(III)	2.654	—	41,228	—	221	—
$A^1\Sigma_u^+$	2.466	[2.50]	43,207	[40,349]	253	[260]
$b^3\Pi_u$(II)	2.561	—	44,669	—	200	—
$^1\Pi_u$	2.535	—	45,327	—	216	—
$^1\Delta_g$	2.747	—	47,668	—	174	—
$^1\Pi_u$(II)	2.74	—	51,294	—	108	—

[a]Theoretical spectroscopic constants are from Reference 159. Experimental values are from Reference 37. Numbers in square brackets are uncertain values.

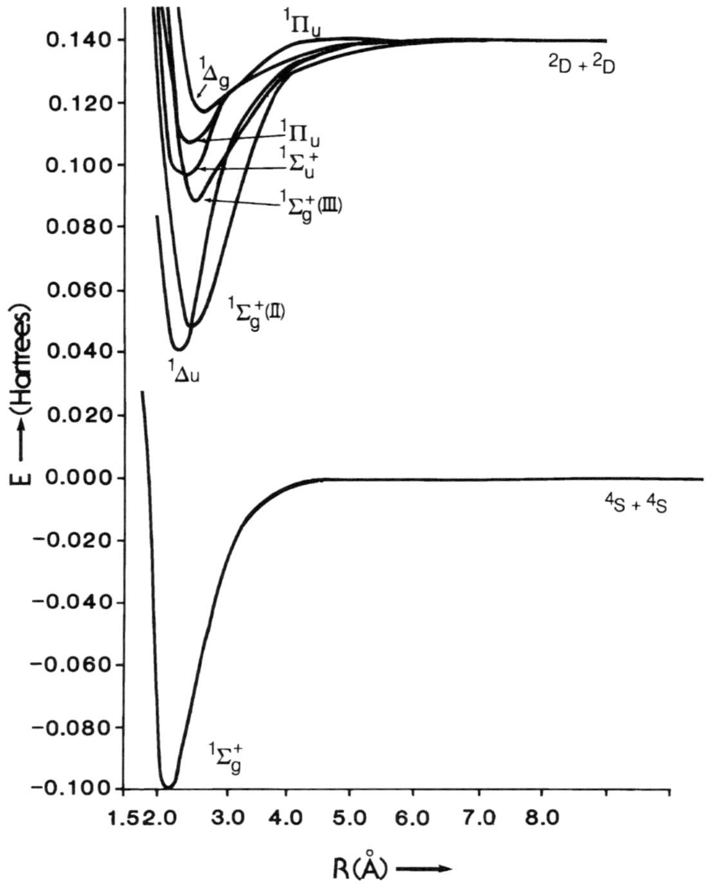

Figure 3.3 Potential-energy curves for the low-lying singlet electronic states of As_2. (Reprinted from Ref. 159; copyright 1987, Academic Press, Inc.)

Heimbrook et al. (156) observed vibronic bands of As_2 in the red, yellow, and blue near-UV regions. These transitions were labeled $c \leftrightarrow X^1\Sigma_g^+$, $e \leftrightarrow X$, and $a \leftrightarrow X$, respectively. These authors (156) also tentatively assigned the a, c, and e states to the $^1\Sigma_g^+$, $^3\Sigma_u^+$, and $^3\Delta_u$ states, respectively. As seen from Table 3.12, theoretical calculations support the assignment of the a, c, and e states to the $^3\Sigma_u^-$, $^3\Sigma_u^+$, and $^3\Delta_u$ states, respectively. The calculated ω_e values of these states are in reasonable agreement with the experimental results.

The $b \rightarrow X$ system was observed by Almy and co-workers (141–143) although no spectroscopic properties have been reported for the b state, since the $b \rightarrow X$ system appears to be weak. A possible candidate for the b state is a $^3\Pi_u$ state. The author's calculations (159) supported the existence of a $^3\Pi_u$ state whose T_e value is 44,669 cm^{-1} and was designated as $^3\Pi_u$(II). The experimen-

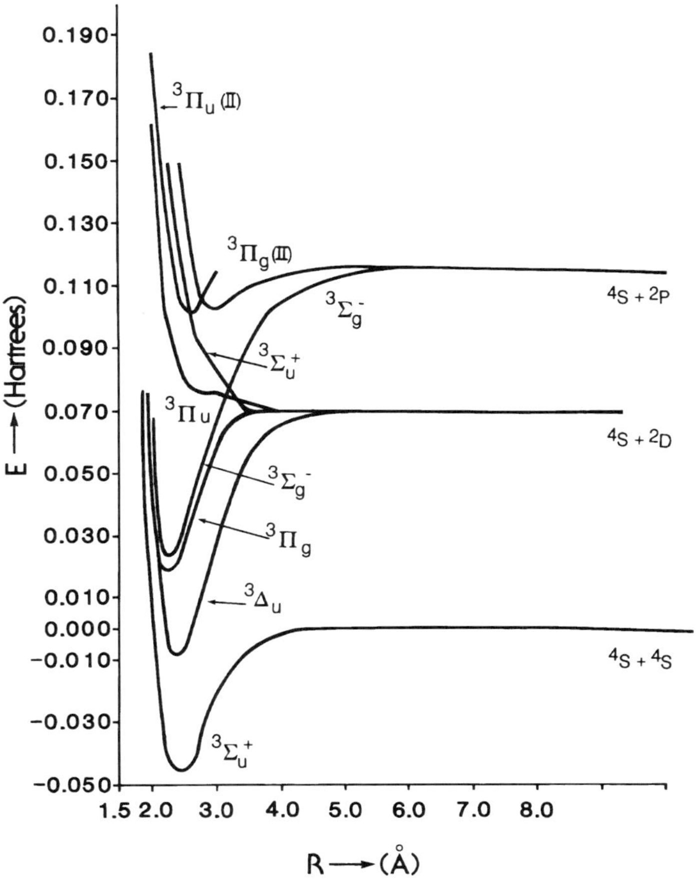

Figure 3.4 Potential-energy curves for the low-lying triplet states of As_2. (Reprinted from Ref. 159; copyright 1987, Academic Press, Inc.)

tally observed bands for the $b \rightarrow X$ systems are consistent with the theoretical T_e value. Although the $^3\Pi_u \rightarrow {}^1\Sigma_g^+$ transition is not allowed in the absence of spin–orbit interaction, the spin–orbit effects seem to be large enough to give rise to the weak $0_u^+ \rightarrow {}^1\Sigma_g^+$ transition, where 0_u^+ arises from $^3\Pi_u$. Alternatively, the $^3\Pi_u(1_u) \rightarrow {}^1\Sigma_g^+$ transition should also have been observable in the perpendicular direction via spin–orbit mixing of $^3\Pi_u$ with $^1\Pi_u$.

Many electronic states are reported in Table 3.12 that have not yet been observed. Specifically, the $^1\Pi_u \leftrightarrow {}^1\Sigma_g^+$ transition is an allowed electric dipole transition. The calculated T_e value of the $^1\Pi_u$ state is 45,327 cm^{-1}. Thus this system should be observable in both absorption and emission. Sibai et al. (150) have observed bands in the 42,400–44,500-cm^{-1} region. The theoretical T_e value of the $^1\Pi_u$ state seems to support the assignment of these bands to the ${}^1\Pi_u \leftarrow X$ transition, although this assignment is not yet confirmed. The ${}^1\Pi_u$(II) $\leftrightarrow X^1\Sigma_g^+$ transition is also an allowed electric dipole transition. From the calculated properties of the ${}^1\Pi_u$(II) state (Table 3.12), this transition lies at 51,300 cm^{-1}. The theoretical ω_e value of the ${}^1\Pi_u$(II) state is 108 cm^{-1}, indicating the broad nature of the potential surface of the ${}^1\Pi_u$(II) state. A number of unclassified bands have been observed (158) in the 47,000–55,000-cm^{-1} region. Thus the absorption bands in this region could tentatively be assigned to the ${}^1\Pi_u$(II) $\rightarrow X^1\Sigma_g^+$ transition.

There are many allowed electronic transitions originating from excited electronic states that appear not to have been observed. The ${}^3\Pi_u \leftrightarrow {}^3\Pi_g$ transition is a strong allowed transition, and could be observed as is the ${}^3\Pi_u \rightarrow {}^3\Pi_g$ emission system. The approximate energy for this transition is 18,616 cm^{-1}. The ${}^3\Delta_g \rightarrow {}^3\Delta_u$ transition occurs at 10,728 cm^{-1}, and the ${}^1\Delta_u \rightarrow {}^1\Delta_g$ transition could be observed in the region of 17,000 cm^{-1}. None of the bound excited ${}^1\Sigma_g^+$ states (Table 3.12) have been characterized experimentally. Emission from the $A^1\Sigma_u^+$ state to all lower ${}^1\Sigma_g^+$ states is allowed. The $A^1\Sigma_u^+ \rightarrow {}^1\Sigma_g^+$(II) emission could be observed at 10,485 cm^{-1}. The ${}^1\Sigma_g^+$(III) and ${}^1\Sigma_g^+$(II) states would be accessible in emission from the $B^1\Sigma_u^+$ state. The ${}^1\Pi_g \rightarrow {}^1\Pi_u$ and ${}^1\Pi_g \rightarrow {}^1\Pi_u$(II) transitions could be observed at 12,065 and 18,032 cm^{-1}, respectively. However, none of these transitions have been observed.

The D_e of the ground state of As_2 was obtained by using the CASSCF/FOCI method (159) as 2.71 eV. The spectroscopic D_e value was obtained by extrapolating the predissociated A–X bands into the ${}^4S + {}^2D$ atomic states (142, 143). The D_e value obtained this way (3.96 eV) (142) should not be regarded as accurate, and usually such extrapolations provide only an upper limit. Nevertheless, second-order correlation corrections could increase the D_e value of the ground state. The D_e of As_2 should, however, be closer to 3.8 eV, and thus CASSCF/FOCI calculations underestimate the D_e of As_2.

Table 3.13 shows the important configurations contributing to the electronic states near the equilibrium geometries of the various states (159). As seen from Table 3.13, the ground state of As_2 was found to be predominantly the triple-bond configuration $1\sigma_g^2 1\sigma_u^2 2\sigma_g^2 1\pi_u^4$, although the $1\sigma_g^2 1\sigma_u^2 2\sigma_g^2 1\pi_u^2 1\pi_g^2$ configurations also make an appreciable contribution, thereby lowering the bond order to 2.64. The $A^1\Sigma_u^+$ state participating in the strong $A \leftrightarrow X$ system is a mixture of $1\sigma_g^2 1\sigma_u^2 2\sigma_g^2 1\pi_u^3 1\pi_g$, $1\sigma_g^2 1\sigma_u^2 2\sigma_g 2\sigma_u 1\pi_u^4$, and other configurations.

3.3.4 Se_2

The earlier studies on the spectroscopy of Se_2 are summarized in Reference 37. The Se_2 dimer has been studied by several investigators (160–172). The O_2 analogs of the Schumann–Runge bands, attributed to the B–X system, are observed for Se_2 in the 26,000-cm^{-1} region. These bands for Se_2 were found to be perturbed by the A state (165). The B–X, A–X, and a–x systems were also observed in a matrix by Bondybey and English (163). Prosser et al. (164)

Table 3.13 Leading Configurations of the Various Electronic States of As_2 at Their Equilibrium Geometries[a]

State	Configuration
$^1\Sigma_g^+$	$1\sigma_g^2 1\sigma_u^2 2\sigma_g^2 1\pi_u^4$ (82%), $1\sigma_g^2 1\sigma_u^2 2\sigma_g^2 1\pi_u^2 1\pi_g^2$ (9%)
$^3\Sigma_u^+$	$1\sigma_g^2 1\sigma_u^2 2\sigma_g^2 1\pi_u^3 1\pi_g$ (75%), $1\sigma_g^2 1\sigma_u^2 2\sigma_g^2 1\pi_u 1\pi_g^3$ (11%)
$^3\Delta_u$	$1\sigma_g^2 1\sigma_u^2 2\sigma_g^2 1\pi_u^3 1\pi_g$ (82%), $1\sigma_g^2 1\sigma_u^2 2\sigma_g^2 1\pi_u 1\pi_g^3$ (7%)
$^3\Pi_g$	$1\sigma_g^2 1\sigma_u^2 2\sigma_g 1\pi_u^4 1\pi_g$ (76%), $1\sigma_g^2 1\sigma_u^2 2\sigma_g 1\pi_u^2 1\pi_g^3$ (10%)
$^3\Sigma_u^-$	$1\sigma_g^2 1\sigma_u^2 2\sigma_g^2 1\pi_u^3 1\pi_g^2$ (87%)
$^3\Delta_g$	$1\sigma_g^2 1\sigma_u^2 2\sigma_g^2 1\pi_u^2 1\pi_g^2$ (86%), $1\sigma_g^2 1\sigma_u^2 2\sigma_u^2 1\pi_u^2 1\pi_g^2$ (4%)
$^1\Delta_u$	$1\sigma_g^2 1\sigma_u^2 2\sigma_g^2 1\pi_u^3 1\pi_g$ (86%), $1\sigma_g^2 1\sigma_u^2 2\sigma_g 2\sigma_u 1\pi_u^2 1\pi_g^2$ (3%)
$^1\Sigma_g^+$(II)	$1\sigma_g^2 1\sigma_u^2 2\sigma_g^2 1\pi_u^2 1\pi_g^2$ (67%), $1\sigma_g^2 1\sigma_u^2 2\sigma_g^2 1\pi_u^4$ (15%)
$^1\Pi_g$	$1\sigma_g^2 1\sigma_u^2 2\sigma_g 1\pi_u^4 1\pi_g$ (82%), $1\sigma_g^2 1\sigma_u^2 2\sigma_g^2 1\pi_u^2 1\pi_g^3$ (5%)
$^1\Sigma_g^+$(III)	$1\sigma_g^2 1\sigma_u^2 2\sigma_g^2 1\pi_u^2 1\pi_g^2$ (89%)
$^1\Sigma_u^+$	$1\sigma_g^2 1\sigma_u^2 2\sigma_g^2 1\pi_u^3 1\pi_g$ (47%), $1\sigma_g^2 1\sigma_u^2 2\sigma_g 2\sigma_u 1\pi_u^4$ (22%), $1\sigma_g^2 1\sigma_u^2 2\sigma_g^2 1\pi_u 1\pi_g^3$ (17%)
$^3\Pi_u$(II)	$1\sigma_g^2 1\sigma_u^2 2\sigma_g 1\pi_u^3 \pi_g^2$ (85%), $1\sigma_g^2 1\sigma_u^2 2\sigma_g 1\pi_u^3 \pi_g^2$ (1.5%)
$^1\Pi_u$	$1\sigma_g^2 1\sigma_u^2 2\sigma_g \pi_u^4 1\pi_g^2$ (81%), $1\sigma_g^2 1\sigma_u^2 2\sigma_g 1\pi_u 1\pi_g^4$ (2%)
$^1\Delta_g$	$1\sigma_g^2 1\sigma_u^2 2\sigma_g 1\pi_u^2 1\pi_g^2$ (61%), $1\sigma_g^2 1\sigma_u^2 2\sigma_g 2\sigma_u 1\pi_g^3 1\pi_g$ (12%), $1\sigma_g^2 1\sigma_u^2 1\sigma_g^2 1\pi_u^4 1\pi_g^2$ (11%)
$^1\Pi_u$(II)	$1\sigma_g^2 1\sigma_u^2 2\sigma_g 1\pi_u^3 1\pi_g^2$ (63%), $1\sigma_g^2 1\sigma_u^2 2\sigma_g 1\pi_u 1\pi_g^4$ (22.4%)

[a]Numbers in parentheses are contributions in percentage.

Source: Reference 159.

observed a $B0_u^+–b^1\Sigma_g^+$ system, by populating the B and A electronic states with a 4131-Å krypton ion laser. In addition, a $C–X$ system was also observed in the 53,000-cm^{-1} region. Below 26,000 cm^{-1}, five λ–*s* states of Se_2 and some of their spin–orbit components have been observed. These states were given the labels B, A, a, b, and X, among which, B, A, b, and X have been tentatively assigned.

The nature of the state perturbing the $B^3\Sigma_u^-$ state is not well known. This has been the subject of many investigations (165, 169, 170). Further, unlike O_2, the nature of the electronic states below the B state is not clearly understood. Theoretical study of the low-lying electronic states of Se_2 was made by the author with the objective of comprehending the nature of the low-lying electronic states of Se_2 (172). He assigned the experimental spectra by a comparison of calculated and observed values. Saxon and Liu (173) have shown that such theoretical calculations yield reasonable spectroscopic constants for O_2 even if errors in D_e may be appreciable. Experimental study of other related group VI and group IV dimers is also quite intense (174–177).

The author (172) carried out CASSCF/FOCI calculations on 20 low-lying electronic states of Se_2 employing ($3s3p1d$) valence Gaussian basis sets and RECPs that included the outer $4s^24p^4$ shells explicitly in the valence space. The internal space consisted of four a_1 orbitals, two b_2 orbitals, and two b_1 orbitals for Se_2. Table 3.14 shows the possible low-lying electronic states of Se_2 in both λ–s and ω–ω couplings. Table 3.15 shows the dissociation relationships for the low-lying electronic states of Se_2. The ground state of Se_2 is evidently an $X^3\Sigma_g^-$ state arising from the $1\sigma_g^2 1\sigma_u^2 2\sigma_g^2 1\pi_u^4 1\pi_g^2$ configuration analogous to O_2.

The spin–orbit interaction was also introduced by the author (172) to some of the electronic states of $Se_2[{}^3\Sigma_g^-, {}^1\Sigma_g^+, {}^3\Pi_u(0_u^+), {}^3\Pi_u(1_u)]$ with the objective of estimating the effect of spin–orbit interaction. This was accomplished through the RCI scheme. A double-ζ STO basis set optimized for the $3P$ ground state of the Se atom was employed in SCF/RCI calculations. The ω–ω states arising from the ${}^3\Sigma_g^-$, ${}^1\Delta_g$, and ${}^1\Sigma_g^+$ states were obtained by using the orbitals generated from SCF calculations of the ${}^3\Sigma_g^-$ state. The ω–ω states arising from the ${}^3\Pi_u$ and ${}^1\Pi_u$ states were calculated by using the ${}^3\Pi_u$ SCF orbitals, since ${}^3\Sigma_g^-$ SCF did not produce a good $2\sigma_u$ orbital, as this is unoccupied in the ${}^3\Sigma_g^-$ state.

The RCI calculations of the 0_g^+ state of Se_2 included single and double excitations from Cartesian reference configurations, arising from $1\sigma_g^2 1\sigma_u^2 2\sigma_g^2 1\pi_u^4 1\pi_g^2({}^3\Sigma_g^-, {}^1\Sigma_g^+)$ and $1\sigma_g^2 1\sigma_u^2 2\sigma_g 1\pi_u^4 1\pi_g^3({}^3\Pi_g)$. The 1_g state in-

Table 3.14 A Few Low-Lying Electronic Configurations of Se_2 and the λ–s and ω–ω States Arising from Them[a]

Electronic Configuration	λ–s States	ω–ω States
$2\sigma_g^2 1\pi_u^4 1\pi_g^2$	${}^3\Sigma_g^-$	$0_g^+, 1_g$
	${}^1\Delta_g$	${}^2g$
	${}^1\Sigma_g^+$	0_g^+
$2\sigma_g^2 1\pi_u^3 1\pi_g^3$	${}^3\Delta_u$	${}^3u, {}^2u, {}^1u$
	${}^3\Sigma_u^-$	$0_u^+, 1_u$
	${}^3\Sigma_u^+$	$0_u^-, 1_u$
	${}^1\Delta_u$	${}^2u$
	${}^1\Sigma_u^+$	0_u^+
	${}^1\Sigma_u^-$	0_u^-
$2\sigma_g^2 2\sigma_u^1 1\pi_u^4 1\pi_g$	${}^3\Pi_u$	$0_u^+, 0_u^-, 1_u, 2_u$
	${}^1\Pi_u$	1_u
$2\sigma_g 2\sigma_u 1\pi_u^4 1\pi_g^2$	${}^5\Sigma_u^-$	$0_u^-, 1_u, 2_u$
	${}^3\Sigma_u^-(2)$	$0_u^+(2), 1_u(2)$
	${}^1\Sigma_u^-$	0_u^-
	${}^3\Delta_u$	$1_u, 2_u, 3_u$
	${}^3\Sigma_u^+$	$0_u^-, 1_u$
	${}^1\Delta_u$	2_u
	${}^1\Sigma_u^+$	0_u^+

[a]The $1\sigma_g^2 1\sigma_u^2$ shell arising from the valence s is not shown.

Table 3.15 Dissociation Relationships for a Few Low-Lying States of Se_2

Molecular States	Atomic States	Energy of Atoms, cm^{-1} [a]
$^1\Sigma_u^+(2)$, $^1\Sigma_u^-$, $^1\Pi_g$, $^1\Pi_u$, $^1\Delta_g$, $^3\Sigma_u^+$, $^3\Sigma_g^-$, $^3\Pi_g$, $^3\Pi_u$, $^3\Delta_u$, $^5\Sigma_g^+(2)$, $^5\Sigma_u^-$, $^5\Pi_g$, $^5\Pi_u$, $^5\Delta_g$	$^3P + {}^3P$	0.0
$^3\Sigma_g^+$, $^3\Sigma_u^+$, $^3\Sigma_g^-(2)$, $^3\Sigma_u^-(2)$, $^3\Pi_g(3)$, $^3\Pi_u(3)$, $^3\Delta_g(2)$, $^3\Delta_u(2)$, $^3\Phi_g$, $^3\Phi_u$	$^3P + {}^1D$	9,576

[a]From Reference 134.

cluded five reference configurations, arising from $^3\Sigma_g^-(1_g)$, $^3\Pi_g(1_g)$, and $^1\Pi_g(1_g)$. The 2_g state included eight reference configurations, arising from $^1\Delta_g$ and $^3\Pi_g$. The author (172) also carried out RCI calculations of the 0_u^+, 1_u, and 2_u components of the $3\Pi_u$ state. The 0_u^+ state included limited single and double excitations from 24 reference configurations, arising from a number of $^3\Pi_u$ references and $^3\Sigma_u^-$ references. The 1_u state calculations included $^3\Pi_u$, $^1\Pi_u$, and $^3\Sigma_u^+$ references, while the 2_u calculations included only $^3\Pi_u$ references.

Table 3.16 shows theoretical and experimental spectroscopic properties (R_e, T_e, and ω_e) of the low-lying electronic states of Se_2. The potential-energy curves of some of these states are shown in Figure 3.5. As seen from Table 3.16, the ground state (X) is the $^3\Sigma_g^-$ state, arising from the $1\sigma_g^2 1\sigma_u^2 2\sigma_g^2 1\pi_u^4 1\pi_g^2$ configuration. The discrepancies in theoretical R_e and ω_e values of the ground state are within 3–6% of the experimental values. The ω_e values of the various states in Table 3.16 were calculated by using a cubic polynomial fit of energies at near-equilibrium geometries. The small discrepancy between calculated and observed values arises partly from the errors in fits (usually within ±115 cm^{-1}) and partly from higher-order correlation corrections and basis-set effects that influence the ω_e values.

The strongest electronic transition originating from the ground state is the $B^3\Sigma_u^-$–$X^3\Sigma_g^-$ system. The theoretical properties of the $^3\Sigma_u^-$ state are in very good agreement with the spectroscopic values, confirming the earlier assignment of this system. There have been a few experimental investigations on electronic states below the B state. Specifically, the A state in the A–X system was identified by Heaven et al. (169). The T_e value of the A state reported by these authors is about 24,111 cm^{-1}. The author's (172) theoretical calculations confirmed the existence of a $^3\Pi_u(0_u^+)$ state with a theoretical T_e of 21,277 cm^{-1} in this region. Thus the A state was found to be most consistent with $^3\Pi_u(0_u^+)$. The $^3\Pi_u(0_u^+)$ spin–orbit state could mix with $^3\Sigma_u^-(0_u^+)$, although the author's RCI calculations revealed this contamination to be negligible for $^3\Pi_u(0_u^+)$ at its equilibrium geometry. Similarly, the mixing of the $^3\Pi_u(1_u)$, $^3\Sigma_u^-(1_u)$, and $^1\Pi_u(1_u)$ states was found to be relatively small for the selenium dimer. The author's (172) theoretical R_e value for the A state did not agree well with the value

Table 3.16 Spectroscopic Properties of As_2[a]

	R_e, Å		T_e, cm^{-1}		ω_e, cm^{-1}	
State	Calc.	Exp.	Calc.	Exp.	Calc.	Exp.
$X^3\Sigma_g^- 0_g^+ X_1$	2.244	2.166	0.0	0.0	357	385
$1_g X_2$	2.240	—	664	512	360	387
$a'^1\Delta_g(2_g)$	2.27	—	4,652	~4,000	341	320
$b^1\Sigma_g^+ 0_g^+$(II)	2.30	—	8,557	7,958	323	355
$a^3\Delta_u$	2.51	—	15,241	16,321	246	268
$^3\Sigma_u^+$	2.52	—	15,651	—	244	—
$A^3\Pi_u(0_u^+)$	(2.71)	2.51	21,277	24,111	197	195
$A^3\Pi_u(1_u)$	(2.72)	2.54	21,961	24,940	194	187
$^3\Pi_u(2_u)$	(2.73)	—	22,837	—	189	—
$B^3\Sigma_u^-$	2.54	2.44	24,822	25,980	232	246
$^3\Sigma_g^-$(II)	3.12	—	30,567	—	133	—
$^3\Pi_g$(II)	3.01	—	31,582	—	122	—
$^3\Pi_u$(II)	3.32	—	33,325	—	—	—
$^3\Delta_u$(II)	3.40	—	33,712	—	—	—
$^1\Delta_u$	2.53	—	36,814	—	233	—
$^1\Pi_u$(II)	2.58	—	39,173	—	325	—
$^1\Delta_u$(II)	2.92	—	41,270	—	141	—
$^1\Pi_g$(III)	2.90	—	42,556	—	159	—

[a]Experimental values for the X_1, X_2, and B states quoted here are from Huber and Herzberg's book (37). Experimental values for the $^1\Delta_g$, $^1\Sigma_g^+(0_u^+)$, $^3\Pi_u(0_u^+)$, and $^3\Pi_u(1_u)$ states are from References 168, 164, 165, and 162, respectively. Theoretical values are from Reference 172.

calculated from the experimental spectroscopic data. A similar deficiency was found in the $^3\Pi_u$ states of Br_2 and I_2. The $^3\Pi_u$ state is weakly bound, and its surface is shallow. The basis set that was used in conjunction with effective core potentials does not appear to be of sufficient quality to compute the spectroscopic properties of weakly bound excited states. Diffuse basis functions and polarization functions could be important for the $^3\Pi_u$ and $^1\Pi_u$ excited states.

Bondybey and English (163) have observed spectroscopic bands with a T_e value of 16,321 cm^{-1}, which they attributed to a weak electronic transition. These authors call this an a–X system, although this designation could also be used for the 4000-cm^{-1} system. To differentiate the two systems, the state with T_e = 4000 cm^{-1} was denoted by the author as a'. Theoretical calculations suggested the existence of a $^3\Delta_u$ state in the neighborhood of the observed electronic bands. The theoretical and experimental ω_e values were also found to be in very good agreement. Further, the $^3\Delta_u \leftrightarrow {}^3\Sigma_g^-$ transition is dipole-forbidden, supporting the experimental observation of weak electronic bands in this system. The above reasonings led the author to the assignment of this system to the $a^3\Delta_u$–$X^3\Sigma_g^-$-forbidden transition.

Prosser et al. (164) and Winger et al. (168) characterized an electronic state

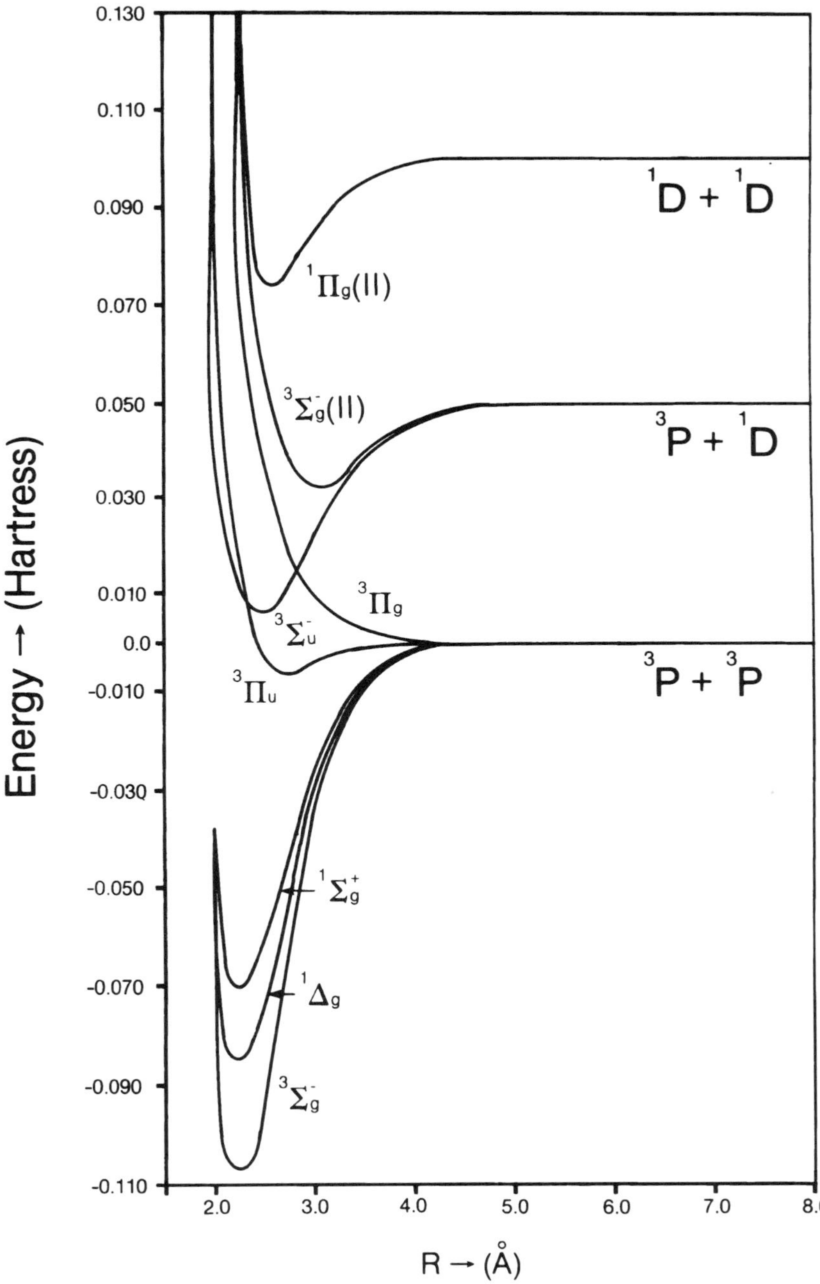

Figure 3.5 Potential-energy curves for the low-lying states of Se_2. (Reprinted from Ref. 172; copyright 1987, American Chemical Society.)

(b) with a T_e value of 7958 cm^{-1}. The b state was tentatively assigned to the $^1\Sigma_g^+$ electronic state. As seen from Table 3.16, the results of theoretical calculations support this assignment. The fact that the theoretical and the experimental T_e values of this state are in excellent agreement, and that there is no other electronic state with this T_e value, led the author (172) to confirm unambiguous assignment of the b state to the $^1\Sigma_g^+$ state, arising from the $1\sigma_g^2 1\sigma_u^2 2\sigma_g^2 1\pi_u^4 1\pi_g^2$ configuration.

Yee and Barrow (170) studied the absorption and fluorescence spectra of gaseous Se_2 in the 5145–4880-Å region. These authors assigned these bands to an $n \leftrightarrow a$ transition, where the upper state is believed by these authors (170) to be a relatively weakly bound 1_u state, which perturbs the $B^3\Sigma_u^-(0_u^+)$ state. The a state is predicted to be approximately 4000 cm^{-1} above the $X^3\Sigma_g^-(0_g^-)$ ground state, and was tentatively assigned to $^1\Delta_g$, based on the lower vibrational frequency of this state compared to that of the ground state ω_e. The theoretical spectroscopic properties for the $^1\Delta_g(2_g)$ state (Table 3.16) certainly support this assignment. The experimental ω_e value (320 cm^{-1}) does not follow the theoretical trend in Table 3.16, which implies that the ω_e value of the $^1\Delta_g(2_g)$ state should be 20 cm^{-1} less than the $^1\Delta_g$ value of the $^3\Sigma_g^-(0_g^+, 1_g)$ states. While the spin–orbit coupling lowers the T_e value of the $^1\Delta_g$ state by 95 cm^{-1}, it does not change the R_e and ω_e values at all. This is primarily a consequence of the fact that the $^1\Delta_g$ state was not found to be contaminated with other λ–s states that give rise to a 2_g ω–ω state such as $3\Pi_g$. It appears that the experimental ω_e of the $^1\Delta_g$ state (320 cm^{-1}) is somewhat low and should be about 368 cm^{-1}; this was arrived at by decreasing the experimental ω_e value of the $^3\Sigma_g^-(1_g)$ state by 19 cm^{-1}, as implied by the theoretical calculations.

Jenouvrier (165) studied the perturbations of the bands in the $B^3\Sigma_u^-(0_u^+)$–X system. Some confusion was introduced by using the label A to designate the state that perturbs the B state and the $A^3\Pi_u(0_u^+)$ state. The A state that was found to perturb the B state was assigned to $A^3\Pi_u(1_u)$ by Jenouvrier (165). It is best to refer to these two states with their Ω quantum number in parentheses. The $A^3\Pi_u(1_u)$ state is 1000 cm^{-1} below the $B0_u^+$ state. There is also another state designated A' assigned to $A[^1\Pi_u(1_u)]$ by Jenouvrier (165), which was also assigned as the state perturbing the B state. The A' state was found to be only 110 cm^{-1} below the B state (165). From Table 3.16 it can be seen that the theoretical 1_u state (predominantly $^3\Pi_u$) is 685 cm^{-1} above the $^3\Pi_u(0_u^+)$ state. The corresponding experimental splitting reported by Jenouvrier (165) is 829 cm^{-1}, and is thus in reasonable agreement with the predicted spin–orbit splitting for this state. However, the larger theoretical R_e value was attributed by the author to the basis set and suggested further extension of basis sets with more diffuse functions and polarization functions, since they appeared to play a significant role, especially for the weakly bound excited states.

As seen from Table 3.16, the theoretical bond length for the $B^3\Sigma_u^-$ state is 0.1 Å longer than the experimental value, primarily because of the limitations of the basis set. The worst case is the $^1\Pi_u$ state, which actually comes out to be repulsive theoretically, while the experimental A' state, which is tentatively assigned to $^1\Pi_u(1_u)$ (165), is bound, although somewhat less so than the A state.

Table 3.16 shows a number of electronic states of Se_2 that are yet to be observed. Specifically, the $^3\Sigma_g^-$(II) state, which has the same symmetry as the ground state, can be observed in the $^3\Sigma_u^-$(II) emission system if the C state can be populated. The $^3\Pi_u$(II) state has a minimum at a long distance. The dipole-allowed $^3\Pi_u(\mathrm{II}) \rightarrow X^3\Sigma_g^-$ transition should be observable in the 33,000-cm^{-1} region.

The theoretical CASSCF/FOCI D_e of the ground state of Se_2 was found to be 2.91 eV (172). A number of experimental D_e values were obtained from the predissociation of $B0_u^+$ bands (3.41, 3.16, and 3.10 eV) (37). Photoionization and thermochemical studies seem to favor the higher value (171). Yee and Barrow (170) obtained an upper bound for D_e as 27,096 + 2λ″, where 2λ″ is the 0_g^+–1_g splitting. They used a somewhat smaller value of 367 cm^{-1} for this splitting compared to the theoretical splitting and the observed splitting (Table 3.16). The author corrected this by taking the 0_g^+–1_g splitting as 512 cm^{-1}. The author obtained an upper value for D_e as 27,608 cm^{-1}, still eliminating the highest of the three possible D_e values considered by these authors (27,700, 25,710, or 25,166 cm^{-1}). Thus, the D_e value should be about 3.19 eV. The theoretical value of 2.91 eV seems to favor this more than the 3.41-eV value.

Table 3.17 shows the weights of the leading configurations in the FOCI wavefunctions of the electronic states of Se_2. As seen from this table, the $^3\Sigma_g^-$, $^1\Delta_g$, and $^1\Sigma_g^+$ states are dominated by the $2\sigma_g^2 1\pi_u^4 1\pi_g^2$ configuration. The $^3\Delta_u$ state observed in the $^3\Delta_u$–$X^3\Sigma_g^-$ system arises from the $2\sigma_g^2 1\pi_u^3 1\pi_g^3$ configuration. The $A^3\Pi_u$ state was found to be a mixture of the $2\sigma_g^2 2\sigma_u 1\pi_u^4 1\pi_g$ and $2\sigma_g^2 2\sigma_u 1\pi_u^2 1\pi_g^3$ configurations. The $B^3\Sigma_u^-$ state was found to be a mixture of the $2\sigma_g^2 1\pi_u^3 1\pi_g^3$, $2\sigma_g 2\sigma_u 1\pi_u^4 1\pi_g^2$, and $2\sigma_g 2\sigma_u 1\pi_u^2 1\pi_g^4$ configurations. This seems to suggest that the C state observed in the $C^3\Sigma_u^-$ system should be dominated by the $2\sigma_g 2\sigma_u 1\pi_u^4 1\pi_g^2$ configuration, although, since this state is observed in the 53,000-cm^{-1} region, contributions from Rydberg configurations could become significant.

The spin–orbit contamination of the 0_g^+ ground state of Se_2 was found to be 81% $^3\Sigma_g^-$ and 8% $^1\Sigma_g^+$. The 0_g^+(II) state is 74% $^1\Sigma_g^+$ and 8% $^3\Sigma_g^-$. The 1_u component of $^3\Pi_u$ was found to be actually 73% $^3\Pi_u(1\sigma_g^2 1\sigma_u^2 2\sigma_g^2 2\sigma_u 1\pi_u^4 1\pi_g)$, 2% $^1\Pi_u(1\sigma_g^2 1\sigma_u^2 2\sigma_g^2 2\sigma_u 1\pi_u^4 1\pi_g)$, 18% $^3\Pi_u(1\sigma_g^2 1\sigma_u^2 2\sigma_g^2 2\sigma_u 1\pi_u^2 1\pi_g^3)$, and 0.2% $^1\Pi_u(1\sigma_g^2 1\sigma_u^2 2\sigma_g^2 2\sigma_u 1\pi_u^2 1\pi_g^3)$.

3.3.5 Br_2 and Br_2^+

Spectroscopic investigations up to 1977 have been summarized in the book by Huber and Herzberg (37), although Brand and Hoy (38) published a subsequent review on the experimental spectra of halogens. The electronic spectra of Br_2 in the region below 60,000 cm^{-1} (37, 38) reveal the existence of the X, A, B, C, D, E, F, G, and H states. Among these, the A, B, C, F, and G states are

Table 3.17 Leading Configuration(s) Contributing to the Low-Lying States of Se_2 at Their Equilibrium Geometries[a]

State	Configuration
$^3\Sigma_g^-$	$2\sigma_g^2 1\pi_u^4 1\pi_g^2$ (89%)
$^1\Delta_g$	$2\sigma_g^2 1\pi_u^4 1\pi_g^2$ (89%), $2\sigma_g^2 1\pi_u^2 1\pi_g^4$ (6%)
$^1\Sigma_g^+$	$2\sigma_g^2 1\pi_u^4 1\pi_g^2$ (82%), $2\sigma_g^2 1\pi_u^2 1\pi_g^4$ (9%)
$^3\Delta_u$	$2\sigma_g^2 1\pi_u^3 1\pi_g^3$ (90%), $2\sigma_u^2 1\pi_u^3 1\pi_g^3$ (3%)
$^3\Sigma_u^+$	$2\sigma_g^2 1\pi_u^3 1\pi_g^3$ (90%), $2\sigma_u^2 1\pi_u^3 1\pi_g^3$ (3%)
$^3\Pi_u$	$2\sigma_g^2 2\sigma_u 1\pi_u^4 1\pi_g$ (66%), $2\sigma_g^2 2\sigma_u 1\pi_u^2 1\pi_g^3$ (22%)
$^3\Sigma_u^-$	$2\sigma_g^2 1\pi_u^3 1\pi_g^3$ (73%), $2\sigma_g 2\sigma_u 1\pi_u^4 1\pi_g^2$ (14%), $2\sigma_g 2\sigma_u 1\pi_u^2 1\pi_g^4$ (4%)
$^3\Sigma_g^-$(II)	$2\sigma_g 2\sigma_u 1\pi_u^3 1\pi_g^3$ (42%), $2\sigma_g^2 1\pi_u^2 1\pi_g^4$ (38%), $2\sigma_g^2 1\pi_u^4 1\pi_g^2$ (8%)
$^3\Pi_g$(II)	$2\sigma_g^2 2\sigma_u 1\pi_u^3 1\pi_g^2$ (78%), $2\sigma_u^2 2\sigma_g 1\pi_u^2 1\pi_g^3$ (7.3%)
$^3\Pi_u$(II)	$2\sigma_g^2 2\sigma_u 1\pi_u^2 1\pi_g^3$ (48%), $2\sigma_g^2 2\sigma_u 1\pi_u^4 1\pi_g$ (23%) $2\sigma_g 2\sigma_u^2 1\pi_u^3 1\pi_g^2$ (20%)
$^3\Delta_u$(II)	$2\sigma_g 2\sigma_u 1\pi_u^4 1\pi_g^2$ (38%), $2\sigma_g 2\sigma_u 1\pi_u^2 1\pi_g^4$ (27%), $2\sigma_g^2 1\pi_u^3 1\pi_g^3$ (24%), $2\sigma_u^2 1\pi_u^3 1\pi_g^3$ (5%)
$^1\Delta_u$	$2\sigma_g^2 1\pi_u^3 1\pi_g^3$ (73%), $2\sigma_g 2\sigma_u 1\pi_u^4 1\pi_g^2$ (14%), $2\sigma_g 2\sigma_u 1\pi_u^2 1\pi_g^4$ (3%)
$^1\Pi_g$(II)	$2\sigma_g 1\pi_u^4 1\pi_g^3$ (84%), $2\sigma_g 2\sigma_u^2 1\pi_u^4 1\pi_g$ (2%)
$^1\Pi_g$(III)	$2\sigma_g^2 2\sigma_u 1\pi_u^3 1\pi_g^2$ (87%)

[a]The $1\sigma_g^2 1\sigma_u^2$ shell is not shown.

Source: Reference 172.

observed in either absorption from the ground state or emission to the ground state. The *D*, *E*, and *H* states are observed in $D \rightarrow B$, $E \leftrightarrow B$, and $H \rightarrow B$ transitions. The ground state *X* is assigned unambiguously to the Σ_g^+ state. The *A* and *B* states are the 1_u and 0_u^+ spin–orbit components of a $^3\Pi_u$ state. The *C* state was assigned to the $^1\Pi_u$ state. The exact nature of these states (electronic configurations contributing to these states) and the other upper states was not clearly established from experiment alone, although the symmetry of the upper states is known in many cases from the intensities of these transitions.

The rotational analysis, Franck–Condon factors, and the long-range potentials have been obtained for the *B* state by Barrow et al. (178). Ishiwata et al. (179) identified an ion-pair state, which they designated as an *f* state by a two-photon excitation technique. A transition from the ground state *X* to *B* followed by *B* to *f* was induced by using a two-photon technique. Subsequently, the fluoresence from the *f* to *B* state was detected. As these authors (179) pointed out, there was some controversy on the nature of some of the ion-pair states of Br_2. Mulliken (180) derived the expressions for the energy levels related to the $^3\Pi$ and $^1\Pi$ states, which enabled interpretation and assignment of the A–X, B–X, and C–X systems.

Theoretical investigations, before the author's calculations (181) were limited to the ground state of Br_2 (182, 183), of Schwerdtfeger et al. (182) employed an SCF/CI method with semiempirical pseudopotentials for the ground state of Br_2. The calculation by Andzelm et al. (183) employed a single-configuration SCF treatment, which tended to give a poor D_e. The author carried out the first theoretical calculations on the excited states of Br_2. Boerrighter et al. (184) performed HF/limited CI calculations on the electronic states of Br_2^+.

The author (181) carried out CASSCF/FOCI calculations on 16 electronic states of Br_2 and 6 low-lying electronic states of Br_2^+. Spin–orbit effects were taken into account by using the relativistic configuration interaction method, which employed a double-ζ STO + polarization basis set for the bromine atom. The CASSCF/FOCI calculations were made by using a valence Gaussian basis set of $(3s3p1d)$ and $(4s4p2d)$ quality. For the ground state, the author also carried out full second-order CI calculations, which included up to 210,880 configurations and tested the effect of spin–orbit interaction on the dissociation energies.

The theoretical spectroscopic properties (R_e, T_e, ω_e) of the bound states of Br_2 calculated by the author are shown in Table 3.18, and those of Br_2^+ are shown in Table 3.19. The potential-energy curves of some of the electronic states dissociating to the ground-state atoms in the absence of spin–orbit interaction are shown in Figure 3.6. The corresponding curves for Br_2^+ obtained with the FOCI method of calculation are shown in Figure 3.7.

As seen from Table 3.18, the $X0_g^+$ ground state of Br_2 has a theoretical FOCI R_e value ($3s3p1d$ basis) 0.1 Å longer than the experimental value of 2.28 Å. The larger basis set at the SOCI level of calculation yielded better results. The calculated FOCI ω_e was found to be 10% lower than the experimental value of 325 cm^{-1}, although the SOCI method improves this.

Spectroscopic bands of Br_2 observed below 60,000 cm^{-1} were assigned to the $A \leftrightarrow X$, $B \leftrightarrow X$, and $C \leftarrow X$ transitions. The A and B states are now well characterized as $^3\Pi_u(1_u)$ and $^3\Pi_u(0_u^+)$ states, respectively. The A and B states observed in the $A \leftrightarrow X$ and $B \leftrightarrow X$ transitions can thus be assigned unambiguously, since the 0_u^+ state should be well above $^3\Pi_u(1_u)$. The experimental T_e values for the B and A states (15,900, 13,900 cm^{-1}) are consistent with the trend suggested by theoretical calculations. The experimental ω_e values (153, 167 cm^{-1}) were found to be in very good agreement with the theoretical ω_e value for the $^3\Pi_u$ state (155 cm^{-1}). The C state was observed in the $C \leftarrow X$ absorption system. It is also considered to be responsible for the predissociation of the B state. Although at the FOCI level of approximation the $^1\Pi_u$ state appeared to be repulsive, higher-order correlation effects and a larger basis set could lead to a bound $^1\Pi_u$ state. Nevertheless, the C state must be relatively weakly bound, since the absorption spectra that corresponded to the $C \leftarrow X$

Table 3.18 Spectroscopic Properties of ABr_2[a]

Assignment	State	R_e, Å Theory	R_e, Å Exp.	T_e, cm^{-1} Theory	T_e, cm^{-1} Exp.	ω_e, cm^{-1} Theory	ω_e, cm^{-1} Exp.
$X0_g^+$	$^1\Sigma_g^+$	2.33	2.28	0	0	321	325
$A'2_u$	$^3\Pi_u(2_u)$	2.81	2.68	10,360	[13,220]	153	[165]
$A1_u$	$^3\Pi_u(1_u)$	2.83	2.70	10,620	13,905	154	153
$B0_u^+$	$^3\Pi_u(0_u^+)$	2.81	2.68	11,931	15,903	155	168
$D'2_g$	$^3\Pi_g$(II)	3.21	3.17	48,227	48,930	157	151
$E0_g^+$	$^3\Pi_g$(II) 0_g^+	—	3.19		49,778		151
$F0_u^+$	$^3\Sigma_u^- 0_u^+$	3.14	3.28	49,530	53,900	165	156
	$^3\Sigma_u^-(1_u)$	3.14	—	49,630	—	165	—
$D0_u^+$	$^3\Pi_u$(II) 0_u^+	3.38	3.17	49,965	49,929	162	134
$f0_g^+$	$^3\Sigma_g^-$(II) 0_g^+	3.23	3.17	50,277	53,102	157	153
$G0_u^+$	$^1\Sigma_u^+$(II) 0_u^+	3.00	—	57,578	56,337	169	[250]
	$^1\Pi_g$(II)	3.15	—	60,674	—	170	—
	$^1\Pi_u$(II)	3.32	—	62,210	—	162	—

[a]Theoretical spectroscopic constants for $X(0_g^+)$ were obtained with the CASSCF/SOCI ($4s4p2d$ basis) method. The theoretical $D_e(Br_2)$ includes spin–orbit effects = 1.87 eV compared to experimental $D_e(Br_2)$ = 1.97 eV. All theoretical constants are from Reference 181. Experimental constants and assignments are from Reference 38. Numbers in square brackets are uncertain values.

system are continuous (37, 38). The D state observed in the $D \rightarrow B$ system is most probably due to the $^3\Pi_g \rightarrow {}^3\Pi_u$ that the 0_u^+ state. It was suggested by the author (181) that the 0_g^+ and 1_g components of the $^3\Pi_g$(II) state are the most probable candidates for this transition. Since the $^3\Pi_g$(II) state arises from more than a half-filled π shell, the 2_g and 1_g states are lower than the 0_g^+ and 0_u^-

Table 3.19 Spectroscopic Properties of Br_2^+[a]

State	r_e, Å Theory	T_e, cm^{-1} Theory	T_e, cm^{-1} Exp.	ω_e, cm^{-1} Theory	ω_e, cm^{-1} Exp.
$X_1\ ^2\Pi_g(\frac{3}{2})$	2.30	0	0	343	376
$X_2\ ^2\Pi_g(\frac{1}{2})$	2.30	3,043	2,820	342	—
$A_1\ ^2\Pi_u(\frac{3}{2})$	2.62	11,017	—	194	190
$A_2\ ^2\Pi_u(\frac{1}{2})$	2.60	14,117	16,414	137	132
$^2\Delta_g$	3.04	20,312	—	128	—
$B^2\Sigma_g^+$	3.05	20,584	—	127	—
$^2\Pi_u$ (II)	3.58	20,718	—	89	—
$^2\Sigma_g^+(\frac{1}{2})$	2.79	25,033	—	199	—

[a]Theoretical constants are from Reference 181.

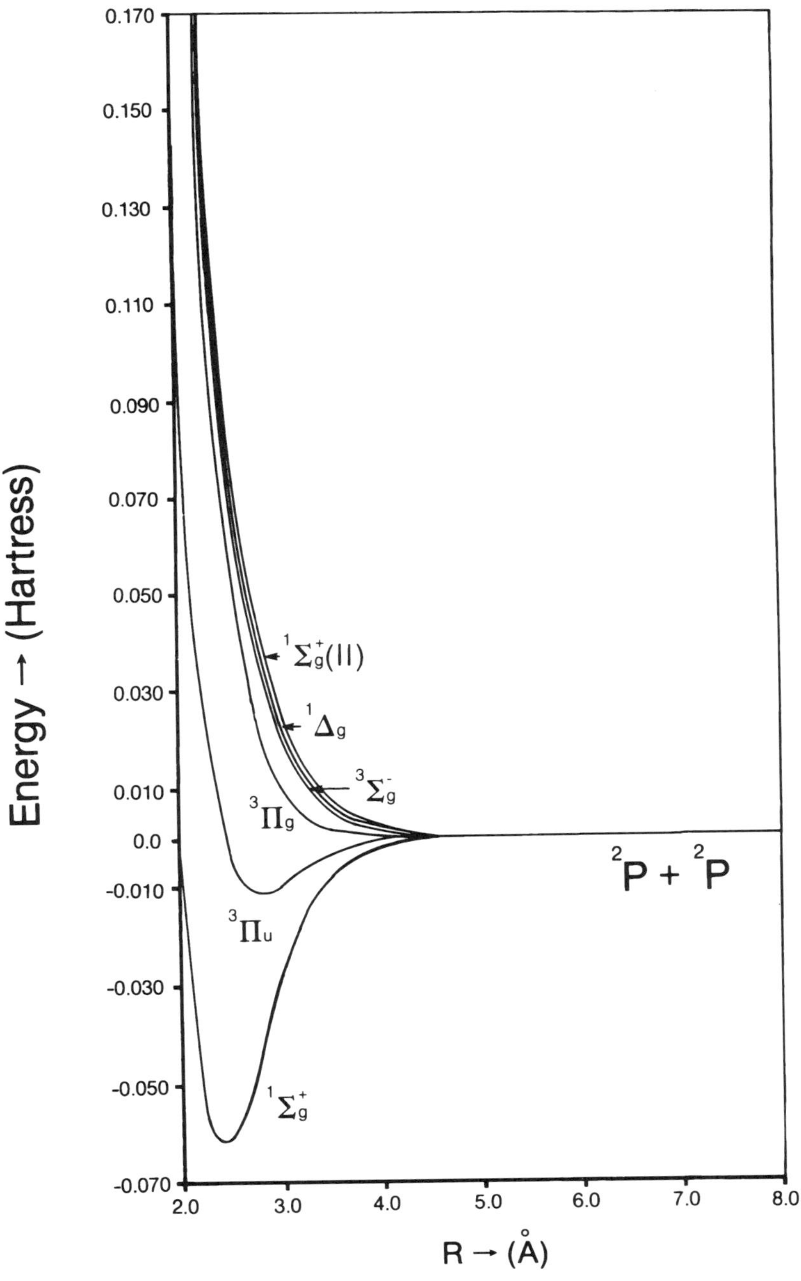

Figure 3.6 Potential-energy curves for the low-lying states of Br_2. (Reprinted from Ref. 181; copyright 1988, Elsevier Science Publishers B.V.)

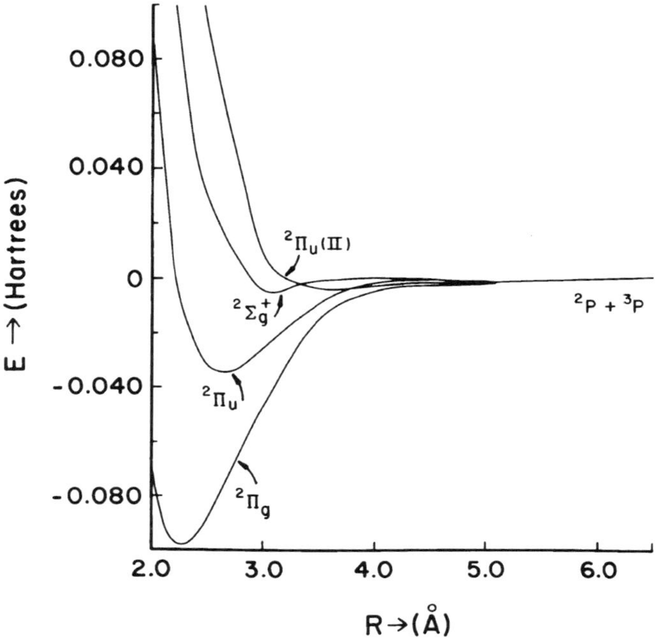

Figure 3.7 Potential-energy curves for the low-lying states of Br_2^+. (Reprinted from Ref. 181; copyright 1988, Elsevier Science Publishers.)

states, and consequently, the D state was assigned to 1_g. The E state observed in the $E \leftrightarrow B$ transition was assigned to the $^3\Pi_g(\text{II})(0_g^+)$ state.

The F state observed in the $F \rightarrow X$ system could be due to the $^3\Sigma_u^-(0_u^+) \rightarrow {}^1\Sigma_g^+(0_g^+)$ or $^3\Pi_u(\text{II})(0_u^+) \rightarrow {}^1\Sigma_g^+(0_g^+)$ transition. The theoretical spin–spin effects reported in Table 3.18 for the $^3\Sigma_u^-$ state as well as the $^1\Sigma_u^+$ state were obtained with a calculation that included only the $^3\Sigma_u^-$ and $^1\Sigma_u^+$ reference configurations, and are thus crude.

Ishiwata et al. (179) studied the $f \rightarrow B$ transition using the $X \rightarrow B \rightarrow f$ two-photon transition. These authors call the f state an ion-pair state, since the calculated R_e value of the f state (3.17 Å) is much longer than the R_e value of the ground state. The f state was tentatively assigned to a $^3\Sigma_g^-(0_g^+)$ state. It can be seen from Table 3.18 that the theoretical R_e and ω_e values of the $^3\Sigma_g^-(\text{II})$ state are in excellent agreement with the experimental values obtained by Ishiwata et al. (179). The observed T_e value of this state is lower than the value reported in Table 3.18, since the $^3\Sigma_g^-(\text{II})$ state is split by spin–orbit interaction. The 0_g^+ component is lower than the unsplit $^3\Sigma_g^-$ state. Thus, the f state participating in the $f \rightarrow B$ system was found to be most consistent with the 0_g^+ component of the $^3\Sigma_g^-(\text{II})$ state. The first $^3\Sigma_g^-$ state was found to be repulsive.

The $G \rightarrow X$ system observed in the 56,337-cm^{-1} region should be due to the $G^1\Sigma_u^1(0_u^+) \rightarrow X^1\Sigma_g^+(0_g^+)$ transition. The calculated T_e values of 57,576 cm^{-1} in the absence of spin–orbit interaction was found to be most consistent with this assignment. The rotational analysis of this system should be carried out to confirm this assignment.

The spectroscopic constants of Br_2^+ are shown in Table 3.19. Ionization of the highest occupied orbital of the $^1\Sigma_g^+$ ground state of Br_2, namely, the $1\pi_g$ orbital, yields the lowest state of Br_2^+ as the $^2\Pi_g$ state arising from the $1\sigma_g^2 1\sigma_u^2 2\sigma_g^2 1\pi_u^4 1\pi_g^3$ configuration. There is another possible low-lying state arising from the $1\sigma_g^2 1\sigma_u^2 2\sigma_g 1\pi_u^4 1\pi_g^4$ configuration ($2\Sigma_g^+$). The author (181) carried out CASSCF/FOCI calculations that employed the ($3s3p1d$) basis set on the $^2\Pi_g$, $^2\Pi_u$, $^2\Delta_g$, $^2\Sigma_g^+$, and $^2\Pi_u$(II) electronic states of Br_2^+, which dissociated into $Br^+(^3P) + Br(^2P)$.

The ground state of Br_2^+ is the $^2\Pi_g(\frac{3}{2})$ electronic state arising primarily from the $1\sigma_g^2 1\sigma_u^2 2\sigma_g^2 1\pi_u^4 1\pi_g^3$ configuration (see Table 3.19). The R_e value of the ground state of Br_2^+ is about 0.08 Å shorter than that of the Br_2 ground state. The vibrational frequency is correspondingly larger (about 40 cm^{-1}). There are no experimental R_e values known for the electronic states of Br_2^+. Since the ($3s3p1d$) basis set tended to yield R_e values 0.1 Å larger than the true values, the author (181) predicted that the true R_e values of the $^2\Pi_g^-(\frac{3}{2}, \frac{1}{2})$ states should be 2.22 Å.

The experimental photoelectron spectra of Br_2 yielded substantially different ω_e values for the two components of the $^2\Pi_u$ excited state (185). The reliability of these analyses was questioned in the literature. The value deduced from the photoelectron spectra (185) for $^2\Pi_u(\frac{3}{2})$ is 190 cm^{-1}, while for the $^2\Pi_u(\frac{1}{2})$ state it is 152 cm^{-1}. Usually, the vibrational frequencies of the spin–orbit components of a λ–s state cannot differ that much, unless there is considerable contamination of one of the components with another λ–s state due to spin-orbit coupling. As seen from Table 3.19, the theoretical vibrational frequencies of the two components of the first excited $^2\Pi_u$ state are 194 and 137 cm^{-1}, supporting the experimental values deduced from the photoelectron spectra (185). The substantial difference in the two vibrational frequencies was puzzling. This difference arises from the interaction of the $^2\Pi_u(\frac{1}{2})$ state with the $^4\Sigma_u^-(\frac{1}{2})$ and $^2\Sigma_u^+(\frac{1}{2})\lambda$–$s$ states in the vicinity of its equilibrium geometry. These two states undergo avoided crossing, influencing the R_e and ω_e values of the $^2\Pi_u(\frac{1}{2})$ state. Note that the R_e is about 0.02 Å shorter than that of the $^2\Pi_u(\frac{3}{2})$ state. The $^2\Pi_u(\frac{3}{2})$ component, however, is less contaminated, and thus its vibrational frequency is not substantially different from that of the $^2\Pi_u$ state obtained without the spin–orbit term.

The photoelectron spectra of Br_2 were recorded by Cornford et al. (185) as well as Potts and Price (186). The photoelectron spectra exhibited two states in the 19,000–21,600-cm^{-1} region and a third state in the 30,000-cm^{-1} region. Since photoelectron spectra typically involve ionization of a neutral molecule, these bands must arise from vertical rather than adiabatic transitions. The theoretical vertical transition energies and the splittings of the photoelectron spectral

Table 3.20 Vertical Transition Energies for Br_2^+[a]

Transition	T_{vert}, cm^{-1}	
	Theory	Experimental
$^2\Pi_g(\frac{3}{2}) \rightarrow {}^2\Pi_g(\frac{1}{2})$	3,030	2,820
$^2\Pi_g(\frac{3}{2}) \rightarrow {}^2\Pi_u(\frac{3}{2})$	18,170	19,290
$^2\Pi_g(\frac{3}{2}) \rightarrow {}^2\Pi_u(\frac{1}{2})$	21,478	21,602
$^2\Pi_g(\frac{3}{2}) \rightarrow {}^2\Sigma_g^+(\frac{1}{2})$	34,600	(30,700)[b]
$^2\Pi_g(\frac{3}{2}) \rightarrow {}^2\Pi_u(\text{II})$	56,752	—

[a]Theoretical values are from Reference 181.
[b]The value is not certain.

peaks are compared in Table 3.20. The agreement between these two values is remarkable, which led the author to reinterpret the photoelectron spectral bands as due to vertical rather than adiabatic transitions.

Table 3.21 shows the spectroscopic constants of the X ground state of Br_2 obtained by the author using various theoretical methods. The simplest calculation employed the FOCI method with a $(3s3p1d)$ basis set, while the most sophisticated calculation used the SOCI method with a $(4s4p2d)$ basis set (210,880 CSFs). The uncorrected D_e is the value obtained from the appropriate calculations without the Davidson correction for unlike clusters and spin–orbit correction. The Davidson correction was estimated from the calculation by Schwerdtfeger et al. (182), which yielded a D_e of 1.33 eV without the Davidson correction using a CISD method and 1.73 eV after the correction. The spin–orbit interaction decreased the D_e by 0.3 eV. Both these corrections were applied by the author (Table 3.21). As seen from Table 3.21, the $(4s4p2d)$ basis set yielded better D_e, R_e, and ω_e values compared to the $(3s3p1d)$ basis set. The improvement in D_e obtained by the SOCI method is marginal, although there are improvements in the R_e and ω_e values. The best D_e (1.88 eV) obtained by the author (181) compared quite favorably with an experimental value of

Table 3.21 Spectroscopic Properties of the Ground State of Br_2 at Various Levels of Theory[a]

Method	Basis Set	R_e, Å	ω_e, cm^{-1}	D_e, eV	
				Uncorrected[a]	Corrected[b]
FOCI	$(3s3p1d)$	2.37	303	1.73	1.79
FOCI	$(4s4p2d)$	2.34	319	1.82	1.88
SOCI	$(4s4p2d)$	2.33	321	1.77	1.87

[a]Without spin–orbit and Davidson's corrections.
[b]With spin–orbit and Davidson's corrections.
Source: Reference 181.

1.97 eV. The same calculation gave an ω_e value of 321 cm^{-1} compared to the experimental values of $\omega_e = 325$ cm^{-1} and $R_e = 2.28$ Å (37).

The theoretical spin–orbit splitting of the bromine atom (181) ($^2P_{3/2}$–$^2P_{1/2}$ splitting) from asymptotic splittings of molecular states of Br_2 was found to be 3915 cm^{-1}, compared to an experimental value (134) of 3685 cm^{-1}. The curves in Figure 3.6, which do not include spin–orbit splittings, are actually split apart at long distances. The spin–orbit contamination of various λ–*s* states became significant in this region in the RCI wavefunctions. For the ground state $^1\Sigma_g^+$ the curve was found to be almost identical with the one reported in Figure 3.6 at near-equilibrium distances in the absence of spin–orbit interaction. At long distances the curve shifted downward to dissociate the $^2P_{3/2} + ^2P_{3/2}$ atoms. The ground-state atoms were found to be more stabilized by the spin–orbit interaction than the ground-state molecule.

The potential-energy curves in Figure 3.7 for Br_2^+ also did not include spin–orbit effects and were actually split apart by spin–orbit interaction. At equilibrium the RCI spin–orbit splitting was found to be 3030 cm^{-1}. The spin–orbit splitting of the Br^+ ion ^3P state into $^3P_2$, $^3P_1$, and $^3P_0$ components was obtained by the author and co-workers (139), using the RCI method in an investigation on the collision of Br^+ with Kr. The $^3P_2$–$^3P_1$ and $^3P_2$–$^3P_0$ splittings of Br^+ were calculated at 3182 and 4675 cm^{-1}, respectively.

A laser spectroscopic investigation of the *A*–*X* system by Hamilton (190) yielded a T_e value of 16,414 cm^{-1} for the *A* state. If the *A* state is assigned to $2\Pi_u(\frac{1}{2})$, the agreement between the calculated *T* and this value is good. Boerrighter et al. (184) carried out HF/limited CI calculation that included spin–orbit matrix elements in an empirical manner for Br_2^+. The author's RCI (181) calculations, which included single and double excitations from many λ–*s* configurations, were in overall agreement with Boerrighter's (184) findings that the vibrational frequencies of the two components of the $^2\Pi_u$ state differ as a result of avoided crossings. However, the author's results differ from those of Boerrighter et al. (184) in the states that induce avoided crossings, in that the author's calculations revealed it to be the $^4\Sigma_u^-(\frac{1}{2})$ state, which undergoes avoided crossing with $^2\Pi_u(\frac{1}{2})$. The $^2\Pi_u(\frac{3}{2})$ state mixes heavily with $^4\Sigma_u^-(\frac{3}{2})$ instead of $^2\Delta_u(\frac{3}{2})$ as suggested by Boerrighter et al. (184), but does not undergo avoided crossing. The bond lengths calculated by Boerrighter et al. (184) for the $^2\Pi_u$ components are also much too long.

Table 3.22 depicts the contributions of the leading configurations in the FOCI wavefunction of the various electronic states in the absence of spin–orbit interaction. As seen from Table 3.22, with the exceptions of $^3\Pi_g$(II), $^3\Pi_u$(II), $^3\Sigma_g^-$(II), $^1\Sigma_u^+$, $^1\Pi_g$(II), and $^1\Pi_u$(II), other states have leading contributions with weights ≥85%.

The 0_g^+ ground state of Br_2 was found to be 93% $^1\Sigma_g^+(2\sigma_g^2 1\pi_u^4 1\pi_g^4)$ and 0.2% $^3\Pi_g(2\sigma_u 1\pi_u^3 1\pi_g^4)$. The $^3\Pi_u(0_u^+)$ state was also found to be pure $^3\Pi_u(0_u^+)$ at near-equilibrium geometries. The $^3\Pi_u(1_u)$ state is 49% $^3\Pi_u(2\sigma_g^2 2\sigma_u 1\pi_u^4 1\pi_g^3)$, 12% $^1\Pi_u(2\sigma_g^2 2\sigma_u 1\pi_u^4 1\pi_g^3)$, and 7% $^3\Sigma_u^+(2\sigma_g^2 2\sigma_u^2 1\pi_u^3 1\pi_g^3)$ at 6.50 bohr. The $^3\Pi_u(2_u)$ state is 62% $^3\Pi_u(2\sigma_g^2 2\sigma_u 1\pi_u^4 1\pi_g^3)$, 30% $^3\Pi_u(2\sigma_g 2\sigma_u^2 1\pi_u^3 1\pi_g^4)$, and 9% $^3\Delta_u(2\sigma_g^2 2\sigma_u^2 1\pi_u^3 1\pi_g^3)$. The 0_u^+(II) state is predominantly $^3\Sigma_u^-(0_u^+)$ (91%); $^1\Sigma_u^+(0_u^+)$

Table 3.22 Contributions of Various Configurations to the Low-Lying Electronic States of Br_2 at Their Near-Equilibrium Geometries[a]

State	Configurations(s)
$^1\Sigma_g^+$	$2\sigma_g^2 1\pi_u^4 1\pi_g^4$ (94%), $2\sigma_u^2 1\pi_u^4 1\pi_g^4$ (3%)
$^3\Pi_u$	$2\sigma_g^2 2\sigma_u 1\pi_u^4 1\pi_g^3$ (89%), $2\sigma_g 2\sigma_u^2 1\pi_u^3 1\pi_g^4$ (8%)
$^1\Pi_u$	$2\sigma_g^2 2\sigma_u 1\pi_u^4 1\pi_g^3$ (90%), $2\sigma_g 2\sigma_u^2 1\pi_u^3 1\pi_g^4$ (7%)
$^3\Pi_g$(II)	$2\sigma_u^2 2\sigma_g 1\pi_u^4 1\pi_g^3$ (64%), $2\sigma_g^2 2\sigma_u 1\pi_u^3 1\pi_g^4$ (26%)
$^3\Sigma_u^-$	$2\sigma_g^2 2\sigma_u^2 1\pi_u^3 1\pi_g^3$ (90%)
$^3\Pi_u$(II)	$2\sigma_g 2\sigma_u^2 1\pi_u^3 1\pi_g^4$ (64%), $2\sigma_g^2 2\sigma_u 1\pi_u^4 1\pi_g^3$ (26%)
$^3\Sigma_g^-$(II)	$2\sigma_g^2 2\sigma_u^2 1\pi_u^2 1\pi_g^4$ (56%), $2\sigma_g^2 2\sigma_u^2 1\pi_u^4 1\pi_g^2$ (34%)
$^1\Sigma_u^+$	$2\sigma_g 2\sigma_u 1\pi_u^4 1\pi_g^4$ (74%), $2\sigma_g^2 2\sigma_u^2 1\pi_u^3 1\pi_g^3$ (15%)
	Rydberg Configurations (3%)
$^1\Pi_g$(II)	$2\sigma_g 2\sigma_u^2 1\pi_u^4 1\pi_g^3$ (58%), $2\sigma_g^2 2\sigma_u 1\pi_u^3 1\pi_g^4$ (33%)
$^1\Pi_u$(II)	$2\sigma_g 2\sigma_u^2 1\pi_u^3 1\pi_g^4$ (64%), $2\sigma_g^2 2\sigma_u 1\pi_u^4 1\pi_g^3$ (25%)

[a]The $1\sigma_g^2 1\sigma_u^2$ shell is omitted. For the $^1\Pi_u$ state the contribution is at 2.50 Å.
Source: Reference 181.

states arising from $2\sigma_g 2\sigma_u 1\pi_u^4 1\pi_g^4$ and $2\sigma_g^2 2\sigma_u^2 1\pi_u^3 1\pi_g^3$ make nonnegligible contributions (3%). The $^3\Pi_1(1_g)$ state was found to be $^3\Pi_g$ (34%) and $^1\Pi_g$ (30%), $^3\Pi_g$ (20%) and $^1\Pi_g$ (15%) arising from $2\sigma_g 2\sigma_u^2 1\pi_u^4 1\pi_g^3$ 6.00 bohr. Consequently, spin–orbit contamination is more significant in excited states than the ground state.

The leading configuration contributing to the FOCI of the $^2\Pi_g$ ground state of Br_2^+ was found to be $(1\sigma_g^2 1\sigma_u^2)2\sigma_g^2 1\pi_u^4 1\pi_g^3$ (0.950) at 2.25 Å. The $^2\Pi_u$ state was predominantly $(1\sigma_g^2 1\sigma_u^2)2\sigma_g^2 1\pi_u^3 1\pi_g^4$ (0.925), while the $^2\Pi_u$(II) state was found to be predominantly $(1\sigma_g^2 1\sigma_u^2)1\pi_u^4 2\sigma_g 1\pi_g^3 2\sigma_u$ (0.902). The $^2\Sigma_g^+$ and $^2\Delta_g$ states arise from the $(1\sigma_g^2 1\sigma_u^2)2\sigma_g^2 2\sigma_u 1\pi_g^3 2\pi_g^3$ configuration (85%) at 3.0 Å. The $^3\Sigma_g^+$(II) state was found to be predominantly $1\sigma_g^2 1\sigma_u^2 2\sigma_g 1\pi_u^4 1\pi_g^4$.

Table 3.23 displays the contributions of various λ–*s* configurations to the

Table 3.23 Contributions of Various λ–*s* Configurations to the Low-Lying States of Br_2^+[a]

State	*R*, a.u.	Contribution
$^2\Pi_g(\frac{3}{2})$	4.75	$2\sigma_g^2 1\pi_u^4 1\pi_g^3{:}^2\Pi_g$ (90), $2\sigma_u^2 1\pi_u^4 1\pi_g^3{:}^2\Pi_g$ (3), $2\sigma_g 2\sigma_u 1\pi_u^3 1\pi_g^4{:}^4\Pi_g$ (2)
$^2\Pi_g(\frac{3}{2})$	6.00	$2\sigma_g^2 1\pi_u^4 1\pi_g^3{:}^2\Pi_g$ (75), $2\sigma_u^2 1\pi_u^4 1\pi_g^3{:}^2\Pi_g$ (6), $2\sigma_g 2\sigma_u 1\pi_u^3 1\pi_g^4{:}^4\Pi_g$ (6)
$^2\Pi_g(\frac{1}{2})$	4.75	$2\sigma_g^2 1\pi_u^4 1\pi_g^3{:}^2\Pi_g$ (89), $2\sigma_u^2 1\pi_u^4 1\pi_g^3{:}^2\Pi_g$(3), $2\sigma_g 2\sigma_u 1\pi_u^4 1\pi_g^3{:}^2\Pi_g$ (5),

Table 3.23 (*Continued*)

State	R, a.u.	Contribution
		$2\sigma_g 2\sigma_u 1\pi_u^4 1\pi_g^3{:}^4\Pi_g$ (0.2), $2\sigma_g 1\pi_u^4 1\pi_g^4{:}^2\Sigma_g^+$ (0.9)
$^2\Pi_g(\frac{1}{2})$	6.00	$2\sigma_g^2 1\pi_u^4 1\pi_g^3{:}^2\Pi_g$ (74), $2\sigma_u^2 1\pi_u^4 1\pi_g^3$ (5), $2\sigma_g 2\sigma_u 1\pi_u^4 1\pi_g^3{:}^2\Pi_g$(11); $^4\Pi_g$ (7)
$(\frac{3}{2})_u$	4.50	$2\sigma_g^2 1\pi_u^3 1\pi_g^4{:}^2\Pi_u$ (82), $2\sigma_g 2\sigma_u 1\pi_u^4 1\pi_g^3{:}^2\Pi_u$ (7); $^4\Pi_u$ (1.3), $2\sigma_u^2 1\pi_u^3 1\pi_g^4{:}^2\Pi_u$ (0.8), $2\sigma_g^2 2\sigma_u 1\pi_u^4 1\pi_g^2{:}^4\Sigma_u^-$ (1.3), $^2\Delta_u$ (0.7)
$(\frac{3}{2})_u$	4.75	$2\sigma_g^2 1\pi_u^3 1\pi_g^4{:}^2\Pi_u$ (76), $2\sigma_g 2\sigma_u 1\pi_u^4 1\pi_g^3{:}^2\Pi_u$ (8); $^4\Pi_u$ (1.6), $2\sigma_u^2 1\pi_u^3 1\pi_g^4{:}^2\Pi_u$ (1.9), $2\sigma_g^2 2\sigma_u 1\pi_u^4 1\pi_g^2{:}^4\Sigma_u^-$ (6), $^2\Delta_u$ (1)
$(\frac{3}{2})_u$	5.00	$2\sigma_g^2 1\pi_u^3 1\pi_g^4{:}^2\Pi_u$ (68), $2\sigma_g 2\sigma_u 1\pi_u^4 1\pi_g^3{:}^2\Pi_u$(7.6); $^4\Pi_u$ (2), $2\sigma_u^2 1\pi_u^3 1\pi_g^4{:}^2\Pi_u$ (2.2), $2\sigma_g^2 2\sigma_u 1\pi_u^4 1\pi_g^2{:}^4\Sigma_u^-$ (11); $^2\Delta_u$ (1.3)
$(\frac{3}{2})_u$	5.25	$2\sigma_g^2 1\pi_u^3 1\pi_g^4{:}^2\Pi_u$ (59), $2\sigma_g 2\sigma_u 1\pi_u^4 1\pi_g^3{:}^2\Pi_u$ (7); $^4\Pi_u$ (2.3), $2\sigma_u^2 1\pi_u^3 1\pi_g^4{:}^2\Pi_u$ (2.4); $2\sigma_g^2 2\sigma_u 1\pi_u^4 1\pi_g^2{:}^4\Sigma_u^-$ (16); $^2\Delta_u$ (1.3)
$(\frac{1}{2})_u$	4.25	$2\sigma_g^2 1\pi_u^3 1\pi_g^4{:}^2\Pi_u$ (87), $2\sigma_g 2\sigma_u 1\pi_u^4 1\pi_g^3{:}^2\Pi_u$ (6.7); $^4\Pi_u$ (0.4), $2\sigma_u^2 1\pi_u^3 1\pi_g^4{:}^2\Pi_u$ (1.2), $2\sigma_g^2 2\sigma_u 1\pi_u^4 1\pi_g^2{:}^4\Sigma_u^-$ (1.4); $^2\Sigma_u^+$ (0.1)
$(\frac{1}{2})_u$	4.50	$2\sigma_g^2 1\pi_u^3 1\pi_g^4{:}^2\Pi_u$ (81), $2\sigma_g 2\sigma_u 1\pi_u^4 1\pi_g^2{:}^2\Pi_u$ (7.2); $^4\Pi_u$ (0.6), $2\sigma_u^2 1\pi_u^3 1\pi_g^4{:}^2\Pi_u$ (1.5); $2\sigma_g^2 2\sigma_u 1\pi_u^4 1\pi_g^2{:}^4\Sigma_u^-$ (6)
$(\frac{1}{2})_u$	4.75	$2\sigma_g^2 1\pi_u^3 1\pi_g^4{:}^2\Pi_g$ (26), $2\sigma_g 2\sigma_u 1\pi_u^4 1\pi_g^3{:}^2\Pi_u$ (3); $^4\Pi_u$ (0.2), $2\sigma_u^2 1\pi_u^3 1\pi_g^4{:}^2\Pi_u$ (0.6), $2\sigma_g^2 2\sigma_u 1\pi_u^4 1\pi_g^2{:}^4\Sigma_u^-$ (50); $^2\Sigma_u^+$ (3.4)
$(\frac{1}{2})_u$	5.00	$2\sigma_g^2 1\pi_u^3 1\pi_g^4{:}^2\Pi_g$ (3), $2\sigma_g^2 2\sigma_u 1\pi_u^4 1\pi_g^2{:}^4\Sigma_u^-$ (61); $^2\Sigma_u^+$ (7.3)
$(\frac{1}{2})_u$	5.25	$2\sigma_g^2 1\pi_u^3 1\pi_g^4{:}^2\Pi_g$ (1.1), $2\sigma_g^2 2\sigma_u 1\pi_u^4 1\pi_g^2{:}^4\Sigma_u^-$ (58); $^2\Sigma_u^+$ (9)
$(\frac{1}{2})_u$	5.50	$2\sigma_g^2 1\pi_u^3 1\pi_g^4{:}^2\Pi_g$ (84), $2\sigma_g 2\sigma_u 1\pi_u^4 1\pi_g^3{:}^2\Pi_u$ (8); $^4\Pi_u$ (1), $2\sigma_u^2 1\pi_u^3 1\pi_g^4{:}^2\Pi_u$ (5)

[a]Note that the $^2\Pi_u(\frac{1}{2})$ state exhibits an avoided crossing. The $1\sigma_g^2 1\sigma_u^2$ shell is omitted. Numbers in parentheses are contributions in percentage.

Source: Reference 181.

low-lying states of Br_2^+. As seen from Table 3.23, the $^2\Pi_g(\frac{3}{2})$ state is predominantly composed of the $2\sigma_g^2 1\pi_u^4 1\pi_g^3$ configuration. At long distances, the $4\Pi_g$ state arising from the $2\sigma_g 2\sigma_u 1\pi_u^3 1\pi_g^4$ configurations makes a substantial contribution. The $^3\Pi_g(\frac{1}{2})$ component also exhibits a parallel behavior.

The compositions of the two spin–orbit components of $^2\Pi_u$ are interesting as a function of internuclear distance. The $\frac{3}{2}$ component of $^2\Pi_u$ was found to be predominantly $2\sigma_g^2 1\pi_u^3 1\pi_g^4(^2\Pi_u)$ up to 4.75 bohr. At 5.00 bohr, the $^4\Sigma_u^-(\frac{3}{2})$ state arising from the $2\sigma_g 2\sigma_u 1\pi_u^4 1\pi_g^2$ configuration makes a substantial contribution. Although this contribution is significant (~11–16%), it is not large enough to induce an avoided crossing. The $(\frac{1}{2})_u$ component, to the contrary, exhibited a different composition. Up to R = 4.50 bohr, this state was found to be predominantly the $^2\Pi_u$ state arising from the $2\sigma_g^2 1\pi_u^3 1\pi_g^4$ configuration. However, at 4.75 bohr, it became predominantly $^4\Sigma_u^-(\frac{1}{2})$ (50%), while $(\frac{1}{2})$ makes a 26% contribution. At 5.00 bohr, the contribution of $^2\Pi_u(\frac{1}{2})$ drops to 3%, while $^4\Sigma_u^-(\frac{1}{2})$ increases to 61% accompanied by an increased $^2\Sigma_u^+(\frac{1}{2})$ contribution (7%), whereas at 5.50 bohr, the $^2\Pi_u(\frac{1}{2})$ dominates once again. This avoided crossing of $^4\Sigma_u^-(\frac{1}{2})$ and $^2\Pi_u(\frac{1}{2})$ explains the large difference in the vibrational frequencies (ω_e values) in the photoelectron spectra of these two states.

3.3.6 In_2

Wajnkrac (191) and Ginter et al. (122) have obtained the emission spectra of In_2 in a King furnace in the 16,800–20,000-cm^{-1} region with bands centered at 18,000 cm^{-1}. Douglas et al. (123) studied the electronic absorption spectra of Al_2, Ga_2, and In_2 in cryogenic matrices. These authors found two systems, one at 13,000 cm^{-1} and the other centered near 27,690 cm^{-1}, for In_2. Douglas et al. (123) assigned the ground state of Al_2, Ga_2, and In_2 to a $^1\Sigma_g^+$ state and the absorption bands to $A^1\Pi_u \rightarrow {}^1\Sigma_g^+$ and $B^1\Sigma_g^+ \rightarrow X^1\Sigma_g^+$ transitions, respectively. The theoretical investigation of the author (133) on Ga_2 revealed these assignments to be incorrect, and showed that the most consistent assignments of the bands observed by Douglas et al. (123) for Ga_2 are $A^3\Pi_g \leftarrow X^3\Pi_u$ and $B^3\Pi_g \leftarrow X^3\Pi_u$ (see Section 3.2.1). The dissociation energy of In_2 was calculated as $D_e \sim 1$ eV using mass-spectrometric techniques (193).

Froben et al. (192) obtained the Raman spectra of matrix-isolated Ga_2, In_2, and Tl_2. They deduced approximate spectroscopic constants for the ground state of In_2 and Tl_2, using the Guggenheimer method, assuming a ground state of $^3\Sigma^-$ for In_2. This estimate was subsequently found to be incorrect by the author and Li (194).

The author and Li (194) carried out CASSCF/FOCI/RCI calculations employing RECPs for the indium atom, which retained the outer $4d^{10}5s^25p^1$ shells explicitly in the molecular calculations. A $(3s3p3d)$ valence Gaussian basis set was employed. The $4d^{10}$ shells of the In atoms were allowed to relax at the CASSCF level, but no excitations from these shells were allowed. The CASSCF active space consisted of four a_1, two b_2, and two b_1 orbitals in the C_{2v} group.

The RCI calculations (194) were carried out to estimate the spin–orbit cor-

Table 3.24 Dissociation Relationships for the Low-Lying States of In_2

	Dissociation Limits		
		Energies, cm^{-1}	
λ–s States	Atomic States	Exp.[a]	Theory[b]
$^3\Pi_u$, $^3\Sigma_g^-$, $^1\Pi_u$, $^1\Delta_g$	$^2P + {}^2P$	0.0	0.0
$^1\Sigma_g^+$ (2), $^1\Sigma_u^-$, $^3\Delta_u$			
$^3\Sigma_u^+$(2), $^3\Pi_g$, $^1\Pi_g$			
$^3\Sigma_g^+$, $^3\Sigma_u^+$, $^3\Pi_g$, $^3\Pi_u$, $^1\Sigma_g^+$, $^1\Sigma_u^+$, $^1\Pi_g$, $^1\Pi_u$	$^2P + {}^2S(5s^26s)$	24,373	26,578
$^3\Sigma_g^+$ (2), $^3\Sigma_u^+$ (2), $^3\Sigma_g^-$, $^3\Sigma_u^-$, $^3\Pi_g$(2), $^3\Pi_u$ (2), $^3\Delta_g$, $^3\Delta_u$	$^2P + {}^2P(5s^26p)$	31,816	32,746
$^1\Sigma_g^+$ (2), $^1\Sigma_u^+$ (2), $^1\Sigma_g^-$, $^1\Sigma_u^-$, $^1\Pi_g$ (2), $^1\Pi_u$ (2), $^1\Delta_g$, $^1\Delta_u$			
$0_g^+, 0_u^-, 1_u$	$^2P_{1/2} + {}^2P_{1/2}$	0.0	0.0
$2_u, 2_g, 1_u(2), 1_g(2)$	$^2P_{1/2} + {}^2P_{3/2}$	2,213	2,195
$0_g^+, 0_u^+, 0_g^-, 0_u^-$,			
$3_u, 2_u, 2_g, 1_u(2)$	$^2P_{3/2} + {}^2P_{3/2}$	4,426	4,389
$1_g, 0_g^+(2), 0_u^-(2)$			

[a]From Reference 134.
[b]From Reference 194.

rections for the spectroscopic properties, dissociation energies, and potential-energy curves. RCI calculations were also carried out omitting the spin–orbit term. The differences in spectroscopic properties and energies obtained with and without the spin–orbit term were then applied as corrections to the corresponding CASSCF/FOCI properties.

Table 3.24 shows the dissociation relationships for the low-lying states of In_2 with and without spin–orbit terms. Also shown in that table are the theoretical and experimental asymptotic splittings (134) of molecular states. As seen from Table 3.24, agreement between the theoretical and experimental in atom splittings is very good.

The spectroscopic properties of 23 electronic states of In_2 are shown in Table 3.25 (194). The spectroscopic properties listed in that table for states without ω–ω components were obtained without the spin–orbit integral. The ground state of In_2 is a $^3\Pi_u(0_u^-)$ state as seen from Table 3.25.

Figures 3.8 and 3.9 show the potential-energy curves of electronic states of In_2 without and with spin–orbit terms, respectively. The curves in Figure 3.8 were obtained by using the CASSCF/FOCI calculations. The curves in Figure 3.9 were obtained by using RCI calculations that included the spin–orbit coupling accurately, but correlations were included to a lesser degree compared to the CASSCF/FOCI method.

As seen from Table 3.25, the calculated R_e, ω_e, and D_e for the ground state

Table 3.25 Spectroscopic Properties of In_2 Calculated with the FOCI/RCI Method

State	r_e (Å)	T_e (cm^{-1})	ω_e, (cm^{-1})	D_e (eV)
$X^3\Pi_u(0_u^-)$	3.14	0	111	0.83
$^3\Pi_u(0_u^+)$	3.12	231	114	1.07
$^3\Pi_u(1_u)$	3.13	625	113	0.75
$^3\Pi_u(2_u)$	3.14	1,397	113	0.93
$^3\Pi_u$	3.13	963	114	1.02
$^3\Sigma_g^-(0_g^+)$	2.89	1,444	132	0.61
$^3\Sigma_g^-(1_g)$	2.88	1,947	141	0.82
$^3\Sigma_g^-$	2.87	2,006	142	0.90
$^1\Pi_u[1_u(\mathrm{II})]$	3.20	4,701	98	0.58
$^1\Pi_u$	3.20	4,629	98	0.57
$^1\Sigma_g^+[0_g^+(\mathrm{II})]$	3.29	5,112	91	0.51
$^1\Delta_g(2_g)$	2.98	5,158	109	0.41
$^1\Delta_g$	2.97	5,274	110	0.49
$^1\Sigma_g^+(\mathrm{II})[0_g^+(\mathrm{III})]$	3.29	8,334	89	—
$^3\Sigma_u^-[0_u^+(\mathrm{II})]$	3.04	19,600	119	2.78
$^3\Sigma_u^-$	3.04	19,635	120	2.77
$B^3\Pi_g(\mathrm{II})[0_g^-(\mathrm{II})]$	3.49	26,104	70	1.14
$^3\Pi_g(\mathrm{II})$	3.49	26,631	70	1.14
$^1\Delta_u$	3.08	27,149	120	1.84
$^3\Delta_u(\mathrm{II})$	2.77	28,582	181	1.66
$^3\Pi_u(\mathrm{II})$	4.67	30,952	47	0.60
$^3\Sigma_u^-(\mathrm{II})$	2.75	31,767	—	—
$^1\Pi_g(\mathrm{II})$	3.56	33,689	54	1.03

Source: Reference 194.

[$^3\Pi_u(0_u^-)$] of In_2 are 3.14 Å, 111 cm^{-1}, and 0.83 eV, respectively. The In–In bond strength in the 0_u^- state is reduced by 19% as a result of spin–orbit interaction. The R_e value is correspondingly increasing, while ω_e is decreased. The theoretical spectroscopic properties of the ground state of In_2 are in conflict with the values reported by Froben et al. (192), who used an approximate Guggenheimer technique and obtained an R_e of 2.8 Å. This value is too short for In_2. However, the vibrational frequency and the dissociation energy of 0.87 eV deduced by Froben et al. agreed with the Balasubramanian–Li (194) theoretical D_e value of 0.83 eV. The mass-spectrometric methods yield a slightly higher D_e of 1 eV. A similar discrepancy in the R_e value of Tl_2 was noted before. Pitzer and co-workers (195, 196) obtained a bond length of 3.54 Å for the ground state of Tl_2, disagreeing with the 3-Å value reported by Froben et al. (192) using the same technique.

The bands of $A \leftarrow X$, $B \leftarrow X$, and the emission systems were centered at 13,000, 27,690, and 18,000 cm^{-1}, respectively (122, 123, 130). Table 3.26 shows both vertical and adiabatic transition energies of allowed dipole tran-

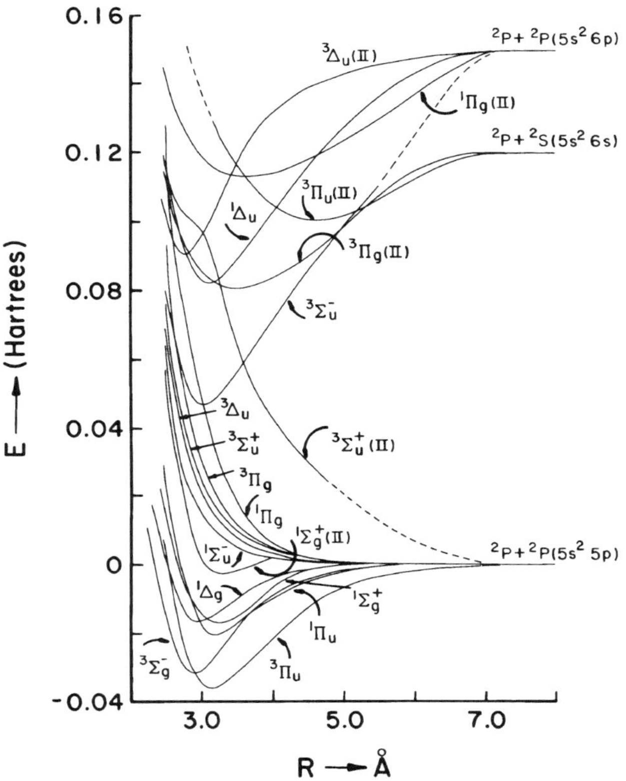

Figure 3.8 Potential-energy curves of In_2 obtained without the spin–orbit term. (Reprinted from Ref. 194; copyright 1988, American Institute of Physics.)

sitions of In_2. As seen from Table 3.26, the most consistent transition for the 13,000-cm^{-1} absorption band is the $A[^3\Pi_g(0_g^-)] \leftarrow X[^3\Pi_u(0_u^-)]$ transition. The theoretical vertical transition energy of 12,423 cm^{-1} is in very good agreement with the experimental band center (13,000 cm^{-1}). Since the $^3\Pi_g$ curve is repulsive, it is seen only in absorption. The $B \leftarrow X$ system was assigned by the author and Li (194) to the $B(^3\Pi_g) \leftarrow X(0_u^-)$ transition. The theoretical transition energy of 26,400 cm^{-1} was found to be in excellent agreement with the experimental value of 27,690 cm^{-1}.

The emission spectra observed by Ginter et al. (122) (18,000-cm^{-1} region) are most consistent with the $^3\Sigma_u^-(0_u^+) \leftrightarrow {}^3\Sigma_g^-(0_g^+)$ transition. The theoretical T_e of 18,156 cm^{-1} found by the author and Li was in remarkable agreement with this value; thus the author and Li assigned this to the $^3\Sigma_u^-(0_u^+) \leftrightarrow {}^3\Sigma_g^-(0_g^+)$ system.

Many transitions listed in Table 3.26 have not yet been observed. Among the transitions listed in Table 3.26, only the above-mentioned three transitions have been observed. Thus there is considerable room for further study in the electronic spectra of In_2. There are no vibrational or rotational analyses done

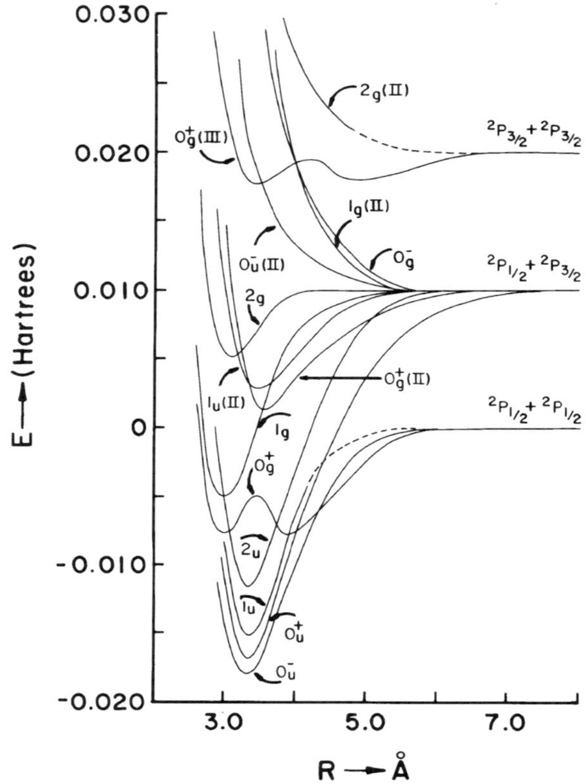

Figure 3.9 Potential-energy curves of In_2 including spin–orbit effects. (Reprinted from Ref. 194; copyright 1988, American Institute of Physics.)

on the observed spectra. Specifically, the vibrational–rotational analysis of the bands in $({}^3\Sigma_u^- - {}^3\Sigma_g^-)$ should be done for a comparison of theoretical R_e and ω_e values of these states.

Spin–orbit effects were found to be significant for In_2, since the atomic splitting of the indium atom is 2213 cm^{-1}. The ${}^3\Pi_u(0_u^-)$–${}^3\Pi_u(2_u)$ splitting of 1400 cm^{-1} was about 64% of the indium ${}^2P_{1/2}$–${}^2P_{3/2}$ atomic splitting. The spin–orbit splitting stabilizes the ${}^3\Pi_u(0_u^+)$ state with respect to the separated atoms, while it destabilizes the ${}^3\Pi_u(0_u^-)$ ground state of In_2, since the ${}^3\Pi_u(0_u^+)$ state dissociates into $In({}^2P_{1/2})$ + $In({}^2P_{3/2})$ atoms, while the ground state (0_u^-) dissociates into $In({}^2P_{1/2})$ + $In({}^2P_{1/2})$. The influence of spin–orbit coupling on the bond lengths and the vibrational frequencies is subtle. The spin–spin splitting of the ${}^3\Sigma_g^-$ state is much smaller (503 cm^{-1}) than that of the ${}^3\Pi_u$ state. Spin–orbit coupling also changes the shapes of the potential-energy curves of some of the excited states of In_2. For example, the double minima in the 0_g^+ and 0_g^+(III) surfaces are attributed to avoided crossings induced by spin–orbit coupling.

Table 3.26 Some Allowed Electric Dipole Transitions and the Transition Energies for In_2[a]

Transition	T, cm^{-1}	Transition	T, cm^{-1}
$^3\Sigma^-_{1_g} \leftrightarrow {}^3\Pi_{0^-_u}$	1,947	$^3\Delta_u$(II) $\leftrightarrow$ $^3\Pi_g$(II)	1,951
$^3\Pi_{0^-_g} \leftarrow {}^3\Pi_{0^-_u}$	12,423	$^3\Pi_u$(II) $\leftrightarrow$ $^3\Pi_g$(II)	4,321
$^3\Pi_g[1_g(\text{II})] \leftarrow {}^3\Pi_{0^-_u}$	128,889	$^3\Pi_g$(II) $\leftarrow$ $^3\Pi_{0^+_u}$	26,400
$^3\Pi_g[2_g(\text{II})] \leftarrow {}^3\Pi_{1_u}$	13,220	$^3\Sigma^-_{0^+_g} \leftrightarrow {}^3\Pi_{1_u}$	819
$^3\Pi_g(\text{II})[0^-_g(\text{II})] \leftrightarrow {}^3\Pi_{0^-_u}$	26,104	$^3\Sigma^-_{1_g} \leftrightarrow {}^3\Pi_{1_u}$	322
$^3\Pi_g(\text{II}) \leftrightarrow {}^3\Pi_{0^-_u}$	26,631	$^3\Sigma^-_u[0^+_u(\text{II})] \leftrightarrow {}^3\Sigma^-_{0^+_g}$	18,156
$^3\Pi_g[1_g(\text{II})] \leftarrow {}^3\Pi_{1_u}$	12,264	$^3\Sigma^-_{1_g} \leftrightarrow {}^3\Pi_{2_u}$	550
$^3\Pi_g[1_g(\text{II})] \leftarrow {}^3\Pi_{0^+_u}$	12,658	$^3\Pi_g(\text{II}) \leftrightarrow {}^3\Pi_{2_u}$	25,234
$^3\Pi_g[1_g(\text{II})] \leftarrow {}^3\Pi_{2_u}$	11,492	$^3\Sigma^-_u(\text{II}) \leftrightarrow {}^3\Sigma^-_{0^+_g}$	30,323
$^3\Pi_{0^-_g} \leftarrow {}^3\Pi_{1_u}$	11,803	$^3\Pi_u(\text{II}) \leftrightarrow {}^3\Sigma^-_{0^+_g}$	29,508
$^3\Pi_g[2_g(\text{II})] \leftarrow {}^3\Pi_{2_u}$	12,448	$^3\Sigma^-_u[0^+_u(\text{II})] \leftrightarrow {}^3\Sigma^-_{1_g}$	17,653
$^3\Pi_g(\text{II})\ [0^-_g(\text{II})] \leftrightarrow {}^3\Pi_{1_u}$	26,006	$^3\Sigma^-_u(\text{II}) \leftrightarrow {}^3\Sigma^-_{1_g}$	29,820
$^3\Sigma^-_{0^+_g} \leftrightarrow {}^3\Pi_{0^+_u}$	1,213	$^3\Pi_u(\text{II}) \leftrightarrow {}^3\Sigma^-_{1_g}$	29,005
$^3\Sigma^-_{1_g} \leftrightarrow {}^3\Pi_{0^+_u}$	1,715	$^1\Sigma^+_g[0^+_g(\text{II})] \leftrightarrow {}^1\Pi_u[1_u(\text{II})]$	411
$^1\Pi_g(\text{II}) \leftrightarrow {}^1\Pi_u[1_u(\text{II})]$	28,988	$^1\Delta_{2_g} \leftrightarrow {}^1\Pi_u[1_u(\text{II})]$	457
$^1\Delta_u \leftrightarrow {}^1\Delta_{2_g}$	21,991	$^1\Sigma^+_g(\text{II})(0^+_g(\text{III})) \leftrightarrow {}^1\Pi_u[1_u(\text{II})]$	3,633
$^1\Pi_g(\text{II}) \leftrightarrow {}^1\Delta_u$	6,540		

[a] A two-way arrow indicates both states are bound. A one-way arrow indicates that the originating state is bound. The transition energies are adiabatic if both states are bound and vertical if one state is bound.

Source: Reference 194.

Table 3.27 shows the contributions of various important configurations to the FOCI wavefunctions of the electronic states of In_2. As seen from Table 3.27, the $^3\Pi_u$ ground state and the $^3\Sigma^-_g$ state are dominated by a single electronic configuration. The $^1\Sigma^+_g$ and $^1\Sigma^+_g$(II) states are mixtures of several electronic configurations. For the $^1\Sigma^+_g$ state, both the $2\sigma_g^2$ and $1\pi_u^2$ configurations make substantial contributions. Thus correlation effects are somewhat significant for the $^1\Sigma^+_g$ states. The shoulder in the $^3\Sigma^+_u$(II) curve (see Fig. 3.8) is due to an avoided crossing as seen from Table 3.27. At long distances this state was found by the author and Li (194) to be predominantly composed of the $2\sigma_g 2\sigma_u$ configuration. At short distances, the contribution from the $1\sigma_u 2\sigma_g 1\pi_u^2$ configurations became important, resulting in an avoided crossing of these two states. The $^3\Sigma^-_u$ state was found to be a mixture of $1\pi_u 1\pi_g$, $1\sigma_u 2\sigma_g 1\pi_u^2$, and other configurations.

Table 3.28 shows the contributions of various λ–s states in the RCI wavefunctions of In_2. The 0^-_u, 0^+_u, and 2_u states are relatively pure ($^3\Pi_u$), while the 1_u state exhibited 2% triplet–singlet mixing. The most significant effect

Table 3.27 Contributions of Important Configurations to the FOCI Wavefunctions of In_2

State	Percentage of Contributions	State	Percentage of Contributions
$^3\Pi_u$	$2\sigma_g 1\pi_u$ (87)	$^3\Sigma_u^+$(II) $R = 3.4$ Å	$2\sigma_g 2\sigma_u$(64), $2\sigma_g 3\sigma_u$(10), $2\sigma_g 4\sigma_u$ (8), $2\sigma_u 4\sigma_g$ (2)
$^3\Sigma_g^-$	$1\pi_u^2$ (85), $1\pi_g^2$ (2)	$R = 3.2$ Å	$2\sigma_g 2\sigma_u$(56), $2\sigma_g 3\sigma_u$(11), $2\sigma_g 4\sigma_u$ (10), $1\sigma_u 2\sigma_g 1\pi_u^2$ (5)
$^1\Pi_u$	$2\sigma_g 1\pi_u$ (86), $1\pi_g 2\sigma_u$ (3)	$R = 3.0$ Å	$1\sigma_u 2\sigma_g 1\pi_u^2$ (45), $2\sigma_g 2\sigma_u$ (22), $2\sigma_g 3\sigma_u$ (6), $2\sigma_g 4\sigma_u$ (4)
$^1\Delta_g$	$1\pi_u^2$ (82), $1\pi_g^2$ (6)	$R = 2.5$ Å	$1\sigma_u 2\sigma_g 1\pi_u^2$ (76), $1\pi_u 1\pi_g$ (8), $1\sigma_u 2\sigma_g 1\pi_g^2$ (4)
$^1\Sigma_g^+$	$2\sigma_g^2$ (51), $1\pi_u^2$ (28), Rydberg (4), $1\pi_g^2$ (2)	$^3\Pi_g$(II)	$1\pi_u 2\sigma_u$ (47), $1\sigma_u 1\pi_u$ (28), $1\pi_u 3\sigma_u$ (3), $1\sigma_g 1\pi_g$ (2), $1\pi_u 4\sigma_u$ (3), $2\sigma_g 1\pi_g$ (2), $1\sigma_u 2\sigma_g 5\sigma_u 1\pi_g$ (1)
$^1\Sigma_g^+$(II)	$1\pi_u^2$ (53), $2\sigma_g^2$ (17), $1\pi_g^2$ (15)	$^1\Delta_u$	$1\pi_u 1\pi_g$(68), $1\sigma_u 2\sigma_g 1\pi_u^2$ (14), $1\pi_u 2\pi_g$ (3), $1\pi_u 3\pi_g$ (2)
$^1\Sigma_u^-$	$1\pi_u 1\pi_g$ (87)	$^3\Delta_u$(II)	$1\sigma_u 2\sigma_g 1\pi_u^2$ (79), $1\pi_u 1\pi_g$ (4), $1\sigma_u 2\sigma_g 1\pi_g^2$ (4), $1\sigma_g 2\sigma_g 1\pi_u 1\pi_g$ (1)
$^3\Delta_u$	$1\pi_u 1\pi_g$ (89)	$^3\Pi_u$(II)	$2\sigma_u 1\pi_g$(61), $2\sigma_g 1\pi_u$ (9), $1\sigma_u 1\pi_g$ (7), $1\sigma_u 2\sigma_g 1\pi_u 2\sigma_u$ (3), $4\sigma_u 1\pi_g$ (1), $1\sigma_g 2\sigma_g 2\sigma_u 1\pi_g$ (1), $3\sigma_u 1\pi_g$ (1), $1\sigma_g 1\pi_u$ (2), $1\sigma_g 1\sigma_u 2\sigma_u 1\pi_u 1\pi_g^2$ (1)
$^3\Sigma_u^+$	$1\pi_u 1\pi_g$ (89)	$^1\Pi_g$(II)	$2\sigma_u 1\pi_u$(50), $2\sigma_g 1\pi_g$ (15), $1\sigma_u 1\pi_u$ (10), $3\sigma_u 1\pi_u$ (4), $4\sigma_u 1\pi_u$ (4)
$^3\Pi_g$	$2\sigma_g 1\pi_g$ (77), $1\pi_u 2\sigma_u$ (4), $1\sigma_u 1\pi_u$ (3)	$^3\Sigma_u^-$(II)	$1\sigma_u 2\sigma_g 1\pi_u^2$ (71), $1\pi_u 1\pi_g$ (10), $1\sigma_u 2\sigma_g 1\pi_g^2 1\pi_g^2$ (1)
$^1\Pi_g$	$2\sigma_g 1\pi_g$ (69), $1\pi_u 2\sigma_u$ (13), $2\sigma_g 2\pi_g$ (3)		
$^3\Sigma_u^-$	$1\pi_u 1\pi_g$ (61), $1\sigma_u 2\sigma_g 1\pi_u^2$(19), $1\sigma_u 2\sigma_g 1\pi_g^2$(2), $1\pi_u 2\pi_g$(3), $1\pi_u 3\pi_g$ (2)		

Source: Reference 194.

Table 3.28 Weights of Various λ–s Configurations in the ω–ω States in In_2

States	R, bohr	Percentage of Contributions	States	R, bohr	Percentage of Contributions
0_u^-	6.0	$2\sigma_g 1\pi_u(^3\Pi_u)$ (93)	2_g	6.0	$1\pi_u^2(^1\Delta_g)$ (84), $1\pi_g^2(^1\Delta_g)$ (11)
0_u^+	6.0	$2\sigma_g 1\pi_u(^3\Pi_u)$ (95)		5.5	$1\pi_u^2(^3\Sigma_g^-)$ (68), $1\sigma_u 2\sigma_u 1\pi_u^2(^3\Sigma_g^-)$ (19)
1_u	6.0	$2\sigma_g 1\pi_u(^3\Pi_u)$ (93), $2\sigma_g 1\pi_u(^1\Pi_u)$ (2)	0_g^+(III)	6.5	$1\pi_u^2(^1\Sigma_g^+)$ (59), $2\sigma_g^2(^1\Sigma_g^+)$ (9), $1\pi_g^2(^1\Sigma_g^+)$ (16), $1\pi_u^2(^3\Sigma_g^-)$ (10)
2_u	6.0	$2\sigma_g 1\pi_u(^3\Pi_u)$ (95)			
0_g^+	5.5	$1\pi_u^2(^3\Sigma_g^-)$ (89), $1\pi_g^2(^3\Sigma_g^-)$ (4)		7.5	$1\pi_u^2(^1\Sigma_g^+)$ (44), $1\pi_u^2(^3\Sigma_g^-)$ (21), $1\pi_g^2(^1\Sigma_g^+)$ (20), $1\pi_g^2(^3\Sigma_g^-)$ (5)
	6.5	$1\pi_u^2(^3\Sigma_g^-)$ (42), $2\sigma_g^2(^1\Sigma_g^+)$ (25), $1\pi_u^2(1\Sigma_g^+)$ (16)	0_g^-	6.0	$2\sigma_g 1\pi_g(^3\Pi_g)$ (92)
	7.5	$2\sigma_g^2(^1\Sigma_g^+)$ (70), $2\sigma_u^2(^1\Sigma_g^+)$ (5), $1\pi_u^2(^1\Sigma_g^+)$ (8)	1_g(II)	6.0	$2\sigma_g 1\pi_g(^3\Pi_g)$ (87), $2\sigma_g 1\pi_g(^1\Pi_g)$ (4)
1_g	5.5	$1\pi_u^2(^3\Sigma_g^-)$ (91), $1\pi_g^2(^3\Sigma_g^-)$ (5)	0_u^-(II)	6.0	$1\pi_u 1\pi_g(^1\Sigma_u^-)$ (65), $1\pi_u 1\pi_g(^3\Sigma_u^+)$ (23)
0_g^+(II)	5.5	$1\pi_u^2(^1\Sigma_g^+)$ (71), $2\sigma_g^2(^1\Sigma_g^+)$ (8), $1\pi_u^2(^3\Sigma_g^-)$ (8), $1\pi_g^2(^3\Sigma_g^-)$ (18)	2_g(II)	6.0	$2\sigma_g 1\pi_g(^3\Pi_g)$ (92)
	6.5	$2\sigma_g^2(^1\Sigma_g^+)$ (57), $1\pi_u^2(^3\Sigma_g^-)$ (33)	0_u^+(II)	5.5	$1\pi_u 1\pi_g(^3\Sigma_u^-)$ (97)
	7.5	$1\pi_u^2(^3\Sigma_g^-)$ (51), $2\sigma_g^2(^1\Sigma_g^+)$ (14), $1\pi_g^2(^3\Sigma_g^-)$ (15), $1\pi_u^2(^1\Sigma_g^+)$ (9), $1\pi_g^2(^1\Sigma_g^+)$ (5)	0_g^-(II)	6.5	$2\sigma_u 1\pi_u(^3\Pi_g)$ (55), $1\sigma_u 1\pi_u(^3\Pi_g)$ (40)
1_u(II)	6.0	$2\sigma_g 1\pi_u(^1\Pi_u)$ (92), $2\sigma_g 1\pi_u(^3\Pi_u)$ (2)			

Source: Reference 194.

of spin–orbit contamination was found to be on the 0_g^+ states. The 0_g^+ curves exhibited avoided crossings. At short distances, the 0_g^+(I) state was found to be predominantly $1\pi_u^2(^3\Sigma_g^-)$, while at intermediate distances, contributions from $1\pi_u^2(^3\Sigma_g^-)$ and $2\sigma_g^2(^1\Sigma_g^+)$ became significant. At longer distances, the two states undergo an avoided crossing, leading to a second minimum in the 0_g^+ surface. This avoided crossing is transferred to the 0_g^+(II) and 0_g^+(III) roots, as seen from the weights of the various λ–*s* states as a function of internuclear distance (Table 3.28).

Table 3.29 shows the Mulliken populations of various electronic states of In_2. As seen from Table 3.29, the $5s^2$ shell is not an inert shell, since Mulliken populations of In are between 1.79 and 1.84. The participation of the $5s^2$ shell in bonding is certainly of interest, since for Tl_2 the corresponding $6s^2$ shell was found to be inert (195, 196). The $5s^2$ shell is not that inert for In_2, since some of the 5*s* electrons are promoted to the 5*p* shell. The In–In bond is thus expected to be stronger than the Tl–Tl bond, which is consistent with their calculated dissociation energies. Some of the excited states, especially the $^3\Sigma_u^-$, $^3\Pi_g$(II), $^3\Delta_u$(II), and $^3\Sigma_u^-$(II) states, arise from excitation of the 5*s* electron into the 5*p*

Table 3.29 Mulliken Population Analysis in In_2[a]

	Gross Population				Overlap
State	In	In(*s*)	In(*p*)	In(*d*)	In–In
$^3\Pi_u$	13.000	1.835	1.133	10.031	0.335
$^3\Sigma_g^-$	13.000	1.786	1.183	10.030	0.544
$^1\Pi_u$	13.000	1.838	1.128	10.033	0.280
$^1\Delta_g$	13.000	1.794	1.179	10.027	0.477
$^1\Sigma_g^+$	13.000	1.859	1.114	10.027	0.256
$^1\Sigma_g^+$(II)	13.000	1.855	1.127	10.018	0.212
$^1\Sigma_u^-$	13.000	1.834	1.143	10.023	−0.120
$^3\Delta_u$	13.000	1.842	1.136	10.022	−0.156
$^3\Sigma_u^+$	13.000	1.842	1.136	10.022	−0.161
$^3\Pi_g$	13.000	1.824	1.146	10.031	−0.341
$^1\Pi_g$	13.000	1.872	1.100	10.028	−0.740
$^3\Sigma_u^-$	13.000	1.645	1.322	10.033	0.146
$^3\Sigma_u^+$(II)	13.000	1.845	1.151	10.005	−2.550
$^3\Pi_g$(II)	13.000	1.649	1.321	10.030	−1.017
$^1\Delta_u$	13.000	1.697	1.266	10.037	0.067
$^3\Delta_u$(II)	13.000	1.369	1.612	10.019	0.755
$^3\Pi_u$(II)	13.000	1.824	1.170	10.005	−0.398
$^1\Pi_g$(II)	13.000	1.772	1.200	10.027	−1.012
$^3\Sigma_u^-$(II)	13.000	1.378	1.590	10.032	0.756

[a]The population analyses were carried out at R_e when the state was bound; otherwise they were carried out at the R_e of the ground state.

Source: Reference 194.

shell as seen from their Mulliken s populations. The s populations for these states are 1.38–1.70. The In–In overlap populations are larger for the $^3\Sigma_g^-$, $^3\Pi_u$, and $^1\Delta g$ states.

3.3.7 Sn_2

Bondybey and co-workers have obtained the electronic spectra of matrix-isolated Sn_2 using laser-induced fluorescence. The spectroscopic constants of the $X0_g^+$ ground state of Sn_2 are the 0_u^+–$X0_g^+$ system at 18,223 cm^{-1}. An emission system in the 12,000–14,000-cm^{-1} region was assigned as terminating in a state 3000–5000 cm^{-1} above the ground state. The spectroscopic properties of some of these states have been calculated from the electronic spectra.

The author and Pitzer (201) used the SRF/RCI techniques including spin–orbit coupling on 10 low-lying electronic states of Sn_2 with a double-ζ STO basis set. The partition function of Sn_2 was also calculated with the objective of correcting the thermodynamic D_0° value of Sn_2. Subsequently, Pacchioni (202) calculated the low-lying electronic states of Sn_2 and Pb_2, using semiempirical pseudopotentials and the MRDCI method.

Table 3.30 shows the dissociation limits for low-lying molecular states of Sn_2 and the energies of separated atoms. The tin dimer has many more low-lying electronic states compared to Ge_2, mainly due to larger spin–orbit coupling. Table 3.31 shows the theoretical spectroscopic constants for Sn_2 calculated by the author and Pitzer (201). Figure 3.10 shows the potential-energy curves of the g states of Sn_2, while Figure 3.11 shows the corresponding curves for the u states of Sn_2.

The vibrational frequency of the ground state of Sn_2 is well known as 188 cm^{-1} in rare-gas matrices (199, 200) and 190 cm^{-1} in the gas phase (197). The theoretical value 197 cm^{-1} agrees well within the estimated uncertainty

Table 3.30 Dissociation Limits of Some Molecular States of Sn_2

Molecular States	Dissociated Atoms	Energies of Dissociated Atoms in cm^{-1} Sn_2 [a]
0_g^+	$^3P_0 + {}^3P_0$	0.0
$0_g^-, 0_u^-, 1_g, 1_u$	$^3P_0 + {}^3P_1$	1,691.8
$0_g^+, 0_u^+, 1_g, 1_u, 2_g, 2_u$	$^3P_0 + {}^3P_2$	3,427.7
$0_g^+(2), 0_u^-, 1_g, 1_u, 2_g$	$^3P_1 + {}^3P_1$	3,383.6
$0_g^+, 0_u^+, 0_g^-(2), 0_u^-(2), 1_g(3),$ $1_u(3), 2_g(2), 2_u(2), 3_g, 3_u$	$^3P_1 + {}^3P_2$	5,119.5
$0_g^+(3), 0_u^-(2), 1_g(2), 1_u(2),$ $2_g(2), 2_u, 3_g, 3_u, 4_g$	$^3P_2 + {}^3P_2$	6,855.4
$0_g^+, 0_u^+, 1_g, 1_u, 2_g, 2_u$	$^3P_0 + {}^1D_2$	10,304.8

[a] Atomic energies from Reference 134.

Table 3.31 Spectroscopic Constants of Several Low-Lying States of Sn_2 and the Corresponding Values Obtained without the Spin–Orbit Term[a]

State	T_e, cm^{-1}	R_e, Å	ω_e cm^{-1}
$X0_g^+$	0	2.76	197
1_g	342	2.75	205
2_u	1,477	2.62	218
1_u	2,509	2.62	220
2_g	7,159	2.81	178
0_u^-	4,084	2.63	219
0_u^+	4,192	2.63	221
1_u(II)	7,260	2.62	232
1_g(II)	9,964	2.95	88
0_g^+(II)	8,002	2.78	116
$C0_u^+$(II)	18,223		
$^3\Sigma_g^-$	1,694	2.75	204
$^3\Pi_u$	3,218	2.62	219
$^1\Sigma_g^+$	7,901	2.81	140

[a]The T_e value of the 0_u^+(II) is an experimental value from Reference 140.
Source: Reference 201.

of 10 cm^{-1}. Most of the excited electronic states that have been observed in Sn_2 lie above the 10,000-cm^{-1} level of the highest state among those calculated theoretically. The first 0_u^+ state of Sn_2 is at 4,192 cm^{-1}, which appears not to have been studied experimentally.

The strongest absorption in Sn_2 is due to a second 0_u^+ state at 18,223 cm^{-1} from the ground state. An emission in the 12,000–14,000-cm^{-1} range for Sn_2 was interpreted (198) as leading to a state about 3,000–5,000 cm^{-1} above the ground state with a harmonic vibration frequency of ~195 cm^{-1}. Both the 0_u^+ and 0_u^- states near 4100 cm^{-1} fit the first criterion, and the 1_u state at 2500 cm^{-1} is probably close enough to be considered. These are all components of the $^3\Pi_u$ term in Λ–S coupling. Their theoretical ω_e values are definitive. The author and Pitzer (201) suggested the possibility that this array of bands contains contributions from more than one transition involving the 0_u^+, 0_u^-, or 1_u states.

The thermodynamic "third law" method (204, 205) of calculation of D_e of Sn_2 from mass-spectrometric data requires knowledge of the partition function of the molecule and the dissociated atoms. The molecular parameters of Sn_2 were estimated in this calculation. Spectroscopic measurements (87, 140, 199, 200) have provided vibrational frequencies in good agreement with the Balasubramanian–Pitzer theoretical calculations. The author and Pitzer (201) corrected the D_e obtained from mass spectrometry using the correct partition function.

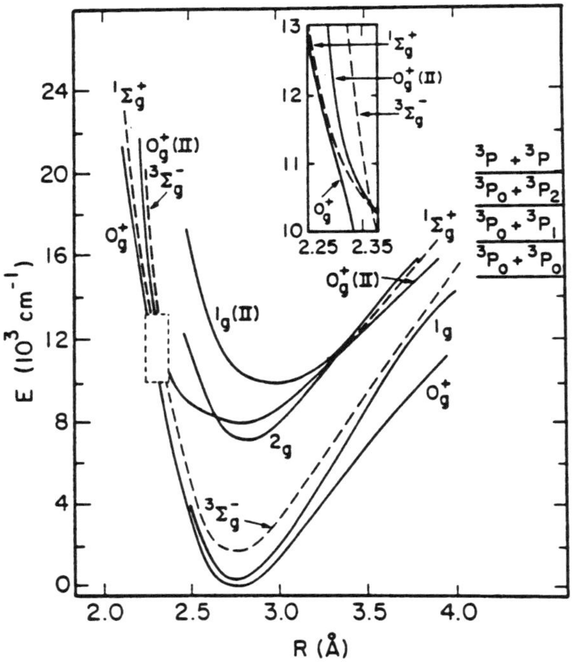

Figure 3.10 Potential-energy curves for the g electronic states of Sn_2. (Reprinted from Ref. 201; copyright 1983, American Institute of Physics.)

The author and Pitzer (201) calculated the partition function for Sn_2 using $\omega_{e,0} = 189$ cm^{-1} and the theoretical spectroscopic constants in Table 3.31 for other states. At 1,600 K the partition function was found to be 2.93, which was only a bit smaller than the value 3 assumed before on the basis of a $^3\Sigma_g^-$ ground state. A much greater change arises, however, from the reduction of $\omega_{e,0}$ from the value of 300 cm^{-1} assumed before (204, 205) to 189 cm^{-1}. The change in R_e is smaller but also significant.

The mass-spectral data (204) were presented in the form of D_0° values for a wide range of temperatures from 1472 to 1769 K calculated on the basis of a partition function with defined parameters. The correction to each D_0° value can be shown to be

$$\delta D_0^\circ = RT \ln\left(\frac{q'}{q''}\right)$$

where q' was the originally assumed partition function and q'' is the corrected value. At 1,600 K, the ratio q''/q' is 1.43, which yields a δD_0° of -1.15

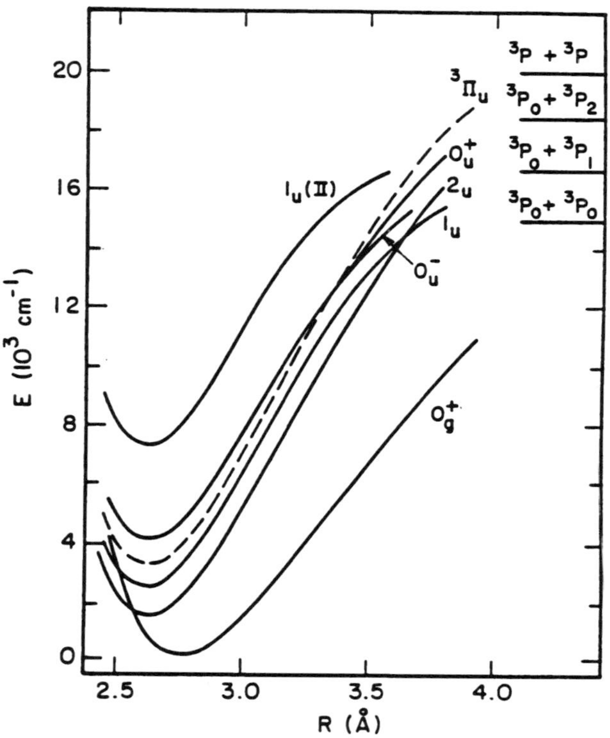

Figure 3.11 Potential-energy curves for the u electronic states of Sn_2. (Reprinted from Ref. 201; copyright 1983, American Institute of Physics.)

kcal/mol. The corrected D_0° is 1.94 eV, in excellent agreement with a theoretical value of 1.86 eV.

3.3.8 Sb_2

Genard (206) obtained the first fluorescence spectrum of antimony vapor, which revealed four bands in the 2,968–3,132-Å region. Naudé (207) observed two bands, which were assigned to the $D \leftarrow X^1\Sigma_g^+$ and $F \leftarrow X^1\Sigma_g^+$ systems. Nakamura and Shidei (208) showed the existence of another system. Almy and Schultz (209) found two other systems, assigned to $A \leftrightarrow X^1\Sigma_g^+$ and $B \leftrightarrow X^1\Sigma_g^+$. Mrozowski and Santaram (210) studied the $D \rightarrow X^1\Sigma_g^+$ system in emission. Sfeila et al. (211) carried out the vibrational and rotational analyses of the $B \rightarrow X$ bands observed in emission, which provided the spectroscopic constants for the B and X states. Topouzkhanian et al. (212, 213) carried out the vibrational and rotational analysis of the $D \leftrightarrow X$ systems.

Gerber and Kuscher (214) as well as Bondybey and co-workers (215) obtained the laser-induced fluorescence spectra in rare-gas matrices, while Sontag and Weber (216, 217) investigated both the laser-induced fluorescence and the Raman spectra of Sb_2 and Sb_4 in rare-gas matrices. Thus, accurate spectroscopic constants of the X, A, and B states are available.

Bondybey et al. (215) observed fluorescence bands in the 15,000 cm^{-1} region, tentatively assigned to a $^3\Sigma_u^+$–$X^1\Sigma_g^+$, based on the similarity of these bands to the Vegard–Kaplan system of N_2. The author and Li (219) showed that this assignment is not consistent with the theoretical separation of the $^3\Sigma_u^+$ state from the ground state and reassigned these bands to the $^3\Delta_u(1_u)$–$X^1\Sigma_g^+$ system. Mass spectroscopic investigations of antimony clusters were also carried out by Kordis and Gingerich (218). The vacuum UV spectra of Sb_2 have been recorded (147).

The author and Li (219) carried out relativistic complete active space MCSCF (CASSCF) calculations followed by large-scale configuration interaction and RCI calculations on 33 electronic states of Sb_2 lying below 44,000 cm^{-1}. Subsequent to this study, Das et al. (405) computed the spectroscopic constants of 22 electronic states lying 40,000 cm^{-1} above the ground state in the absence of spin–orbit coupling. Subsequently spin–orbit effects are also included leading to the spectroscopic constants of 33 electronic states. We shall discuss and compare the results of the two computational studies and the experimental studies after 1990, subsequent to the discussion and comparison of the results obtained by the author and Li (219).

The author and Li (219) employed an extended $(4s4p3d)$ basis set that included polarization functions. The outer $4d^{10}5s^25p^3$ shells (15 electrons) were explicitly retained in these calculations. The rest of the core electrons were replaced by the relativistic effective core potentials. Both CASSCF and CI calculations of all the electronic states were done in the C_{2v} group, and a few states were also calculated in the D_{2h} group to compute the effect of higher-order correlations. Following CASSCF, CI calculations were carried out with the first-order CI (FOCI) approximation.

Although the CASSCF/FOCI calculations appeared to be satisfactory for calculating the R_e and ω_e values of various electronic states, they were found to be somewhat less accurate for energy separations of electronic states of Sb_2. These calculations yielded only 70% or less of the experimental dissociation energies, especially for group V dimers, for which correlation effects were found to be quite severe. Hence, the effects of higher-order excitations not included in the CASSCE/FOCI calculations were studied by the CASSCF/MRSDCI methodology for a few low-lying electronic states of interest. The MRSDCI calculations of Sb_2 included single and double excitations from a set of chosen reference configurations that had a coefficient ≥ 0.07 in the CASSCF. The CASSCF calculations included up to 400 configurations, while the MRSDCI calculations included between 60,000 and 300,000 configurations. The RCI calculations were made to include the spin–orbit effects (219). These calculations included possible low-lying λ–s states that had the same Ω symmetry and potentially mix in the RCI.

Table 3.32 shows the spectroscopic constants of 33 electronic states

Table 3.32 Spectroscopic Properties of Sb_2 Calculated with the FOCI/RCI Method[a]

	R_e, Å		T_e, cm^{-1}		ω_e, cm^{-1}		D_e, eV
State	Theory	Exp.	Theory	Exp.	Theory	Exp.	Theory
$X^1\Sigma_g^+(0_g^+)$	2.59	2.34	0	0	246	270	1.86
$^1\Sigma_g^+$	2.59		836		245		1.85
$^3\Sigma_u^+(1_u)$	2.82		7,823		177		0.93
$^3\Sigma_u^+(0_g^-)$	2.83		7,910		176		0.92
$^3\Sigma_u^+$	2.83		8,409		176		0.91
$^3\Delta_u(2_u)$	2.78		16,036		198		1.63
$A^3\Delta_u(1_u)$	2.78		16,388	14,991	197	217	1.42
$^3\Delta_u(3_u)$	2.78		16,412		199		1.53
$^3\Delta_u$	2.78		16,505		197		1.58
$^3\Pi_g(0_g^-)$	2.67		19,887		217		1.11
$^3\Pi_g(0_g^+)$	2.69		20,439		198		1.07
$^3\Pi_g(1_g)$	2.67		21,052		215		0.80
$^3\Pi_g(2_g)$	2.67		21,425		211		0.70
$^3\Pi_g$	2.67		22,418		214		0.96
$B^3\Sigma_u^-(0_u^+)$	2.76	2.48	22,176	19,068	205	219	1.59
$^3\Sigma_u^-(1_u)$	2.76		22,435		205		1.59
$^3\Sigma_u^-$	2.76		22,260		205		1.58
$^1\Delta_u(2_u)$	2.77		21,370		204		2.44
$^1\Delta_u$	2.76		22,976		206		2.46
$^3\Sigma_g^+$	3.07		23,472		140		0.72
$^3\Delta_g$	3.07		23,496		140		0.72
$^1\Sigma_g^+$(II)	2.93		26,158		193		2.06
$^1\Sigma_u^-$	2.73		26,614		211		2.57
$^1\Pi_g$	2.67		27,570		218		1.89
$^3\Sigma_g^-$	3.05		31,319		143		1.42
$^1\Sigma_g^+$(III)	3.05		33,525		151		1.15
$^1\Delta_g$	3.05		33,583		148		1.14
$K, L\ ^3\Pi_u$(II)	2.96		36,154	31,400	141		0.86
$D^1\Sigma_u^+$	2.89		36,243	32,087	169	209	1.79
$^1\Pi_u$	2.91		37,156		152		0.70
$^1\Sigma_g^-$	3.16		39,048		116		0.47
$^1\Pi_u$(II)	3.14		42,100		86		0.65
$^1\Pi_g$(II)	3.21		42,647		120		0.58

[a]Theoretical constants are from Reference 219. Most of the experimental values are from Reference 137 (see text).

of Sb_2, including spin–orbit effects obtained by the author and Li with the CASSCF/FOCI/RCI methodology. The calculated FOCI potential-energy curves are shown in Figures 3.12 and 3.13 for the singlet and triplet states, respectively. Figure 3.14 shows the potential-energy curves of Sb_2, which included spin–orbit effects. As seen from Table 3.32, the spin–orbit components of a given λ–s state came out in a group. That is, the spin–orbit contamination for Sb_2 is not so large as to cause a change in the relative ordering of the electronic states. The ordering of the spin–orbit components of most of the low-lying states of Sb_2 followed Hund's rule with the exception of the $^3\Delta_u(1_u)$ state. The contamination of $^3\Sigma_u^+(1_u)$ with $^3\Delta_u(1_u)$ lowered $^3\Sigma_u^+(1_u)$ and raised $^3\Delta_u(1_u)$ relative to the $^3\Delta_u$ state. The spin–spin and spin–orbit splittings of the $^3\Sigma_u^+$ and $^3\Sigma_u^-$ states were found to be smaller compared to the $^3\Delta_u$ and $^3\Sigma_u^-$ states. Similarly, spin–orbit mixing of $^3\Delta_u(2_u)$ with $^1\Delta_u(2_u)$ lowered the $^3\Delta_u(2_u)$ state.

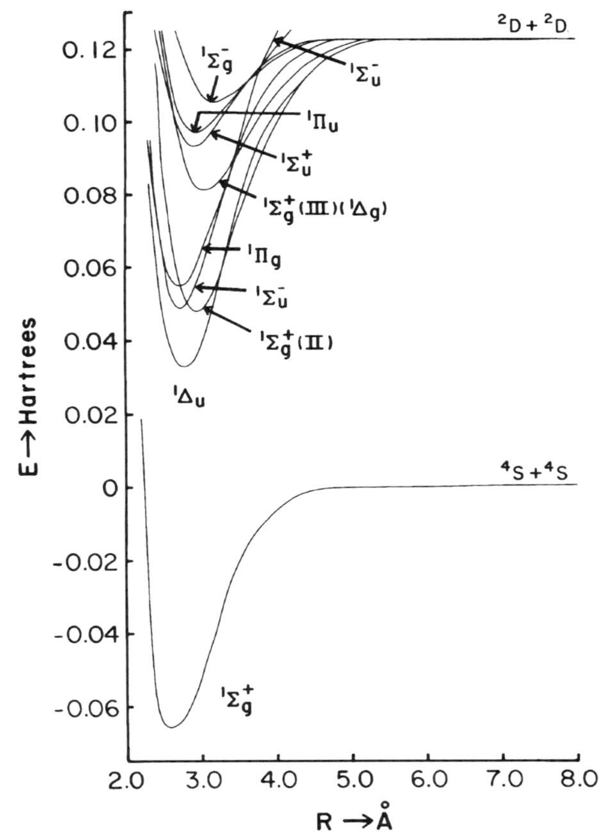

Figure 3.12 Potential-energy curves of the singlet electronic states of Sb_2. (Reprinted from Ref. 219; copyright 1989, Academic Press, Inc.)

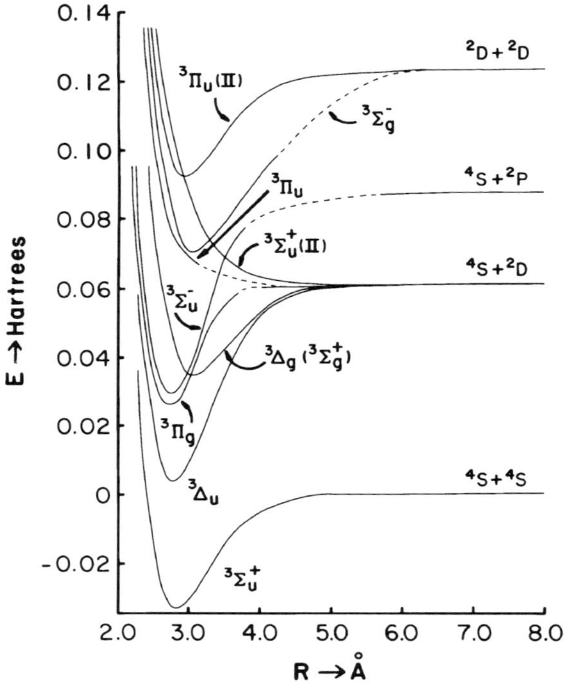

Figure 3.13 Potential-energy curves of the triplet electronic states of Sb_2. (Reprinted from Ref. 219; copyright 1989, Academic Press, Inc.)

Table 3.33 shows the spectroscopic constants of nine low-lying electronic states of Sb_2 obtained with more accurate CASSCF/MRSDCI/RCI calculations (219). In comparing the results of Table 3.32 and 3.33, one can see that R_e shrinks by 0.01 Å as a result of higher-order correlations. The vibrational frequency improved by 4.5% for the ground state. The main effect of higher-order correlations not included in the FOCI was found to be on the dissociation energies. The calculated D_e values increased by 17% as a result of the higher-order correlations. The refined theoretical D_e (2.17 eV) (219) was still about 30% smaller than an experimental (thermodynamic) value of 3.09 eV obtained by the Knudsen effusion mass-spectrometric method (218).

The R_e and ω_e values of the X ground state deduced from laser-induced fluorescence experiments are 2.48 ± 0.02 Å and 269 cm^{-1}. The best levels of calculations (MRSDCI) of the author and Li (219) yielded 2.58 Å and 259 cm^{-1}. The slightly longer bond length is attributable to the effective core potential approximations by the author and Li (219). The longer bond lengths and slightly lower vibrational frequencies appear to be the general trend in the effective core potential calculations and are consistent with other calculations on comparable systems.

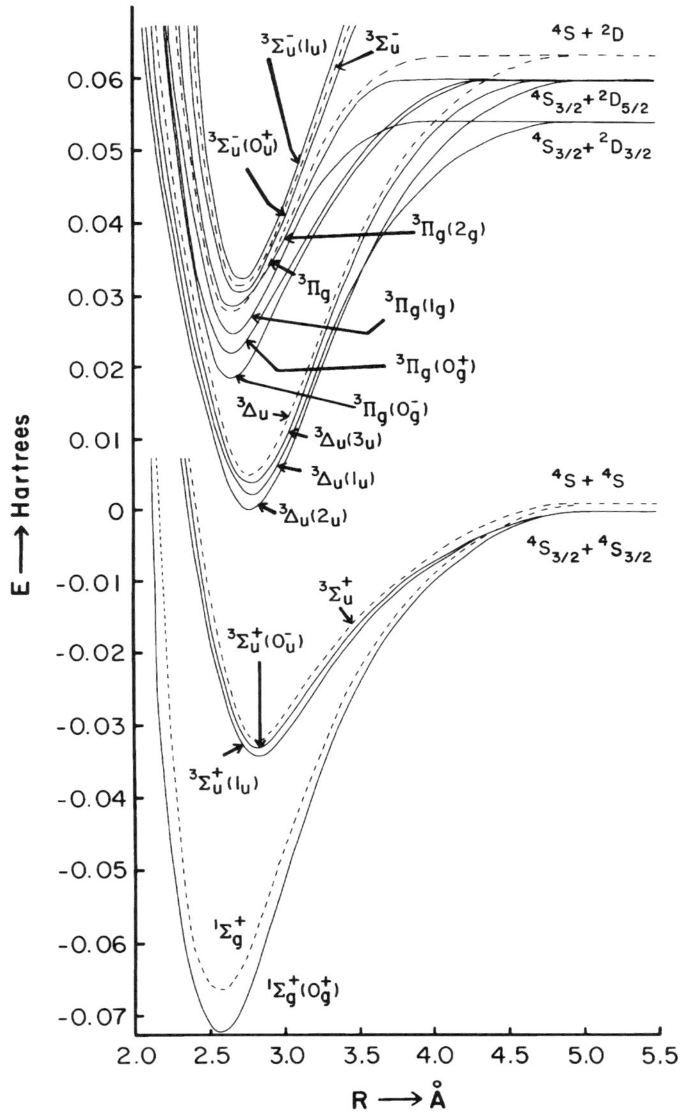

Figure 3.14 Potential-energy curves of the low-lying electronic states of Sb_2 including spin–orbit effects. (Reprinted from Ref. 219; copyright 1989, Academic Press, Inc.)

Two experimental systems labeled *B–X* and *A–X*, where *X* is the $^1\Sigma_g^+(0_g^+)$ ground state, have been studied to a considerable extent. The T_e value of the *B* state deduced from the experimental spectra is 19,000–19,070 cm^{-1}. The author and Li (219) calculated the CASSCF/MRSDCI/RCI separation of $^3\Sigma_u^-(0_u^+)$ from the $X(0_g^+)$ ground state to be 20,255 cm^{-1}, in reasonable agreement with the

Table 3.33 Spectroscopic Constants of Various Low-Lying Electronic States of Sb_2 Calculated with the MRSDCI/RCI Methods

	R_e, Å		T_e, cm^{-1}		ω_e, cm^{-1}		D_e (eV)	
State	Theory	Exp.	Theory	Exp.	Theory	Exp.	Theory	Exp.
$X^1\Sigma_g^+(0_g^+)$	2.58	2.34	0	0	259	270	2.17	3.09
$^1\Sigma_g^+$	2.58		836		258		2.16	
$^3\Sigma_u^+(1_u)$	2.78		8,760		199		1.04	
$^3\Sigma_u^+(0_u^-)$	2.79		8,847		198		1.03	
$^3\Sigma_u^+$	2.79		9,346		198		1.02	
$^3\Delta_u(2_u)$	2.77		15,284		199		1.80	
$A^3\Delta_u(1_u)$	2.77	2.64	15,636	14,991	198	217	1.59	
$^3\Delta_u(3_u)$	2.77		15,660		200		1.73	
$^3\Delta_u$	2.77		15,753		198		1.75	
$B^3\Sigma_u^-(0_u^+)$	2.75	2.48	20,255	19,068	205	219	2.03	
$^3\Sigma_u^-(1_u)$	2.75		20,514		205		2.03	
$^3\Sigma_u^-$	2.75		20,339		205		2.02	
$^1\Delta_u(2_u)$	2.73		23,509		234		—	
$^1\Delta_u$	2.72		25,115		232		—	
$D^1\Sigma_u^+$	2.86		33,945	32,087	180	209	—	
$^3\Pi_u$(II)	2.94		36,296		161		—	

Source: Reference 219.

experimental results. The ω_e value of the B state deduced from experiments was found to be 218 cm^{-1}. The small discrepancy between the theoretical value and the experimental result is about the same for both the ground state and the excited states. The predicted difference in the ω_e values between the X and B states (54 cm^{-1}) was found to be in excellent agreement with the experimental difference of the ω_e values of the X and B states. Likewise, the theoretical difference in the R_e values of the X and B states (0.17 Å) was found to be in full agreement with the experimental difference of 0.17 Å. This appeared to substantiate the assumption that the small discrepancy between the theoretical and experimental results is due mainly to the effective core potential approximation. The calculations of the author and Li (219) confirmed the earlier assignment of the B state to the 0_u^+ component of $^3\Sigma_u^-$ by Bondybey et al. (215).

Bondybey et al. (215) as well as Sontag and Weber (216) studied the A–X system both in the gas phase and in a rare-gas matrix. The T_e and ω_e values of the A state deduced from these experiments are 14,990 and 217 cm^{-1}, respectively. The R_e value of the A state has not yet been obtained from the gas-phase experiments, but in the Ar matrix a value of 2.64 Å has been suggested. As seen from Table 3.33, the most consistent electronic transition for the A–X system with the theoretical calculations of Li and the author is the $^3\Delta_u(1_u)$–$^1\Sigma_g^+(0_g^+)$ transition. The theoretical adiabatic transition energy of 15,636 cm^{-1} is in very good agreement with the observed value of 14,995 cm^{-1}. Further, the theoretical R_e and ω_e values when corrected for the ECP approximation and electron correlation ($\Delta R_e = -0.17$ Å, $\Delta\omega_e = +18$ cm^{-1}) yield 2.60 Å and 216 cm^{-1}. These values are in remarkable agreement with the experimental values of 2.64 Å and 217 cm^{-1}. Further, there is no other allowed electric dipole transition in this region. Bondybey and co-workers (215) noted that the fluorescence in the 15,000-cm^{-1} region exhibited an intensity ratio of $I_{\|}/I_{\perp} = 1.32$. This ratio is close to a $\Delta\Omega = +1$ intensity ratio of $\frac{4}{3} = 1.33$, suggesting the A state to be a 1_u state. However, they assigned the A state to $^3\Sigma_u^+(1_u)$, based on the Begard–Kaplan system of N_2 and As_2. But the separations of the electronic states of N_2 are quite different from those of Sb_2, and thus it seemed that it is not entirely correct to assign the observed transitions of Sb_2 based on N_2 or even As_2. The $^3\Sigma_u^+(1_u)$ state of Sb_2 is much lower ($T_e \sim 8760$ cm^{-1}) in energy and thus did not appear to be a probable candidate for the A state seen in the region of 15,000 cm^{-1} (see Table 3.33). Moreover, the $^3\Sigma_u^+$ state has a slightly longer bond length (2.79 Å) and much lower ω_e value (177 cm^{-1}). Even if these values are corrected for the ECP approximation, the theoretical R_e and ω_e values of the $^3\Sigma_u^+(1_u)$ state are 2.65 and 195 cm^{-1}, respectively. The experimentally observed A state had a higher ω_e value of 217 cm^{-1}. All these reasons together with the energy separation of $^3\Delta_u(1_u)$ led the author and Li (219) to speculate that the A state observed in the A–X system is most probably the 1_u component of the $^3\Delta_u$ state.

The experimental T_e and ω_e values are uncertain for the D–X system observed in the 32,087-cm^{-1} region. Theoretical calculations of the author and Li (219) suggested three candidates for the D state [Π_u(II), $^1\Sigma_u^+$, and $^1\Pi_u$]. The vibrational frequencies of the $^3\Pi_u$ and $^1\Pi_u$ states are somewhat smaller compared to an uncertain experimental ω_e of the D state (209 cm^{-1}) (37). The most probable candidate for the D state was believed to be the $^1\Sigma_u^+$ state, since its ω_e value should increase by at least 15 cm^{-1}, due to higher order correlations (184 cm^{-1}). The predicted theoretical ω_e value of the D state after correction for the ECP approximation (202 cm^{-1}) was found to be in remarkable agreement with an uncertain experimental value of 209 cm^{-1}. The author and Li assigned the observed D–X system tentatively to the $D^1\Sigma_u^+$–$X^1\Sigma_g^+$ transition.

Sontag and Weber (216) identified a new electronic system, which they called K–X, in the 32,000-cm^{-1} region below the D state. The vibrational and rotational constants of the K state have also been determined. The R_e and ω_e values deduced from laser-induced fluorescence experiments are 2.841 Å and 127 cm^{-1}, respectively. From Table 3.33, the most consistent candidate for the k state is $^3\Pi_u(0_u^+)$, arising from the second root of the $^3\Pi_u$ state. The theoretical R_e and ω_e values (respectively 2.96 Å and 141 cm^{-1}) are in reasonable agreement with the experimental values, although the experimental values themselves are not very definitive because of perturbations with nearby electronic states.

Sibai et al. (213) carried out the rotational analysis of the $D \leftarrow X$ system. They found that the vibrational levels of the D state were quite perturbed by another electronic state in this region designated as the L state. As seen from Tables 3.32 and 3.33, there are many probable candidates for the L state. Since the K and L states are close in energy and are both below the $D(^1\Sigma_u^+)$ state, the author and Li (219) argued that the experimental (uncertain) ω_e value of the H state (479 cm^{-1}) is unreasonable for a heavy species such as Sb_2.

Subsequent to the theoretical study of the author and Li, two other theoretical studies appeared on Sb_2 (405–407). An experimental study by Breidohr et al. (407) on Sb_2 will also be considered here for discussion. Toscano and Russo (406), in their theoretical study on Sb_2, neglected the spin–orbit coupling. A study by Das et al. (405) contains a complete comparison of the results that they obtained with the experiment and the previous theoretical study by the author and Li (219).

Das et al. (405) used the RECPs for antimony atoms that included all except the outer $3d^{10}4s^2 5p^3$ shells in the core. These RECPs were used in conjunction with a valence [$3s3p3d$] Gaussian basis set for the antimony atom. A single-configuration SCF approach was used for the generation of the orbitals compared to a CASSCF method for the generation of the orbitals in the study by the author and Li (219). The CASSCF approach, as noted in the previous chapters (this volume and Part A) includes full CI within the valence space, and assures proper dissociation of the potential-energy curves. The CI techniques used by Das et al. (405) is based on the MRDCI method, which invokes configuration selection and energy approximation. The techniques used by the author and Li (219), namely, first-order CI (FOCI) and MRSDCI, do not invoke configuration selection or energy extrapolation. All configurations were fully included in both the CI matrix diagonalization and energy computation in the procedures employed by the author and Li. The spin–orbit effects were incorporated in both the techniques.

Das et al. (405) have considered several more electronic states, especially when spin–orbit coupling is included. They also computed the radiative lifetimes of the electronic states of Sb_2, and discussed the computed values in light of the experimental measurements of the lifetimes of some low-lying electronic states. We shall first discuss the theoretical results of Das et al. compared with those of the author and Li, and then compare with the experiment of Breidohr et al.

Table 3.34 shows the spectroscopic constants of Sb_2 computed by Das et al. (405) together with the corresponding values reported by the author and Li (219) using the CASSCF/FOCI method discussed earlier. In general, the bond lengths of the electronic states obtained by Das et al. are slightly longer, and the vibrational frequencies are correspondingly smaller. Since the basis sets and the RECPs used for the antimony atoms are quite similar, the main difference appears to arise from the level of electron correlation treatment in the two works. In spite of such differences between the two techniques, as seen from

Table 3.34 Comparison of Spectroscopic Constants of Sb_2 by Two Theoretical Methods in Absence of Spin–Orbit Effect

	R_e, Å		T_e, cm^{-1}		ω_e, cm^{-1}	
State	Ref. 219	Ref. 405	Ref. 219	Ref. 405	Ref. 219	Ref. 405
$^1\Sigma_g^+$	2.59	2.57	0	0	245	243
$^3\Sigma_g^+$	2.83	2.796	7,573	7,908	176	178
$^5\Sigma_g^+$	—	3.055	—	13,522	—	—
$^3\Delta_u$	2.78	2.746	15,669	15,326	197	206
$^3\Sigma_u^-$	2.76	2.729	21,424	19,928	205	212
$^1\Delta_u$	2.76	2.703	22,140	23,864	206	220
$^1\Sigma_u^-$	2.73	2.722	25,378	21,037	211	213
$^3\Delta_g$	3.07	3.038	22,660	22,615	140	141
$^3\Pi_g$	2.67	2.663	21,582	20,795	214	201
$^5\Pi_u$		2.933		24,796		132
$^3\Sigma_g^+$	3.07	—	22,636	—	140	—
$^3\Delta_g$	3.07	3.04	22,660	22,615	140	141
$^1\Sigma_g^+$(II)	2.93	2.92	26,158	24,814	193	185
$^1\Pi_g$	2.67	2.667	26,734	26,256	218	206
$^3\Sigma_g^-$	3.05	3.039	30,483	28,787	143	143
$^1\Sigma_g^+$(III)	3.05	3.015	32,689	31,827	151	149
$^1\Delta_g$	3.05	3.075	32,747	37,014	148	161
$^3\Pi_u$(II)	2.96	2.949	35,318	35,006	141	140
$^1\Sigma_u^+$	2.89	2.858	35,407	33,704	169	164
$^1\Pi_u$	2.91	2.886	36,320	35,751	152	150
$^1\Sigma_g^-$	3.16	3.023	38,212	31,804	116	147
$^1\Pi_u$(II)	3.14	3.063	41,264	40,697	86	79
$^1\Pi_g$(II)	3.21	3.12	41,811	38,827	120	121

Table 3.34, the overall agreement in the results reported by the two groups is excellent.

Since the spin–orbit coupling for the antimony atom is nonnegligible, it is important to compare the results including spin–orbit effects. Table 3.35 reproduces a table from the work of Das et al. for this purpose, which contains the results of Das et al., the author and Li, as well the experiment. As seen from this table, the overall agreement between the theories and the experiment is very good, considering that Sb_2 is a very heavy dimer for, which the spectroscopic constants are hard to compute. It is commendable that Das et al. have computed several more electronic states, especially when spin–orbit coupling is included as these data could provide valuable information on the potential new spectra on the dimer, as well as the thermodynamic properties of the system. In general, the vibrational frequencies computed by the author and Li are closer to the experimental results primarily because the electron correlation effects are included to

Table 3.35 Spectroscopic Constants of Some Low-Lying Ω States of Sb_2 (Excitation Energy T_0, Equilibrium Bond Length r_e, and Vibrational Frequency ω_e)

	T_e, cm^{-1}			r_e, Å			ω_e, cm^{-1}		
	Calc.			Calc.			Calc.		
State	Ref. 405	Ref. 219	Exp.	Ref. 405	Ref. 219	Exp.	Ref. 405	Ref. 219	Exp.
$X0_g^+$	0	0	0	2.578	2.58	2.476[a]	240	259	269[a,b]
1_u(I)	7,522	8,760	9,474[b]	2.792	2.78	—	180	199	206[b]
0_u^-(I)	8,153	8,847	—	2.798	2.79	—	177	198	—
0_g^+(II)	13,148	—	—	3.056	—	—	127	—	—
1_g(I)	13,372	—	—	3.057	—	—	126	—	—
2_g(I)	14,046	—	—	3.058	—	—	124	—	—
2_u(I)	14,614	15,284	—	2.742	2.77	—	205	199	—
3_u(I)	15,361	15,660	—	2.754	2.77	—	200	200	—
$A1_u$	15,508	15,636	14,991[c]	2.750	2.77	2.640[d]	206	198	217[c]
0_g^-(I)	19,702	19,887	—	2.664	2.67	—	201	217	—
$B0_u^+$	19,941	20,255	19,067[a,c]	2.748	2.75	2.645[a]	204	205	218[a]
0_g^+(III)	20,191	20,439	—	2.670	2.69	—	192	198	—
1_u(III)	20,435	20,514	—	2.752	2.75	—	202	205	—
1_g(II)	20,934	21,052	—	2.666	2.67	—	200	215	—
0_u^-(II)	21,360	—	—	2.728	—	—	210	—	—

2_g(II)	22,749	—	—	3.001	—	—	156	—	—
2_g(III)	22,879	21,425	—	2.672	2.67	—	196	211	—
3_g(I)	22,999	—	—	3.050	—	—	136	—	—
1_g(III)	23,275	—	—	3.053	—	—	136	—	—
2_u(II)	24,682	—	—	2.935	—	—	135	—	—
3_u(II)	24,683	—	—	2.918	—	—	134	—	—
2_u(III)	24,698	23,509	—	2.715	2.73	—	220	234	—
0_g^+(IV)	24,830	—	—	2.902	—	—	192	—	—
1_u(IV)	25,075	—	—	2.942	—	—	131	—	—
0_u^-(III)	25,841	—	—	2.951	—	—	120	—	—
0_u^+(II)	26,244	—	—	2.906	—	—	149	—	—
1_g(IV)	27,330	—	—	2.668	—	—	197	—	—
1_u(V)	27,581	—	—	2.904	—	—	149	—	—
0_u^+(III)	30,578	—	—	3.253	—	—	78	—	—
2_u(IV)	31,917	—	—	3.010	—	—	129	—	—
0_g^-(II)	32,453	—	—	3.025	—	—	149	—	—
2_u(V)	34,278	—	—	2.993	—	—	132	—	—
$D0_u^+$	34,493	33,945	32,087[d]	2.856	2.86	—	165	180	210[d]

[a]Reference 216.
[b]Reference 407.
[c]Reference 215.
[d]Reference 37.

a higher order in the work of the author and Li through the CASSCF/MRSDCI techniques. Das et al. have suggested that the state observed by Sontag and Weber denoted as the K state by them should actually be the $D(0_u^+)$ state shown in Table 3.35. This is based on the excellent agreement between the computed and experimental T_e values. Thus both D and K states are attributed by these authors to the same 0_u^+(IV) state. This conclusion is also supported by Das et al.'s intensity computations.

Breidohr et al. (407) detected the ${}^3\Sigma_u^+–X^1\Sigma_g^+$ emission process, but the lifetime of this state has not yet been measured. Das et al. computed the radiative lifetimes of several electronic states of Sb_2, reported in Table 3.36. Let us consider the 1_u spin–orbit state of the ${}^3\Sigma_u^+$ state to illustrate the role of relativistic effects on the lifetime. An electronic transition to this state from the closed-shell singlet ground state is spin-forbidden in the absence of spin–orbit coupling. However, the 1_u spin–orbit state has a radiative lifetime of 6.18 ms for transition to the ground state. Clearly, the spin–orbit mixing of the closed-shell ground state with the ${}^3\Pi_g$ state should be responsible for this transition probability.

Table 3.36 Radiative Lifetimes of Some of the Excited States (f) at $v' = 0$ of Sb_2 from which Transition $i \leftarrow f$ Takes Place[a]

	τ_p, s						
$i \backslash f$	1_u(I)	$A1_u$	0_g^-(I)	$B0_u^+$	0_u^+(II)	0_g^-(II)	$D0_u^+$
$X0_g^+$	6.18(−3)	2.98(−4)	—	1.44(−2)	2.03(−5)	—	3.70(−8)
1_u(I)	—	—	9.99(−6)	—	—	7.03(−3)	—
0_u^-(I)	—	—	4.25(−4)	—	—	2.77(−2)	—
0_g^+(II)	—	1.16(+2)	—	3.71(+2)	1.71(−3)	—	1.74(−4)
1_g(I)	—	1.95(+4)	—	8.36	2.52(−2)	—	5.10(−2)
2_g(I)	—	1.85(+5)	—	—	—	—	—
$A1_u$	—	—	—	—	—	1.58(−2)	—
0_g^+(III)	—	—	—	—	1.24(−4)	—	3.28(−5)
1_u(III)	—	—	—	—	—	1.29(−2)	—
1_g(II)	—	—	—	—	9.92(−3)	—	6.13(−4)
0_u^-(II)	—	—	—	—	—	1.85	—
1_g(III)	—	—	—	—	1.08(−2)	—	1.24(−3)
0_g^+(IV)	—	—	—	—	2.23(−1)	—	1.01(−6)
1_u(IV)	—	—	—	—	—	1.41(+1)	—
0_u^-(III)	—	—	—	—	—	4.84	—
1_g(IV)	—	—	—	—	—	—	1.70(−4)
1_u(V)	—	—	—	—	—	3.69(−1)	—
	—	—	—	τ(s)	—	—	—
	6.18(−3)	2.98(−4)	9.76(−6)	1.44(−2)	1.72(−5)	3.10(−3)	3.56(−8)

[a]Values in parentheses denote powers of 10, τ_p and τ are partial and total lifetimes, respectively.
Source: Reproduced from Reference 405.

It is interesting that, although the optical transition from the 0_g^- state to the ground state is forbidden, it has a short radiative lifetime (9.76 μs), due primarily to its transition to the $a1_u$ state. Thus Das et al. (405) note that it is a bit surprising that the corresponding band system has not been observed. The intensity for the B–X system was computed to be very small by Das et al. The experimental radiative lifetimes of the B state have been very uncertain, with values in the range of 2–20 μs. Das et al. have computed the radiative lifetime of the B state ($v' = 0$) to be 14.4 ms, which differs substantially from the measured values. It should also be noted that Bondybey et al. (215) have reported the ratio of the transition strengths of the B–X and A–X systems to be 30, which is a discrepancy by a factor of 1500 compared to the computed transition strengths of these systems. Das et al. compute the ratio of the lifetimes of the B and A states to be 50. They have thus concluded that the collision mechanisms appear to be predominant factors in the depopulations of both B and A states. Das et al. have computed the transition dipole moment for the A–X transition to be 0.066 D at $r = 4.85\ a_0$, compared to 0.0021 for the B–X system.

One of the strongest transitions to the ground state is from the $D0_u^+$ state, which is predominantly $^1\Sigma_u^+$. Thus the D state was computed to have a relatively short lifetime (35.6 ns). At present there are no experimental measurements on the lifetime of this state. Das et al. find the two 0_u^+ states lying below the D and B states to be mixtures of $^3\Pi_u$ and $^5\Pi_u$. The main contribution to the transition moment and the lifetime is from the $^3\Pi_u$–$^3\Pi_g$ transition. Das et al. compute the lifetime of the 0_u^+(II) state to be 17.2 μs.

3.3.9 Te_2

The electronic spectra of Te_2 are of considerable interest (170, 220–235), since they serve as possible wavelength standards. Further, the Te_2 molecule has been considered as a candidate for optically pumped lasers. Theoretically, the Te_2 molecule is considered interesting, as a result of relativistic effects and the complexity of electronic states due to large spin–orbit coupling.

The $B \leftrightarrow X$ and $A \leftrightarrow X$ systems, which are the strongest observed transitions of Te_2, have been observed in both absorption and emission spectra, as well as laser-induced fluorescence, and chemiluminescence spectra. Yee and Barrow (170) observed a weak fluorescence series originating from the perturbed $B(0_u^+)$ terminating to a state $X_2 < 2230\text{cm}^{-1}$ above the X ground state. Laser-induced fluorescence (LIF) investigations by a number of authors (231, 232, 234) later revealed that the X_2 state is 1975–1977 cm^{-1} above the X ground state.

The LIF spectra of matrix-isolated Te_2, in the 16,400-cm^{-1} region, have been obtained by Bondybey and English (231). The lower state involved in this fluorescence was assigned to the $X_2(1_g)$ state, since excitation to the emitting state from the ground state X could not be achieved. The emitting state (A') was tentatively assigned to $^3\Pi_u(2_u)$. Subsequently, Ahmed and Nixon (232) reassigned the A' state to a 1_u state, since they observed another, weaker, fluorescence series in the 14,091-cm^{-1} region. Ahmed and Nixon (232) suggested that the upper state in this fluorescence series is a 2_u state. Verges et al. (235) argued that the A' state is most likely $^3\Sigma_u^+(1_u)$, based on the intensities of the observed bands and a comparison to the analogous O_2 bands for which $^3\Sigma_u^+$–$^3\Sigma_g^-$ system is well characterized.

Effantin et al. (230) observed a b state, which they assigned to $^1\Sigma_g^+$, in the new $B(0_u^+)$–$b(^1\Sigma_g^+)$ system using Fourier transform spectroscopy. The spectroscopic properties of the lower state have been characterized. However, no transitions to a lower $^1\Delta_g$ state have been observed to date, although Verges et al. (235) predicted that such a state should be about 6,500 cm^{-1} above the ground state. Verges et al. (229) also observed an electronic state, which they designated $B(1_u)$, which is very close to the $B(0_u^+)$ state.

The author and Ravimohan (236) carried out CASSCF/FOCI/RCI calculations of 22 low-lying electronic states of Te_2. These authors employed relativistic effective core potentials with the outer $4d^{10}5s^25p^4$ shells of the Te atom retained explicitly in the valence space. A $(3s3p4d/3s3p2d)$ valence Gaussian basis set was employed. Extensive RCI calculations that included all possible reference configurations to yield the desired Ω states were made employing a double-ζ STO basis set.

Table 3.37 depicts the spectroscopic constants for 22 low-lying states of Te_2 obtained by the author and Ravimohan (236) together with available experimental values. The potential-energy curves for several of these states are given in Figure 3.15 (without spin–orbit coupling) and Figure 3.16 (RCI, including spin–orbit coupling).

As seen from Table 3.37, the ground state of Te_2 is an $X(0_g^+)$ state. The 1_g state, which is the other Ω component of the $^3\Sigma_g^-\lambda$–s state, is 2229 cm^{-1} above the 0_g^+ state. The theoretical T_e of the 1_g state was found to be in very good agreement with the value of 1975 cm^{-1} reported in References 13 and 14. The calculated R_e, T_e, and ω_e values were found to be in very good agreement with available experimental data for the electronic states below 10,000 cm^{-1}. The theoretical R_e values of excited states above 10,000 cm^{-1} are much longer than the experimental values. This difference was attributed mainly to the basis-set limitations in the theoretical calculations by the author and Ravimohan (236). For the excited states the basis sets must be extended further with more diffuse functions and polarization functions.

As seen from Table 3.37, theoretical calculations confirmed the earlier assignments of the $A(0_u^+)$–$X(0_g^+)$ and $B(0_u^+)$–$X(0_g^+)$ systems. The $A(0_u^+)$ state was found to be $^3\Pi_u(0_u^+)$, while the $B(0_u^+)$ state was found to be $^3\Sigma_u^-(0_u^+)$. The A'' state observed by Bondybey and English (231) in the weak fluorescence, which was assigned to $^3\Pi_u(2_u)$ by these authors (231), is most consistent with $^3\Delta_u(1_u)$ (Table 3.37). The theoretical T_e value of 17,759 cm^{-1} was found to be in very good agreement with the experimental state. The notation A' was also used for another electronic state with a T_e value of 14,091 cm^{-1} by other authors, and thus caution must be exercised in talking about this state. The state that Bondybey and English (231) designate A' is observed in the $A' \rightarrow X_2(1_g)$ fluorescence. The lower state is assigned to 1_g, since excitation to the A' state could not be

Table 3.37 Spectroscopic Constants for Te_2

State	R_e, Å		T_e, cm^{-1}		ω_e, cm^{-1}	
	Theory[g]	Exp.	Theory	Exp.	Theory	Exp.
$X\,^3\Sigma_g^-\,(0_g^+)$	2.68	2.56[a]	0	0	210	247[a]
$X\,^3\Sigma_g^-\,(1_g)$	2.66	2.55[a]	2,229	1,975[a]	216	250[a]
	—	—	—	1,977[b]	—	—
$^1\Delta_g(2_g)$	2.70	2.55[a]	6,383	[6,500][c]	199	[235][c]
$^3\Delta_u(3_u)$	2.92	—	9,142	—	158	—
$^1\Sigma_g^+(0_g^+(\text{II}))$	2.73	—	10,446	9,600[d]	182	229[d]
	—	—	—	9,591[b]	—	—
$^3\Delta_u(2_u)$	2.97	—	11,193	—	127	—
$A'\,^3\Sigma_u^+(1_u)$	3.00	—	14,369	14,091[b]	127	—
$A''\,^3\Delta_u(1_u)$	3.00	2.8 ± 0.05[e]	17,759	17,789[b]	128	120[b]
$A^3\Pi_u(0_u^+)$	3.25[f]	2.88[a]	17,052	19,451[e]	142	144[e]
"	—	—	—	19,399[b]	—	—
$B^3\Sigma_u^-(0_u^+(\text{II}))$	3.03[f]	2.82[a]	21,606	22,207[e]	131	162[e]
"	—	—	—	22,165[b]	—	—
$B^3\Sigma_u^-(1_u)$	—	—	22,414	22,222[d]	—	150[d]
$^3\Sigma_g^-$	2.65	—	3,218	—	222	—
$^1\Delta_g$	2.68	—	6,900	—	205	—
$^1\Sigma_g^+$	2.73	—	9,318	—	187	—
$^3\Delta_u$	2.97	—	13,900	—	137	—
$^3\Sigma_u^+$	2.98	—	14,342	—	135	—
$^3\Pi_u$	—	—	19,629	—	—	—
$^3\Sigma_u^-$	3.03[f]	—	22,413	—	132	—
$^3\Sigma_g^-(\text{II})$	3.57[f]	—	25,722	—	92	—
$^3\Pi_g(\text{II})$	3.17[f]	—	27,233	—	53	—
$^3\Pi_g(\text{III})$	3.42[f]	—	27,647	—	81	—
$^1\Pi_g(\text{II})$	3.05[f]	—	34,043	—	147	—
$^1\Pi_g(\text{III})$	3.42[f]	—	37,189	—	84	—

[a]Reference 37.
[b]Reference 232.
[c]Predicted value as in Reference 234. This state is yet to be observed.
[d]Reference 230.
[e]Reference 231.
[f]Calculated R_e and ω_e values for these states may not be accurate because of basis set limitations.
[g]All theoretic constants are from Reference 236. Numbers in square brackets are uncertain values.

achieved from the $X(0_g^+)$ ground state. The author and Ravimohan (236) suggested the notation A'' for the $^3\Delta_u(1_u)$ state to avoid confusion with the A' state participating in the 14,091-cm^{-1} bands.

Ahmed and Nixon (232) observed another, weaker fluorescence series in the 14,000-cm^{-1} region, which they assigned to the forbidden $A'(2_u)$–$X(0_g^+)$ transition. However, Verges et al. (235) correctly argued that the A' state should most probably be $^3\Sigma_u^+(1_u)$. As seen from Table 3.37, the theoretical T_e value of the $^3\Sigma_u^+(1_u)$ state (14,369 cm^{-1}) is in remarkable agreement with the experimental

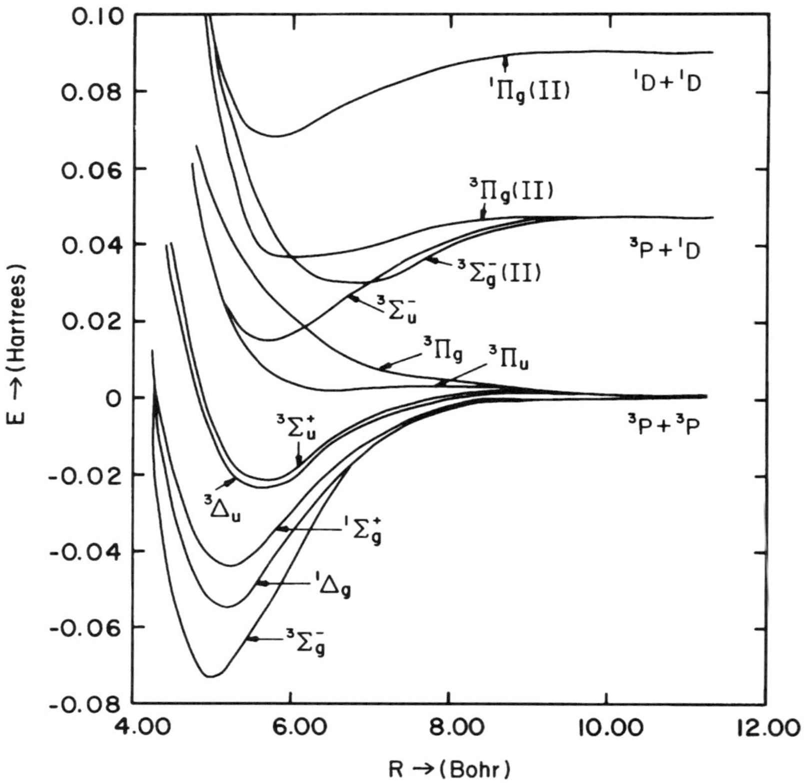

Figure 3.15 Potential-energy curves of some low-lying electronic states of Te_2. (Reprinted from Ref. 236; copyright 1987, Academic Press, Inc.)

T_e value of the A' state (14,091 cm^{-1}). Further, the theoretical and experimental ω_e values agree well if the A' state is assigned to $^3\Sigma_u^+(1_u)$. Consequently, the theoretical calculations of the author and Ravimohan (236) supported Verges et al.'s (235) tentative assignment.

Verges et al. (229) and Effantin et al. (230) observed a $B(1_u)$ state, which was found to be close to $B(0_u^+)$. The author and Ravimohan (236) estimated the separation of this state at the R_e of $^3\Sigma_u^-$ (22,414 cm^{-1}). Although this may not be very accurate, it is close to the experimental T_e of 22,222 cm^{-1}.

Verges et al. (229) identified a singlet state (b) with a T_e value of 9600 cm^{-1}. These authors tentatively assigned the b state to $^1\Sigma_g^+$. As seen from Table 3.37, the calculations of the author and Ravimohan are in agreement with this assignment. Verges et al. (235), in another investigation, predicted that there should be a $^1\Delta_g$ state with an approximate T_e value of 6500 cm^{-1}, although this state is yet to be observed. The theoretical separation of the $^1\Delta_g$ state (6383 cm^{-1}) was found to be in remarkable agreement with this prediction.

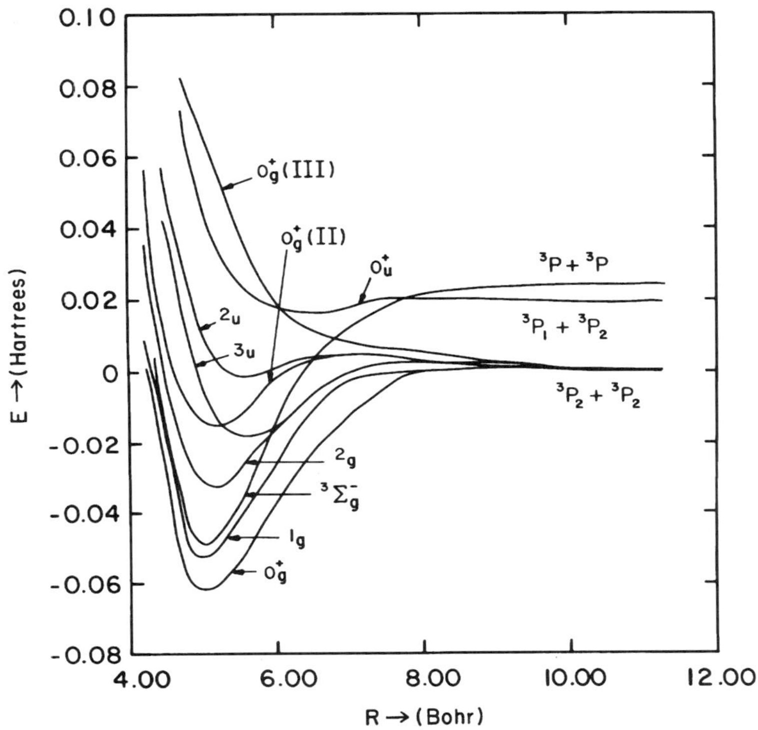

Figure 3.16 Potential-energy curves of low-lying ω–ω states of Te_2 including spin–orbit interaction. (Reprinted from Ref. 236; copyright 1987, Academic Press, Inc.)

Table 3.37 also lists the properties of many electronic states without spin–orbit coupling with T_e values of 25,700 cm^{-1}, none of which have been observed experimentally so far. The theoretical R_e, T_e, and ω_e values for these states, however, should not be considered to be very accurate, since the basis set and level of theory employed by the author and Ravimohan (236) were not adequate to calculate these properties with reliable accuracy.

The theoretical dissociation energy (236) (D_e) obtained with the CASSCF/FOCI/RCI method for the $X(0_g^+)$ state of Te_2 was found to be 1.69 eV. The spin–orbit interaction decreased the D_e value, since the separated atoms are more stabilized by the spin–orbit interaction compared to the molecule. Huber and Herzberg (37) listed a D_0° value of 2.68 eV, based on a weighted mean of a number of values obtained from spectroscopic and thermochemical methods. The difference between this value and the theoretical value is due primarily to limitations of theoretical calculations. As pointed out by Saxon and Liu (173), although D_e values of group IV dimers such as O_2 are not obtained accurately by (modest level) theoretical calculations, the spectroscopic properties near the well should be reasonable.

Table 3.38 Leading Electron Configurations Contributing to Some Low-Lying λ–s States of Te_2 Near-Equilibrium Geometry

$^3\Sigma_g^-$	$2\sigma_g^2 1\pi_u^4 1\pi_g^2$ (90%), $2\sigma_g^2 1\pi_u^2 1\pi_g^4$ (3%)
$^1\Delta_g$	$2\sigma_g^2 1\pi_u^4 1\pi_g^2$ (83%), $2\sigma_g^2 1\pi_u^2 1\pi_g^4$ (5%)
$^1\Sigma_g^+$	$2\sigma_g^2 1\pi_u^4 1\pi_g^2$ (78%), $2\sigma_g^2 1\pi_u^2 1\pi_g^4$ (14%)
$^3\Delta_u$	$2\sigma_g^2 1\pi_u^3 1\pi_g^3$ (94%) —
$^3\Sigma_u^+$	$2\sigma_g^2 1\pi_u^3 1\pi_g^3$ (90%) —
$^3\Sigma_u^-$	$2\sigma_g^2 1\pi_u^3 1\pi_g^3$ (79%), $2\sigma_g 2\sigma_u 1\pi_u^4 1\pi_g^2$ (10%)
$^3\Sigma_g^-$(II)	$2\sigma_g^2 1\pi_u^2 1\pi_g^4$ (80%), $2\sigma_g^2 1\pi_u^4 1\pi_g^2$ (10%)
$^3\Pi_g$(II)	$2\sigma_g 1\pi_u^4 1\pi_g^3$ (68%), $2\sigma_g^2 2\sigma_u 1\pi_u^3 \pi_g^2$ (18%)
$^3\Pi_g$(III)	$2\sigma_g^2 2\sigma_u 1\pi_u^3 1\pi_g^2$ (82%), $2\sigma_g 2\sigma_u^2 1\pi_u^2 \pi_g^3$ (4%)
$^3\Pi_u$(I)	$2\sigma_g^2 2\sigma_u 1\pi_u^4 1\pi_g$ (46%), $2\sigma_g^2 2\sigma_u 1\pi_u^2 1\pi_g^3$ (32%)
$^1\Pi_g$(II)	$2\sigma_g 1\pi_u^4 1\pi_g^3$ (87%) —
$^1\Pi_g$(III)	$2\sigma_g^2 2\sigma_u 1\pi_u^3 1\pi_g^2$ (92%) —

Source: Reference 236.

Table 3.38 shows the leading configurations in the CI wavefunctions of the bound electronic states of Te_2. The nature of the electronic states listed in Table 3.37 could be best understood by using Table 3.38. Spin–orbit coupling makes significant contributions to the electronic states of Te_2. The $X(0_g^+)$ ground state was found to be actually 71% $^3\Sigma_g^-(0_g^+)$ and 19% $^1\Sigma_g^+(0_g^+)$ at its equilibrium geometry. The spin–orbit contamination for other states such as $^1\Delta_g(2_g)$, $^3\Sigma_g^-(1_g)$, $^3\Delta_u(3_u)$, $^3\Delta_u(2_u)$, and $^3\Sigma_u^+(1_u)$ were not particularly significant at their equilibrium geometries, although at longer distances, the spin–orbit contaminations became significant for all electronic states.

3.3.10 I_2 and I_2^+

Huber and Herzberg (37), as well as Brand and Hoy (38), have summarized the spectroscopic properties of I_2 and I_2^+. From empirical evidence and known experimental data at the time, Mulliken and other authors (237–239) obtained potential-energy curves of I_2, together with conclusions and assignments. The electronic spectra of I_2 in the region below 55,000 cm^{-1} indicate the existence of X, A', A, B, B', B'', C, D, D', E, G', F', F, G, H, and I states. The A', A, B, and B' states are respectively the 2_u, 1_u, 0_u^+, and 0_u^- spin–orbit components of the $^3\Pi_u$ state (37, 243). The G, G', E, and F states are assigned to the 2_g, 1_g, and O_g^+ components of the $^3\Pi_u$ state and $^1\Sigma_u^+$, respectively (245). Among these states, the transitions $A \rightarrow X$, $B \leftrightarrow X$, $B'' \leftrightarrow X$, $C \leftarrow X$, $G \leftarrow A'$, $D \leftrightarrow X$, $E \rightarrow B$, $G' \leftarrow A$, $F' \rightarrow X$, $F \leftrightarrow X$, $H \rightarrow B$, $I \rightarrow B$, and $D' \rightarrow A'$ are observed.

The absorption spectra of I_2 in CCl_4 and n-heptane are analyzed in terms of transitions from the ground state to the three electronically excited states

[$^3\Pi(1_u)$, $^1\Pi(1_u)$, $^3\Pi_u(0_u^+)$] (239). King and McLean (247) recorded the three-photon absorption spectrum of I_2 at 16,400–18,300 cm^{-1} and conformed the presence of the $F'(0_u^+)$ state by rotational analysis of the $3P2C$ spectrum. Venkateswarlu et al. (245, 248) studied the photoacoustic spectroscopy of I_2 and found that the structure observed at 20,200–20,750 cm^{-1} is beyond the convergence limit of $X^1\Sigma^+(0_g^+) \rightarrow B^3\Pi(0_u^+)$; this transition was attributed to a two-photon absorption. They also studied the $E \rightarrow B$ transition of I_2 at 4400–4000 Å and discussed the origin of the upper state, $E(0_g^+)$, of this transition. Many weakly bound electronic states and ion-pair states of I_2 have also been detected. Experimental investigations (238) on diffuse and predissociation spectra of I_2 have led to the characterization of electronic states labeled a and another state labeled D', which was earlier called G (37).

The I_2^+ halogen positive ion has been the topic of many investigations (249, 250). Leach (249) has reviewed the various experimental investigations on I_2^+, Br_2^+, and Cl_2^+. Li and the author (251) carried out CASSCF/FOCI/SOCI/RCI calculations on many electronic states of I_2 and I_2^+. They found 30 low-lying electronic states of I_2 and 13 electronic states of I_2^+. These authors employed RECPs that retained the $4d^{10}5s^25p^5$ outer shells of the iodine atom in the calculations, while the rest of the core electrons were replaced by RECPs. A flexible $(4s4p2d)$ valence Gaussian basis set was employed for the iodine atom. The final SOCI calculations included up to 105,004 configurations for I_2 and up to 230,000 configurations for I_2^+. Spin–orbit effects were taken into account by using the relativistic configuration interaction calculations.

Table 3.39 shows the calculated spectroscopic properties (R_e, T_e, ω_e) of 29 electronic states of I_2. Table 3.40 shows the spectroscopic constants of 13 electronic states of I_2^+. In Table 3.39, the experimental spectroscopic constants are shown for comparison. The spectroscopic properties in Tables 3.39 and 3.40 were obtained by using the $(4s4p2d)$ basis set and a combination of FOCI/SOCI/RCI methods. For many of the excited states with energy separations $> 30{,}000$ cm^{-1}, only the FOCI method was used. The T values in these tables are T_e values for the bound states and are the vertical transition energies for the repulsive states.

Figure 3.17 and Figure 3.18 show the potential-energy curves of the low-lying electronic states of I_2 with and without spin–orbit coupling, respectively. Figure 3.19 shows the potential-energy curves of I_2^+. Tables 3.41 and 3.42 depict the contributions of important configurations to the FOCI/RCI wavefunctions of I_2 and I_2^+, respectively.

In Table 3.39, the principal components of the electronic states are shown in Mulliken's $ijkl$ notation, where the set of integers i, j, k, and l designate the electronic configuration $2\sigma_g^i 1\pi_u^j 1\pi_g^k 2\sigma_u^l$. The labels for the various experimentally observed states are as per the review by Brand and Hoy (38). It must be pointed out that some of these labels have changed during the years. For example, the electronic state labeled D' in Table 3.39 was denoted G earlier as seen in Reference 37. The labels with questions marks are tentative assignments. The ensuing discussion should be consulted for the possibilities.

Table 3.39 Experimental and Theoretical Spectroscopic Constants of I_2[a]

State	*ijkl*	T, cm^{-1}		Assignment	R_e, Å		ω_e, cm^{-1}	
		Theory	Exp.		Theory	Exp.	Theory	Exp.
$^1\Sigma_g^+(0_g^+)$	2440	0	0	X	2.70	2.67	210	214.5
$^3\Pi_u(2_u)$	2431	11,500	10,042	A'	3.30	3.08	93	109
$^3\Pi_u(1_u)$	2431	12,522	11,888	A	3.30	—	90	93
$^3\Pi_u(0_u^+)$	2431	16,560	15,725	B	3.25	3.02	110	126
$^3\Pi_u(0_u^-)$	2431	21,000 (repulsive)	—	B'	—	—	—	—
$^1\Pi_u(1_u)$	2431	25,700 (repulsive)	21,000 (max)	B''	—	—	—	—
$^3\Pi_g(1_g)$	2341	27,900 (repulsive)	—	a	—	—	—	—
$^3\Pi_g(2_g)$	2341	30,763 (repulsive)	—	—	—	—	—	—
$^3\Sigma_g^-(0_g^+)$	2422	31,700 (repulsive)	—	a'	—	—	—	—
$^3\Sigma_g^+(1_u)$	1441	32,500 (repulsive)	37,000 (max)	C	—	—	—	—
$^3\Sigma_g^-(1_g)$	2422	34,700 (repulsive)	—	—	—	—	—	—
$^3\Pi_g(0_g^+)$	2341	35,800 (repulsive)	—	—	—	—	—	—
$^3\Pi_g(0_g^-)$	2341	35,869	—	—	—	—	—	—
$^1\Pi_g(1_g)$	2341	36,200	—	—	—	—	—	—

$1_u(^3\Delta, ^3\Sigma^+)_u$	2332	36,750	—	—	—	—	—	—
$^1\Delta_g(2_g)$	2422	37,340	—	—	—	—	—	—
$^3\Sigma_u^-(0_u^+)$	2332	41,800	41,029	D	3.74	3.58	90	95
$^3\Delta_u$	2332	41,940	41,621	$\gamma 1_u$	—	3.67	—	95
		41,789	—	$\delta 2_u$	—	(4.0)	—	100
$^3\Sigma_g^-(\mathrm{II})$	2242	42,500	—	—	—	—	—	—
$^3\Pi_g(\mathrm{II})$	1432	42,711	40,386	$D'(2_g)$	3.85	3.61	90	105
$^3\Pi_g(\mathrm{II})0_g^+$	1432	—	41,412	$E(0_g^+)$	—	3.67	—	101
$^3\Pi_u(\mathrm{II})$	1342	46,014	45,230	F'?	4.02	—	94	93
$^1\Sigma_u^+(0_u^+)$	1441	47,500	47,217	F	3.596	3.61	102	96
$^1\Sigma_g^+(\mathrm{II})$	0442	51,200	47,026	f	—	3.574	103	104
$^1\Delta_u(2_u)$	2332	53,145	—	H?	3.69	—	101	—
$^1\Sigma_g^+(\mathrm{III})$	2422	54,350 (repulsive)	—	—	—	—	—	—
$^3\Pi_g(\mathrm{II})1_g$	1432	55,000	47,559	G	3.528	—	89	—
$^1\Pi_g(\mathrm{II})_g^1$	1432	56,000	52,000	I?	3.84	—	92	107
$^1\Pi_u(\mathrm{II})$	1342	57,200	—	—	3.94	—	95	—
$^1\Sigma_u^+(\mathrm{II})$	2332	65,200	51,706	$F''0_u^+$	3.63	3.48	102	131

[a]Theoretical $D_e(I_2)$ = 1.45 eV including spin–orbit effects. Experimental $D_e(I_2)$ = 1.54 eV. The notation $ijkl$ is used to designate the electronic configuration $2\sigma_g^i 1\pi_u^i 1\pi_g^k 2\sigma_u^1$ following Mulliken (180). Theoretical values are from Reference 251, while experimental values are from Reference 38. For repulsive states, the reported T values are vertical energy separations. For bound states, these are T_e.

Table 3.40 Spectroscopic Properties of I_2^+ Calculated by the FOCI (SOCI)/RCI Method[a]

State		T, cm^{-1} Calc.	Assignment	r_e, Å	ω_e, cm^{-1}	D_e, eV
$^2\Pi_g(\frac{3}{2})_g$		0	X	2.70	213	1.92
	SOCI:	0		2.69	217	2.06
$^2\Pi_g(\frac{1}{2})_g$		5,979	X	2.70	204	1.16
	SOCI:	5,979		2.69	208	1.30
$^4\Sigma_u^-(\frac{1}{2})_u$		7,948		3.17	126	0.92
$^4\Sigma_u^-(\frac{3}{2})_u$		8,258		3.19	125	0.88
$^2\Delta_u$		12,215		3.48	99	0.74
$^2\Pi_u(\frac{3}{2})_u$		9,961	A	3.12	128	—
	SOCI:	10,068	(vertical: 12687)	3.09	132	—
$^2\Pi_u(\frac{1}{2})_u$		15,010	A	3.14	108	—
	SOCI:	15,117	(vertical: 17736)	3.11	112	—
$^2\Sigma_u^+$		14,470		3.52	102	—
$^2\Delta_g$		17,739		3.61	71	—
$^2\Sigma_g^+$		19,361	B	3.62	72	—
$^2\Pi_g$(II)		40,358 (repulsive)		—	—	—
$^2\Pi_u$(II)		44,162 (repulsive)		—	—	—
$^4\Sigma_u^-$(II)		44,462 (repulsive)		—	—	—

[a]All values are from Reference 251. Reported T values are adiabatic for bound states; otherwise they are vertical separations.

The theoretical ground-state spectroscopic constants (251) are R_e = 2.70 (SOCI/RCI) and 2.77 Å (FOCI/RCI), ω_e = 210 cm^{-1} (FOCI/RCI), and D_e = 1.45 eV, which includes the spin–orbit effects. From Table 3.41 it is seen that the main configuration contributing to the ground state is $2\sigma_g^2$ (28%) and $1\pi_u^3 2\sigma_u^1$ (5%) become significant. The I–I bond in the $^1\Sigma_g^+(0_g^+)$ state is weakened by the spin–orbit interaction. The experimental spectroscopic data (37) are R_e = 2.67 Å, ω_e = 214.5 cm^{-1}, and D_0° = 1.54 eV, which are in excellent agreement with the calculation results of the author and Li.

Skorko (252) found a strong continuous absorption region at higher temperatures and pressures in the spectrum of I_2 at λ 3427 Å, and a similar but weaker region with a long-wavelength edge at λ 3263 Å. Mulliken (237) assumed that the λ 3427-Å and λ 3263-Å regions can be identified with absorptions from the $^3\Pi_u(2_u)$ and $^3\Pi_u(1_u)$ states terminating to $^3\Pi_g(2_g)$ and $^3\Pi_g(1_g)$, respectively, and estimated a T_e value of 10,100 cm^{-1} for $^3\Pi(2_u)$ and 11,888 cm^{-1} for $^3\Pi_u(1_u)$. Ashby et al. (241) photographed the $A^3\Pi_u(1_u)$–$X^1\Sigma_g^+$ system of I_2

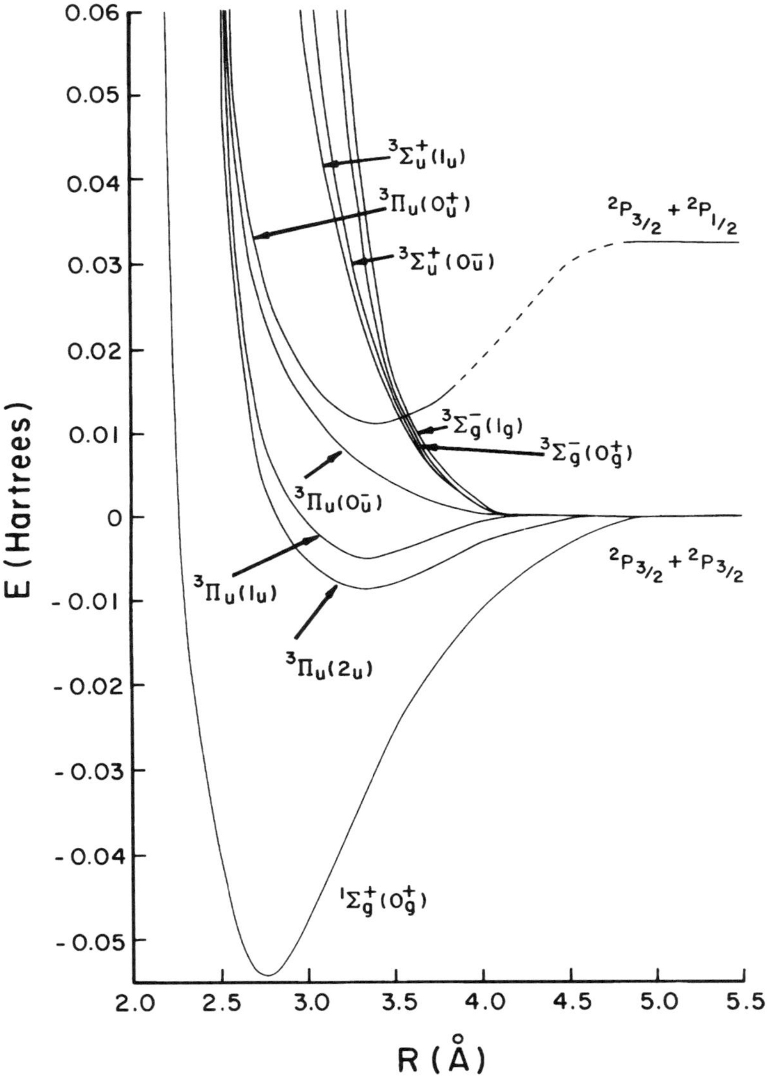

Figure 3.17 Potential-energy curves of the low-lying electronic states of I_2 including spin–orbit effects. (Reprinted from Ref. 251; copyright 1989, Academic Press, Inc.)

at 8,000–13,400 Å in absorption, which facilitated calculation of constants for the $A^3\Pi_u(1_u)$ state: T_e = 10,906 cm^{-1}, D_e = 1641 cm^{-1}, and ω_e = 92.5 cm^{-1}. Gerstenkorn et al. (242) determined the IR absorption spectrum of the $A^3\Pi_u(1_u)$ state at 7,000–12,500 cm^{-1} using Fourier spectroscopy and found the separation between the $X^1\Sigma_g^+$ and $A^3\Pi_u(1_u)$ states to be 10,846 cm^{-1}; the D_e of the A state was calculated as 1639.77 cm^{-1}. Gerstenkorn et al. (253) obtained some

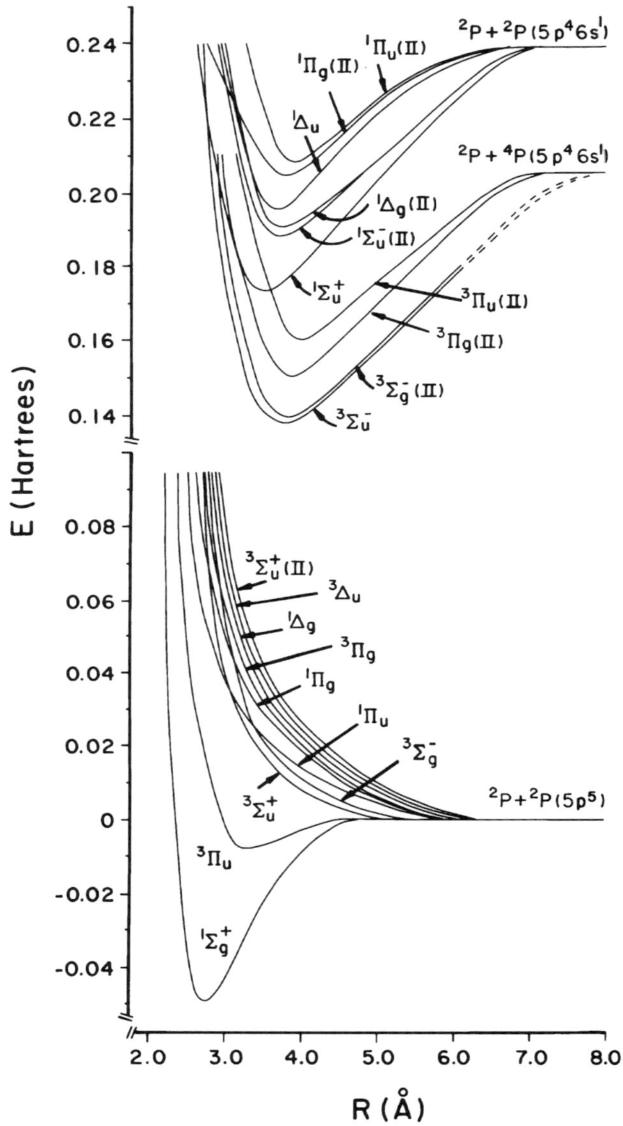

Figure 3.18 Potential-energy curves of the electronic states of I_2 without spin–orbit effects. (Reprinted from Ref. 251; copyright 1989, Academic Press, Inc.)

visible absorption bands of I_2 by the same technique. The separation between the $X^1\Sigma_g^+$ and $B^3\Pi_u(0_u^+)$ states deduced by these authors is $T_{0,0} = 15724.586$ cm^{-1}.

The spin–orbit components of the $^3\Pi_u$ state of I_2 are not as strongly bound as those of the ground state. The basis sets and levels of electron correlation

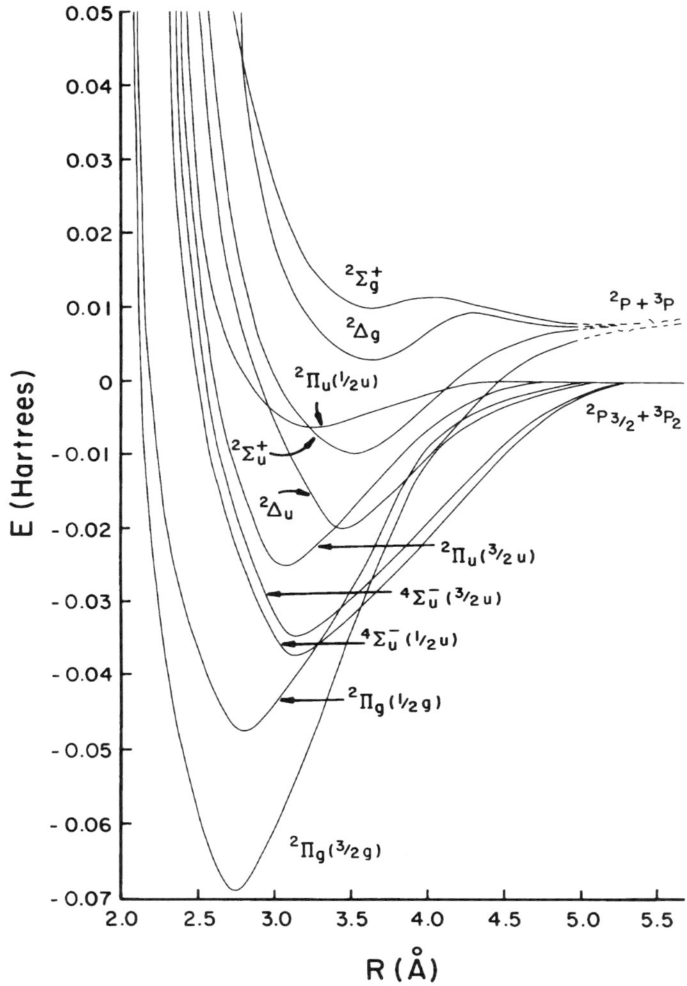

Figure 3.19 Potential-energy curves of the electronic states of I_2^+. (Reprinted from Ref. 251; copyright 1989, Academic Press, Inc.)

treatments employed by Li and the author (251) are not likely to be as accurate for less bound excited states compared to the ground state. Nevertheless, a full SOCI treatment that included up to 105,000 configurations in the D_{2h} symmetry cannot be regarded as a poor approximation. Thus the theoretical R_e and ω_e values of the $B(0_u^+)$ state are not in as good agreement with the experimental values as the ground state is.

The $B'^3\Pi_u(0_u^-)$ and $B''^1\Pi_u(1_u)$ states were both assigned as repulsive states (37) from the $^2P_{3/2}$ + $^2P_{3/2}$ atomic states at long distance; the B'' state is also the state responsible for the absorption continuum with a maximum at 20,050

Table 3.41 Contributions of Important Configurations to the FOCI/RCI Wavefunctions of I_2

State	r (Å)	Percentage of Contributions	State	r, Å	Percentage of Contributions
${}^1\Sigma_g^+(0_g^+)$	2.75	$2\sigma_g^2 1\pi_u^4 1\pi_g^4({}^1\Sigma_g^+)$ (95), $1\pi_u^4 1\pi_g^4 2\sigma_u^2(1)({}^1\Sigma_g^+)$ (4),	${}^3\Pi_g(1_g)$	2.75	$2\sigma_g^2 1\pi_u^3 1\pi_g^4 2\sigma_u^1({}^3\Pi_g)$ (41), $2\sigma_g^1 1\pi_u^4 1\pi_g^3 2\sigma_u^2({}^3\Pi_g)$ (51)
		$2\sigma_g^2 1\pi_u^3 1\pi_g^4 2\sigma_u^1({}^3\Pi_g)$ (0.6), $2\sigma_g^2 1\pi_u^4 1\pi_g^2 2\sigma_u^2({}^3\Sigma_g^-)$ (0.3)	${}^3\Sigma_g^-(0_g^+)$	2.75	$2\sigma_g^2 1\pi_u^4 1\pi_g^2 2\sigma_u^2({}^3\Sigma_g^-)$ (93), $2\sigma_g^2 1\pi_u^2 1\pi_g^4 2\sigma_u^2({}^3\Sigma_g^-)$ (2)
	4.00	$2\sigma_g^2 1\pi_u^4 1\pi_g^4({}^1\Sigma_g^+)$ (67), $1\pi_u^4 1\pi_g^4 2\sigma_u^2({}^1\Sigma_g^+)$ (28),	${}^3\Sigma_g^-(1_g)$	2.75	$2\sigma_g^2 1\pi_u^4 1\pi_g^2 2\sigma_u^2({}^3\Sigma_g^-)$ (94)
		$2\sigma_g^2 1\pi_u^3 1\pi_g^4 2\sigma_u^1({}^3\Pi_g)$ (5)	${}^1\Delta_g(2_g)$	2.75	$2\sigma_g^2 1\pi_u^4 1\pi_g^2 2\sigma_u^2({}^1\Delta_g)$ (92), $2\sigma_g^2 1\pi_u^2 1\pi_g^4 2\sigma_u^2({}^1\Delta_g)$ (4)
${}^3\Pi_u(2_u)$	3.25	$2\sigma_g^2 1\pi_u^4 1\pi_g^3 2\sigma_u^1({}^3\Pi_u)$ (97)	${}^3\Pi_g(\mathrm{II})(2_g)$	3.75	$2\sigma_g^1 1\pi_u^4 1\pi_g^3 2\sigma_u^2({}^3\Pi_g)$ (62), $2\sigma_g^2 1\pi_u^3 1\pi_g^4 2\sigma_u^1({}^3\Pi_g)$ (33)
	3.50	$2\sigma_g^2 1\pi_u^4 1\pi_g^3 2\sigma_u^1({}^3\Pi_u)$ (97)	${}^3\Pi_g(\mathrm{II})(0_g^+)$	3.75	$2\sigma_g^1 1\pi_u^4 1\pi_g^3 2\sigma_u^2({}^3\Pi_g)$ (62), $2\sigma_g^2 1\pi_u^3 1\pi_g^4 2\sigma_u^1({}^3\Pi_g)$ (33)
	4.00	$2\sigma_g^2 1\pi_u^4 1\pi_g^3 2\sigma_u^1({}^3\Pi_u)$ (67), $2\sigma_g^1 1\pi_u^3 1\pi_g^4 2\sigma_u^2({}^3\Pi_u)$ (30)	${}^3\Pi_g(\mathrm{II})(0_g^-)$	3.75	$2\sigma_g^1 1\pi_u^4 1\pi_g^3 2\sigma_u^2({}^3\Pi_g)$ (62), $2\sigma_g^2 1\pi_u^3 1\pi_g^4 2\sigma_u^1({}^3\Pi_g)$ (33)
${}^3\Pi_u(1_u)$	3.25	$2\sigma_g^2 1\pi_u^4 1\pi_g^3 2\sigma_u^1({}^3\Pi_u)$ (57), $2\sigma_g^2 1\pi_u^4 1\pi_g^3 2\sigma_u^1({}^1\Pi_u)$ (3),	${}^3\Pi_g(\mathrm{II})(1_g)$	3.75	$2\sigma_g^1 1\pi_u^4 1\pi_g^3 2\sigma_u^2({}^3\Pi_g)$ (58), $2\sigma_g^1 1\pi_u^4 1\pi_g^3 2\sigma_u^2({}^1\Pi_g)$ (0.3)
		$2\sigma_g^1 1\pi_u^4 1\pi_g^4 2\sigma_u^1({}^3\Sigma_u^+)$ (39)			$2\sigma_g^2 1\pi_u^3 1\pi_g^4 2\sigma_u^1({}^3\Pi_g)$ (38), $2\sigma_g^2 1\pi_u^3 1\pi_g^4 2\sigma_u^1({}^1\Pi_g)$ (0.3)
	3.50	$2\sigma_g^2 1\pi_u^4 1\pi_g^3 2\sigma_u^1({}^3\Pi_u)$ (11), $2\sigma_g^1 1\pi_u^4 1\pi_g^4 2\sigma_u^1({}^3\Sigma_u^+)$ (86),	${}^1\Sigma_u^-$	2.75	$2\sigma_g^2 1\pi_u^3 1\pi_g^3 2\sigma_u^2({}^1\Sigma_u^-)$ (99)
		$2\sigma^2 1\pi_u^4 1\pi_g^3 2\sigma_u^1({}^1\Pi_u)$ (1)	${}^3\Sigma_u^-$	3.75	$2\sigma_g^2 1\pi_u^3 1\pi_g^3 2\sigma_u^2({}^3\Sigma_u^-)$ (88)

	4.00	$2\sigma_g^2 1\pi_u^4 1\pi_g^3 2\sigma_u^1 (^3\Pi_u)$ (31),	$1_u\ {}^3\Delta_u\ {}^3\Sigma_u^+(\mathrm{II})$	2.75	$2\sigma_g^2 1\pi_u^3 1\pi_g^3 2\sigma_u^2 (^3\Delta_u)$ (49),
		$2\sigma_g^1 1\pi_u^4 1\pi_g^4 2\sigma_u^1 (^3\Sigma_u^+)$ (47),			$2\sigma_g^2 1\pi_u^3 1\pi_g^3 2\sigma_u^2 (^3\Sigma_u^+)$ (48)
		$2\sigma_g^2 1\pi_u^3 1\pi_g^4 2\sigma_u^1 (^3\Pi_u)$ (13),	$^3\Delta_u$	2.75	$2\sigma_g^2 1\pi_u^3 1\pi_g^3 2\sigma_u^2 (^3\Delta_u)$ (98)
		$2\sigma_g^1 1\pi_u^4 1\pi_g^3 2\sigma_u^1 (^1\Pi_u)$ (2)			
$^3\Pi_u(0_u^+)$	3.25	$2\sigma_g^2 1\pi_u^4 1\pi_g^3 2\sigma_u^1 (^3\Pi_u)$ (96),	$^3\Sigma_u^-(\mathrm{II})$	3.75	$2\sigma_g^2 1\pi_u^2 1\pi_g^4 2\sigma_u^2 (^3\Sigma_u^-)$ (55),
		$2\sigma_g^1 1\pi_u^4 1\pi_g^4 2\sigma_u^1 (^1\Sigma_u^+)$ (1)			$2\sigma_g^2 1\pi_u^4 1\pi_g^2 2\sigma_u^2 (^3\Sigma_g^-)$ (34)
	3.50	$2\sigma_g^2 1\pi_u^4 1\pi_g^3 2\sigma_u^1 (^3\Pi_u)$ (96),	$^3\Sigma_u^+(\mathrm{II})$	2.75	$2\sigma_g^2 1\pi_u^3 1\pi_g^3 2\sigma_u^2 (^3\Sigma_u^+)$ (97)
		$2\sigma_g^1 1\pi_u^4 1\pi_g^4 2\sigma_u^1 (^1\Sigma_u^+)$ (1)			
	4.00	$2\sigma_g^2 1\pi_u^4 1\pi_g^3 2\sigma_u^1 (^3\Pi_u)$ (67),	$^3\Pi_u(\mathrm{II})$	4.00	$2\sigma_g^1 1\pi_u^3 1\pi_g^4 2\sigma_u^2 (^3\Pi_u)$ (63),
		$2\sigma_g^1 1\pi_u^3 1\pi_g^4 2\sigma_u^2 (^3\Pi_u)$ (31)			$2\sigma_g^2 1\pi_u^4 1\pi_g^3 2\sigma_u^1 (^3\Pi_u)$ (25)
$^3\Pi_u(0_u^-)$	2.75	$2\sigma_g^2 1\pi_u^4 1\pi_g^3 2\sigma_u^1 (^3\Pi_u)$ (84),	$^1\Sigma_u^+$	3.50	$2\sigma_g^1 1\pi_u^4 1\pi_g^4 2\sigma_u^1 (^1\Sigma_u^+)$ (78),
		$2\sigma_g^1 1\pi_u^4 1\pi_g^4 2\sigma_u^1 (^3\Sigma_u^+)$ (10)			$2\sigma_g^2 1\pi_u^3 1\pi_g^3 2\sigma_u^1 (^1\Sigma_u^+)$ (14)
	3.25	$2\sigma_g^2 1\pi_u^4 1\pi_g^3 2\sigma_u^1 (^3\Pi_u)$ (32),	$^1\Sigma_g^+(\mathrm{II})$	4.00	$1\pi_u^4 1\pi_g^4 2\sigma_u^2 (^1\Sigma_g^+)$ (50),
		$2\sigma_g^1 1\pi_u^4 1\pi_g^4 2\sigma_u^1 (^3\Sigma_u^+)$ (66)			$2\sigma_g^2 1\pi_u^4 1\pi_g^4 (^1\Sigma_g^+)$ (16),
	4.00	$2\sigma_g^2 1\pi_u^4 1\pi_g^3 2\sigma_u^1 (^3\Pi_u)$ (28),			$2\sigma_g^2 1\pi_u^2 1\pi_g^4 2\sigma_g^2 (^1\Sigma_g^+)$ (12),
		$2\sigma_g^1 1\pi_u^4 1\pi_g^4 2\sigma_u^1 (^3\Sigma_u^+)$ (55)			$2\sigma_g^2 1\pi_u^4 1\pi_g^2 2\sigma_u^2 (^1\Sigma_u^-)$ (10)
		$2\sigma_g^1 1\pi_u^3 1\pi_g^4 2\sigma_u^2 (^3\Pi_u)$ (15)	$^1\Sigma_u^+(\mathrm{II})$	3.75	$2\sigma_g^2 1\pi_u^3 1\pi_g^3 2\sigma_u^2 (^1\Sigma_u^-)$ (89)
$^1\Pi_u(1_u)$	2.75	$2\sigma_g^2 1\pi_u^4 1\pi_g^3 2\sigma_u^1 (^1\Pi_u)$ (87),	$^1\Delta_g(\mathrm{II})$	3.75	$2\sigma_g^2 1\pi_u^2 1\pi_g^4 2\sigma_u^2 (^1\Delta_g)$ (51),
		$2\sigma_g^2 1\pi_u^4 1\pi_g^3 2\sigma_u^1 (^3\Pi_u)$ (12)			$2\sigma_g^2 1\pi_u^4 1\pi_g^2 2\sigma_u^2 (^1\Delta_g)$ (38)
$^1\Pi_g(1_g)$	2.75	$2\sigma_g^2 1\pi_u^3 1\pi_g^4 2\sigma_u^1 (^1\Pi_g)$ (50),	$^1\Sigma_g^+(\mathrm{III})$	2.75	$2\sigma_g^2 1\pi_u^4 1\pi_g^2 2\sigma_u^2 (^1\Sigma_g^+)$ (85),
		$2\sigma_g^1 1\pi_u^4 1\pi_g^3 2\sigma_u^2 (^1\Pi_g)$ (48)			$2\sigma_g^2 1\pi_u^2 1\pi_g^4 2\sigma_u^2 (^1\Sigma_g^+)$ (4)
$^3\Sigma_u^+(1_u)$	2.75	$2\sigma_g^1 1\pi_u^4 1\pi_g^4 2\sigma_u^1 (^3\Sigma_u^+)$ (98)	$^1\Delta_u$	3.75	$2\sigma_g^2 1\pi_u^3 1\pi_g^3 2\sigma_u^2 (^1\Delta_u)$ (88)
$^3\Sigma_u^+(0_u^-)$	2.75	$2\sigma_g^1 1\pi_u^4 1\pi_g^4 2\sigma_u^1 (^3\Sigma_u^+)$ (98)	$^1\Pi_g(\mathrm{II})$	3.75	$2\sigma_g^1 1\pi_u^4 1\pi_g^3 2\sigma_u^2 (^1\Pi_g)$ (67),
					$2\sigma_g^2 1\pi_u^3 1\pi_g^3 2\sigma_u^1 (^1\Pi_g)$ (23)

Table 3.41 (*Continued*)

State	r (Å)	Percentage of Contributions	State	r, Å	Percentage of Contributions
$^3\Pi_g(2_g)$	2.75	$2\sigma_g^2 1\pi_u^3 1\pi_g^4 2\sigma_u^1 (^3\Pi_g)$ (51), $2\sigma_g^1 1\pi_u^4 1\pi_g^3 2\sigma_u^2 (^3\Pi_g)$ (46)	$^1\Pi_u$(II)	4.00	$2\sigma_g^1 1\pi_u^3 1\pi_g^4 2\sigma_u^2 (^1\Pi_u)$ (57), $2\sigma_g^2 1\pi_u^4 1\pi_g^3 2\sigma_u^1 (^1\Pi_u)$ (31)
$^3\Pi_g(0_g^+)$	2.75	$2\sigma_g^2 1\pi_u^3 1\pi_g^4 2\sigma_u^1 (^3\Pi_g)$ (53), $2\sigma_g^1 1\pi_u^4 1\pi_g^3 2\sigma_u^2 (^3\Pi_g)$ (45)	$^3\Pi_g$(III)	3.00	$2\sigma_g^2 1\pi_u^4 1\pi_g^3 3\sigma_g^1 (^3\Pi_g)$ (83), $2\sigma_g^1 1\pi_u^3 1\pi_g^4 2\sigma_u^1 3\sigma_g^1 (^3\Pi_g)$ (5)
$^3\Pi_g(0_u^-)$	2.75	$2\sigma_g^2 1\pi_u^3 1\pi_g^4 2\sigma_u^1 (^3\Pi_g)$ (53), $2\sigma_g^1 1\pi_u^4 1\pi_g^3 2\sigma_u^2 (^3\Pi_g)$ (45)	$^1\Pi_u$(III)	3.00	$2\sigma_g^1 1\pi_u^3 1\pi_g^4 2\sigma_u^2 (^1\Pi_u)$ (42), $2\sigma_g^2 1\pi_u^4 1\pi_g^3 3\sigma_u^1 (^1\Pi_u)$ (33) $2\sigma_g^2 1\pi_u^3 1\pi_g^4 3\sigma_u^1 (^1\Pi_u)$ (8), $2\sigma_g^2 1\pi_u^4 1\pi_g^3 2\sigma_u^1 (^1\Pi_u)$ (5)
$^1\Pi_g$(III)	3.00	$2\sigma_g^2 1\pi_u^4 1\pi_g^3 3\sigma_g^1 (^1\pi_g)$ (84), $2\sigma_g^1 1\pi_u^3 1\pi_g^4 2\sigma_u^1 3\sigma_g^1 (^1\Pi_g)$ (5),			
$^1\Sigma_u^+$(II)	3.75	$2\sigma_g^2 1\pi_u^3 1\pi_g^3 2\sigma_u^2 (^1\Sigma_u^+)$ (73), $2\sigma_g^1 1\pi_u^4 1\pi_g^4 2\sigma_u^1 (^1\Sigma_u^+)$ (15)	$^1\Sigma_u^+$(III)	3.25	$2\sigma_g^2 1\pi_u^4 1\pi_g^2 2\sigma_u^1 3\sigma_g^1 (^1\Sigma_u^+)$ (50), $2\sigma_g^2 1\pi_u^2 1\pi_g^4 2\sigma_u^1 3\sigma_g^1 (^1\Sigma_u^+)$ (32) $2\sigma_g^1 1\pi_u^3 1\pi_g^3 2\sigma_u^2 3\sigma_g^1 (^1\Sigma_u^+)$ (4)
$^3\Sigma_u^+$(III)	3.50	$2\sigma_g^2 1\pi_u^4 1\pi_g^2 2\sigma_u^1 3\sigma_g^1 (^3\Sigma_u^+)$ (50), $2\sigma_g^2 1\pi_u^2 1\pi_g^4 2\sigma_u^1 3\sigma_g^1 (^3\Sigma_u^+)$ (32) $2\sigma_g^1 1\pi_u^3 1\pi_g^3 2\sigma_u^2 3\sigma_g^1 (^3\Sigma_u^+)$ (4)	$^3\Sigma_g^+$	3.25	$2\sigma_g^1 1\pi_u^4 1\pi_g^4 3\sigma_g^1 (^3\Sigma_g^+)$ (90)
$^3\Pi_u$(III)	3.00	$2\sigma_g^2 1\pi_u^4 1\pi_g^3 3\sigma_u^1 (^3\Pi_u)$ (43), $2\sigma_g^2 1\pi_u^3 1\pi_g^4 3\sigma_g^1 (^3\Pi_u)$ (39), $2\sigma_g^1 1\pi_u^3 1\pi_g^4 2\sigma_u^2 (^3\Pi_u)$ (4)	$^3\Sigma_g^+$(II)	3.00	$2\sigma_g^2 1\pi_u^4 1\pi_g^3 2\pi_g^1 (^3\Sigma_g^+)$ (72), $2\sigma_g^1 1\pi_u^4 1\pi_g^4 3\sigma_g^1 (^3\Sigma_g^+)$ (15)

Source: Reference 251.

Table 3.42 Contributions of Important Configurations to the FOCI/RCI Wavefunctions of I_2^+

State	r, Å	Percentage of Contributions	State	r, Å	Percentage of Contributions
$^2\Pi_g(\frac{3}{2})_g$	2.75	$2\sigma_g^2 1\pi_u^4 1\pi_g^3 (^2\pi_g)$ (93), $1\pi_u^4 1\pi_g^3 2\sigma_u^2 (^2\pi_g)$ (3)		3.50	$2\sigma_g^2 1\pi_u^3 1\pi_g^3 2\sigma_u^1 (^2\Delta_g)$(87), $2\sigma_g^1 1\pi_u^4 1\pi_g^2 2\sigma_u^2 (^2\Delta_g)$ (6)
$^2\Pi_g(\frac{1}{2})_g$	2.75	$2\sigma_g^2 1\pi_u^4 1\pi_g^3 (^2\pi_g)$ (92), $1\pi_u^4 1\pi_g^3 2\sigma_u^2 (2\pi_g)$ (3)		5.00	$2\sigma_g^1 1\pi_u^4 \pi_g^2 2\sigma_u^2 (^2\Delta_g)$(32), $2\sigma_g^1 1\pi_u^2 1\pi_g^4 2\sigma_u^2 (^2\Delta_g)$ (30), $2\sigma_g^2 1\pi_u^3 1\pi_g^3 2\sigma_u^1 (^2\Delta_g)$ (34)
$^4\Sigma_u^-(\frac{1}{2})_u$	3.25	$2\sigma_g^2 1\pi_u^4 1\pi_g^2 2\sigma_u^1 (^4\Sigma_u^-)$ (65), $2\sigma_g^2 1\pi_u^2 1\pi_g^4 2\sigma_u^1 (^4\Sigma_u^-)$ (31) $2\sigma_g^2 1\pi_u^4 1\pi_g^2 2\sigma_u^1 (^2\Sigma_u^+)$ (2)	$^2\Sigma_g^+$	2.75	$2\sigma_g^1 1\pi_u^4 1\pi_g^4 (^2\Sigma_g^+)$(87), $2\sigma_g^2 1\pi_u^3 1\pi_g^3 2\sigma_u^1 (^2\Sigma_g^+)$ (6)
$^4\Sigma_u^-(\frac{3}{2})_u$	3.25	$2\sigma_g^2 1\pi_u^4 1\pi_g^2 2\sigma_u^1 (^4\Sigma_u^-)$ (64), $2\sigma_g^2 1\pi_u^2 1\pi_g^4 2\sigma_u^1 (^4\Sigma_u^-)$ (33)		3.50	$2\sigma_g^2 1\pi_u^3 1\pi_g^3 2\sigma_u^1 (^2\Sigma_g^+)$ (87), $2\sigma_g^1 1\pi_u^4 1\pi_g^2 2\sigma_u^2 (^2\Sigma_g^+)$ (5)
$^2\Delta_u$	3.50	$2\sigma_g^2 1\pi_u^4 1\pi_g^2 2\sigma_u^1 (^2\Delta_u)$ (58), $2\sigma_g^2 1\pi_u^2 1\pi_g^4 2\sigma_u^1 (^2\Delta_u)$ (35)		5.00	$2\sigma_g^1 1\pi_u^4 1\pi_g^2 2\sigma_u^2 (^2\Sigma_g^+)$ (32), $2\sigma_g^1 1\pi_u^2 1\pi_g^4 2\sigma_u^2$ (30), $2\sigma_1^2 1\pi_u^3 1\pi_g^3 2\sigma_u^1 (^2\Sigma_g^+)$ (32)
$^2\Pi_u(\frac{3}{2})_u$	3.00	$2\sigma_g^2 1\pi_u^3 1\pi_g^4 (^2\Pi_u)$ (78), $2\sigma_g^1 1\pi_u^4 1\pi_g^3 2\sigma_u^1 (^2\Pi_u)$ (14)	$^2\Pi_g$(II)	2.75	$2\sigma_g^2 1\pi_u^4 1\pi_g^1 2\sigma_u^2 (^2\Pi_g)$ (72), $2\sigma_g^2 1\pi_u^2 1\pi_g^3 2\sigma_u^2 (^2\Pi_g)$ (19)
$^2\Pi_u(\frac{1}{2})_u$	3.00	$2\sigma_g^2 1\pi_u^3 1\pi_g^4 (^2\Pi_u)$ (57), $2\sigma_g^1 1\pi_u^4 1\pi_g^3 2\sigma_u^1 (^2\Pi_u)$ (19) $2\sigma_g^2 1\pi_u^4 1\pi_g^2 2\sigma_u^1 (^4\Sigma_u^-)$ (9)	$^2\Pi_u$(II)	2.75	$2\sigma_g^1 1\pi_u^4 1\pi_g^3 2\sigma_u^1 (^2\Pi_u)$ (91)
$^2\Sigma_u^+$	3.50	$2\sigma_g^2 1\pi_u^4 1\pi_g^2 2\sigma_u^1 (^2\Sigma_u^+)$ (57), $2\sigma_g^2 1\pi_u^2 1\pi_g^4 2\sigma_u^1 (^2\Sigma_u^+)$ (36)	$^4\Sigma_u^-$(II)	2.75	$2\sigma_g^2 1\pi_u^2 1\pi_g^4 2\sigma_u^1 (^4\Sigma_u^-)$ (78), $2\sigma_g^1 1\pi_u^3 1\pi_g^3 2\sigma_u^2 (^4\Sigma_u^-)$ (13)
$^2\Delta_g$	2.75	$2\sigma_g^2 1\pi_u^3 1\pi_g^3 2\sigma_u^1 (^2\Delta_g)$ (94)			

Source: Reference 251.

cm^{-1} and for the predissociation of $B^3\Pi_u(0_u^+)$. As can be seen from Figure 3.17, the $^3\Pi_u(0_u^+)$ curve crosses with the $^3\Sigma_u^+(1_u)$, $^3\Sigma_u^+(0_u^-)$, $^3\Sigma_g^-(0_g^+)$, $^3\Sigma_g^-(1_g)$, $^1\Pi_g(1_g)$, and $^1\Pi_u(1_u)$ curves. Hence the $^3\Pi_u(0_u^+)$ state is predissociated. This was noted before (238) on the basis of experimental findings. There has been some doubt about the $B'^3\Pi_u(0_u^-)$ state being responsible for the magnetic-field-induced predissociation of the $^3\Pi_u(0_u^+)$ state (see Ref. 37). The theoretical calculations of Li and the author (251) show (Fig. 3.17) that the repulsive state $^3\Pi_u(0_u^-)$ lies below the $^3\Pi_u(0_u^+)$ state and thus may not be responsible for the predissociation. The theoretical $B''^1\Pi_u(1_u)$ state is repulsive, and its vertical energy from the ground state is 22% higher than the data from the absorption spectrum. The vertical energy separations of most of the states were not calculated exactly at the R_e of the ground state. A distance close to the R_e for which the calculational value existed was used. Part of the difference could be attributed to this, although the basis sets and the levels of electron correlations used by Li and the author are not fully adequate for accurate determination of the energy separations of the excited states.

The C state of I_2 has been studied by many authors. Mathieson and Rees (255) assigned the C state to 2332 $^3\Sigma_u^+(\text{II})(0_u^-)$. Mulliken (237) and Clear and Wilson (254) assigned the C state to 1441 $^3\Sigma_u^+(1_u)$. The experimental data indicate that the C state has a weak but broad absorption continuum with a maximum at 2700 Å (37,000 cm^{-1}) and dissociates into $^2P_{2/3}$ + $^2P_{1/2}$ atomic states. The theoretical calculations of Li and the author (251) revealed (see Fig. 3.17 and Table 3.39) that the 1441 $^3\Sigma_u^+(1_u)$ state dissociates into $^2P_{3/2}$ + $^2P_{3/2}$ at long distances; the theoretical vertical energy of the C state was found to be 32,510 cm^{-1}, which is lower than the experimental value. However, the 1_u state, which is a mixture of the 2332 $^3\Delta u(1_u)$ and 2332 $^3\Sigma_u^+(\text{II})(1_u)$ states, dissociates into $^2P_{3/2}$ +^{2} $P_{1/2}$ and has a vertical energy (36,751 cm^{-1}) near the experimental value. The mixing between $^3\Sigma_u^+(1_u)$ and $^3\Delta_u(1_u)$ could provide the necessary transition moment for the $C \leftarrow X$ transition if the C state were assigned to this state. Thus the possibility of C being this state should not be ruled out. In both cases, theoretical calculations of Li and the author (251) supported Mulliken's argument that C should be a 1_u state.

Chen et al. (246) observed the $D \leftarrow X$ (resonance-enhanced) three-photon excitation spectrum of I_2 by a continuous-wave (CW) intracavity absorption method. The spectroscopic constants for the D state obtained by these authors are ν_{00} = 40,998 cm^{-1}, ω_0' = 113 cm^{-1}, and $\omega_0' x_0$ = 0.045 cm^{-1}. Other authors (37,237) assigned the D state to the $^1\Sigma_u^+$ state with T_e = 40,679 cm^{-1} and ω_e = 104.4 cm^{-1}. On the basis of the strong intensity of the D–X emission system, Mulliken (237) suggested two possible assignments for the D state. He argued that the D–X system could be 2332 $^3\Sigma_u^-(0_u^+) \leftarrow X$, 1441 $^1\Sigma_u^+$–X, or 1342 $^3\Pi_u(0_u^+) \leftarrow X$. The last one was excluded based on the low intensity and higher frequency. The 2332 $^3\Sigma_u^-(0_u^+)$–X transition needs excitation of two electrons from the ground state, and consequently, would be less probable compared to the 1441 $^1\Sigma_u^+$ system. However, as Mulliken correctly argued, the spin–orbit mixing between 2332 $^3\Sigma_u^-(0_u^+)$ and 1441 $^1\Sigma_u^+(0_u^+)$ would be significant and thus could provide a considerable transition moment for the 2332 $^3\Sigma_u^-(0_u^+)$–X system. In any event, the $^3\Sigma_u^-(0_u^+)$ state should be observable for I_2, since less intense systems have been characterized for I_2.

The theoretical energy separations (251) in Table 3.39 tended to substantiate the assignment of the D state to a mixture of $^3\Sigma_u^-(0_u^+)$ and $^1\Sigma_u^+(0_u^+)$. If this assignment is correct, then the intense F state observed in the F–X system (experimental T_e of 41,000 cm^{-1}) could be assigned to 1441 $^1\Sigma_u^+$. Furthermore, theoretical calculations of Li and the author (251) placed the 2332 $^1\Sigma_u^+$ state, which was thought of as a candidate for the experimental F state ($T_e \sim 41,000$ cm^{-1}) 65,000 cm^{-1} above the ground state. Theoretical calculations of Li and the author (251) were most consistent with the assignment of D to a mixture of $^3\Sigma_u^-(0_u^+)$ and $^1\Sigma_u^+(0_u^+)$ and the assignment of F to 1441 $^1\Sigma_u^+(0_u^+)$. The theoretical energy separation of the F state is higher than the experimental value as expected.

The Li–Balasubramanian (251) theoretical R_e and ω_e values of the 1441 $^1\Sigma_u^+$ state are respectively 3.55 Å and 102 cm^{-1}, in very good agreement with the experimental R_e and ω_e values of the F state (3.61 Å, 96 cm^{-1}); this agreement provides further confidence in the assignment of the F state to 1441 $^1\Sigma_u^+$.

The electronic states labeled a and a' observed from the diffuse and predissociation spectra (238) were assigned before to 2341 $^3\Pi_g(1_g)$. The a' state was assigned to a 0_g^+ state, although the exact principal configuration contributing to this state was not known. As seen from Table 3.39, the theoretical calculations confirm the assignment of a to 2341 $^3\Pi_g(1_g)$ and suggest that the a' state would be 2422 $^3\Sigma_g^-(0_g^+)$. The a state is responsible for the predissociation of B–X systems. As seen from Figure 3.17, both the $^3\Sigma_g^-(1_g)$ and $^3\Sigma_g^-(0_g^+)$ curves cross the 0_u^+ state. However, $^3\Pi_g(1_g)$ is lower than $^3\Sigma_g^-(0_g^+)$ (Table 3.39) and thus would cross with the $B(0_u^+)$ state at a shorter distance compared to $^3\Sigma_g^-(0_g^+)$. Thus the earlier assignment of the state causing predissociation to $a(1_g)$ rather than $a'(0_g^+)$ is in agreement with the results of the theoretical calculations of Li and the author (251).

The presence of the $F'(0_u^+)$ state has been confirmed by King and McLean (247) with the three-photon absorption spectrum of I_2. Spectroscopic constants reported in Reference 37 for this state are $T_e = 42{,}230$ cm^{-1} and $\omega_e = 93.4$ cm^{-1}. The calculated data in Table 3.39 are most consistent with the assignment of the F' state to $^3\Pi_u$(II)(0_u^+). The theoretical ω_e value is in excellent agreement with the experimental results.

Brand et al. (256) studied the resonant two-photon $E(0_g^+) \leftarrow B(0_u^+) \leftarrow X(0_g^+)$ transition of I_2 vapor by polarization spectroscopy. Huber and Herzberg (37) assigned the E state to the $^3\Pi_g(0_g^+)$ state with $T_e = 41411.4$ cm^{-1}, $\omega_e = 101.59$ cm^{-1}, and $R_e = 3.65$ Å. Theoretical calculations (251) confirmed this assignment. The $E(0_g^+)$ state is thus a 1432 $^3\Pi_g(0_g^+)$ state. Viswanathan and Tellinghuisen (257) examined an emission system of I_2 in Ar at 2830–2890 Å and interpreted this system as a charge-transfer transition originating from an ion-pair state near $T_e = 47{,}000$ cm^{-1} and $\omega_e = 105.7$ cm^{-1} and terminating in a weakly bound state that dissociates into two ground-state atoms. The transition is tentatively assigned to the $0_u^- \rightarrow 2431\ 0_u^-(^3\Pi)$ transition. Table 3.39 shows that the 0_g^- state should be the 1432 $^3\Pi_g$ state of the theoretical calculations. The theoretical calculations predicted that the 2431 $0_u^-(^3\Pi_u)$ state is a repulsive state (251). This prediction is consistent with a somewhat repulsive $^3\Pi_u(0_u^-)$, constructed from experimental results with a weak vibrational structure in the long-range part of the well.

Kawasaki et al. (258) studied the two- and three-photon absorption spectra of I_2 in the gas phase. The emission band at ~340 nm is attributed to the 1432 $^3\Pi_g(2_g) \rightarrow 2431\ ^3\Pi_u(2_u)$ transition, and the 385-nm band is attributed to the $^3\Pi_g(1_g) \rightarrow {}^3\Pi_u(1_u)$ transition. Huber and Herzberg (37) assigned the G state to the $^3\Pi_g(2_g)$ state with T_e = 42,300 cm^{-1}, although later the label D' was given instead of G. Theoretical calculations of Li and the author (251) supported these assignments. Tellinghuisen and co-workers (238, 259) studied the emission spectrum $D'(2_g) \rightarrow 2332\ ^3\Delta_u(2_u)$ and the detailed vibrational system for the $D'(2_g) \rightarrow A'[^3\Pi_u(2_u)]$ transition. The experimental spectroscopic constants of the D' state are T_e = 40,388.24 cm^{-1}, D_e = 31,794 cm^{-1}, R_e = 3.61 Å, and ω_e = 103.953 cm^{-1}. As seen from Table 3.39, theoretical calculations confirm the assignment of the D' (called G earlier) state.

The emission bands in the 3460–3015-Å region in the presence of foreign gases were assigned to the $H \rightarrow B$ transition (37). The H state, later denoted as f', has the constants T_e = 46,063 cm^{-1} and ω_e = 103.7 cm^{-1}. The weak emission bands (2785–2731 Å) were found and attributed to the $I \rightarrow B$ transition (37). The spectroscopic constants of the I state are T_e = 51,973 cm^{-1} and ω_e = 112.4 cm^{-1}. These assignments are both based on the assumption that the lower state is the B state, but it is not entirely certain whether the lower state has been correctly identified as the B state (37). As seen from Table 3.39, there are two bound states, $^1\Sigma_g^+$(II) and $^1\Pi_g$(II), whose T_e values are both around 50,000 cm^{-1}. Thus the assignments of the f' state as the $^1\Sigma_g^+$(I) state and the I state as the $^1\Sigma_g$(II) state seem reasonable.

The spectroscopic properties of the low-lying states if I_2^+ can be discussed in terms of ionization of the highest occupied orbitals of the $^1\Sigma_g^+$ ground state of I_2. On ionizing the $1\sigma_g$ orbital, one would obtain the lowest state of I_2^+ as the $^2\Pi_g$ state arising from the $1\sigma_g^2 1\sigma_u^2 2\sigma_g^2 1\pi_u^4 1\pi_g^3$ configuration. There are other possible low-lying states arising from the configurations $1\sigma_g^2 1\sigma_u^2 2\sigma_g^2 1\pi_u^4 1\pi_g^2 2\sigma_u^1$, $1\sigma_g^2 1\sigma_u^2 2\sigma_g^2 1\pi_u^3 1\pi_g^4$, and $1\sigma_g^2 1\sigma_u^2 2\sigma_g^1 1\pi_u^4 1\pi_g^4$.

Li and the author (251) carried out CASSCF/FOCI/(SOCI)/RCI calculations on 13 electronic states (Table 3.40) of I_2^+. It can be seen from Table 3.40 that the ground state of I_2^+ is the $^2\Pi_g(\frac{3}{2})_g$ state arising primarily from the $1\sigma_g^2 1\sigma_u^2 2\sigma_g^2 1\pi_u^4 1\pi_g^3$ configuration. The first excited state is $^2\Pi_g(\frac{1}{2})_g$ from the same configuration.

The photoelectron spectrum of I_2 (260, 261) reveals the IP of I_2 is 9.311 eV for the $^1\Sigma_g^+(I_2) \rightarrow {}^2\Pi_g(\frac{3}{2})_g(I_2^+)$ ionization, and IP = 9.953 eV for the $^1\Sigma_g^+(I_2) \rightarrow {}^2\Pi_g(\frac{1}{2})_g(I_2^+)$. The theoretical adiabatic ionization potentials by the SOCI/RCI methods are IP = 8.62 eV [to $^2\Pi_g(\frac{3}{2})_g$] and IP = 9.38 eV [to $^2\Pi_g(\frac{1}{2})_g$].

The experimental spectroscopic constants (37, 260) are $D_0^\circ = 2.683$ eV and $\omega_e = 240$ cm^{-1} for the $^2\Pi_g(\frac{1}{2})_g$ state. For the $^2\Pi_g(\frac{3}{2})_g$ state, the theoretical constants obtained by SOCI/RCI are $D_e = 2.06$ eV and $\omega_e = 217$ cm^{-1}; and for the $^2\Pi_g(\frac{1}{2})_g$ state, $T_e = 5979$ cm^{-1} and $\omega_e = 208$ cm^{-1}. As seen from Table 3.40, the $^2\Pi_g(\frac{3}{2})_g$ and $^2\Pi_g(\frac{1}{2})_g$ states are quite pure. Hund's rule should apply in this case; specifically, the energy level of the $^2\Pi_g(\frac{3}{2})_g$ state should be lower than that of the $^2\Pi_g(\frac{1}{2})_g$ state. No experimental R_e values have been reported for the electronic states I_2^+. Since the $(4s4p2d)$ basis-set yields an R_e value about 0.06 Å longer, Li and the author (251) predicted that the experimental R_e values of the $^2\Pi_g(\frac{3}{2}, \frac{1}{2})$ states should both be around 2.63 Å.

The A state of I_2^+ has been assigned to the $^2\Pi_u(\frac{1}{2}, \frac{3}{2})$ states by photoelectron spectra (37, 261, 262) with $T_e = 12{,}420$ cm^{-1}[$^2\Pi_u(\frac{3}{2})_u$] and $T_e = 18{,}950$ cm^{-1}[$^2\Pi_u(\frac{1}{2})_u$)]. The calculated T_e values and the vertical transition energies from the ground state to the $^2\Pi_u(\frac{3}{2})_u$ and $^2\Pi_u(\frac{1}{2})_u$ states are listed in Table 3.40. Table 3.42 shows that the $^2\Pi_u(\frac{3}{2})_u$ component is relatively pure, while the $^2\Pi_u(\frac{1}{2})_u$ state exhibits considerable contamination with the $^4\Sigma_u^-(\frac{1}{2})$ and $^2\Sigma_u^+(\frac{1}{2})$ near the equilibrium geometry (181).

The $^2\Pi_u(\frac{1}{2})$ state of I_2^+ also behaves qualitatively similar to Br_2^+ in that it also mixes with $^4\Sigma_u^-(\frac{1}{2})$, which results in a lower ω_e for the $^2\Pi_u(\frac{1}{2})$ component. The actual amount of mixing is smaller for I_2^+ compared to Br_2^+. Actually, for Br_2^+ there is an avoided crossing of $^2\Pi_u(\frac{1}{2})$ and $^4\Sigma_u^-(\frac{1}{2})$ (181). The $^2\Pi_u(\frac{1}{2})$ state of Br_2^+ is designated as the A state and was observed by Hamilton (190) in a laser spectroscopic investigation. Direct laser spectroscopic investigation of the electronic states of I_2^+ is yet to be undertaken.

An electronic state labeled B and assigned to a $^2\Sigma_g^+$ state (37, 249, 250, 261, 262) was observed with $T = 27{,}900$ cm^{-1}. Eland (261) suggested that the $B^2\Sigma_g^+$ state of I_2^+ is fully dissociated to the ground state $I^+ + I$. The calculated $^2\Sigma_g^+$ state is a weakly bound state with $T_e = 19{,}361$ cm^{-1}, $R_e = 3.62$ Å, and $\omega_e = 72$ cm^{-1}. Figure 3.19 shows that the $B^2\Sigma_g^+$ state is dissociated to the ground state $I_+ + I$. Leach (249) and McLoughlin et al. (250) predicted that the $B^2\Sigma_g^+$-state curve should have a maximum at a long distance. The theoretical potential curve of the B state obtained by Li and the author (Fig. 3.19) confirmed this prediction of a maximum near 4 Å. This maximum is due to an avoided crossing of configurations contributing to the CI wave functions (Table 3.42). This state is dominated by mainly $1\sigma_g^2 1\sigma_u^2 2\sigma_g^2 1\pi_u^3 1\pi_g^3 2\sigma_u^1$ at 3.5 Å; at long distances (>4 Å), the state is a heavy mixture of $1\sigma_g^2 1\sigma_u^2 2\sigma_g^1 1\pi_u^4 1\pi_g^2 2\sigma_u^2$ and $1\sigma_g^2 1\sigma_u^2 2\sigma_g^1 1\pi_g^4 2\sigma_u^2$ configurations (see Table 3.42). This avoided crossing resulted in the maximum in the potential-energy curve of the $B^2\sigma_g^+$ state. At very short distances there is another avoided crossing.

Table 3.42 shows the weights of the various configurations contributing to the RCI wavefunctions of the electronic states of I_2^+. As seen from Table 3.42, most of the electronic states are relatively pure. The contribution of $^4\Sigma_u^-(\frac{1}{2})$ to $^2\Pi_u(\frac{1}{2})$ is noticeable (9%) as seen from Table 3.42. The $^4\Sigma_u^-(\frac{3}{2})$ state is somewhat purer at its equilibrium geometry.

3.3.11 Tl_2 and Tl_2^+

There are a few experimental studies on Tl_2 concerning the determination of the dissociation energies. Drowart and Honig (263) suggested an upper limit of 0.9 eV for D_0° of Tl_2, based on failure to observe Tl_2 in Tl vapor above 900 K (264). Ginter et al. (122) observed emission bands in the 15,300–16,000-cm^{-1} region and absorption bands in the 23,000-cm^{-1} region. It is not clear whether these bands arise from Tl_2 or another cluster of thallium atoms. The related Tl_2^+ ion has been observed by Berkowitz and Walter (265), but no experimental spectroscopic constants have been reported for Tl_2. Balducci and Piacente (266) have carried out a mass-spectrometric investigation of Tl_2 and obtained a D_0° value of 0.63 eV using $\omega_e = 136$ cm^{-1} for Tl_2.

Christiansen and Pitzer (195) used a small MCSCF-spinor method for the electronic states of Tl_2, which, however, failed to yield a strongly bound ground state for Tl_2. This study also considered the potential energy curves of the $(\frac{1}{2})_g$ and $(\frac{1}{2})_u$ states of Tl_2^+.

Nine low-lying electronic states of Tl_2 were considered by Christiansen (196). The potential-energy curves obtained by him for the electronic states dissociating to the ground-state atoms and the excited $^2P_{1/2} + {}^2P_{3/2}$ limits, including spin–orbit coupling, are shown in Figure 3.20. The theoretical spectroscopic constants are shown in Table 3.43. The potential-energy curves of the $(\frac{1}{2})_g$ and $(\frac{1}{2})_u$ states of Tl_2^+ are shown in Figure 3.21. As seen from Figure 3.20, the 0_u^- ground state of Tl_2 is only weakly bound. The Tl_2^+ ion, in fact, is a bit more bound ($D_e = 0.58$ eV) than Tl_2 (Fig. 3.20) (195).

Christiansen's (196) RCI calculations failed to yield an appreciably bound 0_u^- state ($D_e = 0.16$ eV), in conflict with an experimental mass-spectral value obtained by Balducci and Piacente (266) (0.6 ± 0.15 eV). However, Christiansen recalculated the D_0° value by using a more accurate partition function derived from his molecular calculations, as 0.37 eV. The revised value is somewhat improved, yet it is in disagreement with the RCI value of 0.16 eV. Electron correlation and spin–orbit effects are expected to be large, and coupled for Tl_2 (267).

3.3.12 Pb_2

The first vibrational analysis of the observed spectrum of Pb_2 was carried out by Shawhan (268). Subsequently, the electronic spectra of Pb_2 in both the gas phase and rare-gas matrices have been obtained by several investigators (87, 197, 269–273, 275, 278, 279). There are many mass spectrometric studies of Pb_2 also (263, 274, 276). Johnson et al. (272) using the laser-induced fluorescence (LIF) technique obtained superior spectra, showed that the early analysis of Shawhan cannot be correct. Bondybey and English (87, 197) studied the lead dimer LIF spectra extensively in rare-gas matrices.

Pitzer and the author (277) employed the SCF/RCI calculations for 10 electronic states of Pb_2 and their potential-energy curves. Table 3.44 gives the molecular states of Pb_2 that dissociate to atoms with total energies up to 22,000

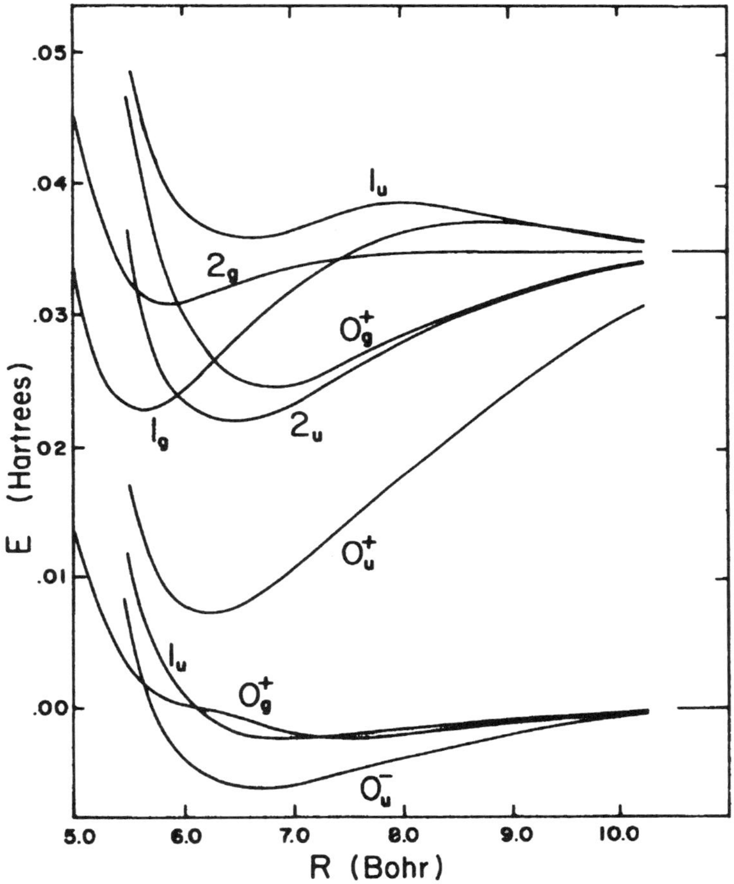

Figure 3.20 Potential-energy curves of nine low-lying states of Tl_2 including spin–orbit effects. (Reprinted from Ref. 196; copyright 1983, American Institute of Physics.)

cm^{-1} above the lowest states $^3P_0$ + $^3P_0$. Pitzer and the author (277) limited their calculations to those terms expected to be relatively low in energy (within about 15,000 cm^{-1} of the lowest 0_g^+ state).

Table 3.45 shows the theoretical spectroscopic constants of Pb_2 obtained by Pitzer and the author (277) together with available experimental data. The potential-energy curves are shown in Figure 3.22 for the g terms, and in Figure 3.23 for the u terms; terms with the lowest 0_g^+ states shown in both figures. These potential-energy curves were obtained by Pitzer and the author (277).

As seen from Figure 3.22, the spin–orbit (SO) greatly lowers the 0_g^+ component state of $^3\Sigma_g^-(\sigma_g^2\pi_u^2)$ near the potential minimum. At shorter distances this 0_g^+ state undergoes an avoided crossing with the $^1\Sigma_g^+(\pi_u^4)$ state (also 0_g^+), yielding a marked shoulder in the repulsive side of the curve. The other component of $^3\Sigma_g^-$, the 1_g term, is lowered less by the SO effect and has a normal curve shape.

Table 3.43 Spectroscopic Constants for the Low-Lying States of Tl_2[a]

State	T_e, cm^{-1}	R_e, Å	ω_e, cm^{-1}
$X0_u^-$	0	3.54	39
1_u(I)	814	3.71	30
0_g^+(I)	860	3.97	25
0_u^+	2,900	3.30	85
2_u	6,200	3.41	54
1_g	6,370	3.00	87
0_g^+(II)	6,780	3.61	54
2_g	8,130	3.14	61
1_u(II)	9,280	3.49	39

[a]All values are theoretical constants from Reference 196.

Likewise, the 2_g curve ($^1\Delta_g$ in type *a*) is quite normal. The $^1\Sigma_g^+$ state (without SO) has a second minimum at longer distances; in this region it is primarily $\sigma_g^2\pi_u^2$ rather than π_u^4.

The $^3\Pi_u$ state splits into 2_u, 1_u, 0_u^-, and 0_u^+ components in increasing energy near their minima. But the 2_u term dissociates to higher-energy atoms than 1_u and thus the curves cross near 3.3 Å. The energy differences in these *u* terms can be understood best by considering first the π_u^3 component, which yields a lower $^2\Pi_{3/2}$ term if a high-energy $\pi_{3/2}$ spinor is vacant. If a lower-energy $\pi_{1/2}$ spinor is vacant, a higher-energy $^2\Pi_{1/2}$ term results. Then, when the spin of the σ_g electron is coupled to the π_u^3 group, there is a smaller splitting of $^2\Pi_{3/2}$ to 2_u and 1_u, with 2_u lowest, in agreement with Hund's third rule. From the $^2\Pi_{1/2}$

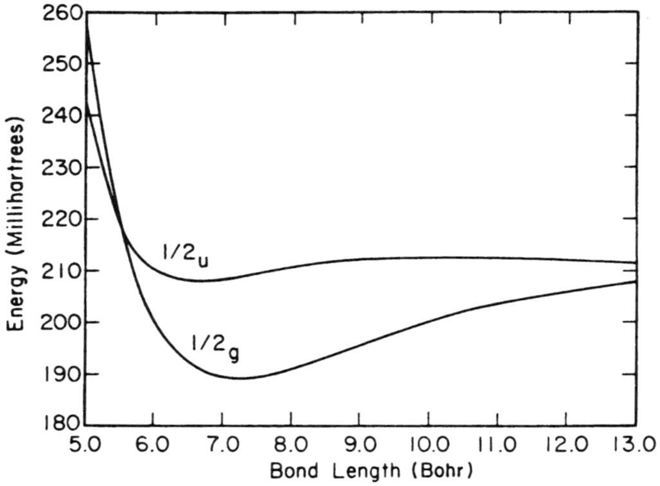

Figure 3.21 Potential-energy curves of two low-lying states of Tl_2^+. (Reprinted from Ref. 195; copyright 1981, American Institute of Physics.)

Table 3.44 Molecular States of Pb_2 Related to Atoms in Several Low-Energy States and Their Energies[a]

Dissociation	Limit	Molecular States
$^3P_0 + {}^3P_0$	0.0	0_g^+
$^3P_0 + {}^3P_1$	7,819.4	$0_g^-, 0_u^-, 1_g, 1_u$
$^3P_0 + {}^3P_2$	10,650.5	$0_g^+, 0_u^+, 1_g, 1_u, 2_g, 2_u$
$^3P_1 + {}^3P_1$	15,638.7	$0_g^+(2), 0_u^-, 1_g, 1_u, 2_g$
$^3P_1 + {}^3P_2$	18,469.8	$0_g^+, 0_u^+, 0_g^-(2), 0_u^-(2), 1_g(3), 1_u(3), 2_g(2), 2_u(2), 3_g, 3_u$
$^3P_2 + {}^3P_2$	21,300.9	$0_g^+(3), 0_u^-(2), 1_g(2), 1_u(2), 2_g(2), 2_u, 3_g, 3_u, 4_g$
$^3P_0 + {}^1P_2$	21,457.9	$0_g^+, 0_u^+, 1_g, 1_u, 2_g, 2_u$

[a]The number of states of a given symmetry is given in parentheses. The energies of the dissociation atoms are in cm^{-1} and are from Reference 134.

term for π_u^3 there arises the 0_u^- and 0_u^+ terms and the second 1_u term, which is $^1\Pi_u$ in type *a* coupling.

The ground 0_g^+ state is only about half as strongly bound as the $^3\Sigma_g^-$ state without the spin–orbit effect. Among the states arising from $^3\Pi_u$ on introduction of the spin–orbit term, the dissociation energy of 2_u is almost unchanged, whereas that for 0_u^- is considerably reduced. Some of these differences are due to different dissociation limits.

The strongest absorption (near 19,800 cm^{-1}) is from $X(0_g^+)$ to the second 0_u^+ of *F* state denoted by $F \leftarrow X$ (275). An energy curve for this state was added by Pitzer and the author (277) to Figure 3.23 at the experimental energy. Since

Table 3.45 Spectroscopic Parameters for Pb_2[a]

	T_e, cm^{-1}		ω_e, cm^{-1}		R_e, Å	
Term	Theory	Exp.	Theory	Exp.	Theory	Exp.
$X0_g^+$	0	0	103	110	2.9_7	2.93
$A1_g$	4,150	$a \leq 5,500$	124	122	2.9_4	—
2_u	6,670		119	—	2.7_0	—
1_u	7,570		119	—	2.7_1	—
2_g	10,130		105	—	3.0_3	
$B0_u^-$	12,920	12,457	106	—	2.7_5	—
1_g(II)	13,320		100	—	3.0_7	—
0_g^+(II)	13,640		74	—	3.1_8	—
$C0_u^+$	14,130	15,314	115	129	2.7_4	—
D —	—	$a + 13,433$	—	—	—	—
E —	—	$a + 14,500$	—	—	—	—
1_u(II)	15,820	—	123	—	2.7_3	—
$F0_u^+$	—	19,800	—	157	—	—

[a]The *X*, *A*, *B*, and *C* labels are those of Bondybey and English (87, 197). The theoretical constants are from Reference 277.

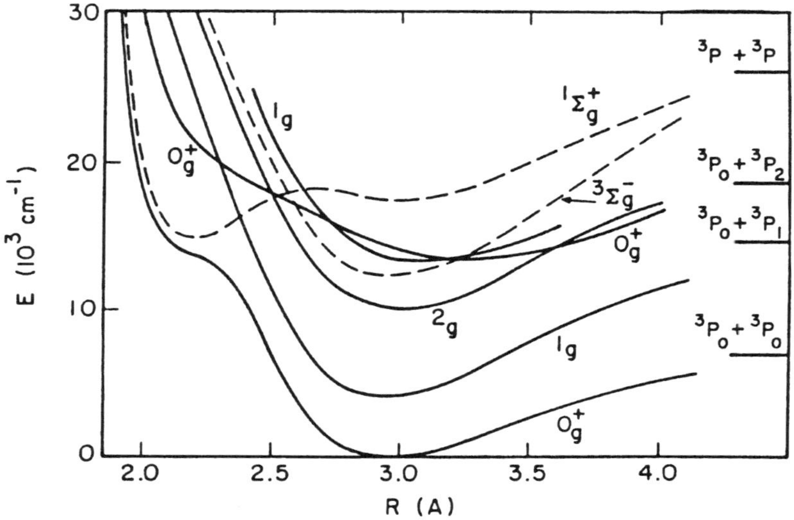

Figure 3.22 Potential-energy curves of the g states of Pb_2. (Reprinted from Ref. 277; copyright 1982, American Chemical Society.)

this state arises primarily from the configuration $\sigma_g^2\pi_u\pi_g$, it is expected to have a larger R_e than that for $\sigma_g^2\pi_u^2$; this was found to be in agreement with the red shading of the experimental bands. The transition probability from 0_g^+ to the lower 0_g^+ state was found to be experimentally much smaller, in agreement with the nature of these states. Theoretical spectroscopic properties obtained by

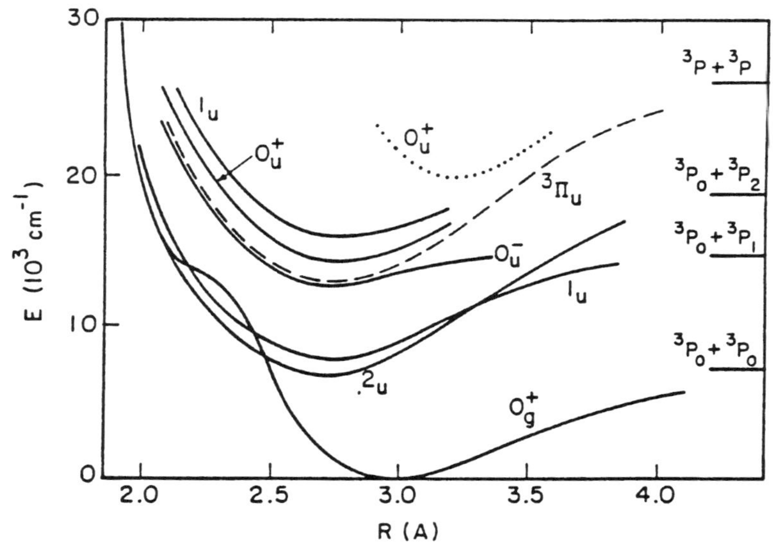

Figure 3.23 Potential-energy curves for the u states of Pb_2. (Reprinted from Ref. 277; copyright 1982, American Chemical Society.)

Pitzer and the author for this state agreed reasonably well with those measured, in particular the large anharmonicity and low dissociation energy. Also the theoretical R_e value for the lowest 0_u^+ is smaller than that for 0_g^+, in agreement with the blue shading of the observed bands (87, 197).

Bondybey and English (87, 197) interpreted a set of emission bands for Pb_2 in an inert matrix as arising by internal conversion from the F state to the long-lived B state at 12,457 cm^{-1}. Although this state is not connected to the ground state by a dipole-allowed transition, it was found to radiate slowly as a result of matrix distortion or higher-order effects. Bondybey and English (197) suggested the 2_u state, but its energy is much too low, as seen from Table 3.45. However, theoretical calculations of Pitzer and the author (277) indicated this to be the 0_u^- state.

Matrix spectra exhibited emission bands near 13,400 cm^{-1} connecting two new states. Since the upper D state arises by internal conversion from the F state, the lower or A state cannot be more than 6540 cm^{-1} above the ground state. Pitzer and the author argued that the most probable assignment for the A state is 1_g. The theoretical T_e of the 1_g state was found to be 4150 cm^{-1}. The 2_u state cannot be completely eliminated from consideration, since the computed (277) theoretical energy of 6670 cm^{-1} might be in error by several hundred reciprocal centimeters. The theoretical ω_e of the 1_g term was found to agree quite well with the experimental value. The D state at an energy of 17,500 cm^{-1} (or 19,800 cm^{-1} if the A state is 2_u) could not be calculated with reliable accuracy (277).

The computed D_e for the 0_g^+ ground state of Pb_2 of 0.88 eV (277) was almost in exact agreement with an experimental value of 0.86 ± 0.01 eV obtained by Gingerich et al. (276). Although this agreement was considered fortuitous by Pitzer and Balasubramanian (277), the agreement of such magnitude on a very heavy dimer such as Pb_2, for which both spin–orbit effects and relativistic effects are large, is quite good.

Sontag and Weber (278) obtained the R_e value of the ground state of Pb_2 from experimental spectra after the publication of the theoretical calculations of Pb_2 by Pitzer and the author (277). In general, R_e values obtained by RECP/RCI calculations tend to be a bit longer, and thus this agreement should be considered excellent.

The blue-green system that Bondybey and English (87) called the $F \leftarrow X$ system was studied further by Berg et al. (279) in laser-induced fluoresence. From these spectra the spectroscopic constants of the X and F [Berg et al. (279) call the F state B] states were obtained by these authors (279). The constants thus obtained were close to the values obtained by Bondybey and English.

Pitzer (203) corrected the D_0° value of Pb_2 obtained from mass-spectrometric measurements. This correction was obtained by recalculating the partition function of Pb_2. The corrected D_0° value of Pb_2 was found to be in excellent agreement with the direct theoretical D_e value.

Pacchioni (202) completed calculations on the low-lying states of Sn_2 and Pb_2 after the appearance of the paper by Pitzer and the author (277). Pacchioni (202) used the Hafner–Schwarz model potentials (63) compared to the *ab ini-*

tio RECPs derived from DHF calculations, which were used by the author and Pitzer (277). Pacchioni ignored spin–orbit effects in his computation of the potential-energy curves (PEC) of Pb_2 and Sn_2. As demonstrated by the calculations of Pitzer and the author (277), the PEC obtained for Pb_2 without spin–orbit effects have little resemblance to the actual PEC, which included spin–orbit effects. Consequently, Pacchioni obtained a D_e approximately twice the experimental value, which he subsequently corrected using a semiempirical procedure to arrive at a D_e close to the value obtained by the author and Pitzer (277). For dimers of the sixth-row atoms, spin–orbit effects are so large that calculations without inclusion of spin–orbit terms have very little to do with the real molecule.

Figure 3.24 shows the weights of various λ–*s* configurations in the Ω states of Pb_2 and Sn_2. As seen from this figure, for Pb_2 the λ–*s* populations change dramatically as a function of internuclear distances. At short distances, there is an avoided crossing of $\pi_u^4(^1\Sigma_g^+)$ with $\sigma_g^2\pi_g^2(^3\Sigma_g^-)$. This results in the shoulder of the $X(0_g^+)$ curve of Pb_2. At long distances many configurations contribute, leading to dissociation into $^3P_0 + {}^3P_0$ atoms instead of $^3P + {}^3P$.

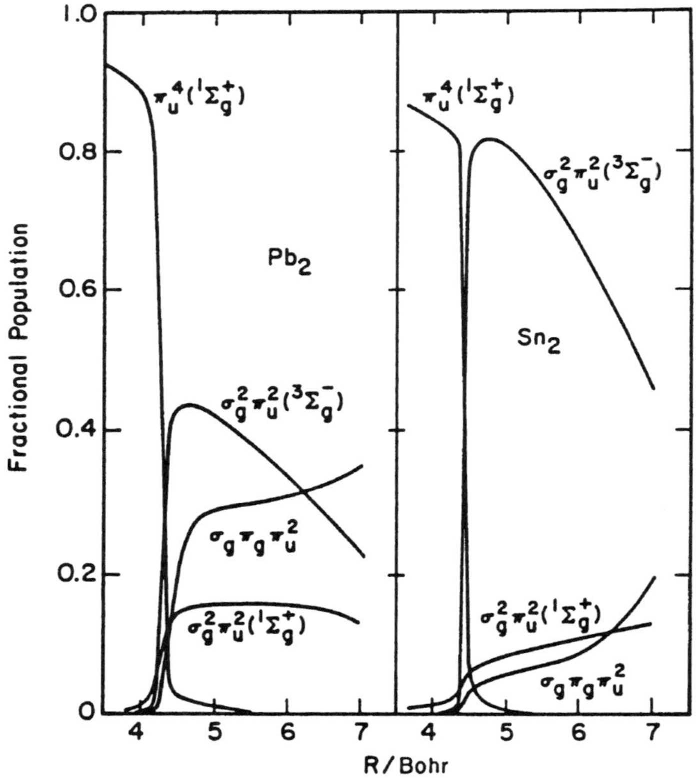

Figure 3.24 Fractional λ–*s* populations with low-lying states of Pb_2 and Sn_2. (Reprinted from Ref. 201; copyright 1983, American Institute of Physics.)

3.3.13 Bi_2

Among the sixth-row *p*-block dimers, Bi_2 is perhaps the most intensely studied system. There are several experimental spectroscopic investigations (280–291, 408–414) on Bi_2 and the related Bi_4 clusters. Polak et al. (413) obtained the negative-ion photodetachment spectra of Bi_2^-, Bi_3^-, and Bi_4^-. This study has provided new information on some of the low-lying excited electronic states that could not be observed earlier in the ordinary optical spectra. Fink et al. (414) have observed the band spectra of Bi_2 in the near-infrared region. Although weak band sequences have been observed in the 7620–7800-cm^{-1} region, no definitive assignments were made by these authors.

Among the experimentally observed systems, the $A(0_u^+)$–$X(0_g^+)$ system is the most widely studied. Gerber et al. (287, 288) made a comprehensive study of the spectra of Bi_2 and obtained the potential-energy curves of the low-lying electronic states using the Rydberg–Klein–Rees (RKR) method. The lifetime of and transition moments of the *A*–*X* system have also been measured by Ehret and Gerber (281). The transition moments deduced from the experiment for the *A*–*X* system do not appear to change for internuclear distances of 2.6–3.3 Å. The lifetimes of the vibrational levels were found to be in the range of 50–600 ns. Gerber et al. (287, 288) had earlier suggested that there exists a new X' state below the previously known ground state of Bi_2. However, a subsequent laser-induced fluorescence (LIF) study of the matrix-isolated Bi_2 by Bondybey and English (290) revealed that this lower state is really due to Bi_4, thus establishing that $X(0_g^+)$ is the ground state of Bi_2. Fabre et al. (291) as well as Effantin et al. (409) observed five electronic transitions in the infrared–visible region. Another experimental study on Bi_2 is due to Briedhor et al. (416).

There have been four independent theoretical studies on the electronic states of Bi_2. The earliest study was due to Christiansen (292), who carried out generalized valence bond (GVB) calculations followed by RCI calculations in conjunction with a double-ζ basis set. A subsequent study by Wang et al. (412) used a more extensive technique to treat the electron correlation and spin–orbit effects through CASSCF/MRSDCI/RCI techniques that employed 15-electron RECPs for Bi in conjunction with a (4*s*4*p*5*d*) valence Gaussian basis set. This was also restricted to the ground state.

The first theoretical study on several excited electronic states of Bi_2 is due to the author and Liao (415), who employed CASSCF/FOCI/SOCI/RCI techniques to study several electronic states of Bi_2 and Bi_2^-. The SOCI method included full second-order configuration interaction. Spectroscopic constants of 26 electronic states of Bi_2 and six electronic states of Bi_2^- were obtained in that study. The potential-energy curves of the electronic states of Bi_2 dissociating into several excited-state asymptotes of the separated atoms were computed together with the electronic states arising from the ground states of the Bi atoms. The experimentally observed lifetimes and transition moments were qualitatively explained by the author and Liao on the basis of the computed potential-energy curves.

Subsequent to the study by the author and Liao (415), Das et al. (417) carried out a single-configuration SCF followed by the relativistic CI (RCI) study on the low-lying electronic states of Bi_2. The RCI treatment by these authors is based on the previous MRDCI technique, which invokes configuration selection and energy extrapolation. These authors have computed the spectroscopic constants of 23 electronic states of Bi_2. These authors have also computed the transition probabilities and the radiative lifetimes involving the low-lying electronic states. We shall consider the results of the studies by the author and Liao (415) as well as Das et al. (417) together with the known experimental data on the low-lying electronic states.

The results of the CASSCF/FOCI/RCI computations of the author and Liao (415) are shown in Table 3.46. Figures 3.25 and 3.26 show the potential-energy curves computed by the author and Liao (415) without and with spin–orbit coupling, respectively. As seen from Table 3.46, the spectroscopic constants computed using a more extended basis sets and 15-electron RECPs by Wang et al. (412) earlier are also included in that table. The vibrational constants are slightly improved if one utilized this extended basis and 15-electron RECPs. Table 3.47 shows the spectroscopic constants computed by the author and Liao (415), including spin–orbit effects together with the experimental data known on the low-lying electronic states of Bi_2. As seen from Table 3.47, although theoretical results do not have the accuracies within 0.02 Å in bond distances and 10 cm^{-1} in vibrational frequencies, the relative trend is reproduced by the theoretical study of the author and Liao. This is, however, understandable in view of the large number of electrons, electron correlation effects, and relativistic effects present in Bi_2, which collectively make this species one of the most difficult to study among the main-group dimers. In spite of the understandable differences between theoretical and experimental results, assignments of the observed spectra could be made. Detailed discussions and comparison with the observed spectra could be found in the original paper of the author and Liao. In this chapter we restrict ourselves to the salient points pertinent of the spectra.

The author and Liao (415) reassigned a state observed by Gerber and Broida (288), which they assigned to $B(0^-)$ with $T_e = 6500$ cm^{-1} with an error of 1000 cm^{-1}. Subsequently Polak et al. (413) also observed an electronic state in this region with a T_e value of 5460 cm^{-1}. Both of these states were assigned to the 1_u component of $B^3\Sigma_u^+$ state by the author and Liao (415) (see Table 3.47). The other electronic state, labeled $A'(0)$ by Gerber and Broida (288) with $T_e = 9500$ cm^{-1}, was also observed by Polak et al. in their negative ion photodetachment spectra with a T_e of 8220 cm^{-1}. This state was assigned by the author and Liao (415) to the 0_u^- component of the $B^3\Sigma_u^+$ state. The a state observed by Polak et al. (413) with T_e was assigned to the 2_u spin–orbit component of the ^3D_u state. The author and Liao have argued that this state is the same as the one observed by Teichmann and Nixon (289).

Effantin et al. (409) have used the UV lines of Ar^+ and Kr^+ lasers to excite a new system of Bi_2, which they assign to the $V(0_u^+)$–$X(0_g^+)$ system. It was

Table 3.46 Theoretical Spectroscopic Properties of Bi_2[a]

State	R_e, Å	T_e, cm^{-1}	ω_e, cm^{-1}
$^1\Sigma_g^+(0_g^+)$	2.76 (2.84)	0.0	151 (171)
$^3\Sigma_u^+(1_u)$	3.00	6,651	114
$^3\Sigma_u^+(0_u^-)$	3.03	11,780	112
$^3\Delta_u(2_u)$	2.97	12,833	133
$^5\Sigma_g^+(0_g^+)$	3.22	13,700	111
$^5\Sigma_g^+(1_g)$	3.24	14,424	96
$^3\Pi_g(0_g^+)$	3.00	18,892	99
$^3\Delta_u(3_u)$	2.93	18,905	138
$^3\Pi_g(0_g^-)$	2.90	21,062	139
$A^3\Sigma_u^-$	2.87	22,300	140
$^3\Pi_g(2_g)$	2.97	23,657	122
$^1\Sigma_g^+$(II)	3.09	22,170	137
$^1\Delta_u$	2.91	22,481	145
$^1\Pi_g$	2.92	25,948	123
$^1\Sigma_u^-$	2.89	26,308	152
$^3\Delta_u$(II)	3.22	28,786	96
$^1\Sigma_g^+$(III)	3.22	30,855	88
$^3\Pi_g$(II)	3.38	31,815	75
$^1\Pi_u$	3.16	34,107	99
$^1\Sigma_u^+$	3.05	36,208	119
$^1\Pi_g$(II)	3.33	37,931	92
$^1\Pi_u$(II)	3.34	37,995	71
$^3\Pi_g$(III)	3.34	40,179	97
$^3\Delta_u$(II)	3.32	42,575	107
$^1\Delta_u$(II)	3.29	45,004	112
$^1\Sigma_u^+$(II)	3.33	51,657	105

[a] T_e values without Ω designation in parentheses are splittings relative to $^1\Sigma_g^+$ in the absence of spin–orbit coupling. Numbers in parentheses obtained using 15 electron RECPs, (4*s*4*p*4*d*) basis set, and MRSDCI method (Ref. 412). All other numbers are from Reference 415.

observed that the *V* state radiated to another new state designated by these authors as $B'0_g^+$. Indeed, the theoretical computations of the author and Liao (415) found such a state, which corresponds to the 0_g^+ spin–orbit component of the $B^5\Sigma_g^+$ state. As seen from Table 3.47, the computed T_e value for this state supports the experimentally observed state.

Teichmann and Nixon (289) observed a state labeled *b* with a T_e of 15,246

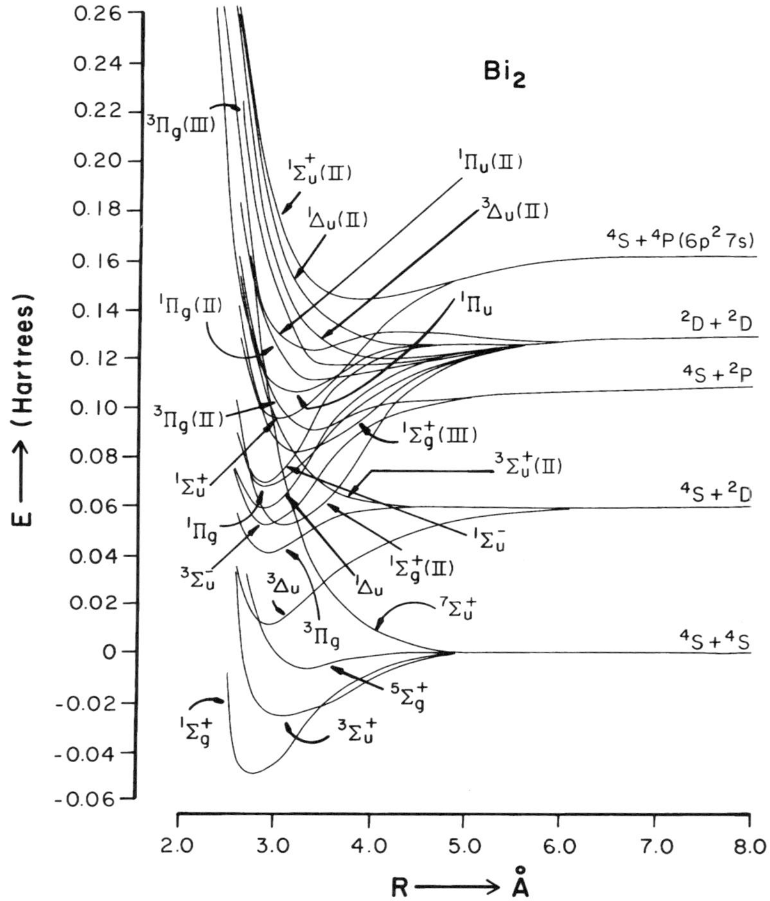

Figure 3.25 Potential-energy curves of electronic states of Bi_2 without spin–orbit coupling. (Reproduced from Ref. 415; copyright 1991, American Institute of Physics.)

cm^{-1}. On the basis of the computed energy separation, the author and Liao were inclined to argue that it could be probably assigned to the forbidden $^3\Delta_u(3_u)$–X system. However, Das et al. (417), in a subsequent study, argued that this is probably due to the $^3\Delta_u(1_u)$–X system on the basis of the closer agreement between the *b*–*a* energy separation of the computed and experimental results.

The author and Liao (415) have assigned an electronic state observed by Polak et al. with $T_e = 16{,}850\ cm^{-1}$ deduced from the negative-ion photodetachment spectra to the $^3\Pi_g(0_g^+)$ state on the basis of the computed energy separation for this state. As mentioned before, the *A*–*X* system has been characterized quite well experimentally. Although the T_e value computed by the author and Liao (415) is somewhat higher than the observed value, since the spin–orbit effect could not be included very accurately, assignment of this state is unambiguous,

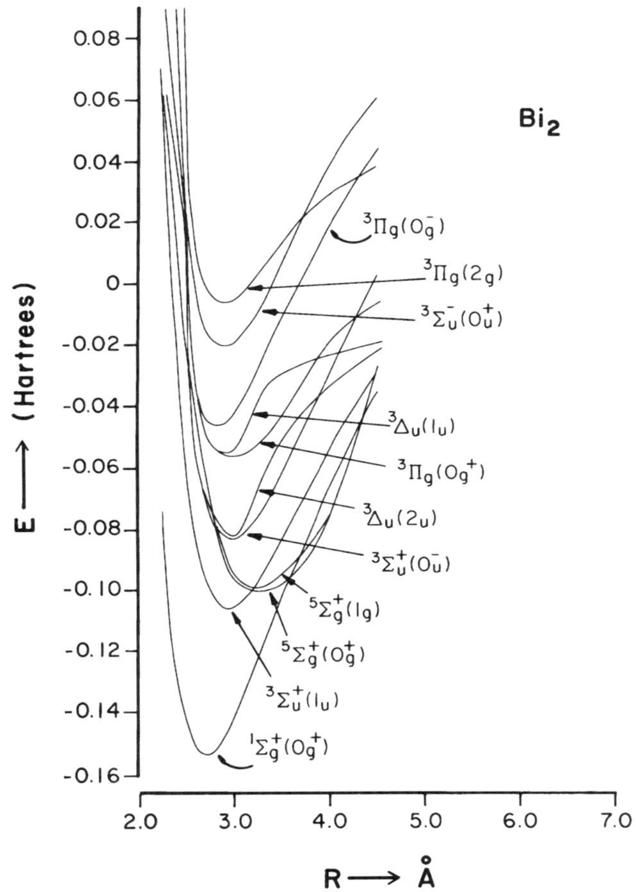

Figure 3.26 Potential-energy curves of electronic states of Bi_2 with spin–orbit coupling. (Reproduced from Ref. 415; copyright 1991, American Institute of Physics.)

since this is one of the most strongly allowed transitions, and it has been studied quite extensively.

Gerber and Broida (288) have observed a doublet system labeled the *E–B* system by these authors, where the *B* state is the $^3\Sigma_u^+(1_u)$ state. The *E* state was tentatively assigned by the author and Liao to either 0_g^- or 0_g^+ component of the $^3\Pi_g$ state. Another system labeled *N–G* was observed by Gerber et al. (287) in photoluminescence where the lower *G* state was presumed to be identical to the one observed by Reddy and Ali (283) in the *G–A* system. The *G* state is probably of *g* symmetry. The author and Liao (415) argued that this state should be probably an excited $^1\Sigma_g^+$ state with a computed T_e value of 30,855 cm^{-1}. An emission system observed by Reddy and Ali (283) called the *H–A* system was attributed to the $H[^1P_g(\mathrm{II})]$–$A(0_u^+)$ by the author and Liao (415). The computed

Table 3.47 Assignment of Experimentally Observed Systems Consistent with Theoretical Constants

	T_e, cm^{-1}		ω_e, cm^{-1}		
State	Exp.	Theory	Exp.	Theory[g]	Footnote
$X0_g^+$	0	0	173	151 (171)	—[a,i]
$B^3\Sigma_u^+$	5,460 (6,500)	6,651	134	114	—[a,h]
$A'\,^3\Sigma_u^+(0_u^-)$	8,225	11,780	132	112	—[a,g]
$a^3\Delta_u(2_u)$	9,875 (12,395)	12,833	125	133	—[a,c]
$B'\,^5\Sigma_g^+(0_g^+)$	10,826	13,700	91 (106)	111	—[a,b]
$b^3\Delta_u(3_u)$	15,246	18,905	132	138	—[c]
$^3\Pi_g(0_g^+)$	16,850	18,892	143	145	—[d]
$cB''\,^1\Delta_u(2_u)$	17,336	22,481	145	140	—[d]
$A^3\Sigma_u^-(0_u^+)$	17,739	22,300	132	140	—[e]
$E^3\Pi_g(0_g^\pm)$	(26,505)	31,815	64	75	—[e]
$G^1\Sigma_g^+$	(29,609)	30,855	107	88	—[e]
$V^5\Sigma_u^+(0_u^+)$	30,172	—	32	—	—[b]
$H^1\Pi_g(\text{II})$	(32,657)	37,931	—	92	—[e]
$I^3\Pi_g(\text{III})$	32,167	40,179	156	97	—[e]
$C^1\Sigma_u^+$	36,456	36,208	155	119	—[e]
$R^1\Sigma_u^+(\text{II})$	53,200	51,657	140	105	—[f]

[a]Experimental constants from Reference 413.
[b]The experimental ω_e of 106 cm^{-1} was obtained by Effantin et al. (409) for this state.
[c]Experimental T_e of 12,395 cm^{-1} in the Ar matrix from Reference 289.
[d]Experimental constants in the Ar matrix from Reference 289.
[e]Experimental values from Reference 37.
[f]Experimental constants from Reference 408.
[g]All theoretical values from Reference 415.
[h]An earlier experimental value of 6500 ± 1000 cm^{-1} is reported in Reference 286.
[i]The number in parentheses obtained using 15*e* RECPs and a (4*s*4*p*4*d*) basis set (Reference 412).

results were found to be in good agreement with the experiment. Damany et al. (408) observed several systems for Bi_2 that led to *C*, *F*, *Q*, *R*, *S*, and *T* states of Bi_2. Although assignment of all these states is not definitive at present, the *C* state was assigned by the author and Liao (415) to a $^1\Sigma_u^+$ state, which was computed in the 36,000-cm^{-1} region above the ground state.

The author and Liao (415) computed the D_e of Bi_2 as 1.88 eV using the CASSCF/SOCI/RCI techniques. This value is somewhat lower than an experimental value of 2.03 eV deduced by Rovner et al. (282) from the mass spectrum. Ehret and Gerber (281) have deduced a spectroscopic dissociation energy of 2.08 eV from the convergence of one of the spectral systems.

As mentioned before, Das et al. (417) use the SCF/RCI techniques using the MRDCI procedure for configuration selection and energy extrapolation. The spectroscopic constants that they have obtained together with their comparison with the results of the author and Liao (415) and the experiment are shown in Table 3.48. As seen from this table, overall agreement between the results computed by the two groups is quite good, considering the differences in the two techniques. For example, Das et al. use a single-configuration SCF to obtain orbitals for the RCI, while the author and Liao have used the full complete active space MCSCF (CASSCF) method for the CI computations. A problem that still remains with both techniques is that the computed bond lengths are too long compared to the experiment. It appears that typically the RECPs employed in both techniques yield bond lengths longer than the experiment.

Das et al. (417) computed the transition moments and the lifetimes of some of the electronic states of Bi_2. Table 3.49 shows the radiative lifetimes of some of the excited states of Bi_2 as computed by Das et al. (417). Experimental data on the radiative lifetimes are scarce at present. Since the computed bond lengths are too long compared to the experiment, it is anticipated that the computed lifetimes will also be in larger error. Ehret and Gerber (281) obtained the radiative lifetimes for the various vibrational and rotational levels of the A state, which has a T_e of 17,740 cm^{-1}. Table 3.49 shows both the partial lifetimes for a number of transitions and the total lifetimes for seven electronic states of Bi_2.

Some striking contrasts between the electronic states of Bi_2 and Sb_2 are in the lifetimes of the excited electronic states. The spin–orbit effect plays an important role in enhancing the transition moments of some of the electronic transitions that are nonrelativistically forbidden. For example, consider the ${}^3\Sigma_u^+$–$X^1\Sigma_g^+$ system. This transition is nonrelativistically spin-forbidden, and is thus expected to have small transition moments for the lighter group (V) dimers. For Bi_2, as evidenced from Table 3.49, the $B1_u$ state has a radiative lifetime of 2.29 ms. Since the ground state of Bi_2 acquires considerable ${}^3\Pi_g$ character, and ${}^3\Pi_g$–${}^3\Sigma_u^+$ transition is allowed, the transition moment for the ${}^3\Sigma_u^+(1_u)$–$X0_g^+$ becomes nonzero, thereby decreasing the lifetime of the $B1_u$ state. On the other hand, the lifetime of the $A'O_u^-$ state is longer, since the transitions to lower states are forbidden after the inclusion of spin–orbit interaction. Nonrelativistically, both of these states would have the same lifetimes. On the basis of the lifetime of the ${}^3\Sigma_u(1_u)$ state, Das et al. (417) suggested that the assignment of the b state observed by Teichmann and Nixon (289) should be due to this state rather than the 3_u component suggested by the author and Liao (415). However, Das et al. (417) have suggested that the a state could very well by due to 3_u, since the A state is not computed to have a strong transition moment, and thus Das et al. (417) considered the possibility of A being the 1_u component of ${}^3\Delta_u$. At this time these assignments still remain very tentative.

Das et al. (417) find only weak transitions from $A0_u^+$ to the $X0_g^+$ state. However, Ehret and Gerber (281) find that the radiative lifetimes of the $A0_u^+$ state vary from 0.595 μs ($v' = 1$) to 40 ns ($v' = 34$). These values are in conflict with the results of Das et al., who compute a radiative lifetime of 72 μs for the $v' = 0$ state of the $A0_u^+$ state. Das et al. consider this disagreement to be somewhat unusual, since typically their computed lifetimes differ by a factor of 2–3 compared to experiment, while a factor of 120 discrepancy is found

Table 3.48 Calculated and Experimental Spectroscopic Constants of Some Low-Lying Ω States of Bi_2 (Transition Energies T_e, Bond Lengths r_e, and Vibrational Frequencies ω_e)

	T_e, cm^{-1}			r_e, Å			ω_e, cm^{-1}		
	Calculated			Calculated			Calculated		
State	Ref. 417	Ref. 415	Exp.[a]	Ref. 417	Ref. 415	Exp.[a]	Ref. 417	Ref. 415	Exp.[a]
$X0_g^+$	0	0	0	2.768	2.76	2.660	153	151	173
$B1_u$	5,139	6,651	5,480	2.961	3.00	2.863	113	114	134
	—	—	5,510[b]	—	—	2.860[b]	—	—	134[b]
$A'0_u^-$	8,896	11,780	8,245	2.988	3.03	2.860	114	112	132
$B'0_g^+$	9,867	13,700	10,826	3.231	3.22	3.110	92	111	106
1_g	11,428	14,424	—	3.296	3.24	—	75	96	—
$a2_u$	11,507	12,833	9,900	2.904	2.97	2.905	127	133	125
$b3_u$	14,507	18,905	15,246[c]	2.945	2.93	—	121	138	132[c]
1_u(II)	17,278	—	—	2.903	—	—	158	—	—
0_g^-	19,674	21,062	—	2.903	2.90	—	108	139	—
$A0_u^+$	20,239	22,300	17,741[d]	2.966	2.87	2.863	122	140	132
1_g(II)	22,363	—	—	2.904	—	—	97	—	—
0_u^+(II)	22,962	—	—	2.902	—	—	136	—	—
0_g^+(III)	23,055	18,892	—	3.073	3.00	—	102	99	—
1_u(III)	23,160	—	—	2.993	—	—	108	—	—
3_g	23,300	—	—	3.286	—	—	58	—	—
2_g(II)	24,729	—	—	3.267	—	—	84	—	—
1_u(IV)	25,692	—	—	3.057	—	—	148	—	—
2_u(II)	26,200	22,481	—	4.127	—	—	103	140	—
0_g^+(IV)	26,734	—	—	3.013	—	—	127	—	—
2_u(III)	27,829	—	—	2.963	—	—	139	—	—
3_u(II)	28,302	—	—	3.164	—	—	80	—	—
2_g(III)	29,291	—	—	3.013	—	—	59	—	—
0_g^+(V)	30,725	—	—	3.430	—	—	105	—	—

[a]Reference 413 unless otherwise indicated.
[b]Reference 416.
[c]Reference 409.
[d]Reference 20.

Source: Reproduced from Reference 417.

Table 3.49 Radiative Lifetimes of Some of the Excited States ($v = 0$) of Bi_2[a]

	Partial Lifetimes τ_p (s) for Transitions to								Total Lifetime
State	$X0_g^+$	$B1_u$	$A'0_u^-$	$B'0_g^+$	1_g	1_u(II)	0_g^-	$A0_u^+$	τ, s
$B1_u$	2.29(−3)	—	—	—	—	—	—	—	2.29(−3)
1_u(II)	1.28(−5)	—	—	1.68(+1)	—	—	—	—	1.28(−5)
0_g^-	—	1.85(−5)	2.72(−5)	—	—	—	—	—	1.10(−5)
$A0_u^+$	7.20(−5)	—	—	6.04	8.21(+2)	—	—	—	7.20(−5)
1_g(II)	—	3.19(−5)	4.86(−5)	—	—	7.51(−4)	—	—	1.18(−5)
0_u^-(II)	—	—	—	—	—	—	1.69(+2)	—	1.69(+2)
0_g^+(III)	—	5.48(−5)	—	—	—	2.16(−3)	—	1.91(−2)	5.33(−5)

[a]Values in parentheses denote powers of 10.

Source: Reproduced from Reference 417.

for the computed lifetime of the $A0_u^+$ state. It is not clear whether this discrepancy is due to defects in theory such as longer computed bond lengths or lack of inclusion of electron correlation effects, to difficulties in the experimental measurements. Das et al. have suggested that self-quenching in the experimental work could become important or a different assignment could be feasible for the A state. Then one must wonder as to which experimental state the 0_u^+ state should correspond, since this state should definitely be observable.

3.4 ELECTRONIC STRUCTURE OF MIXED CLUSTERS

Several theoretical and experimental studies have been devoted to the electronic structure of mixed main-group clusters. Of special importance among these are the Ga_xAs_y, In_xP_y, semiconductors that are of immense technological value as they are useful in constructing fast devices. A number of investigators have studied the properties of GaAs surfaces (293–300). A pioneering study by Smalley and co-workers (97) on supersonic beams of semiconductor clusters of formula Ga_xAs_y has revealed that the relative abundance of the larger clusters revealed a binomial distribution, while the abundance of smaller clusters deviated considerably from the binomial distribution. In particular, the diatomic clusters were found to be mostly GaAs and As_2 with very little Ga_2. Further, photoionization with an ArF excimer laser revealed even–odd alternation in the photoionization cross sections.

As the topic of lighter clusters is not the main theme of this book, we mainly provide references to this topic. The author (301) has made the first complete active space MCSCF calculations followed by first-order CI calculations of potential-energy curves and spectroscopic properties of the low-lying electronic states of GaAs and $GaAs^+$. He considered 12 electronic states of GaAs and 4 electronic states of $GaAs^+$. Valence Gaussian ($3s3p1d$) basis sets for Ga and As atoms in conjunction with RECPs that retained the outer $4s^24p^1$ and $4s^24p^3$ shells of Ga and As, respectively, were used in these calculations. After these calculations were published, Lemire et al. (303) reported the electronic spectrum of GaAs. The bond lengths calculated by the author were longer than the experimental values of Morse. This motivated the author to reinvestigate GaAs. An error was found in one of the ECP parameters, which was subsequently corrected (305). A more detailed study (404) with corrected ECP parameters included the spectroscopic constants of several electronic states of GaAs, $GaAs^-$, and $GaAs^+$. The corrected spectroscopic constants for GaAs were, however, found to be in good agreement with both experiment and all-electron calculations.

Following the theoretical calculations of the author (301), Knight and Petty (302) observed the ESR spectrum of $GaAs^+$ by laser-evaporated GaAs followed by photoionization. The experimental spectrum confirmed the $^4\Sigma^-$ ground state for $GaAs^+$ predicted by the author.

Lemire et al. (303) studied the optical spectra of jet-cooled GaAs generated by laser vaporization of a GaAs crystal. These authors observed bands in the 23,000–24,000-cm^{-1} region. All the bands except the $^3\Pi(0^+)$–$X^3\Sigma^-(0^+)$ bands were found to be predissociated above the $v = 1$ level. Basically, the results of these calculations confirmed the author's prediction (301) of a $^3\Sigma^-$ ground state for GaAs.

Meier et al. (403) published an all-electron study of the electronic states of GaAs. They also found a $^3\Sigma^-$ ground state for GaAs and a very low-lying $^3\Pi$ state (T_e ~ 0.17–0.24 eV) above the ground state using the Hartree–Fock/MRDCI method. Most of the discrepancies between the author's earlier results (301) and Meier et al.'s all-electron results (403) are due to an error in one of the ECP parameters mentioned before. The bond lengths listed in Table 3.49 are within 0.04–0.05 Å of the all-electron results for these states. The revised ω_e values are also in much better agreement with the all-electron results. It is, however, worth noting that the $^1\Delta$ state has a larger ω_e if $^1A_1$ CASSCF orbitals are used in the calculations, while the $^1A_2$ orbitals yield a more reasonable ω_e. Hence the larger ω_e for the $^1\Delta$ state obtained before (301) is due to the change of $^1A_1$ orbitals, due to avoided crossing of the $1\sigma^2 2\sigma^2 1\pi^4$ and $1\sigma^2 2\sigma^2 3\sigma^2 \pi^2$ configurations in the CASSCF calculations. The results of the author and Meier et al. have been compared and discussed in considerable detail elsewhere. Interested readers are referred to these references.

A theoretical study in the author's laboratory (418) considered the electronic states of the Ga_3As_2 and Ga_2As_3 clusters using the CASSCF/MRSDCI techniques in conjunction with the quasi-Newton–Raphson geometry optimization at the CASSCF level. Similar studies are being considered on Ga_xP_y and In_xP_y clusters as well. For the Ga_3As_2 cluster, we found two nearly degenerate Jahn–Teller distorted electronic states, while Ga_2As_3 exhibits D_{3h} structure. While the geometries computed by Smalley and co-workers (419) using the density function method agrees quite well for the Ga_2As_3, and one of the minima of the Ga_3As_2 clusters, substantial differences were found in the equilibrium geometry of the other Jahn–Teller distorted minimum of Ga_3As_2. These theoretical studies are in accord with the observed ESR spectra for Ga_2As_3 (420), which predict a double ground state.

Figure 3.27 shows the geometries considered for the Ga_3As_2 and Ga_2As_3 clusters. Table 3.50 shows the geometries and energy separations of the two nearly degenerate Jahn–Teller distorted minima for the Ga_3As_2 cluster, while Table 3.51 shows the energy separations of the undistorted D_{3h} minima. Table 3.52 shows the geometries and energy separations of the electronic states of the Ga_2As_3 cluster.

Among other studied heteronuclear diatomics, from the standpoint of relativistic effects and spectroscopic constants, are rare-gas/halide ion systems and interhalogen compounds. The interest in rare-gas/halogen (RgX) neutral and ionic compounds arises partly because they are potential candidates for chemical lasers (189, 306–331). The generation and deexcitation of neutral RgX

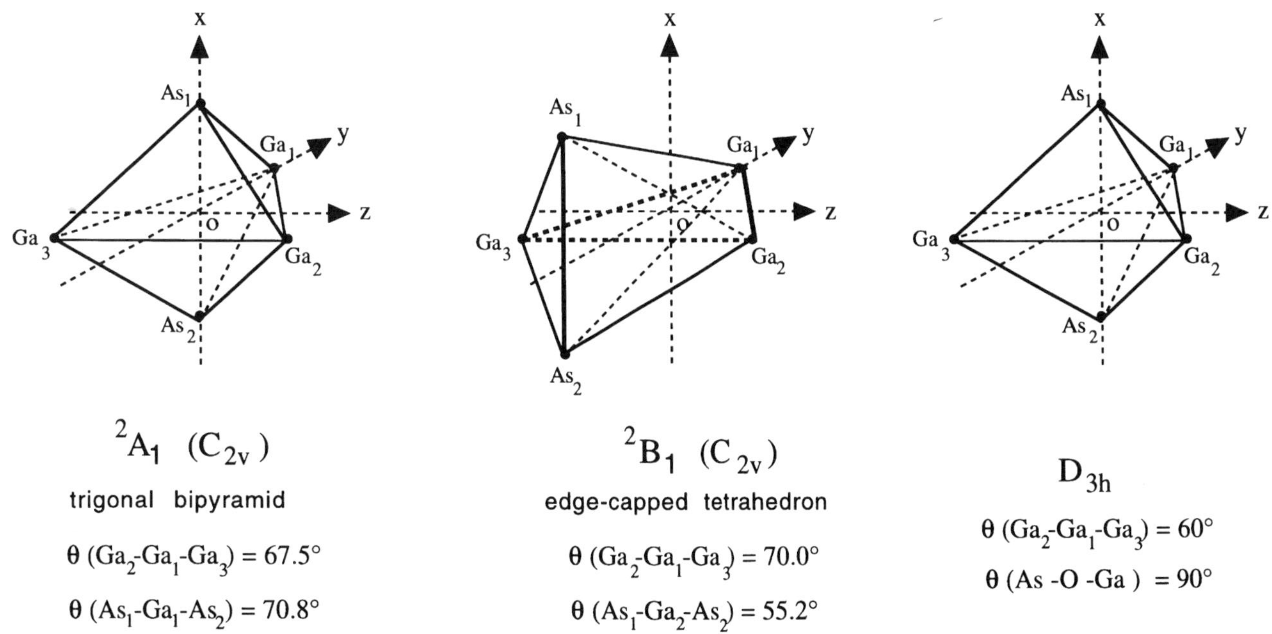

Figure 3.27 Geometrics considered for Ga_2As_3 and Ga_3As_2. (Reproduced from Ref. 417; copyright 1995, Elsevier Science Publishers, B.V.)

Table 3.50 Geometries and Energy Separations for the Electronic States of Ga_3As_2 with C_{2v} Structure

	$^2A_1$		$^2B_1$	
State	Ref. 418	Ref. 419[b]	Ref. 418	Ref. 419[b]
Ga_2–Ga_1–Ga_3 (deg)[a]	67.5	—	70.0	94.9
As_1–Ga_1–As_2 (deg)	70.8	67.9	52.6	—
As_1–Ga_2–As_2 (deg)	61.4	—	55.2	—
Ga_1–Ga_2 (Å)	3.702	3.72	3.986	2.62
Ga_2–Ga_3 (Å)	4.114	—	4.572	—
Ga_1–As_1 (Å)	2.401	2.41	2.829	—
Ga_2–As_1 (Å)	2.725	2.59	2.707	2.28
As_1–As_2 (Å)	2.782	2.70	2.507	2.24
E(CAS) (eV)	0.00	—	0.03	—
E(MRSDCI) (eV)	0.00	—	0.02	—
E(MRSDCI+Q) (eV)	0.00	—	0.005	—

[a]The labels of all atoms are defined in Figure 3.28.
[b]Energy separation between the two states found in Reference 419 is 0.01 eV at the density function level of theory.

species have also been extensively studied. Rare-gas oxides have also been the topics of many investigations, since they are potential candidates for optical pumping reactions.

The author and co-workers (326) have carried out SCF/RCI calculations on nine low-lying Ω state of $KrBr^+$ including spin–orbit effects. Some λ–*s* electronic states without spin–orbit effects were also studied by the same method. A valence double-ζ + *d* polarization STO basis set was employed together with RECPs for Kr and Br^+. The author and co-workers (326) investigated all low-lying states dissociating below 11,500 cm^{-1}. Experimental and theoretical

Table 3.51 Geometries and Energy Separations for the Electronic States of Ga_3As_2 with D_{3h} Structure

State		CASSCF			MRSDCI		
C_{2v}	D_{3h}	Ga–Ga, Å	Ga–As, Å	E, eV	Ga–Ga, Å	Ga–As, Å	E, eV[a]
$^2A_1$, $^2B_2$	$^2E'$	3.828	2.606	0.35	3.812	2.591	0.26(0.19)
$^2B_1$, $^2A_2$	$^2E''$	4.228	2.746	0.16	4.184	2.724	0.19(0.21)
$^4B_2$	$^4A_2'$	2.830	2.530	1.12	2.828	2.517	1.00(1.02)
$^4A_2$	$^4A_1''$	3.310	2.569	1.31	3.274	2.538	1.10(1.02)
$^4A_1$	$^4E'$	3.691	2.737	2.25	3.637	2.685	1.91(1.89)
$^4B_1$	$^4E''$	3.149	2.685	2.50	3.093	2.646	2.10(2.01)

[a]The values in parentheses are the Davidson corrected energies.
Source: Reproduced from Reference 418.

Table 3.52 Geometries and Energy Separations for the Electronic States of Ga_2As_3 with D_{3h} Structure

State		CASSCF			MRSDCI		
C_{2v}	D_{3h}	As–As, Å	Ga–As, Å	E, eV	As–As, Å	Ga–As, Å	E, eV[a]
$^2B_1$	$^2A_2''$ [b]	2.509	2.591	0.00	2.595	2.594	0.00(0.00)
$^2A_2$	$^2A_1''$	2.481	2.884	0.29	2.481	2.876	0.87(1.01)
$^2B_2$	$^2A_2'$	2.542	2.847	1.65	2.542	2.851	2.27(2.44)
$^2A_1$	$^2A_1'$	2.605	2.600	2.76	2.581	2.607	2.87(2.82)

[a]The values in parentheses are the Davidson corrected energies.
[b]The density function theory yielded $^2A_2''$ ground state with As–As and Ga–As distances of 2.62 and 2.65 Å, respectively, from Reference 419.
Source: Reproduced from Reference 418.

investigations of Ar–Cl^+ collisions (328, 329) as well as Ne–F^+ energy transfers (327) have also been carried out. Theoretical calculations on related species can be found in References 323, 329, and 331.

Interhalogen diatomics have been the topics of numerous studies for many years (187, 188, 332–354). The IF, ICl, and IBr species have been some of the most studied species in this group. Brand and Hoy (38) reviewed the spectroscopic properties and ion-pair states of homonuclear and heteronuclear halogens. The spectroscopic properties of the ICl diatomic have been obtained by a number of techniques such as absorption spectra (332, 333), emission spectra (345, 349) vacuum UV spectra (341), time-resolved fluorescence (337), chemiluminescence of the I + Cl reaction (336), two-step excited fluorescence (340), and three-photon resonance (348). The photoelectron spectra of ICl have been recorded, and the spectra reveal the existence of four low-lying states of ICl^+ (350). The author (353) carried out RCI calculations of the spectroscopic properties of five low-lying states of ICl [0^+(I),2,1,0^-,0^+(II)], and spectral assignments of various transitions among these states were accomplished or confirmed. The continuous absorption spectra as well as diffuse emission spectra of ICl in the region below 45,000 cm^{-1} were interpreted, and assigned to appropriate electronic transitions. In a subsequent work, the author (354) extended the calculations of ICl to the 0^-(II), 0^+(III), and 1(II) states as well as the corresponding states without spin–orbit interaction. He also carried out RCI calculations on the $\frac{3}{2}$, $\frac{1}{2}$, $\frac{3}{2}$(II), $\frac{1}{2}$(II), $\frac{5}{2}$, $^2\Pi$, $^2\Sigma^+$, and $^4\Pi$ states of ICl^+ (354).

3.5 SPECTROSCOPIC PROPERTIES AND POTENTIAL-ENERGY SURFACES OF TRIMERS

3.5.1 Ga_3

There are no experimental spectroscopic studies on Ga_3 except the one done by Smalley and co-workers (97), who detected Ga_3 in a laser-evaporated GaAs supersonic jet beam of composition Ga_xAs_y. The author and Feng (355) carried out complete active space MCSCF followed by CI calculations that included up to 32,000 configurations. They employed the same RECPs and basis sets used in the earlier calculations on Ga_2 (133). Those calculations were carried out on several low-lying electronic states of Ga_3 and the linear $^1A_1$ closed-shell state of Ga_3^+.

Isoelectronic Al_3 was investigated by Basch (356) theoretically, by Howard et al. (357) experimentally using the ESR spectroscopic method, and subsequently by Fu et al. (358), who obtained the spectra of jet-cooled Al_3 and found a $^2A_1$ ground state (near-equilateral-triangle geometry) with a low-lying $^4A_2$ state only 5 kcal/mol above $^2A_1$. To the contrary, the ESR spectra of Al_3 in a rare-gas matrix favored a quartet ground state. A discrete band system in the 5200–6100-Å region was observed together with an absorption continuum that exhibited a lifetime of 24–35 μs. These authors (358) deduced $D_0(Al_2\text{–}Al) \leq$ 2.40 eV from the predissociation of the continuous absorption bands.

Table 3.53 shows the optimized equilibrium geometries of seven electronic states of Ga_3 and two electronic states of Ga_3^+ using both the CASSCF and CI method (355). Figure 3.28 shows the CI geometries of the three lowest-lying electronic states of Ga_3. As seen from Table 3.53, the ground state of Ga_3 is a $^2A_1$ state with an equilibrium geometry of a near-equilateral triangle. The apex angle reduces by 1° on inclusion of high-order correlations in the CI calculations. The bond length of the base of the triangle is about 2.63 Å at the CI level, while the equal sides are a bit shorter (2.58 Å). It is expected that the Ga–Ga bond lengths should be slightly longer than the actual bond lengths, since for Ga_2 and GaAs, the 3e RECPs of Ga gave slightly longer bond lengths. The $^2A_1$ state was found to consist of predominantly $1a_1^2 2a_1^2 3a_1 1b_2^2 1b_1^2$ configuration.

Table 3.53 Geometries and Energies of the Electronic States of Ga_3[a]

		CASSCF			POLCI		
Species	State	θ, deg	r (Å)[b]	E,[c] eV	θ, deg	r, Å[b]	E,[c] eV
Ga_3	$^2A_1$	62.4	2.51	0.0	61.2	2.58	0.0
Ga_3	$^4A_2$	76.4	2.62	0.24	72.6	2.57	0.24
Ga_3	$^2B_1$	50.6	2.92	0.27	61.2	2.60	0.28
Ga_3	$^4B_1$	54.5	2.88	0.29	56.5	2.74	0.32
Ga_3	$^2B_2$	68.8	2.54	0.67	65.5	2.58	0.69
Ga_3	$^4B_2$	58.3	3.05	0.78	61.8	2.93	1.27
Ga_3	$^4A_1$	180	2.80	1.18	180	2.79	1.38
Ga_3^+	$^1A_1$	180	3.07	4.92	180	3.00	5.99
Ga_3^+	$^3B_2$	180	2.90	5.57	180	2.78	6.36

[a]All constants are from theoretical calculations in Reference 355.
[b]The two equal sides of the isosceles triangle; θ = apex angle.
[c]Zero energy for CASSCF is −6.086778 hartree; zero energy for POLCI is −6.159135 hartree.

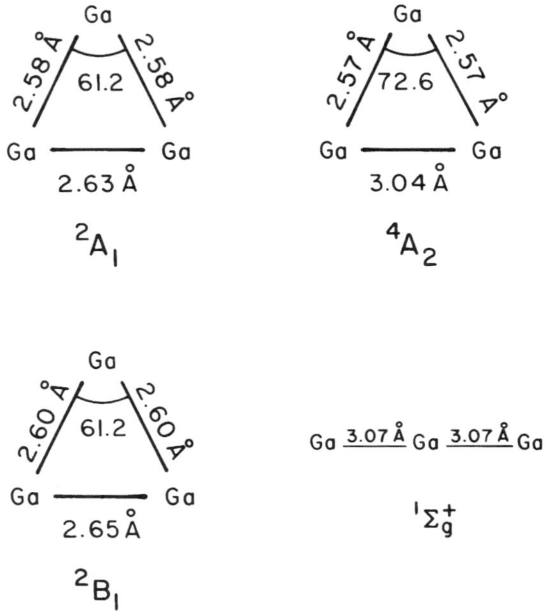

Figure 3.28 Equilibrium geometries of low-lying electronic states of Ga_3. (Reproduced from Ref. 355; copyright 1988, Elsevier Science Publishers, B.V.)

Two very low-lying excited electronic states—$^4A_2$ and $^2B_1$—were found by the author and Feng (355). The $^4A_2$ state was found to be 5.5 kcal/mol above the $^2A_1$ state, while the T_e of the $^2B_1$ state was 6.5 kcal/mol. The $^4A_2$ state arises from the $1a_1^2 2a_1^2 3a_1 1b_2^2 2b_2 1b_1$ configuration. The equilibrium geometries of the $^4A_2$ and $^2B_1$ states are isosceles triangles.

The $^4B_1$ state ($1a_1^2 2a_1^2 3a_1 4a_1 1b_2^2 1b_1$) has a very acute triangular equilibrium geometry. The two equal sides of the isosceles triangle have longer bond lengths (2.74 Å). The base of the triangle has a shorter bond length (2.59 Å), which suggests enhanced bonding along the base of the triangle. Two of the other three electronic states—$^2B_2$ and $^4B_2$—have isosceles triangular geometries, while the $^4A_1$ state has a linear equilibrium geometry.

The geometric effect of higher-order electron correlations is not substantial, as seen from Table 3.53 by comparing the CASSCF and CI geometries for most of the states except the $^2B_1$ state, for which CASSCF calculations yield a very acute triangle structure, which is corrected to a near-equilateral-triangular structure by CI. The energy separations are, in general, more sensitive to electron correlation effects.

The CASSCF atomization energy was calculated by carrying out a long-distance CASSCF calculation for the linear $^2A_1$ geometry ($R = 8.0$ Å). The atomization energy calculated in this way was found to be 2.32 eV (355). Higher-order correlation effects not included in the CASSCF can certainly increase the atomization energy. The corresponding CASSCF D_e of the Ga_2 dimer calculated earlier by the author (133) is 0.92 eV. Consequently, Ga_3 is at least 1.4 eV more stable than the dimer. The corresponding calculations on Al_3 by Basch (357) revealed that the aluminum trimer is about 1.93 eV more stable than the dimer, while the experimental results of Fu et al. (358) imply $D_0(Al_2–Al) < 2.4$ eV.

As seen from Table 3.53, there are two low-lying electronic states for the Ga_3^+ positive ion. The ground state is a closed-shell $^1A_1$ state with a linear geometry. The adiabatic ionization energy of Ga_3 is 5.99 eV. The $^1A_1$–$^3B_2$ splitting of the Ga_3^+ positive ion is 0.65 and 0.37 eV at the CASSCF and CI levels of theory, respectively. Consequently, this energy separation is influenced to a large extent by higher-order correlations. More ionization occurs at the base atoms of the triangle than the apex atom, as evidenced from the Mulliken population analysis. The adiabatic ionization potential of the Ga_3 cluster (5.99 eV) is quite comparable to the experimental ionization potential of the Ga atom (134), which is 6.0 eV. It is interesting to note that both the Ga atom and Ga_3 clusters form a stable closed-shell singlet electronic state on ionization, leading to lower adiabatic ionization energies compared to even clusters. Hence odd–even alternations are anticipated in ionization energies.

The lowest $1a_1$ orbital of Ga_3 is the totally symmetric bonding combination of the $4s$ orbitals of the three Ga atoms. The $2a_1$ orbital is also composed predominantly of the $4s$ orbitals on the three Ga atoms with the combination $-4s(1)+4s(2)+4s(3)$, where the label 1 is given for the central apex atom. Consequently, the $2a_1$ orbital is antibonding along the sides, and bonding along the base of the triangle. The $1b_2$ orbital was found to be predominantly the $4p_y(1) + 4s(2) - 4s(3)$ orbital. Hence this orbital is antibonding along the base, and bonding along the sides. Consequently, the $1b_2$ orbital is stabilized if the sides are shorter than the base of the triangle. Conversely, the $2a_1$ orbital is stabilized if the sides are longer than the base of the triangle. The $1b_1$ orbital is a symmetric combination of the $4p_x$ orbitals on the three Ga atoms perpendicular to the plane of the cluster, and thus is "π-like" in character. The $3a_1$ orbital is a bonding orbital resulting from the mixing of the $4p_z$ orbital of the central atom with the $4p_y$ and $4p_z$ orbitals of the base of the triangle. The $4s$ orbitals on the three atoms also participate in this orbital. Similarly, the $2b_2$ orbital is a bonding orbital resulting from the interaction of the $4p_y$ orbital of the central atom with the antisymmetric combination of the $4p_y$ and $4p_z$ orbitals of the base atoms with appropriate signs. The $2b_1$, $4a_1$, and $1a_2$ orbitals are antibonding orbitals.

The Mulliken population analysis reveals that the total s populations of the three metal atoms in all the electronic states are between 4.90 and 5.33. The $^2B_2$ and $^2A_1$ states have the smallest s populations, while the $^4B_2$ and $^4A_1$ states have the largest s populations. The corresponding p populations are between 3.36 and 3.75. The $^2A_1$ and $^2B_2$ states have the largest p populations, while the $^4B_2$ and $^4A_1$ states have the smallest p populations. Hence there is considerable $4s$ and $4p$ electron transfer in the electronic states of Ga_3. The d populations are about

0.3–0.35 for all the electronic states of Ga_3. Consequently, participation of the d polarization functions is quite important for all the electronic states of Ga_3. A comparison of the total gross populations of the central and side atoms of Ga_3 and $Ga_3^+(^2A_1, ^1A_1)$ revealed that most of the electron removed in the ionization process comes from the base atom, rather than the central atom. Of course, there is considerable rearrangement of this geometry from a near-equilateral-triangular to a linear structure on ionization. Comparison of the total s, p, and d populations of the ground states of Ga_3 and Ga_3^+ shows that the electron is removed from the $4p$ orbital, and that there is considerable $4p \rightarrow 4s$ electron transfer on ionization.

The electronic states of Al_3 show considerable similarities to the corresponding state of Ga_3. Basch (356) found the ground state of Al_3 to be a $^2A_1$ state with an isosceles triangular geometry (apex angle = 56.2°). The $^2B_1$ and $^4A_2$ states respectively were found to be 0.22 and 0.23 eV above the ground state. The energy separations for Al_3 are thus comparable to the corresponding electronic states of Ga_3. The geometries are, however, somewhat different as expected. The Al_3^+ ion has a $^1A_1$ closed-shell ground state. The $^1A_1$–$^3B_2$ splitting for Ga_3^+ obtained by the author and Feng (355) (0.37 eV) is quite comparable to the corresponding splitting for Al_3^+.

3.5.2 In_3

Feng and the author (369) carried out complete active space MCSCF followed by multireference singles + doubles CI calculations, which included up to 177,000 configurations on seven low-lying electronic states of In_3 (194). They employed a valence ($3s3p1d$) basis set used earlier for In_2 together with $3e$ RECPs for the In atom. In addition, spin–orbit effects for the low-lying states of In_3 were obtained. Complete bending potential-energy surfaces as a function of bending angle were also obtained (369).

Table 3.54 shows the equilibrium geometries of the bent states of In_3 and

Table 3.54 Geometries of Energies of the Bent Electronic States of In_3

	CASSCF				MRSDCI		
State	θ, deg	r,[a] Å	E,[b] eV	State	θ, deg	r,[a] Å	E,[b] eV
$^4A_2$	75.2	2.97	0.0	$^4A_2$	71.04	2.95	0.0
$^4B_1$	54.8	3.25	0.03	$^4B_1$	55.2	3.12	0.01
$^2B_1$	47.6	3.37	0.07	$^2B_1$	48.9	3.31	0.11
$^2A_1$	49.6	3.60	0.28	$^4B_2$	60.0	3.29	0.35
$^4B_2$	58.4	3.39	0.35	$^2A_1$	51.0	3.49	0.45
$^2B_2$	74.3	3.17	0.37	$^2B_2$	74.4	3.04	0.45

[a]The two equal sides of the isosceles triangle; θ = apex angle.
[b]Zero energy for CASSCF is −5.513007 hartrees; zero energy for MRSDCI is −5.609780 hartrees. The zero energy is for the $^4A_2$ ground state at the reported geometry.
Source: Reference 369.

their energy separations at both the CASSCF and MRSDCI levels of theory. Figure 3.29 shows the actual bending potential-energy surfaces for several electronic states obtained with the CASSCF method. As seen from Table 3.54, the equilibrium geometries predicted by CASSCF calculations are not substantially different from the MRSDCI geometries. The bond lengths changed by ≤ 0.13 Å. The bond angles changed at most by 4° at the MRSDCI level of theory compared to the zeroth-order CASSCF level. Table 3.55 shows the equilibrium geometries and the energy separations of the linear electronic states of In_3.

The MRSDCI method predicted that two nearly degenerate states of ${}^4A_2$ and ${}^4B_1$ are isosceles triangles, although ${}^4B_1$ is very acute. Note that the linear limit of the ${}^4A_2({}^4\Sigma_u^-)$ state is only 0.07 eV above the ground state.

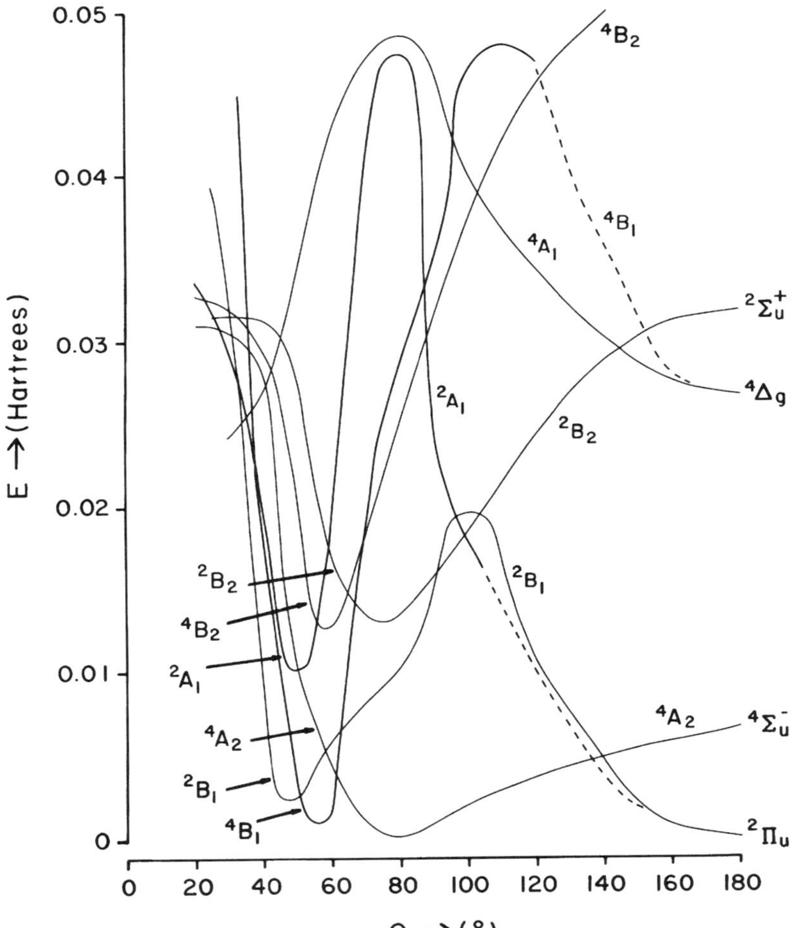

Figure 3.29 Bending potential-energy surfaces of In_3. (Reproduced from Ref. 369; copyright 1989, Elsevier Science Publishers B.V.)

Table 3.55 Properties of the Linear Electronic States of In_3

	CASSCF			MRSDCI	
State	R, Å	E, eV	State	R, Å	E, eV
$^2\Pi_u$	3.07	−0.02	$^4\Sigma_u^-$	2.90	0.07
$^4\Sigma_u^-$	2.90	0.18	$^2\Pi_u$	3.05	0.37
$^4\Delta_g$	3.14	0.73	$^4\Delta_g$	3.13	0.77
$^2\Sigma_u^+$	2.95	0.87	$^2\Sigma_u^+$	2.92	0.85
$^4\Pi_g$	2.85	1.29	$^4\Pi_g$	2.90	1.13

Source: Reference 369.

The geometry of the $^2B_1$ state is acute, and thus the In–In bond lengths for the two sides of the isosceles triangle are large. Both $^4B_1$ and $^2B_1$ have short In–In bond lengths at the base of the isosceles triangle. Hence the $^4B_1$, $^2B_1$, and $^2A_1$ electronic states exhibited enhanced bonding at the base, while for the other states, bond strengths at the equal sides of the isosceles triangles are greater.

The energy separations of the various electronic states are quite sensitive to higher-order correlations. For example, the $^2A_1$ and $^4B_2$ states switch order between CASSCF and MRSDCI calculations. The energy separations of most of the excited states increase at the MRSDCI level of theory, implying stabilization of the ground state.

The atomization energy, specifically, the energy for the process $In_3 \rightarrow 3In$, was calculated through a long-distance calculation ($R = 8.0$ Å) for the linear $^4A_2$ state. The MRSDCI calculations yielded an atomization energy of 48 kcal/mol for the ground state of In_3 (369).

The $1a_1$ orbital of the ground state is the totally symmetric linear combination of the $5s$ orbitals on the three atoms. The $2a_1$ orbital is $In_1(5s) + In_2(5s) + In_3(5s)$ for the $^4A_2$ state, where In_1 represents the apex atom of the isosceles triangle. For the $^2A_1$ state, the $2a_1$ orbital was found to be predominantly $In_2(5s) + In_3(5s)$, since the equilibrium geometry of this state is a very acute triangle ($\theta_e \sim 51°$). The $3a_1$ orbital for the $^4A_2$ state was found to be composed predominantly of the $5s$, $5p_y$, and $5p_z$ orbitals of the base atoms. The $1b_2$ orbital for the $^4A_2$ state was found to be predominantly the $In_2(5s)$–$In_3(5s)$ antibonding orbital for both the $^4A_2$ and $^2A_1$ states. The $2b_2$ orbital of the $^4A_2$ state was found to be predominantly the $In(5p_y)$–$In_2(5p_z) + In_3(5p_z)$ bonding orbital. The $1b_1$ orbital of the $^4A_2$ state was found to be a $In(5p_x) + In_2(5p_x) + In_3(5p_x)$ bonding orbital in the plane perpendicular to the plane of the trimer.

The Mulliken population analysis of the electronic states reveals that the apex In atom has a gross population >3.0 for the $^4A_2$ and $^2B_2$ states, while it is <3.0 for the other states. This trend is generally consistent with the dipole moments for these states. The s populations of the metal atom are somewhat smaller for all the electronic states. The gross p populations for most of the electronic states are larger than their natural atomic occupations, which suggested some mixing of $5s^25p^1$ and $5s^15p^2$ configurations, especially in the $^4A_2$, $^4B_2$ $^2B_2$, and $^4A_1$ states.

As noted in Section (3.5.1), the author and Feng (355) studied the electronic states of Ga_3. The $^2A_1$ state was found to be the ground state of Ga_3, while the $^4A_2$ state was found to be 0.24 eV above the ground state. For In_3, however, the $^4A_2$ and $^4B_1$ states are considerably lower than the $^2A_1$ state. The $^2A_1$ state, in fact, was found to be 0.35 eV above the $^4A_2$ state. For both the species, the $^4A_1$ state of linear geometry is the highest in energy. For Ga_3, the $^4A_1$ state was found to be 1.38 eV above the ground state, while the corresponding state for In_3 was found to be 0.77 eV above the ground state. The electronic states were found to be more closely packed, as expected for In_3 compared to Ga_3. The difference in the properties of Ga_3 and In_3 seems to arise from both relativistic effects and the fact that the atomic energy separations are lower for In compared to Ga (134). The ^4P state arising from $4s4p^2$ is 38,000 cm^{-1} above for Ga, while the corresponding state ($5s5p^2$) is 35,000 cm^{-1} above the ground state for In (134).

A major difference between the electronic states of Ga_3 and In_3 is in the ionicity of the metal–metal bonding. The dipole moments of the electronic states of In_3 were found to be considerably larger compared to the corresponding states of Ga_3. For example, the dipole moment of the $^2A_1$ state of Ga_3 was found to be 0.02 D, while the corresponding state of In_3 has a dipole moment of 0.80 D. Similarly, the dipole moment of the $^4A_2$ state of Ga_3 is 0.17 D, while the corresponding state of In_3 has a dipole moment of -0.48 eV. This trend is consistent with the greater metallic character of the (more electropositive) In atom compared to the gallium atom.

3.5.3 Ge_3

Although the electronic states of Si_3 have been the topic of several experimental and theoretical studies (421–426), this is not the case with Ge_3. In a fascinating study, Grev and Schaefer (424) found that there are two nearly degenerate electronic states of $^1A_1(C_{2v})$ and $^3A_2'(D_{3h})$ symmetries for Si_3. At the highest level of theory the $^1A_1$ state with isosceles triangular geometry appears to prevail (423). Theoretical calculations have predicted several excited electronic states of Si_3 (423). Some of these states were observed by Neumark and co-workers (426) using negative photodetachment spectroscopy of Si_3^-. Winstead et al. (427) have observed the laser-induced plasma spectrum of cooled Si_3. They observed transitions from an upper state E to both the X ground state and the A state, which is predicted to be 0.45 eV above the ground state. It is interesting to note the excellent agreement in the vibrational frequency of the E state in Reference 427 and the previous work of Weltner and McLeod (428).

Pacchioni and Koutecký (429) performed single-point D_{3h} SCF/CI calculations on the singlet state of Ge_3. Dixon and Gole (430) considered the $^1A_1(C_{2v})$ and $^3A_2'(D_{3h})$ structures using the local density function (LDF) approximation. Dai et al. (431) considered the potential-energy surfaces and the spectroscopic constants of several electronic states of Ge_3 using the CASSCF/MRSDCI techniques in conjunction with RECPs. The results discussed in this section are primarily from this work, although comparisons are also made with the previous theoretical studies and the experimental data on the related Ge_3.

Dai et al. (431) used a restricted CAS-MCSCF method to generate the orbitals for higher-order CI calculations and to compute the entire bending potential-energy surfaces of eight electronic states of Ge_3. Higher-order MRSDCI calculations were carried out following the CAS-MCSCF at selected regions in the potential-energy surfaces. All calculations were made in the C_{2v} point group orienting the trimer on the yz plane. Relativistic effective core potentials for Ge were used in the Dai et al. study. A $(3s3p1d)$ valence Gaussian basis set was employed by these authors.

Dai et al. (431) performed MRSDCI calculations following the CAS-MCSCF method, since the higher-order electron correlation effects not included in the CASSCF could be important. The configurations generated by allowing single + double excitation from all configurations in the CASSCF with coefficients ≥ 0.07 were included in the MRSDCI.

Figure 3.30 shows the bending potential-energy surfaces of eight electronic states of Ge_3 obtained by Dai et al. (431). As evidenced from Figure 3.30, the $^1A_1$ state and $^3B_2(^3A_2$ at $\theta = 60°)$ states are nearly degenerate, analogous to Si_3 (423). As evidenced from Figure 3.30, for all eight states, the potential minima correspond to bent isosceles triangular geometries. The linear electronic states are considerably higher in energy.

Table 3.56 shows the equilibrium geometries of all eight bend electronic states. As evidenced from Table 3.56, at the CASSCF level the $^1A_1$ state ($\theta_e = 83.4°$) prevails, while at the MRSDCI level the $^1A_1$–$^3A_2'$ energy separation decreases. At the MRSDCI+Q level of theory the $^3A_2'$ state is slightly more stabilized relative to $^1A_1$. A previous all-electron study on Si_3 (423), which employed a large McLean–Chandler basis set augmented by two sets of d-polarization functions, revealed that at the highest level of theory the $^1A_1$ state is the ground state of Si_3. At the highest MRSDCI level of theory the $^1A_1$–$^3A_2'$ energy separation for Si_3 was computed as 3.5 kcal/mol using a $TZ+2d$ basis set. For Ge_3 the picture is more complicated, since the basis set extension as well as the inclusion of $3d$ correlation effects could alter the relative ordering of these two states separated by only a small energy. As seen from Table 3.56 at the MRSDCI level the $^1A_1$ state is the ground state, while the $^3A_2'$ state is 1 kcal/mol above. However, at the MRSDCI+Q level of theory the $3A_2'$ state is 0.2 kcal/mol below the $^1A_1$ state. Thus it seems that the ground state of Ge_3 cannot be established unambiguously.

The equilibrium geometry of the $^1A_1$ state is an isosceles triangle with an apex angle of 83.3° and Ge–Ge equal bond lengths of 2.32 Å. As seen from Table 3.56, the equilibrium geometries of the $^1A_1$ and $^3A_2'$ states do not change much between the CASSCF and MRSDCI levels of approximation. The $^3B_2$ state forms an equilateral triangular minimum and corresponds to a $^3A_2'$ state in the D_{3h} group.

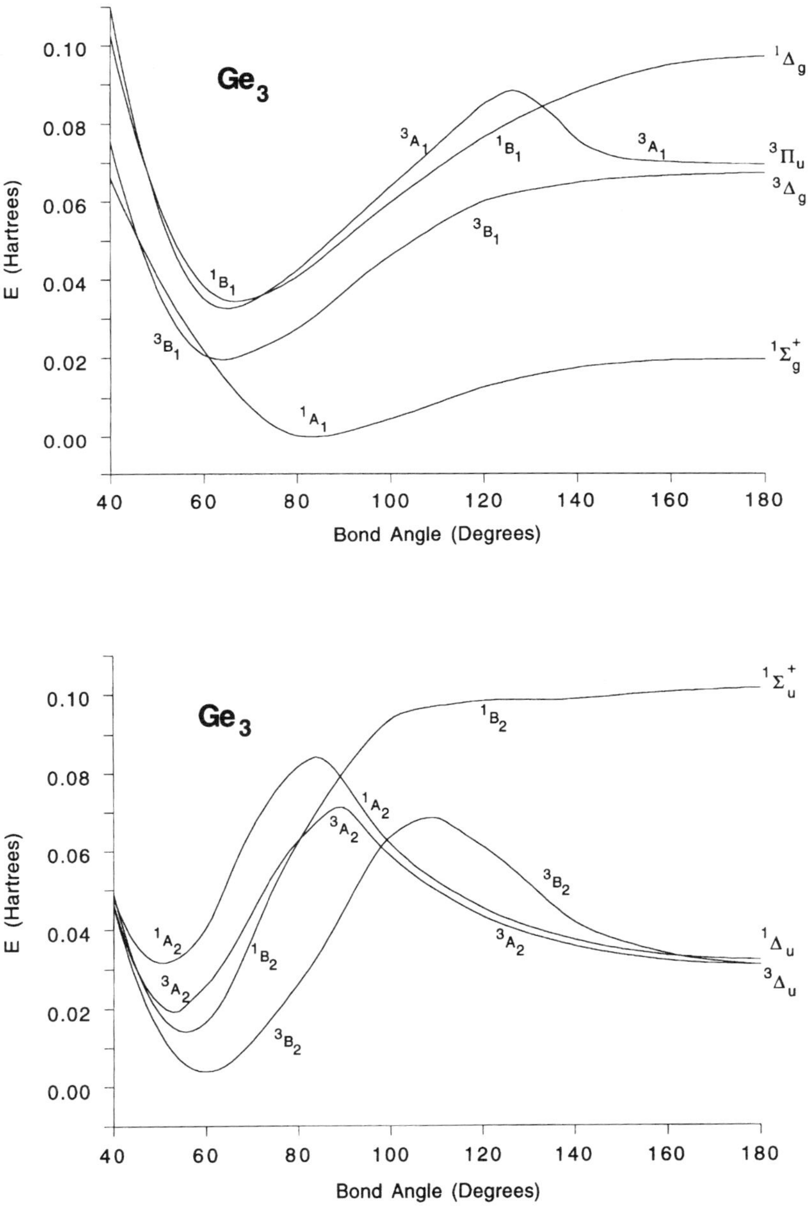

Figure 3.30 Bending potential-energy surfaces of Ge_3. (Reproduced from Ref. 431; copyright 1992, Elsevier Science Publishers B.V.)

Table 3.56 Equilibrium Properties of the Bent Electronic States of Ge_3

	CASSCF			MRSDCI		
State	θ_e, deg[a]	R_e, Å[a]	E, eV[a]	θ_e, deg[a]	R_e, Å[a]	E, eV[c,d]
$^1A_1$	83.4	2.294	0.00	83.3	2.320	0.00 (0.00)
$^3A_2'(^3B_2)$	60.0	2.457	0.11	60.0	2.452	0.05 (−0.01)
$^1B_2$	55.6	2.514	0.39	57.3	2.484	0.44 (0.48)
$^3B_1$	64.1	2.528	0.53	66.9	2.507	0.70 (0.69)
$^3A_2$	52.9	2.694	0.53	54.2	2.679	0.73 (0.84)
$^1A_2$	50.7	2.767	0.86	52.6	2.725	1.06 (1.17)
$^3A_1$	65.1	2.421	0.89	66.2	2.428	0.88 (0.81)
$^1B_1$	66.9	2.499	0.94	68.2	2.504	1.07 (1.13)

[a] θ_e refers to the apex angle of the isosceles triangle, while R_e refers to two equal sides of the isosceles triangle.
[b] The zero CASSCF energy is for the $^1A_1$ bent minimum, and is −11.13072 hartrees.
[c] The zero MRSDCI energy is for the $^1A_1$ bent minimum, and is −11.33007 hartrees.
[d] The values in parentheses are the Davidson corrected energies.

Dixon and Gole (430) computed the $^1A_1$ and $^3A_2'$ states of Ge_3 using the LDF theory, which led to two nearly degenerate states for Ge_3. Their calculated Ge–Ge equal side bond lengths for the $^1A_1$ and $^3A_2'$ states are 2.287 and 2.436 Å compared to Dai et al.'s values of 2.32 and 2.452 Å, respectively. Their equilibrium apex angle for the $^1A_1$ state is 85°, compared to the Dai et al. result of 83.3°.

Among the excited states listed for Ge_3 in Table 3.56, electronic transitions to $^1B_2$ and $^1B_1$ states from the $^1A_1$ state are likely to be strong. These states are predicted to lie 0.44 (0.48) and 1.07 (1.13) eV above the $^1A_1$ state. Neumark and co-workers (426) have recorded the negative-ion photodetachment spectra for Si_3. They find that the A^1B_2 state is 0.45 eV above the $^1A_1$ state of Si_3. The previous single-point ($\theta = 60°$) computation placed the A^1B_2 state 0.41 eV above the X^1A_1 state (423). The equilibrium geometry of the A^1B_2 state for Ge_3 is not too far from $\theta = 60°$, and Figure 3.30 evidently suggests that the energy separation at $\theta = 60°$ and $\theta_e = 57°$ are not too far off. Winstead et al. (427) also observed the A state at 0.45 eV above the ground state, and concluded that the most probable candidate for the A state is the $^1B_2$ state.

The $^3B_1$ state of Ge_3 is theoretically computed at 0.7 eV (0.69 including the Davidson correction) above the ground state. The θ_e of this state is ~60° (>60°). Neumark and co-workers (426) observed the B state of Si_3 at 0.89 eV above the ground state. The Dai et al. (431) computations on Ge_3 appear to confirm the previous theoretical study on Si_3 (423) and the assignment of the A state to $^1B_2$ for Si_3. It should be noted that the assignment of the B state is complicated by the presence of the $^3A_2$ and $^3A_1$ states, which are only 0.07 and 0.18 eV, respectively, above the $^3B_1$ state. In fact, Rohlfing and Raghavachari (422) favor the assignment of the B state of Si_3 to $^3A_1$, but assign the C state to

$^3B_1$. The assignment of the A state to $^1B_2$ is more definitive than the assignment of the B state to $^3B_1$ or $^3A_1$ or $^3A_2$.

The $^3B_2$ state at its minimum (60°) is described as being predominantly $1a_1^2 2a_1^2 3a_1^2 4a_1^1 1b_2^2 2b_2^1 1b_1^2$ (84%). At θ = 100° the weight of this component increases to 85%, while at θ = 110° is composed of $1a_1^2 2a_1^2 3a_1^1 1b_2^2 2b_2^2 3b_2^1 1b_1^2$ (85%). Therefore the saddle point and the shape of the PES of the $^3B_2$ state are due to this avoided crossing, which is similar to Si_3 (423). The $^3A_1$ state is predominantly described by $1a_1^2 2a_1^2 3a_1^1 4a_1^1 1b_2^2 2b_2^2 1b_1^2$ (91~94%) for θ between 60° and 120°, while between θ = 130° and 150° it is predominantly $1a_1^2 2a_1^2 3a_1^2 1b_2^2 2b_2^1 3b_2^1 1b_1^2$ (86%). Hence the saddle point (θ = 126°) of the $^3A_1$ state is due to this avoided crossing. The $^3A_2$ and $^1A_2$ states have the same leading configuration $1a_1^2 2a_1^2 3a_1^2 4a_1^2 1b_2^2 2b_2^1 1b_1^1$ (83%) between θ = 50° and 80°, while at θ = 100° they are composed of $1a_1^2 2a_1^2 3a_1^2 1b_2^2 2b_2^2 3b_2^1 1b_1^1$ (82%). The saddle points of both states just appear between θ = 80° and 100°.

Most electronic states of Ge_3 have gross Mulliken populations with the apex Ge atom > 4.0, implying that the polarity of the $\mathrm{Ge}\diagup\overset{\mathrm{Ge}^-}{}\diagdown\mathrm{Ge}^+$ is $\mathrm{Ge}^+\diagup\overset{\mathrm{Ge}^-}{}\diagdown\mathrm{Ge}^+$ with the exception of the $^3B_2$ state, which has the same gross population (4.0) for three Ge atoms. This explains the isosceles triangle geometry, rather than an equilateral triangular geometry. The $^3A_2$ and $^1A_2$ states have $\mathrm{Ge}^-\diagup\overset{\mathrm{Ge}^+}{}\diagdown\mathrm{Ge}^-$ polarity. All states exhibit deviation from the atomic $4s^2 4p^2$ population of Ge. The $4d$ populations are between 0.26 and 0.30 for all bent states due to $4d$ polarization. The $4p$ population of the apex atom is 2.41 for the $^1A_1$ state and 2.32–2.59 for all linear states suggesting mixing of the $4s^2 4p^2$ configuration with the $4s^1 4p^3$ configuration, especially for the apex Ge atom. In the $^1A_1$ ground state, the s, p, and d populations are 1.44, 2.41, 0.29 for the apex atom and 1.67, 2.00, and 0.26 for each base atom, respectively.

3.5.4 Sn_3

Tin clusters have been a topic of a few studies (370, 371). At room temperature in the solid state Sn is metallic, but at low temperature it exhibits a semiconducting crystallographic modification with Ge structure (370). The mass spectra of Sn_n clusters have been recorded by Martin and Schaber (370). The mass spectrum of Sn_x^+ exhibited a gradual decrease in intensity for X = 1–13, and the x = 14 peak was almost absent; however, for x = 15, 16, the intensity increases. For the Ge_x^+ clusters, the peak with x = 13 is the low-intensity peak. All the clusters exhibited enhanced stabilities for x = 6,10. Further, Martin and Schaber (370) showed that Ge_{14}^+ is quite stable, while Sn_{14}^+ is unstable.

The author (371) studied the low-lying electronic states of equilateral triangular, isosceles triangular, and linear configurations of Sn_3 employing a complete active space MCSCF followed by multireference singles and doubles CI calculations that included up to 228,250 configurations. He employed relativistic effective potentials for the tin atom with the $d^{10}s^2p^2$ outer shell as

Table 3.57 Properties of the Low-Lying Electronic States of Sn_3

State	Method	r_e, Å[a]	θ, deg	E[b]
$^1A_1$	5410-CAS	2.757	85	−204.080192
$^1A_1$	4310-CAS	2.728	85	−204.038657
$^1A_1$	MRSDCI	2.757	85	−204.158123
$^1A_1$	POLCI	2.757	85	−204.124963
$^3A_{2'}(^3B_2)$	5410-CAS	2.96	60	−204.074416
$^3A_{2'}(^3B_2)$	4310-CAS	3.077	60	−204.035962
$^3A_{2'}(^3B_2)$	MRSDCI	2.927	60	−204.148786
$^3A_2$	4310-CAS	3.077	60	−204.029980
$^1B_2$	4310-CAS	2.927	60	−204.022106
$^1A_2$	4310-CAS	3.088	60	−204.014320
$^3B_1$	4310-CAS	3.073	60	−204.028423
$^1B_1$	4310-CAS	3.084	60	−204.012746
$^1\Sigma_g^+(^1A_1)$	5410-CAS	2.714	180	−204.067007
$^3B_2$	5410-CAS	2.792	180	−204.048230
3 Sn(^3P)	SCF	—	—	−203.986118
3 Sn(^3P)	SDCI	—	—	−204.045293
$^1A_1$	5410-CAS	7.500	180	−203.991871
$^1A_1$	POLCI	7.500	180	−204.000095

[a]Refers to the two equal distances of the isosceles triangle for the triangular structure. For the linear structure, this refers to the Sn–Sn bond length. MRSDCI calculations for the $^1A_1$ and $^3B_2$ states were carried out at the 5410-CAS-optimized geometries.

[b]Energies are in hartrees (1 hartree = 27.21 eV).

Source: Reference 371.

the valence shell. A valence + polarization Gaussian ($3s3p2d$) basis set was employed for the tin atom.

Table 3.57 depicts the properties of several electronic states of Sn_3 obtained with many methods employing a ($3s3p1d$) basis set (371). As seen from Table 3.57, two nearly degenerate structures exist for Sn_3. The first structure is an isosceles triangle with an apex angle of 85° and is the minimum of the $^1A_1$ state. The other structure is an equilateral triangular structure, which is a minimum of the $^3B_2$ state. The splitting of these two states was calculated as 3.6 kcal/mol at the MRSDCI level (371).

The energy separation between the bent $^1A_1$ structure and the $^1\Sigma_g^+$ linear structure was found to be about 8.56 kcal/mol (371). Consequently, Sn_3 resembles Si_3 and Ge_3 in possessing two nearly degenerate equilibrium geometries (see Section 3.5.3).

The bending potential-energy surfaces of the two nearly degenerate states, (viz., $^1A_1$ and $^3B_2$) are shown in Figure 3.31. As seen from Figure 3.31, the $^3B_2$ state has a barrier at about $\theta = 100°$ and a weakly bound second minimum at near-linear geometry. This was attributed to an avoided crossing by the author (371). This is reminiscent of the potential-energy surfaces of Si_3 and Ge_3 (see Section 3.5.3).

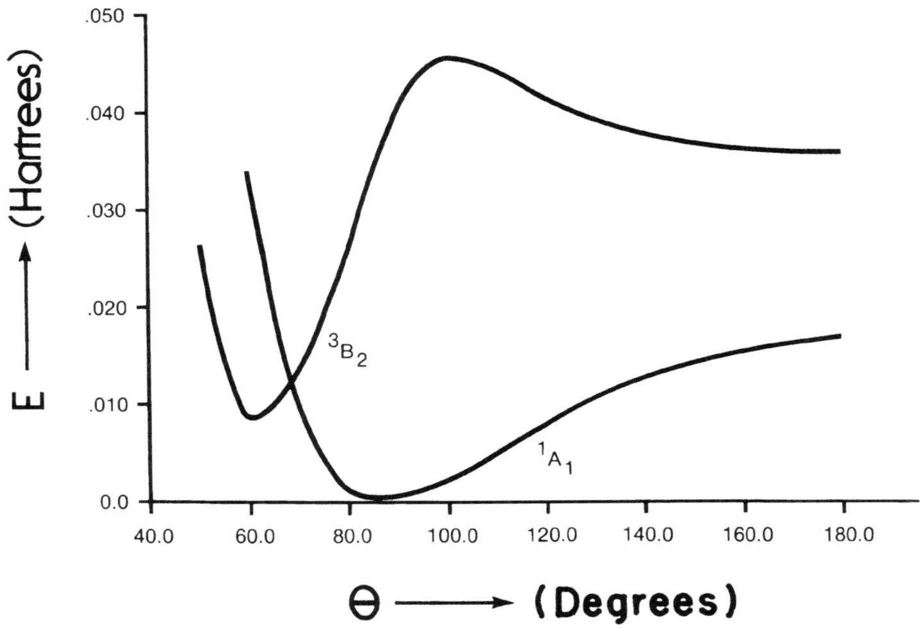

Figure 3.31 Bending potential-energy surfaces of Sn_3. (Reproduced from Ref. 371; copyright 1986, American Institute of Physics.)

The properties of the $^1A_1$ and $^3B_2$ states were also calculated with a $(3s3p2d)$ basis. The uncontraction of the most diffuse d function shrinks the bond length for the $^1A_1$ and $^3B_2$ states by about 0.04 Å. The $^1A_1$–$^3A_2'$ splitting in the larger basis set was calculated as 2.3 kcal/mol at the CASSCF and 3.9 kcal/mol at the MRSDCI levels. Consequently, this splitting was found to decrease in the larger basis set.

The atomization energy of Sn_3 (i.e., the energy for $Sn_3 \rightarrow 3Sn$) was calculated by the MRSDCI method in the absence of spin–orbit interaction as 3.07 eV or 71 kcal/mol. CI calculations carried out by setting the Sn–Sn bond lengths to 7.5 Å gave an atomization energy of 3.4 eV. The CI value should be considered more accurate, since in the MRSDCI method, the energy of the individual atoms (SDCI) was compared with the MRSDCI energy of the molecule. Thus the value of 71 kcal/mol is not correlation consistent and is only a lower bound. The spin–orbit interaction should decrease the atomization energy of Sn_3. This is because the contribution of spin–orbit interaction for the closed-shell $^1A_1$ isosceles triangular structure is zero. Thus, applying a correction of 0.99 eV to the atomization energy, the author (371) obtained a lower bound of 2.4 eV or 55.3 kcal/mol. The dissociation energy of Sn_2 was calculated (201) as 1.86 eV in the presence of spin–orbit interaction. It is thus evident that Sn_3 is at least 0.5 eV more stable than the dimer.

The effect of spin–orbit interaction should be interesting on the PES of Sn_3. It would split the $^3B_2$ state into A_1, B_1, and A_2 components in the C_{2v}^2 double group. For the equilateral triangle, $3A_2'$ is split into A_1' and E'' representations in the D_{3h}^2 double group. The $^1A_1$ isosceles structure correlates with the A_1 representations in the double group. As seen from Figure 3.31, the bending potential-energy surfaces of the $^1A_1$ and $^3B_2$ states cross at $\theta = 70°$. Thus the corresponding A_1 components in the presence of spin–orbit interaction would undergo avoided crossing. In the presence of spin–orbit interaction, the PES of the A_1 component may thus contain shoulder and minimum or double minima (one at $\theta = 60°$ and the other at $\theta = 85°$) separated by a small barrier.

The electronic structure of Sn_3 resembles that of Si_3 and Ge_3 in many ways. For Si_3 (see Section 3.5.3), also, there are two nearly degenerate structures ($^1A_1$ and $^3A_2'$). The splitting between these two states was found to be about 3.6 kcal/mol. The atomization energy of Si_3 is larger as expected. Relativistic effects are, however, negligible for Si_3.

The bond lengths of the two equal sides of the isosceles triangle (2.723 Å) are shorter than the bond lengths for the $^3A_2'$ structure. This implies double bonds along the sides of the isosceles triangle. The linear ($^1\Sigma_g^+$) structure was found to exhibit multiple bonding. The $^3A_2'$ structure can be regarded as a resonance hybrid of three equivalent structures, each having one double bond. Thus the bond lengths for the $^3A_2'$ equilateral structure are a bit longer than the two equal sides of the isosceles triangle. The bond lengths obtained with the two-d function (2d) basis set are about 0.03 Å shorter than the corresponding values obtained using a one-d function (1d) basis set. This was attributed by the author (371) primarily to a polarization effect resulting from the uncontraction of the most diffuse *d* function.

3.5.5 Group V Trimers P_3–Bi_3

Group V trimers form an interesting class of clusters from the standpoint of Jahn–Teller distortion, since the ground states of these trimers in equilateral triangular forms belong to degenerate irreducible representation, and are thus candidates for Jahn–Teller distortion. Experimental studies on the group V trimers have been mainly on the lighter clusters such as P_3, As_3, and Sb_3 (432–440). A striking contrast between N_3 and the heavier group V trimers is that N_3 exhibits a linear ground state, while the heavier trimers exhibit bent ground states. Andrews and Mielke (440) studied the vibronic spectra of P_3 and P_{4+} in solid argon. A broad absorption in the 427-nm region was assigned to P_3. There are several theoretical studies, especially on the lighter group V clusters (441–449).

Matrix-isolation techniques have been used for the characterization of Sb_3 by Sontag and Weber (435). Cabaud et al. (434) observed Sb_3 in high-temperature mass spectrometric study. Sullivan et al. (436) measured the heats of sublimation of Sb_3 and Bi_3. Rovner et al. (438) characterized Bi_3 in the vapor of pure bismuth and Pb + Bi alloy. The dissociation energy of Bi_3 has also been measured by these authors by partial pressure measurements (438, 439).

There are a few theoretical studies on these trimers (441–448). Earlier work of Murrel et al. (441) focused on a comparative theoretical study of N_3 and P_3. While P_3 was found to exhibit small Jahn–Teller distortion (4°), the Jahn–Teller distortion opens up the N_3 trimer all the way to a linear structure. Hamilton and Schaefer (446) studied the P_3^- negative ion, which exhibits two nearly degenerate states. Jones and Hohl (447) employed the molecular dynamics study to investigate P_2–P_8.

The author, Sumathi, and Dai (448) conducted an extensive theoretical study of the potential-energy surfaces of the electronic states of group V trimers (P_3–Bi_3). They employed CASSCF/MRSDCI/RCI computational schemes that included up to 800,000 configurations to characterize the ground and excited electronic states of these species. All these trimers exhibit $^2E''$ ground state for the equilateral triangular geometry, which undergoes distortion, resulting in $^2A_2$ and $^2B_1$ electronic states in the C_{2v} group. The positive ions of these species were also considered, and it was found that these ions belong to $^1A_1'(D_{3h})$ ground states. The results of this study (448) explained the experimentally observed contrast (450) in the photofragmentation patterns of antimony and bismuth clusters. In the ensuing paragraphs, we provide the tables of geometries and energies, and discuss the electronic properties of these species. The potential-energy surfaces are also provided for relative comparison. Relativistic effects are discussed in relation to the periodic trends within the group, and it is shown that Bi_3 behaves differently compared to its lighter analogs mainly because of relativistic effects.

Figure 3.32–3.35 show the potential-energy surfaces for the bending motions for two lowest electronic states of the neutral trimers and the ground state of the positive ion. The potential-energy curves have the common feature that the two curves corresponding to the neutral trimer cross at $\theta = 60°$, resulting in the degenerate $2E''$ state. However, since the positive ion gives rise to a closed-shell ground state, it has an equilateral triangular minimum. The Jahn–Teller distortion from the ideal D_{3h} structure is about 5–7° for the P_3 neutral cluster. Table 3.58 shows the geometries and energy separations of the low-lying electronic states of the neutral P_3. The two C_{2v} Jahn–Teller components are separated by only 0.22 eV, as seen from this table. The adiabatic ionization energy of P_3 is computed as 7.31 eV at the MRSDCI+Q level compared to an experimental value of 7.85 eV estimated from the threshold for the formation of P_3^+.

Table 3.59 shows the computed geometries and energy separations for the low-lying electronic states of As_3 and As_3^+. As seen from this table, the author and co-worker (448) find several quartet and doublet excited states within 2.7 eV of the ground state of As_3. The energy separation between the C_{2v} Jahn–Teller components is only 0.05 eV. From the computed dissociation energy in Table 3.59 and the experimental D_e of As_2 (3.96 eV), the author and coworkers (448) obtained the atomization energy of As_3 to be 5.3 eV. This value (122 kcal/mol) is, however, considerably lower than the experimental value of 147.7 kcal/mol deduced by Bennet et al. (433). The author and co-workers (448) noted the intensity of electron correlation effects by comparing the MRSDCI treatment with a full p-SOCI and suggested an improvement of at least 0.2 eV

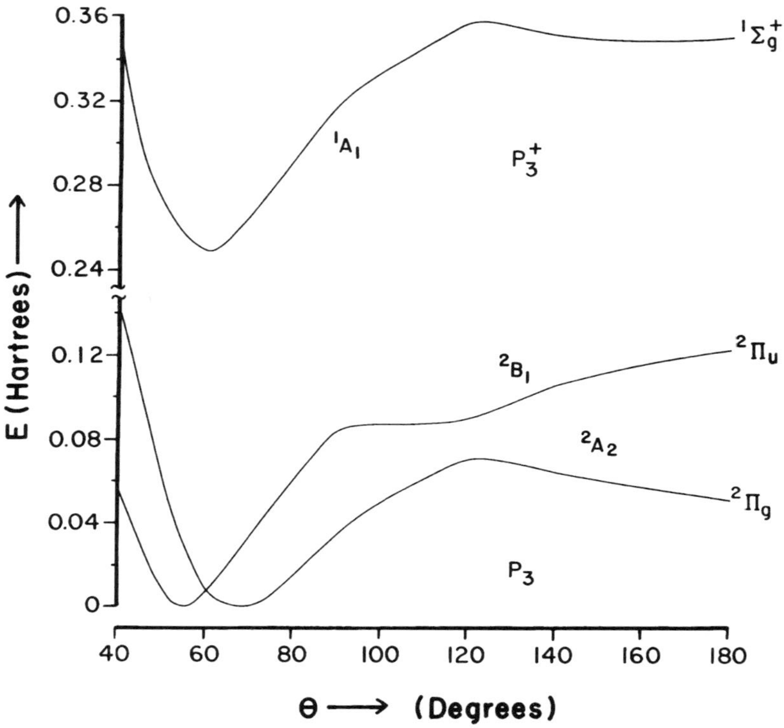

Figure 3.32 Bending potential-energy surfaces of P_3 and P_3^+. (Reproduced from Ref. 448; copyright 1991, American Institute of Physics.)

in the computed dissociation energy. Even the corrected value thus obtained was lower than the experimental value.

Table 3.60 shows geometries and energy separations reported by the author and co-workers (448) for Sb_3. A similar set of electronic states is found for Sb_3. The energy separations of the excited electronic states for Sb_3 are somewhat lower than those for As_3. The dissociation energy of Sb_3 is a topic of some controversy. Sullivan et al. (436) have obtained the heat of formation for the reaction $Sb_3(g) = 3Sb(g)$ as 133 kcal/mol assuming a linear ground state for Sb_3 with Sb–Sb bold lengths of 2.82 Å. Although this assumption is not supported by the computations of the author and co-workers (448), the correction to AE due to this incorrect geometry should be only 3–5 kcal/mol in thermochemical second or third law computations. On the other hand, Cabaud et al. (434) reported an AE of 143 kcal/mol, which appears to be too high. These authors find metastable peaks at masses 274.1 and 60.5, which are attributed to fragmentation of the tetrameter to yield trimer and dimer to yield monomer, respectively. Duncan and co-workers (450) also found that the tetrameric anti-

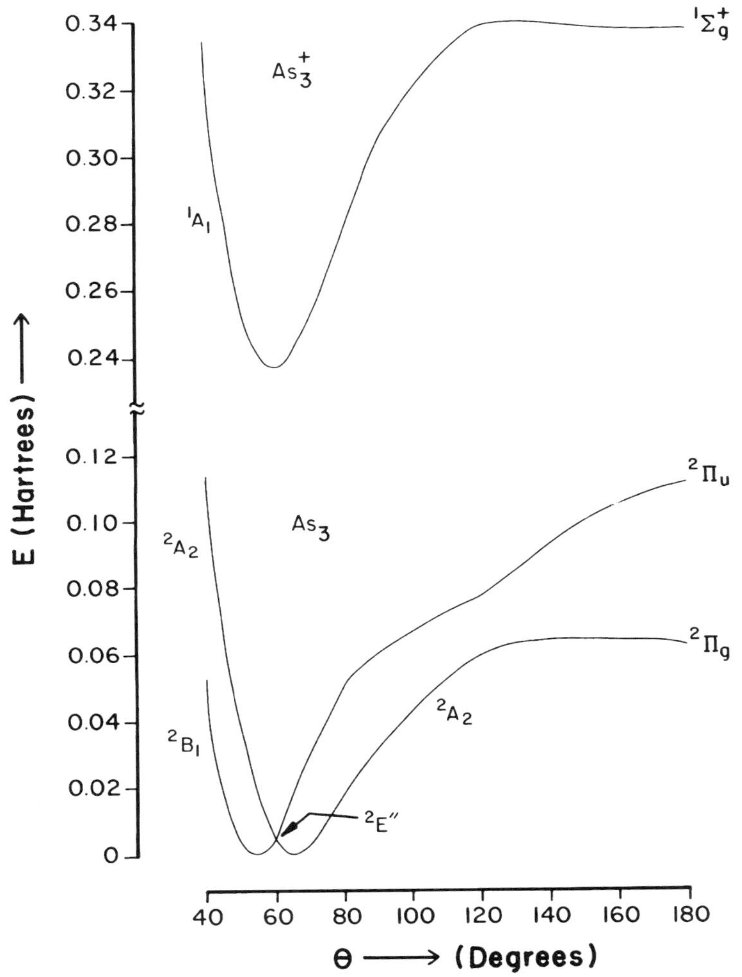

Figure 3.33 Bending potential-energy surfaces of As_3 and As_3^+. (Reproduced from Ref. 448; copyright 1991, American Institute of Physics.)

mony cluster fragments mostly into trimer. This is supported by the computations of the author and co-workers (448), who find that the trimeric ion forms a closed-shell ground state, and is thus quite stable.

Table 3.61 shows the geometries and energy separations of Bi_3 and Bi_3^+ computed by the author and co-workers (448). The spin–orbit effect lowered the ground state near its equilibrium geometry by 0.3 eV. The doublet–quartet mixing in the ground state was seen when spin–orbit effects were included in that the RCI wavefunction of the E(I) state in the double groups consists of 82%

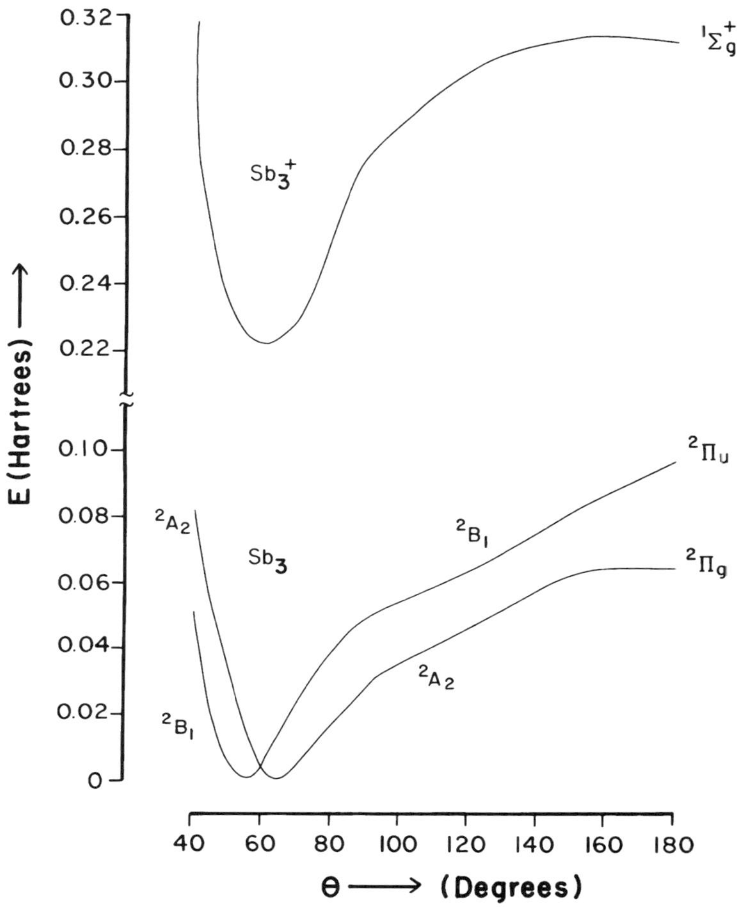

Figure 3.34 Bending potential-energy surfaces of Sb_3 and Sb_3^+. (Reproduced from Ref. 448; copyright 1991, American Institute of Physics.)

$^2A_2$ and 7% $^4A_2$. Since the spin–orbit stabilization in the ground state was found to be larger than the separation between the C_{2v} Jahn–Teller components, the author and co-workers (448) argued that the spin–orbit effect will quench the Jahn–Teller distortion in the case of Bi_3 trimer. This is clearly a consequence of relativistic effect.

The spin–orbit corrected atomization energy of Bi_3 was computed as 3 eV using the experimental D_e of Bi_2. This value is in good agreement with the experimental result (438) obtained from partial pressure measurements (3.7 eV). Although negative ion photoelectron spectra of Bi_3^- have been obtained by Polak et al. (451), the spectra were found to be too perturbed for any definitive assignments.

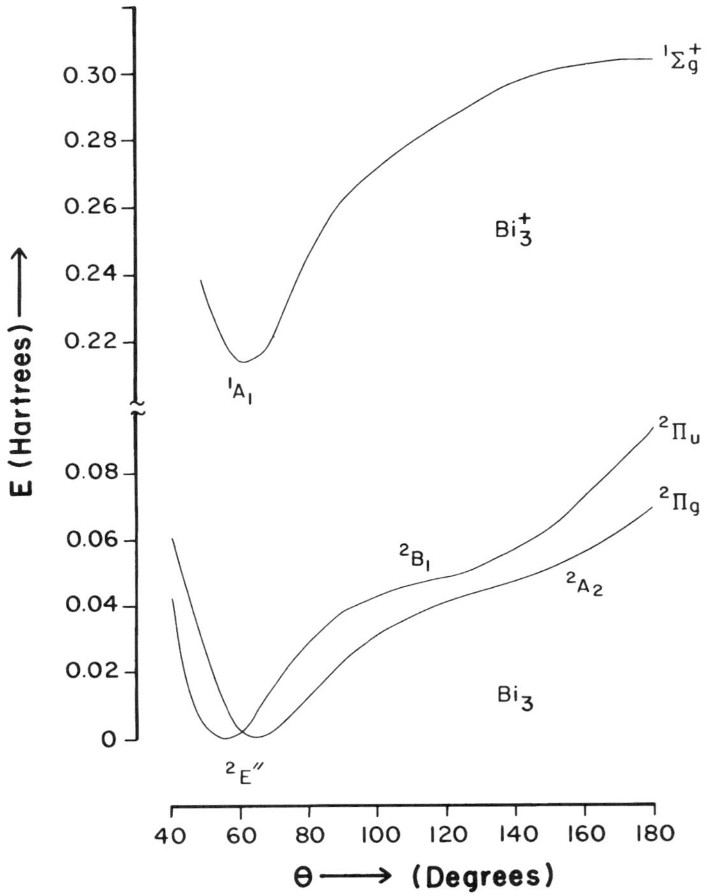

Figure 3.35 Bending potential-energy surfaces of Bi_3 and Bi_3^+. (Reproduced from Ref. 448; copyright 1991, American Institute of Physics.)

Geusic et al. (450) found in their photofragmentation spectra that Bi_3^+ fragments mostly into Bi^+ (73%) in contrast with Sb_3^+, which fragments mostly into Sb_2^+. The author and co-workers (448) attribute this anomaly to relativistic effects. The adiabatic IPs of Bi_2 and Bi are 7.4 and 7.29 eV, respectively, compared with the IPs of 8.5 and 8.64 eV, respectively, for Sb_2 and Sb. This means that Bi^+ is more readily formed, since the neutral Bi atom has lower IP compared to the dimer, while the opposite trend is noted for Sb_2 and Sb. The smaller IP of the Bi atom compared to its dimer was attributed by the author and co-workers (448) to relativistic effects. In the case of Bi, ionization results in the $^3P_0$ ground state, while the ionization of the closed shell Bi_2 leads to the open-shell $(\frac{3}{2})_u$ ground state for the ion. The spin–orbit stabilization of the

Table 3.58 Geometries of Electronic States of P_3 and P_3^+ at CASSCF and MRSDCI Levels of Theory

		CASSCF			MRCI		
System	State	r_e, Å	θ_e, (deg)	E, eV	r_e, Å	θ_e, deg	E, eV
P_3	$^2A_2$	2.13	66.4	0.0	2.10	66.7	0.0[a]
P_3	$^2B_1$	2.25	55.0	0.006	2.21	55.0	0.22
P_3^+	$^1A_1$	2.14	60.0	6.64	2.12	60.0	7.26(7.31)
$P(^4S) + P_2(^1\Sigma_g^+)$	—	—	—	0.90	—	—	1.24(1.44)

[a]Zero energy for the $^2A_2$ ground state at the CASSCF level is −19.174272 hartrees. Zero energy for the $^2A_2$ ground state at the MRCI level is −19.430804 hartrees. Numbers in parentheses include Davidson's correction.

Source: Reproduced from Reference 448.

atom and dimer with respect to ionization is thus different leading to lower IP for Bi compared to the dimer.

The Mulliken population analysis of the various electronic states of all group V trimers reveals that Bi_3 is quite different from its lighter analogs. For example, the $^2A_2$ state of P_3 has the population $3s^{1.8}3p^{2.93}3d^{0.22}$, while the same state of Bi_3 has the population $6s^{2.0}6p^{2.97}$. In all the electronic states the Bi_3 has enhanced 6s population, which was attributed by the author and co-workers (448) to the relativistic stabilization of the 6s orbital of the bismuth atom. The base atoms of the trimers were found to shed greater electronic density compared to the apex atom. Of course, in the final ground state of the positive ion, all three atoms have the same populations due to the D_{3h} symmetry.

Table 3.59 Geometries of Electronic States of As_3 and As_3^+

		CASSCF			MRCI		
System	State	r_e, Å	θ_e, deg	E, eV	r_e, Å	θ_e, deg	E, eV
As_3	$^2A_2$	2.37	65.4	0.009	2.35	65.5	0.0
As_3	$^2B_1$	2.50	55.0	0.0	2.48	55.0	0.05
As_3	$^4A_2''$	2.58	60.0	1.72	2.56	60.0	1.89
As_3	$^4E'$	2.50	60.0	2.40	2.46	60.0	2.19
As_3	$^2E'$	2.55	60.0	2.77	2.48	60.0	2.73
As_3^+	$^1A_1'$	2.39	60.0	6.47	2.37	60.0	7.1
$As(^4S) + As_2(^1\Sigma_g^+)$	—	—	—	0.73	—	—	1.02(1.25)[a]

[a]Number in parentheses includes Davidson's correction.

Source: Reproduced from Reference 448.

Table 3.60 Geometries and Energy Separations of Electronic States of Sb_3 and Sb_3^{+a}

		CASSCF			MRCI		
System	State	r_e, Å	θ_e, deg	E, eV	r_e, Å	θ_e, deg	E, eV
Sb_3	$^2A_2$	2.77	64.9	0.007	2.76	65.8	0.0
Sb_3	$^2B_1$	2.95	55.8	0.0	2.90	55.6	0.04
Sb_3	$^4A_2''$	3.00	60.0	1.18	2.95	60.0	1.31
Sb_3	$^4E'$	2.92	60.0	1.94	2.87	60.0	1.76
Sb_3	$^2E'$	2.97	60.0	2.18	2.88	60.0	2.21
Sb_3^+	$^1A_1'$	2.81	60.0	6.06	2.79	60.0	6.38(6.44)
$Sb(^4S) + Sb_2(^1\Sigma_g^+)$	—	—	—	0.69	—	—	0.98(1.16)

[a]Number in parentheses includes Davidson's correction. The total energy corresponding to the CASSCF zero is −16.006562, while the MRCI is −16.162763 a.u.

Source: Reproduced from Reference 448.

3.5.6 Se_3, Te_3, and Po_3

Among the clusters of the group VI elements, sulfur clusters appear to have been the most studied species and are known to exist predominantly as rings (452, 453). Relatively less is known about the heavier analogs (453). Although selenium exhibits a variety of allotropes (453) akin to sulfur, it is documented to be different from sulfur in the three phases. Group VI clusters have been the topic of several studies (454–467).

While structures of many of the sulfur clusters, S_n ($n \leq 20$), are now characterized, this is not the case with Se_n or Te_n clusters, although small Se_n ring

Table 3.61 Geometries and Energy Separations of Electronic States of Bi_3 and Bi_3^{+a}

		CASSCF			MRCI		
System	State	r_e, Å	θ_e, deg	E, eV	r_e, Å	θ_e, deg	E, eV
Bi_3	$^2A_2$	3.0	64.5	0.005	2.94	65.6	0.0
Bi_3	$^2B_1$	3.14	55.3	0.0	3.08	55.8	0.04
Bi_3	$^4A_2''$	3.21	60.0	0.88	3.12	60.0	0.87
Bi_3	$^4E'$	3.13	60.0	1.73	3.05	60.0	1.61
Bi_3	$^2E'$	3.18	60.0	1.91	3.08	60.0	1.97
Bi_3^+	$^1A_1'$	3.0	60.0	5.83	2.97	60.0	6.12(6.16)
$Bi(^4S) + Bi_2(^1\Sigma_g^+)$	—	—	—	0.56	—	—	0.75(0.94)

[a]Number in parentheses includes Davidson's correction. The total energy corresponding to MRCI zero is −16.178151 a.u.

Source: Reproduced from Reference 448.

fragments have been characterized as part of infinite chains. Crystalline phases of Se_6 and Se_8 have been studied and mass-spectroscopic studies have led to the identification of Se_n for $n = 2$–9. Tellurium exists primarily as molecular chains. Two allotropic forms of Te—amorphous and metallic tellurium—have been characterized. Although larger clusters of tellurium such as Te_n ($n = 3$–7) have now been identified, Te_2 appears to be the most studied spectroscopically among the Te clusters. Polonium appears to be the least studied of group VI clusters, although the $A \rightarrow X$ spectroscopic band of $^{210}Po_2$ dimer has been identified by Charles et al. (460). The analysis has led to the spectroscopic constants of the X and A states and an approximate $D_0^\circ = 1.9$ eV for the ground state of Po_2.

The resonance Raman spectrum of the selenium trimer has been obtained by Schnöckel et al. (456, 457) in argon and nitrogen matrices. This was accomplished by reacting superheated selenium vapor with Se_2 on the surface of the matrix, which resulted in a series of fluorescence bands with band origin at ~17,000 cm^{-1}. Brabson and Andrews (458) obtained the infrared and visible spectra of Se_n ($n = 2$–4) in solid argon in the 480–575-nm region attributed to Se_3 and 710–850 nm band system assigned to Se_4. In addition, the IR band at 350 cm^{-1} was attributed to Se_3. Interestingly, the 480–575-nm bands disappeared in the superheated Knudsen cell experiments. It should also be noted that the visible bands of Se_3 overlapped with the red emission system of Se_3 at 509–568 nm.

Schnöckel (457) has studied the resonance Raman spectrum of Te_3. Hassanzadeh et al. (459) studied the absorption spectra of tellurium clusters in solid argon, which led to the determination of a structured absorption band centered at 665 nm assigned to Te_3. In addition, infrared bands attributed to Te_4 were observed by these authors.

There are several theoretical studies on sulfur clusters, especially S_3 and S_4, for which high-level *ab initio* calculations have been done (460–466). The structures of Se_n clusters ($n = 3$–8) have been studied using the molecular dynamics combined with density function methods in their ground states (464–466). Von Niessen et al. (464) obtained the valence electron affinities of Se_3 and Te_3 using the Green function method. For S_3 two nearly degenerate states with D_{3h} and C_{2v} geometries are known to exist.

Dai and the author (467) carried out *ab initio* relativistic complete active space multiconfiguration self-consistent field (CASMCSCF) calculations followed by multireference singles + doubles configuration interaction (MRSDCI) computations in conjunction with RECPs. In addition the spin–orbit effects are considered using the relativistic configuration interaction (RCI) method. The entire potential-energy surfaces of several electronic states were considered, and the observed spectra of Se_3 and Te_3 were assigned.

Dai and the author (467) used the relativistic effective core potentials (RECPs) for the Se, Te, and Po atoms generated by Ross et al. (468) with the outer ns^2np^4 shells included in the valence space (n is the principal quantum number of the valence s and p orbitals). It was noted by these authors that the $5d^{10}6s^2 6p^4$ RECPs for the polonium atom were erroneous, as they gave unreasonable results. Although the RECPs for Po that included only the $6s^2 6p^4$ shells gave qualitatively reasonable results, Christiansen and co-workers (468), subsequently published an erratum, which showed that both the RECPs for the polonium atom were in error. Recently, Dai and the author (469) recomputed the potential-energy surfaces and the equilibrium geometries using the corrected RECPs reported by Christiansen and co-workers, and an erratum was submitted by the author and Dai. In this chapter, we discuss the results of the correct RECPs for the polonium atom. The optimized (valence $3s3p$) Gaussian basis sets for the selenium, tellurium, and polonium atoms were augmented this with one set of diffuse s, p and one set of six component d-type polarization functions. The exponents of the $3d$-type polarization functions were optimized for the $^1A_1$ states of the trimers near their equilibrium geometries at the CASMCSCF level. The final basis sets for Se, Te, and Po atoms are uniformly of the $(4s4p1d)$ quality.

Dai and the author (467) carried out separate CASMCSCF (complete active space multiconfiguration self-consistent field) calculations for each electronic state of different spin multiplicity (singlet, triplet, and quintet) and spatial symmetry in the C_{2v} group for all three trimers. Then Dai and the author made the multireferences singles + doubles configuration interaction calculations (MRSDCI) including the Davidson correction (MRSDCI+Q) for the low-lying states of these trimers.

Dai and the author used the RCI method to include the spin-orbit effects. The RCI calculation of the A_1 states included all low-lying reference configurations in the C_{2v}^2 double group, namely, the $^1A_1(A_1)$, $^3B_2(A_1)$, $^3A_2(A_1)$, and $^3B_1(A_1)$ relativistic states. The RCI computations included 20 reference configurations. Single + double excitation from these reference configuration resulted in a total of 9621 determinants. The MRSDCI calculations of Dai and the author included up to 1,660,000 CSFs.

3.5.6.1 Se$_3$ Figure 3.36 shows the potential energy surface of 12 electronic states of Se_3 as a function of the Se–Se–Se apex angle θ, while Table 3.62 shows the actual equilibrium geometries and their energy separations. As seen from Figure 3.36, all singlet and most triplet states exhibit isosceles or equilateral triangular geometries. For quintet states and the $^3A_1$ state, the linear electronic geometries are lower or almost degenerate with the bent minima.

The $^1A_1$ ground state exhibits double minima separated by a barrier (Fig. 3.36). These correspond to the $D_{3h}(1A_1')$ closed-ring structure and the $C_{2v}(^1A_1)$ open-chain structure. The apex angle of the C_{2v} minimum is ~115.5° at the CASSCF level, while the Se–Se bond lengths corresponding to two equal sides of the isosceles triangle are 2.27 Å. The existence of double minima for S_3 and Se_3 is documented by several theoretical studies before (462, 464). At the CASSCF level of theory the C_{2v} minimum was found to be 0.2 eV below the

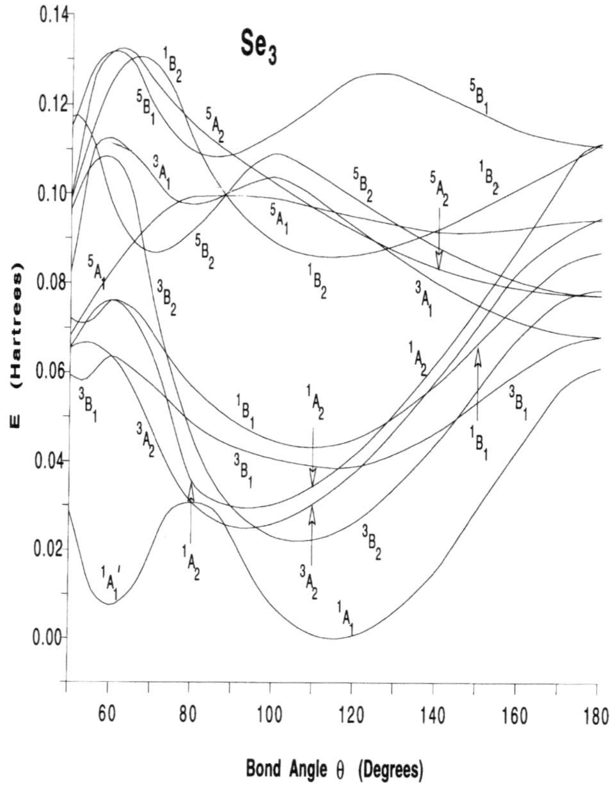

Figure 3.36 Bending potential-energy surfaces of Se_3. (Reproduced from Ref. 467; copyright 1995, American Institute of Physics.)

equilateral triangular (D_{3h}) minimum. The two minima are separated by a saddle point at ~80° with a CASSCF barrier height of ~19 kcal/mole required to convert the C_{2v} minimum to the D_{3h} minimum.

Schnöckel et al. (456) obtained the resonance Raman spectra of Se_3 in Ar and nitrogen matrices, which consisted of 10 fluorescence bands with the origin at 17,000 cm^{-1}. Brabson and Andrews (458) found a well-structured band system in the 480–575-nm region, attributed to Se_3. Their assignment to Se_3 was based on the fact that these bands appeared only in experiments with the microwave source and disappeared in the superheated Knudsen cell experiments. This is primarily because the required concentration of selenium atoms is not accomplished in the thermal experiments so as to produce Se_3.

A significant portion of the 480–575-nm system attributed to Se_3 was found to overlap with the red emission system of Se_2. The ω_e'' values of this system and Se_3 were found to be sufficiently different to distinguish the two systems. Brabson and Andrews (458) attributed their Se_3 bands to be the companion of

Table 3.62 Geometries and Energy Separations of Electronic States of Se_3[a]

		CASSCF			MRSDCI+Q		
Geometry	State	θ_e, deg	R_e, Å	E, eV	θ_e, deg	R_e, Å	E, eV
Bent	$^1A_1$	115.5	2.274	0.00	114.2	2.253	0.00(0.00)
	$^1A_1'$	60.0	2.436	0.20	60.0	2.402	0.14(0.13)
	$^3B_2$	106.7	2.361	0.60	104.9	2.335	0.50(0.62)
	$^3A_2$	93.5	2.346	0.68	91.4	2.308	0.58(0.54)
	$^1A_2$	94.0	2.355	0.80	92.0	2.313	0.71(0.69)
	$^3B_1$	115.5	2.361	1.05	113.3	2.317	1.02(0.99)
		52.6	2.674	1.58	—	—	—
	$^1B_1$	109.4	2.392	1.18	107.5	2.346	1.13(1.13)
		52.9	2.685	1.94	—	—	—
	$^1B_2$	114.3	2.416	2.33	108.7	2.357	2.13(2.05)
	$^5B_2$	70.9	2.602	2.36	—	—	—
	$^5A_1$	146.3	2.602	2.48	—	—	—
	$^3A_1$	80.1	2.573	2.65	—	—	—
	$^5B_1$	85.8	2.673	2.95	—	—	—
Linear	$^1\Sigma_g^+$	180.0	2.434	1.66	180.0	2.373	1.62(1.60)
	$^3\Pi_u$	180.0	2.391	1.85	180.0	2.349	1.57(1.61)
	$^5\Pi_g$	180.0	2.491	2.10	—	—	—
Se + Se_2	$^3P + {}^3\Sigma_g^-$	5.0	2.229	1.16	5.0	2.216	1.08(1.31)
3Se	$3(^3P)$	180.0	10.0	3.57	180.0	10.0	3.77(4.17)

[a]In this table θ_e refers to the apex angle of the isosceles triangle while R_e refers to two equal sides of the isosceles triangle. For the diatomic, it is simply the Se–Se distance. The values in parentheses are the Davidson corrected energies.

Source: Reference 467.

the fluorescence band system composed of 10 bands observed by Schnöckel et al. (456). Brabson and Andrews (458) noted that the electronic transition in question must involve a large change in the equilibrium internuclear distance, and thus the origin of the transition could not be observed both in absorption and fluorescence. Assuming that the origin of the Schnöckel et al. system lies at 17,000 cm^{-1}, Brabson and Andrews deduced the T_{00} for the excited state as 17,005 (2.11 eV) cm^{-1} in absorption and 16,819 cm^{-1} (2.09 eV) in fluorescence. They deduced the ω_e value to be between 294 and 318 cm^{-1}.

As seen from Table 3.62, the most consistent electronic transition for the 480–575-nm band system of Brabson and Andrews as well as the fluorescence system of Schnöckel et al. is the $^1B_2$–X^1A_1 electronic transition. This transition is dipole-allowed, and adiabatic MRSDCI and MRSDCI+Q transition energies of 2.13 and 2.05 eV, respectively, computed by Dai et al. (467) fit with the experimental fluorescence T_{00} value of 2.09 eV. Since the fluorescence is due to an adiabatic transition, Dai et al. suggested that the T_{00} from fluorescence should correspond to the theoretically computed T_e = 2.05 eV. The zero-point corrections in the ground and $^1B_2$ states are similar and should approximately cancel out.

As seen from Table 3.62 at the MRSDCI level the Se–Se side bond lengths change from 2.25 Å in the $^1A_1$ state to 2.357 Å in the $^1B_2$ state. The apex angle changes from 114° to 108.7°. This means that the Se–Se base bond length changes from 3.78 to 3.83 Å. All these evidences are consistent with the suggestion of Brabson and Andrews that there is a large change in the equilibrium Se–Se distance in the excited state. This, combined with the fact that there is no other excited electronic state to which the transition is dipole-allowed from the ground state, leads to unequivocal assignment of the 480–575-nm band system to the $^1B_2$–X^1A_1 electronic transition.

Schnöckel et al. (457) deduced an apex bond angle of 110–120° from the C_{2v} ground state of Se_3 based on their $2\nu_1$ resonance Raman band with resolved spectra of mixed isotopic molecule (67% ^{76}Se and 38% ^{82}Se). Brabson and Andrews (458) estimated the IR band width of Se_3 to be 7 cm^{-1} assuming a bond angle of 115° for the ground state of Se_3. The MRSDCI bond angle of 114.2° computed by Dai et al. (467) is in excellent agreement with this assumption. A previous MD/DF method yielded somewhat larger bond angle of 118° compared to the value reported by Dai et al. Hohl et al. (466) obtained the Se–Se side bond lengths as 2.20 Å compared to our value of 2.25 Å. Von Niessen et al. (464) reported the ionization energies and electron affinities with assumed geometries for Se_3. At the SCF level they find that the D_{3h} structure is 23 kcal/mol more stable than the C_{2v} structure. To the contrary, Dai et al. (467) find that at the CASSCF and MRSDCI levels of theories the $^1A_1(C_{2v})$ structure to be more stable, with the C_{2v} structure 3.5 kcal/mol more stable than the D_{3h} structure at the MRSDCI+Q level. The observed spectra as well as high-level theoretical computations on S_3 support our finding that the $C_{2v}(^1A_1)$ structure is more stable.

The dissociation energy for the process

$$Se_3(^1A_1) \longrightarrow Se(^3P) + Se_2(^3\Sigma_g^-)$$

was computed by Dai et al. (467) as 1.31 eV at the MRSDCI+Q level. The atomization energy to dissociate Se_3 into three selenium atoms (^3P) was computed by Dai et al. (467) as 4.17 eV at the MRSDCI+Q level. From the computed dissociation and atomization energies of Se_3, Dai et al. (467) deduced that the dissociation energy D_e of the selenium dimer as 2.86 eV. This result is consistent with a direct CASSCF/FOCI result of 2.91 eV for the D_e of Se_2 obtained by the author. There are three experimental D_e values for Se_2, but the author recommended a value of 3.19 eV, which is in reasonable agreement with the Dai et al. (467) result. On the basis of 0.3 eV error in the MRSDCI+Q D_e of Se_2, the atomization energy computed by Dai et al. for Se_3 should be at least 0.6 eV in error, and thus a lower bound AE of 4.8 eV was recommended for Se_3.

3.5.6.2 *Electronic States and Geometries of Te_3* Table 3.63 shows the optimized geometries and energy separations of the electronic states of Te_3 and the dissociation limit (Te + Te_2, and 3Te) obtained by using the CASMCSCF and the MRSDCI+Q calculations. Figure 3.37 shows the bending potential-energy surfaces of Te_3 as a function of the Te–Te–Te apex bond angle θ. Analogous to Se_3, the $^1A_1$ state of Te_3 exhibits a double minimum. The two minima are nearly degenerate. The narrow bent minima for $^3B_1$ and $^1B_1$ near $\theta \sim 60$ and the curve crossings exhibited by Te_3 are similar to the corresponding states of Se_3. The barrier for the conversion of the C_{2v} structure to the D_{3h} structure is smaller for Te_3 as expected for a heavier system.

As seen from Table 3.63, at the CASMCSCF and MRSDCI levels, the $^1A_1(C_{2v})$ state is 0.01 and 0.04 eV above the $^1A_1'(D_{3h})$ state, respectively, but at the MRSDCI+Q level the $^1A_1(C_{2v})$ state is 0.04 eV lower than the $^1A_1'$ state. The low-lying bent $^3B_2$, $^3A_2$, $^1A_2$, $^3B_1$ (obtuse bond angle minimum), and $^1B_1$ (obtuse bond angle minimum) states are 0.40–0.86 eV above the ground state at the CASMCSCF level. For the acute bond angle minima of the $^3B_1$ and $^1B_1$ states, the energy separations are up to 0.5 eV lower than the corresponding energy separations of Se_3.

Table 3.63 Geometries and Energy Separations of Electronic States of Te_3

		CASSCF			MRSDCI+Q		
Geometry	State	θ_e, deg	R_e, Å[a]	E, eV	θ_e, deg	R_e, Å[a]	E, eV
Bent	$^1A_1$	113.2	2.673	0.01	111.5	2.637	0.04(−0.04)
	$^1A_1'$	60.0	2.827	0.00	60.0	2.775	0.00(0.00)
	$^3B_2$	105.4	2.750	0.40	103.6	2.711	0.33(0.37)
	$^3A_2$	92.0	2.736	0.49	90.5	2.683	0.45(0.33)
	$^1A_2$	92.7	2.748	0.58	90.8	2.693	0.54(0.45)
	$^3B_1$	110.4	2.767	0.78	105.6	2.720	0.78(0.69)
		54.7	3.031	1.13	—	—	—
	$^1B_1$	106.2	2.790	0.86	103.4	2.734	0.84(0.78)
		54.3	3.061	1.42	—	—	—
	$^1B_2$	115.0	2.816	1.88	108.1	2.736	1.77(1.62)
Linear	$^1\Sigma_g^+$	180.0	2.844	1.34	180.0	2.777	1.31(1.09)
	$^3\Pi_u$	180.0	2.792	1.54	180.0	2.725	1.29(1.15)
	$^3\Sigma_u^+$	180.0	2.934	1.66	180.0	2.851	1.72(1.42)
	$^1\Pi_u$	180.0	2.866	1.95	—	—	—
	$^3\Pi_g$	180.0	2.973	2.04	—	—	—
Te + Te_2	^3P + $^3\Sigma_g^-$	5.0	2.615	0.96	5.0	2.589	0.86(0.97)
3Te	$3(^3P)$	180.0	10.0	2.90	180.0	10.0	2.97(3.24)

[a]For θ_e and R_e, see Table 3.62.
[b]The values in parenthesis are the Davidson corrected energies.
Source: Reference 467.

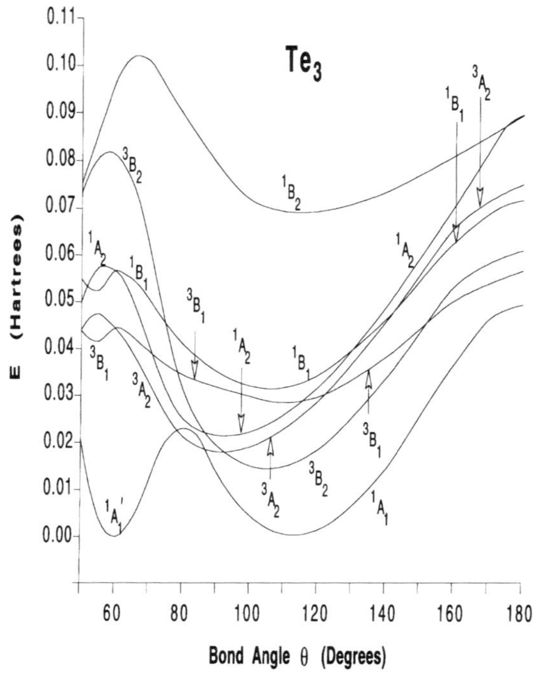

Figure 3.37 Bending potential-energy surfaces of Te_3. (Reproduced from Ref. 467; copyright 1995, American Institute of Physics.)

Schnöckel et al. (456, 457) obtained the resonance Raman spectra of the matrix-isolated Te_3 and reported the vibrational frequency for the symmetry stretching as 206.33 cm^{-1} for $^{26}Te_3$ and 203.1 cm^{-1} for $^{130}Te_3$. The laser excitation of Te_3 in nitrogen matrix at 647 nm produced a series of resonance Raman bands with the origin ~15,454 cm^{-1}. Hassanzadeh et al. (459) studied the absorption spectra of tellurium clusters in solid argon. They found Te_2, Te_3, and two structural isomers for Te_4. They found seven absorption bands attributed to Te_3 in the 650–705-nm region. The origin of the absorption band was located at ~1.87 eV. The structural features in the absorption spectra of Te_3 were found to be similar to Se_3. The vibrational fundamentals for Te_3 supported a C_{2v} structure for Te_3.

As seen from Figure 3.37, Dai and the author (467) find that the potential-energy surface of the $^1A_1$ state contains double minima. However, the $^1B_2$ excited state does not exhibit a minimum near $\theta \sim 60°$, and it exhibits a saddle point. Consequently, although the $^1A_1$ state contains two nearly degenerate double minima, the observed spectrum is most consistent with the $^1B_2$–X^1A_1 system analogous to the assignment for Se_3.

Dai and the author (467) favored the assignment of the 650–705-nm absorption band system of Hassanzadeh et al. as well as the resonance Raman spectrum of Schnöckel with origin at 15,454 cm^{-1} to the $^1B_2$–X^1A_1 dipole-allowed transition. Their computed adiabatic MRSDCI and MRSDCI+Q transition energies of 1.77 and 1.62 eV, respectively, were found to be close to the absorption band center at 1.87 eV and the fluoresence spectrum near 1.91 eV. Furthermore, there are no other bent minima in this region. Dai and the author noted that the Te–Te side bond lengths change from 2.637 to 2.736 Å for the $^1B_2$–X^1A_1 electronic transition. The corresponding Te–Te base bond length increases from 4.36 to 4.43 Å. The geometry changes in this transition are in accord with the observed spectral features and the corresponding state of Se_3.

The energy to dissociate Te_3 into the tellurium atom (^3P) and Te_2 dimer ($^3\Sigma_g^-$) was calculated as 1.01 eV at the MRSDCI+Q level, 0.30 eV lower than the dissociation energy of Se_3. The atomization energy of Te_3 was computed as 3.28 eV at the MRSDCI+Q level, which is 0.89 eV lower than the AE of Se_3. From the dissociation energy to form Te_2 + Te and the AE of Te_3, Dai and the author deduced the dissociation energy D_e of Te_2 as 2.27 eV, in reasonable agreement with the experimental value of 2.68 eV. Dai and the author attributed the discrepancy of 0.4 eV to basis-set limitations.

The average MRSDCI+Q atomization energies per atom for Te_3 and Te_2 are 1.08 and 1.135 eV, respectively, suggesting that Te_2 has a slightly more average bond energy compared to Te_3. Furthermore, the energy for the $Te_3(^1A_1) \rightarrow 3Te(^3P)$ process was computed by Dai and the author as 2.27 eV at the MRSDCI+Q level. This means that larger clusters of tellurium would dissociate mostly into Te_2 as opposed to Te_3. Furthermore, Te_3 should be formed primarily through the $Te_2 + Te \rightarrow Te_3$ process. Hassanzadeh and Andrews (455) have suggested this process as the growth mechanism for Te_3 and larger clusters. This is also consistent with the absence of Te_3 in the spectrum on annealing, since recombination of tellurium atoms to yield Te_2 is more favorable than formalation of Te_3 by adding to Te_2. Te_n clusters were found to decompose mostly into Te_2 in the superheating experiments.

3.5.6.3 Electronic States and Geometries of Po$_3$ Dai and the author (467) carried out CASSCF/MRSDCI computations for the electronic states of Po_3 as well as the ground state of the dimer. They noted in their paper that both polonium RECPs (the ones that retained the $6s^26p^4$ shells in the valence space and the ones that retained the $5d^{10}6s^26p^4$ shells in the valence space) had some problems. One of them gave reasonable Po–Po bond lengths, but unreasonable results, when spin–orbit effects were included, while the other set of RECPs gave unreasonable Po–Po bond lengths but reasonable spin–orbit effects. After publication of the work of Dai and the author on Po_3, Christiansen and co-workers published an erratum (468) correcting the RECPs of the polonium atom. Dai and the author (469) investigated the electronic states of Po_3 using the corrected RECP parameters reported by Christiansen and co-workers. In this section all results reported are the corrected results employing the corrected RECPs and the same CASSCF/MRSDCI techniques employed before for the electronic states of Po_3.

Table 3.64 shows the optimized geometries and energy separations of the electronic states of Po_3 and the dissociation limits (Po + Po_2, and 3Po) obtained by using the CASMCSCF and the MRSDCI(+Q) methods using the corrected RECPs. Table 3.65 shows the properties of the low-lying electronic states of Po_3 including spin-orbit effects. Figure 3.38 shows the bending potential-energy surfaces of Po_3 as a function of the Po–Po–Po apex bond angle θ. As seen from Figure 3.38, the $^1A_1$ ground state exhibits double minima similar to Se_3 and Po_3. The apex bond angle $\theta = 111.2°$ is also similar to Se_3 and Te_3 for the C_{2v} minimum. The Po–Po side bond lengths for the two equal sides of the isoceles triangle are shorter for the C_{2v} minimum compared to the D_{3h} minimum. Although the C_{2v} and D_{3h} minima are nearly degenerate at the CASSCF level, the D_{3h} minimum prevails at the MRSDCI level.

There are several low-lying with bent geometries, as seen from Table 3.64. Among these the $^1B_1$ and $^1B_2$ states are of special interest, as transitions to these states are optically allowed. These transitions are predicted to take place in the 1.17- and 1.68-eV regions by the MRSDCI and CASSCF computations, respectively. The geometries of these two states are closer to the C_{2v} minimum of the $^1A_1$ state. The $^3B_2$, $^3A_2$, $^1A_2$, and $^3B_1$ (obtuse-angle minimum) are

Table 3.64 The Geometries and Energy Separations of Electronic States of Po_3

		CASSCF			MRSDCI+Q		
Geometry	State	θ_e, deg[a]	R_e, (Å)[a]	E (eV)	θ_e (deg)[a]	R_e (Å)[a]	E, eV[b]
Bent	$^1A_1'$	60.0	2.976	0.00	60.0	2.940	0.00(0.00)
	$^1A_1$	112.8	2.832	0.07	111.2	2.797	0.31(0.21)
	$^3B_2$	109.1	2.893	0.40	105.0	2.874	0.55(0.49)
	$^3A_2$	90.7	2.886	0.47	87.5	2.842	0.63(0.53)
	$^1A_2$	91.4	2.899	0.56	87.7	2.861	0.70(0.59)
	$^3B_1$	108.8	2.927	0.74	102.5	2.895	1.13(1.09)
		55.2	3.179	1.02	—	—	—
	$^1B_1$	105.4	2.945	0.81	101.9	2.902	1.17(1.14)
		54.7	3.201	1.30	—	—	—
	$^1B_2$	118.7	2.976	1.68	—	—	—
	$^3A_1$	77.1	3.114	1.79	—	—	—
Linear	$^1\Sigma_g^+$	180.0	2.970	1.04	180.0	2.900	1.39(1.20)
	$^3\Delta_u$	180.0	3.050	1.34	180.0	3.016	1.60(1.31)
	$^3\Pi_u$	180.0	2.939	1.37	180.0	2.873	1.51(1.40)
	$^1\Pi_u$	180.0	2.994	1.68	—	—	—
	$^3\Sigma_u^-$	180.0	3.066	1.77	—	—	—
	$^1\Sigma_u^-$	180.0	3.074	2.05	—	—	—
	$^1\Sigma_u^+$	180.0	3.011	2.14	—	—	—
Po + Po_2	$^3P + {}^3\Sigma_g^-$	5.0	2.766	0.96	5.0	2.754	1.22(1.18)
3Po	$3(^3P)$	180.0	10.0	2.69	180.0	10.0	3.30(3.59)

[a] θ_e and R_e, see Table 3.62.
[b] The values in parentheses are the Davidson corrected energies.
Source: Reference 467.

Table 3.65 Properties of Low-Lying Electronic States of Po_3 Including the Spin–Orbit Effect

State	θ_e, deg	R_e, Å	E, eV	Weight, %			
				$^1A_1$	$^3B_2$	$^3A_2$	$^3B_1$
$^1A_1'$	60	2.940	0.00	100	—	—	—
$^1A_1'(A_1')$	60	2.971	−0.81	71	1	12	12
$^1A_1$	111.2	2.797	0.31	100	—	—	—
$^1A_1(A_1)$	111.6	2.787	−0.55	70	1	11	14

Source: Reproduced from Reference 469.

0.50–1.13 eV above the ground state. These energy separations are comparable to the corresponding energy separations of Te_3.

The energy to dissociate Po_3 into the Polonium atom (^3P) and Po_2 dimer ($^3\Sigma_g^-$) for the $Po_3(^1A_1) \rightarrow Po(^3P) + Po_2(^3\Sigma_g^-)$ process was computed as 1.22 eV at the MRSDCI+Q level. The atomization energy for the $Po_3(^1A_1) \rightarrow 3Po(^3P)$

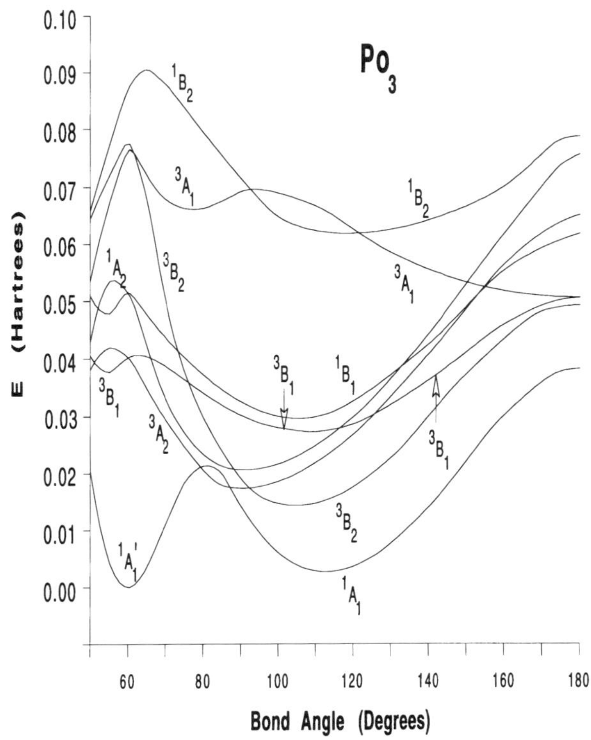

Figure 3.38 Bending potential-energy surfaces of Po_3. (Reproduced from Ref. 467; copyright 1995, American Institute of Physics.)

process was computed as 3.30 eV at the MRSDCI+Q level. However, the spin–orbit stabilization for the Po atom is expected to be larger than for the trimer, and we expect the atomization energy to be lower for the Po_3 trimer, due to spin–orbit effects.

3.5.6.4 Relative Comparison of Group VI Trimers The $^1A_1'$ states of Se_3, Te_3, and Po_3 are represented predominantly by the $1a_1'^2 2a_1'^2 1a_1''^2 1e'^4 2e'^4 1e''^4$ configuration in the D_{3h} symmetry ($1a_1^2 2a_1^2 3a_1^2 4a_1^2 1b_2^2 2b_2^2 1b_1^2 2b_1^2 1a_2^2$ in C_{2v} symmetry), with the coefficients > 0.91 for Se_3, Te_3, and Po_3. Promotion of two electrons from $2b_1$ ($1e''$ in D_{3h}) to the $3b_2$ orbital results in the configuration ($1a_1^2 2a_1^2 3a_1^2 4a_1^2 1b_2^2 2b_2^2 3b_2^2 1b_1^2 1a_2^2$) of the obtuse minimum of the $^1A_1$ state (the coefficients are between 0.84 and 0.87 for Se_3, Te_3, and Po_3). The $4a_1^2 3b_2^2 2b_1^2$ configuration forms the second leading configuration with the coefficients of 0.30–0.36 for Se_3, Te_3, and Po_3. Therefore, the saddle points at $\theta = 79$–$81°$ and the shapes of the PES of the $^1A_1$ states of Se_3, Te_3, and Po_3 are due to the avoided crossing of $(\cdots)\ 4a_1^2 1b_2^2 2b_2^2 1b_1^2 2b_1^2 1a_2^2$ with $(\cdots)\ 4a_1^2 1b_2^2 2b_2^2 3b_2^2 1b_1^2 1a_2^2$.

The low-lying $^3B_2$ state arises from $4a_1^2 3b_2^2 2b_1^1 1a_2^1$ configuration or relative to the $^1A_1$ state it involves an electronic excitation from $1a_2$ to $2b_1$. For the $^3A_2$ and $^1A_2$ states Dai and the author found the same configuration of $4a_1^2 3b_2^1 2b_1^1 1a_2^2$ (the coefficients are between 0.86 and 0.88 for Se_3, Te_3, and Po_3). Dai and the author found the same leading configuration of $4a_1^1 3b_2^2 2b_1^1 1a_2^2$ for the obtuse-angle minima of the $^3B_1$ and $^1B_1$ states with the coefficients between 0.72 and 0.78 for Se_3, Te_3, and Po_3. Also, Dai and the author found the same leading configuration of $4a_1^2 3b_2^1 2b_1^2 1a_2^1$ for the acute-angle minima and the saddle points of these two states with the coefficients about 0.90 for Se_3, Te_3, and Po_3.

For the $^1A_1$ states of Se_3, Te_3, and Po_3, the highest occupied orbitals are composed of

$$\psi[4a_1] = p_{z1} - (p_{z2} + p_{z3}) + (p_{y2} + p_{y3}) + s_1 + (s_2 + s_3)$$
$$\psi[3b_2] = -p_{y1} + (p_{y2} + p_{y3}) + (s_2 - s_3)$$
$$\psi[1b_1] = p_{x1} + (p_{x2} + p_{x3})$$

The $1b_1$ orbital is perpendicular to the molecular plane, while the $4a_1$ and $3b_2$ orbitals are in-plane orbitals.

The dipole moments (μ_e) of the $^1A_1$ states of Se_3, Te_3, and Po_3 with C_{2v} geometries are 0.59, 0.66, and 0.85 D, respectively, with (apex atom) + (base atoms) polarities, suggestive of depletion of electronic density at the apex atom. The dipole moments of the excited $^3A_2$ and $^1A_2$ states are found larger than other states, implying the stronger ionicities of the bond between apex atom and base atoms for the $^3A_2$ and $^1A_2$ states. The excess electronic charge at the base atoms leads to a much longer bond length between the base atoms due to repulsion. This explains the formation of obtuse angle minima for the three trimers.

The Mulliken population analyses of the low-lying electronic states of Se_3, Te_3, and Po_3 reveal that for almost all states the apex atom has an effective Mulliken population of 5.73–5.96, while each base atom has an effective Mulliken population of 6.02–6.14, implying that the same polarity of the bond between the apex atom and each base atom as inferred from the dipole moments.

For the $^1A_1C_{2v}$ states, the apex atom and each base atom have an effective Mulliken population of $4s^{1.82}4p^{3.69}4d^{0.32}$ and $4s^{1.92}4p^{4.04}4d^{0.12}$ for Se_3, $5s^{1.87}5p^{3.75}5d^{0.23}$ and $5s^{1.93}5p^{4.02}5d^{0.12}$ for Te_3, $6s^{1.93}6p^{3.60}6d^{0.28}$ and $6s^{1.96}6p^{3.73}6d^{0.13}$ for Po_3, respectively. The increased 6*s* population in the case of Po_3 compared to Te_3 could be attributed to the relativistic mass–velocity effects. For the $^1A_1'$ states all three atoms have the same effective Mulliken populations of $4s^{1.91}4p^{3.90}3d^{0.19}$, $5s^{1.93}5p^{3.91}4d^{0.16}$, and $6s^{1.99}6p^{3.90}5d^{0.07}$ for Se_3, Te_3, and Po_3, respectively.

All states of Se_3, Te_3, and Po_3 exhibit deviation from the atomic ns^2np^4 population. The *nd* populations are 0.16–0.32 for the apex atoms and 0.16–0.19 for each base atom due to *d* polarization. The *np* populations are 3.42–3.91 for the apex atoms and 3.87–4.05 for each base atom. The *ns* populations are 1.82–1.98 for the apex atoms and 1.91–1.98 for each base atom.

Figures 3.39*a–f* contain the overall comprehensive trends for Se_3, Te_3, and Po_3. Figures 3.39*a,b* depict the bond energy trends. Dai et al. find that both the dissociation or atomization energies are the largest for Se_3 and smallest for Po_3 as expected. However, note that the spin–orbit effects were not included in computing the AEs of Po_3.

Figure 3.39*c* shows the dipole moment of three selected states of Se_3, Te_3, and Po_3. Regardless of the $^1A_1$ ground states, in the $^3A_2$ (or $^1A_2$) states with the higher μ_e values, or the $^3B_2$ states with the lower μ_e values, the μ_e values of Te_3 are found to be the largest, while the μ_e values of Po_3 are found to be the smallest. For the overlap population trend as seen from Figure 3.39*d*, the reverse trends of the μ_e are found. Figures 3.39*e, f* show population analyses for the $^1A_1$ states of Se_3, Te_3, and Po_3. The total populations of the apex atoms of the $^1A_1$ states of Se_3, Te_3, and Po_3 are all below the dashed lines, which indicate transfer of electronic charge from the apex to the base atoms. Both Figures 3.39*c,e* indicate that the strongest ionicities of the bonds between the apex atom and each base atom.

3.6 COMPARISON OF THE PROPERTIES AND PERIODIC TRENDS AMONG DIMERS

In this section we compare the spectroscopic properties of the low-lying electronic states of dimers of some of the groups for which complete sets of data are available. Before comparison of the properties of dimers is made, it is relevant to compare a few atomic properties within some of the main-group elements considered here. Figure 3.40 shows the $^2P(ns^2np^1)$–$^4P(ns^1np^2)$ energy separations of B–Tl and C^+–Pb^+ as obtained from experimental atomic spectral data. This comparison reveals an interesting trend. In going from In to Tl as well as Sn^+ to Pb^+, we see a dramatic rise in this energy separation. This increase is

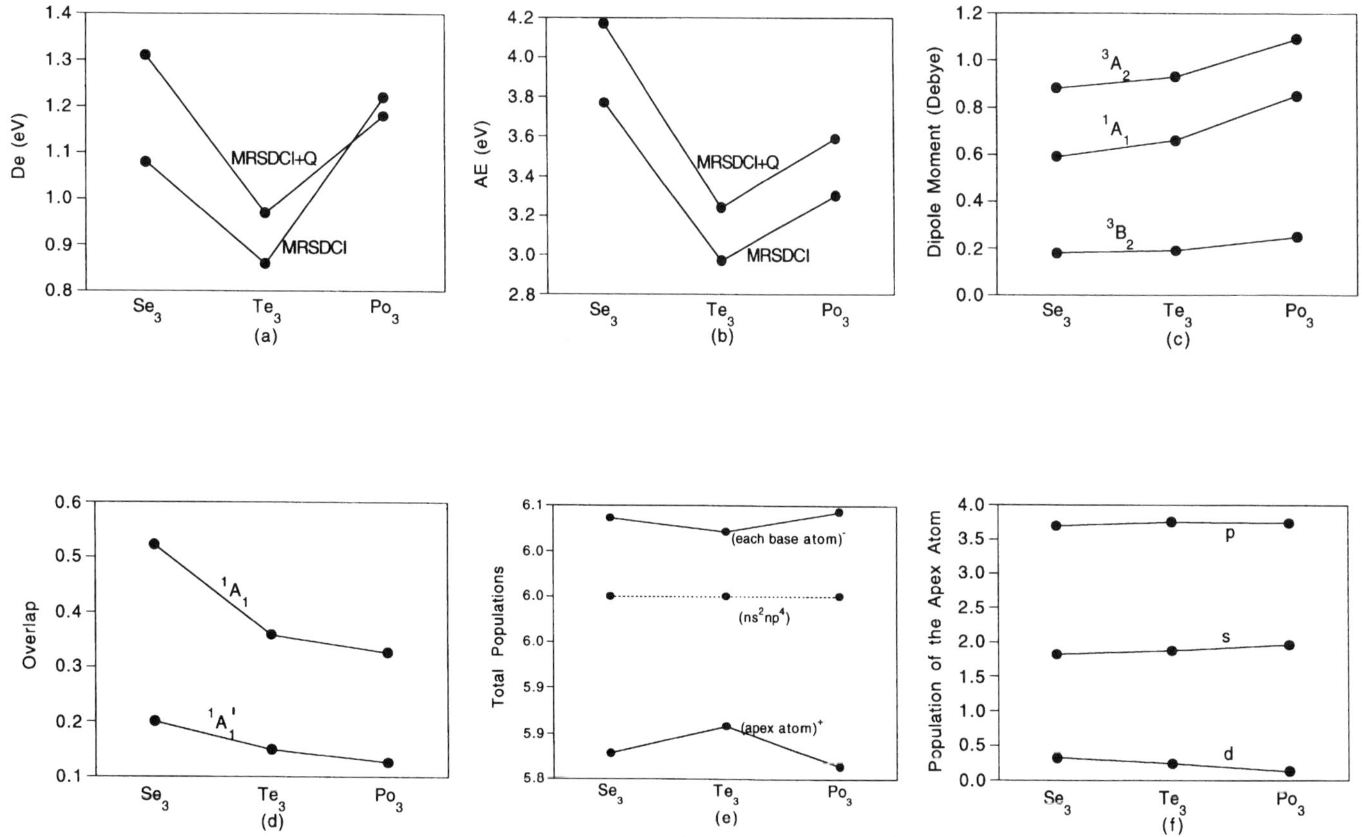

Figure 3.39 (*a–f*) Comprehensive periodic bends for the various properties of Se_3–Po_3. (Reproduced from Ref. 467; copyright 1995, American Institute of Physics.)

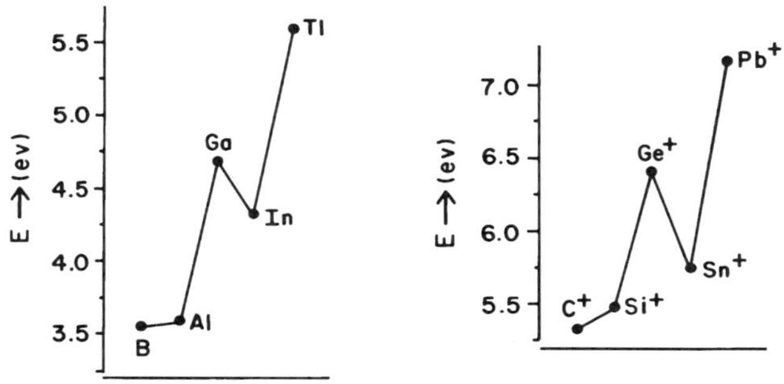

Figure 3.40 ^2P–^4P energy separations of B to Al and C^+ to Pb^+. (Reprinted from Ref. 52; copyright 1989, American Chemical Society.)

due primarily to a phenomenon referred to as the relativistic inert-pair effect. The relativistic mass–velocity contraction stabilizes the $6s$ orbital of the atoms Au to At. As a result, the $6s^2$ shell is quite stabilized and does not participate in bonding. Consequently, the $6s$–$6p$ promotion energy dramatically increases in comparison to the corresponding $5s$–$5p$ promotion energy as evidenced from Figure 3.40.

The other important relativistic effect that has substantial impact on the bonding energies and spectroscopic properties of very heavy dimers is the spin–orbit effect, which is also a relativistic effect. This is illustrated in Figure 3.41 for the Pb and Sn atoms. Figure 3.42 shows the periodic trend for $^3P_0$–$^3P_2$ splittings of atoms C to Pb of group IV elements. As seen from these figures, the spin–orbit splitting is significantly larger for Pb than for Sn. This changes the coupling of the Pb atom to intermediate in that the ground state of the Pb atom is 88% $^3P_0$ and 12% $^1S_0$. Similarly, the $J = 2$ state of Pb is 70% $^3P_2$ and 30% $^1P_2$.

Figure 3.43 compares the dissociation energies of the heavy dimers of three of the groups considered in this investigation. The dashed line below the solid line is the theoretical curve. As seen from Figure 3.43, the theoretical trend mimics the experimental trend closely. With the exception of Pb_2, the theoretical D_e is always lower than the corresponding experimental value. As seen from Figure 3.43, as one goes down a column of the periodic table, D_e decreases. The decreases in the D_e are especially large in comparing In_2 to Tl_2 or Sn_2 to Pb_2. The As_2–Bi_2 curve is almost collinear. The experimental D_e (266) of Tl_2 is a bit uncertain, although it has been corrected (267). The sharp drop in the D_e in moving down from In to Tl is primarily a consequence of the inert-pair effect and the spin–orbit effect. As mentioned above, the $6s^2$ shell of Tl is inert. The spin–orbit effects further weaken the bonding, since the atoms are more stabilized compared to the dimer. The destabilization of the bond in Pb_2 compared to Sn_2 is due primarily to the spin–orbit coupling, since the D_e obtained without the spin–orbit integrals is approximately twice the corresponding value with the spin–orbit term.

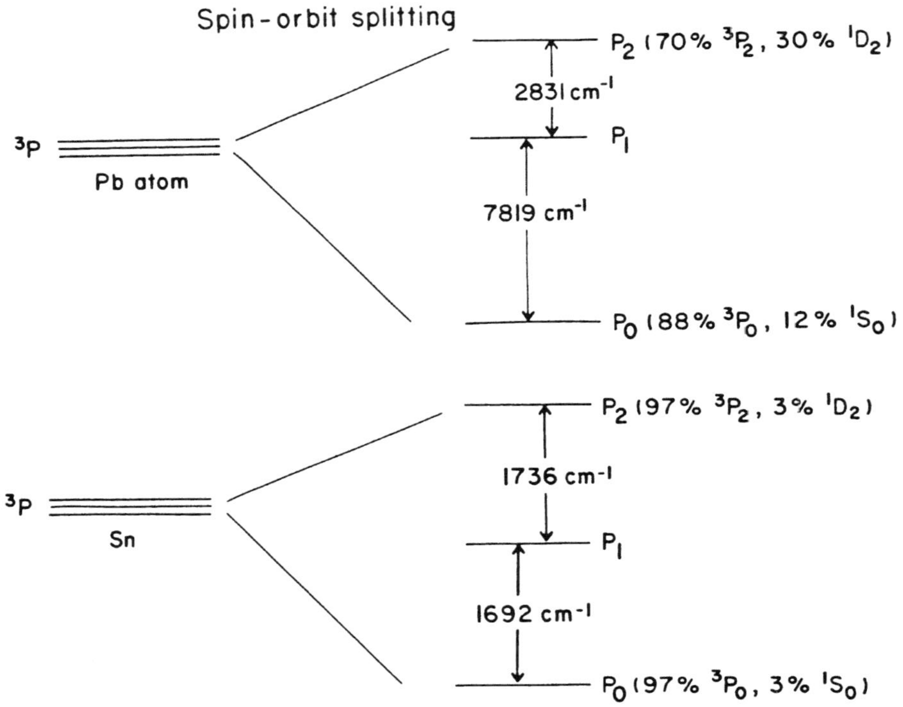

Figure 3.41 Spin–orbit splittings of Sn and Pb atoms. (Reprinted from Ref. 52; copyright 1989, American Chemical Society.)

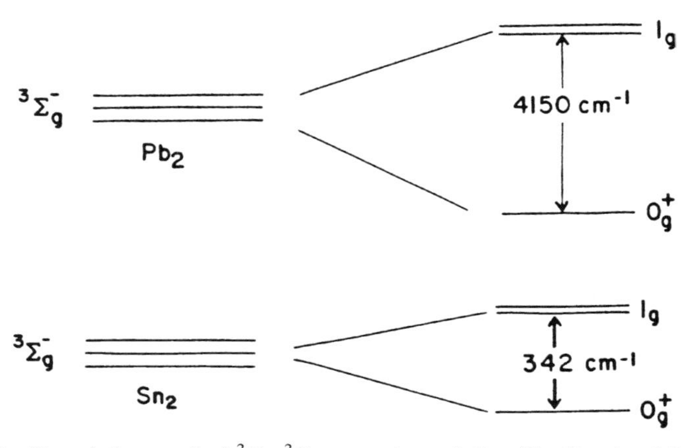

Figure 3.42 The relative trend of $^3P_0$–$^3P_2$ separations of C to Pb. (Reprinted from Ref. 52; copyright 1989, American Chemical Society.)

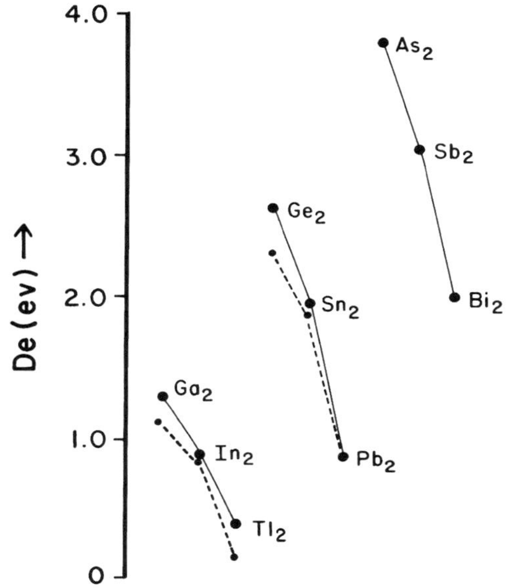

Figure 3.43 Comparison of the dissociation energies of Ga_2 to Bi_2. The dashed line below the solid line (experimental values) is the theoretical curve. (Reproduced from Ref. 2; copyright 1990, American Chemical Society.)

Figure 3.44 compares the equilibrium bond lengths (R_e values) of Ga_2 to Bi_2. As seen from Figure 3.44, the bond length increases as one goes down the periodic table within a column. The increase is especially large in moving from In to Tl due to both the inert-pair effect and the spin–orbit effect. The bond length decreases as one moves toward the right of the periodic table (i.e., from Ga_2 to As_2 and from In_2 to Sb_2). This is primarily because of increases in the bond order as one moves to the right until one reaches group V. The dimers of As and Sb form almost a triple bond, while the bond order of Bi_2 decreases to 2.2 mainly because of spin–orbit coupling of ${}^1\Sigma_g^+(0_g^+)$ with ${}^3\Pi_g(0_g^+)$. Thus, Ga_2, Ge_2, and As_2 form approximately single, double, and triple bonds, respectively, while the bond orders for Tl_2, Pb_2, and Bi_2 are much smaller. The ratio of bond orders as obtained from the ratio of Mulliken overlap populations for In_2 and Sb_2 is 2.62.

Next we compare the energy separations of some spectroscopically interesting states of the various dimers. Figure 3.45 compares the experimental transition energy for the ${}^3\Sigma_u^-(0_u^+)$–$X^1\Sigma_g^+(0_g^+)$ transition of As_2 to Bi_2. As seen from Figure 3.45, this transition energy decreases as one goes down the periodic table. This is the usual trend except that the decrease is smaller for Bi_2 primarily because of spin–orbit coupling.

Figure 3.47 compares the ${}^3\Pi_u(0_u^+)$–$X^3\Sigma_g^-(0_g^+)$ energy separations of Ge_2 to Pb_2. The spin–orbit effects (Fig. 3.46) are substantially larger for both the states

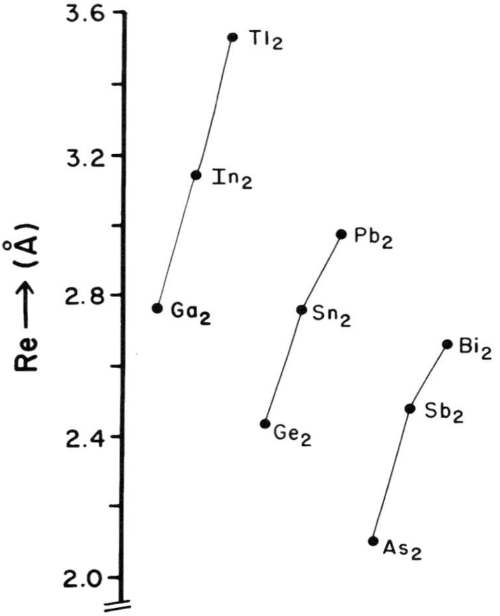

Figure 3.44 Comparison of the R_e values of the ground states of Ga_2 to Bi_2. (Reproduced from Ref. 2; copyright 1990, American Chemical Society.)

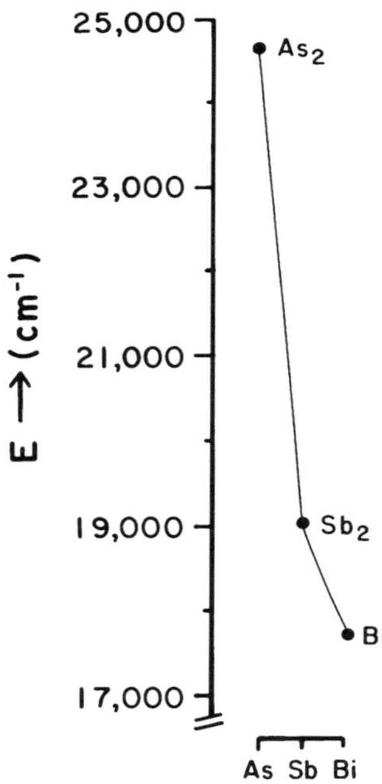

Figure 3.45 Comparison of ${}^3\Sigma_u^-(0_u^+)$–$X\,{}^1\Sigma_g^+(0_g^+)$ energy separations of As_2, Sb_2, and Bi_2. (Reproduced from Ref. 2; copyright 1990, American Chemical Society.)

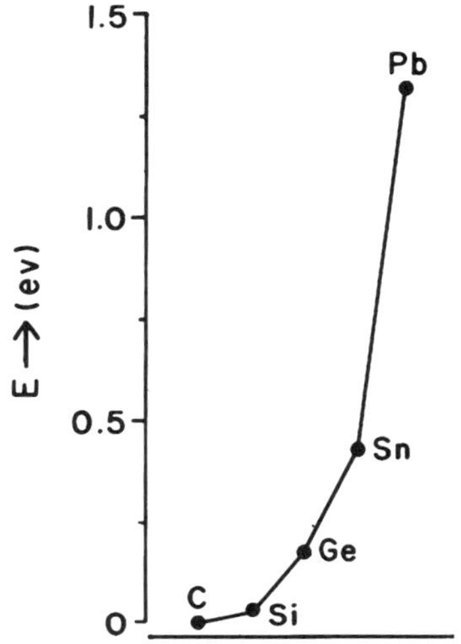

Figure 3.46 Spin–orbit splittings of C–Pb.

of group IV dimers. Note the dramatic increase in the energy separation for Pb_2. This is primarily a consequence of the large spin–orbit splitting of the Pb atom, as seen in Figure 3.46. For both Ge_2 and Si_2 the ${}^3\Pi_u$ state is very low-lying and has a very small spin–orbit splitting. For Si_2, it is, in fact, speculated that ${}^3\Pi_u$ could be the ground state [37].

Figure 3.48 shows the periodic trend for the ${}^3\Sigma_u^-(0_u^+)$–$X^3\Sigma_g^-(0_g^+)$ separation of Ge_2, Sn_2, and Pb_2. This exhibits a very interesting trend as seen from Figure 3.48. The tin dimer follows the trend of decrease in the energy separation as one goes down the periodic table. However, this energy separation increases

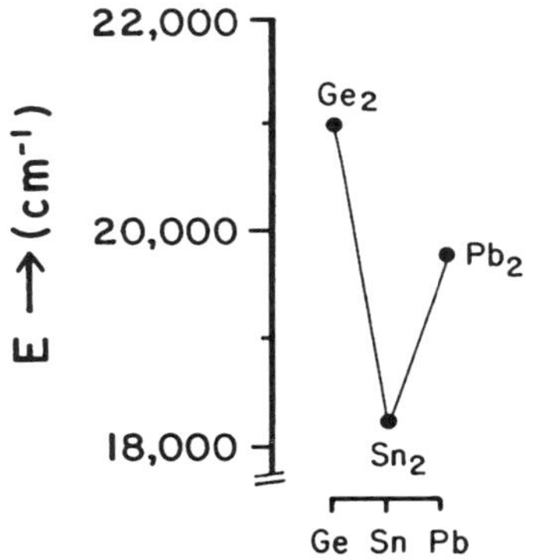

Figure 3.47 Comparison of ${}^3\Pi_u(0_u^+)$–$X^3\Sigma_g^-(0_g^+)$ energy separations of Ge_2, Sn_2, and Pb_2. (Reproduced from Ref. 2; copyright 1990, American Chemical Society.)

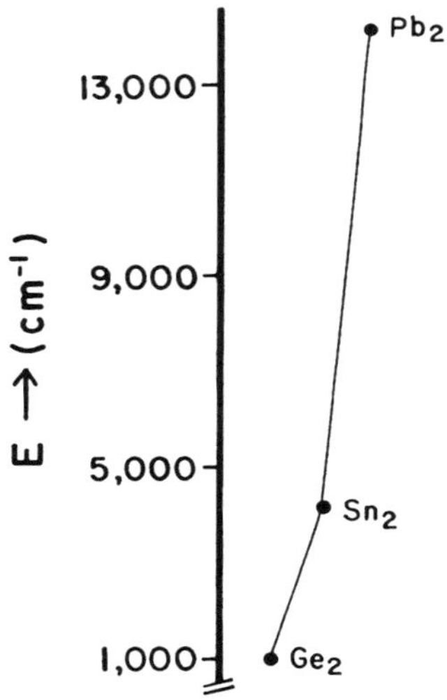

Figure 3.48 Comparison of ${}^3\Sigma_u^-(0_u^+)$–$X^3\Sigma_g^-(0_g^+)$ energy separations of Ge_2, Sn_2, and Pb_2. (Reproduced from Ref. 2; copyright 1990, American Chemical Society.)

in comparing Sn_2 with Pb_2 primarily because of spin–orbit effects (Fig. 3.46). Consequently, relativistic effects seem to play an important role in the spectroscopic properties of the sixth-row dimers.

REFERENCES

1. M. L. Mandich, W. D. Reents, and V. E. Bondybey, "Main Group Clusters: A Review," in *Atomic and Molecular Clusters*, E. Bernstein, ed., Elsevier, 1990.
2. K. Balasubramanian, *Chem. Rev.* **90,** 93, (1990).
3. M. D. Morse, *Chem. Rev.* **86,** 1049 (1986).
4. J. Koutecky and P. Fantucci, *Chem. Rev.* **86,** 539 (1986).
5. M. F. Jarrold, in *Modern Inorganic Chemistry: Gas Phase Inorganic Chemistry*, vol. 4, D. H. Russell, ed., Plenum, New York, 1990.
6. J. D. Corbett, *Chem. Rev.* **85**, 383 (1985); *Prog. Inorg. Chem.* **21,** 129 (1976).
7. M. Grätzel, *Acc. Chem. Res.* **14,** 376 (1981).
8. R. C. Baetzold and J. F. Hamilton, *Prog. Solid State Chem.* **15,** 1 (1983).
9. R. E. Smalley, "Supersonic Carbon Cluster Beams," in *Cluster Spectroscopy*, E. R. Bernstein, ed., Elsevier, 1990.
10. R. B. King, in *Physics and Chemistry in Small Clusters*, P. Jena, B. K. Rao, S. N. Khanna, eds., Plenum, New York, 1987.
11. M. Kappe, *Chem. Rev.* **88,** 369 (1988).
12. S. Sagano, Y. Nishina, and S. Ohinishi, eds., *Microclusters*, Springer-Verlag, Berlin, 1987.

13. M. Moskovits, ed., *Metal Clusters*, Wiley, New York, 1986.
14. R. Beuhler and L. Friedman, *Chem. Rev.* **86,** 521 (1986).
15. J. C. Phillips, *Chem. Rev.* **86,** 619 (1986).
16. J. M. Alford, P. E. Williams, D. J. Trevor, and R. E. Smalley, *Internatl. J. Mass Spectrom. Ion Phys.* **72,** 33 (1986).
17. M. Comisarow, in *Advances in Mass Spectrometry*, N. R. Daly, ed., Heyden, London, 1978.
18. T. P. Martin, ed., *Elemental and Molecular Clusters*, Springer, New York, 1987.
19. J. L. Margrave, ed., *The Characterization of High-Temperature Vapors*, Wiley, New York, 1967.
20. R. E. Honig, *Advances in Mass Spectrometry*, R. M. Elliott, ed., Pergamon, London, 1963.
21. J. L. Beauchamp, *Annu. Rev. Phys. Chem.* **22,** 571 (1971).
22. K. A. Gingerich, in *Mass Spectroscopy in Inorganic Chemistry*, R. F. Gould, ed., Advances in Chemistry Series 72, American Chemical Society: Washington, DC, 1968, pp. 291–300.
23. A. Henglein, *Pure Appl. Chem.* **56,** 1215 (1984).
24. L. Brus, *J. Phys. Chem.* **90,** 2555 (1986).
25. E. S. Robbins, R. E. Leckenby, and P. Willis, *Adv. Phys.* **16,** 739 (1967).
26. M. T. Bowers, ed., *Gas Phase Ion Chemistry*; Academic Press, New York, 1979.
27. R. Steudel, *Curr. Top. Chem.* **102,** 149 (1982).
28. R. Steudel, in *Studies in Inorganic Chemistry*, A. Müller and B. Kerbs, eds., Elsevier: Amsterdam, 1984.
29. G. A. Ozin, *Catal. Rev.* **16,** 191 (1977).
30. E. L. Muetterties, *Science* **196,** 839 (1977).
31. R. E. Smalley, in *Comparison of Ab Initio Quantum Chemistry with Experiment*, R. J. Bartlett, ed., Reidel, Dordrecht, 1985.
32. A. J. Bares, A. Muller, W. J. Orville-Thomas, and R. Gaufreys. *Matrix-Isolation Spectroscopy*, Reidel, Dordrecht, 1981.
33. M. F. Geusic, M. D. Morse, S. C. O'Brien, and R. E. Smalley, *Rev. Sci. Instrum.* **56,** 2123 (1985).
34. O. Cheshnovsky, S. H. Yang, C. L. Pettiette, and M. J. Craycraft, *Rev. Sci. Instrum.* **58,** 2131 (1987).
35. W. Weltner, Jr. and R. J. Van Zee, *Annu. Rev. Phys. Chem.* **35,** 291 (1984).
36. O. Cheshnovsky, C. L. Pettiette, and R. E. Smalley, in *Ion and Cluster Ion Spectroscopy*, P. Maier, ed., Elsevier, Amsterdam, 1988.
37. K. P. Huber and G. Herzberg, *Molecular Spectra and Molecular Structure.* IV. *Constants of Diatomic Molecules*, Van Nostrand Reinhold, New York, 1979.
38. J. C. D. Brand and A. R. Hoy, *Spectrosc. Rev.* **23,** 285 (1987).
39. Ch. A. Brau, in *Excimer Lasers*, 2nd ed., Ch. K. Rhodes, ed., Springer, Berlin, 1984.
40. J. Tellinghuisen, *Applied Atomic Collision Physics*, Vol. 3, Academic Press, New York, 1982.

41. K. S. Pitzer, *Acc. Chem. Res.* **12,** 271 (1979).
42. P. Pyykkö and J. P. Desclaux, *Acc. Chem. Res.* **12,** 276 (1979).
43. K. S. Pitzer, *Internatl. J. Quantum Chem.* **25,** 131 (1984).
44. M. Kraus and W. J. Stevens, *Annu. Rev. Phys. Chem.* **35,** 357 (1984).
45. P. Pyykkö, *Adv. Quantum Chem.* **11,** 353 (1978).
46. P. Pyykkö, ed., "Proceedings of the Symposium on Relativistic Effects in Quantum Chemistry," *Internatl. J. Quantum Chem.* **25** (1984).
47. K. Balasubramanian and K. S. Pitzer, *Adv. Chem. Phys.* **67,** 287 (1987).
48. P. Pyykkö, *Chem. Rev.* **88,** 563 (1988).
49. W. H. E. Schwarz, in *Theoretical Models of Chemical Bonding*, Z. Maksić, ed., Springer-Verlag, Berlin, 1990.
50. G. Malli, *Stud. Phys. Theor. Chem.* **21,** 199 (1982).
51. P. Pyykkö, *Relativistic Theory of Atoms and Molecules*, Springer-Verlag, Berlin, New York, 1986.
52. K. Balasubramanian, "Feature Article," *J. Phys. Chem.* **93,** 6585 (1989).
53. K. Balasubramanian, *J. Molec. Struc.* **202,** 291 (1989).
54. K. Balasubramanian, *Chem. Rev.* **89,** 1801 (1989).
55. Y. S. Lee, W. C. Ermler, and K. S. Pitzer, *J. Chem. Phys.* **67,** 5961 (1977).
56. Y. S. Lee, W. C. Ermler, and K. S. Pitzer, *J. Chem. Phys.* **67,** 360 (1980).
57. P. A. Christiansen, Y. S. Lee, and K. S. Pitzer, *J. Chem. Phys.* **71,** 4445 (1979).
58. Y. S. Lee, W. C. Ermler, K. S. Pitzer, and A. D. McLean, *J. Chem. Phys.* **70,** 288 (1979).
59. L. Kahn, P. Baybutt, and D. G. Truhlar, *J. Chem. Phys.* **65,** 3826 (1976).
60. W. C. Ermler, Y. S. Lee, P. A. Christiansen, and K. S. Pitzer, *Chem. Phys. Lett.* **71,** 70 (1981).
61. M. M. Hurley, L. F. Pacios, P. A. Christiansen, R. B. Ross, and W. C. Ermler, *J. Chem. Phys.* **84,** 6840 (1986).
62. L. A. LaJohn, P. A. Christiansen, R. B. Ross, T. Atashroo, and W. C. Ermler, *J. Chem. Phys.* **87**, 2812 (1987).
63. P. Hafner and W. H. E. Schwarz, *Chem. Phys. Lett.* **65**, 537 (1979).
64. P. J. Hay and W. R. Wadt, *J. Chem. Phys.* **82,** 270 (1985).
65. W. R. Wadt and P. J. Hay, *J. Chem. Phys.* **82,** 284 (1985).
66. P. J. Hay and W. R. Wadt, *J. Chem. Phys.* **82,** 299 (1985).
67. P. A. Christiansen, K. Balasubramanian, and K. S. Pitzer, *J. Chem. Phys.* **76,** 5087 (1982).
68. K. Balasubramanian, *J. Chem. Phys.* **89,** 5731 (1988).
69. R. M. Pitzer and N. W. Winter, *J. Chem. Phys.* **92,** 3061 (1988).
70. The major authors of ALCHEMY II are B. Liu, B. Lengsfield, and M. Yoshimine.
71. G. E. Gibson and A. MacFarlane, *Phys. Rev.* **46,** 1059 (1934).
72. G. Herzberg, *Ann Phys.* **15,** 677 (1932).
73. G. M. Almy and F. M. Sparks, *Phys. Rev.* **44,** 365 (1933).
74. S. M. Naude, *Phys. Rev.* **45,** 280 (1934).

75. W. A. Chupka and M. G. Ingram, *J. Chem. Phys.* **21,** 1313 (1953).
76. J. Drowart, R. P. Burns, G. DeMaria, and M. G. Ingram, *J. Chem. Phys.* **31,** 1131 (1959).
77. T. P. Martin, *J. Chem. Phys.* **83,** 78 (1985).
78. T. G. Dietz, M. A. Duncan, D. E. Powers, and R. E. Smalley, *J. Chem. Phys.* **74,** 6511 (1981).
79. D. E. Powers, S. G. Hansen, M. E. Geusic, D. L. Michalopoulos, and R. E. Smalley, *J. Chem. Phys.* **78,** 2866 (1983).
80. D. E. Powers, S. G. Hansen, M. E. Geusic, A. C. Puiu, J. B. Hopkins, T. G. Dietz, M. A. Duncan, P. R. R. Langridge-Smith, and R. E. Smalley, *J. Phys. Chem.* **86,** 2556 (1982).
81. J. R. Heath, Y. Liu, S. C. O'Brien, Q.-L. Zhang, R. F. Curl, F. K. Tittel, and R. E. Smalley, *J. Chem. Phys.* **83,** 5520 (1985).
82. R. G. Wheeler, K. LaiHing, L. L. Wilson, and M. A. Duncan, *Chem. Phys. Lett.* **131,** 8 (1986).
83. K. LaiHing, R. G. Wheeler, W. L. Wilson, and M. A. Duncan, *J. Chem. Phys.* **87,** 3401 (1987).
84. R. G. Wheeler, K. LaiHing, W. L. Wilson, J. D. Allen, R. B. King, and M. A. Duncan, *J. Am. Chem. Soc.* **108,** 8101 (1986).
85. L. A. Bloomfield, R. R. Freeman, and W. L. Brown, *Phys. Rev. Lett.* **54,** 2246 (1985).
86. E. A. Rohlfing, D. M. Cos, and A. Kaldor, *J. Chem. Phys.* **81,** 3322 (1984).
87. V. E. Bondybey and J. H. English, *J. Chem. Phys.* **67,** 3405 (1977).
88. V. E. Bondybey, *J. Chem. Phys.* **66,** 995 (1977).
89. V. E. Bondybey and J. H. English, *J. Chem. Phys.* **79,** 4746 (1983).
90. H. M. Rosenstock, K. Draxl, B. W. Steiner, and J. T. Herron, *J. Phys. Chem. Ref. Data* **6,** Suppl. 1 (1977).
91. D. B. Jacobson and B. S. Freiser, *J. Am. Chem. Soc.* **108,** 27 (1986).
92. M. F. Jarrold, A. J. Illies, and M. T. Bowers, *J. Am. Chem. Soc.* **107,** 7339 (1985).
93. P. A. Montano, G. K. Shenoy, E. E. Alp, W. Schulze, and J. Urban, *Phys. Rev. Lett.* **56,** 2076 (1986).
94. J. F. Hamilton and P. C. Longel, *Thin Solid Films* **16,** 49 (1973).
95. M. J. Yacaman, K. Heinemann, C. Y. Yang, and H. Poppa, *J. Cryst. Growth* **47,** 182 (1979).
96. G. Apai, J. F. Hamilton, J. Stohr, and A. Thomson, *Phys. Rev. Lett.* **43,** 165 (1979).
97. S. C. O'Brien, Y. Liu, Q. Zhang, J. R. Heath, F. K. Tittel, R. F. Curl, and R. E. Smalley, *J. Chem. Phys.* **84,** 4074 (1986).
98. J. R. Heath, Y. Liu, S. C. O'Brien, Q. Zhang, R. F. Curl, F. K. Tittel, and R. E. Smalley, *J. Chem. Phys.* **83,** 5520 (1985).
99. Y. Liu, Q. Zhang, F. K. Tittel, R. F. Curl, and R. E. Smalley, *J. Chem. Phys.* **85,** 7434 (1986).
100. Q. Zhang, Y. Liu, R. F. Curl, F. K. Tittel, and R. E. Smalley, *J. Chem. Phys.* **88,** 1670 (1988).
101. O. Cheshnovsky, S. H. Yang, C. L. Pettiette, M. J. Craycraft, Y. Liu, and R. E. Smalley, *Chem. Phys. Lett.* **138,** 119 (1987).

102. D. J. Trevor, D. M. Cox, K. C. Reichmann, R. O. Brickman, and A. Kaldor, *J. Phys. Chem.* **91,** 2598 (1987).

103. L. A. Bloomfield, M. E. Geusic, R. Freeman, and W. L. Brown, *Chem. Phys. Lett.* **121,** 33 (1985).

104. V. E. Bondybey, W. D. Meents, and M. L. Mandich, unpublished results.

105. J. L. Elkind, J. M. Alford, F. D. Weiss, R. T. Laaksonen, and R. E. Smalley, *J. Chem. Phys.* **84,** 2397 (1987).

106. M. Rasanen, L. A. Heimbrook, G. P. Schwartz, and V. E. Bondybey, *J. Chem. Phys.* **85,** 86 (1986).

107. J. C. Phillips, *J. Chem. Phys.* **87,** 1712 (1987).

108. S. Beck, *J. Chem. Phys.* **87,** 4233 (1987).

109. M. L. Steigerwald, A. P. Alivisatos, J. M. Gibson, T. D. Harris, K. Kortan, A. J. Mullar, A. M. Theyer, T. M. Duncan, D. C. Douglass, and L. E. Brus, *J. Am. Chem. Soc.* **110,** 3046 (1988).

110. K. LaiHing, P. Y. Chen, and M. A. Duncan, *J. Phys. Chem.* **91,** 6521 (1987).

111. R. G. Wheeler, K. LaiHing, W. L. Wilson, and M. A. Duncan, *J. Chem. Phys.* **88,** 2831 (1988).

112. M. E. Geusic, R. R. Freeman, and M. A. Duncan, *J. Chem. Phys.*, (1990).

113. M. E. Geusic, R. R. Freeman, and M. A. Duncan, *J. Chem. Phys.* **88,** 163 (1988).

114. S. Taylor, G. W. Lemire, Y. Hamrick, Z. Fu, and M. D. Morse, *J. Chem. Phys.* **89,** 5517 (1988).

115. E. Dörenburg and H. Z. Hintenberger, *Naturforsch.* **14A,** 765 (1959).

116. C. E. Richter and M. Trapp, *Internatl. J. Mass Spectrom. Ion Phys.* **38,** 21 (1981).

117. M. L. Mandich, A. M. Desantolo, and K. A. Keating, unpublished results.

118. K. M. Ervin, J. Ho, and W. C. Lineberger, *J. Chem. Phys.* **89,** 4514 (1988).

119. D. G. Leopold, J. Almlöf, W. C. Lineberger, and P. R. Taylor, *J. Chem. Phys.* **86,** 3780 (1988).

120. D. G. Leopold, J. Ho, and W. C. Lineberger, *J. Chem. Phys.* **86,** 1715 (1987).

121. C. L. Pettiette, S. H. Yang, M. J. Craycraft, J. Conceicao, R. T. Laaksonen, O. Cheshnovsky, and R. E. Smalley, *J. Chem. Phys.* **88,** 5377 (1988).

122. D. S. Ginter, M. L. Ginter, and K. K. Innes, *J. Chem. Phys.* **69,** 480 (1965).

123. M. A. Douglas, R. Hauge, and J. L. Margrave, *J. Phys. Chem.* **87,** 2945 (1983).

124. H. Basch, W. J. Stevens, and M. Krauss, *Chem. Phys. Lett.* **109,** 212 (1984).

125. G. R. Mohlmann, *Laser Chem.* **5,** 1 (1984).

126. C. W. Bauschlicher and S. R. Langhoff, *J. Chem. Phys.* **109,** 4627 (1989).

127. A. E. Douglas and G. Herzberg, *Phys. Rev.* **57,** 742 (1940).

128. W. R. M. Graham and W. Weltner, Jr., *J. Chem. Phys.* **65,** 1516 (1976).

129. M. DuPuis and B. Liu, *J. Chem. Phys.* **68,** 2092 (1973).

130. D. S. Ginter, M. L. Ginter, and K. K. Innes, *Astrophys. J.* **139,** 365 (1963).

131. W. A. Chupka, J. Berkowitz, C. F. Giese, and M. G. Inghram, *J. Phys. Chem.* **62,** 611 (1958).

132. J. Drowart and R. E. Honig, *J. Phys. Chem.* **61,** 980 (1957).

133. K. Balasubramanian, *J. Phys. Chem.* **90,** 6786 (1986); "Erratum," *J. Phys. Chem.* **93,** 8388 (1989).
134. C. E. Moore, *Atomic Energy Levels*, National Bureau of Standards, Washington, DC, 1971.
135. F. W. Froben and W. Schulze, *Surf. Sci.* **156,** 765 (1983).
136. J. E. Kingcade, Jr., U. V. Choudary, and K. A. Gingerich, *Inorg. Chem.* **18,** 3094 (1979).
137. I. Shim, H. M. Nagarathna-Naik, and K. A. Gingerich, *Internatl. J. Quantum Chem.* **29,** 975 (1986).
138. G. Pacchioni, *Molec. Phys.* **49,** 727 (1983).
139. K. Balasubramanian, *J. Molec. Spectrosc.* **123,** 228 (1987).
140. V. E. Bondybey, M. Heaven, and T. A. Miller, *J. Chem. Phys.* **78,** 3593 (1983).
141. G. M. Almy and G. D. Kinzer, *Phys. Rev.* **47,** 721 (1935).
142. G. D. Kinzer and G. M. Almy, *Phys. Rev.* **52,** 814 (1937).
143. G. M. Almy, *J. Phys. Chem.* **41,** 47 (1937).
144. S. Mrozowski and C. Santaram, *J. Opt. Soc. Am.* **57,** 522 (1967).
145. P. Perdigon, F. Martin, and J. D'Incan, *J. Molec Spectrosc.* **36,** 341 (1970).
146. P. Perdigon and J. D'Incan, *Can. J. Phys.* **48,** 1140 (1970).
147. R. J. Donovan and P. Strachan, *Trans. Faraday Soc.* **67,** 3407 (1971).
148. S. L. Bennett and J. L. Margrave, *J. Chem. Phys.* **59,** 5814 (1973).
149. F. Martin, P. Perdigon, and J. D'Incan, *J. Molec. Spectrosc.* **50,** 45 (1974).
150. A. M. Sibai, P. Perdigon, and A. Topouzkhanian, *Naturforsch.* **29A,** 429 (1974).
151. F. Martin, J. Figuet, and P. Perdigon, *J. Molec. Spectrosc.* **57,** 319 (1975).
152. P. Perdigon and F. Martin, *Molec. Spectrosc.* **83,** 40 (1980).
153. S. Ebel, H. T. Dieck, and H. Walthner, *Inorg. Chim. Acta* **53,** L101 (1981).
154. G. Wannous, C. Effantin, F. Martin, and J. D'Incan, *J. Molec. Spectrosc.* **91,** 1 (1982).
155. R. A. Kok and M. B. Hall, *Inorg. Chem.* **22**, 728 (1983).
156. L. A. Heimbrook, N. Chestnoy, M. Rasanen, G. Schwartz, and V. E. Bondybey, *J. Chem. Phys.* **83,** 6091 (1985).
157. Y. Watanabe, Y. Sakai, and H. Kashiwagi, *Chem. Phys. Lett.* **120,** 363 (1985).
158. A. Topouzkhanian and A. M. Sibai, *Spectrochim. Acta* **28A,** 2197 (1972).
159. K. Balasubramanian, *J. Molec. Spectrosc.* **121,** 465 (1987).
160. R. F. Barrow, G. C. Chandler, and C. B. Meyer, *Philos. Trans. Roy. Soc.* **A260,** 395 (1966).
161. G. Gouedard and J. C. Lehmann, *J. Phys.* **B9,** 2113 (1976).
162. K. G. Ibbs and A. J. McCaffery, *J. Chem. Soc., Faraday Trans. 2* **77,** 631, 637 (1981).
163. V. E. Bondybey and J. H. English, *J. Chem. Phys.* **72,** 6479 (1980).
164. S. J. Prosser, R. F. Barrow, J. Verges, C. Effantin, and J. D'Incan, *J. Phys.* **B13,** L547 (1980).
165. A. Jenouvrier, *Can. J. Phys.* **61,** 1531 (1983).
166. F. Ahmed and E. R. Nixon, *J. Molec. Spectrosc.* **83,** 64 (1980).

167. S. L. N. G. Krishnamachari and T. V. Venkatachalam, *Chem. Phys. Lett.* **67,** 69 (1979).
168. R. Winger, I. Barnes, E. H. Fink, J. Wildt, and F. Zabel, *Chem. Phys. Lett.* **73,** 297 (1980).
169. M. Heaven, T. A. Miller, J. H. English, and V. E. Bondybey, *Chem. Phys. Lett.* **91,** 251 (1982).
170. K. K. Yee and R. F. Barrow, *J. Chem. Soc., Faraday Trans. 2* **68,** 1181 (1972).
171. J. Drowart and S. Smones, *J. Chem. Soc., Faraday Trans. 2* **73,** 1755 (1977).
172. K. Balasubramanian, *J. Phys. Chem.* **91,** 5166 (1987).
173. R. P. Saxon and B. Liu, *J. Chem. Phys.* **67,** 5432 (1977).
174. V. E. Bondybey and J. H. English, *J. Chem. Phys.* **68,** 1865 (1978).
175. V. E. Bondybey and J. H. English, *J. Chem. Phys.* **72,** 3113 (1980).
176. J. M. Ricks and R. F. Barrow, *Can. J. Phys.* **47,** 2423 (1969).
177. M. C. Heaven, T. A. Miller, and V. E. Bondybey, *J. Phys. Chem.* **87,** 2072 (1983).
178. R. F. Barrow, T. C. Clark, J. A. Coxan, and K. K. Yee, *J. Molec. Spectrosc.* **51,** 428 (1974).
179. T. Ishiwata, M. Ohtoshi, and I. Tanaka, *J. Chem. Phys.* **81,** 2300 (1984).
180. R. S. Mulliken, *Phys. Rev.* **57,** 500 (1940).
181. K. Balasubramanian, *Chem. Phys.* **119,** 41 (1988).
182. P. Schwerdtfeger, L. V. Szentpaly, K. Vogel, H. Silberbach, H. Stoll, and H. Preuss, *J. Chem. Phys.* **84,** 1606 (1986).
183. J. Andzelm, M. Klobukowski, and E. Radzio-Andzelm, *J. Comput. Chem.* **5,** 146 (1984).
184. P. M. Boerrighter, M. A. Buijse, and J. G. Snijders, *Chem. Phys.* **111,** 47 (1987).
185. A. B. Cornford, D. C. Frost, C. A. McDowell, J. L. Ragle, and J. A. Stenhouse, *J. Chem. Phys.* **54,** 2651 (1971).
186. A. W. Potts and W. C. Price, *Trans. Faraday Soc.* **67,** 1242 (1971).
187. V. R. Dibeler, J. A. Walker, and K. E. McCulloh, *J. Chem. Phys.* **53,** 4715 (1970).
188. P. Venkateswarlu, *Can. J. Phys.* **48,** 1055 (1970).
189. K. Balasubramanian, J. J. Kaufman, P. C. Hariharan, and W. S. Koski, *Chem. Phys. Lett.* **129,** 169 (1986).
190. P. A. Hamilton, *Chem. Phys. Lett.* **140,** 591 (1987).
191. R. Wajnkrac, *Z. Phys.* **104,** 122 (1937).
192. F. W. Froben, W. Schulze, and U. Kloss, *Chem. Phys. Lett.* **99,** 500 (1983).
193. G. DeMaria, J. Drowart, and M. G. Ingram, *J. Chem. Phys.* **31,** 1076 (1959).
194. K. Balasubramanian and J. Q. Li, *Chem. Phys.* **88,** 4979 (1988).
195. P. A. Christiansen and K. S. Pitzer, *J. Chem. Phys.* **74,** 1162 (1981).
196. P. A. Christiansen, *J. Chem. Phys.* **79,** 2928 (1983).
197. V. E. Bondybey and J. H. English, *J. Chem. Phys.* **76,** 2165 (1982).
198. M. A. Epting, M. T. McKenzie, Jr., and E. R. Nixon, *J. Chem. Phys.* **73,** 134 (1980).
199. V. E. Bondybey and J. H. English, *J. Molec. Spectrosc.* **84,** 388 (1980).
200. R. A. Teichmann, III, M. Epting, and E. R. Nixon, *J. Chem. Phys.* **68,** 336 (1978).

201. K. Balasubramanian, K. S. Pitzer, *J. Chem. Phys.* **78,** 321 (1983); "Erratum," *J. Chem. Phys.* **80,** 592 (1984).
202. G. Pacchioni, *Molec. Phys.* **55,** 211 (1985).
203. K. S. Pitzer, *J. Chem. Phys.* **74,** 3078 (1981).
204. M. Ackerman, J. Drowart, F. E. Stafford, and G. Verhaegen, *Chem. Phys.* **36,** 1557 (1962).
205. J. Drowart and R. E. Honig, *J. Phys. Chem.* **61,** 980 (1957).
206. J. Genard, *Phys. Rev.* **44,** 468 (1933).
207. S. M. Naudé, *Phys. Rev.* **45,** 280 (1934).
208. G. Nakamura and T. Shidei, *Jpn. J. Phys.* **10,** 11 (1935).
209. G. M. Almy and H. A. Schultz, *Phys. Rev.* **51,** 62 (1937).
210. S. Mrozowski and C. Santaram, *J. Opt. Soc. Am.* **57,** 522 (1967).
211. J. Sfeila, P. Perdigon, F. Martin, and B. Femelat, *J. Molec. Spectrosc.* **42,** 239 (1972).
212. A. Topouzkhanian, A. M. Sibai, and J. D'Incan, *Z. Naturforsch.* **29A,** 436 (1974).
213. A. M. Sibai, A. Topouzkhanian, and P. Perdigonl, *Z. Naturforsch.* **31A,** 145 (1976).
214. G. Gerber and G. Kuscher, *Chem. Phys.* **60,** 119 (1981).
215. V. E. Bondybey, G. P. Schwarz, and J. E. Griffiths, *J. Molec. Spectrosc.* **89,** 328 (1981).
216. H. Sontag and R. Weber, *J. Molec. Spectrosc.* **91,** 72 (1982).
217. H. Sontag and R. Weber, *Chem. Phys.* **70,** 23 (1982).
218. J. Kordis and K. A. Gingerich, *J. Chem. Phys.* **55,** 5141 (1973).
219. K. Balasubramanian and J. Q. Li, *Molec. Spectrosc.* **135,** 169 (1989).
220. B. Rosen, *Naturwissen.* **14,** 978 (1926).
221. R. F. Barrow and R. P. Duparq, *Proc. Roy. Soc. (London)* **A237,** 279 (1972).
222. J. Berkowitz and W. A. Chupka, *J. Chem. Phys.* **50,** 4245 (1969).
223. E. O. Degenkolb, H. Mayforth, and J. I. Steinfield, *Chem. Phys. Lett.* **8,** 288 (1971).
224. T. J. Stone and R. F. Barrow, *Can. J. Phys.* **53,** 1976 (1975).
225. J. Berkowitz, *J. Chem. Phys.* **62,** 4074 (1975).
226. S. T. Lee, S. Suzer, and D. A. Shirley, *Chem. Phys. Lett.* **41,** 25 (1976).
227. K. K. Lin, L. C. Balling, and J. J. Wright, *Chem. Phys. Lett.* **123,** 37 (1986).
228. W. G. Thorpe, W. R. Carper, and S. J. Davis, *Chem. Phys. Lett.* **123,** 493 (1976).
229. J. Verges, J. D'Incan, C. Effantin, D. J. Greenwood, and R. F. Barrow, *J. Phys.* **12,** L301 (1979).
230. C. Effantin, J. D'Incan, J. Verges, M. T. MacPherson, and R. F. Barrow, *Chem. Phys. Lett.* **70,** 560 (1980).
231. V. E. Bondybey and J. H. English, *J. Chem. Phys.* **72,** 6478 (1980).
232. F. Ahmed and E. R. Nixon, *J. Molec. Spectrosc.* **87,** 101 (1981).
233. A. Pardo, J. M. L. Poyato, and J. Basulto, *J. Molec. Spectrosc.* **93,** 245 (1982).
234. R. Winter, I. Barnes, E. H. Fink, J. Wildt, and F. Zabel, *Chem. Phys. Lett.* **86,** 118 (1982).

235. J. Verges, C. Effantin, O. Babaky, J. D'Incan, S. J. Prosser, and R. F. Barrow, *Phys. Scripta* **25,** 338 (1982).
236. K. Balasubramanian and Ch. Ravimohan, *J. Molec. Spectrosc.* **126**, 220 (1987).
237. R. S. Mulliken, *Phys. Rev.* **46,** 549 (1934); *J. Chem. Phys.* **55,** 288 (1971).
238. J. Tellinghuisen, *J. Chem. Phys.* **82,** 4012 (1985); *J. Phys. Chem.* **87,** 5136 (1983); *J. Chem. Phys.* **78,** 2374 (1983).
239. M. Geilhoupt and Th. Dorfmüller, *Chem. Phys.* **76,** 443 (1983).
240. R. A. Ashby, *Can. J. Phys.* **57,** 698 (1979).
241. R. A. Ashby and C. W. Johnson, *J. Molec. Spectrosc.* **84,** 41 (1980).
242. S. Gerstenkorn, P. Luc, and J. Verges, *J. Phys.* **B14**, L193 (1981).
243. A. Beckman, H. Fietz, P. Baierl, W. Kiefer, *Chem. Phys. Lett.* **86,** 140 (1982).
244. S. Gerstenkorn and P. Luc, *J. Phys.* (Les Ulis, France) **46,** 867 (1985).
245. P. Venkateswarlu, G. Chakrapani, M. C. George, Y. V. Rao, and C. Okafor, *Pramana* **29,** 261 (1987).
246. K. M. Chen, L. E. Steenhoek, and E. S. Yeung, *Chem. Phys. Lett.* **59,** 222 (1978).
247. G. W. King and T. D. McLean, *Chem. Phys. Lett.* **121,** 57 (1985).
248. P. Venkateswarlu, T. Putcha, and R. Y. Vankateswara, *Spectrochim. Acta, Part A* **42A,** 285 (1986).
249. S. Leach, *J. Phys. Chem.* **92,** 5373 (1988).
250. R. G. McLoughlin, J. D. Morrison, and D. L. Smith, *Internatl. J. Mass Spectrom. Ion Proc.* **58,** 201 (1984).
251. J. Q. Li and K. Balasubramanian, *J. Molec. Spectrosc.* **138,** 162 (1989).
252. E. Skorko, *Nature* **131,** 366 (1933); *Acta Phys. Polon.* **3,** 191 (1934).
253. S. Gerstenkorn, P. Luc, and J. Sinzell, *J. Phys.* (Orsay, France) **41,** 1419 (1980).
254. R. D. Clear and K. R. Wilson, *J. Molec. Spectrosc.* **101,** 39 (1973).
255. L. Mathieson and A. L. G. Rees, *J. Chem. Phys.* **25,** 753 (1956).
256. J. C. D. Brand, A. K. Kalukar, and A. B. Yamashita, *Opt. Commun.* **39,** 235 (1981).
257. K. S. Viswanathan and J. Tellinghuisen, *J. Molec. Spectrosc.* **101,** 285 (1983).
258. M. Kawasaki, K. Tsukiyama, M. Kuwana, K. Obi, and I. Tanaka, *Chem. Phys. Lett.* **67,** 365 (1979).
259. A. L. Guy, K. S. Viswanathan, A. Sur, and J. Tellinghuisen, *Chem. Phys. Lett.* **73,** 582 (1980). J. Tellinghuisen, *J. Molec. Spectrosc.* **94,** 231 (1982).
260. B. R. Higginson, D. R. Lloyd, and P. L. Roberts, *Chem. Phys. Lett.* **19,** 480 (1973).
261. J. H. D. Eland, *J. Chem. Phys.* **70,** 2926 (1979).
262. A. B. Cornford, D. C. Frost, C. A. McDowell, J. L. Ragle, and I. A. Stenhouse, *J. Chem. Phys.* **54,** 2651 (1971).
263. J. Drowart and R. E. Honig, *J. Phys. Chem.* **61,** 980 (1957).
264. R. C. Miller and P. Kush, *Phys. Rev.* **99,** 1314 (1955).
265. J. Berkowitz and T. A. Walter, *J. Chem. Phys.* **49,** 1184 (1968).
266. G. Balducci and V. Piacente, *J. Chem. Soc., Chem. Commun.*, 1287 (1980).
267. K. Balasubramanian, unpublished results.
268. E. N. Shawhan, *Phys. Rev.* **48,** 343 (1935).

269. S. S. Puri and H. Mohan, *Indian J. Pure Appl. Phys.* **13,** 206 (1975).
270. L. Natanson, *Acta Phys. Polon.* **7,** 275 (1937).
271. S. Weniger, *J. Phys.* **28,** 595 (1967).
272. S. E. Johnson, D. Cannell, J. Lunacek, and H. P. Broida, *J. Chem. Phys.* **56,** 5723 (1972).
273. J. Lochet, *Phys.* **B11**, 1735 (1978).
274. L. Brewer and C. A. Chang, *J. Chem. Phys.* **56,** 1728 (1972).
275. R. A. Teichman, III and E. R. Nixon, *J. Molec. Spectrosc.* **59,** 299 (1976).
276. K. A. Gingerich, D. L. Cooke, and F. Miller, *Chem. Phys.* **64,** 4027 (1976).
277. K. S. Pitzer and K. Balasubramanian, *J. Phys. Chem.* **86,** 3068 (1982).
278. H. Sontag and R. Weber, *J. Molec. Spectrosc.* **100,** 75 (1983).
279. L.-E. Berg, S. Ismail, L. Klynning, H. Martin, A. Pereira, and P. Royen, *Phys. Scripta* **23,** 1047 (1981).
280. N. Aslund, R. F. Barrow, W. G. Richards, and D. N. Travis, *Ark. Fys.* **30,** 171 (1965).
281. G. Ehret and G. Gerber, *Chem. Phys.* **66,** 27 (1982).
282. L. Rovner, A. Drowart, and J. Drowart, *Trans. Faraday Soc.* **63,** 2906 (1967).
283. S. P. Reddy and M. K. Ali, *J. Molec. Spectrosc.* **35,** 285 (1970).
284. B. Wellegehausen, D. Friede, and G. Steger, *Opt. Commun.* **26,** 391 (1978).
285. W. P. West and H. P. Broida, *Chem. Phys. Lett.* **56,** 283 (1978).
286. S. Drosch and G. Gerber, *J. Chem. Phys.* **77,** 123 (1982).
287. G. Gerber, K. Sakurai, and H. P. Broida, *J. Chem. Phys.* **64,** 3410 (1976).
288. G. Gerber and H. P. Broida, *J. Chem. Phys.* **64,** 3423 (1976).
289. R. A. Teichman and E. R. Nixon, *J. Chem. Phys.* **67,** 2470 (1977).
290. V. E. Bondybey and J. H. English, *J. Chem. Phys.* **73,** 42 (1980).
291. G. Farbe, J. P. Bacci, C. Athenour, R. Stringat, and P. Bernath, *Can. J. Phys.* **60,** 73 (1982).
292. P. A. Christiansen, *Chem. Phys. Lett.* **109,** 145 (1984).
293. C. A. Swarts, T. C. McGill, and W. A. Goddard, III, *Surf. Sci.* **110,** 400 (1981).
294. C. A. Swarts, W. A. Goddard, III, and T. C. McGill, *J. Vacuum Sci. Technol.* **19,** 360, 551 (1981).
295. B. K. Agarwal, *Phys. Rev.* **B23,** 2995 (1981).
296. N. Nishida, *Solid State Commun.* **28,** 551 (1979); *Surf. Sci.* **72,** 589 (1978).
297. G. Demeyer, R. Hoogewijs, W. Lambrecht, J. Vennik, and G. Dalma, *Surf. Sci.* **106,** 498 (1981).
298. R. A. Hoepful, J. Shah, D. Block, and A. C. Gossard, *Appl. Phys. Lett.* **48,** 148 (1986).
299. A. C. Marsh and J. C. Inkson, *J. Phys.* **C19,** 43 (1986).
300. G. W. Lemire, G. A. Bishea, S. Heidecke, and M. D. Morse, *J. Chem. Phys.* **92,** 121 (1990).
301. K. Balasubramanian, *J. Chem. Phys.* **86,** 3410 (1987).
302. L. B. Knight and P. T. Petty, *J. Chem. Phys.* **88,** 481 (1988).

303. G. W. Lemire, G. A. Bishea, S. Heidecke, and M. D. Morse, *J. Chem. Phys.* **92,** 121 (1990).

304. I. Shim, private communications.

305. K. Balasubramanian, "Erratum," *J. Chem. Phys.* **92,** 2123 (1990).

306. L. G. Piper, J. E. Velazco, and D. W. Setser, *J. Chem. Phys.* **59,** 3323 (1973).

307. J. E. Velazco, J. H. Kolts, and D. W. Setser, *J. Chem. Phys.* **65,** 3468 (1973).

308. J. Tellinghuisen, A. K. Hays, J. M. Hoffman, and G. C. Tisome, *J. Chem. Phys.* **65,** 4473 (1976).

309. D. L. King, L. G. Piper, and D. W. Setser, *J. Chem. Soc., Faraday Trans. 2* **73,** 177 (1977).

310. M. Rokni, J. H. Jacobs, and J. A. Mangano, *Phys. Rev.* **A16,** 2216 (1977).

311. R. E. Olson, B. Liu, *Phys. Rev.* **A17,** 1568 (1978).

312. T. H. Dunning, Jr. and P. J. Hay, *J. Chem. Phys.* **66,** 3767 (1977) and references cited therein.

313. J. S. Cohen, W. Wadt, and P. J. Hay, *J. Chem. Phys.* **71,** 2955 (1979) and references cited therein.

314. H. P. Watkins, N. E. Sondergaard, and W. S. Koski, *Radiochim. Acta* **29,** 87 (1981).

315. J. H. Holloway, *Nobel Gas Chemistry*, Methuen, London, 1968.

316. A. Henglein and G. A. Muccini, *Angew. Chem.* **72,** 630 (1960).

317. I. Kuen and F. Howorka, *J. Chem. Phys.* **70,** 595 (1979).

318. J. Berkowitz and W. A. Chupka, *Chem. Phys. Lett.* **7,** 447 (1970).

319. J. Berkowitz, W. A. Chupka, P. W. Guyon, J. H. Holloway, and R. Sphor, *J. Chem. Phys.* **75,** 1461 (1971).

320. H. P. Watkins and W. S. Koski, *Chem. Phys. Lett.* **77,** 470 (1981).

321. M. A. Gardner, A. M. Karo, and A. C. Wahl, *J. Chem. Phys.* **65,** 1222 (1976).

322. D. H. Liskow, H. F. Schaefer, III, P. S. Bagus, and B. Liu, *J. Am. Chem. Phys.* **95,** 4056 (1973).

323. B. Liu and H. F. Schaefer, III, *J. Chem. Phys.* **55,** 2369 (1971).

324. K. Wendell, C. A. Jones, J. J. Kaufman, and W. S. Koski, *J. Chem. Phys.* **63,** 750 (1975).

325. R. B. Sharma and W. S. Koski, unpublished results.

326. K. Balasubramanian, J. J. Kaufman, P. C. Hariharan, and W. S. Koski, *Chem. Phys. Lett.* **129,** 165 (1986).

327. M. Hotoka, B. Roos, J. B. Delos, R. Srivastava, R. B. Sharma, and W. S. Koski, *Phys. Rev.* **A35,** 4515 (1987).

328. K. Balasubramanian, P. Feng, J. J. Kaufman, P. C. Hariharan, and W. S. Koski, *Phys. Rev.* **A37,** 3204 (1988).

329. D. A. Chapman, K. Balasubramanian, S. H. Lin, J. J. Kaufman, P. C. Hariharan, and W. S. Koski, *Phys. Rev.* **A39,** 4428 (1989).

330. M. Hotoka, B. Roos, K. Balasubramanian, N. Semo, R. B. Sharma, and W. S. Koski, in *Advances in Chemical Reaction Dynamics*, R. Rencsepis, C. Capellos, eds., Reidel, Dordrecht, 1986, pp. 135–143.

331. P. Y. Feng, K. Balasubramanian, J. J. Kaufman, P. C. Hariharan, and W. S. Koski, *Phys. Rev.* **42,** 6428 (1990).

332. W. G. Brown and G. E. Gibson, *Phys. Rev.* **40,** 529 (1932).

333. W. G. Brown, *Phys. Rev.* **40,** 1040 (1932).

334. W. V. F. Brooks and B. C. Crawford, *J. Chem. Phys.* **23,** 363 (1955).

335. E. Hulten, N. Jarlaster, and L. Koffman, *Ark. Fys.* **18,** 479 (1960).

336. M. A. A. Clyne and J. A. Coxon, *Proc. Roy. Soc. (London)* **298,** 424 (1967).

337. G. W. Holleman and J. I. Steinfield, *Chem. Phys. Lett.* **12,** 431 (1971).

338. M. S. Child and R. B. Bernstein, *J. Chem. Phys.* **59,** 5916 (1973).

339. F. E. Cummings and W. Klemperer, *J. Chem. Phys.* **60,** 2035 (1974).

340. R. H. Barnes, C. E. Moeller, J. F. Kircher, and C. M. Verber, *Appl. Phys. Lett.* **24,** 610 (1974).

341. P. Venkateswarlu, *Can. J. Phys.* **53,** 812 (1975).

342. C. D. Olson and K. K. Innes, *J. Chem. Phys.* **64,** 2405 (1976).

343. G. W. King and R. G. McFadden, *Chem. Phys. Lett.* **58,** 119 (1978).

344. R. D. Gordon and K. K. Innes, *J. Molec. Spectrosc.* **78,** 350 (1979).

345. J. A. Coxon and M. A. Wickramaaratchi, *J. Molec. Spectrosc.* **79,** 380 (1980).

346. J. A. Coxon, R. M. Gordon, and M. A. Wickramaaratchi, *J. Molec. Spectrosc.* **79,** 363 (1980).

347. J. C. D. Brand, V. D. Deshpande, A. R. Hoy, and S. M. Jaywant, *J. Molec. Spectrosc.* **100,** 416 (1983).

348. J. C. D. Brand, D. Bissieres, A. R. Hoy, and S. M. Jaywant, *Chem. Phys. Lett.* **109,** 101 (1984).

349. J. D. Spivey, J. G. Ashmore, and J. Tellinghuisen, *Chem. Phys. Lett.* **109,** 456 (1984).

350. A. W. Potts and W. C. Price, *Trans. Faraday Soc.* **67,** 1242 (1971).

351. J. C. D. Brand, A. R. Hoy, and S. M. Jaywant, *J. Molec. Spectrosc.* **106,** 188 (1984).

352. A. B. Cornford, Ph.D. thesis, University of British Columbia, Vancouver, BC (as referred to in Ref. 341).

353. K. Balasubramanian, *J. Molec. Spectrosc.* **110,** 339 (1985).

354. K. Balasubramanian, *Chem. Phys.* **95,** 225 (1985).

355. K. Balasubramanian and P. Y. Feng, *Chem. Phys. Lett.* **146,** 155 (1988).

356. H. Basch, *Chem. Phys. Lett.* **136,** 289 (1987).

357. J. A. Howard, R. Sutcliffe, T. S. Tse, H. Dahamane, and B. Mile, *J. Phys. Chem.* **89,** 3595 (1985).

358. Z. Fu, G. W. Lemire, Y. M. Hamrick, S. Taylor, J.-C. Shui, and M. D. Morse, *J. Chem. Phys.* **88,** 3524 (1988).

359. K. Balasubramanian, *J. Chem. Phys.* **87,** 3518 (1989).

360. W. D. Reents, *J. Chem. Phys.* **90,** 4258 (1989).

361. G. Pacchioni and J. Koutecký, *J. Chem. Phys.* **84,** 3301 (1986).

362. J. R. Sabin, J. Oddershede, G. H. F. Diercksen, and N. E. Grüner, *J. Chem. Phys.* **84,** 354 (1986).

363. K. Raghavachari, *J. Chem. Phys.* **83,** 3525 (1985).
364. K. Raghavachari, *J. Chem. Phys.* **84,** 5672 (1986).
365. R. S. Grev and H. F. Schaefer, III, *Chem. Phys. Lett.* **119,** 111 (1985).
366. G. H. F. Diercksen, N. E. Grüner, J. Oddershede, and J. R. Sabin, *Chem. Phys. Lett.* **117,** 29 (1985).
367. K. Balasubramanian, *Chem. Phys. Lett.* **125,** 400 (1986).
368. K. Balasubramanian, *Chem. Phys. Lett.* **135,** 288 (1987).
369. P. Y. Feng and K. Balasubramanian, *Chem. Phys.* **138,** 89 (1989).
370. T. P. Martin and H. Schaber, *J. Chem. Phys.* **83,** 855 (1985).
371. K. Balasubramanian, *J. Chem. Phys.* **85,** 3401 (1986).
372. K. Balasubramanian, *Chem. Phys. Lett.* **150,** 71 (1988).
373. K. Balasubramanian, *Chem. Rev.* **85,** 599 (1985).
374. K. Balasubramanian, *Theor. Chim. Acta* **51,** 37 (1979).
375. K. Balasubramanian, *Indian J. Chem.* **16B,** 1094 (1978).
376. K. Balasubramanian, *Theor. Chim. Acta* **53,** 129 (1979).
377. K. Balasubramanian, *Ann. NY Acad. Sci.* **319,** 576 (1979).
378. G. Pólya, *Acta Math.* **65,** 145 (1937).
379. R. B. King, *J. Math. Chem.* **1,** 15 (1987).
380. R. B. King, *Stud. Phys. Theor. Chem.* **28,** 99 (1983).
381. R. B. King and D. H. Rouvray, *Theor. Chim. Acta* **48,** 207 (1978).
382. R. B. King, *Inorg. Chem.* **27,** 1941 (1988).
383. R. B. King, in *Physics and Chemistry of Small Clusters*, P. Jena, B. K. Rao, and S. N. Khanna, eds., Plenum, New York, 1987.
384. R. B. King, *Inorg. Chem.* **24,** 1716 (1985).
385. E. L. Muetterties, *J. Am. Chem. Soc.* **91,** 1636 (1969).
386. E. L. Muetterties, *J. Am. Chem. Soc.* **90,** 5097 (1968).
387. W. G. Klemperer, *J. Am. Chem. Soc.* **94,** 6940 (1972).
388. K. Balasubramanian, *Internatl. J. Quantum Chem.* **21,** 411 (1982).
389. K. Balasubramanian, *Internatl. J. Quantum Chem.* **22,** 385 (1982).
390. K. Balasubramanian, *J. Phys. Chem.* **22,** 1013 (1982).
391. K. Balasubramanian, *J. Phys. Chem.* **86,** 4668 (1982).
392. W. N. Lipscomb, *Science* **153,** 373 (1966).
393. R. B. King, *Inorg. Chem.* **25,** 506 (1986).
394. R. B. King, in *New J. Chem.* **12,** 493 (1988).
395. K. Balasubramanian, *Stud. Phys. Theor. Chem.* **23,** 149 (1983).
396. K. Balasubramanian, *J. Comput. Chem.* **4,** 302 (1983).
397. K. Balasubramanian, *Croat. Chim. Acta* **57,** 1465 (1984).
398. K. Balasubramanian and X. Y. Liu, *J. Comput. Chem.* **9,** 406 (1988).
399. K. Balasubramanian and X. Y. Liu, *Internatl. J. Quantum Chem.* **22,** 319 (1988).
400. M. Randić, D. O. Oakland, and D. J. Klein, *J. Comput. Chem.* **7,** 35 (1986).
401. M. Randić, *Chem. Phys. Lett.* **42,** 383 (1976).

402. K. Balasubramanian, *J. Phys. Chem.* **94,** 7764 (1990).
403. U. Meier, S. D. Peyerimhoff, P. S. Bruna, and F. Grein, *J. Mol. Spectrosc.* **134,** 259 (1989).
404. K. Balasubramanian, *J. Molec. Spectrosc.*, **139,** 405 (1990).
405. K. K. Das, H.-P. Libermann, G. Hersch, and R. J. Buenker, *J. Chem. Phys.* **102,** 8462 (1995).
406. M. Toscano and N. Russo, *Z. Phys. D.* **D22,** 683 (1992).
407. R. Breidohr, O. Shestakov, and E. H. Fink, *Chem. Phys. Lett.* **218,** 12 (1994).
408. N. Damany, J. Figuet, and A. Topouzkhanian, *Chem. Phys.* **63,** 157 (1982).
409. C. Effantin, A. Topouzkhanian, J. Figuet, J. D'Incan, R. F. Barrow, and J. Verger, *J. Phys.* **B15,** 3829 (1982).
410. J. H. Glownia, J. A. Misewich, and P. P. Soronkin, *J. Chem. Phys.* **92,** 3375 (1990).
411. G. Ehret, and G. Gerber, *Springer Ser. Opt. Sci.* **49,** 140 (1985).
412. L. S. Wang, Y. T. Lee, D. A. Shirley, K. Balasubramanian, and P. Feng, *J. Chem. Phys.* **93,** 6310 (1990).
413. M. Polak, J. Ho, G. Gerber, and W. C. Lineberger, *J. Chem. Phys.* **95,** 3053 (1991).
414. E. H. Fink, K. D. Setzer, D. A. Ramsay, and M. Vervloet, *Chem. Phys. Lett.* **179,** 95 (1991).
415. K. Balasubramanian, and D. W. Liao, *J. Chem. Phys.* **95,** 3064 (1991).
416. R. Briedhor, K. D. Setzer, O. Shestakov, E. H. Fink, and W. Zyrnicki, *J. Molec. Spectrosc.* **166,** 251 (1994).
417. K. K. Das, H-P. Libermann, and R. J. Buenker, *J. Chem. Phys.* **102,** 4518 (1995).
418. M. Z. Liao, D. Dai, and K. Balasubramanian, *Chem. Phys. Lett.* **239,** 124 (1995).
419. L. Lou, L. Wang, L. P. F. Chibante, R. T. Laaksonen, P. Nordlanden, and R. E. Smalley, *J. Chem. Phys.* **94,** 8015 (1991); L. Lou, P. Nordlander, and R. E. Smalley, *J. Chem. Phys.* **97,** 1858 (1992).
420. R. J. Van Zee, S. Li, and W. Weltner, Jr., *J. Chem. Phys.* **98,** 433 (1993).
421. D. W. Liao and K. Balasubramanian, *J. Chem. Phys.* **96,** 8938 (1992).
422. K. Raghavachari, *J. Chem. Phys.* **84,** 5672 (1986); K. Raghavachari and C. M. Rohlfing, *J. Chem. Phys.* **93,** 2114 (1992).
423. K. Balasubramanian, *Chem. Phys. Lett.* **125,** 400 (1986).
424. R. S. Grev and H. F. Schaefer, III, *Chem. Phys. Lett.* **119,** 111 (1985).
425. K. Balasubramanian, *Chem. Phys. Lett.* **135,** 287 (1987).
426. T. N. Kistopoulos, C. J. Chick, A. Weaver, and D. M. Neumark, *J. Chem. Phys.* **93,** 6108 (1990).
427. G. B. Winstead, K. X. He, D. Grantier, T. Hammond, and J. L. Gole, *Chem. Phys. Lett.* **181,** 222 (1991).
428. W. Weltner, Jr., and D. McLeod, *J. Chem. Phys.* **41,** 235 (1964).
429. G. Pacchioni and J. Koutecký, *J. Chem. Phys.* **84,** 3301 (1986); G. Pacchioni and J. Koutecký, *Ber. Bunsengs. Phys. Chem.* **88,** 242 (1984).
430. D. A. Dixon and J. L. Gole, *Chem. Phys. Lett.* **188,** 560 (1992).
431. D. G. Dai, K. Sumathi, and K. Balasubramanian, *Chem. Phys. Lett.* **193,** 251 (1992).

432. J. D. Carette and L. Kerwin, *Can. J. Phys.* **39,** 1300 (1961).
433. S. L. Bennett, J. L. Margrave, and J. F. Franklin, *J. Chem. Phys.*, **61,** 1647 (1974).
434. B. Cabaud, A. Hareau, P. Nounov, and R. Uzan, *Internatl. J. Mass Spectrosc. Ion Phys.* **59,** 5814 (1973).
435. H. Sontag and R. Weber, *Chem. Phys.* **70,** 23 (1982).
436. C. L. Sullivan, J. E. Prusaczyk, and K. D. Carlson, *High Temp. Sci.* **4,** 212 (1972).
437. V. E. Bondybey, G. P. Schwartz, J. E. Griffiths, and J. H. English, *Chem. Phys. Lett.* **76,** 30 (1980).
438. L. Rovner, A. Drowart, and J. Drowart, *Trans. Faraday* **C63,** 2906 (1967).
439. B. Eberle, H. Sontag, and R. Weber, *Surf. Sci.* **156,** 751 (1985).
440. L. Andrews and Z. Mielke, *J. Phys. Chem.* **94,** 2348 (1990).
441. J. N. Murrel, O. Novaro, S. Castillo, and V. Saunders, *Chem. Phys. Lett.* **90,** 421 (1982).
442. O. Novarao and S. Castillo, *Internatl. J. Quantum Chem.* **26,** 411 (1984).
443. J. N. Murrel, *Chem. Phys. Lett.* **55,** 1 (1978).
444. J. K. Burdett and C. J. Marsden, *New. J. Chem.* **12,** 797 (1988).
445. K. Raghavachari, R. C. Haddon, and J. S. Binkley, *Chem. Phys. Lett.* **122,** 219 (1985).
446. T. J. Hamilton and H. F. Schaefer, III, *Chem. Phys. Lett.* **166,** 303 (1990).
447. R. O. Jones and D. Hohl, *J. Chem. Phys.* **92,** 6713 (1990).
448. K. Balasubramanian, K. Sumathi, and D. Dai, *J. Chem. Phys.* **95,** 3494 (1991).
449. H. Zhang and K. Balasubramanian, *J. Chem. Phys.* **97,** 3437 (1992).
450. M. E. Geusic, R. R. Freeman, and M. A. Duncan, *J. Chem. Phys.* **89,** 223 (1988).
451. M. Polak, G. Gerber, and W. C. Lineberger, *J. Chem. Phys.* **95,** 3053 (1991).
452. J. Donohue, *The Structures of the Elements*, Wiley, New York, 1974.
453. G. Lucovsky, in *The Physics and Chemistry of Selenium and Tellurium*, E. Gerlach and P. Grosse, eds., Springer, Berlin, 1979.
454. G. D. Brabson, Z. Mielke, and L. Andrews, *J. Phys. Chem.* **95,** 79 (1991).
455. P. Hassanzadeh and L. Andrews, *J. Phys. Chem.* **96,** 7679 (1992).
456. H. Schnöckel, H. J. Gocke, and R. Elsper, *Z. Anorg. Allg. Chem.* **78,** 494 (1982).
457. H. Schnöckel, *Z. Anorg. Allg. Chem.* **72,** 510 (1984).
458. G. D. Brabson and L. Andrews, *J. Phys. Chem.* **96,** 9172 (1992).
459. P. Hassanzadeh, C. Thompson, and L. Andrews, *J. Phys. Chem.* **96,** 8246 (1992).
460. G. W. Charles, D. Litimma, D. J. Hunt, and G. Pish, *J. Opt. Soc. Am.* **47,** 291 (1957).
461. J. E. Rice, R. D. Amos, N. C. Handy, T. J. Lee, and H. F. Schaefer, III, *J. Chem. Phys.* **85,** 963 (1986).
462. G. E. Quelch, H. F. Schaefer, III, and C. J. Marsden, *J. Am. Chem. Soc.* **112,** 8719 (1990).
463. K. Raghavachari, C. M. Rohlfing, and J. S. Binkley, *J. Chem. Phys.* **93,** 5862 (1990).
464. W. Von Niessen, L. S. Cederbaum, and F. Tarantelli, *J. Chem. Phys.* **91,** 3582 (1989).

465. W. Von Niessen, *J. Chem. Phys.* **95,** 8301 (1991).
466. D. Hohl, R. O. Jones, R. Carr, and M. Parrinello, *Chem. Phys. Lett.* **139,** 540 (1987).
467. D. Dai and K. Balasubramanian, *J. Chem. Phys.* **99,** 5239 (1993).
468. R. B. Ross, J. M. Powers, T. Atashroo, W. C. Ermler, L. A. LaJohn, and P. A. Christiansen, "Erratum," *J. Chem. Phys.* **101,** 10198(E) (1994).
469. D. Dai and K. Balasubramanian, "Erratum," *J. Chem. Phys.* submitted.

4

RELATIVISTIC EFFECTS ON MOLECULES CONTAINING LANTHANIDES AND ACTINIDES

Relativistic effects are expected to play an important role in the chemistry of molecules containing lanthanide and actinide elements, due to the large number of protons in the nucleus of these elements. The discovery of several superheavy actinide elements at Berkeley by Seaborg and co-workers was indubitably one of the motivating factors for the growth of relativistic quantum chemistry. An article by Hoffman (1) outlines the connection between relativity and superheavy elements. The current author (2) reviewed the connection between relativity and the electronic properties of molecules containing lanthanides and actinides. This chapter is derived principally from this review by the author (2). However, numerous editorial changes were made to the references and the text so as to fit the current format of the book.

The electronic structures of molecules containing lanthanide and actinide atoms are extremely interesting because of the complex array of electronic states resulting from open f-shell electronic configurations. There are seven possible real projections for the $4f$ orbitals, which could accommodate up to 14 electrons. As a result, numerous possible electronic states of varied spin multiplicities and spatial symmetries result even for a simple diatomic molecule consisting of a lanthanide or actinide element. A wealth of spectroscopic data has been accumulated up to now on several diatomics containing f-block elements. The spectra are considerably complex, and definitive assignments are not always feasible.

The $4f$-shell electrons in ionic compounds such as lanthanide oxides and halides are relatively easier to deal with. In other compounds they are more challenging to deal with, since they may be involved in the chemical bonds. The book by Salahub and Zerner (3) as well as the introductory remarks by Atkins (4) entitled "f for facts, factors and fallacies" summarize this varied and rather intriguing behavior of $4f$ electrons.

Pitzer and co-workers (5) showed that not all of the "lanthanide contraction" is due to incomplete screening of the 4*f*-shell electrons. In fact, the magnitude of the 4*f*-shell effect and relativistic effects were found to be comparable and hence indeed nearly half of "lanthanide contraction" itself is due to relativistic effects. In this chapter, this point is further exemplified by considering the electronic states of HfH, HfH_2, and Pt- and Au-containing molecules, and comparing them with their lighter analogs (ZrH and ZrH_2). The lanthanide contraction plays a very important role in the third-row transition-metal compounds containing Hf–At as well as in the chemistry of molecules containing Hf–At. The third-row transition-metal elements (Hf–Au) form stronger chemical bonds compared to the second-row elements because of the 4*f*-shell incomplete screening and the relativistic mass–velocity effect, both of which lead to the contraction of the 6*s* orbital of Hf–Au. For the same reason, the $6s^2$ shell of Hg, Tl, Pb, Bi, and other elements is considerably lower in energy, and hence the shell is relatively inert, leading to the "inert-pair effect" phenomenon.

The electronic spectra of rare-earth diatomics such as LaO, CeO, and PrO are notoriously complex as noted by Field and co-workers (6, 7). Since an array of molecular electronic states result from incomplete 4*f* shells, several electronic states have nearly the same properties. Furthermore, spectroscopic bands arising from one set of states can be highly perturbed by another set of states in the near proximity. Moreover, there are several isotopes for lanthanides with large nuclear spins, which lead to complex hyperfine structures.

In spite of the above-mentioned complexities, progress on the electronic structure computations of lanthanide and actinide molecules as well as spectroscopy of lanthanide molecules in general and oxides in particular, has been impressive. The fact that molecules such as $C_{60}La$ (8) and $U(C_8H_8)_2$ (9, 10) can be computed is a testimonial to the progress in this area, and the impact that large-scale *ab initio* computations has made in the chemistry of molecules containing lanthanides and actinides.

The objective of this chapter is to outline the importance of relativistic effects for the electronic structure and spectroscopy of molecules containing lanthanide and actinide elements. First, we review the developments in both theoretical computations and experimental spectra of molecules containing lanthanides and actinides. It is not intended to provide a comprehensive set of available data on these molecules. A few molecules of both experimental and theoretical interest are identified and discussed. The importance of relativistic effects in lanthanide, actinide chemistry, and postlanthanide chemistry is considered.

4.1 RELATIVITY AND THE LANTHANIDE CONTRACTION

Bagus et al. (5) studied the effects of relativity and the lanthanide contraction (gradual decrease of atomic radii from La to Lu) on Hf to Bi atoms, which follow the lanthanides. It is important to know how much of the contraction on the 6*s* valence orbital is caused by relativity as opposed to incomplete screening of the $4f$ shells. To accomplish this goal, Bagus et al. (5) performed Hartree–Fock (HF) calculations on "pseudoatoms" corresponding to Hf, Re, Au, Hg, Tl, Pb, and Bi without $4f$ electrons, but with the atomic number reduced by 14. The orbital energies of the pseudoatoms were compared with Dirac–Fock orbital energies of the real atoms by Bagus et al. Of course, this technique would thus separate the contraction effect due to $4f$ shells and relativity.

Typically the seven projections of the $4f$ shells occupy a small volume and thus fail to screen the nucleus. This results in an incomplete screening of the $4f$-shell electrons. Consequently, for example, the $6s$ orbital of the Hf atom is not fully shielded by the $4f$-shell electrons. This, in turn, leads to a contraction of the outer $6s$ orbital. The contraction of the radii of the outer electrons is more commonly referred to as the "lanthanide contraction." However, not all the orbital radii contraction arises from the incomplete screening of the $4f$ shells. As a matter of fact, nearly half of the contraction is due to relativity.

Table 4.1a reproduces the results of Bagus et al., showing a critical comparison of the orbital energies in hartree atomic units for both the HF pseudoatom

Table 4.1a Orbital Binding Energies (in a.u.) of Atoms Hf–Bi[a]

		$5d$		$6s$		$6p$	
	$5d_{3/2}$		$5d_{5/2}$	$(6s_{1/2})$	$6p_{1/2}$		$6p_{3/2}$
Pseudo-Hf		0.3192		0.1805		—	
Hf(HF)		0.2992		0.2104		—	
Hf(DHF)	0.2473		0.2355	0.2397	—		—
Pseudo-Re		0.4660		0.2031		—	
Re(HF)		0.4538		0.2347		—	
Re(DHF)	0.3972		0.3661	0.2783	—		—
Pseudo-Au		0.5372		0.1905		—	
Au(HF)		0.5210		0.2208		—	
Au(DHF)	0.4935		0.4287	0.2917	—		—
Pseudo-Hg		0.7191		0.2288		—	
Hg(HF)		0.7142		0.2610		—	
Hg(DHF)	0.6501		0.5746	0.3280	—		—
Pseudo-Tl		0.9472		0.3162		0.1836	
Tl(HF)		0.9683		0.3611		0.1924	
Tl(DHF)	0.8945		0.8062	0.4492	0.2114		0.1765
Pseudo-Pb		1.1772		0.4025		0.2268	
Pb(HF)		1.2245		0.4589		0.2398	
Pb(DHF)	1.1388		1.0360	0.5665	0.2751		0.2199
Pseudo-Bi		1.4131		0.4906		0.2693	
Bi(HF)		1.4874		0.5582		0.2862	
Bi(DHF)	1.1389		1.2710	0.6862	0.3385		0.2612

[a]Pseudo-, relativistic, and nonrelativistic atoms are compared.

Source: Reproduced from Reference 5.

(without $4f$ shells) and the actual atom, for which relativistic Dirac–Fock energies are listed. Note that the actual orbital energies are negative and for convenience the magnitudes of these energies are shown in Table 4.1a. The difference in the orbital energies of the pseudoatom and HF atom measures the effect of the $4f$ shell. The difference between the DHF orbital energy and HF orbital energy measures the effect of relativity.

As seen from Table 4.1a, the difference in the $6s$ orbital energy of the pseudo-Hf and Hf (HF) is 0.0299 hartree, while the difference between Hf (HF) and Hf (DHF) energy is 0.0293. The former energy of stabilization is due to the incomplete screening of the $4f^{14}$ shell, while the latter energy of stabilization is due to relativity. This means that the effect of the lanthanide contraction is comparable to the effect of relativity.

Figure 4.1 shows systematically the magnitude of orbital energies of Cu, Ag, and Au as well as the Zn–Cd–Hg triad. The plot joining the pseudoatom (without $4f^{14}$ shells and without relativity) is almost collinear (which is the expected trend). The dotted line is the actual Dirac–Fock trend including relativistic effects and $4f^{14}$ shells, while the solid line is the HF line without rel-

Table 4.1b Comparison of Radial Expectation Values (R) for Hf, Re, Au, Hg, Tl, Pb, and Bi

		$5d$		$6s$		$6p$	
	$d_{3/2}$		$d_{5/2}$		$p_{1/2}$		$p_{3/2}$
Pseudo-Hf		2.5048		4.6934		—	
Hf(HF)		2.2277		4.0684		—	
Hf(DHF)	2.3376		2.4198	3.6939	—		—
Pseudo-Re		2.0326		4.2162		—	
Re(HF)		1.7999		3.6942		—	
Re(DHF)	1.8301		1.9047	3.2770	—		—
Pseudo-Au		1.7228		4.2230		—	
Au(HF)		1.5433		3.7006		—	
Au(DHF)	1.5359		1.6185	3.0609	—		—
Pseudo-Hg		1.6040		3.7500		—	
Hg(HF)		1.4327		3.3284		—	
Hg(DHF)	1.4312		1.4987	2.8434	—		—
Pseudo-Tl		1.5042		3.3294		4.2434	
Tl(HF)		1.3412		2.9669		3.9262	
Tl(DHF)	1.3387		1.3940	2.5792	3.5166		4.0123
Pseudo-Pb		1.4214		3.0475		3.7532	
Pb(HF)		1.2671		2.7242		3.4569	
Pb(DHF)	1.2641		1.3119	2.3916	3.0739		3.5162
Pseudo-Bi		1.3506		2.8336		3.4116	
Bi(HF)		1.2046		2.5939		3.1366	
Bi(DHF)	1.2012		1.2439	2.2429	2.7802		3.1862

Source: Reproduced from Reference 5.

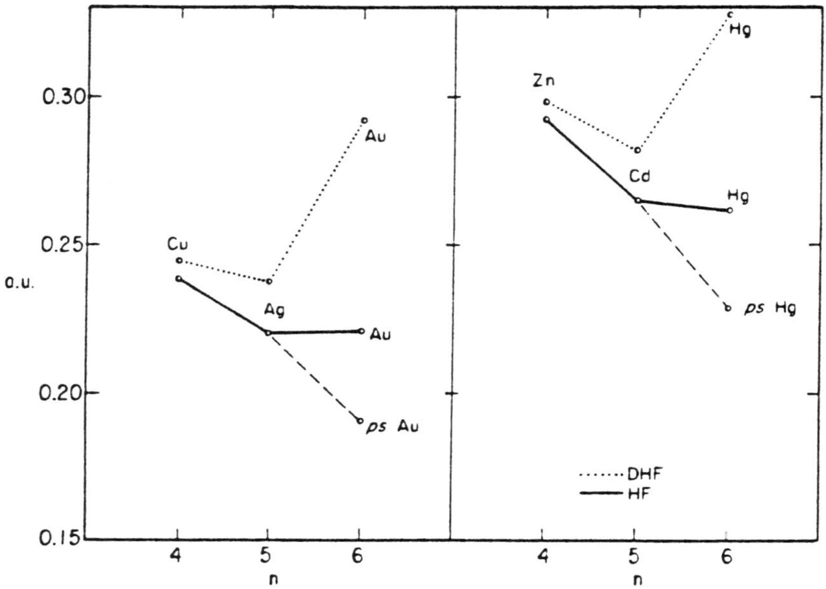

(a)

Figure 4.1 Illustration of lanthanide contraction. (4.1.a) Magnitude of orbital energies of valence *s* electrons of Cu, Ag, and An as well as Zn, Cd and Hg. Also shown in 4.1.b are the expectation values of orbital radii for Cu–Au. The label 6*s* is the pseudoatom without $4f^{14}$ shells. (Reproduced from Ref. 5.)

ativistic effects. It is clear that the dramatic change in trend for Au and Hg is due to both incomplete screening of the $4f^{14}$ shell and the effect of relativity.

Table 4.1b shows the orbital radii expectation values for the pseudo-, HF and DHF atoms. Figure 4.1*b* shows the orbital radii expectation values obtained by Bagus et al. As seen from both Table 4.1b and Figure 4.1*b*, the $6s_{1/2}$ orbitals of Hf–Bi are contracted both due to lanthanide contraction and relativity. For the gold atom (Fig. 4.1*b*), the contraction due to lanthanide contraction is 0.523 a.u. (atomic unit). The relativistic contraction is 0.64 a.u. Consequently, the relativistic contraction effect is larger than the lanthanide contraction effect for the gold atom. For the Hf atom, the contraction is 0.625 a.u., while the relativistic contraction is only 0.3745 a.u. Hence the lanthanide contraction is larger than the relativistic contraction for hafnium.

This effect of relativistic and lanthanide contraction has a dramatic impact in the chemical and spectroscopic compounds of molecules containing Hf to Bi. We have already discussed the impact of this for the properties of the third-row transition-metal compounds (Chapter 1), and in particular the difference in the properties of the third-row coinage-metal compared to the second- and first-row metal atoms. The trend in the ionization potentials of the Cu, Ag, and

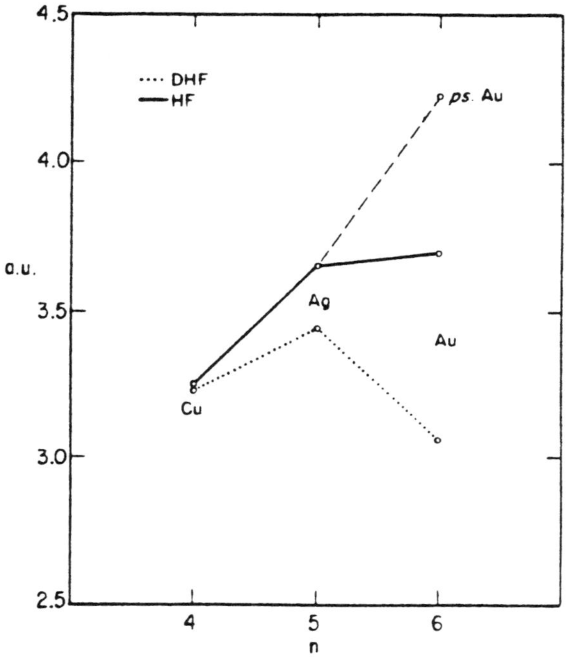

(*b*)

Figure 4.1 (*Continued*)

Au atoms discussed in Chapter 1 is akin to the DHF orbital energies of Cu, Ag, and Au computed by Bagus et al. (Fig. 4.1*a*). It is clear that nearly half of the anomalous behavior of the gold atom is due to relativity, while the other half is due to the lanthanide contraction.

The lanthanide and relativistic contractions have other important consequences on molecular and chemical properties. Figure 4.2 compares the equilibrium bond distances of NiH, PdH, and PtH. Note that the PtH diatomic has significantly shorter bond distance compared to PdH, which is attributed to the contractions of the outer 6*s* orbital of the Pt atom arising from both lanthanide contraction and relativity.

Atoms such as Hg, Tl, Pb, and Bi form considerably weaker bonds compared to their lighter members in the same group. This, too, is a consequence of the contraction of the 6*s* orbital of these elements. Note that for Hg–Bi the $6s^2$ shell is complete. The unusual stability of the 6*s* shell leaves the filled $6s^2$ shell inert in that it does not actively participate in bonding. Hence, two Hg atoms form only a van der Waals dimer Hg_2, while Cd_2 and Zn_2 are strongly bound. For the same reason Hg_2^{2+} is unusually stable (11).

The formation of Hg–Hg bonds would require promotion of an electron

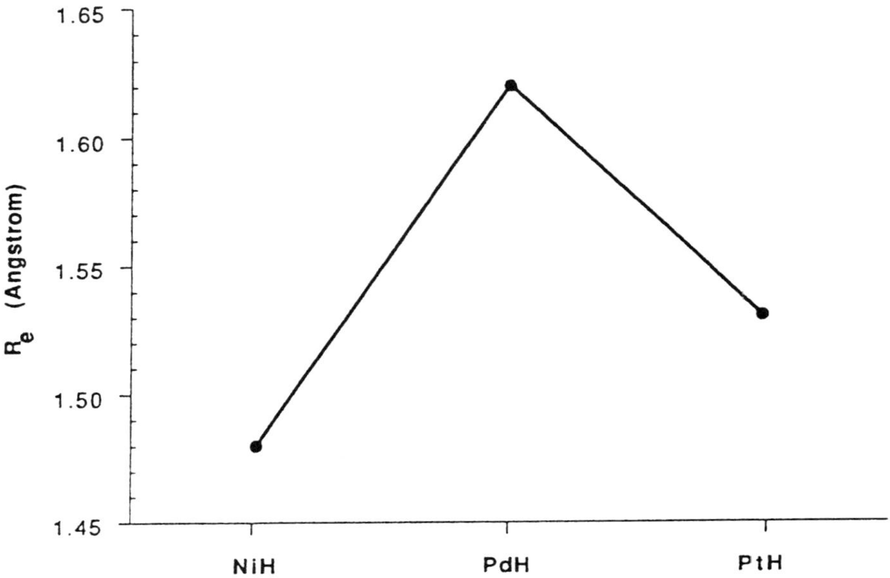

Figure 4.2 The r_e values of the $^2\Delta(\frac{3}{2})$ states of NiH, PdH, and PtH diatomics.

from the 6*s* orbital to the 6*p* orbital. Figure 4.3 compares the *ns*–*np* promotion energy by comparing the $(ns^2)^1S$-$(nsnp)^1P$ energy separations of Zn, Cd, and Hg. As evidenced from Figure 4.3, mercury usually has a larger promotion energy resulting from stabilization of the 6*s* orbital due to the lanthanide and relativistic contractions. Consequently, these effects are extremely significant and often dramatically alter the spectroscopic properties and the reactivities of the third-row transition-metal compounds as well as compounds of Hg–Bi.

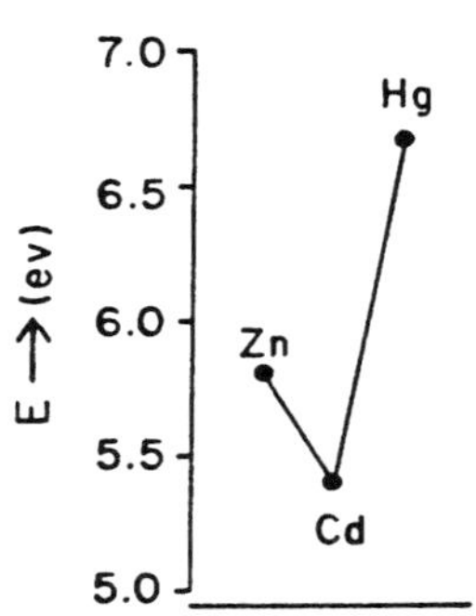

Figure 4.3 The $ns^2(^1s)$–$nsnp(^1p)$ energy separations for Zn, Cd, and Hg. (Reproduced from Ref. 14.)

4.2 *AB INITIO* THEORETICAL TECHNIQUES FOR LANTHANIDE- AND ACTINIDE-CONTAINING MOLECULES

In addition to the *ab initio* relativistic effective core potential (RECP) techniques (12–45) and the other relativistic techniques discussed before for general molecules (see Part A, Chapter 4), a few techniques, especially applicable to lanthanide- and actinide-containing molecules, are worth mentioning in this section. The RECP technique has been applied extensively for several heavy molecules (12–45) including the lanthanide- and actinide-containing molecules (8–10, 46–50).

Preuss, Stoll, and co-workers (51–60) at Stuttgart adapted several other ECP methods, which include core polarization effects with considerable success in molecular calculations. Dolg et al. (60) have also reviewed the theoretical and experimental developments pertinent to the electronic structure molecules composed of lanthanides and actinides. These authors have generated complete sets of ECPs for all elements in the periodic table, including lanthanides and actinides (53). Preuss, Stoll, and co-workers have in particular applied these ECPs for computing the spectroscopic constants and potential-energy curves of lanthanide oxides such as GdO, EuO, and YbO. The results of the specific calculations on these species are described subsequently.

Pyykkö and Desclaux (61–64) have used the one-center Dirac–Fock method to study diatomic hydrides containing very heavy atoms. In particular, the work of Pyykkö (65) on several lanthanide hydrides is worth citing. Pyykkö (66, 67) has given an impressive compilation of references pertinent to the topics of relativistic effects. It should be stated that the one-center DF method is applicable only to diatomic hydrides. A critical comparison of the properties computed by this method with the state-of-the-art *ab initio* method suggests that the properties obtained by the one-center method are not always in good agreement with more accurate procedures. Also, the properties of the excited states have not been computed using this method. Pyykkö and co-workers (68–72) have also studied several actinide-containing molecules using theoretical schemes that have addressed the relativistic effects.

There have been several all-electron relativistic calculations on very heavy diatomic hydrides such as AgH and AuH (73, 74). The calculations of Lee and McLean (73, 74) employed four-component STO (Slater-type orbital) spinors as basis sets for all-electron calculations of AuH, AgH, Ag_2 and other molecules. Ramos et al. (75) used this method to study the sixth-row hydrides (see also Refs. 76 and 77). These results have been compared with other techniques (78).

Sadlej and co-workers (79–81) have had considerable success in using the Breit–Pauli Hamiltonian in all-electron relativistic calculations, which included electron correlation effects. Techniques that include electron correlation effects such as the coupled cluster method or the Moller–Plesset (MP) perturbation method have been adapted to include relativistic effects. The authors have carried out such computations on molecules such as AgH and AuH.

The coupled (electron) pair formalism (CPF) (82), modified coupled pair formalism (MCPF), and more averaged coupled pair formalism (ACPF) have been used by Bauschlicher and co-workers (83–90) for related compounds that contain a metal atom with ligands. The coupled cluster methods (91, 92) have been used mostly for lighter molecules and have not yet been used for molecules containing lanthanides or actinides. However, the MCPF method has been used for La-containing species by Bauschlicher and co-workers (83–90).

4.3 APPROXIMATE METHODS FOR LANTHANIDE AND ACTINIDE MOLECULES

4.3.1 Ligand Field Theory

Field (6, 7) has combined the well known molecular-orbital pictures of *s* and *p* orbitals with the ligand field theoretical treatment of the *d* and *f* orbitals. This approach has provided very simple and yet promising interpretation for lanthanide oxides. Lanthanide oxides such as CeO and PrO are highly ionic. The conventional MO theory for the AB diatomic molecule by itself ignores the A^+B^- or $A^{2+}B^{2-}$ ionic character of such compounds. Although indirectly from MO theory, this can be rationalized as well, but the amount of effort is usually more significant using the LCAO-MO approach.

The ligand field theory combines the electrostatic model of the AB molecule with the MO picture. It tends to rationalize how the metal ion's energy levels are affected by the surrounding field generated by the ligand. The simplest form of the ligand field theory is the crystal field theory and is described in textbooks (93). The more accurate forms of ligand field theory take into consideration the electronic structures from an MO standpoint. This theory considers the overlap of the metal orbitals with the ligand orbitals. There are many variations of this theme.

The $(n-1)d$ and $(n-2)f$ orbitals occupy a small volume as discussed before, and thus especially the overlap of the $(n-2)f$ orbital with ligand orbitals would be rather small. Consequently, the potential-energy curves of states belonging to a specific electronic configuration wth metal $4f$ orbitals and σ and π MOs derived from *s* and *p* with ligand orbitals are similar. This is the underlying philosophy of the implementation of ligand field theory to interpret complex spectra of lanthanide oxides.

4.3.2 Local Density Functional (LDF) Method

Accurate determination of many electronic wavefunctions, which include electron correlation effects to a high order, is an extremely difficult many-body problem requiring a multimillion CI expansion (94–97). Considerable progress has been made with the advent of graphic-unitary group approach (94, 96) and the symbolic CI expansion technique (97). Indeed, the billion configuration CI computation has become possible with such techniques (95). Yet one needs to seek alternative approximate techniques to address the problem of electron correlation. The philosophy behind the density functional method (98–102) is that the properties of the ground state of a molecule can be determined if its charge density is known even though the electronic wavefunction may not be known. Thus if the density can be computed directly without the knowledge of wavefunctions, then one can circumvent the tedious problem of having to determine the wavefunction. The density functional method seeks to replace the Hartree–Fock nonlocal exchange potential with other forms of potentials. The first form of the density functional theory is the Slater local exchange approximation. The Hobenberg–Kohn theorem (98) is the basis of subsequent LDF methods (99–102), which basically state that knowledge of the charge density suffices to determine the ground state properties. The multiple scattering $X\alpha$ method or muffin-tin method and its relativistic analog (102) is one way of implementing the LDF method. There are, of course, several basic set expansion techniques. The major advantage of the LDF method is that the replacement of the conventional HF nonlocal potential with some form of local potential, which includes both correlation effects and exchange effects leading to simple Hartree–Fock like equations. There are no CI diagonalizations involving matrices of the order of a million $\times$ million. Instead computationally inexpensive HF-like equations are solved. The LDF method has been extended to include relativistic effects through complex exchange potentials [see, e.g., Becke (100)]. The readers are also referred to the excellent book by Parr and Yang (101) on density function theory.

Kohn and Sham (99) showed that the exact ground-state energy E_0 is given by

$$E_0 = -\frac{1}{2}\sum_i \langle\psi_i(1)|\nabla_i^2|\psi_i(1)\rangle - \sum_\mu \frac{Z_\mu \rho(1)}{r_{1\mu}}\,d\tau_1 + \frac{1}{2}\int\int \frac{\rho(1)\rho(2)}{r_{12}}\,d\tau_1 d\tau_2 + E_{xc}(\rho)$$

where ρ is the density, $\psi_i(1)$ terms are Kohn–Sham orbitals, and E_{xc} (exchange correlation energy) is a function of the density ρ, which can be expressed as the sum of the probability associated with ψ_i. The Kohn–Sham orbitals are obtained variationally solving Hartree–Fock-like equations

$$\hat{F}(1)\psi_i(1) = \varepsilon_i \psi_i(1)$$

where

$$\hat{F}(1) = -\frac{1}{2}\nabla_1^2 - \sum_\mu \frac{Z_\mu}{r_{1\mu}} + \sum_j J_j(1) + V_{xc}(1)$$

where $V_{xc}(1)$ is the exchange correlation functional given by

$$V_{xc} = \frac{\delta E_{xc}(\rho)}{\delta \rho}$$

This Fock-like equation can thus be solved iteratively leading to self-consistency if the functional $E_{xc}(\rho)$ is known. Several approximations have been used in the literature for E_{xc} [see Parr and Yang (101)]. In the local density functional method E_{xc} is approximated as

$$E_{xc} = \int \rho_{(x,y,z)} \varepsilon_{xc}(\rho)\, d\tau$$

where ε_{xc} is the exchange correlation energy per atom in a homogeneous electron gas. The practical implementation of the LDF method uses an initial guess for ρ to find E_{xc}, V_{xc}, and then the Fock-like equation is iterated. The Kohn–Sham orbitals can be expressed as a linear combination of Gaussians or localized muffin-tin orbitals in the muffin-tin approximation. The main problem with the DF method is in the choice of E_{xc}, for which several approximations are made, including the $X\alpha$ approximation.

The LDF method is generally applied for the ground state, although Salahub (102) used it to obtain excited-state properties. There is also a question as to whether the density function method can treat electronic states of different spin multiplicities size consistently and whether it produces correct dissociation behavior. Often the dissociated atoms are above their true energies, which leads to dissociation energies higher than the experimental values. It should also be stated that the assignment of electronic states arising from a complex set of open-shell *d* and *f* electronic configurations from Kohn–Sham orbital occupancies is not always unambiguous. There appears to be difficulty in differentiating electronic states that have similar occupancies in the Kohn–Sham orbitals. Hence, at present, it appears that all electronic states derived from complex open-shell *d* and *f* electronic configurations are not always well differentiated by the LDF method. Finally in the LDF method, one seeks to replace complex many-body nature of the electron correlation problem by one set of local potentials. It is not evident that this could always be accomplished in practice, although in principle such a function should exist.

4.3.3 Neglect of Differential Overlap/Spin–Orbit CI Method (INDO/S-CI)

Rösch, Zerner, and co-workers (103–110) have extended the previous intermediate neglect of differential overlap for spectroscopy INDO/S method (109, 110) to include spin–orbit interaction through a spin–orbit interaction enhanced INDO/S-CI method. In the INDO/S method one-center core integrals are taken from experimental IPs. The two-center electron integrals are parametrized using the Weiss–Matya–Nishimoto formula (a variation of the integral parametrization used in the Paraisen–Parr–Pople method). All one-center integrals are fully accounted for explicitly for d and f electrons. The one-center spin–orbit operator $H_{so} = \zeta(r)\vec{l} \cdot \vec{s}$ is used to evaluate spin–orbit integrals. The radial ζ integrals are taken from atomic spectroscopy. The spin–orbit CI part of the INDO/S-CI is an adaptation of the Pitzer double group CI (111) approach, which we have already described. This method has been applied by Zerner, Rösch, and co-workers for several lanthanide compounds such as LaO, CeO, GdO, LuO, and PrO.

4.4 ELECTRONIC STRUCTURE OF LANTHANIDE HYDRIDES

First we consider the electronic states of lanthanide hydrides. Although YH and SCH are strictly not lanthanide or rare-earth compounds, it is important to investigate these compounds as well as compare and contrast them with their periodic analog—namely, LaH. The results of HfH and HfH_2 and comparison with ZrH and ZrH_2 are also included to demonstrate the effect of the lanthanide contraction.

4.4.1 YH and ScH

The author and Wang (21) carried out CASSCF followed by full second-order CI (SOCI) calculations, which included single + double excitations from all CASSCF configurations. They used a $(3s3p3d1f)$ valence Gaussian basis set together with RECPs, which retained $4d^1 5s^2$ shells of the Y atom in the valence space. For the hydrogen atom, van de Duijneveldt's $(5s1p/3s1p)$ basis set was used. The author and Wang studied 29 electronic states of YH. In addition, spin–orbit coupling was included through the RCI method. Langhoff et al. (87) studied four electronic states of YH using the SCF/MCPF method comparing Hay–Wadt ECPs (33, 34), which included $4s^2 4p^6 4d^1 5s^8$ shells of Y. The spin–orbit coupling was neglected in this work.

Table 4.2 shows the spectroscopic constants obtained by the author and Wang (21) for the electronic states of YH, while Figure 4.4 shows the computed potential-energy curves. Both Langhoff et al. (87) and the author and Wang (21) predicted the ground state of YH to be $^1\Sigma^+$, in noticeable disagreement with Bernard and Bacis (112, 113), who had carried out the rotational analysis of a system, which they designated as a $^3\Phi$–$^3\Delta$ system. It now appears that this assignment is incorrect. The most consistent assignment for this system appears to be a $^3\Delta$–$^3\Pi$ system.

Bernard and Bacis (113) observed six electronic systems ascribed to YH. They also carried out the rotational analysis of these systems and identified the transitions as $^3\Phi \rightarrow {}^3\Delta$, ($^1\Pi$, $^1\Sigma$)-$^1\Sigma$, and ($^1\Sigma$, $^1\Delta$) $\rightarrow {}^1\Pi$. These assignments appear to be incorrect and in any case are not consistent with the theoretically

Table 4.2 Spectroscopic Properties of Low-Lying Electronic States of YH

State	R_e, Å	T_e, cm^{-1}	ω_e, cm^{-1}
$^1\Sigma^+(0^+)$	1.865	0	1510
$^1\Sigma^+$	1.865	11	1510
$^3\Pi(0^-)$	1.907	7,884	1488
$^3\Pi(0^+)$	1.907	7,905	1488
$^3\Pi(1)$	1.907	8,187	1489
$^3\Pi(2)$	1.907	8,458	1489
$^3\Pi$	1.907	8,179	1489
$^3\Delta(3)$	1.944	9,062	1432
$^3\Delta(2)$	1.944	9,000	1431
$^3\Delta$	1.944	9,002	1432
$^3\Delta(1)$	1.944	9,047	1433
$^3\Sigma^+(1)$	1.944	10,164	1337
$^3\Sigma^+(0^-)$	1.944	10,171	1342
$^3\Sigma^+$	1.944	10,175	1334
$^1\Delta(2)$	1.950	10,991	1366
$^1\Delta$	1.950	10,992	1366
$^1\Pi(1)$	1.944	12,687	1380
$^1\Pi$	1.944	12,685	1380
$^1\Sigma^+(\text{II})(0^+)$	2.034	15,049	—
$^1\Sigma^+(\text{II})$	2.034	15,050	—
$^3\Pi(\text{II})(2)$	1.965	20,226	1418
$^3\Pi(\text{II})(1)$	1.965	20,478	1418
$^3\Pi(\text{II})(0^-)$	1.965	20,789	1417
$^3\Pi(\text{II})(0^+)$	1.965	20,789	1417
$^3\Pi(\text{II})$	1.965	20,510	1418
$^1\Pi(\text{II})(1)$	1.955	21,990	1386
$^1\Pi(\text{II})$	1.955	21,991	1386
$^1\Gamma(4)$	1.988	30,051	1298
$^1\Gamma$	1.988	30,051	1298

Source: Reproduced from Reference 21.

predicted $^1\Sigma^+$ ground state for YH. Their $B'^1\Sigma^+$–$A'^1\Sigma^+$, when reassigned as a $B^1\Sigma^+$–$^1\Sigma^+$ system, agrees with the computed T_e of the $^1\Sigma^+(\text{II})$ state of the author and Wang (21).

The ν_{00} value of the experimental system is ~14,300 cm^{-1}, while the computed SOCI value for the $^1\Sigma^+(\text{II})$ state is 15,049 cm^{-1}, including spin–orbit coupling. The $B^1\Delta$–$A^1\Pi$ system of Bernard and Bacis (112, 113) has a ν_{00} value of 11,378 cm^{-1}. This system corresponds to the $^1\Pi$–$X^1\Sigma^+$ system whose computed transition energy is 12,687 cm^{-1} (the author and Wang). Likewise, the $^3\Phi$–$^3\Delta$ system of Bernard and Bacis should also be reassigned.

The Balasubramanian–Wang computed dissociation energy of YH (3.05 eV) agreed with a value of 2.95 eV obtained by Langhoff et al. (87). However, the

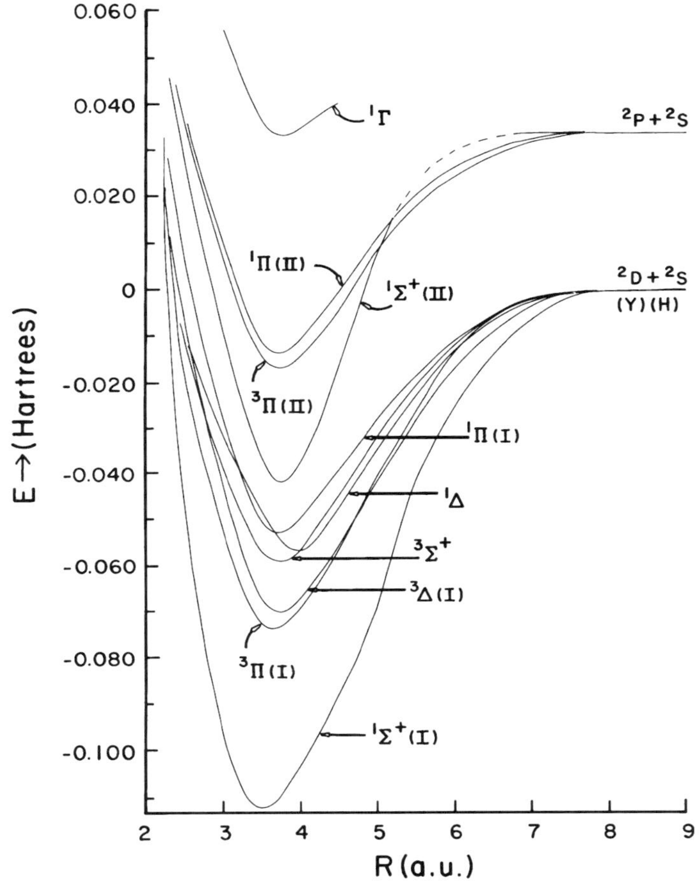

Figure 4.4 Potential-energy curves of the selected states of YH. (Reproduced from Ref. 21.)

dipole moment computed by the author (7) using ($4s^24p^64d^15s^2$) RECPs with a large basis set was found to be 0.3 D lower than the value of Langhoff et al.

Following the theoretical calculations, Simard and co-workers (114) obtained $C^1\Sigma^+$–$X^1\Sigma^+$ spectroscopic system of YH(YD) as well as a system in the 19,600-cm^{-1} region. These spectra are shown in Figures 4.5 and 4.6, respectively. As seen from the assignment of the observed spectra, Simard and co-workers confirm the theoretically predicted $X^1\Sigma^+$ ground state of YH.

Figure 4.7 shows the calculated energy levels of YH and the experimentally observed levels. As seen from Figure 4.7, the computed values are in excellent agreement with experiment considering the approximations introduced in the theoretical procedures.

Table 4.3 shows the contributions of important configurations in the SOCI of the electronic states of YH. In Table 4.3, the π and δ orbitals are nonbond-

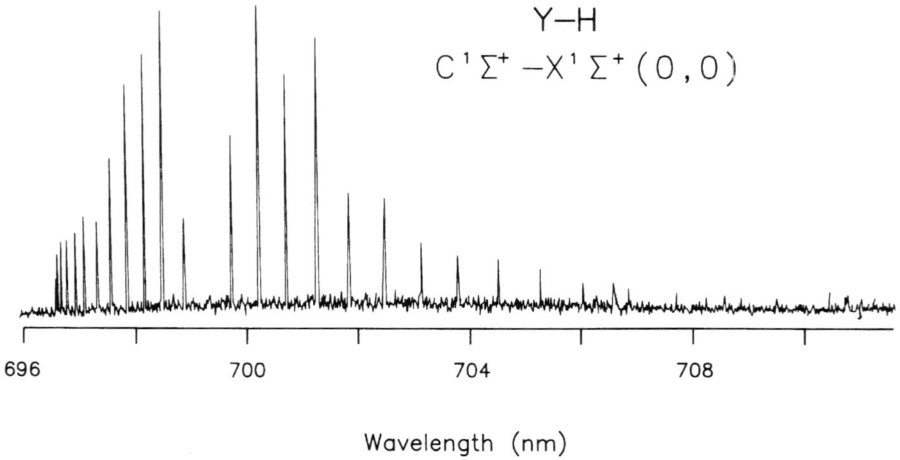

Figure 4.5 The *C–X* system of YH. (Reproduced from Ref. 114.)

ing yttrium orbitals. The contribution from the 5*p* orbital of Y is very noticeable. Most of the other electronic states are dominated by a single configuration except the $^1\Sigma^+$(II) state, which is a mixture of several configurations.

Table 4.4 shows the Mulliken populations of various electronic states of YH. As one can see from that table, the gross yttrium populations of all the electronic states are considerably smaller than 3.0, indicating the ionic character of the YH bond with the Y^+H^- polarity. Note that the 5*p* population is between

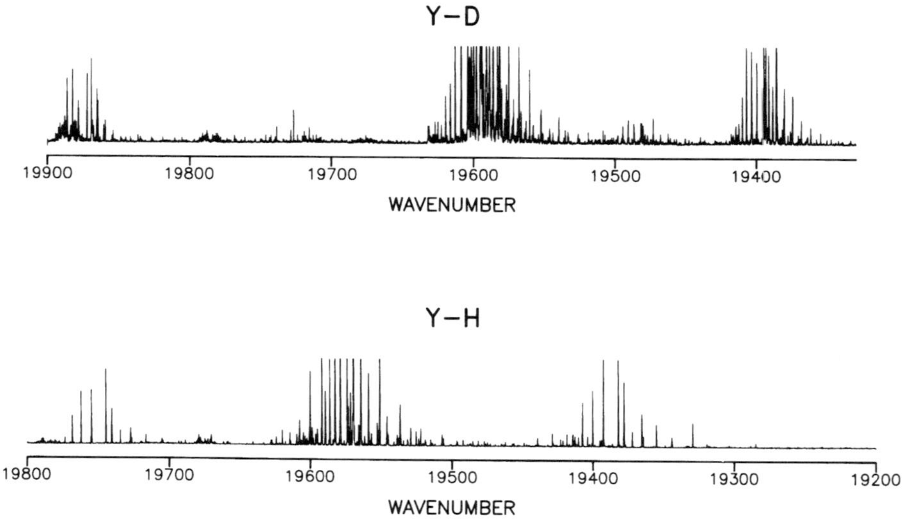

Figure 4.6 The spectra of YD and YH in the 19,600-cm^{-1} region. (Reproduced from Ref. 114.)

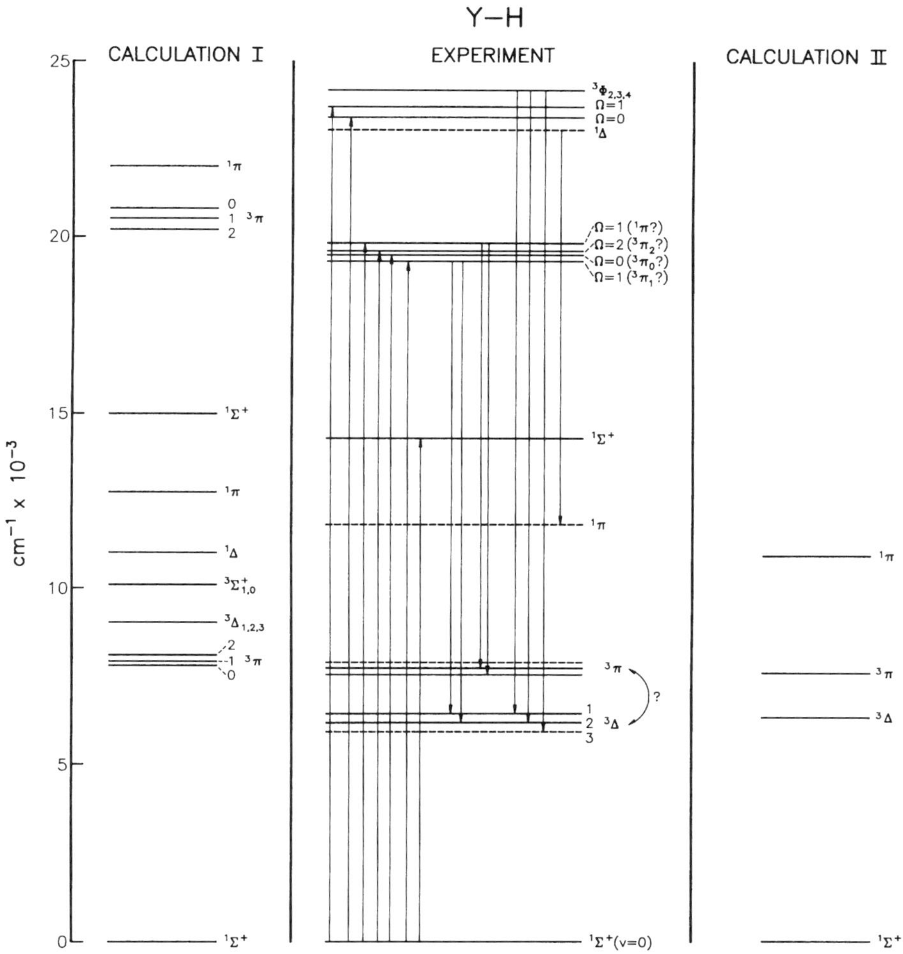

Figure 4.7 Predicted and observed energy levels of YH. Calculation I is from Ref. 21, while calculation II is from Ref. 87. (Reproduced from Ref. 114.)

0.2 and 0.986 in various states. This indicates considerable *dsp* hybridization. This hybridization is minimum for the $^1\Gamma$ state (0.126) and maximum for the $^1\Pi$(II) (0.986), $^3\Pi$(II) (0.61), $^3\Sigma^+$ (0.751), $^3\Pi$ (.603), and $^1\Delta$ (.610) states. The *d* population of the ground state is 0.690, and the *d* populations of many other states are between 0.96 and 1.5, with the exception of the $^1\Gamma$ and $^3\Pi$(II) states.

Bauschlicher and Walch (83) considered full-valence MCSCF/CI calculations on the lowest six electronic states of ScH. Subsequently, Chong et al. (89) studied the dipole moments of the first-row hydrides using the MCPF and CASSCF/MRCI techniques.

Table 4.5 shows the CASSCF and FOCI results for ScH obtained by

Table 4.3 Contributions of Important Configurations in SOCI Wavefunctions of YH

State	Percentage of Contributions
$^1\Sigma^+$	$1\sigma^2 2\sigma^2$ (86), $1\sigma^2 1\pi^2$ (8)
$^3\Pi$	$1\sigma^2 2\sigma^1 1\pi^1$ (94)
$^3\Delta$	$1\sigma^2 2\sigma^1 1\delta^1$ (96)
$^3\Sigma^+$	$1\sigma^2 2\sigma^1 3\sigma^1$ (81)
$^1\Delta$	$1\sigma^2 2\sigma^1 1\delta^1$ (81)
$^1\Pi$	$1\sigma^2 2\sigma^1 1\pi^1$ (80)
$^1\Sigma^+$(II)	$1\sigma^2 2\sigma^1 3\sigma^1$ (58), $1\sigma^2 1\pi 2\pi$ (7), $1\sigma^2 1\delta^2$ (6), $1\sigma^2 2\sigma 4\sigma$ (5), $1\sigma^2 3\sigma^2$ (5)
$^3\Pi$(II)	$1\sigma^2 1\pi^1 1\delta^1$ (89)
$^1\Pi$(II)	$1\sigma^2 2\sigma^1 2\pi^1$ (85)
$^1\Gamma$	$1\sigma^2 1\delta^2$ (89)

Source: Reproduced from Reference 21.

Bauschlicher and Walch (83). We note that they too obtain a $^1\Sigma^+$ ground state for ScH analogous to YH. An initial comparison of Table 4.5 with Table 4.2 reveals that the energy separations of the excited states of ScH are below the corresponding values of YH.

The ground state of ScH was found to be composed mainly of Sc(d) and H(s) σ bonds, while the excited states of ScH exhibited *sp* hybrid bonds. The dipole moment of ScH in its $^1\Sigma^+$ state was computed as 1.374 D using the MCPF method if four electrons are correlated and 1.641 D if 12 electrons are correlated.

The analysis of the contour plots of the highest occupied σ and π orbitals of

Table 4.4 Mulliken Population Analyses of Electronic States of YH

	Gross Population							Overlap
State	Y	Y(s)	Y(p)	Y(d)	H	H(s)	H(p)	Y–H
$^1\Sigma^+$	2.694	1.550	0.454	0.690	1.306	1.296	0.010	0.639
$^3\Pi$	2.696	0.947	0.663	1.086	1.304	1.295	0.009	0.690
$^3\Delta$	2.709	0.962	0.368	1.380	1.291	1.283	0.008	0.672
$^3\Sigma^+$	0.735	1.025	0.751	0.959	1.265	1.258	0.007	0.700
$^1\Delta$	2.700	0.791	0.610	1.300	1.300	1.292	0.008	0.680
$^1\Pi$	2.769	0.918	0.350	1.500	1.231	1.224	0.007	0.653
$^1\Sigma^+$(II)	2.782	0.900	0.522	1.361	1.218	1.211	0.006	0.670
$^3\Pi$(II)	2.730	0.293	0.681	1.757	1.270	1.262	0.007	0.680
$^1\Pi$(II)	2.827	0.632	0.986	1.209	1.173	1.168	0.005	0.652
$^1\Gamma$	2.758	0.345	0.126	2.287	1.242	1.235	0.007	0.664

Source: Reproduced from Reference 21.

Table 4.5 Summary of FV-MCSCF and First-Order CI for ScH (R_e in bohr; T_e in cm^{-1})

	MCSCF		FO	
State	R_e	T_e	R_e	T_e
$^1\Sigma^+$	3.39	0	3.37	0
$^3\Delta$	3.67	1,412	3.65	2,328
$^3\Delta^a$	3.67	2,098	3.65	3,023
$^3\Pi^a$	3.68	4,502	3.61	4,995
$^3\Sigma^{+\,a}$	3.91	5,875	3.90	6,317

[a]Core taken from $^1\Sigma^+$ SCF.

Source: Reproduced from Reference 83.

the $^1\Sigma^+$ ground state of ScH obtained by Bauschlicher and Walch (83) reveals that the 7σ orbital is mainly Sc(4s), but is polarized outward perpendicular to the bond axis. The 3π orbital is a nonbonding Sc(4p) orbital. The 6σ bonding orbital is mainly Sc($3d\sigma$) + H(1s) orbital.

Following the Bauschlicher–Walch (83) calculation, the author (12) studied ScH using RECPs and a CASSCF/FOCI method. Table 4.6 compares the all-electron result with the ECP result. As seen from Table 4.6, the agreement between the ECP and all-electron result is good.

4.4.2 YH_2 and ScH_2

The author and Ravimohan (20) computed the potential-energy surfaces of three electronic states of YH_2 of $^2A_1$, $^2B_1$, and $^2A_2$ symmetries. They employed a ($3s2p4d/3s2p3d$) valence Gaussian basis set in conjunction with RECPs, which retained the outer $4d^1 5s^2$ shells of the Y atom in the valence space. A full CASSCF was followed by MRSDCI calculations, which retained all configurations in the CASSCF with coefficients ≥0.05 as reference configurations.

Table 4.7 shows the computed equilibrium geometries and energy separations of the electronic states of YH_2, while Figure 4.8 shows the potential-

Table 4.6 Comparison of ECP and All-Electron Results for ScH

	R_e, Å		T_e, cm^{-1}	
State	ECP	All-Electron[a]	ECP	All-Electron
$^1\Sigma^+$	1.744	1.78	0	0
$^3\Delta$	1.884	1.93	2,050	2,328

[a]From Reference 83.

Source: Reproduced from Reference 12.

Table 4.7 Energies and Equilibrium Structures for Low-Lying States of YH_2

	CASSCF			MRSDCI		
State	R, Å	θ, deg	E, eV	R, Å	θ, deg	E, eV
$^2A_1$	1.95	123	0.0^a	1.94	123	0.0^a
$^2\Sigma_g^+$	2.02	180	0.37	2.01	180	0.36
$^2B_1$	2.04	133	1.11	2.02	132	1.04
$^2\Pi_g$	2.10	180	1.18	2.08	180	1.16
$^2A_2$	2.04	130	1.66	2.02	126	1.56
$^2\Delta_g$	2.10	180	1.74	2.08	180	1.71

[a]The absolute CASSCF valence energy of $^2A_1$ is −2.550578 hartrees, and the MRSDCI valence energy of $^2A_1$ is −2.573984 hartrees.

Source: Reproduced from Reference 20.

energy surfaces of three electronic states of YH_2 obtained by the author and Ravimohan (20).

It is evident from Table 4.7 and Figure 4.8 that $^2A_1$ is the ground state of YH_2 with a bent equilibrium geometry (r_e = 1.94 Å, θ_e = 123°). This state is more stable than $^2\Sigma_g^+$ by 0.36 eV. The author and Ravimohan (20) have found that the equilibrium structures in all the cases are bent rather than linear. How-

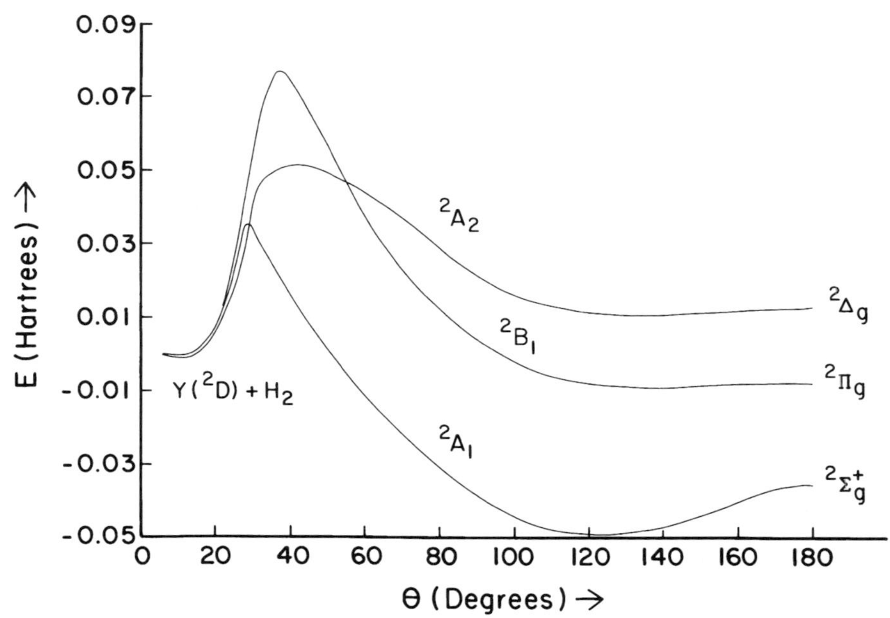

Figure 4.8 Potential-energy surfaces of YH_2. (Reproduced from Ref. 20.)

ever, unlike in the $^2A_1$ state, $^2A_1$ and $^2B_1$ are very close to the linear states in energy. The stability of the $^2A_1$ ground state with respect to Y + H_2 dissociation was found to be 31 kcal/mol at the CASSCF level of theory. It should be mentioned that the author and Ravimohan inadvertently compared their computed dissociation energy with the enthalpy of the solid Y + H_2 reaction.

In Figure 4.8 for each θ = H–Y–H bond angle, Y–H bond lengths were optimized. All three states have barriers for the insertion of the yttrium atom into H_2 to form YH_2. The saddle points occur near $28°, 37°$, and $40°$ for $^2A_1$, $^2B_1$, and $^2A_2$ states, respectively. The saddle points were obtained as the superposition of the surfaces obtained by starting from the CASSCF wavefunction of the linear structure and the CASSCF wavefunction of the dissociated structure (Y + H_2). The barrier height of the $^2A_1$ surface (0.98 eV) is lower than the corresponding values of $^2B_1$ (2.09 eV) and $^2A_2$ (1.41 eV) states.

Table 4.8 shows the dipole moments of the $^2A_1$, $^2B_1$, and $^2A_2$ states at their bent equilibrium geometries obtained from the MRSDCI density matrices. All three states have Y^+H^- polarities. The dipole moment of the $^2A_1$ state is the smallest, while that of $^2A_2$ state is the largest. The large dipole moments of these states reveal the ionic character of Y–H bonds.

Analyses of the leading configurations near equilibrium geometries of the three states of YH_2 revealed that they are dominated by their leading configuration with coefficients ≥0.98. The $^2A_1$ ground state is composed of the $1a_1^2 2a_1 1b_2^2$ configuration, wherein the $1a_1$ and $2a_1$ orbitals are mixtures of $Y(5s)+H(1s)+H_2(1s)$ and $Y(4d)+H_1(1s)+H_2(1s)$, respectively. The $1b_2$ orbital is composed of $Y(5p_y) + H_1(1s)–H_2(1s)$.

Table 4.9 shows the Mulliken population analyses of the electronic states of YH_2. The Y population strongly deviates from its neutral 3.0 population, suggesting transfer of up to 0.6 electron to the hydrogen atoms. We also note significant deviation from the neutral $4d^15^2$ electronic configuration near the equilibrium geometries. The Y–H overlaps are also quite significant.

The author (12) has studied the insertion of $Sc(^2D)$ into H_2 in the $^2A_1$ state. In addition, the linear $^2\Sigma_g^+$, $^2\Pi_g$, and $^2\Delta_g$ states of the ScH_2 molecule were studied. He used a complete active space MCSCF method followed by full second-order

Table 4.8 Dipole Moments for Low-Lying States of YH_2.[a]

State	μ, D[a]
$^2A_1$	1.41
$^2B_1$	3.81
$^2A_2$	6.01

[a]Polarity Y^+H^-.

Source: Reproduced from Reference 20.

Table 4.9 Mulliken Population Analysis for Low-Lying State of YH_2[a]

	Gross Populations					Y–H
State	Y (total)	H (total)	Y(s)	Y(d)	Y(p)	Overlaps
$^2A_1$	2.36	2.64	0.83	1.01	0.52	1.35
$^2\Sigma_g^+$	2.40	2.60	0.98	0.92	0.50	1.36
$^2B_1$	2.34	2.66	0.41	1.47	0.45	1.34
$^2\Pi_g$	2.38	2.62	0.44	1.46	0.48	1.34
$^2A_2$	2.47	2.53	0.54	1.55	0.38	1.33
$^2\Delta_g$	2.51	2.49	0.53	1.42	0.56	1.34

[a]Populations of the two H atoms are combined together.

Source: Reproduced from Reference 20.

CI (SOCI). For the Sc atom, RECPs that retained the outer $3d^14s^2$ shells were retained in the valence space. The $(4s5d/3s3d)$ valence Gaussian basis set augmented by two sets of p functions suggested by Wachters was used for the Sc atom.

Figure 4.9 shows the bending potential-energy surface of ScH_2. Again θ is the $\mathrm{H}\overset{\mathrm{Sc}}{\diagup\;\diagdown}\mathrm{H}$ bond angle as in Figure 4.9. For each θ, Sc–H bond lengths were optimized. Table 4.10 shows the geometries and energy separations of three linear electronic states of ScH_2 at the CASSCF and SOCI levels of theory.

As seen from both Figure 4.9 and Table 4.10, the Sc atom has to surpass a large barrier of 1.57 eV (SOCI) to insert into H_2 to form the $^2\Sigma_g^+$ state. The $^2\Sigma_g^+$ state thus formed is only 0.22 eV more stable than the dissociated Sc + H_2 species. As seen from Figure 4.9, there exists a shallow obtuse angle bent minimum for ScH_2. At the CASSCF level, this minimum has r_e = 1.81 Å and θ_e = 133.4°. At this level, this minimum was found to be 0.04 eV below $^2\Sigma_g^+$. The author thus suggested that the possibilities of an obtuse-angle $^2A_1$ ground state for ScH_2 cannot be ruled out, but such a minimum formed only a shallow potential well.

Table 4.11 shows the Mulliken populations of the SOCI natural orbitals of ScH_2 obtained by the author (12). The leading configuration of the $^2\Sigma_g^+$ state of ScH_2 is $1\sigma_g^2 1\sigma_u^2 1\sigma_g^1$ (−0.967). The $1\sigma_g$ orbital is primarily a bonding orbital of Sc($4s$) and H($1s$) [Sc($3d$) makes nonnegligible contributions]. The $2\sigma_g$ orbital is dominantly Sc($3d$) and H($1s$), but is slightly antibonding with respect to hydrogen $1s$ orbitals. The $1\sigma_u$ is primarily Sc($4p$) + H($1s$)–H_2($1s$). As seen from Table 4.11, the Sc($4s$) net population is smaller, but the $4p$ population is not zero, suggesting considerable $4p$ mixing in the $1\sigma_u$ orbital.

While the bending potential-energy surface of the $^2A_1$ state of ScH_2 has a shallow bent minimum (θ_e = 133.4°, r_e = 1.81 Å), the $^2A_1$ surface YH_2 has a well-bound minimum (θ_e = 123°, r_e = 1.95 Å). The $^2A_1$–$^2\Sigma_g^+$ splitting is very

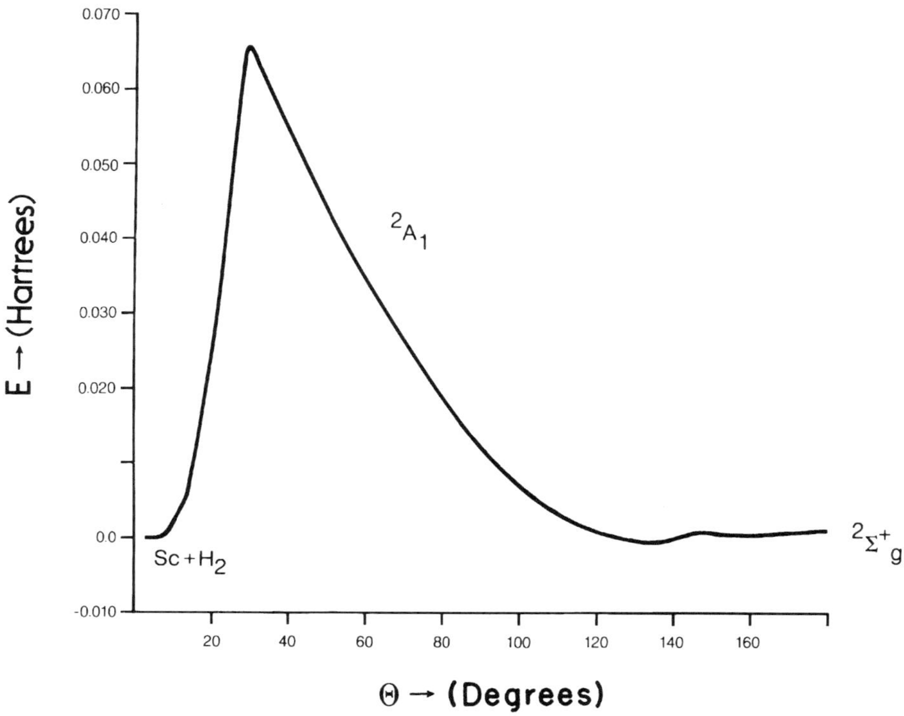

Figure 4.9 Potential-energy surface of the $^2A_1$ state of ScH_2. (Reproduced from Ref. 12.)

Table 4.10 Geometries and Energies of Low-Lying States of ScH_2

Method	State	r_e, Å	θ_e, deg	E, eV
SOCI	$^2\Sigma_g^+$	1.84	180.0	0.0[a]
SOCI	$^2\Pi_g$	1.91	180.0	0.16
SOCI	$^2\Delta_g$	1.91	180.0	0.62
SOCI	Sc + H_2	—	—	0.22
SOCI	$^2A_1$ (barrier)	—	—	1.57[b]
CASSCF	$^2\Sigma_g^+$	1.86	180.0	0.0[c]
CASSCF	$^2\Pi_g$	1.94	180.0	0.19
CASSCF	$^2\Delta_g$	1.94	180.0	0.68
CASSCF	$^2A_1$	1.83	29.0	1.81[d]

[a] SOCI energies reported with respect to the energy of the $^2\Sigma_g^+$ state at its equilibrium geometry.
[b] The energy splitting between the linear geometry and the barrier evaluated at the CAS geometries.
[c] Energies reported with respect to the $^2\Sigma_g^+$ CASSCF energy.
[d] The energy of the barrier at the estimated CASSCF geometry with respect to the $^2\Sigma_g^+$ minimum.
Source: Reproduced from Reference 12.

Table 4.11 Mulliken Population Analysis (Net and Overlap) of Low-Lying States of ScH_2 at their Equilibrium Geometries[a]

	Net Population							
State	Sc	H	Sc(*s*)	Sc(*d*)	Sc(*p*)	H(*s*)	H(*p*)	Overlap
$^2\Sigma_g^+$	1.77	1.90	0.75	0.81	0.22	1.89	0.01	1.33
$^2\Pi_g$	1.67	1.99	0.41	1.16	0.21	1.98	0.01	1.35
$^2\Delta_g$	1.74	1.92	0.47	1.14	0.25	1.91	0.01	1.35

[a]Populations of both hydrogen atoms are grouped together.
Source: Reproduced from Reference 12.

small for ScH_2, while it is somewhat large for YH_2. The barrier to insert the Sc atom into H_2 is about 36 kcal/mol in the $^2A_1$ surface, while for the Y atom the corresponding barrier is about 23 kcal/mol, which is much smaller. Thus the Y atom is found to be more reactive with H_2 than Sc. The $^2A_1$ bent YH_2 state is more stable with respect to Y + H_2 by about 31 kcal/mol, while the $^2\Sigma_g^+$ state of ScH_2 is only 5 kcal/mol more stable than Sc + H_2. Thus bonding in YH_2 is much stronger than in ScH_2.

The total gross population of Sc in the $^2\Sigma_g^+$ state of ScH_2 is 2.44, while the corresponding population is 2.40 in YH_2. This suggests greater charge transfer from Y to H in YH_2 compared to ScH_2. The Y(5*p*) population of all electronic states of YH_2 (cf. Table 4.9) are considerably larger than the corresponding Sc(4*p*) populations. This suggests greater involvement of the Y($5p_y$) orbital in the $1b_2$ orbital of YH_2.

4.4.3 Electronic States of LaH

Das and the author (27) have computed the spectroscopic constants and potential-energy curves of 20 electronic states of LaH. They used RECPs with $5s^25p^65d^16s^2$ valence shells for the La atom together with a (5*s*5*p*3*d*1*f*) valence Gaussian basis set. CASSCF followed by the FOCI method of calculation was used to compute the potential-energy curves, while the spectroscopic constants were obtained using the SOCI method. In addition, the spin–orbit effects were introduced using the RCI method.

Table 4.12 shows the computed energy separation for the atomic states of La compared with experimental values together with the possible molecular states of LaH. As evidenced from Table 4.12, there are several possible low-lying electronic states for LaH.

Figure 4.10 shows the potential-energy curves for LaH obtained using the CASSCF/FOCI method, while Table 4.13 shows the SOCI spectroscopic constants of LaH. As seen from Table 4.13, the ground state of LaH is a $X^1\Sigma^+$ state with $r_e = 2.08$ Å and $\omega_e = 1433\ \text{cm}^{-1}$.

Bernard and Bacis (113) observed $B \leftrightarrow A$, $C \rightarrow A$, and $b \rightarrow a$ systems of

Table 4.12 Dissociation Relationships for Some Low-Lying Electronic States of LaH

Molecular State	Atomic State (La + H)	Energy, cm^{-1} SOCI	Exp.[a]
$^1\Sigma^+$, $^1\Pi$, $^1\Delta$, $^3\Sigma^+$, $^3\Pi$, $^3\Delta$	$a^2D(d^1s^2) + {}^2S$	0	0
$^1\Sigma^-$, $^1\Pi$, $^1\Delta$, $^1\Phi$, $^3\Sigma^-$, $^3\Pi$, $^3\Delta$, $^3\Phi$	$a^2F(d^2s^1) + {}^2S$	5,706	6,961
$^1\Sigma^+$, $^1\Pi$, $^1\Delta$, $^3\Sigma^+$, $^3\Pi$, $^3\Delta$	$b^2D(d^2s^1) + {}^2S$	9,437	8,249
$^1\Sigma^+$, $^1\Pi$, $^1\Delta$, $^1\Phi$, $^1\Gamma$, $^3\Sigma^+$, $^3\Pi$, $^3\Delta$, $^3\Phi$, $^3\Gamma$	$a^2G(d^2s^1) + {}^2S$	11,193	9,280

[a]Averaged over *J*.

Source: Reproduced from Reference 27.

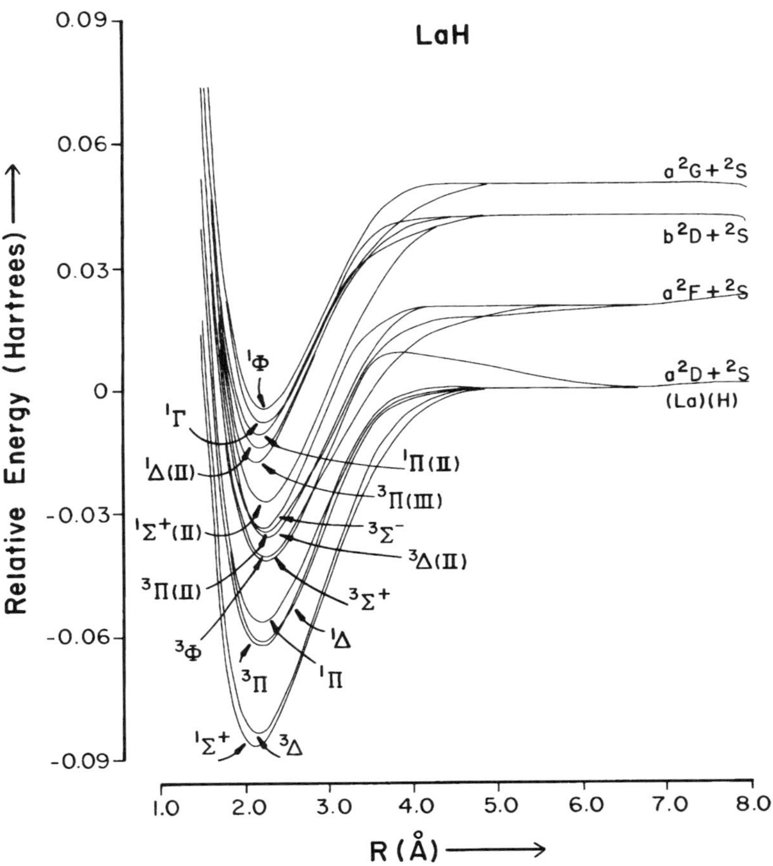

Figure 4.10 Potential-energy curves for LaH. (Reproduced from Ref. 27.)

Table 4.13 Spectroscopic Properties of LaH

State	r_e, Å	ω_e, cm^{-1}	T_e, cm^{-1}	D_e, eV	μ_e, D
$X\,^1\Sigma^+$	2.08	1,433	0	2.60	2.42
$^3\Delta$	2.13	1,352	2,805	2.24	3.63
$a^3\Pi$	2.12	1,341	5,147	1.92	2.32
$^1\Pi$	2.13	1,309	6,226	1.80	3.95
$^1\Delta$	2.16	1,299	6,510	1.77	0.25
$^3\Phi$	2.19	1,240	10,612	2.00	4.89
$^3\Sigma^+$	2.20	1,203	11,794	1.20	−0.49
$^3\Pi$(II)	2.19	1,228	11,956	1.83	4.24
$^3\Sigma^-$	2.18	1,247	12,035	1.86	5.40
$A\,^1\Sigma^+$(II)	2.20	1,230	13,025	2.17	2.17
$^3\Delta$(II)	2.24	1,166	14,020	1.61	1.63
$B^1\Pi$(II)	2.16	1,293	15,729	1.77	3.24
$^3\Pi$(III)	2.09	1,377	15,880	1.81	3.14
$^3\Sigma^-$(II)	2.17	1,251	16,107	1.35	4.17
$^1\Gamma$	2.18	1,260	16,838	1.88	6.51
$^1\Phi$	2.18	1,259	17,333	1.79	4.46
$^1\Delta$(II)	2.18	1,234	17,427	1.60	5.49
$^1\Delta$(III)	2.20	1,226	20,109	—	2.21
$C\,^1\Pi$(III)	2.19	1,342	20,170	—	6.15
$b^3\Delta$(III)	2.10	1,364	23,256	—	4.48

Source: Reproduced from Reference 27.

LaH. They performed a rotational analysis of the observed systems and assigned these three systems to $^1\Delta$–$^1\Pi$, $^1\Sigma$–$^1\Pi$, and $^3\Phi$–$^3\Pi$ systems, respectively.

It is evident from Table 4.13 that the $^1\Pi$ state is not the ground state of LaH. Hence, Das and the author (DB) argued that the *A* state of Bernard and Bacis should be reassigned to the $X^1\Sigma^+$ ground state of LaH. Das and the author suggested reassignment of the Bernard–Bacis assignment in accordance with Table 4.13. These suggested assignments are also consistent with LaF (115, 116). The ground state of LaF is now well known to be a $X^1\Sigma^+$ state. For LaF, excited $A^1\Sigma^+$, $B^1\Pi$, $C^1\Pi$, $D^1\Sigma^+$, and $E^1\Sigma^+$ states have been observed and assigned through the analysis of $A \leftarrow X$, $B \leftarrow X$, $C \leftarrow X$, $D \leftarrow X$, and $E \leftarrow X$ systems. The ν_{00} values for these transitions are 11,662, 16,184, 20,960, 22,485, and 22,574 cm^{-1}, respectively. The well-characterized spectra of LaF supported the assignment in Table 4.12.

The $A^1\Sigma^+ \leftrightarrow X^1\Sigma^+$ system predicted at 13,000 cm^{-1} has not yet been observed. The corresponding transition for LaF has been observed at 11,662 cm^{-1}. The $B \leftrightarrow X$ system with a theoretically predicted T_e value of 15,730 cm^{-1} is consistent with the ν_{00} band head of the experimentally observed $B \leftrightarrow A$ system (15,619 cm^{-1}). It is also consistent with the rotational analysis of Bernard and Bacis, which suggested that $\Delta\lambda = +1$ for this transition. For LaF this system appears with $\nu_{00} = 16{,}184$ cm^{-1}.

The $C \rightarrow A$ system of Bernard and Bacis with a $\nu_{00} = 18{,}595\ \text{cm}^{-1}$ was found to be most consistent with the $C^1\Pi \rightarrow X^1\Sigma^+$ system, by DB with a theoretical $T_e = 20{,}170\ \text{cm}^{-1}$. Likewise, the triplet–triplet transition observed with $\nu_{00} = 16{,}000\ \text{cm}^{-1}$ was assigned to $b^3\Delta \leftrightarrow a^3\Pi$ by DB with a theoretical $\nu_{00} = 18{,}100\ \text{cm}^{-1}$. The differences between theoretical and experimental values were found to be within the error bars of calculations for excited states.

The dipole moments in Table 4.13 are large and positive, suggesting La^+H^- polarities. Some of the excited states have very large dipole moments showing significant La^+H^- charge separations.

Table 4.14 shows the weights of the leading configurations for the electronic states of LaH. Even the ground state of LaH is a mixture of $1\sigma^2 2\sigma^2$, $1\sigma^2\delta^2$, and $1\sigma^2\pi^2$ configurations. The excited electronic states are considerably more complex, as evidenced from Table 4.14.

We now compare the properties of ScH, YH, and LaH. There is a sharp rise in the bond length of YH compared to that of ScH. However, the bond lengths of YH (1.95 Å) and LaH (2.08 Å) are quite similar. This is primarily because of the sharp rise in the $5p_y$ orbital participation in YH compared to ScH. Although the $6p$ population decreases for LaH, the $6s$ population rises slightly. The La($5d$) population is enhanced in the $^1\Sigma^+$ state.

Figure 4.11 compares the D_e values of ScH, YH, and LaH. We note that the M–H bond strength is significantly enhanced in going from ScH to YH, but the D_e decreases in going from YH to LaH.

Another quantity of interest is the μ_e/r_e ratio, shown in Figure 4.12. This

Table 4.14 Contributions of Various Leading Configurations to SOCI Wavefunctions at Equilibrium Geometries of Electronic States of LaH

State	Electronic Configurations (Percentage)
$^1\Sigma^+$	$1\sigma^2 2\sigma^2$ (86), $1\sigma^2 1\delta^2$ (6), $1\sigma^2 1\pi^2$ (4)
$^3\Delta$; $^1\Delta$	$1\sigma^2 2\sigma 1\delta$ (97); $1\sigma^2 2\sigma 1\delta$ (87), $1\sigma^2 1\pi^2$ (10)
$^3\Pi$; $^1\Pi$	$1\sigma^2 2\sigma 1\pi$ (95); $1\sigma^2 2\sigma 1\pi$ (74), $1\sigma^2 1\delta 1\pi$ (20)
$^3\Phi$; $^1\Phi$	$1\sigma^2 1\delta 1\pi$ (95; 90)
$^3\Sigma^+$	$1\sigma^2 2\sigma 3\sigma$ (52), $1\sigma^2 2\sigma 4\sigma$ (24), $1\sigma^2 2\sigma 5\sigma$ (12)
$^3\Sigma^-$	$1\sigma^2 1\delta^2$ (60), $1\sigma^2 1\pi 2\pi$ (4), $1\sigma^2 2\pi 3\pi$ (4), $1\sigma^2 1\pi^2$ (3), $1\sigma^2 1\delta 2\delta$ (3)
$^3\Pi$(II); $^1\Pi$(II)	$1\sigma^2 1\delta 1\pi$ (86; 56), $1\sigma^2 2\sigma 1\pi$ (3; 16)
$^1\Sigma^+$(II)	$1\sigma^2 2\sigma 3\sigma$ (23), $1\sigma^2 1\delta^2$ (32), $1\sigma^2 2\sigma 4\sigma$ (11), $1\sigma^2 2\sigma 5\sigma$ (10)
$^3\Delta$(II)	$1\sigma^2 2\sigma 1\delta$ (42), $1\sigma^2 3\sigma 1\delta$ (17), $1\sigma^2 4\sigma 1\delta$ (14), $1\sigma^2 5\sigma 1\delta$ (10), $1\sigma^2 6\sigma 1\delta$ (7)
$^3\Sigma^-$(II)	$1\sigma^2 1\delta^2$ (30), $1\sigma^2 1\pi 2\pi$ (12), $1\sigma^2 1\pi 3\pi$ (12), $1\sigma^2 1\pi^2$ (5), $1\sigma^2 2\pi 3\pi$ (5)
$^3\Pi$(III); $^1\Pi$(III)	$1\sigma^2 2\sigma 1\pi$ (55), $1\sigma^2 2\sigma 2\pi$ (22); $1\sigma^2 1\delta 1\pi$ (39), $1\sigma^2 2\sigma 1\pi$ (17), $1\sigma^2 1\delta 2\pi$ (16)
$^1\Gamma$	$1\sigma^2 1\delta^2$ (95)
$^1\Delta$(II)	$1\sigma^2 2\sigma 1\delta$ (40), $1\sigma^2 1\pi^2$ (11), $1\sigma^2 3\sigma 1\delta$ (8), $1\sigma^2 1\pi 2\pi$ (8), $1\sigma^2 4\sigma 1\delta$ (7)
$^1\Delta$(III)	$1\sigma^2 1\pi^2$ (36), $1\sigma^2 2\sigma 1\delta$ (22), $1\sigma^2 1\pi 2\pi$ (13), $1\sigma^2 1\pi 3\pi$ (9)
$^3\Delta$(III)	$1\sigma^2 1\pi 2\pi$ (41), $1\sigma^2 1\pi 3\pi$ (17), $1\sigma^2 1\pi 4\pi$ (12), $1\sigma^2 2\pi 5\pi$ (9)

Source: Reproduced from Reference 27.

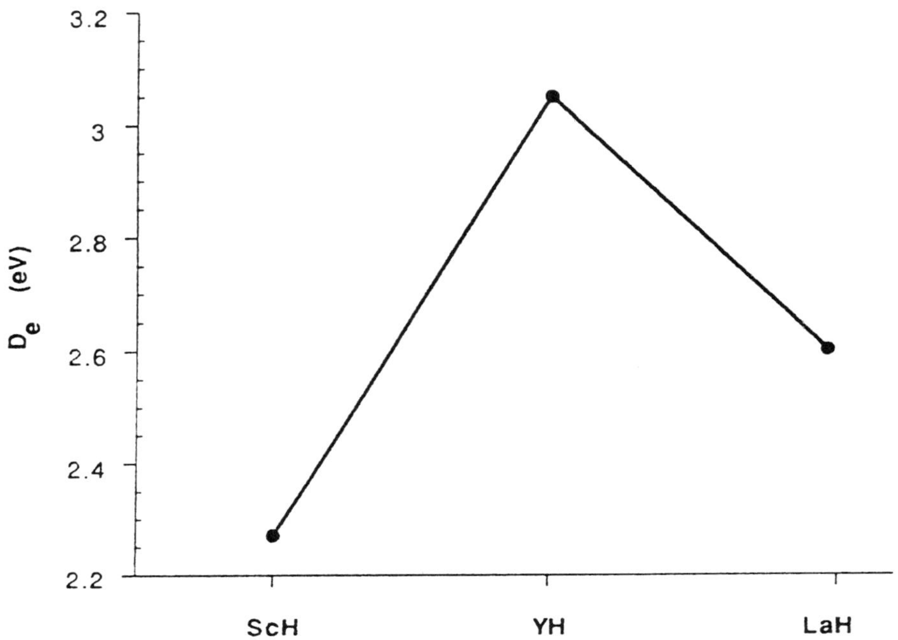

Figure 4.11 Dissociation energies of ScH, YH and LaH. (Reproduced from Ref. 2.)

quantity compares the extent of charge transfer in the three species. As seen from Figure 4.12, the μ_e/r_e values are similar for ScH and YH, but dramatically increase for LaH. This suggests significant ionic character of the La–H bond.

The 1σ valence orbital of LaH is composed of La($6s$) + H($1s$), while the 2σ orbital is composed of La($5d$), La($6p_z$) and H($1s$). Of course, the π and δ orbitals are nonbonding. The Mulliken population analysis of the $^1\Sigma^+$ ground state of LaH revealed that the La population is $6s^{1.52}6p^{0.22}5d^{0.997}$

4.4.4 LaH⁺

Das and the author (30) have computed the spectroscopic constants and potential-energy curves of LaH^+. They used a CASSCF method followed by full second-order CI calculations in conjugation with $5s^25p^65d6s^2$ RECPs of Ross et al. (196). The basis set used was ($5s5p3d$) valence Gaussian basis set. At both CASSCF and SOCI levels, excitations from the $5s^25p^6$ shells were not allowed. The CASSCF calculations included four a_1, one b_2, one b_1, and one a_2 orbitals in the active space. The spin–orbit effects were included using the RCI method.

Elkind et al. (117–119) have investigated the reactions of metal ions with H_2. Among several systems, they have studied the periodic trends in the reactivity

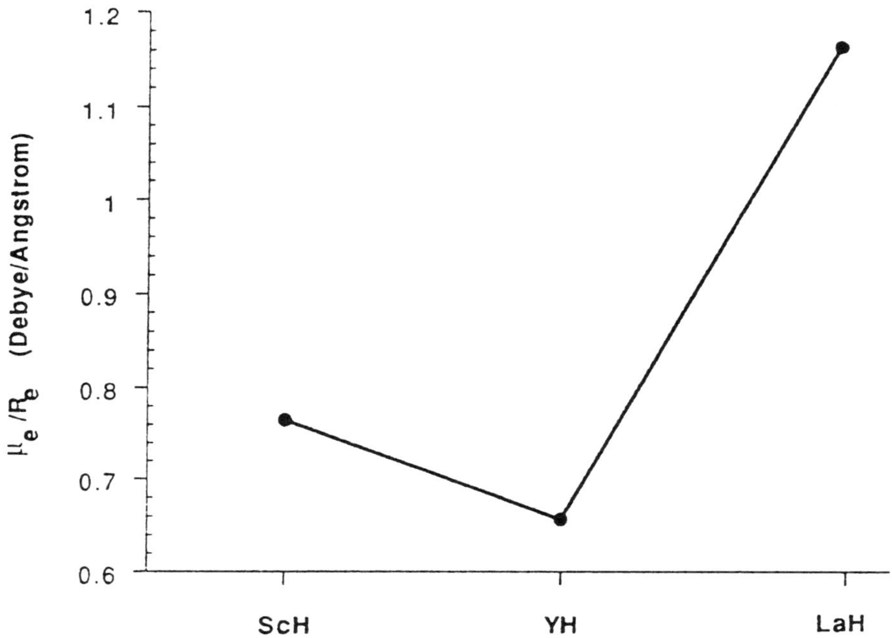

Figure 4.12 Dipole moment per unit M–H distance for ScH, YH, and LaH. (Reproduced from Ref. 2.)

of Sc^+, Y^+, La^+, and Lu^+ with H_2 using ion–molecule collisions of the type

$$La^+ + H_2 \rightarrow LaH^+ + H.$$

They measured the D_0° value of LaH^+ as 2.48 ± 0.09 eV. Elkind et al. proposed several mechanisms for the formation of LaH^+. The theoretical calculations of Das and the author not only provided an unambiguous mechanism for the formation of LaH^+ but also explained the observed branching ratios of LaD^+ and LaH^+ in the La^+ + HD reaction. There are similar studies by Elkind et al. (117–119) on the reactivities of other transition-metal ions with H_2. [See also Armentrout and Beauchamp (120), Martinho and Beauchamp (121), and Schilling et al. (122) for related works.]

Table 4.15 compares the computed dissociation limits of the electronic states of LaH^+ by Das and the author (30) together with experimental atomic data from Moore's book (123). The experimental values are averaged with respect to J, permitting comparison of energies in the absence of spin–orbit coupling. As seen from Table 4.15, although the ground state of La^+ is a ^3F state arising from the $5d^2$ configuration the excited electronic state of ^1D and ^3D symmetries arising from the $5d^16s^1$ configuration are quite closer. Although the ground

Table 4.15 Dissociation Relationships of Low-Lying States of LaH⁺

Atomic States	Molecular States	Energy, cm^{-1}	
		Theory	Exp.[a]
$a^3F(5d^2) + {}^2S$	$^2\Sigma^-, {}^2\Pi, {}^2\Delta, {}^2\Phi,$ $^4\Sigma^-, {}^4\Pi, {}^4\Delta, {}^4\Phi$	0	0
$a^1D(5d^16s^1) + {}^2S$	$^2\Sigma^+, {}^2\Pi, {}^2\Delta$	288	179
$a^3D(5d^16s^1) + {}^2S$	$^2\Sigma^+, {}^2\Pi, {}^2\Delta,$ $^4\Sigma^+, {}^4\Pi, {}^4\Delta$	1,604	1,590

[a]Experimental values are J-averaged from Reference 123.

Source: Reproduced from Reference 30.

state of the La atom is ^2D arising from $5d^16s^2$, it is quite interesting that La^+ has a $5d^2$ ground state, rather than $5d^16s^1$ or $6s^2$ ground states. Since the lanthanide contraction is experienced only by Hf onward, the 6*s* orbital of La has no contraction from this effect, since the 4*f* shells have not yet started filling up. The effect of relativity is also rather small. Consequently, the removal of a 6*s* electron is favored for La^+. Subsequent to the removal of a 6*s* electron, the

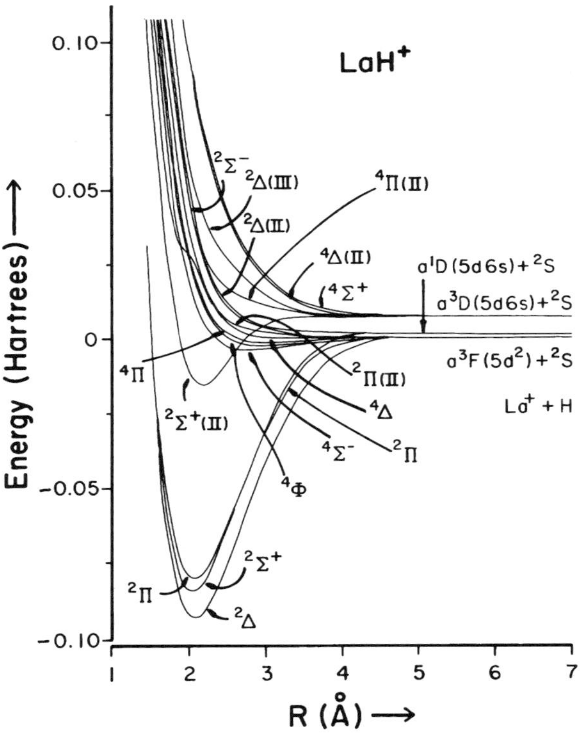

Figure 4.13 Potential-energy curves for LaH⁺. (Reproduced from Ref. 30.)

Table 4.16 Spectroscopic Constants of Low-Lying States of LaH^+

State	r_e, Å	ω_e, cm^{-1}	T_e, cm^{-1}	μ_e, D	D_e, eV Theory	D_e, eV Exp.[a]
$^2\Delta$	2.074	1480	0	4.41	2.54	2.57 ± 0.09
$^2\Sigma^+$	2.040	1517	1,948	3.60	2.33	
$^2\Pi$	2.064	1462	2,855	4.08	2.19	
$^2\Sigma^+$(II)	2.184	1287	17,082	—	—	
$^4\Sigma^-$	2.774	407	19,597	—	—	
$^4\Phi$	2.972	346	19,874	—	—	
$^4\Pi$	3.087	229	20,127	—	—	

[a]The experimental D° from Reference 118 was corrected for zero-point correction using our calculated ω_e to obtain experimental D_e.

Source: Reproduced from Reference 30.

electronic configuration rearranges to form a ^3F state for La^+. Note that Hf^+ has a $5d6s^2$ (^2D) ground state rather than $5d^26s$, since removal of a $6s$ electron becomes considerably more difficult for Hf due primarily to lanthanide contraction and to a less extent because of relativity.

As evidenced from Table 4.15, there are several doublet and quartet low-lying electronic states of LaH^+. Figure 4.13 shows the potential-energy curves for several electronic states of LaH^+ computed by Das and the author (30). Table 4.16 shows the spectroscopic constants of bound electronic states of LaH^+. As seen in Figure 4.13, doublet states dissociating into $La^+(^3F)$ + H are bound. The $^4\Sigma^-$, $^4\Phi$, and $^4\Pi$ states have only long-range minima.

The quartet states of LaH^+ do not form very deep potential wells, since La^+ in these states do not hybridize much and remain in a configuration closer to the $La^+(5d^2)$ configuration. The low-spin states of La^+, to the contrary, undergo significant $5d$–$6s$ as well as $5d6s6p$ hybridization, so that pronounced minima are formed.

As seen from Table 4.16, the ground state of LaH^+ is $^2\Delta$. Schilling et al. (122), Alvarado-Swaisgood et al. (124), and Aglada et al. (125) have studied ScH^+. They find that ScH^+ too has a $^2\Delta$ ground state. The first excited $^2\Sigma^+$ state of LaH^+ is only ~1950 cm^{-1} above the ground state. The r_e values of all doublet states are uniformly between 2.04 and 2.18 Å. The quartet states to the contrary form longer r_e values consistent with the high-spin characters of these states.

The $^4\Pi$-$^2\Delta$, $^4\Phi$-$^2\Delta$ weak spin–change transitions are predicted in the 20,000-cm^{-1} region. The $^2\Sigma^+$(II)–$^2\Sigma^+$ transition occurs in the 15,000-cm^{-1} region. Since the dipole moments of charged species are gauge-dependent, all μ_e values listed in Table 4.16 were obtained by placing the dipole operator on the La atom. The positive polarity of the dipole moment of the doublet states indicates that there is electronic charge transfer from La^+ to H. Since the IP of La is much smaller than that of H, most of the ionization takes place on La.

As we stated before, Elkind et al. (118) have measured the $D°(LaH^+)$ as 2.48 eV using guided-ion mass spectrometry. The experimental D_e value including zero-point correction is 2.57 eV. The CASSCF/SOCI value of 2.54 eV computed by Das and the author (30) was in excellent agreement with the experiment.

The ground state of LaH is a $^1\Sigma^+$ state with $r_e = 2.08$ Å and $\omega_e = 1433\,\text{cm}^{-1}$ (see Section 4.4.3). The adiabatic ionization energy for the process

$$\text{LaH}(^1\Sigma^+) \longrightarrow \text{LaH}^+(^2\Delta) + e^-$$

was computed as 9.5.33 eV by Das and the author. They also computed the SOCI IP of the La atom as 9.5.3 eV in reasonable agreement with the experimental value of 9.5.61 eV (123). Therefore, Das and the author predicted theoretical adiabatic IP of LaH to be roughly 0.3 eV below its true value and thus 9.5.65 ± 0.1 eV.

Das and the author have computed the spin–orbit effects for low-lying states of LaH^+. They found that the spin–orbit splittings of the low-lying doublet electronic states of LaH^+ to be between 590 and 694 cm^{-1}. The $^2\Pi_{1/2}$–$^2\Pi_{3/2}$ spin–orbit splitting was found to be 694 cm^{-1}, while the ground-state splitting was calculated as 590 cm^{-1}. The spin–orbit mixing of different states was found to be rather small for all states except the $\frac{3}{2}$ states, for which $^2\Delta_{3/2}$ and $^2\Pi_{3/2}$ mixing was roughly 2%. Since the ground state of LaH^+ is formed through promotion of some of the 5*d* electronic density to the 6*s* orbital, the spin–orbit separation reduces near the potential well of LaH^+.

Table 4.17 shows the leading configuration in the SOCI wavefunctions of LaH^+ computed by Das and the author (30). The $^2\Delta$ ground state of LaH^+ is predominantly $1\sigma^2 1\delta$. The 1σ orbital was found to be a mixture of $La^+(d\sigma)$ and $La^+(6s) + H(1s)$ orbital, while the 1δ is a nonbonding $La^+(d\delta)$ orbital. The $^2\Sigma^+$ state is somewhat more complex in that it is a mixture of $1\sigma^2 2\sigma$, $1\sigma^2 3\sigma$, and $1\sigma^2 4\sigma$ configurations. The La(6*s*) orbital makes significant contribution to the 2σ orbital. The $^2\Pi$ state is analogous to the $^2\Delta$ state. Two quartet states

Table 4.17 Contributions of Important Configurations in SOCI Wavefunctions of LaH^+

State	Configuration (Percentage)
$^2\Delta$	$1\sigma^2 1\delta$ (97)
$^2\Sigma^+$	$1\sigma^2 2\sigma$ (73), $1\sigma^2 3\sigma$ (10), $1\sigma^2 4\sigma$ (5)
$^2\Pi$	$1\sigma^2 1\pi$ (97)
$^2\Sigma^+$(II)	$1\sigma^2 2\sigma$ (34), $1\sigma^2 3\sigma$ (27), $1\sigma^2 4\sigma$ (10), $1\sigma^2 5\sigma$ (9), $1\sigma^2 6\sigma$ (6)
$^2\Sigma^-$	$1\sigma 1\pi^2$ (85), $1\sigma 1\delta^2$ (15)
$^4\Phi$	$1\sigma 1\pi 1\delta$ (99)
$^4\Pi$	$1\sigma 1\pi 1\delta$ (58), $1\sigma 2\sigma 1\pi$ (36)

Source: Reproduced from Reference 30.

Table 4.18 Mulliken Population Analyses of Low-Lying States of LaH^{+a}

	Gross Population						Overlap
State	La	H	La(s)	La(p)	La(d)	H(s)	Population
$^2\Delta$	1.76	1.24	0.15	0.04	1.56	1.24	0.53
$^2\Sigma^+$	1.79	1.21	0.87	0.04	0.88	1.20	0.53
$^2\Pi$	1.79	1.21	0.18	0.06	1.55	1.20	0.53
$^2\Sigma^+$(II)	1.70	1.30	0.32	0.57	0.82	1.29	0.47
$^4\Sigma^-$	2.07	0.93	0.03	0.03	2.01	0.93	0.09
$^4\Phi$	2.06	0.94	0.02	0.02	2.02	0.94	0.09
$^4\Pi$	2.06	0.94	0.12	0.02	1.92	0.94	0.08

[a]The $5s^2 5p^6$ core populations are subtracted from the Mulliken populations in this table.
Source: Reproduced from Reference 30.

of LaH^+, the $^4\Sigma^-$ and $^4\Pi$ states, are especially complex, while the $^4\Phi$ state is relatively simple. Since 2σ participates in the second leading configuration of $^4\Pi$ (36%), the La(6s) orbital makes nonnegligible contribution to the $^4\Pi$ state. The $^4\Sigma^-$ state is a mixture of $1\sigma 1\pi^2$ and $1\sigma 1\delta^2$.

Table 4.18 shows Mulliken population analyses of electronic states of LaH^+ obtained by Das and the author (30). There is a transfer of 0.24e from La^+ to H in the $^2\Delta$ ground state. Note that the $^2\Sigma^+$ state has a $5d^{0.88}6s^{0.87}$ population, since 2σ orbital is singly occupied in the leading configuration of the $^2\Sigma^+$ state. The $^2\Sigma^+$(II) state has larger $La^+(6s)$ population compared to other states. The La(6p) population is relatively small in all, but the $^2\Sigma^+$(II) state. In the ground state, the La population is $5d^{1.56}6s^{0.15}$, and hence 6s population is rather small. Note that La experiences the effects of neither lanthanide contraction nor relativity as much as Hf and ensuing elements.

4.4.5 LaH_2^+

Das and the author (30) have computed the bending potential-energy surfaces of eight electronic states of LaH_2^+ arising from the insertion of $La^+(^3F; 5d^2)$ and $La^+(^1D; 5d^1 6s^1)$ into H_2. They used the CASSCF/SOCI methods together with the same RECPs and basis sets described in Section 4.4.4.

Figure 4.14 shows the computed potential-energy surfaces of eight electronic states of LaH_2^+. In Figure 4.14, θ is the H–La–H bond angle. For each θ, the La–H bond lengths were optimized and the optimized energies are plotted in Figure 4.14. Separate CASSCF calculations were made for each state. The CASSCF active space included four a_1, two b_2, one b_1, and one a_2 orbitals in the C_{2v} symmetry. For very small bond angles ($\theta \sim 5^\circ$) the optimized La–H bond lengths become very long, hence leading to $La^+ + H_2$ dissociation. Consequently, the bending potential-energy surfaces model insertion of La^+ into H_2.

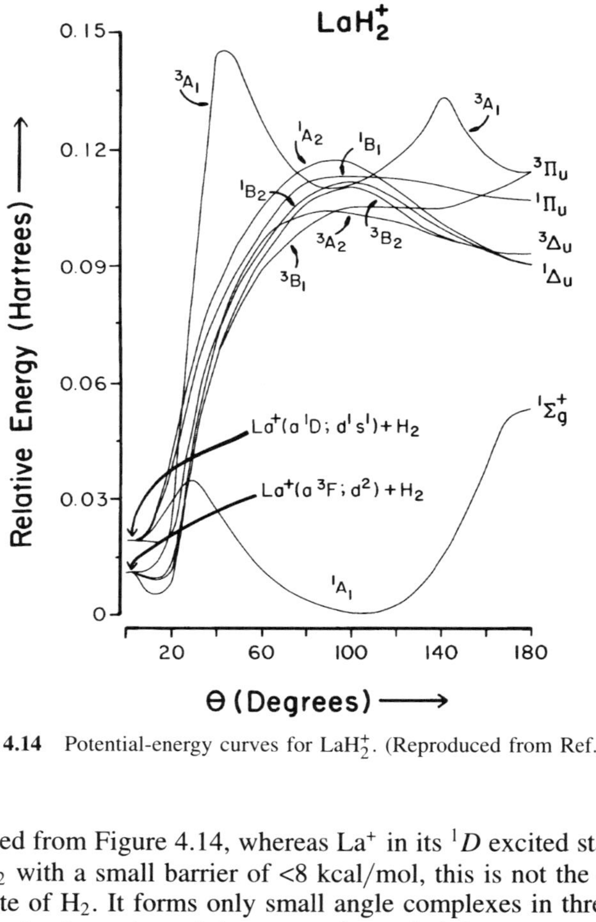

Figure 4.14 Potential-energy curves for LaH_2^+. (Reproduced from Ref. 30.)

As evidenced from Figure 4.14, whereas La^+ in its ^1D excited state $(5d^16s^1)$ inserts into H_2 with a small barrier of <8 kcal/mol, this is not the case for the ^3F ground state of H_2. It forms only small angle complexes in three channels. Subsequent to the formation of these complexes La^+ has to surmount a high barrier to insert into H_2 in all channels for the triplet state. The intersection of the $^1A_1$ surface with the triplet surfaces could facilitate nonadiabatic channels for the reactivity of ^3F state of La^+. However, spin–orbit coupling, which is one possible source of interaction, was found to be rather small. Also, the La–H bond lengths at the θ intersection of the two states are quite different. However, since $La^+(^1D)$ is only a few hundred reciprocal centimeters above $La^+(^3F)$, the initial beam could consist of $La^+(^1D)$ in large percentage and then could undergo insertion into H_2.

The insertion of ^1D state of La^+ into H_2 leads to several states, among which $^1A_1$ is decisively the only favorable state. The ground state of LaH_2^+ is $^1A_1$ and has a bent equilibrium geometry. Table 4.19 shows the equilibrium geometries of the low-lying electronic states of LaH_2^+, from which it can be seen that the ground state of LaH_2^+ is a $^1A_1$ state with $r_e = 2.057$ Å and $\theta_e = 106°$. We note that $^3A_1$ is the only other state exhibiting a bent molecular geometry.

Table 4.19 Geometries and Energies of Low-Lying States of LaH_2^+

	CASSCF			SOCI		
State	R_e, Å	θ_e, deg	E, eV	R_e, Å	θ_e, deg	E, eV
$^1A_1$	2.070	105.0	0.0	2.057	106.0	0.0
$^1\Sigma_g^+$	2.205	180.0	1.44	2.193	180.0	1.42
$^1\Delta_u$	2.277	180.0	2.49	2.263	180.0	1.42
$^3\Delta_u$	2.274	180.0	2.53	2.263	180.0	2.68
$^3\Pi_u$	2.258	180.0	3.10	2.219	180.0	2.87
$^1\Pi_u$	2.289	180.0	2.90	2.265	180.0	3.00
$^3A_1$	2.280	92.3	3.00	2.260	92.6	3.11

Source: Reproduced from Reference 30.

The $^3B_1$, $^3B_2$, and $^3A_2$ states contain small-angle minima corresponding to loose complexes of La^+ with H_2. These weak complexes are 2–3 kcal/mol more stable than $La^+(^3F) + H_2$. Hence, they are not likely to have sufficiently long lifetimes to be observed in hot molecular beams.

Alvarado-Swaisgood and Harrison (124) found that ScH_2^+ has a $^1A_1$ ground state with $r_e = 1.745$ Å and $\theta_e = 109.6.7°$ compared to LaH_2^+, which also has a θ_e of 106°. The longer La–H bond lengths are expected, since La is subjected to neither lanthanide contraction nor relativistic effects as much as Hf–Au.

Table 4.20 shows the geometries of the saddle points in the LaH_2^+ bending potential-energy surfaces. As seen from this table, the $^1A_1$ state exhibits a relatively weak saddle point at $\theta = 29°$ (a barrier height of 8 kcal/mol). The $^3A_1$

Table 4.20 Saddle Points in Potential-Energy Surfaces of LaH_2^{+a}

State	R, Å	θ, deg	Barrier Height, kcal/mol[b]
$^1A_1$	2.10	29.4	8
$^1B_1$	2.32	118.5	61
$^3A_1$	2.33	140.3	67
$^3A_1$	2.43	45.5	77
$^3B_1$	2.27	122.0	57
$^1B_2$	2.35	99.0	60
$^1A_2$	2.34	92.4	61
$^3B_2$	2.33	98.6	61
$^3A_2$	2.28	90.0	56

[a]Bond lengths were optimized at the SOCI level, while bond angles were obtained from CASSCF results.

[a]Barrier heights were calculated with respect to the corresponding dissociated limits at the SOCI level.

Source: Reproduced from Reference 30.

surface contains a very acute θ saddle point (barrier height = 77 kcal/mol). The barrier in the $^3A_1$ surface arises from an avoided crossing. Three other triplet states contain saddle points with obtuse bond angles ($\theta \geq 90°$) with larger barrier heights (>56 kcal/mol). The La–H bond lengths at the saddle point are longer than the bond lengths at minima. Hence, for the complete insertion of La^+ into H_2, to form a LaH_2^+ molecule, the $La^+(^3F)$ ion has to surmount a barrier of at least 56 kcal/mol.

The dissociation energy for the process

$$LaH_2^+(^1A_1) \rightarrow La^+(^3F; d^2) + H_2$$

was computed by Das and the author (30) as 11.3 kcal/mol at the SOCI level of theory. The D_e relative to $La^+(^1D; 5d6s)$ was found to be 12.5 kcal/mol. The D_e ($HLa–H^+$) was computed by these authors as 58.6 kcal/mol at the SOCI level. The D_e ($HLa^+–H$) is almost identical to the D_e ($La^+–H$). Alvarado-Swaisgood and Harrison (124) found the first and second bond energies of ScH_2^+ to be 59.4.7 and 51.7 kcal/mol.

Elkind et al.'s (118) experiment on the reactivities of La^+, Sc^+, Y^+, and Lu^+ with H_2 using guided-ion-beam mass spectrometry revealed that La^+ reacted with H_2 primarily through an insertion mechanism. Their surface ionization (SI) beam of La^+ was found to be 69% $^3F(d^2)$, 12% $^1D(ds)$, and 15% $^3D(ds)$. Since the La^+ (^3F; d^2) forms only loose complexes with H_2 [2–3 kcal/mol more stable than La^+ (^3F)+H_2], at experimental translational energies >4 eV, it is not expected that these loose complexes will be long-lived. Nor do their collisions result in LaH^+ + H unless substantial barriers are surpassed.

As seen from Figure 4.14, the La^+ [$^1D(ds)$] is an attractive candidate for insertion into H_2 as it forms a stable LaH_2^+. Hence the primary mechanism for the reactivity of La^+ with H_2 to produce LaH^+ is the insertion of $La^+[^1D(ds)]$ into H_2 to produce $LaH_2^+(^1A_1)$. The possibility suggested by Elkind et al. (118), specifically, through the insertion of $La^+(^3F; d^2)$ was not supported by the calculations of Das and the author (30), since the triplet surfaces have to surpass a substantial barrier. The experimentally observed statistical behavior of the branching ratio in the La^+ + HD reaction was also explainable from theory.

Table 4.21 shows the important configurations in the SOCI wavefunctions for the minima of LaH_2^+. All states of LaH_2^+ with the exception of $^3A_1$ were found to be dominated by a single configuration. The $1a_1$ orbital of $^1A_1$ near its θ_e was composed of La(5d), smaller amount of La(6s) and $H_1(1s)$ + $H_2(1s)$ orbitals. The $1b_2$ orbital was found to be a mixture of La($d\pi$) and $H_2(\sigma_u)$, $H_1(1s)$–$H_2(1s)$ orbitals. The contribution of $La^+(6p)$ was found to be rather small. However, in the $^1\Sigma_g^+$ state the La(6p) orbital makes an enhnaced contribution to the $1\sigma_u$ orbital.

The $^3A_1$ double hump in the surface arises from two avoided crossings. For $\theta \ll 60°$, this state was found to be predominantly $1a_1^2 2a_1 3a_1$ (90%). For larger θs ($60 < \theta < 140$), this state was found to be dominated by the $1a_1 2a_1 1b_2^2$.

Table 4.21 Leading Configurations of Low-Lying States of LaH_2^+

State	Configuration (Percentage)
$^1A_1$	$1a_1^2 1b_2^2$ (95)
$^1\Sigma_g^+$	$1\sigma_g^2 1\sigma_u^2$ (96)
$^1\Delta_u$	$1\sigma_g^2 1\sigma_u 1\delta_g$ (93) $1\sigma_g 2\sigma_g 1\sigma_u 1\delta_g$ (4)
$^3\Delta_u$	$1\sigma_g^2 1\sigma_u 1\delta_g$ (93) $1\sigma_g 2\sigma_g 1\sigma_u 1\delta_g$ (4)
$^3\Pi_u$	$1\sigma_g^2 1\sigma_u 1\pi_g$ (92) $1\sigma_g 2\sigma_g 1\sigma_u 1\pi_g$ (4)
$^1\Pi_u$	$1\sigma_g^2 1\sigma_u 1\pi_g$ (92) $1\sigma_g 2\sigma_g 1\sigma_u 1\pi_g$ (5)
$^3A_1$	$1a_1 2a_1 1b_2^2$ (88), $1a_1^2 2a_1 3a_1$ (3), $1a_1 2a_1 1b_2 2b_2$ (3)

Source: Reproduced from Reference 30.

The overlap of La($d\pi$) and La(6π) with the $H_2(1\sigma_u^*)$ orbital becomes larger as θ increases, leading to larger occupancy of the $1b_2$ orbital. The minimum at $\theta = 92°$ therefore arises from this avoided crossing. As $\theta > 140°$, the $1a_1^2 1b_2 2b_2$ configuration (90%) dominates. The second barrier near $\theta = 140°$ in the $^3A_1$ surface is due to the avoided crossing of $1a_1 2a_1 1b_2^2$ with $1a_1^2 1b_2 2b_2$. The saddle point in the $^1B_1$ surface ($\theta = 119°$) was found to arise from the avoided crossing of $1a_1^2 1b_2 1a_2$ and $1a_1 1b_2^2 1b_1$. At $\theta = 120°$ the $^1B_1$ state was found to be 56% $1a_1^2 1b_2 1a_2$ and 32% $1a_1 1b_2^2 1b_1$, while at $\theta = 100°$ it was found to be 59% $1a_1^2 1b_2^2 1b_1$ and 29% $1a_1^2 1b_2 1a_2$.

The saddle point in the $^3B_1$ surface ($\theta = 122°$) was found to arise from the avoided crossing of $1a_1^2 1b_2 1a_2$ and $1a_1 1b_2^2 1b_1$. At $\theta = 110°$ the $^3B_1$ state was found to be 56% $1a_1^2 1b_2 1a_2$ and 34% $1a_1 1b_2^2 1b_1$. The $^3A_2$ surface near its saddle point ($\theta = 90°$) was found to be 52% $1a_1^2 1b_2 1b_1$ and 38% $1a_1 1b_2^2 1a_2$. For all larger angles the $1a_1^2 1b_2 1b_1$ configuration was found to dominate.

Table 4.22 contains the Mulliken populations of the electronic states of

Table 4.22 Mulliken Population Analyses for Low-Lying States of LaH_2^+

	Gross Population						
State	La	H	La(s)	La(p)	La(d)	H(s)	Overlap Population
$^1A_1$	1.52	2.48	0.27	0.09	1.16	2.46	1.05
$^1\Sigma_g^+$	1.42	2.58	0.28	0.43	0.71	2.57	1.07
$^1\Delta_u$	1.88	2.12	0.20	0.07	1.61	2.11	0.63
$^3\Delta_u$	1.88	2.12	0.21	0.07	1.60	2.11	0.64
$^3\Pi_u$	1.93	2.07	0.23	0.09	1.61	2.06	0.65
$^1\Pi_u$	1.92	2.08	0.26	0.07	1.59	2.07	0.59
$^3A_1$	1.88	2.12	0.43	0.07	1.38	2.11	0.62

Source: Reproduced from Reference 30.

LaH_2^+. The equilibrium bond angle (106°) for the $^1A_1$ state was found to be determined by the La($5d^{1.16}6s^{0.27}$) hybridization. The $5d$ population is attributed to both ($5d\pi$) orbital mixing in the $1b_2$ orbital and $5d(\sigma/\delta)$ mixing in the $1a_1$ orbital. The $6p$ population of the linear $^1\Sigma_g^+$ state was found to be larger (0.43). All triplet and excited singlet electronic states except $^3A_1$ were found to have similar populations [La($6s$) = 0.20–0.26, La($5d$) = 1.59–1.61]. The $^3A_1$ state differed from this trend (see Table 4.22). The La^+–H overlap population for each bond was found to be 0.525 for the $^1A_1$ ground state. This is identical to the La–H^+ overlap population of 0.53 (see Table 4.18).

4.4.6 LaH_2

Das and the author (28) studied the bending potential-energy surfaces of four doublet and four quartet electronic states of the LaH_2 molecule. They used a CASSCF method followed by full second-order CI calculations. They used ($5s5p3d1f$) as well as ($5s5p3d$) valence Gaussian basis sets in conjugation with an active space comprising four a_1, two b_2, one b_1, and one a_2 orbitals.

Figure 4.15 shows the bending potential-energy surfaces for LaH_2 arising from ^2D and ^4F electronic states of LaH_2. The ground state of La is $^2D(5d^16s^2)$,

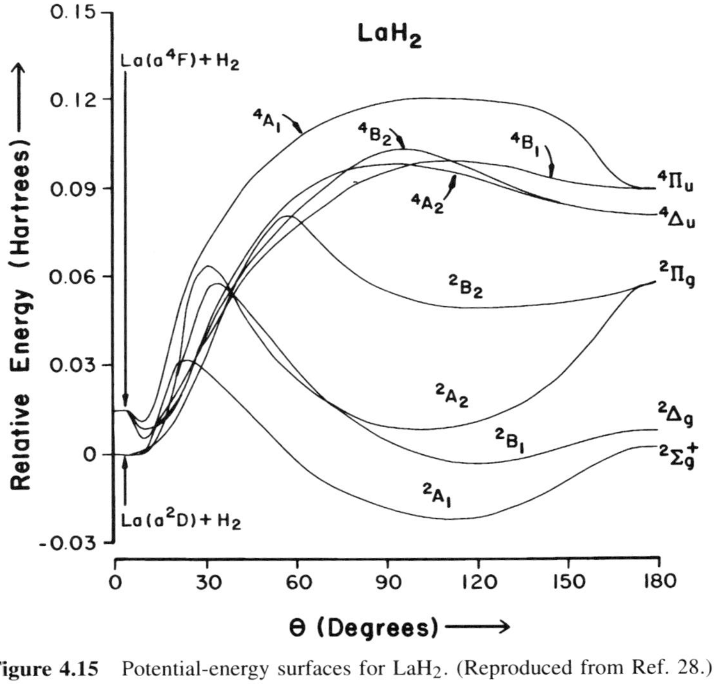

Figure 4.15 Potential-energy surfaces for LaH_2. (Reproduced from Ref. 28.)

while there are several excited states arising from $5d^26s^1$. We considered the ^4F state with the objective of comparing and contrasting high-spin and low-spin reactivities of the La atom.

As evidenced from Figure 4.15, the ^2D ground state of La atom inserts into H_2 relatively easily compared to the excited high-spin ^4F. This is reminiscent of LaH_2^+ discussed in Section 4.4.5. The ^2D state of La also forms other bent doublet electronic states, but these are somewhat higher in energy and also need to surmount larger barriers for insertion into H_2 (see Fig. 4.15). All quartet states have to surpass substantial barriers for insertion into H_2.

Table 4.23 shows the geometries of the bent and linear states of LaH_2 computed by Das and the author (28) at both CASSCF and SOCI level of theories. That table also lists the results with and without $4f$ functions. It is easily seen that $4f$ functions do not play an important role for LaH_2.

The ground state of LaH_2 is of $^2A_1$ symmetry with r_e = 2.14 Å and θ_e = 111.9°. The higher-order correlation effects in the SOCI do not change the geometries much compared to the CASSCF. The ground-state geometry of LaH_2 is similar to YH_2 computed by the author and Ravimohan (20). YH_2 also has a $^2A_1$ ground state (see Section 4.4.2) with r_e = 1.94 Å and θ_e = 123°. The dissociation energy for $LaH_2(^2A_1)$ to yield $La(^2D) + H_2$ was computed as 11 kcal/mol at the SOCI level.

Table 4.24 reports the geometries of the saddle points and barrier heights relative to the dissociation limits. All saddle points have longer La–H bond lengths compared to minima. The saddle-point θs are also similar to YH_2. For YH_2 the $^2A_1$, $^2B_1$, and $^2A_2$ states contain saddle points near θ = 20°, 37°, and 40°, respectively.

Table 4.23 Geometries and Energy of Low-Lying States of LaH_2

	CASSCF			SOCI			
			E, eV			E, eV	
State	r_e, Å	θ_e, deg	($5s5p3d$)	r_e, Å	θ_e, deg	($5s5p3d$)	($5s5p3d1f$)
$^2A_1$	2.158	111.4	0.0	2.140	111.9	0.0	0.0
$^2B_1$	2.225	120.5	0.52	2.190	119.7	0.48	0.48
$^2\Sigma_g^+$	2.277	180.0	0.66	2.260	180.0	0.69	0.68
$^2A_2$	2.200	102.2	0.83	2.174	102.8	0.76	0.75
$^2\Delta_g$	2.339	180.0	0.81	2.312	180.0	0.84	0.88
$^2B_2$	2.207	122.4	1.94	2.195	119.8	1.32	1.30
$^2\Pi_g$	2.322	180.0	2.17	2.310	180.0	1.50	1.49
$^4\Delta_u$	2.314	180.0	2.80	2.298	180.0	2.94	3.01
$^4\Pi_u$	2.279	180.0	3.03	2.259	180.0	3.11	3.12

Source: Reproduced from Reference 28.

Table 4.24 Saddle Points in Various Potential-Energy Curves of LaH_2[a]

State	r, Å	θ, deg	Barrier Height, kcal/mol
$^2A_1$	2.21	24	21
$^2B_1$	2.16	33	36
$^2A_2$	2.17	31	40
$^2B_2$	2.11	57	51
$^4A_2$	2.42	93	52
$^4B_1$	2.40	110	53
$^4B_2$	2.44	97	55
$^4A_1$	2.45	108	66

[a]Barrier heights were calculated with respect to the corresponding dissociation limits.

Source: Reproduced from Reference 28.

The $^2A_1$ surface of LaH_2 intersects the $^4B_1$, $^4B_2$, and $^4A_2$ surfaces at $\theta \approx 25°$, while the $^2A_2$ and $^2B_1$ surfaces cross these quartet surfaces near 40°. These intersections may provide nonadiabatic channels for the quartet-state atom to hop to the doublet-state surfaces. The spin–orbit couplings among electronic states of different spin multiplicities near intersections are negligible. The RCI calculations suggest small SO coupling.

Table 4.25 shows the important configurations for the low-lying states of

Table 4.25 Leading Configurations for Electronic States of LaH_2[a]

State	Configuration
$^2A_1$	$1a_1^2 2a_1 1b_2^2$ (95)
$^2B_1$	$1a_1^2 1b_2^2 1b_1$ (95)
$^2\Sigma_g^+$	$1\sigma_g^2 2\sigma_g 1\sigma_u^2$ (95)
$^2A_2$	$1a_1^2 1b_2^2 1a_2$ (94)
$^2\Delta_g$	$1\sigma_g^2 1\sigma_u^2 1\delta_g$ (96)
$^2B_2$	$1a_1^2 1b_2^2 2b_2$ (94)
$^2\Pi_g$	$1\sigma_g^2 1\sigma_u^2 1\pi_g$ (95)
$^4\Delta_u$	$1\sigma_g^2 2\sigma_g 1\sigma_u 1\delta_g$ (92), $1\sigma_g 2\sigma_g 3\sigma_g 1\sigma_u 1\delta_g$ (3)
$^4\Pi_u$	$1\sigma_g^2 2\sigma_g 1\sigma_u 1\pi_g$ (88), $1\sigma_g 2\sigma_g 3\sigma_g 1\sigma_u 1\pi_g$ (3)

[a]Numbers in parentheses are percentages.

Source: Reproduced from Reference 28.

LaH_2. The $^2A_1$ ground state was found to be predominantly $1a_1^2 2a_1 1b_2^2$ (95%). The $^2B_1$ and $^2A_2$ excited states were composed of $1a_1^2 1b_2^2 1b_1$ and $1a_1^2 1b_2^2 1a_2$ configurations, respectively. The low-lying $^2\Sigma_g^+$, is composed of $1\sigma_g^2 2\sigma_g 1\sigma_u^2$ (95%). The highly excited quartet linear states were found to be considerably pure. The $^4\Delta_u$ and $^4\Pi_u$ states could be represented mainly by $1\sigma_g^2 2\sigma_g 1\sigma_u 1\delta_g$ (92%) and $1\sigma_g^2 2\sigma_g 1\sigma_u 1\pi_g$ (88%) configurations, respectively. The saddle points in the doublet potential-energy surfaces arise from avoided curve crossings. The saddle point at $\theta = 24°$ of the $^2A_1$ state arises as a result of the avoided crossing of the $1a_1^2 2a_1 1b_2^2$ configuration with the $1a_1^2 2a_1^2 3a_1$ configuration. For $\theta < 30°$, the $^2A_1$ state is dominated by $1a_1^2 2a_1^2 3a_1$, while for $\theta > 30°$, the state is described predominantly by the $1a_1^2 2a_1 1b_2^2$ configuration. At the saddle point, these two configurations mix strongly. Similar avoided crossings occur for the $^2B_1$, $^2B_2$, and $^2A_2$ states. Near $\theta = 30°$, the potential-energy curve of the $^2B_1$ state arising from the $1a_1^2 1b_2^2 1b_1$ configuration undergoes an avoided crossing with the $^2B_1$ ($1a_1^2 1a_1^2 1b_1$) surface. The saddle point in the $^2A_2$ state surface is due to the avoided crossing of $^2A_2$ ($1a_1^2 1b_2^2 1a^2$) and $^2A_2$ ($1a_1^2 2a_1^2 1a_2$) configurations. For the $^2B_2$ state, the dominant configuration for $\theta > 60°$ is $1a_1^2 1b_2^2 2b_2$ (90%), while for $\theta < 60°$, the state is predominantly $1a_1^2 2a_1^2 1b_2$ (90%).

Table 4.26 shows the Mulliken population analyses of LaH_2. The total gross population of the lanthanum atom is less than 3.0 for all electronic states of LaH_2, indicating La^+H^- polarity of La–H bonds. Except for the excited $^4\Delta_u$ and $^4\Pi_u$ states, 0.6 electronic population is transferred from the lanthanum atom to the two hydrogen atoms of LaH_2. There is a significant loss of the electronic density from the 6*s* orbital of the lanthanum atom. The La(6*p*) populations are large for most of the low-lying states. The 5*d* orbitals of the lanthanum atom also gain significant amounts of electron density. In the $^2A_1$ ground state, there is significant electronic charge transfer from the 6*s* orbital to the 6*p* and 5*d*

Table 4.26 Mulliken Population Analyses for Electronic States of LaH_2

State	Gross Populations						Overlap Population
	La(total)	H(total)	La(*s*)	La(*p*)	La(*d*)	H(*s*)	
$^2A_1$	2.38	2.62	0.74	0.27	1.37	2.16	1.09
$^2B_1$	2.37	2.63	0.14	0.28	1.95	2.62	1.13
$^2\Sigma_g^+$	2.39	2.61	0.89	0.49	1.01	2.60	1.14
$^2A_2$	2.46	2.54	0.21	0.20	2.05	2.53	1.13
$^2\Delta_g$	2.36	2.64	0.22	0.48	1.66	2.62	1.15
$^2B_2$	2.45	2.55	0.18	0.70	1.58	2.53	1.15
$^2\Pi_g$	2.49	2.51	0.27	0.56	1.66	2.50	1.17
$^4\Delta_u$	2.82	2.18	0.87	0.10	1.85	2.17	0.62
$^4\Pi_u$	2.88	2.13	0.85	0.16	1.86	2.11	0.64

Source: Reproduced from Reference 28.

orbitals. The La–H bond is composed of 31% La(6s), 11% La(6p), and 58% La(5d) at the equilibrium geometry.

The first excited $^2B_1$ state has a rather low La(6s) population. The La(5d) population of this state is nearly 2.0, while the La(6p) population is comparable to that of the ground state. Thus the mixing of different configurations of La is very large. For the low-lying linear state ($^2\Sigma_g^+$), the La(6p) population is 1.01. Hence the participation of the 6p orbital in the La–H bonding is significant. The La–H bonds for this state have 37% La(6s), 21% La(6p), and 42% La(5d) characters. The La(6p) population of the $^2\Pi_g$ state is also large. For these two doublet linear states, the overlap of the La(6p_y) orbital with $H_2(1\sigma_u)$ along the internuclear axis is stronger. The excited $^4\Delta_u$ and $^4\Pi_u$ states have relatively low La(6p) populations. The La–H bonds for these two states have about 30% La(6s), 4% La(6p), and 66% La(5d) characters. Consequently, the La atom almost retains the $5d^2 6s^1$ configuration in these two quartet states. The overlap populations of all the doublet states are above 1.0, while the two quartet linear states have overlaps of 0.63.

The high-spin states of LaH_2 form linear structures, but the low-spin states form bent minima. This is primarily because the low-spin state of the La atom can form bond pairs with the H(1s) electrons, while the high-spin state is primarily nonbonding. Hence the bent minima are formed through the La(d), La($d_{x^2-z^2}$), La($d_{x^2-2y^2+z^2}$), La(6p), and La(6s) hybridization in the low-spin states, while effective orbital overlap in the high-spin states is achieved primarily through the La($d\sigma$), La(6s), and H(1s) orbitals. The most effective overlap of La($d\sigma$), La(6s), and H(1s) is achieved for the linear geometry; consequently, high-spin states are linear, while many low-spin states are bent. The La–H bond lengths of the linear high-spin states are longer than the La–H bond lengths in the bent states. For example, the $^4\Delta_u$ linear state has a $r_e \sim 2.298$ Å, while the $^2A_1$ ground state has an $r_e \sim 2.14$ Å. There are several factors that contrast the bond lengths. The primary factor is that La($d\sigma$) + La(6s) overlap with H(1s) is considerably smaller in the linear geometry compared to the La($d\delta$) + La($d\sigma$) + La(6s) overlap with H(1s) in the bent geometry. This is consistent with the La–H overlap Mulliken populations in Table 4.26. For example, the La–H overlap is 1.09 in the $^2A_1$ state compared to 0.62 for the linear $^4\Delta_u$ state. The La–H overlaps are also much smaller for the $^4\Pi_u$ state.

Table 4.27 shows the computed dipole moments of doublet states of LaH_2 at their equilibrium geometries. All four states exhibit strong La^+H^- polarities. The strong La^+H^- polarities of the La–H bonds are quite consistent with the Mulliken populations shown in Table 4.26.

4.4.7 HfH, ZrH, and TiH

The author and Das (24) compared the potential-energy curves of 14 electronic states of HfH using the CASMCSCF followed by the SOCI method. They also included spin–orbit effects using the RCI method. The author and Wang (22)

Table 4.27 Dipole Moments for Bent Minima of LaH_2

State	μ, D
$^2A_1$	3.79
$^2B_1$	5.42
$^2B_2$	4.72
$^2A_2$	7.42

Source: Reproduced from Reference 28.

computed the potential-energy curves and spectroscopic constants of 30 electronic states of ZrH using comparable CASSCF/SOCI methods. Chong et al. (89) computed the properties of two electronic states of TiH. Chen et al. (126) (1991) considered the gas-phase thermochemistry of TiH. The experimental D_0° (Ti–H) was determined using guided ion-beam mass spectrometry as 2.12 eV in agreement with Chong et al.'s computed CASSCF/CI D_e of 2.12 eV. Of course, zero-point correction to the experimental value yields a D_e of 2.21 ± 0.09 eV. We discuss the results of all these computations and compare them primarily to show the effect of lanthanide contraction on HfH.

4.4.7.1 HfH The potential-energy curves of 14 electronic states of HfH computed by the author and Das (29) are shown in Figure 4.16. Table 4.28 shows the computed spectroscopic constants of HfH in the absence of spin–orbit coupling. The author and Das predicted $^2\Pi$(II)–$X^2\Delta$, $^2\Delta$(II)–$X^2\Delta$, and $^2\Pi$(III)–$X^2\Delta$ transitions of HfH to occur in the 15,000-, 17,000-, and 19,000-cm^{-1} regions, respectively, in the absence of spin–orbit effect.

Table 4.29 contains the spectroscopic constants of HfH, including spin–orbit effects. It is evident from Table 4.29 that spin–orbit coupling strongly mixes several electronic states of HfH, leading to an intermediate coupling description of electronic states. The T_e values also change up to 1500 cm^{-1}.

The author and Das (29) computed the D_e of HfH without spin–orbit coupling as 2.87 eV. The correction to D_e due to spin–orbit coupling was, however, found to be only 0.01 eV, which yielded spin–orbit corrected D_e of 2.88 eV.

Tables 4.29 and 4.30 show the composition of electronic states of HfH with and without spin–orbit coupling, respectively. Although $^2\Delta$, $^2\Pi$, $^4\Pi$, $^4\Delta$, and $^2\Phi$ states are relatively simple in that weights of the leading configurations are larger than 71%, this is not the case with other states of HfH such as $^4\Sigma^-$, $^2\Sigma^+$, and $^2\Pi$(II) (see Table 4.29).

The spin–orbit coupling has noticeable effects on some of the electronic states of HfH (see Table 4.30). For example, the $\frac{1}{2}$ state is a strong mixture of $^2\Pi$, $^2\Sigma^+$, and $^4\Sigma^-$. Likewise, the $\frac{3}{2}$(II) state is a mixture of $^2\Pi$, $^4\Phi$, and $^4\Sigma^-$. Even the ground state of HfH is contaminated by $^2\Pi$ (8%).

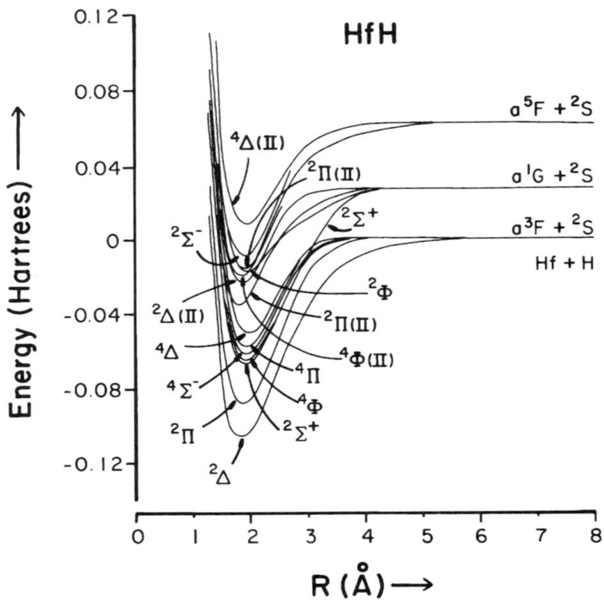

Figure 4.16 Potential-energy surfaces for HfH. (Reproduced from Ref. 29.)

Table 4.28 Spectroscopic Constants of Low-Lying Electronic States of HfH

State	r_e, Å	T_e, cm^{-1}	ω_e, cm^{-1}	μ_e, D	D_e, eV
$^2\Delta$	1.852	0	1,702	0.71	2.87
$^2\Pi$	1.841	1,964	1,682	0.86	2.65
$^4\Phi$	1.882	8,078	1,617	1.41	1.86
$^4\Sigma^-$	1.882	8,807	1,609	1.71	1.76
$^2\Sigma^+$	1.900	9,071	1,540	−0.29	2.47
$^4\Pi$	1.887	9,995	1,596	1.25	1.64
$^2\Pi$(II)	1.838	15,133	1,612	1.85	—
$^4\Delta$	1.942	15,857	1,451	0.06	—
$^2\Phi$	1.907	16,973	1,449	2.41	—
$^2\Delta$(II)	1.875	17,093	1,593	1.80	—
$^2\Sigma^-$	1.870	18,414	1,506	1.70	—
$^2\Pi$(III)	1.837	18,803	1,672	3.14	—
$^4\Phi$(II)	1.832	21,411	1,682	3.32	—
$^4\Pi$(II)	1.924	22,524	1,441	−0.08	—

Source: Reproduced from Reference 29.

Table 4.29 Spectroscopic Constants of Low-Lying ω–ω States of HfH

State	λ–s Description	r_e, Å	ω_e, cm^{-1}	T_e, cm^{-1}	μ_e, D
$\frac{3}{2}$	$^2\Delta + {}^2\Pi$	1.854	1,704	0	0.66
$\frac{1}{2}$	$^2\Pi + {}^2\Sigma^+ + {}^4\Sigma^-$	1.853	1,651	3,807	0.81
$\frac{3}{2}$(II)	$^2\Pi + {}^4\Phi + {}^4\Sigma^-$	1.855	1,654	3,921	0.89
$\frac{5}{2}$	$^2\Delta + {}^4\Phi$	1.853	1,701	4,742	0.67
$\frac{3}{2}$(III)	$^4\Phi + {}^2\Pi$	1.884	1,611	8,989	1.38
$\frac{5}{2}$(II)	$^4\Phi + {}^2\Delta$	1.884	1,616	10,252	1.39
$\frac{1}{2}$(II)	$^4\Sigma^- + {}^2\Pi + {}^2\Sigma^+$	1.881	1,625	11,261	1.59
$\frac{7}{2}$	$^4\Phi$	1.881	1,618	11,788	1.42
$\frac{3}{2}$(IV)	$^4\Sigma^- + {}^2\Pi$	1.882	1,607	12,581	—
$\frac{9}{2}$	$^4\Phi$	1.888	1,608	12,929	1.40

Source: Reproduced from Reference 29.

Table 4.31 shows the Mulliken populations of the electronic states of HfH. The Hf(6*s*) population is larger in $^2\Delta$ and $^2\Pi$ states compared to quartet states. The Hf(6*p*) populations are uniformly larger. The Hf–H overlap populations are larger than 0.5.

Table 4.30 Contributions of Various λ–s States to ω–ω Low-Lying States of HfH

ω–ω State	Contribution
$\frac{3}{2}$	82% $^2\Delta$, 8% $^2\Pi$
$\frac{1}{2}$	72% $^2\Pi$, 9% $^2\Sigma^+$, 4% $^4\Sigma^-$
$\frac{3}{2}$(II)	68% $^2\Pi$, 20% $^4\Phi$, 4% $^4\Sigma^-$
$\frac{5}{2}$	90% $^2\Delta$, 2% $^4\Phi$
$\frac{3}{2}$(III)	95% $^4\Phi$, 1% $^2\Pi$
$\frac{5}{2}$(II)	97% $^4\Phi$, 0.2% $^2\Delta$
$\frac{1}{2}$(II)	85% $^4\Sigma^-$, 6% $^2\Pi$, 5% $^2\Sigma^+$
$\frac{7}{2}$	98% $^4\Phi$
$\frac{3}{2}$(IV)	92% $^4\Sigma^-$, 5% $^2\Pi$
$\frac{9}{2}$	96% $^4\Phi$

Source: Reproduced from Reference 29.

Table 4.31 Mulliken Population of Low-Lying States of HfH[a]

State	Gross Population					Overlap Hf-H
	Hf	H	Hf(*s*)	Hf(*p*)	Hf(*d*)	
$^2\Delta$	3.72	1.28	1.68	0.44	1.59	0.50
$^2\Pi$	3.74	1.26	1.60	0.41	1.73	0.54
$^4\Phi$	3.74	1.26	1.08	0.42	2.24	0.59
$^4\Sigma^-$	3.75	1.25	1.09	0.37	2.29	0.59
$^2\Sigma^+$	3.72	1.28	1.56	0.69	1.47	0.55
$^4\Pi$	3.75	1.25	1.08	0.44	2.23	0.59

[a]The $5s^2 5p^6$ core electrons are omitted for convenience in accounting for Mulliken populations.
Source: Reproduced from Reference 29.

4.4.7.2 ZrH The author and Wang (22) computed the spectroscopic constants of 36 electronic states of ZrH using the CASSCF/SOCI method together with RECPs, which retained the $4d^2 5s^2$ outer shells in the valence space and $(3s2p4d1f/3s2p3d1f)$ basis sets. A subsequent study by the author (17) also compared the properties obtained using $4e$ and $12e$ RECPs for ZrH. Langhoff et al. (87) studied two electronic states of ZrH using the SCF/SDCI/MCPF methods.

Table 4.32 shows the spectroscopic constants of ZrH obtained by the author and Wang (22). Figure 4.17 shows the actual potential-energy curves of some of these electronic states of ZrH. Note that the potential-energy curves in Figure 4.17 were obtained using the inferior FOCI method, while the results in Table 4.32 were obtained using the superior SOCI method.

The spin–orbit splitting of the ground state of ZrH is 546 cm^{-1}, while the $^4\Phi_{3/2}$–$^4\Phi_{4/2}$ excited-state spin–orbit separation is 870 cm^{-1}. Since the spin–orbit contamination of different electronic states is rather small, the author and Wang did not find significant changes in the r_e and ω_e values due to the spin–orbit coupling effect.

The ground state predicted by the author and Wang agreed with the Langhoff et al. (87) SCF/SDCI/MCPF computation. In a subsequent study, the author (17) computed the properties of eight electronic states of ZrH using a superior basis set and $(4s^2 4p^6 5d^2 5s^2)$ RECPs. The spectroscopic constants obtained using either procedure were comparable ($\Delta r_e = 0.04$ Å, $\Delta\omega_e = 40\,\text{cm}^{-1}$, $\Delta D_e =$ 0.17 eV).

Table 4.33 shows the contribution of the important configurations to the electronic states of ZrH. In Table 4.33 the $^2\Delta$, $^4\Phi$, $^4\Pi$, and $^4\Delta$(II) states have leading configurations with weights >76%. All other states are heavily mixed. Table 4.34 shows the Mulliken population analyses of the electronic states of ZrH. The $5p$ contribution is quite significant for all states of ZrH. The $4d$ population

Table 4.32 Spectroscopic Properties of 36 Electronic States of ZrH

State	r_e, Å	T_e, cm^{-1}	ω_e, cm^{-1}
$^2\Delta(\frac{3}{2})$	1.77	0	1,777
$^2\Delta(\frac{5}{2})$	1.77	546	1,779
$^2\Delta$	1.77	292	1,743
$^4\Phi(\frac{3}{2})$	1.82	1,279	1,604
$^4\Phi(\frac{5}{2})$	1.82	1,571	1,605
$^4\Phi(\frac{7}{2})$	1.82	1,857	1,606
$^4\Phi(\frac{9}{2})$	1.82	2,148	1,606
$^4\Phi$	1.82	1,720	1,605
$^2\Pi(\frac{1}{2})$	1.78	2,337	1,740
$^2\Pi(\frac{3}{2})$	1.78	2,637	1,740
$^2\Pi$	1.78	2,526	1,742
$^4\Sigma^-(\frac{1}{2})$	1.82	2,533	1,613
$^4\Sigma^-(\frac{3}{2})$	1.82	2,533	1,613
$^4\Sigma^-$	1.82	2,533	1,613
$^4\Pi(\frac{1}{2})$	1.82	3,521	1,585
$^4\Pi(\frac{3}{2})$	1.82	3,678	1,585
$^4\Pi(\frac{5}{2})$	1.82	3,841	1,585
$^4\Pi$	1.82	3,683	1,585
$^2\Sigma^+(\frac{1}{2})$	1.85	5,227	1,598
$^2\Sigma^+$	1.85	5,221	1,597
$^4\Delta(\frac{1}{2})$	1.92	9,731	1,413
$^4\Delta(\frac{3}{2})$	1.92	9,928	1,413
$^4\Delta(\frac{5}{2})$	1.92	10,136	1,413
$^4\Delta(\frac{7}{2})$	1.92	10,353	1,414
$^4\Delta$	1.92	10,045	1,413
$^2\Pi(\text{II})(\frac{1}{2})$	1.85	10,615	1,565
$^2\Pi(\text{II})(\frac{3}{2})$	1.85	10,758	1,567
$^2\Pi(\text{II})$	1.85	10,687	1,564
$^2\Delta(\text{II})(\frac{3}{2})$	1.88	10,764	1,534
$^2\Delta(\text{II})(\frac{5}{2})$	1.86	11,149	1,551
$^2\Delta(\text{II})$	1.87	10,960	1,548
$^4\Delta(\text{II})(\frac{1}{2})$	1.88	12,940	1,512
$^4\Delta(\text{II})(\frac{3}{2})$	1.88	13,104	1,511
$^4\Delta(\text{II})(\frac{5}{2})$	1.88	13,278	1,512
$^4\Delta(\text{II})(\frac{7}{2})$	1.88	13,467	1,513
$^4\Delta(\text{II})$	1.88	13,205	1,512

Source: Reproduced from Reference 22.

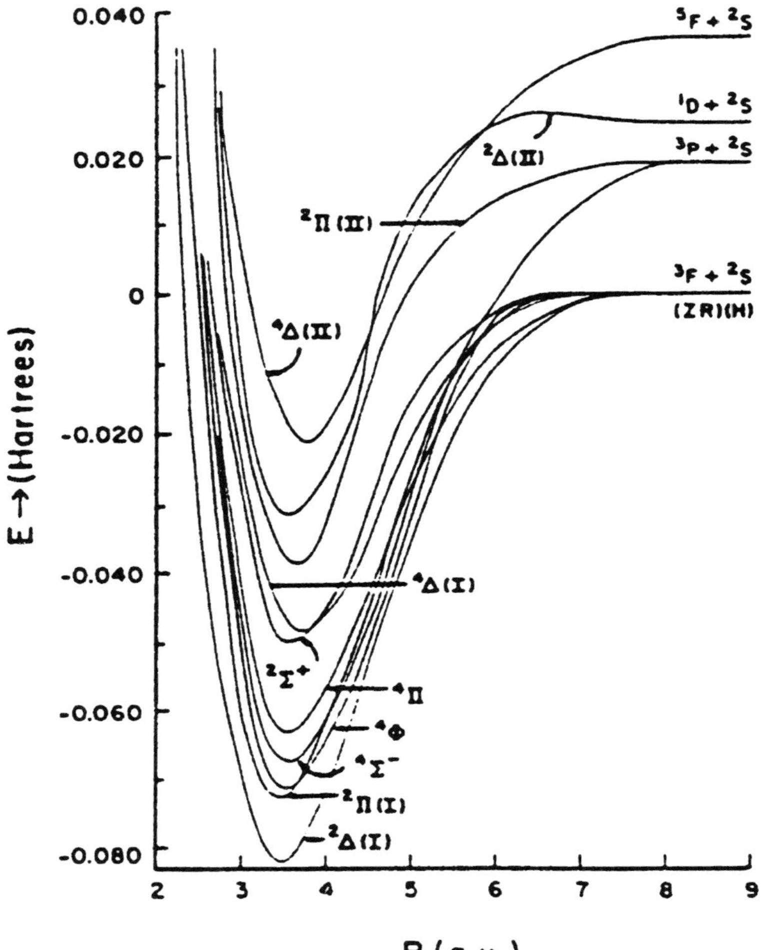

Figure 4.17 Potential-energy surfaces for ZrH. (Reproduced from Ref. 22.)

of the $^2\Delta$ state is smaller in comparison to other states. The total gross populations of the Zr atom are less than 9.4.0 in all the states, indicating that these states are ionic with the polarity Zr^+H^-.

The author (17) computed the dipole moment of ZrH (μ_e) as 1.24 D in the $^2\Delta$ state with Zr^+H^- polarity. The $^4\Phi$ excited state has a significantly larger dipole moment (2.13 D). Langhoff et al. (87) obtain μ_e values of 1.23 and 2.42 D, respectively, using the SCF/MCPF method. Both the calculations suggest that ZrH is ionic with Zr^+H^- polarity.

Table 4.33 Contributions of Important Configurations in SOCI Wavefunctions of ZrH

States	Percentage of Contributions
$^2\Delta$	$1\sigma^2 2\sigma^2 1\delta^1$ (75), $1\sigma^2 2\sigma^1 3\sigma^1 1\delta^1$ (6)
$^4\Phi$	$1\sigma^2 2\sigma^1 1\delta^1 1\pi^1$ (96)
$^2\Pi$	$1\sigma^2 2\sigma^2 1\pi^1$ (33), $1\sigma^2 2\sigma^2 2\pi^1$ (28), $1\sigma^2 2\sigma^1 1\pi^1 1\delta^1$ (7)
$^4\Sigma^-$	$1\sigma^2 2\sigma^1 1\pi^2$ (48), $1\sigma^2 2\sigma^1 1\delta^2$ (46)
$^4\Pi$	$1\sigma^2 2\sigma^1 1\delta^1 1\pi^1$ (87)
$^2\Sigma^+$	$1\sigma^2 2\sigma^2 3\sigma^1$ (59), $1\sigma^2 2\sigma^1 1\pi^2$ (20)
$^4\Delta$	$1\sigma^2 2\sigma^1 3\sigma^1 1\delta^1$ (90)
$^2\Pi$(II)	$1\sigma^2 2\sigma^1 1\pi^1 1\delta^1$ (51), $1\sigma^2 2\sigma^1 2\pi^1 1\delta^1$ (23)
$^2\Delta$(II)	$1\sigma^2 2\sigma^1 3\sigma^1 1\delta^1$ (67), $1\sigma^2 2\sigma^1 1\pi^2$ (10), $1\sigma^2 1\pi^2 1\delta^1$ (9)
$^4\Delta$(II)	$1\sigma^2 1\pi^1 1\delta^2$ (83)

Source: Reproduced from Reference 22.

4.4.7.3 TiH There are several studies on TiH, but the ones by Chong et al. (89) as well as an earlier study of Bauschlicher and Walch (84) provide accurate data for comparison with ZrH and HfH. In the former study, Chong et al. used all electron SCF/SOCI/CPF/MCPF methods in conjugation with [$8s6p4d3f$] Gaussian basis set for Ti and the ($4s3p$) basis set for the hydrogen atom. Their computed spectroscopic constants of the $^4\Phi$ and $^2\Delta$ states of TiH are shown in Table 4.35.

At all levels of theory (SOCI, CPF, MCPF), the high-spin $^4\Phi$ state is the ground state of TiH in contrast with both ZrH and HfH. The $3d$ population is ~0.3 electron larger in the $^4\Phi$ state of TiH compared to $^2\Delta$.

The dipole moment of the $^4\Phi$ state is 2.12 D at the highest level of theory, while the corresponding value is 1.71 D for the $^2\Delta$ state. The r_e values of the $^4\Phi$ and $^2\Delta$ states are 1.781 and 1.73 Å, respectively.

The $3d^2 4s^2$–$3d^3 4s^1$ energy separation of the Ti atom is 0.81 eV, and thus the $^4\Phi$ state is favored via spin exchange stabilization energy. Note that the $3d$ population of the $^4\Phi$ state is between 2.0 and 3.0, indicating considerable mixing of the two configurations. The dipole moment of TiH is increased compared to ScH and VH (89).

4.4.7.4 Effect of Lanthanide Contraction on the TiH–ZrH–HfH Triad The $^2\Delta$ electronic states of the three species have r_e values of 1.78, 1.811, and 1.85 Å, respectively, while the D_e values of the three molecules in the $^2\Delta$ state are 1.77, 2.47, and 2.87 eV. The effect of lanthanide contraction is not fully manifested in r_e values of the three species compared to the D_e values. The stronger Hf–H bond compared to the Zr–H bond is attributed partly to the lanthanide contraction. The μ_e value of TiH, ZrH, and HfH in the $^2\Delta$ state are 1.92, 1.24, and 0.71 D, respectively. We note a steep decrease in the HfH μ_e values.

Table 4.34 Mulliken Population Analyses of ZrH

	Net Population							Gross Population							Overlap
State	Zr	Zr(*s*)	Zr(*p*)	Zr(*d*)	H	H(*s*)	H(*p*)	Zr	Zr(*s*)	Zr(*s*)	Zr(*p*)	H	H(*s*)	H(*p*)	Zr–H
$^2\Delta$	3.441	1.323	0.346	1.551	0.972	0.966	0.006	3.734	1.481	0.419	1.834	1.266	1.255	0.011	0.587
$^2\Pi$	3.428	1.121	0.350	1.720	0.939	0.936	0.003	3.745	1.334	0.412	1.999	1.255	1.247	0.008	0.632
$^4\Phi$	3.399	0.849	0.285	2.177	0.955	0.950	0.005	3.722	1.001	0.363	2.358	1.278	1.268	0.010	0.646
$^4\Sigma^-$	3.398	0.831	0.249	2.219	0.957	0.952	0.005	3.721	0.989	0.321	2.411	1.279	1.270	0.010	0.645
$^4\Pi$	3.388	0.833	0.270	2.189	0.965	0.959	0.005	3.712	0.990	0.342	2.379	1.288	1.278	0.010	0.647
$^2\Sigma^+$	3.394	1.048	0.439	1.704	0.983	0.978	0.005	3.705	1.203	0.593	1.910	1.295	1.285	0.009	0.623
$^4\Delta$	3.342	0.838	0.474	1.953	0.993	0.992	0.001	3.674	0.974	0.639	2.061	1.326	1.322	0.003	0.665
$^2\Pi$(II)	3.430	1.081	0.348	1.717	0.947	0.946	0.001	3.741	1.321	0.419	2.001	1.259	1.252	0.006	0.522
$^2\Delta$(II)	3.402	0.710	0.489	2.211	0.983	0.982	0.002	3.709	0.761	0.610	2.339	1.291	1.286	0.005	0.615
$^4\Delta$(II)	3.435	0.282	0.282	2.862	0.935	0.930	0.004	3.750	0.447	0.346	2.957	1.250	1.242	0.008	0.631

Source: Reproduced from Reference 22.

Table 4.35 Spectroscopic Constants and Other Properties for $^4\Phi$ and $^2\Delta$ States of TiH

	SDCI	CPF	MCPF	CASSCF/CI
	$^4\Phi$ Ground State—5 Electrons Correlated			
E (hartree) + 840	−8.99683	−9.00155	−9.00126	−9.01689
r_e (bohr)	3.439	3.440	3.440	3.443
ω_e (cm^{-1})	1506	1592	1548	1444
D_e (eV)	1.99	2.07	2.06	2.12
μ(D)	1.899	2.308	2.185	—
$d\mu/dr$ (D/bohr)	1.373	1.246	1.283	—
$3d$ population	2.27	2.33	2.30	2.26
	$^4\Phi$ Ground State—13 Electrons Correlated			
E (hartree) + 840	−9.27128	−9.29987	−9.29979	—
r_e (bohr)	3.397	3.347	3.365	—
ω_e (cm^{-1})	1515	1327	1407	—
D_e (eV)	2.06	2.07	2.05	—
μ (D)	1.777	2.307	2.122	—
$d\mu/dr$ (D/bohr)	1.523	0.589	1.064	—
$3d$ population	2.29	2.45	2.41	—
	$^2\Delta$ Excited State—5 Electrons Correlated			
E (hartree) + 840	−8.98095	−8.99095	−8.99102	—
T_e (eV)	0.432	0.289	0.279	—
r_e (bohr)	3.264	3.369	3.352	—
ω_e (cm^{-1})	1323	1025	1705	—
D_e (eV)	1.55	1.78	1.78	—
μ (D)	1.916	1.585	1.557	—
$d\mu/dr$ (D/bohr)	1.288	0.821	0.928	—
$3d$ population	2.04	2.04	2.05	—
	$^2\Delta$ Excited State—13 Electrons Correlated			
E (hartree) + 840	−9.24561	−9.28862	−9.28957	—
T_e (eV)	0.698	0.306	0.278	—
r_e (bohr)	3.198	3.270	3.271	—
ω_e (cm^{-1})	1694	1561	1594	—
D_e (eV)	1.36	1.76	1.77	—
μ (D)	2.057	1.828	1.709	—
$d\mu/dr$ (D/bohr)	1.687	0.706	0.752	—
$3d$ population	2.09	2.17	2.16	—

Source: Reproduced from Reference 89.

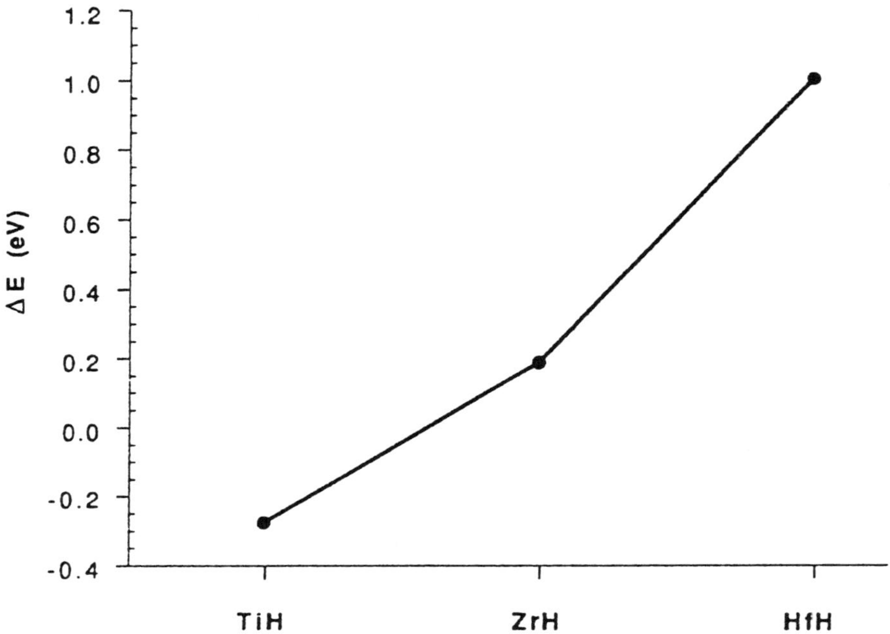

Figure 4.18 Relative trend in the $^2\Delta$–$^4\Phi$ energy separations of TiH, ZrH, and HfH. (Reproduced from Ref. 2.)

The most dramatic manifestation of the lanthanide contraction is in the $^2\Delta$–$^4\Phi$ energy separation, which is plotted in Figure 4.18 for the three species. For Ti, we noted before the $3d^24s^2$–$3d^34s^1$ energy separation is small, and hence the $^4\Phi$ state exhibits a mixed character. The spin exchange energy overcomes the electron correlation energy in the $^4\Phi$ state, resulting in a $^4\Phi$ ground state for TiH. The $^2\Delta$–$^4\Phi$ energy separation therefore increases from TiH to ZrH. But the $^2\Delta$–$^4\Phi$ energy separation is substantially larger for HfH as seen from Figure 4.18, that is, the ΔE values of the $^4\Phi$ state, with $^2\Delta$ taken as 0.0 are −0.28, 0.19, and 1.0 eV for TiH, ZrH and HfH, respectively. The steep rise in the energy of the $^4\Phi$ state for HfH is due to the lanthanide contraction. The $^4\Phi$ state involves enhanced *ns* character. The 6*s*–5*d* promotion energy is considerably larger for Hf, since the 6*s* orbital is stabilized and contracted, due primarily to lanthanide contraction and secondarily to relativity.

This trend can be explained by comparing the $(nd^2ns^2)^3F$–$(nd^3ns^1)^5F$ energy separation of Ti, Zr, and Hf (Fig. 4.19). It is clear from Figure 4.19 that this separation is dramatically larger for Hf. As noted before, lanthanide contraction is somewhat more important for Hf compared to relativity, while the opposite trend is true for the latter third row transition methods such as Pt and Au.

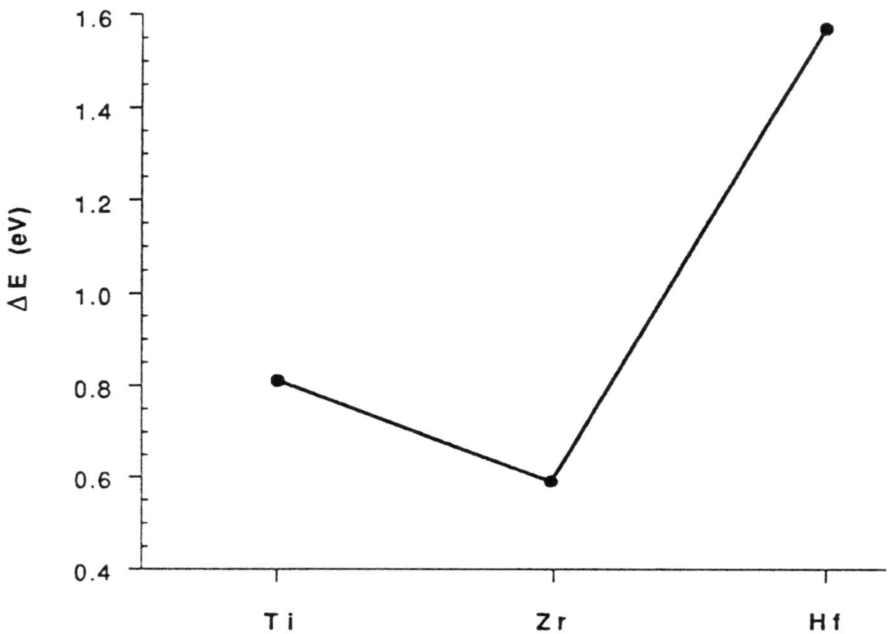

Figure 4.19 $(nd^2ns^2)(^3F)$–$(nd^3ns^1)^5F$ energy separations of Ti, Zr, and Hf. (Reproduced from Ref. 2.)

4.4.8 HfH_2 and ZrH_2

4.4.8.1 HfH_2 Das and the author (31) obtained the potential-energy surfaces of 12 electronic states of HfH_2. They employed CASSCF followed by full second-order CI (SOCI) using RECPs, which retained the outer $(5s^25p^65d^26s^2)$ shells of Hf in the valence space. The computed potential-energy surfaces are shown in Figure 4.20. As seen from Figure 4.20, the excited $^1\Delta$ state of Hf atom $(5d^26s^2)$ inserts into H_2 spontaneously to yield the $^1A_1$ ground state of the HfH_2 molecule. The ^3F ground state of Hf has to surmount barriers for insertion into H_2. The smallest barrier is for the $^3A_2$ state, and it was computed as 17 kcal/mol.

The crossing of $^1A_1$ with triplet surfaces could result in surface hopping from triplet to singlet surfaces. The primary mechanism for this would be spin–orbit coupling. Das and the author (31) used the RCI method to study spin–orbit coupling at θ intersections and found this effect to be small, but nonnegligible. The Hf atom in the $^5F(5d^36s^1)$ state exhibits a different behavior. It forms small-angle complexes with H_2 and subsequently forms linear minima.

Tables 4.36 and 4.37 show the geometries and energy separations of bent and linear electronic states of HfH_2, respectively. The ground state of HfH_2

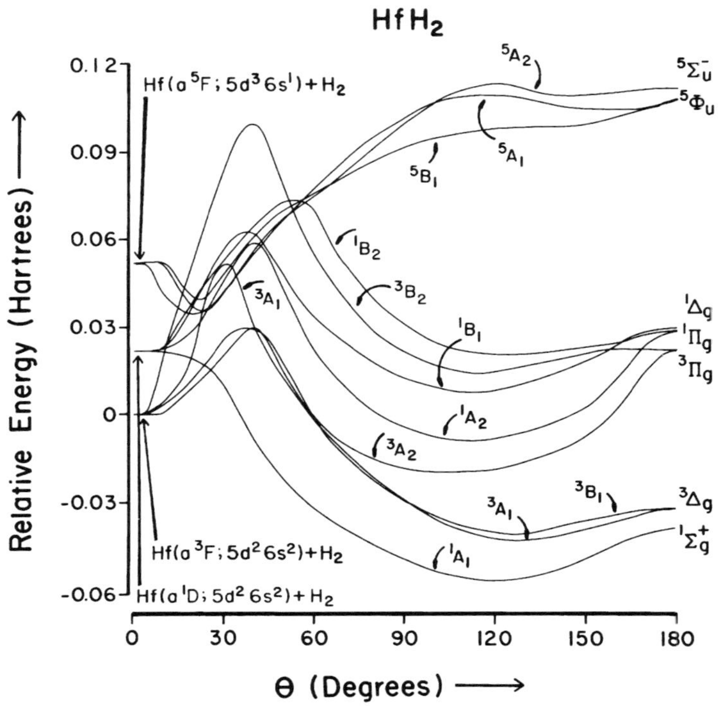

Figure 4.20 Potential-energy surfaces of HfH_2. (Reproduced from Ref. 31.)

Table 4.36 Geometries and Energy Separations of Bent Low-Lying States of HfH_2

	CASSCF			SOCI			
						E, eV	
State	r_e, Å	θ_e, deg	E, eV	r_e, Å	θ_e, deg	Without f	With f
$^1A_1$	1.862	120.0	0.0	1.842	126.7	0.0	0.0
$^3B_1$	1.900	125.5	0.44	1.875	125.7	0.52	0.51
$^3A_1$	1.896	127.1	0.38	1.877	127.8	0.54	0.53
$^3A_2$	1.878	105.6	1.01	1.855	109.3	1.05	1.03
$^1A_1$	1.891	111.6	1.31	1.867	112.6	1.19	1.18
$^1B_2$	1.888	123.6	2.11	1.880	121.5	1.56	1.59
$^1B_1$	1.885	113.1	1.76	1.859	113.2	1.66	1.60
$^3B_2$	1.924	114.5	1.94	1.894	116.8	1.94	1.91
$^5A_2$	2.073	23.3	2.48	2.020	24.4	2.77	2.73
$^5A_1$	2.088	22.7	2.45	2.022	24.3	2.79	2.75
$^5B_1$	2.153	21.7	2.61	2.074	23.4	2.06	2.96

Source: Reproduced from Reference 31.

Table 4.37 Geometries and Energies of Linear States of HfH_2[a]

State	CASSCF		SOCI		
				E, eV	
	r_e, Å	E, eV	r_e, Å	Without f	With f
$^1\Sigma_g^+$	1.934	0.49	1.902	0.34	0.41
$^3\Delta_g$	1.958	0.62	1.937	0.81	0.84
$^3\Pi_g$	1.947	2.14	1.936	1.64	1.67
$^1\Pi_g$	1.953	2.28	1.942	1.76	1.81
$^1\Delta_g$	1.957	2.35	1.934	2.19	2.10
$^5\Sigma_u^-$	2.024	4.56	1.956	4.31	4.22
$^5\Phi_u$	2.011	4.48	1.981	4.27	4.25

[a]Energies were calculated from the $^1A_1$ ground-state energy.
Source: Reproduced from Reference 31.

is a $^1A_1$ state with $r_e = 1.842$ Å and $\theta_e = 129.6.7°$ (Table 4.36). The linear electronic states in Table 4.37 have longer bond lengths compared to the bent states primarily because of enhanced $6p$ participation in the $1\sigma_u$ orbital of the linear state.

Table 4.38 shows the saddle points in the potential-energy surfaces of HfH_2. Only the $^5A_2$ state exhibits an obtuse-angle saddle point. All other very acute-angle saddle points of HfH_2 are due to avoided crossings.

Das and the author (31) have computed the D_e for $HfH_2(^1A_1) \rightarrow Hf(^1\Delta) + H_2$ as 46 kcal/mol and 30 kcal/mol relative to $Hf(^3F)$ ground state. Hence, HfH_2 is considerably more stable than Hf + H_2.

Table 4.39 shows the effect of spin–orbit coupling on the electronic states of HfH_2, while Table 4.40 shows the composition of the RCI wavefunctions obtained including spin–orbit coupling. It is evident from Tables 4.39 and 4.40

Table 4.38 Saddle Points in Potential-Energy Surfaces of HfH_2[a]

State	r, Å	θ, deg	Barrier Height[b] kcal/mol	State	r, Å	θ, deg	kcal/mol
$^1B_1$	1.85	38	27	$^3B_1$	1.86	37	25
$^1A_2$	1.88	40	23	$^3A_2$	1.86	40	17
$^1B_2$	1.84	55	34	$^3B_2$	1.89	40	62
$^3A_1$	1.91	31	32	$^5A_2$	2.06	120	38

[a]Geometries of the saddle points were estimated at the CASSCF level of calculation.
[b]Barrier heights were calculated with respect to the corresponding dissociation limits at the CASSCF level.
Source: Reproduced from Reference 31.

Table 4.39 Effect of Spin–Orbit Contamination on Geometries and Energies of Low-Lying States of HfH_2

State	r_e, Å	θ, deg	E, eV	State	r_e, Å	θ, deg	E, eV
$^1A_1(A_1)$	1.842	126.3	0	$^3A_1$	1.877	127.8	0.55
$^1A_1$	1.842	126.7	0.01	$^3B_1(A_1)$	1.875	125.8	0.51
$^3A_1(A_2)$	1.875	127.3	0.40	$^3B_1(B_2)$	1.878	126.6	0.59
$^3A_1(B_2)$	1.875	127.3	0.44	$^3B_1(A_2)$	1.877	126.5	0.64
$^3A_1(B_1)$	1.878	127.9	0.55	$^3B_1$	1.875	125.7	0.52

Source: Reproduced from Reference 31.

that, while the spin–orbit effect is rather small for the $^1A_1$ state of HfH_2, this is not the case for the triplet states. The $^3A_1$ and $^3B_2$ states are heavily mixed because of the near-degeneracy of the two states, leading to substantial spin–orbit mixing of the two states. Consequently, the spin–orbit effects are important for nearly degenerate states.

Table 4.41 shows the computed dipole moments of the electronic states of HfH_2, which means that although the $^1A_1$ ground state of HfH_2 has only a small dipole moment (0.31 D), this is not the case for the excited states. The negative dipole moment of the $^5B_1$ state is consistent with electronic charge transfer from the hydrogens to Hf in the $^5B_1$ weak complex.

Table 4.42 shows the Mulliken population analyses of the electronic states of HfH_2. The bent $^1A_1$ ground state of HfH_2 has a Hf population of $5d^{1.47}6s^{1.46}6p^{0.52}$. The participation of the $6p$ orbital is thus noticeable in this state. Note that the Hf atom has populations smaller than 4.0 in all low-spin states, suggesting transfer of electronic charge to the hydrogen atom. The substantial difference in the behavior of the $^3B_2$ state compared to other states is explained by the differences in the Mulliken population.

4.4.8.2 ZrH_2 Wang et al. (44) have computed the potential-energy surfaces of ZrH_2. They used the CASSCF/MRSDCI methods together with RECPs,

Table 4.40 Weights of Different Low-Lying States of HfH_2 in RCI Wavefunctions

State	Contribution (%)	State	Contribution (%)
$^1A_1(A_1)$	$^1A_1(97)$, $^3A_2(0.3)$	$^3B_1(A_1)$	$^3B_1(98)$, $^1A_1$ (0.8), $^3A_2$ (0.1)
$^3A_1(A_2)$	$^3A_1(65)$, $^3B_1(32)$	$^3B_1(B_2)$	$^3B_1(64)$, $^3A_1$ (33)
$^3A_1(B_2)$	$^3A_1(64)$, $^3B_1(33)$	$^3B_1(A_2)$	$^3B_1(66)$, $^3A_1$ (33)
$^3A_1(B_1)$	$^3A_1(98)$, $^3A_2(0.1)$		

Source: Reproduced from Reference 31.

Table 4.41 Dipole Moments of Low-Lying States of HfH_2[a]

State	μ, D	State	μ, D
$^1A_1$	0.31	$^1B_1$	1.60
$^3B_1$	1.28	$^3B_2$	3.67
$^3A_1$	1.02	$^5A_2$	0.26
$^3A_2$	1.91	$^5A_1$	0.48
$^1A_2$	0.90	$^5B_1$	−1.48
$^1B_2$	1.96		

[a]Positive values of μ indicate Hf^+H^- polarity of the bonds.

Source: Reproduced from Reference 31.

which retained the outer $4d^2 5s^2$ shells of Zr. They used an extended $(4s4p4d1f)$ valence Gaussian basis set for the Zr atom. It should be noted that the state-averaging procedure was used by Wang et al. (44) to ensure proper dissociation. The MRSDCI included all configurations in the CASSCF with coefficients ≥0.07.

Figures 4.21 and 4.22 show the computed potential-energy surfaces of low-spin and high-spin states of ZrH_2, respectively. The ^1D excited state of the Zr atom inserts spontaneously into H_2 to form the bent $^1A_1$ state of the ZrH_2 molecule. The high-spin ^3F state, on the contrary, needs to surpass barriers for insertion into H_2.

Table 4.42 Mulliken Population Analyses of Low-Lying States of HfH_2

	Gross Populations					Total
State	Hf	H	Hf(s)	Hf(p)	Hf(d)	Overlap Population
$^1A_1$	3.46	2.54	1.46	0.52	1.47	1.11
$^3B_1$	3.44	2.56	1.03	0.49	1.93	1.19
$^3A_1$	3.44	2.56	1.05	0.51	1.88	1.20
$^3A_2$	3.51	2.49	1.04	0.54	1.93	1.17
$^1A_2$	3.50	2.50	0.97	0.63	1.90	1.17
$^1B_2$	3.53	2.47	1.00	0.67	1.86	1.23
$^1B_1$	3.51	2.49	1.06	0.62	1.84	1.12
$^3B_2$	3.53	2.47	0.69	0.45	2.39	1.23
$^5A_2$	3.94	2.06	0.97	0.37	2.59	0.34
$^5A_1$	3.94	2.06	0.97	0.32	2.65	0.33
$^5B_1$	3.94	2.06	1.00	0.59	2.34	0.25
$^1\Sigma_g^+$	3.49	2.51	1.48	0.56	1.45	1.25
$^3\Delta_g$	3.48	2.52	1.06	0.54	1.89	1.26
$^3\Pi_g$	3.56	2.44	1.08	0.58	1.90	1.28
$^1\Pi_g$	3.56	2.44	1.06	0.58	1.92	1.27
$^1\Delta_g$	3.50	2.50	1.08	0.58	1.84	1.22

Source: Reproduced from Reference 31.

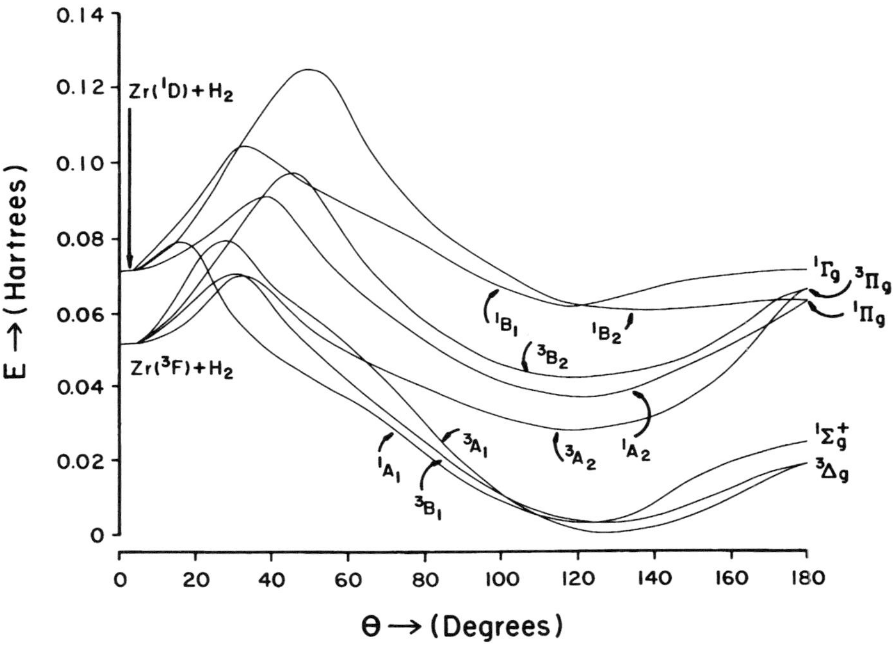

Figure 4.21 Potential-energy surfaces of triplet and singlet states of ZrH_2. (Reproduced from Ref. 44.)

Figure 4.22 shows small-angle complexes of $Zr(^5F)$ with H_2. The formation of these complexes is facilitated by exchange of electronic density from hydrogens to the half-filled $5s$ orbital of Zr in the $^5F(4d^35s^1)$ state.

Table 4.43 shows the computed geometries and energy separations of electronic states of ZrH_2 (bent). It is noted that several electronic states of ZrH_2 are nearly degenerate ($^1A_1$, $^3B_1$, $^3A_1$). In fact, the CASSCF method predicts $^3A_1$ ground state, while MRSDCI predicts $^1A_1$ ground state. Wang et al. computed the D_e of ZrH_2 ($^1A_1$) to yield $Zr(^3F) + H_2$ as 1.11 eV or 26 kcal/mol.

Table 4.44 shows the dipole moments of electronic states of ZrH_2, while Table 4.45 shows the Mulliken populations. The ground state and $^1B_1$ are the least ionic, while all other states of ZrH_2 are very ionic with Zr^+H^- polarities of bond.

The $^1A_1$ ground state of ZrH_2 was found to be 69% $1a_1^22a_1^21b_2^2$, 13% $1a_1^21b_2^21b_1^2$, and 12% $1a_1^23a_1^21b_2^2$ near its r_e and θ_e. As seen from Table 4.45, the $^1A_1$ state has $4d^{1.907}5s^{1.108}5p^{0.422}$ Mulliken population. The $4d$ populations of the excited states are larger. The Zr–H overlaps are between 1.29 and 1.35.

4.4.8.3 Effect of Lanthanide Contraction on the Difference in Properties of HfH_2 and ZrH_2

The most noticeable difference between HfH_2 and ZrH_2 attributed mostly to lanthanide contraction and less to relativity is the

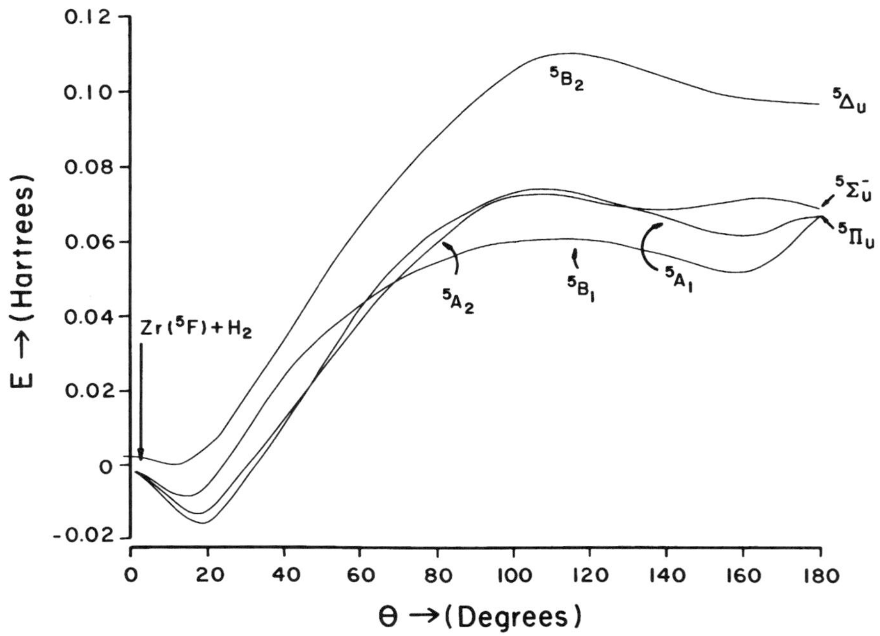

Figure 4.22 Potential-energy surfaces of the quintet states of ZrH_2. (Reproduced from Ref. 44.)

enhanced stability of HfH_2. We note that the D_e of HfH_2 is 30 kcal/mol, while the D_e of ZrH_2 is 24 kcal/mol.

The 6*s* population of Hf in the $^1A_1$ state of HfH_2 is 0.35*e* larger than the 5*s* population of the corresponding state of ZrH_2. The 5*d* population of Hf is correspondingly smaller than Zr (1.47 vs. 1.91). The 6*p* population of Hf is also increased (0.1) compared to Zr, which leads to a larger bond angle for HfH_2.

Table 4.43 Geometries and Energy Separations of Bent Electronic States of ZrH_2

	CASSCF				MRSDCI			
State	R_e, Å	θ_e, deg	E, eV	State	R_e, Å	θ_e, deg	E, eV	D_e, eV
$^3A_1$	1.865	127	0	$^1A_1$	1.806	123	0	1.11
$^3B_1$	1.866	125	0.061	$^3B_1$	1.837	125	0.086	1.02
$^1A_1$	1.839	121	0.072	$^3A_1$	1.837	127	0.145	0.965
$^3A_2$	1.864	121	0.747	$^3A_2$	1.832	121	0.725	0.39
$^1A_2$	1.865	120	0.981	$^1A_2$	1.836	120	0.794	0.32
$^3B_2$	1.924	120	1.144	$^1B_2$	1.887	135	1.074	0.04
$^1B_2$	1.908	136	1.641	$^3B_2$	1.885	119	1.104	—
$^1B_1$	1.828	116	1.648	$^1B_1$	1.787	124	1.481	—

Source: Reproduced from Reference 44.

Table 4.44 Dipole Moments of Bent Electronic States of ZrH_2[a]

State	μ_e, D
$^1A_1$	0.738
$^3B_1$	1.499
$^3A_1$	1.335
$^3A_2$	1.271
$^1A_2$	0.940
$^3B_2$	4.635
$^1B_1$	0.681
$^1B_2$	2.419

[a] Positive polarity means Zr^+H^- polarity of bonds.

Source: Reproduced from Reference 44.

The larger 6*s* populations and enhanced stability of HfH_2 is evidently due to the lanthanide contraction and to a lesser extent to relativity. We note that the atomic 5*d*–6*p* promotion energy is smaller for Hf compared to Zr, resulting in enhanced 6*p* character in all states of HfH_2 (compare Tables 4.42 and 4.45).

4.4.9 UH, UH$^+$, and UH$^-$

Krauss and Stevens (36) used the RECP method and a single configuration SCF treatment to study the high-spin high-Λ states of UH and UH$^+$. They found that bonding in UH is amazingly similar to SrH. This is not surprising, since the radial maximum of the 7*s* orbital of U occurs at 3.66 bohr (127), while the radial maxima of the 6*p*, 6*d*, and 5*f* orbitals of the U atom occur at substantially shorter distances. This means that in the valence configuration of U

Table 4.45 Mulliken Population Analyses of Bent States of ZrH_2

State	Gross Population: Zr	H	Zr(*s*)	Zr(*p*)	Zr(*d*)	Overlaps Zr–H
$^1A_1$	3.437	2.562	1.108	0.422	1.907	1.294
$^3B_1$	3.402	2.598	0.870	0.387	2.144	1.304
$^3A_1$	3.400	2.600	0.887	0.402	2.110	1.316
$^3A_2$	3.473	2.528	0.798	0.467	2.208	1.320
$^1A_2$	3.465	2.534	0.765	0.526	2.175	1.328
$^3B_2$	3.460	2.540	0.544	0.341	2.575	1.296
$^1B_1$	3.452	2.548	0.861	0.461	2.131	1.258
$^1B_2$	3.461	2.540	0.480	0.546	2.147	1.350

Source: Reproduced from Reference 44.

Table 4.46 Relativistic and Nonrelativistic Bond Lengths of LaH, AcH, TmH, LuH, and LrH

	r_e, Å		
Molecule	Nonrelativistic	Relativistic	Exp.
LaH	2.210	2.2222	2.101
AcH	—	2.371	—
TmH	—	2.056	—
LuH	2.019	2.013	1.422
LrH	—	2.041	—

Source: Reproduced with modification from Reference 68.

($6s^2 6p^6 5f^3 6d7s^2$, ^5L), the $6p$, $6d$, and $5f$ orbitals will exhibit atomlike behavior and bonding will be primarily between $7s$ and H. However, it was noted by Krauss and Stevens (36) that the $6p$, $6d$, and $5f$ do extend to the bonding region and were distorted by overlap repulsion and electrostatic effects. Consequently, bonding in UH was found to be similar to SrH.

The $UH^{\pm}$ ions were found to exhibit a more complicated structure, due to the presence of $6s^2 6p^2 5f^3 6d7s$, ^6L configuration only a few hundred reciprocal centimeters above $6s^2 6p^6 5f^3 7s^2$, ^4I state. The role of $6p$, $5f$, and $6d$ orbitals in all actinide compounds needs to be understood. As mentioned in the introduction, the ligand field model works well for the majority of ionic compounds of lanthanides because of the atomic nature of the $4f$ shells in these compounds.

Krauss and Stevens (36) considered the $^6\Lambda$ state correlating into $U(^5L)$ + $H(^2S)$ at the SCF level with the configuration $1\sigma^2[U(6s)]2\sigma^2[U6(p)]3\sigma^2[H(1s)]$-$4\sigma[U(7s)]1\pi^4[U(6p)]2\pi[U(5f)]1\delta[U(5f)]2\delta[U(6d)]1\phi[U(5f)]$. At long distances, the SCF configuration used involved transfer of an electron from hydrogen to U(7s), so that it dissociates into $U(^5L) + H(^2S)$. The Morse curve fit for the energies listed in Table 4.46 yielded R_e = 2.16 Å, ω_e = 1357 cm^{-1}, and $\omega_e x_e$ = 18 cm^{-1} for UH. The D_e was computed as 1.93 eV.

Analysis of the amplitude contour plot of the 3σ, 4σ, and 2σ orbitals of UH shows that the 2σ orbital is almost entirely $U(6p_\sigma)$, while the 4σ orbital is a polarized $7s$ orbital on U. The 3σ orbital has predominant change in s distribution centered on H. The H^- charge distribution appears to envelop nonbonding U valence orbitals.

The similarity between UH and SrH is rather striking. The SrH has an r_e = 2.146 Å compared to UH, which has an r_e = 2.16 Å. The ω_e values of UH and SrH are 1357 and 1206 cm^{-1}, respectively. These constants support the atomlike behavior of $6p$, $6d$, and $5f$ orbitals of U in UH.

4.4.10 AcH, TmH, LuH, and LrH

Pyykkö (68) has performed Dirac–Fock one center (DFOC) calculations on AcH, TmH, LuH, and LrH as well as ScH, YH, and LaH. He considered only the $^1\Sigma^+$ states of these species to gain insight into the nature of bonding in these species. For YH, LaH, and LuH, the DFOC bond lengths are 5% longer than experiment. Pyykkö has provided very insightful comparisons of the orbital energies and hybridizations across the periodic table that are quite interesting.

Table 4.46 shows the computed relativistic and nonrelativistic bond lengths of AcH, TmH, LuH, and LrH computed by Pyykkö. Pyykkö (68) defined lanthanide and actinide contraction as

$$\Delta_{\text{lan}} = r_e(\text{LaH}) - r_e(\text{LuH})$$
$$\Delta_{\text{act}} = r_e(\text{AcH}) - r_e(\text{LrH})$$

The values computed this way are 0.210 and 0.329 Å, respectively. Nonrelativistically, the lanthanide contraction is only 0.191 Å. One could also compare the experimental Δ_{lan} as 0.189 Å, in good agreement with the value of Pyykkö. The relative energies and Morse potentials obtained by Pyykkö for LaH, LuH, AcH, and so on demonstrate the lanthanide and actinide contraction.

Pyykkö (68) noted that nonrelativistic *ns* orbital energies increase monotonically from Sc(4*s*) to Ac(7*s*). On the contrary, the relativistic 7*s* orbital energy of Ac is lower than La(6*s*) for the free atoms as well as for AcH and LaH.

Since LaH and LuH differ by the $4f^{14}$ shells, a comparison of the two molecules is in order. Pyykkö shows the plot of the spherically averaged radial electron densities (Fig. 4.23). As seen from Figure 4.23, the LuH 6*s* electron distribution is shifted to the left with a maximum occurring to the left of the corresponding maximum of LaH. This means that the 6*s* orbital of Lu is contracted,

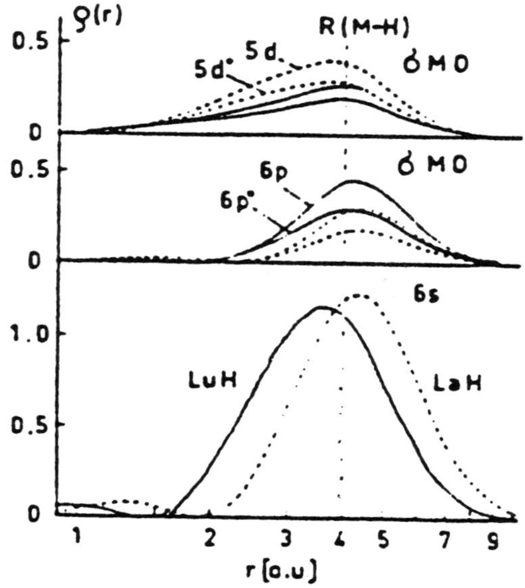

Figure 4.23 Spherically averaged electron densities for the valence orbitals of LaH (---) and LuH (—) at $R = 3.988$ a.u. (Reproduced from Ref. 68.)

since the $4f^{14}$ shells incompletely shield the nucleus, but the nuclear charge is increased by 14 for Lu compared to La. This is the effect of the lanthanide contraction. However, the LuH $5d$ electron distribution is more extended both inward and outward compared to LaH. The inward extension is attributed to the lanthanide contraction of the $4d$ core shell. Pyykkö attributed the outward extension to the attraction of the proton.

4.4.11 Ground-State Properties of Lanthanide Hydrides

Dolg and Stoll (53) have generated nonrelativistic and quasirelativistic energy-adjusted pseudopotentials for the rare-earth elements La–Lu. They have tested their ECPs by studying energetically low-lying superconfigurations of all lanthanide hydrides, fluorides, and oxides.

Table 4.47 shows the spectroscopic constants of [r_e (Å), D_e (eV)] of the lanthanide hydrides computed by Dolg and Stoll at various levels of theory. The column that corresponds to $Q = 11$ represents the results of calculations with $11e$ pseudopotential with the superconfiguration $(4f^{n+1})\sigma^1$, while $Q = 10$ corresponds to the $(4f^n)\sigma^2$ superconfiguration. Thus the Q values are the core charges in the two ECPs. There is very little experimental information on the lanthanide hydrides. The experimental r_e values for YbH and LuH listed in Huber and Herzberg (128) are 2.053 and 1.912 Å, respectively. Note that Dolg and Stoll's r_e for the LaH ground state at the SDCI+Q level (2.075 Å) is in excellent agreement with the value computed by Das and the author (27) (2.08 Å) at the CASSCF/SOCI level of theory. The D_e computed by Dolg and Stoll for LaH is 2.42 eV at the SDCI+Q level compared to a value of 2.60 eV obtained by Das and the author at the SOCI level.

Dolg and Stoll (53) also report the vibrational frequencies and the dipole moments of various lanthanide hydrides. The readers are referred to their original paper for these data. For LaH, their best ω_e of 1450 cm^{-1} agrees well with the value of 1433 cm^{-1} obtained by Das and the author (27). The dipole moment they report for LaH are 2.86, 2.30, and 2.20 D at three levels of theory. The last two values are in agreement with a value of 2.42 D obtained by Das and the author.

An interesting outcome of the Dolg–Stoll computation is the manifestation of lanthanide contraction in the computed r_e values of LaH–LuH. As seen from Table 4.47, the bond lengths for a given level of theory and Q gradually decrease in going from LaH to LuH. For example, at the best level of theory the r_e values of LaH and LuH are 2.079 and 1.949 Å, respectively. Likewise, the D_e values of the two species are 2.61 and 3.18 eV, respectively.

Dolg et al. (57) used the CASSCF/CPF/MRCI/CIPSI methods in conjunction with quasirelativistic effective core potentials to study YbH. They found that the ground state of YbH is a $\Omega = \frac{1}{2}$ state arising from the $4f^{14}\sigma^2\sigma\,(^2\Sigma^+)$ configuration with theoretical $r_e = 2.074$ Å, $D_0 = 1.43$ eV, and $\omega_e = 1276\,cm^{-1}$, in excellent agreement with experimental values of $r_e = 2.053$ Å, $D_0 \leq 1.93$ or ≤ 1.55 eV, and $\omega_e = 1249\,cm^{-1}$. They found an $A^2\Sigma^+$ excited state with an energy separation of 1.0–2.0 eV, depending on the level of theory.

Table 4.47 Spectroscopic Constants [r_e(Å), D_e (eV)] of Lanthanide Hydrides at Various Levels of Theory

		r_e, Å				D_e, eV				
		$Q = 11$		$Q = 10$		$Q = 11$			$Q = 10$	
LnH		CI	+Q	CI	+Q	CI	+Q	+AC	CI	+Q
LaH	*a, d*	2.066	2.075	—	—	2.35	2.42	2.42	—	—
	b, d	2.082	2.089	—	—	2.50	2.57	2.57	—	—
	c, d	2.077	2.084	—	—	2.51	2.58	2.58	—	—
	b, e	2.079	2.085	—	—	2.61	2.71	2.71	—	—
	c, e	2.057	2.062	—	—	3.02	3.18	3.18	—	—
CeH	*a, d*	2.047	2.056	—	—	2.35	2.43	2.43	—	—
PrH	*a, d*	2.032	2.042	—	—	2.36	2.43	1.88	—	—
NdH	*a, d*	2.020	2.028	2.253	2.257	2.37	2.45	1.61	1.71	1.63
PmH	*a, d*	2.011	2.020	2.242	2.247	2.39	2.46	—	1.67	1.59
SmH	*a, d*	2.001	2.010	2.236	2.241	2.40	2.48	0.24	1.64	1.57
EuH	*a, d*	1.990	1.999	2.230	2.235	2.42	2.50	−0.83	1.62	1.55
	b, d	2.010	2.017	2.225	2.232	2.66	2.74	−0.59	1.28	1.23
	b, e	1.997	2.002	2.199	2.203	2.72	2.82	−0.51	1.60	1.59
GdH	*a, d*	1.981	1.989	—	—	2.44	2.52	2.52	—	—
TbH	*a, d*	1.978	1.986	—	—	2.47	2.55	2.52	—	—
DyH	*a, d*	1.972	1.980	—	—	2.50	2.59	1.65	—	—
HoH	*a, d*	1.967	1.975	—	—	2.54	2.62	1.58	—	—
ErH	*a, d*	1.962	1.969	—	—	2.58	2.66	1.77	—	—
TmH	*a, d*	1.958	1.965	2.164	2.169	2.62	2.70	1.07	1.43	1.37
YbH	*a, d*	1.955	1.962	2.173	2.179	2.67	2.74	−0.13	1.42	1.35
	b, d	1.976	1.981	2.148	2.156	3.12	3.18	0.30	1.04	1.01
	b, e	1.958	1.963	2.126	2.132	3.09	3.19	0.31	1.35	1.35
LuH	*a, d*	1.951	1.958	—	—	2.71	2.79	2.79	—	—
	b, d	1.968	1.973	—	—	3.18	3.24	3.24	—	—
	b, e	1.949	1.954	—	—	3.14	3.24	3.24	—	—

[a] MEFIT, HF pseudopotential.
[b] MEFIT, WB pseudopotential.
[c] As footnote *b*, but the *f* pseudopotential is adjusted to La^{10+} $4f^1$ and $5f^1$ ^2F.
[d] Ln $(7s6p5d)/[5s4p3d]$, H $(5s1p)/[3s1p]$.
[e] Ln $(7s6p5d2f)/[5s4p3d2f]$, H $(10s5p1d)/[5s3p1d]$.

Source: Reproduced from Reference 53.

4.5 ELECTRONIC STATES OF DIATOMIC LANTHANIDE AND ACTINIDE HALIDES

In this section we summarize the results of experimental and theoretical studies on diatomic lanthanide and actinide halides. It is pointed out that there exists a large set of spectroscopic data on these species. These are already summarized in Huber and Herzberg (128). The present section focuses primarily on those diatomic molecules for which extensive studies have been made. The readers are referred to the excellent book of Huber and Herzberg (128) for previous studies. A paper by Gotkis (129) provides theoretical insight and systematic comparison of bonding in lanthanide halides.

4.5.1 LaF

The electronic states of LaF are qualitatively similar to LaH, although LaF is significantly more ionic than LaH. The ground state of LaF is also of $^1\Sigma^+$ symmetry and is very ionic. Among others, Barrow et al. (115) have studied the diatomic LaF molecule. They have analyzed the $A \leftarrow X$, $B \leftarrow X$, $C \leftarrow X$, $D \leftarrow X$, and $E \leftarrow X$ systems with ν_{00} values of 11,662, 16,184, 20,960, 22,485 and 22,574 cm^{-1}, respectively. The A, B, C, D, and E states of LaF are assigned to $A^1\Sigma^+$, $B^1\Pi$, $C^1\Pi$, $D^1\Sigma^+$, and $E^1\Sigma^+$ states; of course, the X ground state is of $^1\Sigma^+$ symmetry.

Schall et al. (116) studied LaF using resolved fluorescence laser spectroscopy. Tunable dye lasers were used to excite known systems of LaF. However, the resolved fluorescence spectra subsequent to excitation of $B^1\Pi$–$X^1\Sigma^+$ and $C^1\Pi$–$X^1\Sigma^+$ bands also showed transitions to $X^1\Sigma^+$ and $a^3\Delta$ states. They found yet another $\Omega = 2$ state, which was tentatively assigned to $^1\Delta_2$. These high-resolution spectral bands facilitated accurate determination of the spectroscopic constants of these states. Schall et al. (116) also used the ligand field theory to rationalize their assignments.

Table 4.48 shows the spectroscopic constants of LaF. The values for the

Table 4.48 Spectroscopic Constants of LaF[a]

State	r_e, Å	ω_e, cm^{-1}	T_2, cm^{-1}
$X\,^1\Sigma^+$	2.0265	570	0
$a_1\,^3\Delta_1$	2.05994	537.14	1,432(3)
$a_2\,^3\Delta_2$	2.0543	537.65	1,808
$^1\Delta_2$	—	528	5,478
$A0^+(^1\Sigma^+)$	2.0929	[489]	16,637.96
$B^1\Pi$	2.0971	475.91	16,184
$C^1\Pi$	[2.0671]	[549]	20,959
$D^1\Sigma^+$	[2.04]	—	22,485
$E^1\Sigma^+$	[2.10]	[421]	22,574
$c_1\,^3\Delta_3$	—	—	—
$c_2\,^3\Delta_2$	2.0548	—	—
$c_3\,^3\Delta_1$	—	—	—
$d^3\Phi$	—	—	—
$e^3\Phi$	—	—	—

[a]Numbers in square brackets are uncertain values.

Source: Prepared from Reference 116 and Huber and Herzberg (128).

X, a_1, a_2, $^1\Delta_2$, 0^+, and $B^1\Pi$ states are improved and from Schall et al.'s work (116). Other constants are from Huber and Herzberg (128). Figure 4.24 summarizes the electronic states of LaF observed up to now.

There are several similarities between the electronic states of LaH studied by Das and the author (27) theoretically and LaF (compare Tables 4.13 and 4.48). The ground state of LaH is also $X^1\Sigma^+$ with $r_e = 2.08$ Å and $\omega_e = 1433\,\text{cm}^{-1}$. The shorter LaF bond length is due primarily to large ionicity of the La–F bond. The $^3\Delta_1$, $^3\Delta_3$, and $^3\Delta_2$ states of LaH are computed at 2580, 2588, and

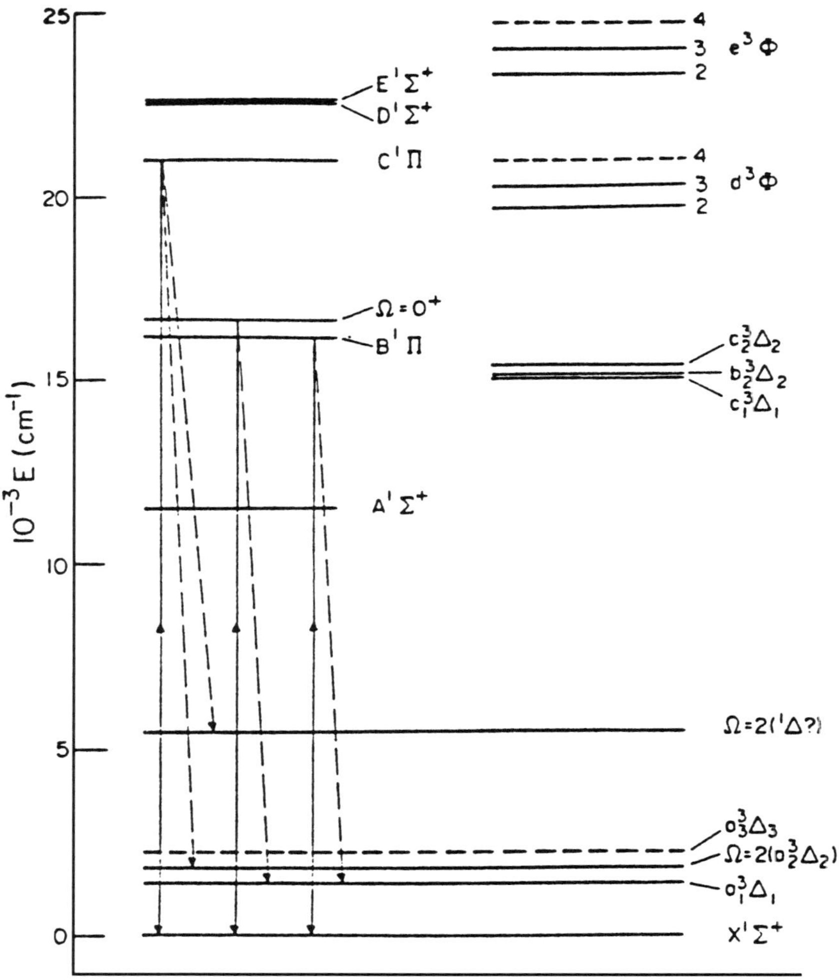

Figure 4.24 Energy levels of LaF known up to now. The dashed lines are uncertain in an absolute sense but are placed correctly relative to other states. The solid vertical lines stand for laser-induced transitions studied by Schall et al. (116), while the dashed vertical lines are observed fluorescence transitions. (Reproduced from Ref. 116.)

2789 cm^{-1}, respectively, while the $^3\Delta_1$ and $^3\Delta_2$ LaF are at 1432 and 1808 cm^{-1}, respectively. The $^1\Delta_2$ states of LaH has a T_e of 6567 cm^{-1}, while the corresponding state of LaF has a T_v of 5478 cm^{-1}.

The $A^1\Sigma^+$(II) state of LaH has a T_e of 13,025 cm^{-1}, while the corresponding state of LaF has a T_e of 16,638 cm^{-1}. The $B^1\Pi$ state of LaH is also close to the A state. The $C^1\Pi$ state of LaH has a T_e of 20,170 cm^{-1}, while the corresponding state of LaF has a T_e of 20,959 cm^{-1}. Consequently, there are many similarities between LaH and LaF. This similarity appears to reveal that these excited states arise primarily from the excitations of the La electrons to upper orbitals composed predominantly of La.

4.5.2 YF and YCl

The diatomic yttrium halides have been the topic of both *ab initio* and experimental studies. Fischell et al. (130) studied the excitation spectra of the YCl diatomic molecule using the laser-induced fluorescence (LIF) method. Xin et al. (131) have studied the $B^1\Pi$–$X^1\Sigma^+$ system of YCl in high resolution. The rotational analysis of the observed bands has yielded very accurate molecular constants for the X and B states of YCl. Shirley et al. (132) studied the molecular-beam optical Stark spectrum of the $B^1\Pi(v = 0)$–$X^1\Pi^+(v = 0)$ band system of YF. The permanent dipole moment μ and the magnetic hyperfine parameter a for the $B^1\Pi$ state have been determined as 2.96(4) D and 149.6.8(3) MHz, respectively. The dipole moment of the $X^1\Sigma^+$ state was determined as 1.82(8) D. Shirley et al. (133) employed the molecular-beam millimeter-wave optical pump-probe spectroscopy to study pure rotational transitions of the YF $^1\Sigma^+$ ground state. This study has yielded improved ground-state rotational constants as $B = 8683.65(1)$ MHz and $D = 0.0079(2)$ MHz, respectively.

Langhoff et al. (88) computed the spectroscopic constants of scandium and yttrium halides using the *ab initio* method. In particular, the YCl diatomic molecule has been studied using CASSCF followed by MRCI and MRCI+Q methods. These authors have computed the spectroscopic constants of 18 electronic states of YCl and reassigned the observed spectra of Fischell et al. (130). We discuss this in this section.

Table 4.49 shows the MRCI and MRCI+Q results of Langhoff et al. (88) for YCl together with the experimental data from Huber and Herzberg (128). Fischell et al. (130) observed the excitation spectra of YCl using LIF, but the two observed bands were not assigned.

Langhoff et al. (88) assigned to the excitation spectra of YCl at 27,116 cm^{-1} to the $J^1\Pi$–$X^1\Sigma^+$ transition. They noted that their computed lifetime (13 ns) is lower than the experimental value of 21 ns. They found that this system contained significant component of $Y^+(5s \rightarrow 5p)$ atomic excitation.

The spectra observed by Fischell et al. with $T_e \sim 22,787$ cm^{-1} were assigned by Langhoff et al. (88) to the $D^1\Pi$–$X^1\Sigma^+$ system. The computed lifetime of 23 ns was found to be in excellent agreement with the experimental value of 28 ns, further supporting the $D^1\Pi$–$X^1\Sigma^+$ assignment.

Table 4.49 Spectroscopic Constants for Excited States of YCl[a]

	MRCI			MRCI+Q			Exp.[b]		
State	r_e, Å	ω_e, cm^{-1}	T_e, cm^{-1}	r_e, Å	ω_e, cm^{-1}	T_e, cm^{-1}	r_e, Å	ω_e, cm^{-1}	T_e, cm^{-1}
$X\,^1\Sigma^+$	4.592	365	0	4.590	364	0	4.547	381	0
$A\,^1\Delta$	4.733	335	9,594	4.729	335	9,557	—	—	—
$B^1\Pi$	4.775	315	12,417	4.770	315	12,284	4.667	330.6	12,128
$C\,^1\Sigma^+$	4.744	314	15,260	4.741	316	15,161	4.694	325	14,908
$D^1\Pi$	4.723	391	23,147	4.704	337	22,765	4.654	345	22,787
$E\,^1\Delta$	4.723	326	23,916	4.715	327	23,792	—	—	—
$F\,^1\Gamma$	4.760	312	25,841	4.743	308	25,460	—	—	—
$G\,^1\Sigma^+$	4.729	325	25,985	4.723	341	25,425	—	—	—
$H\,^1\Phi$	4.749	342	26,624	4.729	322	26,023	—	—	—
$I\,^1\Delta$	4.801	313	28,599	4.775	314	27,722	—	—	—
$J\,^1\Pi$	4.758	315	28,770	4.727	318	27,747	4.668	335	27,116
$a^3\Delta$	4.712	339	6,524	4.707	339	6,673	—	—	—
$b^3\Pi$	4.722	331	8,753	4.716	332	8,788	—	—	—
$c^3\Sigma^+$	4.732	316	10,154	4.724	323	10,255	—	—	—
$d\,^3\Phi$	4.771	309	19,165	4.765	310	19,305	—	—	—
$e^3\Pi$	4.718	325	20,076	4.701	339	20,001	—	—	—
$g^3\Delta^c$	4.786	306	20,332	4.779	307	20,459	—	—	—
$h^3\Pi$	4.756	347	21,027	4.763	340	21,018	—	—	—

[a] In this table, r_e is reported in a_0 and T_e and ω_e are given in cm^{-1}.
[b] Huber and Herzberg (128), except the $B^1\Pi$ state, whose constants are from Reference 131.
[c] The letter f is reserved for the lowest $^3\Sigma^-$ state, which is expected to lie in this region—see the text.
Source: Reproduced from Reference 88.

As seen from Table 4.49, A and B states were assigned to $^1\Delta$ and $^1\Pi$, respectively, by Langhoff et al. The B–X system was also analyzed by Xin et al. (131). The lifetime of the $B^1\Pi$ state was computed as ~291 ns by Langhoff et al. Xin et al. (131) confirmed Langhoff et al.'s computed properties of the $B^1\Pi$ state. The experimental T_e for the B state obtained by Xin et al. (12,128 cm^{-1}) is in excellent agreement with the computed value of Langhoff et al. (12,284 cm^{-1}).

As seen from Table 4.49, there are numerous allowed dipole transitions predicted for YCI, many of which are yet to be observed. Hence there is further scope for experimental studies on YCl.

4.5.3 UF and UF$^-$

Krauss and Stevens (36) have studied the diatomic UF and its ions. The bonding in UF was found to be similar to alkaline-earth fluorides. These authors studied the $^6\Lambda$ state of UF. The SCF configuration used near r_e by these authors is

$$1\sigma^2[F(1s)]2\sigma^2[U(6s)]3\sigma^2[F(2s)]4\sigma^2[U(6p)]5\sigma^2[F(2p)]6\sigma[U(7s)]1\pi^4[U(6p)] \cdot 2\pi^4[F(2p)]3\pi[U(5f)]1\delta[U(5f)]2\delta[U(6d)]1\phi[U(5f)]$$

The UF molecule was found to be ionic with a $U(7s)$ electron transferring to the $F(2p)$ at long distance. The spectroscopic constants of the $^6\Lambda$ ground state of UF were computed as $r_e = 4.12$ bohr, $\omega_e = 511\,\text{cm}^{-1}$, $\omega_e x_e = 2.0\,\text{cm}^{-1}$, and $D_e = 4.92$ eV relative to the neutral separated atoms.

Analysis of the highest occupied orbitals of UF reveals that the 5σ orbital is predominantly $F(2p\sigma)$, although the lobe toward U is compressed by overlap with uranium orbitals. On the other hand, 6σ orbital is composed mainly of $U(7s)$. Krauss and Stevens (36) considered the $^5\Lambda$ state of UF^-. The spectroscopic constants of this state at the SCF level were $r_e = 9.4.30$ bohr, $\omega_e = 429\,\text{cm}^{-1}$, $\omega_e x_e = 1.7\,\text{cm}^{-1}$, and $D_e = 3.1$ eV. The larger D_e and smaller r_e for UF^- compared to UH^- was rationalized by Krauss and Stevens on the basis of smaller size of F^- compared to H^-.

4.5.4 Ground States of LaF–LuF

Dolg and Stoll (53) have computed the ground-state properties of LaF–LuF using different ECPs and SCF/SDCI+Q levels of theory. Tables 4.50 and 4.51 show the r_e and D_e values computed by Dolg and Stoll for LaF–LuF. The computed r_e values of LaF–LuF also show the lanthanide contraction trend. The experimental r_e values for LaF, TbF, HoF, YbF, and LuF are 2.027, 1.96, 1.94, 2.016, and 1.917 Å, respectively. The ground-state results of Dolg and Stoll are thus in good agreement with experiment.

The SDCI+Q method based on a single-reference configuration underestimates the dissociation energies uniformly as seen from Table 4.51. This is expected in view of a single-reference and SCF treatments for the MOs, yet one could obtain a meaningful relative trend, as seen from Table 4.51.

Gotkis (129) developed a simple electron structure model to investigate bonding in lanthanide halides. He investigated the ionization energies for various lanthanides and obtained systematic trends. Gotkis (129) showed that deviations from monotonic trend due to a field-stimulated restructuring of the lanthanide cation core, which involves promotion of an electron from $4f$ to an out-of-core extended σ_{6s} orbital. He grouped lanthanide fluorides into two groups: one with an f^n core (PrF, NdF, SmF, EuF, YbF) and the other with an f^{n-1} core (LaF, GaF, TbF, HoF, ErF, LuF). The former group was found to be similar to alkaline-earth monohalides, while the latter group consisted of Ln^{3+} $(4f^{n-1})$ and F^- and a lone pair $(\sigma_{6s})^2$ localized on the lanthanide atom. Gotkis (129) found that the energies of the two configurations are close for DyF and TmH. He predicted the r_e and ω_e values for the ground state and excited states of these halides. He also discussed the Johnson rule for the variation of properties of lanthanide halides. For the actual tables of molecular constants that Gotkis predicts, the readers are referred to his paper.

Table 4.50 Bond Lengths of Rare-Earth Monofluorides (Å) from SCF and CI (SD) Calculations [Including Davidson's Correction (+Q)] and Experiment

		$Q = 11^g$		$Q = 10^g$	
LnF		CI	+Q	CI	+Q
LaF	*a, d*	2.178	2.181	—	—
	b, d	2.179	2.183	—	—
	c, d	2.169	2.172	—	—
	b, e	2.115	2.116	—	—
	c, e	2.070	2.072	—	—
	b, f	2.116	2.118	—	—
	c, f	2.068	2.070	—	—
CeF	*a, d*	2.156	2.159	—	—
PrF	*a, d*	2.142	2.145	2.249	2.252
NdF	*a, d*	2.127	2.130	2.240	2.242
PmF	*a, d*	2.112	2.115	2.226	2.229
SmF	*a, d*	2.099	2.103	2.214	2.216
EuF	*a, d*	2.087	2.090	2.203	2.206
	b, d	2.084	2.088	2.204	2.208
	b, e	2.020	2.022	2.144	2.145
	b, f	2.017	2.018	2.138	2.138
GdF	*a, d*	2.076	2.079	—	—
TbF	*a, d*	2.065	2.067	—	—
DyF	*a, d*	2.055	2.059	—	—
HoF	*a, d*	2.045	2.049	—	—
ErF	*a, d*	2.037	2.041	—	—
TmF	*a, d*	2.029	2.033	2.146	2.148
YbF	*a, d*	2.021	2.025	2.143	2.146
	b, d	2.013	2.017	2.136	2.140
	b, e	1.953	1.955	2.079	2.080
	b, f	1.944	1.946	2.066	2.067
LuF	*a, d*	2.014	2.018	—	—
	b, d	2.002	2.007	—	—
	b, e	1.944	1.946	—	—
	b, f	1.934	1.936	—	—

[a] MEFIT, HF pseudopotential.
[b] MEFIT, WB pseudopotential.
[c] As footnote *b*, but the *f*-pseudopotential is adjusted to $La^{10+}4f^1$ and $5f^1\ ^2F$.
[d] Ln (7*s*6*p*5*d*)/[5*s*4*p*3*d*], F (9*s*6*p*)/[4*s*3*p*].
[e] Ln (7*s*6*p*5*d*1*f*)/[5*s*4*p*3*d*1*f*], F (9*s*6*p*1*d*)/[4*s*3*p*1*d*].
[f] Ln (7*s*6*p*5*d*2*f*)/[5*s*4*p*3*d*2*f*], F (9*s*6*p*2*d*)/[4*s*3*p*2*d*].
[g] The core charges $Q = 11$ and $Q = 10$ denote the pseudopotentials for a $(4f^n)\ \sigma^2$ and a $(4f^{n+1})\ \sigma^1$ superconfiguration, respectively.

Source: Reproduced from Reference 53.

Table 4.51 Dissociation Energies of Rare-Earth Monofluorides (eV)[a]

		$Q = 11$			$Q = 10$		
LnF		CI	+Q	+AC[b]	CI	+Q	Exp.
LaF	*a, d*	5.14	5.45	5.45	—	—	6.20
	b, d	5.46	5.77	5.77	—	—	—
	c, d	5.50	5.81	5.81	—	—	—
	b, e	5.55	5.98	5.98	—	—	—
	c, e	5.68	6.11	6.11	—	—	—
	b, f	5.56	6.02	6.02	—	—	—
	c, f	5.68	6.15	6.15	—	—	—
CeF	*a, d*	5.17	5.48	5.48	—	—	—
PrF	*a, d*	5.19	5.50	4.95	4.84	4.99	—
NdF	*a, d*	5.22	5.54	4.70	4.83	4.98	5.87
PmF	*a, d*	5.26	5.58		4.79	4.94	—
SmF	*a, d*	5.31	5.63	3.39	4.77	4.91	5.46 5.81
EuF	*a, d*	5.36	5.68	2.35	4.75	4.89	5.42
	b, d	5.77	6.09	2.76	4.38	4.51	5.60
	b, e	5.88	6.32	2.98	4.58	4.87	—
	b, f	5.90	6.37	3.04	4.68	5.02	—
GdF	*a, d*	5.41	5.73	5.73	—	—	6.08
TbF	*a, d*	5.46	5.78	5.75	—	—	—
DyF	*a, d*	5.53	5.85	4.91	—	—	5.46
HoF	*a, d*	5.60	5.92	4.87	—	—	5.57
ErF	*a, d*	5.66	5.98	5.09	—	—	5.83
TmF	*a, d*	5.72	6.04	4.41	4.57	4.72	5.25
YbF	*a, d*	5.80	6.12	3.24	4.52	4.67	4.80
	b, d	6.38	6.69	3.81	4.03	4.20	5.00
	b, e	6.47	6.91	4.03	4.25	4.54	—
	b, f	6.50	6.97	4.09	4.36	4.70	—
LuF	*a, d*	5.87	6.18	6.18	—	—	—
	b, d	6.45	6.76	6.76	—	—	—
	b, e	6.56	6.99	6.99	—	—	—
	b, f	6.57	7.04	7.04	—	—	—

[a]See Table 4.47 for footnotes *a–f*. Most of the experimental constants are from Reference 128.
[b]AC denotes the CI(SD)+Q results after the energy correction to the experimentally observed atomic ground state of the rare-earth atom was applied.
Source: Reproduced from Reference 53.

Dolg et al. (57) have used the CASSCF/MRCI/CPF method in conjunction with quasirelativistic pseudopotentials to study both YbF and YbH, including spin–orbit coupling. For YbF, they computed the molecular constants as r_e = 2.045 Å, D_0 = 4.87 eV, and ω_e = 492 cm^{-1}, in excellent agreement with the experimental values of r_e = 2.016 Å, D_0 = 4.80 eV, or >5.36 eV, ω_e = 502

cm^{-1}. The ground state of YbF was found to be a $\Omega = \frac{1}{2}$ state arising from the $4f^{14}\sigma^2\sigma^2\pi^4\sigma^2$ ($^2\Sigma^+$) configuration. An excited $A^2\Sigma^+$ state of YbF was also found by Dolg et al. (57) with $T_e = 0.44$–1.51 eV depending on the level of treatment. The predicted ground state of YbF was in accord with the ESR studies of Van Zee et al. (134).

4.6 LANTHANIDE OXIDES

Among diatomic lanthanide molecules, lanthanide oxides have been the most studied experimentally (128), since there is a large wealth of experimental information on the lanthanide oxides. The electronic spectra of lanthanide oxides are very complex due to numerous possible electronic states of varied spin multiplicities. Analysis of the spectra and unambiguous assignment are difficult, since many of these states have similar spectroscopic constants and since the 4*f* orbitals do not take part in bonding. The ligand field theory and *ab initio* and INDO/S-CI techniques have aided the assignment of the observed spectra.

We start with the theoretical SDCI+Q computation of Dolg and Stoll (53) on the ground states of lanthanide oxides. These authors have computed the r_e, D_e, μ_e, and ω_e values for the ground states of LaO–LuO using various levels of theory. At present, experimental bond lengths and dissociation energies of several lanthanide oxides are known (128). The experimental r_e values of LaO, CeO, PrO, GdO, TbO, DyO, HoO, YbO, and LuO are 1.826, 1.820, 1.801, 1.803, 1.812, 1.809, 1.814, 1.796, 1.799, 1.807, and 1.790 Å, respectively. The computed SDCI+Q bond lengths are too long for many of these oxides compared to the experimental values despite the fact that large basis sets have been employed. It should be noted that the SDCI method even after inclusion of quadruple correction for a single reference underestimates the dissociation energies of all these oxides uniformly. Although the actinide oxides are not discussed, the work of Krauss and Stevens (36) who have computed the properties of UO and NdO, is cited.

Carette and Hocquet (135) applied the ligand field theory to study the lower energy lines of the lanthanide monoxides. The spin–orbit parameters were also included in the ligand field model. We shall discuss their results on CeO, PrO, and SmO in separate subsections.

In the ensuing subsections, experimental spectra and theoretical studies on individual oxides are discussed. We restrict ourselves to those species that were studied extensively either experimentally or theoretically.

4.6.1 LaO

Earlier experimental studies and known electronic states of LaO are summarized in Huber and Herzberg (128). The ground state of LaO is well established as the $X^2\Sigma^+$ state with $r_e = 1.825$ Å and $\omega_e = 813$ cm^{-1}. There are several excited states of $A^2\Delta$, $A^2\Pi$, $B^2\Sigma^+$, $C^2\Pi$, $D^2\Sigma$, and $F^2\Sigma$ symmetries (128, 135). Experimental studies on LaO and other oxides include the pulsed Fourier transform microwave spectroscopy of YO, LaO, ZrO, and HfO by Suneram et al. (136). The related YO has also been studied extensively (see, e.g., Refs. 137 and 138).

There are several theoretical studies on LaO and related lanthanide oxides. We have already mentioned the ligand field theory model calculations of Field (6, 7) as well as Carette and Hocquet (135). Kotzian and Rösch (104) applied the INDO technique (139) extended to include spin–orbit coupling (see also Ref. 103) to lanthanide oxides (LaO, CeO, GdO, and LuO). The authors call it the "INDO/S-CI method." The INDO parameters were derived from atomic spectra, model Dirac–Fock calculations on lanthanide atoms and ions to derive ionization potentials, Slater–Condon factors, and basis sets. The spin–orbit parameter (ξ) is derived from atomic spectra in this method.

The INDO/S-CI computations were made on LaO by Kotzian et al. (104) at experimental bond distance of 1.826 Å. The ground state configuration of LaO is $[1\sigma/O(2s)]^2[1\pi/O(2p)]^4[2\sigma/O(2p)]^2[3\sigma/La(6s)]^1[1\delta/La(5d)]^0[2\pi/La(6p)]^0$-$[4\sigma/La(6p)]^0[3\pi/LaU(f)]^0[1\phi/La(4f)]^0[5\sigma/La(4f)]^0[2\delta/La(4f)]^0 - [4\pi/La \cdot (5d)]^0[6\sigma/La(5d)]^0$.

Table 4.52 compares the INDO/SCI energy separations with experiment for

Table 4.52 Calculated and Experimental States of LaO[a]

	INDO/S-CI				Experiment	
State	Configuration	T_e	A	State	T_e	A
$1^2\Sigma^+_{1/2}$	3σ 100%	0	—	$X^2\Sigma^+$	0	—
$1^2\Delta_{3/2}$	1δ 100%	8,488	353	$A'^2\Delta_r$	7,468.9	350.5
$1^2\Delta_{5/2}$	1δ 100%	9,195	—	—	8,168.4	—
$1^2\Pi_{1/2}$	2π 99%	13,856	869	$A^2\Pi_r$	12,635.7	862.7
$1^2\Pi_{3/2}$	2π 100%	14,724	—	—	13,497.6	—
$2^2\Sigma^+_{1/2}$	4σ 93%	19,715	—	$B^2\Sigma^+$	17,837.8	—
$2^2\Pi_{1/2}$	3π 94%	22,986	—	—	22,618.9	—
$2^2\Pi_{3/2}$	3π 79%	23,241	255	$C^2\Pi_r$	22,839.6	221.4
	2δ 20%	—	—	—	—	—
$1^2\Phi_{5/2}$	1ϕ 93%	23,310	560	—	—	—
$1^2\Phi_{7/2}$	1ϕ 100%	24,990	—	—	—	—
$2^2\Delta_{3/2}$	2δ 79%	24,788	—	—	—	—
	3π 21%	—	451	—	—	—
$2^2\Delta_{5/2}$	2δ 93%	25,691	—	—	—	—
$3^2\Sigma^+_{1/2}$	5σ 90%	26,007	—	$D(^2\Sigma)$	26,959.0	—
$3^2\Pi_{1/2}$	4π 99%	30,180	408	$F(^2\Sigma)$	28,049.0	—
$3^2\Pi_{3/2}$	4π 99%	30,588	—	—	—	—
$4^2\Sigma^+_{1/2}$	6σ 100%	41,616	—	—	—	—

[a]State energies T_e are given in cm^{-1}. The molecular spin–orbit constant A is defined by the relation $\langle J\Omega\Lambda S\Sigma|H_{SO}|J\Omega\Lambda S\Sigma\rangle = A\Lambda\Sigma$.

Source: Reproduced from Reference 106.

several electronic states of LaO. It is evident from Table 4.52 that the INDO/S-CI method predicts the energy separations for these compounds reasonably well. Kotzian et al. also computed the transition moments for the observed $A \leftrightarrow C$, $B \leftrightarrow X$, $C \leftrightarrow X$, $D \leftrightarrow X$, and $F \leftrightarrow X$ systems.

4.6.2 CeO

There are several experimental studies on the CeO diatomic molecule. Schall et al. (140) studied CeO using sub-Doppler Zeeman spectroscopy. Again, the ligand field model is successful in explaining the observed spectra due to the ionic nature of the diatomic lanthanide oxide. Linton et al. (141) studied the electronic spectrum of CeO using absorption and emission, as well as the laser spectroscopic method. There are many 0–0 bands for CeO arising from electronic transitions between various Ω components of excited states and the Ω states arising from $\cdots (4f)(6s)$ electronic configurations. The CeO molecule is also the simplest lanthanide oxide containing a single $4f$ electron.

The electronic states of CeO are best rationalized in terms of the electronic states of Ce^{2+} ion, since the bonding in CeO is best described as $Ce^{2+}O^{2-}$ ionic bond. Figure 4.25 shows the energy-level diagram of Ce^{+2} $4f6s$ configuration and the corresponding state of CeO. It is evident from Figure 4.25 that there are 16 Ω states arising from the Ce^{2+} $4f6s$ configuration itself.

Table 4.53 shows the spectroscopic constants of CeO derived by Linton et al. (141) from their spectra. As seen from Table 4.53, the X_1 ground state with $\Omega = 2$ and X_2 state with $\Omega = 3$ are very close to each other. Note that X_3 and X_4

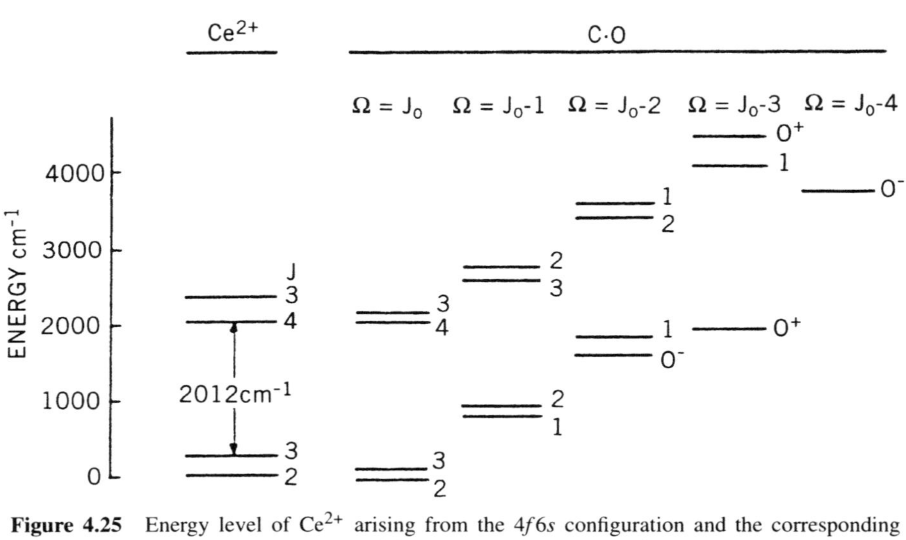

Figure 4.25 Energy level of Ce^{2+} arising from the $4f6s$ configuration and the corresponding states of the diatomic CeO. All states of CeO are arranged by their Ω values. (Reproduced from Ref. 140.)

Table 4.53 Constants for Low-Lying States of CeO (cm^{-1})

State	J_a	J_c	T_0	B_0	$10^7 D_0$	$\Delta G^b_{1/2}$	Footnote
$U_3 0^+$	3	$\frac{7}{2}$	4457.7[a] (30)[b]	0.367	—	—	
$U_2 1$	4	$\frac{7}{2}$	4133[a] (5)	—	—	—	*d*
$T_1 0^-$	4	$\frac{7}{2}$	3821.5 (30)	0.375	—	—	
$V_4 1$	3	$\frac{7}{2}$	3642[a] (5)	—	—	—	*d*
$V_3 2$	4	$\frac{7}{2}$	3462.6 (25)	0.355	—	820.7	
$W_4 2$	3	$\frac{7}{2}$	2771.7 (15)	0.35999 (1)	2.21 (1)	823.4	*e*
$W_3 3$	4	$\frac{7}{2}$	2617.3 (21)	~0.356	—	824.7	*f*
$X_4 3$	3	$\frac{7}{2}$	2140.6 (15)	0.35658 (2)	2.72 (2)	824.1	*g*
$X_3 4$	4	$\frac{7}{2}$	2039.8 (21)	0.35327 (1)	2.24 (1)	822.1	*h*
$U_1 0^+$	3	$\frac{5}{2}$	1931.8[a] (30)	0.377	—	—	
$V_2 1$	3	$\frac{5}{2}$	1869.7[a] (30)	0.343	—	—	
$V_1 0^-$	2	$\frac{5}{2}$	1679.4 (25)	0.35788 (1)	2.52	—	
$W_2 2$	3	$\frac{5}{2}$	912.2 (15)	0.36139 (6)	7.3 (4)	823.0	*i*
$W_1 1$	2	$\frac{5}{2}$	811.6 (25)	~0.361	—	—	*j*
$X_2 3$	3	$\frac{5}{2}$	80.3 (15)	0.35692 (1)	2.71 (1)	822.8	*k*
$X_1 2$	2	$\frac{5}{2}$	0	0.35454 (1)	2.46 (1)	824.3	

[a]Levels of the correct Ω closest to the predicted positions.
[b]Numbers in parentheses represent estimates of the standard deviation in the last two digits.
[c]With the exceptions of the X_1 and X_2 states, uncertainties in $\Delta G_{1/2}$ are ~2 cm^{-1}.
[d]These states have each been observed only once, in fluorescence, at J ~25: their B values are not known.
[e]Originally labeled $v(2)$.
[f]Originally labeled $w(3)$.
[f]$10^3\alpha = 1.4$
[h]$10^3\alpha = 1.27$
[i]Originally labeled $u(2)$.
[j]Originally labeled $y(1)$.
[k]$10^3\alpha = 1.9$

Source: Reproduced from Reference 141.

states arise from the same principal λ–s state, but the splitting between (X_1, X_2) duo with (X_3–X_4) duo is roughly the atomic Ce^{2+} splitting between $J = 2$, 3 and $J = 3$, 4 pairs (see Fig. 4.25). The energy levels in Table 4.53 approximately correspond to the levels in Figure 4.25.

A combination of laser-induced fluorescence with classic absorption and emission spectroscopy has yielded a wealth of data on the excited electronic states of CeO. However, as noted by Linton et al., the spectroscopic data of upper states is less complete and are shown in Table 4.54. Linton et al. (141) used empirical notations taken from atomic spectroscopy to designate these states.

Table 4.54 Constants for Excited States of CeO (cm^{-1})

State	Pre–Ref. 141[a]	T_0	B_0	10^7D_0	$\Delta G_{1/2}$	Transitions[b]
[28.6]1		28,596.1	(e)0.36122 (f)0.36079	3.04	—	$[28.6]^c1-V_10^-$
[26.7]4	E_3	26,715.2	0.35294	4.9	795.5	$[26.7]4-X_34$
[26.2]3	H_2	26,200.0	0.3514	2.2	732.54	$[26.2]3-X_23$
[26.1]2	F_1	26,067.5	0.3522	3.75	—	$[26.1]2-X_12$
[25.3]3	G_2	25,292.3	0.3502	3.03	733.90	$[25.3]3 \leftarrow X_23$
[25.0]2	E_1	25,011.5	0.34863	2.81	744.3	$[25.0]2-X_12$
$[23.7]0^-$		23,674.3	0.35348	0.48	—	$[23.7]0^- -V_10^-$
$[22.7]0^-$		22,722.8	0.35407	2.63	—	$[22.7]0^- -V_10^-$
[22.6]5	D_3	22,556.0	0.35069	2.61	786.0	$[22.6]5-X_34$
[22.5]1		22,505.8	(f)0.3576	6.2	—	$[22.5]1 \rightarrow V_10^-$
[22.5]3	h_2	22,498.4	0.3568	5.0	792.3	$[22.5]3 \rightarrow W_22$ F.0$X_23, -X_43, -W_22, -W_42$
[22.0]4	A_4	22,012.3	0.35443	2.81	768.0	$[22.0]4-X_43$
[21.7]4		21,713.2	0.35353	2.14	—	$[21.7]4-X_23$
[21.4]1	D_1	21,379.2	0.35241 0.35203	2.75 2.59	—	$[21.4]1-X_12$
[21.1]2	g_2	21,061.9	0.35662	3.98	—	$[21.1]2-X_23; [21.1]2-W_22$ F.$-X_23, -X_43, -W_22, -W_42$
[20.9]4	F_2	20,914.5	0.35294	2.97	—	$[20.9]4-X_23$
[20.3]3	C_1	20,273.8	0.34988	2.99	783.6	$[20.3]3-X_12$
[19.9]1		19,926	—	—	—	$[19.9]1-X_12$ F.$-X_12, -W_42, -V_10^-, -V_21, -V_32, -U_10^+, -U_30^+, -T_10^{-1}$
[19.3]3		19,287.5	0.3485	0.4	—	$[19.3]3 \rightarrow X_43$
[18.4]4	C_3	18,386.2	0.34176	2.3	750.7	$[18.4]4 \leftarrow X_34$ F.$-X_34, -X_23, -X_43, -W_33$

[17.2]3	e_2	17,169.5	0.35037	2.44	765.0	[17.2]3 $\leftarrow X_2 3$ F.-X_1 2, $-X_2 3, -X_3 4, -X_4 3, -W_2 2, -W_4 2$
[16.7]1		16,714.8	(e)0.35616 (f)0.35643	4.28 3.45	—	[16.7]1 $\leftarrow V_1 0^-$ F.$-X_1 2, -W_1 1, -W_2 2, -W_4 2, V_1 0^-, -V_2 1, -V_3 2, -U_1 0^+, -U_3 O^+, -T_1 0^-$
[16.5]2	B_1	16,524.2	0.3436	—	720.7	[16.5]2 $\leftarrow X_1 2$ F.$-X_1 2, -X_2 3, -X_4 3, -W_1 1, -W_2 2, -W_3 3, -W_4 2$
[16.5]4	E_2	16,495.4	0.35363	5.1	753.9	[16.5]4 $\leftarrow X_2 3$ F.$-X_2 3, -X_3 4, -X_4 3, -W_2 2, -W_3 3$
[16.0]3	D_2	16,039.2	0.35953	5.3	—	[16.0]3 $\leftarrow X_2 3$
[15.8]4	C_2	15,813.8	0.35216	3.03	—	[15.8]4 $\leftarrow X_2 3$
[15.5]5	B_3	15,489.4	0.34534	2.45	—	[15.5]5 $\leftarrow X_3 4$
[14.7]3		14,701.6	0.35255	2.51	—	[14.7]3 $\leftarrow W_2 2$, [14.7]3 $\leftarrow W_4 2$
[14.2]3	A_3	14,201.8	0.36024	4.48	—	[14.2]3 $\leftarrow X_3 4$
[14.2]0^-		14,197.0	0.35216	7.8	—	[14.2]0^- $\leftarrow V_1 0^-$
[13.9]4	B_2	13,884.3	0.34711	2.86	747.39	[13.9]4 $\leftarrow X_2 3$, [13.9]4 $\leftarrow X_4 3$
			0.35974	4.97	—	
[13.2]1	a_1	13,201.6	0.35921	5.51	—	[13.2]1 $\leftrightarrow X_1 2$
[12.8]2	A_2	12,768.0	0.3535	4.5	—	[12.8]2 $\leftrightarrow X_2 3$
[12.6]3	A_1	12,595.8	0.34672	2.90	—	[12.6]3 $\leftrightarrow X_1 2$

[a]Designations of states prior to Reference 141 are given in the second column.
[b]F. indicates a transition observed in fluorescence.
[c]Numbers in square brackets designate the electronic states.
Source: Reproduced from Reference 141.

The vibrational frequencies ($G_{1/2}$, ω_e) are of the order of 800 cm^{-1} for the low-lying states of CeO. Linton et al. (141) have observed several laser-induced-fluorescence transitions of CeO. They have provided a rather detailed and extensive analysis of their fluorescence excitation spectra. The readers are referred to their original paper for further discussion of their spectra.

Linton et al. (141) noted that the $4f6s$ configuration of Ce^{2+} exhibits a j–j rather than l–s coupling. This is expected in view of the fact that the $4f$ orbital spin–orbit coupling is larger than the $^1F_3$–$^3F_3$ energy separation. Hence they conclude that lanthanide oxide configurations including $4f$ electrons would tend toward Hund Case(c) coupling.

It should be noted that although Ce^{2+} ion has a $4f^2$ configuration, the ligand field calculations of Linton et al. (141) predict reordering of free-ion configurations for the oxides:

$$4f6s \ll 4\delta^2 < 4f5d < 4f6p$$

Hence Linton et al. conclude that for all lanthanide oxides except EuO and YbO, the lowest states should arise from $(4f)^n6s$ configuration rather than $(4f)^{n+1}$.

Schall et al. (140) deduced the electronic g value for four lower states of CeO and compared them with the values derived from the ligand field theory. The agreement was found to be very good.

Kotzian et al. (106, 107) as well as Kotzian and Rösch (104, 105) have applied the INDO/S-CI method extended to include spin–orbit coupling on the diatomic CeO. They have studied 16 electronic states of CeO arising from the $4f6s$ configuration of Ce^{2+} at the experimental r_e of 1.820 Å. The results of their computations are shown in Table 4.55. Note that Kotzian et al. have also compared the INDO/S-CI results with ligand field theories LFT1 and LFT2. As seen from Table 4.55, agreement between the experimental values and the INDO/S-CI as well as ligand field models is quite good.

Kotzian et al. also find that low-lying states of CeO arise from the $4f6s$ configuration. The charge transfer excitations O $\rightarrow$ Ce were omitted in the Kotzian et al. INDO/S-Ci, since the energy for this excitation is >55,000 cm^{-1}. Overall, the experimental trend is reproduced by the INDO/S-CI except that two pairs of states (U_1O^+/V_21 and U_21/U^3O^+) have reverse order in the INDO/S-CI calculation compared to the experiment. Mulliken population analyses of Kotzian et al. revealed that the oxygen atom has $O(2s^{1.923}2p\sigma^{1.731}2p\pi^{3.046})$ population in the ground state. The corresponding Ce population is Ce $4f\sigma^{1.81}4f\pi^{0.329}4f\sigma^{0.284}4f\phi^{0.286}6s^{0.920}5d\sigma^{0.308}5d\pi^{0.882}$.

4.6.3 PrO

PrO was first studied at complete low resolution by Shenyavskaya et al. (142). These authors reported vibrational analyses of 22 electronic transitions of PrO in the 500–1120-nm region. Shenyavskaya et al. (142) designated the observed

Table 4.55 Calculated and Experimental States of CeO of Configuration 4f6s[a]

State	Λ	Ω	J_a	Exp.	LFT1	LFT2	INDO/S-CI Energy	Configuration	
X_1	3	2	2	0.0	0	0	0	$1\phi4\sigma$ 91%	—
X_2	3	3	3	80.3	121.6	197	71	$1\phi4\sigma$ 90%	—
W_1	2	1	2	811.6	805.6	1026	857	$1\delta4\sigma$ 84%	—
W_2	2	2	3	912.2	910.8	1160	918	$1\delta4\sigma$ 82%	$2\pi4\sigma$ 17%
V_1	1	0^-	2	1679.4	1777.7	1756	1,812	$2\pi4\sigma$ 73%	$3\sigma4\sigma$ 25%
V_2	1	1	3	1869.7	1878.4	1919	1,937	$2\pi4\sigma$ 70%	$3\sigma4\sigma$ 27%
U_1	1	0^+	3	1931.8	1866.3	1978	1,850	$2\pi4\sigma$ 75%	$3\sigma4\sigma$ 22%
X_3	3	4	4	2039.8	2021.8	2495	2,084	$1\phi4\sigma$ 99%	—
X_4	3	3	3	2140.6	2185.4	2146	2,154	$1\phi4\sigma$ 98%	—
W_3	2	3	4	2617.3	2632.2	3268	2,726	$1\delta4\sigma$ 84%	—
W_4	2	2	3	2771.7	2762.7	2994	2,791	$1\delta4\sigma$ 84%	—
V_3	1	2	4	3462.6	3501.5	3944	3,453	$2\pi4\sigma$ 80%	$1\delta4\sigma$ 17%
V_4	1	1	3	3642	3600.9	3724	3,562	$2\pi4\sigma$ 81%	$1\delta4\sigma$ 14%
T_1	0	0^-	4	3821.5	4035.2	4109	4,176	$3\sigma4\sigma$ 72%	$2\pi4\sigma$ 25%
U_2	0	1	4	4133	4101.9	4391	4,234	$3\sigma4\sigma$ 68%	$2\pi4\sigma$ 29%
U_3	0	0^+	3	4457.7	4262.8	4476	4,217	$3\sigma4\sigma$ 70%	$2\pi4\sigma$ 23%

[a]Energies are given in cm^{-1}. In the first column the experimental labeling is given. Λ is the projection of the total orbital angular momentum; Ω, the projection of the total angular momentum in the molecular system; and J_a, the value of the atomic total angular momentum from which the molecular state originates. The values of Λ are J_a are taken from the leading determinant.

Source: Reproduced from Reference 106.

systems I through XXII. Delaval et al. (143a) have studied the systems VII and X, but no definitive assignment of the observed electronic transitions came forth. Beaufils et al. (143b) observed the first lines in $\Delta\Omega = +1$ XVII and XX transitions. This led to the assignments of $\Omega'' = 3.5$ and $\Omega' = 9.4.5$ for system XVII and $\Omega'' = 9.4.5$ and $\Omega' = 9.5.5$ for system XX. As pointed out by Dulick and Field (144), some of the earlier assignments of the observed spectra of PrO were either incorrect or speculative.

Dulick and Field (144) provided extensive analyses of the observed systems. They studied the (0, 0) bands of nine prominent systems in the 500–800-nm region. For the first time definitive Ω assignments for the upper and lower electronic states participating in these transitions were provided by Dulick and Field.

Dulick et al. (145) as well as Carette and Hocquet (135) used the ligand field theory to compute the energy levels of PrO. Kotzian and Rösch (104, 105) used the INDO/S-CI method to investigate the electronic states of PrO and TmO. They computed the properties of 33 electronic states of PrO and compared their computed energies with experimental and ligand field theory. All of these studies have provided significant insight into the low-lying electronic states of PrO.

Table 4.56 shows the Dulick–Field assignment of the observed systems of PrO designated by their Ω quantum numbers. Note that Dulick and Field suggest that the lowest state has a Ω quantum number of 3.5. A diagrammatic interpretation of the observed systems and the energy levels of PrO was provided by these authors, which is reproduced in Figure 4.26. Table 4.57 shows the comprehensive assignment of systems I–XXII for PrO observed by Shenyavskaya et al. (142). Tables 4.56 and 4.57 together yield the experimentally known states of PrO. We now proceed to theoretical calculations and insight into the nature of electronic states of PrO.

As mentioned before, there are INDO/S-CI as well as ligand field theoretical studies on PrO. The work of Kotzian and Rösch (104, 105) provides a beautiful summary of all known information on PrO up to now. We will use this work and the paper of Dulick and Field as basis for discussion of the electronic states of PrO. Table 4.58 shows the Ω values of several electronic states of PrO, their energy separations obtained using the INDO/S-CI method of Kotzian and Rösch (105), as well as the ligand field theoretical values of Dulick et al. (145) and Carette and Hocquet (135).

The electronic states of PrO are best rationalized using the energy levels of the Pr^{2+} ion, which has the configuration $4f^26s$. Figure 4.27 shows the energy levels arising from $4f^2(^3H)6s$ and $4f^2(^3H)6p_{1/2}$ configurations of the Pr^{2+}. As seen from Figure 4.27, the coupling is j–j. The spin–orbit coupling of the $4f^2$ shell (^3H) is roughly 2000 cm^{-1}. The PrO diatomic states are rationalized as arising from $Pr^{2+}O^{2-}$. The INDO/SCI yields a ground-state electronic configuration of $[1\sigma/\mathrm{O}(2s)]^2[1\pi/\mathrm{O}(2p)]^4[2\sigma/\mathrm{O}(2p)]^2[1\phi/\mathrm{Pr}(4f)]^{4/7}$-$[1\delta/\mathrm{Pr}(4f)]^{4/7}[2\pi/\mathrm{Pr}(4f)]^{4/7}[3\sigma/\mathrm{Pr}(4f)]^{2/7}[4\sigma/\mathrm{Pr}(6s)]^1$. This configuration is,

Table 4.56 Twelve Observed Electronic Transitions of PrO and Their Suggested Assignments

System	Ω'	T'_0, cm^{-1}	Ω''	T''_0, cm^{-1}
VI	5.5	11,102	4.5	2,157
IX	4.5	16,597	3.5	3,887
X	5.5	13,259	4.5	220
XI	5.5	13,865	4.5	220
XIV	5.5	16,595	4.5	2,157
XVI	5.5	19,169	4.5	3,720
XVII	4.5	16,597	3.5	0
XVIII	7.5	21,321	6.5	3,965
XIX	6.5	19,687	5.5	2,111
XX	5.5	18,069	4.5	220
XXI	4.5	18,885	4.5	220
XXII	5.5	19,169	4.5	220

Source: Reproduced from Reference 144.

of course, consistent with the $Pr^{2+}O^{2-}$ picture. The INDO calculations revealed that the $4f^2 4\sigma$ configuration is the lowest, while $4f^2 2\delta$ and $4f^3$ configurations are respectively 8200 and 18,300 cm^{-1} above the $4f^2 4\sigma$ configuration. Other configurations are much higher (energy > 41,850 cm^{-1}); hence the INDO-SCI studies focused on the electronic states arising from $4f^2 4\sigma$. As seen from Table 4.58, the INDO-SCI model as well as ligand field models reproduce the energies of the experimentally known states ($\Omega \geq 3.5$).

4.6.4 SmO

The spectra of samarium monoxide (SmO) have been studied by Dekock and Weltner (146), Hannigan (147), Dickson and Zare (148), Linton et al. (149), and Bujin and Linton (150, 151). Bujin and Linton (151) summarized the known electronic states of SmO up to now. Carette and Hocquet (135) used the ligand field theory to calculate the lower electronic energy levels of SmO. These authors found 11 electronic states of SmO below 2280 cm^{-1}. The ground state of SmO was calculated as the $X0^-$ electronic state.

Table 4.59 shows the spectroscopic data on the energy separations of the electronic states of SmO. Both experiment and theory seem to agree on the $X0^-$ ground state of SmO. The two ligand field calculations are in reasonable agreement with the experimental energy separations. The calculation of Dulick et al. (145) yields more energy levels than known experimentally. The Carette–Hocquet ligand field calculation is slightly in better agreement with experiment. Linton et al. (149) attributes this to differences in the choice of Sm^{+2} atomic orbital basis sets the Carette–Hocquet eigenfunctions are considered probably more realistic.

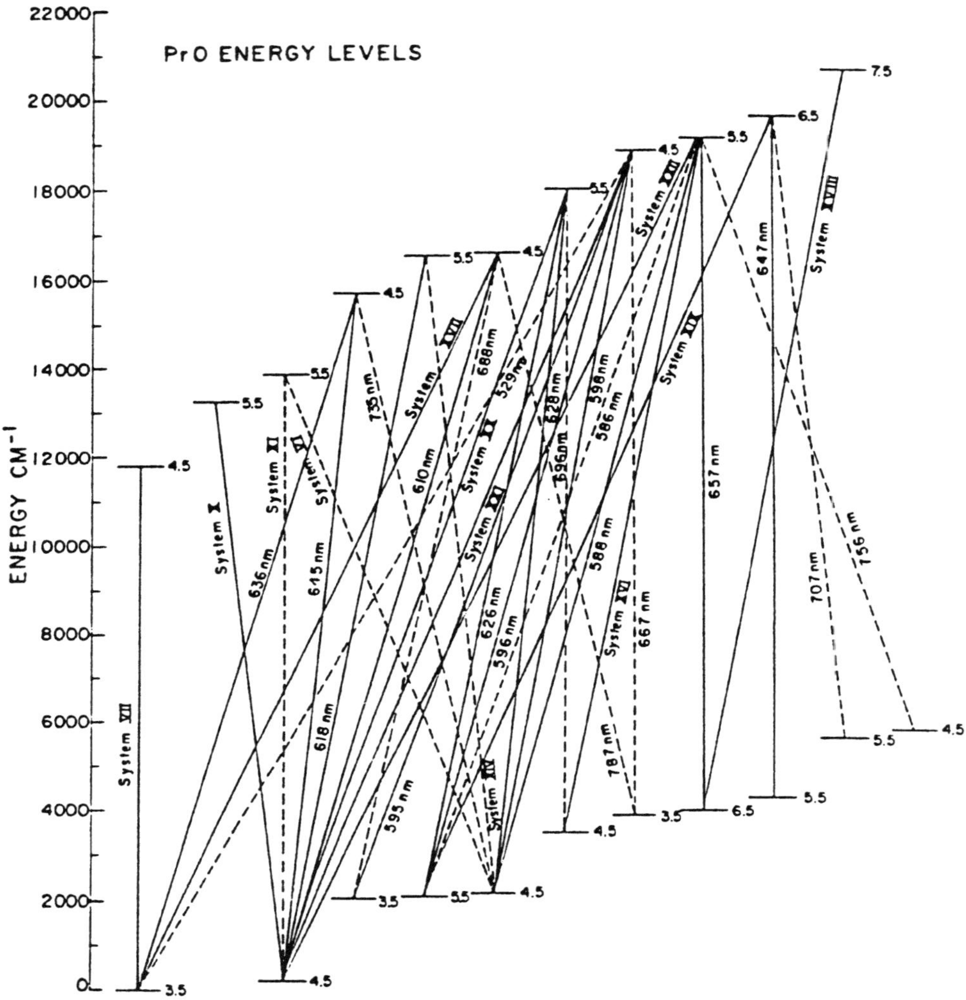

Figure 4.26 Energy-level diagram and the electronic transitions of PrO. The solid line designated electronic transition examined under high resolution. (Reproduced from Ref. 144.)

The spectra of SmO exhibited significant Ω doublings for the lowest states. This was anticipated by calculations, although calculations were not found to be in accord with the observed values. Bujin and Linton (151) attribute this is to incorrect ordering of the *e* and *f* levels in calculations.

The electronic states of SmO can be rationalized from the Sm^{2+} $4f^2 6s(^6H)$ supermultiplet. The overall ordering of the electronic states of SmO closely resembles the low-lying states of Sm^{2+} derived from the $(4f^2 6s)^6H$ electronic supermultiplet. This is further supported by the remarkable agreement of the observed energy levels with the values obtained from ligand field models.

Table 4.57 Electronic Assignments of PrO for Systems I–XXII of Shenyavskaya et al. (142) Suggested by Dulick and Field (144)

System	Bandhead[a] ν, cm^{-1}	Ω'[b]	Ω''
I	9,233	?	?
II	9,996	?	?
III	10,205	?	?
IV	10,438	5.5	4.5 (XX)[c]
V	10,976	?	?
VI	11,109	5.5 (X)	4.5[d]
VII	11,777	4.5	3.5 (XVII)[c,e]
VIII	11,922	2.5	3.5 (XVII)[c]
IX	12,711	4.5 (XVII)	3.5[d]
X	13,047	5.5	4.5 (XX)[c,e]
XI	13,657	5.5	4.5 (XX)
XII	14,006	?	?
XIII	14,090	?	?
XIV	14,384	5.5 (XX)	4.5[d]
XV	14,439	3.5	4.5 (XX)[c]
XVI	15,442	5.5 (XXII)	4.5
XVII	16,610	4.5	3.5
XVIII	17,346	7.5	6.5
XIX	17,567	6.5	5.5
XX	17,863	5.5	4.5
XXI	18,680	5.5	4.5 (XX)
XXII	18,961	5.5	4.5 (XX)

[a]Shenyavskaya et al. (142).
[b]System designations enclosed in parentheses share the indicated upper or lower electronic state.
[c]$\Delta\Omega$ assignments taken from Reference 142.
[d]Assignments suggested by Reference 144.
[e]Assignments taken from Reference 143b.

4.6.5 EuO

Experimental data on EuO are rather limited at the present time. Dulick et al. (145) derived the dissociation energy of EuO as 9.4.92 ± 0.1 eV from a thermodynamic study. They also discussed the electronic structure of EuO using the ligand field model. The $4f^7$–$4f^6\sigma$ (excited) superconfiguration energy separation was found to be 0.60 ± 0.1 eV from thermodynamic studies and 0.41 eV from ligand field theory, although Carette and Hocquet (135) have found a different value of 0.98 eV for this energy separation.

Gabelnick et al. (152) obtained the vibrational frequency (ω_e) of Ar-isolated EuO as 672 cm^{-1}. McDonald (153) obtained a ω_e of 688 cm^{-1} confirming that the ground state arises from a $4f^7$ superconfiguration. An r_e of 1.891 Å was estimated from the rotational constant of ^{151}EuO.

Table 4.58 Energies of Molecular States of PrO (in cm^{-1}) Arising Mainly from Atomic States of Pr(III)

	INDO/S-CI			LFT[b]	LST[c]
Ω	Energy	Composition[a]		Energy	Energy
3.5	0	$\vert H\ 4\ 3.5 > 53\%$	$\vert H\ 5\ 4.5 > 25\%$	0	0
		$\vert H\ 4\ 4.5 > 11\%$			
4.5	235	$\vert H\ 4\ 4.5 > 57\%$	$\vert H\ 5\ 5.5 > 35\%$	233	225
2.5	1,678	$\vert H\ 4\ 3.5 > 71\%$	$\vert H\ 4\ 4.5 > 14\%$	1,867	—
3.5	1,985	$\vert H\ 4\ 4.5 > 66\%$	$\vert H\ 4\ 3.5 > 18\%$	2,074	1,696
5.5	2,136	$\vert H\ 5\ 5.5 > 59\%$	$\vert H\ 6\ 6.5 > 39\%$	2,094	1,995
0.5	2,155	$\vert H\ 6\ 6.5 > 30\%$	$\vert P\ 2\ 2.5 > 23\%$	2,866	—
4.5	2,197	$\vert H\ 5\ 4.5 > 60\%$	$\vert H\ 6\ 5.5 > 30\%$	2,144	2,059
1.5	2,395	$\vert H\ 4\ 3.5 > 21\%$	$\vert H\ 5\ 5.5 > 18\%$	2,924	—
		$\vert P\ 2\ 2.5 > 14\%$	$\vert F\ 4\ 4.5 > 11\%$	—	—
0.5	2,519	$\vert P\ 2\ 1.5 > 17\%$	$\vert H\ 5\ 4.5 > 15\%$	3,107	—
		$\vert H\ 6\ 5.5 > 14\%$	—	—	—
1.5	2,774	$\vert H\ 4\ 3.5 > 40\%$	$\vert H\ 4\ 4.5 > 27\%$	3,125	—
2.5	2,835	$\vert H\ 4\ 4.5 > 62\%$	$\vert G\ 4\ 4.5 > 15\%$	3,099	—
		$\vert H\ 4\ 3.5 > 14\%$	—	—	—
4.5	3,618	$\vert H\ 5\ 5.5 > 47\%$	$\vert H\ 4\ 4.5 > 41\%$	3,767	3,285
3.5	3,753	$\vert H\ 5\ 4.5 > 48\%$	$\vert H\ 4\ 3.5 > 26\%$	3,824	3,357
		$\vert H\ 5\ 5.5 > 11\%$	—	—	—
6.5	4,033	$\vert H\ 6\ 6.5 > 100\%$	—	3,960	3,998
0.5	4,326	$\vert F\ 3\ 2.5 > 19\%$	$\vert H\ 5\ 5.5 > 17\%$	4,941	—
		$\vert F\ 3\ 3.5 > 17\%$	$\vert P\ 0\ 0.5 > 17\%$	—	—
		$\vert P\ 1\ 0.5 > 12\%$	—	—	—
5.5	4,329	$\vert H\ 6\ 5.5 > 97\%$	—	4,246	4,313
1.5	4,439	$\vert H\ 4\ 4.5 > 28\%$	$\vert H\ 6\ 6.5 > 21\%$	4,946	—
		$\vert P\ 1\ 1.5 > 10\%$	—	—	—
0.5	4,456	$\vert H\ 6\ 5.5 > 21\%$	$\vert H\ 4\ 3.5 > 19\%$	5,022	—
		$\vert H\ 4\ 4.5 > 15\%$	$\vert H\ 6\ 6.5 > 14\%$	—	—
		$\vert P\ 2\ 1.5 > 10\%$	—	—	—
2.5	4,508	$\vert H\ 5\ 5.5 > 59\%$	$\vert P\ 2\ 2.5 > 14\%$	4,853	—
3.5	4,568	$\vert H\ 5\ 5.5 > 72\%$	$\vert H\ 5\ 4.5 > 21\%$	4,786	4,195
1.5	4,652	$\vert H\ 5\ 4.5 > 32\%$	$\vert F\ 4\ 3.5 > 13\%$	5,070	—
		$\vert H\ 5\ 5.5 > 11\%$	—	—	—
2.5	4,811	$\vert H\ 5\ 4.5 > 81\%$	—	4,992	—
5.5	5,363	$\vert H\ 6\ 6.5 > 57\%$	$\vert H\ 5\ 5.5 > 41\%$	5,548	5,210
4.5	5,779	$\vert H\ 6\ 5.5 > 46\%$	$\vert H\ 5\ 4.5 > 28\%$	5,812	5,490
		$\vert H\ 6\ 6.5 > 18\%$	—	—	—
0.5	6,050	$\vert P\ 0\ 0.5 > 31\%$	$\vert H\ 4\ 4.5 > 15\%$	6,560	—
		$\vert F\ 2\ 1.5 > 13\%$	—	—	—
1.5	6,213	$\vert H\ 5\ 5.5 > 29\%$	$\vert P\ 1\ 1.5 > 23\%$	6,679	—
		$\vert H\ 6\ 6.5 > 12\%$	—	—	—
2.5	6,422	$\vert H\ 6\ 6.5 > 54\%$	$\vert F\ 2\ 2.5 > 12\%$	6,849	—
0.5	6,427	$\vert P\ 1\ 0.5 > 24\%$	$\vert H\ 5\ 4.5 > 17\%$	6,866	—
		$\vert F\ 2\ 2.5 > 12\%$	$\vert H\ 5\ 5.5 > 10\%$	—	—

Table 4.58 (*Continued*)

Ω	INDO/S-CI Energy	Composition[a]		LFT[b] Energy	LST[c] Energy
4.5	6,540	$\|H\ 6\ 6.5>$ 68%	$\|H\ 6\ 5.5>$ 23%	6,702	6,170
3.5	6,576	$\|H\ 6\ 6.5>$ 88%	—	6,848	6,377
1.5	6,737	$\|H\ 6\ 5.5>$ 29%	$\|H\ 5\ 4.5>$ 12%	7,069	—
		$\|P\ 2\ 1.5>$ 10%	—	—	—
3.5	7,044	$\|H\ 6\ 5.5>$ 90%	—	7,085	6,781
2.5	7,069	$\|H\ 6\ 5.5>$ 73%	—	7,233	—

[a]The atomic states are labeled according to $|L_f\, J_f\, J_a>$. Only states contributing more than 10% are listed.
[b]Reference 145.
[c]Reference 135.
Source: Reference 105.

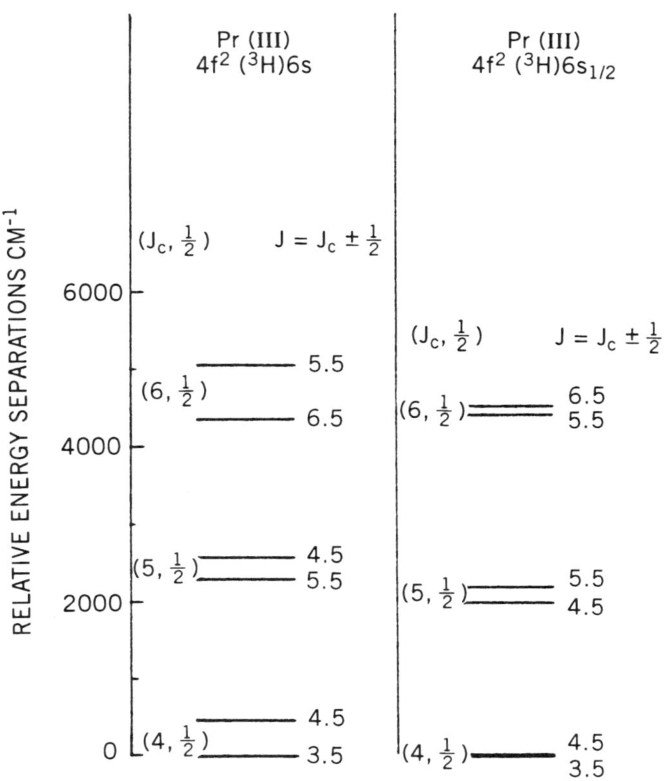

Figure 4.27 Energy levels of Pr^{2+} from $4f^2(^3H)6s$ and $4f^2(^3H)6p$. (Reproduced from Ref. 144.)

Table 4.59 Energies (in cm^{-1}) of the Observed Low-Lying States of SmO

State	Experimental	Calculated(1)[a]	Calculated(2)[b]
$X0^-$	0.0	0.0	0.0
(1)1	146.98	115.0	130.1
(1)2	566.77	498.0	549.0
$(2)0^+$	582.25	640.0	982.0
(2)1	879.30	873.0	1167.3
(1)3	1280.49	1189.0	1299.7
$(3)0^-$	1546.36	1359.0	2018.4
(2)2	1604.25	1517.0	1741.6
$(4)0^+$	1661.0	1603.0	2030.0
(3)1	1661.39	1472.0	2182.9
(3)2	2239.91	1998.0	2839.5
(4)1	2013.5	1963.0	2317.4
(1)4	2286.57	2226.0	2417.1
$(5)0^+$	2520.0	2603.0	2990.0
(5)1	2867.3	2718.0	3272.9

[a]From Reference 135.
[b]From Reference 145.
Source: Reproduced from Reference 151.

Dolg et al. (54, 55) studied the $^8\Sigma^-$ ground state and the first excited $^8\Sigma^-$ state of EuO using *ab initio* SCF and SDCI methods in conjunction with nonrelativistic and quasirelativistic effective core potentials. They obtained the spectroscopic constants of these two states and their potential-energy curves. They employed the Wood–Boring (WB) quasirelativistic ECPs as well as ECPs derived from nonrelativistic Hartree–Fock (HF) calculations. They used an energy-optimized GTO basis set of $(7s6p5d2f)/[5s4p3d2f]$ and $(12s11p9df)/[9s8p6d5f]$ types for Eu. For the oxygen atom, the $(9s5p)/(4s2p)$ basis set of Dunning was used together with one set of d functions. They also include the effect of unlinked quadruple clusters using the Davidson method. The spin–orbit effects were estimated using ECPs, which included spin–orbit operators. The spin–orbit effects were obtained using the double group CI method of Pitzer and Winter (111).

Table 4.60 shows the spectroscopic constants of two electronic states of EuO ($X^8\Sigma^-$ and $A^8\Sigma^-$). The theoretical constants were obtained by Dolg et al. (55) using CISD and CISD+Q levels of theories. The experimental values are from Dulick et al. (145), McDonald (153), and Gabelnick et al. (152). As seen from Table 4.60, experimental data on EuO are restricted to the X ground state, which arises from the $4f^7$ superconfiguration. The $A^8\Sigma^-$ state arises from the $4f^6\sigma^1$ superconfigurations and it appears that no experimental spectroscopic data exist on this state with the exception of an estimate of T_e. Evidently this estimate differs significantly from the CISD+Q value of only 0.04 eV and hence further studies are warranted on the $A^8\Sigma^-$ state.

Table 4.60 Spectroscopic Constants of Two Electronic States of EuO (CISD and CISD+Q Values from Reference 55)

	r_e, Å			D_e/T_e, eV			ω_e, cm^{-1}			μ_e, D
State	CISD	+Q	Exp.	CISD	+Q	Exp.	CISD	+Q	Exp.	CISD
$X^8\Sigma^{-a}$	1.912	1.919	~1.89	3.573	3.916	4.88	716	702	672	9.175
$X^8\Sigma^{-b}$	1.940	1.953	—	3.56	4.13	—	699	676	—	9.67
$A^8\Sigma^{-a}$	1.807	1.814	—	0.002	0.041	0.60	806	834	—	3.987

[a] $4f$ orbitals in the valence space.
[b] $4f$ orbitals in the core.

The spin–orbit contribution estimated from quasirelativistic CI is −0.77 eV for the ground state of EuO. Dolg et al. found that the $A^8\Sigma^-$ excited state of EuO, which is split into Ω = 0.5, 1.5, 2.5, and 3.5 states, is significantly contaminated by the $^8\Pi$ state. All Ω components of the $^8\Sigma^-$ state were found to be within 1400 cm^{-1}.

The Mulliken population analysis of the SCI wavefunction yields Eu with a positive charge of 1.04 with a population of $4f^{9.7.04}5d^{0.84}6s^{0.04}6p^{0.04}$, while the oxygen atom has a $O(2s^{1.9}2p^{9.5.14})$ population. Dolg et al. rationalize that the $X^8\Sigma^-$ ground state arises from a $4f^7$ supermultiplet, while the $A^8\Sigma^-$ state arises from the $4f^6\sigma$ supermultiplet. Dolg et al. found that 99.3% of the unpaired electronic population is in the 4f orbital in the $X^8\Sigma^-$ ground state. The doubly occupied σ orbital of EuO is composed of 21% Eu and 79% O orbitals. However, in the $A^8\Sigma^-$ state 99.7% of the six unpaired electrons are on Eu 4f orbitals, while the seventh unpaired electron is 98% on Eu (6s 80.6%, 6p 19.6.6%, 5d 3.0%).

4.6.6 GdO

Carette et al. (154) studied the spectra of GdO. Energy separations of six electronic states of GdO could be derived from their spectra. These authors have also used the ligand field model to derive the energy levels of eight electronic states of GdO. Dolg et al. (54, 55) studied the $^9\Sigma^-$ and $^7\Sigma^-$ states of GdO using *ab initio* SCF/CISD methods in conjunction with quasirelativistic ECPs. Kotzian et al. (106, 107) used the INDO/S-CI method to compute the energy separations of 45 Ω electronic states of GdO. We use the Kotzian et al. (106, 107) and the Dolg et al. (54, 55) works as the basis for discussion of the electronic states of GdO.

Table 4.61 shows the Dolg et al. (54, 55) CISD and CISD+Q results on GdO for two electronic states. As seen from Table 4.61, agreement between the computed SDCI, SDCI+Q properties with experiment is quite good with the exception of the D_e. It is, however, well known that the dissociation energies cannot be computed very accurately using the SDCI+Q level of theory as this method ignores electron correlation effects from other reference configurations. This method is also not a size-consistent method. Dolg et al. found that relativistic effects lead to a bond expansion of 0.016 Å and lowering of D_e by 0.44 eV. Relativistic effects also increase the dipole moment significantly as a consequence of bond expansion. The Mulliken population analysis of the SCF orbitals by Dolg et al. revealed that seven of the eight unpaired electrons in GdO occupy the 4f orbitals of Gd (99%). Van Zee et al. (155) studied the ESR spectra of GdO, which revealed that the singly occupied σ orbital is mainly the Gd 5d in conflict with the ligand field calculations of Carette et al., which revealed that the singly occupied σ orbital is 77% 6s, 21% 6p, and only 2% 5d. Dolg et al. from the Mulliken population analysis found that 81% of the singly occupied σ orbital is composed of Gd 6s. The charge distribution derived from Mulliken population generally supported the Gd^+O^- form rather than $Gd^{2+}O^{2-}$ used in the ligand field model.

Table 4.61 Spectroscopic Constants of Two Electronic States of GdO (Quasirelativistic CISD, +Q Results From Reference 55)

	r_e, Å			D_e/T_e, eV			ω_e, cm^{-1}			μ_e, D
State	CISD	+Q	Exp.	CISD	+Q	Exp.	CISD	+Q	Exp.	CISD
$^9\Sigma^-$	1.814	1.822	1.812	863	845	824	6.07	6.37	7.39	4.585
$^7\Sigma^-$	1.812	1.820	—	866	849	—	0.23	0.23	0.22	4.364

Table 4.62 INDO/S-CI Energy Separation and Experimental Separations of GdO[a]

	INDO/S-CI		Experiment		LFT
Ω	T_e	Composition	State	T_e	T_e
	0	$1^9\Sigma^-$ 100%	$X^9\Sigma^-$	0	0
	1,166	$1^7\Sigma^-$ 100%	$Y^7\Sigma^-$	1,840	1,999
2	8,505	$1^9\Delta$ 100%			
1	8,612	$1^9\Delta$ 99.7%			
0	8,723	$1^9\Delta$ 99.5%			
1	8,839	$1^9\Delta$ 99.3%			
2	8,959	$1^9\Delta$ 99.2%			11,594
3	9,084	$1^9\Delta$ 99.2%			
4	9,216	$1^9\Delta$ 99.3%			
5	9,355	$1^9\Delta$ 99.5%			
6	9,503	$1^9\Delta$ 100%			
5	14,177	$1^7\Delta$ 99.4%			
4	14,317	$1^7\Delta$ 99.2%			
3	14,449	$1^7\Delta$ 99.1%			
2	14,575	$1^7\Delta$ 99.1%			18,188
1	14,695	$1^7\Delta$ 99.3%			
0	14,810	$1^7\Delta$ 99.5%			
1	14,922	$1^7\Delta$ 99.7%			
3	17,194	$1^9\Pi$ 96.1%			
2	17,283	$1^9\Pi$ 89.5% $1^7\Pi$ 6.1%			
1	17,372	$1^9\Pi$ 81.8% $1^7\Pi$ 12.8%			
0^-	17,430	$1^9\Pi$ 74.5% $1^7\Pi$ 18.8%			
0^+	17,562	$1^9\Pi$ 77.9% $1^7\Pi$ 18.9%			17,647
1	17,630	$1^9\Pi$ 69.4% $1^7\Pi$ 25.9%	$^9\Pi_1$	18,181	
2	17,750	$1^9\Pi$ 58.2% $1^7\Pi$ 35.8%			
3	17,892	$1^9\Pi$ 45.0% $1^7\Pi$ 47.9%			
4	18,060	$1^9\Pi$ 27.5% $1^7\Pi$ 64.2%	$A^9\Pi_4$	18,472	
5	19,846	$1^9\Pi$ 99.9%			
4	20,092	$1^7\Pi$ 27.1% $1^9\Pi$ 71.9%			
3	20,294	$1^7\Pi$ 44.5% $1^9\Pi$ 54.2%			
2	20,471	$1^7\Pi$ 57.7% $1^9\Pi$ 41.0%			
1	20,630	$1^7\Pi$ 68.5% $1^9\Pi$ 30.4%			19,956
0^-	20,767	$1^7\Pi$ 76.5% $1^9\Pi$ 22.3%			
0^+	20,784	$1^7\Pi$ 79.3% $1^9\Pi$ 20.3%			
1	20,910	$1^7\Pi$ 86.2% $1^9\Pi$ 13.3%			
2	21,035	$1^7\Pi$ 93.5% $1^9\Pi$ 6.3%			
0^-	23,507	$2^9\Sigma^-$ 92.1%			
1	23,510	$2^9\Sigma^-$ 92.0%			
2	23,510	$2^9\Sigma^-$ 91.8% $1^7\Pi$ 5.7%	$B^9\Sigma^-$	21,647	26,317
3	23,533	$2^9\Sigma^-$ 91.3% $1^7\Pi$ 6.9%			

Table 4.62 (*Continued*)

	INDO/S-CI		Experiment		LFT
Ω	T_e	Composition	State	T_e	T_e
4	23,554	$2^9\Sigma^-$ 90.8% $1^7\Pi$ 8.7%			
3	26,039	$2^7\Sigma^-$ 96.8%			
2	26,047	$2^7\Sigma^-$ 96.6%	$B_1\Sigma^-$	22,261	25,394
1	26,051	$2^7\Sigma^-$ 96.4%			
0^+	26,053	$2^7\Sigma^-$ 96.4%			

[a]State energies T_e are given in cm^{-1}. The composition of states is given in reference to the nonrelativistic states that most resemble the pure configurations.

Source: Reproduced from Reference 107.

Table 4.62 shows the results of INDO/S-CI computation of Kotzian et al. (106, 107) together with known experimental data from Carette et al. (154). All INDO/S-CI energy separations were computed at single experimental bond length of 1.812 Å. Table 4.62 also shows the overall composition of the Ω states of GdO. As seen from Table 4.62, agreement of the energy levels predicted by the INDO/S-CI method with known experimental T_e values for the first three excited electronic states is good but somewhat less accurate for the $B^9\Sigma^-$ and $B_1^7\Sigma^-$ states. The ligand field model predicts reverse ordering for the $B^9\Sigma^-$ and $B_1^7\Sigma^-$ states compared to the INDO/S-CI model and experiment.

The composition of the Ω states of GdO in Table 4.62 evidently suggests that for the excited states, especially low Ω states arising from $^9\Pi$, $^7\Pi$, and $^9\Sigma^-$ the spin–orbit contaminations are quite significant. For example, the excited $\Omega = 1$ state is 69% $^9\Pi$ and 26% $^7\Pi$. Likewise, the excited $\Omega = 3$ state is 45% $^9\Pi$ and 48% $^7\Pi$, suggesting that these two states undergo an avoided crossing due to spin–orbit coupling.

According to Kotzian et al. (106, 107), the ground-state electronic configuration of GdO is $[1\sigma/O(2s)]^2[1\pi/O(2p)]^4[2\sigma/O(2p)]^2[1\phi/Gd(4f)]^2$-$[1\delta/Gd(4f)]^2[2\pi/Gd(4f)]^2[3\sigma/Gd(4f)]^1[4\sigma/Gd(6s)]^1$. Hence Kotzian et al.'s calculation also supports the theory that most of the singly occupied 4σ orbital consists of Gd 6*s*. The low-energy state of GdO arises from $4f^74\sigma\,(1^9\Sigma^-, 1^7\Sigma^-)$, $4f^72\delta\,(1^9\Delta, 1^7\Delta)$, $4f^73\pi\,(1^9\Pi, 1^7\Pi)$, and $4f^75\sigma\,(2^9\Sigma^-, 2^7\Sigma^-)$ configurations.

Kotzian et al. (106, 107) found evidence from strong transitions with $\Delta\Omega = 0$ between the states of the same multiplicities of $1\Sigma^-$ and $2\Sigma^-$ symmetry, since the computed oscillator strengths were about 0.35 (nonent) and 0.39 (septet). The strongest transitions ($\Delta\Omega = \pm 1$) were calculated between $1^9\Sigma^-$ and $1^{(9,7)}\Pi$ as well as $1^7\Sigma^- \rightarrow 1^{(9,7)}\Pi$. The band system between 16,004 and 17,669 cm^{-1} observed by Yadav et al. (156) could not be assigned by Kotzian et al. (106, 107), since at the present time there exists no experimental information on the electronic quantum numbers.

4.6.7 DyO

Linton et al. (157) studied DyO using the laser spectroscopic method. They used high-resolution wavelength selected excitation and resolved fluorescence spectroscopy to study the electronic, vibrational, and rotational structures. These studies have provided information on five low-lying electronic states and three high-lying states of DyO. The Ω quantum numbers of these states were assigned. The ligand field model was used to gain insight into the electronic structure of DyO, which shows that the observed states can be described as arising from $Dy^{2+}O^{2-}$, in which Dy^{2+} has $4f^9(^6H)6s$ electronic configuration. The ground state of DyO was found to be a $\Omega = 8$ electronic state arising from this configuration.

Table 4.63 shows the spectroscopic constants of states observed by Linton et al. (157), who have also compared their B_0 constants with those of Kaledin and Shenyavskaya (158). Table 4.63 also includes ligand field theoretical values (numbers under column theory). As evidenced from Table 4.63, the ground state of DyO is a $X_{\Omega=8}$ state. The $A_{\Omega=7}$ state is only 770 cm^{-1} above the X state. The $B_{\Omega=6}$, $C_{\Omega=7}$, and $D_{\Omega=6}$ states are also very close in energy. The ligand field model seems to fit the observed spectra quite well (Table 4.63).

Assignment of the observed states is also consistent with the ligand field model, which predicts one $\Omega = 8$ ground state and two $\Omega = 7$, $\Omega = 6$ states close to the ground state all arising from the $4f^9(^6H_{7/5})6s$ configuration of Dy^{2+}. There are, of course, several other electronic state arising from the $4f^96s$ configuration, which are yet to be observed. It appears that such experiments are in progress.

Table 4.63 Energies and Constants (cm^{-1}) of Observed States of DyO[a]

	T_0		B_0		
State	Exp.	Theory	Footnote *b*	Footnote *c*	$\Delta G_{1/2}$
[19.0]9	18,994	—	—	0.3570	—
[18.5]9	18,502	—	—	0.3562	—
[18.3]8	18,300	—	—	—	—
[18.2]6	18,203	—	—	—	—
[18.0]9	18,017	—	—	0.3505	780
[17.1]7	17,070	—	—	—	—
*D*6	2,420	2,382	—	—	—
*C*7	1,683	1,428	—	—	—
*B*6	1,630	1,641	0.35967	—	830
*A*7	770	845	0.35862	—	828.5
*X*8	0	0	0.35867	0.3588	842

[a]Some of the upper states without alphabetic labels designate that these states were not observed in Reference 157.
[b]From Reference 157.
[c]From Reference 158.

Source: Reproduced from Reference 157.

4.6.8 HoO

There are three experimental studies on HoO by Kaledin and Shenyavskaya (158), Liu et al. (159), and Linton and Liu (160). The paper by Liu et al. (159), which deals primarily with the rotational and hyperfine structure in holmium oxide, provides a comprehensive table of known electronic states of HoO. They have compared the observed and computed hyperfine constants of X_1 8.5, X_2 9.7.5, and W_2 9.6.5 Ω states of HoO.

Liu et al. (159) employed laser spectroscopic techniques to study electronic vibrational, rotational, and hyperfine structure of HoO. High-resolution excitation spectra of HoO were obtained, revealing the existence of four low-lying states and four higher electronic states of HoO. These authors have also assigned the Ω values of these electronic states. The ground state of HoO was found to be a X_1 8.5 state. The ligand field model suggests an ionic $Ho^{2+}O^{2-}$ description of these states arising from Ho^{2+} configuration of $4f^{10}(^5I)6s$. The ligand field model fits very well the observed spectra and the assignment of the observed electronic states. Liu et al. (159) also estimated the relative contribution of the 4f and 6s electrons to the hyperfine structure.

Table 4.64 shows the spectroscopic constants of HoO deduced by Liu et al. (159) together with the constants obtained by Kaledin and Shenyavskaya (158). As seen from Table 4.64, the ground state of HoO is assigned to the X_1 8.5 state. The X_2 state is only 609 cm^{-1} above the ground state. The W_1 and W_2 states are relatively low-lying, while all other upper states have $T_0 > 17,500$ cm^{-1}. The observed electronic states fit an ionic $Ho^{2+}O^{2-}$ description of HoO, in which the Ho^{2+} ion arises from the $4f^{10}(^5I)6s$ electronic configuration. The ground-state configuration of HoO can therefore be best described as $4f^{10}\sigma$, where the σ orbital is composed of Ho 6s and O 2p orbitals. Analysis of the hyperfine pattern is indicated that the 4f electrons contribute to a greater extent to the hyperfine pattern compared to the 6s by a factor of 2–9.4. This is in contrast with PrO, for which the f and s contributions were found to be similar. Consequently, Liu et al. (159) concluded that the 4f-electron contribution to hyperfine structure increases as the number of 4f electrons (or holes if 4f is less than half-full) increases.

4.6.9 TmO

Kotzian and Rösch (104, 105) studied the low-lying electronic states of TmO with energy separations of less than 13,528 cm^{-1} relative to the ground state. They have also compared the results of their INDO/S-CI technique with the ligand field theoretical study of Dulick et al. (145) as well as Carette and Hocquet (135). The tables of Kotzin and Rösch (104, 105) are used here as a basis for the discussion on the electronic states of TmO.

Table 4.65 shows the energies of the molecular states of TmO obtained by Kotzian and Rösch (104, 105) together with two ligand field model energy separations. We note that the INDO/S-CI method yields a different ground state of $\Omega = 3.5$, while the LFT models predict a $\Omega = 0.5$ ground state. Kotzian

Table 4.64 Spectroscopic Constants (cm^{-1}) for the Observed States of HoO

		B_0			Hyperfine Width	
State	T_0	Footnote *a*	Footnote *b*	$\Delta G_{1/2}$	(Low J)	d^c
[22.4]8.5[d]	22,371.56	—	0.3448	—	—	—
[20.2]8.5[d]	20,168.63	—	0.3508	—	—	—
[19.4]9.5[d]	19,386.01	—	0.3565	—	—	—
[19.0]7.5	19,010	—	—	—	1.0	0.15
[18.4]6.5	18,361.53	0.3485	0.3485	—	1.1	0.19
[18.3]8.5	18,300	—	—	—	—	—
[17.6]7.5	17,552.72	0.3559	—	787	1.2	0.18
W_2 6.5	1861.1	0.3502	—	—	0.8	0.14
W_1 7.5	1130	—	—	841	2.0	0.28
X_2 7.5	608.78	0.3560	0.3561	—	1.4	0.19
X_1 8.5	0	0.3575	0.3573	841.4	2.4	0.35

Band origins: $[17.6]7.5 - X_1 8.5(0,0)$: 17,552.72
$(0,1)$: 16,711.34
$[17.6]7.5 - X_2 7.5(0,0)$: 16,943.94
$[18.4]6.5 - W_2 6.5(0,0)$: 16,500,51

[a]Values of Reference 159.
[b]Values of Reference 158.
[c]For [17.6]7.5 and X_1 8.5, *d* was calculated from combination differences. For all other states, *d* was estimated from hyperfine width. Both hyperfine width and *d* are very rough approximation.
[d]Not observed in the experiments of Reference 159.

Source: Reproduced from Reference 159.

and Rösch (104, 105) have argued that their INDO/S-CI treatment is consistent with the nonrelativistic $^4\Delta$ ground state for TmO, which would yield a Ω state of 3.5 when projected. Overall, as seen from Table 4.65, the ligand field and INDO/S-CI methods agree reasonably well, when one considers the fact that the $\Omega = 1.5$ state is only ~100 cm^{-1} above the 0.5 state in the LFT methods.

Table 4.65 Energies of Molecular States of TmO (in cm^{-1}) Arising Mainly from Atomic States of H and F of Tm(III)

	INDO/S-CI				
Ω	Energy[a]	Composition		LFT Energy[b]	LST Energy[c]
3.5	0	$\|H\ 6\ 6.5 > 90\%$	$\|H\ 6\ 5.5 > 9\%$	905	731
2.5	24	$\|H\ 6\ 6.5 > 95\%$	—	413	323
1.5	187	$\|H\ 6\ 6.5 > 73\%$	$\|H\ 6\ 5.5 > 25\%$	128	97
2.5	320	$\|H\ 6\ 5.5 > 95\%$	—	1,375	951
4.5	416	$\|H\ 6\ 6.5 > 87\%$	$\|H\ 6\ 6.5 > 11\%$	1,649	1,381

Table 4.65 (*Continued*)

	INDO/S-CI			LFT	LST
Ω	Energy[a]	Composition		Energy[b]	Energy[c]
0.5	512	$\vert H\ 6\ 6.5>$ 85%	$\vert H\ 6\ 6.5>$ 11%	0	0
1.5	629	$\vert H\ 6\ 5.5>$ 72%	$\vert H\ 6\ 6.5>$ 24%	881	536
3.5	668	$\vert H\ 6\ 5.5>$ 89%	$\vert H\ 6\ 6.5>$ 9%	2,150	1,625
0.5	859	$\vert H\ 6\ 5.5>$ 85%	$\vert H\ 6\ 6.5>$ 10%	643	336
5.5	1,671	$\vert H\ 6\ 6.5>$ 95%	—	2,653	2,331
4.5	1,914	$\vert H\ 6\ 5.5>$ 88%	$\vert H\ 6\ 6.5>$ 11%	3,194	2,602
6.5	2,461	$\vert H\ 6\ 6.5>$ 100%	—	3,899	3,583
5.5	2,655	$\vert H\ 6\ 5.5>$ 95%	—	4,485	3,881
3.5	5,201	$\vert F\ 4\ 4.5>$ 83%	—	6,982	—
2.5	5,261	$\vert F\ 4\ 3.5>$ 68%	$\vert F\ 4\ 4.5>$ 23%	7,115	—
2.5	6,289	$\vert F\ 4\ 4.5>$ 68%	$\vert F\ 4\ 3.5>$ 22%	7,396	—
1.5	6,355	$\vert F\ 4\ 3.5>$ 47%	$\vert F\ 4\ 4.5>$ 45%	7,246	—
0.5	6,387	$\vert F\ 4\ 4.5>$ 44%	$\vert F\ 4\ 3.5>$ 12%	7,313	—
—	—	—	$\vert P\ 2\ 2.5>$ 9%	—	—
4.5	6,457	$\vert F\ 4\ 4.5>$ 97%	—	6,870	—
3.5	6,630	$\vert F\ 4\ 3.5>$ 89%	—	7,325	—
0.5	6,637	$\vert H\ 4\ 3.5>$ 54%	$\vert F\ 4\ 4.5>$ 19%	7,588	—
1.5	6,644	$\vert F\ 4\ 3.5>$ 38%	$\vert F\ 4\ 4.5>$ 36%	7,514	—
1.5	7,911	$\vert H\ 5\ 4.5>$ 46%	$\vert F\ 5\ 5.5>$ 44%	8,682	7,771
2.5	7,928	$\vert H\ 5\ 5.5>$ 74%	$\vert H\ 5\ 4.5>$ 16%	9,147	8,093
1.5	8,039	$\vert H\ 5\ 5.5>$ 49%	$\vert H\ 5\ 4.5>$ 46%	9,161	8,119
0.5	8,040	$\vert H\ 5\ 5.5>$ 90%	—	8,502	7,669
0.5	8,065	$\vert H\ 5\ 4.5>$ 90%	—	8,682	7,790
3.5	8,318	$\vert H\ 5\ 5.5>$ 78%	$\vert H\ 5\ 4.5>$ 12%	9,858	8,691
2.5	8,382	$\vert H\ 5\ 4.5>$ 74%	$\vert H\ 5\ 5.5>$ 16%	9,894	8,722
4.5	8,892	$\vert H\ 5\ 5.5>$ 86%	$\vert H\ 5\ 4.5>$ 10%	—	9,648
3.5	8,930	$\vert H\ 5\ 4.5>$ 83%	$\vert H\ 5\ 5.5>$ 13%	—	9,678
5.5	10,038	$\vert H\ 5\ 5.5>$ 100%	—	—	11,010
4.5	10,070	$\vert H\ 5\ 4.5>$ 89%	$\vert H\ 5\ 5.5>$ 11%	—	11,034
1.5	11,636	$\vert H\ 4\ 3.5>$ 67%	$\vert H\ 4\ 4.5>$ 23%	—	14,048
2.5	11,728	$\vert H\ 4\ 4.5>$ 72%	$\vert H\ 4\ 3.5>$ 18%	—	14,649
0.5	12,076	$\vert H\ 4\ 3.5>$ 48%	$\vert H\ 4\ 4.5>$ 29%	—	13,798
1.5	12,100	$\vert H\ 4\ 4.5>$ 59%	$\vert H\ 4\ 3.5>$ 17%	—	14,546
0.5	12,688	$\vert H\ 4\ 4.5>$ 47%	$\vert H\ 4\ 3.5>$ 28%	—	14,094
3.5	12,849	$\vert H\ 4\ 4.5>$ 84%	$\vert H\ 4\ 3.5>$ 12%	—	15,752
2.5	12,859	$\vert H\ 4\ 3.5>$ 78%	$\vert H\ 4\ 4.5>$ 19%	—	15,553
3.5	12,869	$\vert H\ 4\ 3.5>$ 83%	$\vert H\ 4\ 4.5>$ 13%	—	17,122
4.5	13,528	$\vert H\ 4\ 4.5>$ 97%	—	—	17,385

[a]The atomic states are labeled according to $\vert L_f J_f J_a >$. Only states contributing more than 9% are listed.

[b]Reference 145.

[c]Reference 135.

Source: Reference 105.

Kotzian and Rösch did other model calculations by changing the INDO bonding parameter β_f and elongating the bond distance. Both changes were found to yield a linear dependence of k on the corresponding energies. Kotzian and Rösch concluded that the rather small discrepancy in energies and different ground states in the two methods should probably be due to the crude point-charge approximation used in the LFT model on the Tm 6s orbital. However, since at present there are no experimental data on TmO, it is hard to conclude as to which model predicts the correct ground state and energies for TmO. It would certainly be interesting to obtain the spectra of TmO.

4.6.10 YbO

McDonald et al. (161) have studied YbO using a laser spectroscopic method. Linton et al. (141, 162) have studied YbO before and analyzed the $O^+ – {}^1\Sigma^+$ transitions of YbO. McDonald et al.'s work (161) seems to provide evidence of linkage of the Yb^{2+} $f^{13}s$ and f^{14} configurations. McDonald et al. (161) find that the $X^1\Sigma^+$ ground state of YbO arises for the Yb^{2+} $4f^{14}$ configuration, while five excited states with $\Omega = 0, 1, 2$; $\Omega = 1$; and $\Omega = 3$ arise from the $4f^{13}6s$ configuration of Yb^{2+}. This is not quite consistent with the ground-state calculations by Dolg et al. (52, 53).

In a subsequent study Dolg et al. (57) considered YbO together with YbH and YbF using SCF, CASSCF, large-scale CPF, MRCI as well as CIPSI calculations employing quasirelativistic effective core potentials (ECP). Dolg et al. (57) use both nonrelativistic and quasirelativistic energy-adjusted *ab initio* pseudopotentials for Yb with extended Gaussian basis sets. They computed the spectroscopic constants of 13 low-lying electronic states of YbO. Spin–orbit effects were incorporated using energy-adjusted *ab initio* spin–orbit operators. They find 25 Ω states for YbO. In particular, their computed spectroscopic constants (r_e = 1.895 Å, D_0 = 2.47 eV, ω_e = 653 cm^{-1}) are not in complete agreement with the constants deduced by McDonald et al. (161) for the $4f^{14}\sigma^2\sigma^2\pi^4\ {}^1\Sigma^+(0^+)$ state (R_e = 1.807 Å, D_0 = 9.4.29 eV, ω_e = 499 cm^{-1}).

It should also be mentioned that at the best level of theory Dolg et al. (57) obtain the $\Omega = 0^-$ state arising from the $4f^{13}\sigma^2\sigma^2\pi^4\sigma$ configuration 0.93 eV lower than the experimentally assigned $\Omega = 0^+$ ground state arising from the $4f^{14}\sigma^2\sigma^2\pi^4$ configuration. Dolg et al. (57) note that differential correlation effects and relativistic effects are quite significant for YbO.

Dolg et al. (57) also note that their computed r_e = 1.895 Å for the ${}^1\Sigma^+$ is too long compared to a value of 1.807 Å derived by Linton et al. (141, 162), while the computed r_e for the ${}^3\Sigma^+$ state (1.768 Å) is closer to the experimentally deduced value. The major disagreement between the work of Dolg et al. (57) and McDonald et al. (161) is probably in the energy separation of $4f^{14}$ and $4f^{13}\sigma$ configurations. Whereas McDonald et al. (161) deduced this energy separation as 910 cm^{-1} (with $4f^{14}$ lower), Dolg et al. (57) computed an energy separation of 6200 cm^{-1}, with $4f^{13}\sigma$ lower in the absence of spin–orbit coupling.

As noted from Dolg et al.'s work, the spin–orbit coupling between the 0^+

states causes further complication. When spin–orbit coupling is included, the ground state appears to change from $4f^{13}\sigma$ to $4f^{14}$ in the bonding region. This demonstrates strong coupling of the 0^+ states arising from these two electronic configurations. This could, in turn, lead to different spectroscopic constants in the presence of the spin–orbit operator. Thus YbO is certainly an interesting candidate from the standpoint of relativistic effects and electronic correlation effects. The readers are referred to Dolg et al. (57) for further details.

Table 4.66 shows the spectroscopic constants of YbO as derived from experimental spectra as well as three ligand field models. The $X^1\Sigma^+$ state arising from Yb^{2+} $(4f^{14})$ is assigned as the ground state of YbO in Table 4.66. The $4f^{13}6s$ configuration of Yb^{2+} yields several Ω states, among which the $\Omega = 0^-$ and $\Omega = 1$ states are very close to each other. The $\Omega = 0^+$ state from the $4f^{13}6s$ configuration of Yb^{2+} has not yet been observed. The $\Omega = 2, 1, 3$ states are higher than the $\Omega = 0^+$ state of YbO. The state at 4637 cm^{-1} is tentatively assigned to the $4f^{13}5d\,(0^+)$ state by McDonald et al. (161).

The assignment by McDonald et al. (161) of the observed states is based on

Table 4.66 Spectroscopic Constants of Low-Lying YbO States (in cm^{-1})

State	Exp.[a]	LFT Fit[b]	LFT Calc.[c]	LFT Calc.[d]
(J_a, Ω)				
$f^{14}(0, 0^+)$	0	−11	276	−11
$f^{13}s(4, 0^-)$	910	910[c]	910[c]	910[c]
$f^{13}s(4, 1)$	1,015	1,125	997	1,013
$f^{13}s(3, 0^+)$	...	1,628	1,262	1,289
$f^{13}s(4, 2)$	2,408	2,330	1,697	1,646
$f^{13}s(3, 1)$	2,702	2,688	1,895	1,807
$f^{13}s(4, 3)$	4,287	4,319	2,950	2,840
$f^{13}d\,?(?, 0^+)$	4,637	...	...	...
$f^{13}s(3, 2)$	...	4,818	3,197	3,021
$f^{13}s(4, 4)$	...	6,868	4,597	4,676
$f^{13}s(3, 3)$	...	7,559	4,894	4,467
$f^{13}s(2, 0^-)$	...	11,341	...	11,208
$f^{13}s(3, 1)$	...	11,675	...	11,415
$f^{13}s(3, 0^+)$	...	11,981	...	11,562
$f^{13}s(2, 1)$	...	13,807	...	12,559
$f^{13}s(3, 2)$	...	13,973	...	12,665
$f^{13}s(2, 2)$	...	16,882	...	14,508
$f^{13}s(3, 3)$	...	17,305	...	14,697

[a]Experimental values from McDonald et al. (161).
[b]Fit of five experimental $f^{13}s$ energies, varying only G_3 and B_0^2 in the ligand field effective Hamiltonian by McDonald et al. (161).
[c]Ligand field calculation with no adjustable parameters. [See Dulick et al. (145).]
[d]From Carette and Hocquet (135).
[e]Calculated energies shifted to coincide with lowest energy observed $f^{13}s$ state by McDonald et al. (161).

the fact that the vibrational frequencies associated with states arising from $4f^{14}$ and $4f^{13}6s$ configurations are expected to be very different near the $Yb^{2+}O^{2-}$ atomic-ion-in molecule limit. They argued that the $4f^{14}\,{}^1\Sigma^+$ will have a ω_e close to that of BaO whose ground state ω_e is 609.5.7 cm^{-1}. Since the ground state ω_e of YbO was deduced as 683 cm^{-1} by McDonald et al. (161), they suggest the assignment of the ground state as $X^1\Sigma^+(4f^{14})$. The $4f^{13}6s$ states have larger ω_e values. It should be noted [as done by McDonald et al. (161)] that the vibrational frequencies were deduced by Yb isotope shift and thus may not be reliable for multiply perturbed upper states arising from Yb^+O^- states. McDonald et al. (161) noted that the electronic states arising from f^Ns configurations of other species such as CeO, PrO, SnO, DyO, and HoO have larger ω_e values compared to the f^{N+1} configuration. The LFT calculations also yield results in reasonable agreement with the experimentally deduced energies. The readers are referred to the latest paper by Dolg et al. (57) for the information concerning the ground state of YbO and the effect of spin–orbit coupling. The spin–orbit coupling of $4f^{14}$, $4f^{13}6s$ states can certainly lead to different spectroscopic constants in the bonding region.

4.7 ELECTRONIC STRUCTURE OF SELECTED LANTHANIDE AND ACTINIDE POLYATOMICS

There are numerous studies on polyatomics containing lanthanides and actinides in the literature. Comprehensive review of such species is beyond the scope of this chapter. Nevertheless, a review of theoretical and related experimental studies on polyatomics containing lanthanides and actinides can shed significant light on the nature of chemical bonds in such systems. In this connection, we point out two reviews on the topic that provide a comprehensive summary of the literature on the topic. A review by Pyykkö (65) on developments in the theory of f-element molecules, especially polyatomics, should be consulted. This review of Pyykkö contains the references and a table of relativistic calculations on molecules containing f elements. To summarize, in addition to the molecules already reviewed, Pyykkö lists polyatomic hydrides such as CeH_4, ThH_4, UH_6, MH^+, MH_2 (M = Yb,No) UO_2^{2+}, AnX_n (An = U–Pu, X = F–I for several n), AnO_n^{m-} (for several n, m, An = actinide), $An(Cn)_n$ (Cn = C_nH_n), $An(BH_4)_4$, and other inorganic multicenter complexes. The level of theory included extended Hückel, Dirac–Slater discrete variation, Dirac–Slater multiple scattering, quasirelativistic multiple scattering, pseudopotential, perturbative Hartree–Fock, relativistic extended Hückel, CNDO, and INDO. The readers are referred to the literature cited by Pyykkö (65) for the actual details on these computations.

The second review is due to Pepper and Bursten (47), which focus on the electronic structure of actinide-containing molecules and the details of calculations on actinide hydrides, actinide halides, actinide oxides, cydopentadienyl–actinide complexes, actinocenes, metal–metal bonding in actinide systems, and miscellaneous other actinide systems. This review also consists of descriptions of theoretical techniques employed to study the actinide-containing molecules.

We have identified a few lanthanide- and actinide-containing polyatomics that we think provide a basis to gain insight into relativistic effects and bonding in some of these systems. We select only those lanthanide- or actinide-containing polyatomics that are topical and interesting, or those that provide a basis to gain significant insight into the nature of bonding in such systems.

4.7.1 Polyatomic Oxides and Complexes

Pyykkö and Zhao (72) investigated the uranium–oxygen clusters such as UO_2^{m+}, UO_2^{2-}, UO_6^{2-}, UO_6^{6-}, and $(UOU^+)_2$ using quasirelativistic pseudopotential studies at Hartree–Fock or MP2 level of theory. They found that the U–O distances for OUO^{2+}, UO_4^{2-}, and cubic UO_6^{6-} are comparable to the crystal data, and that the theoretical calculations of the large variation in U–O_{ax} distance as a function of U–O_{eq} distance for UO_6^{6-} (D_{4h}) followed the experimental curve. These variations were interpreted by Pyykkö and Zhao as a frozen soft e_g mode. Apparently the large uranyl bond length is a classic problem as illustrated in the handbook of Dennig (163).

Pyykkö and Zhao (72) used the quasirelativistic effective core potential of Hay et al. (33, 34) and the Gaussian 90 codes to study these uranium–oxygen clusters. Table 4.67 shows the results of the Pyykkö–Zhao calculations for various UO_n^{m+} and UN_2 clusters. As seen from Table 4.67, the agreement between the Pyykkö–Zhao calculation with the experiment is quite reasonable for the vibrational–vibrational frequencies.

Pyykkö and Jové (71) have made relativistic extended Hückel calculations to interpret ^{237}Np nuclear quadrupole coupling and isomer shifts in neptunyl compounds. They also reviewed actinide compounds containing multiple bonds. They noted that the diffuse weakly bound 7*s* and 7*p* orbitals are relatively unimportant in the actinide bonding. The 6*d* and 5*f* orbitals appear to be more important for the covalent bonding of the early actinides (Th–Am) and the 6*p* semicore orbital. The possibility of the unusual $0 = An = O^{n+}$ and An = Pa–Am bonding over a large domain by varying the equilateral ligands was discussed by Pyykkö and Jové.

Multiple bonding in actinides is indeed quite interesting. Katz et al. (165) as well as Gmelin (166) list examples of such compounds. For example, the $O = U = O^{2+}$ molecule with a U–O bond length of 1.50–2.08 Å is known. Likewise, Wadt (42) found O = Th = O multiple bonding in thorium dioxide, while Marian et al. (167) and Heaven et al. (168) found multiple bonding in Th≡O and U≡O, respectively.

The other interesting species, although not polyatomic, is the uranium dimer U_2, which exhibits interesting metal–metal bonding. Pepper and Bursten (46) have studied the U_2 molecule. Although actinide–actinide bonding has been a topic of several synthetic and computational studies, as noted by Pepper and Bursten (46), no discrete molecules containing a direct bond between two actinide atoms has been isolated, although U_2 appears to have been observed

Table 4.67 Calculated Relativistic U–O Bond Lengths *R* (in pm) and Breathing Frequencies $\nu_1(a_1)$ (in cm^{-1})

Species	Method	R	ν_1
UO_2^{2+}	HF	162.5	1,216
	MP2	173.2	922
	HF	163	—
	HFS	170.0	—
	Exp.	—	~830
UO_2^+	HF	171.2	1,031
	MP2	177.0	1,074
	Exp.	—	808
UO_2	HF	181.8	—
	MP2	179.7	—
	HFS	187.3	—
	HFS	183.4	—
	Exp.	—	765
UO_4^{2-}	HF	187.3	872
	Exp.	199	—
UO_6^{6-}	HF	214.8	592
	Exp.	208–209	—
	Exp.	—	797
UN_2	HF	181.8	783
	MP2	191.1	816

Source: Reproduced from Reference 72.

in the gas phase by Gingerich (169). Pepper and Bursten (46), in their U_2 relativistic ECP-CASSCF computation, reviewed the previous literature. They have identified two distinct sets of states with potential energy minima. The first set of states ($^{13}A_g$) have approximately 3.0 Å r_e values; these states essentially comprise atomic 5*f* and bonding 6*d* and 7*s* orbitals. The second set of states (^5A_g) has relatively shorter r_e (2.2. Å); the bonds in these states are composed of 5*f*, 6*d*, and 7*s* orbitals of the two atoms. There are other studies on An–An bonding. Ortiz (170) and Makhyoun et al. (171) have investigated $Th(\eta^5\text{-}Cp)_2\text{-}(\mu\text{-}PH_2)_2Ni(Co)_2$ complex using EHT and QR-MS levels of theories, respectively. Ortiz (170) reported in the Th–Ni overlap populations as a function of bond distance. The overlaps were negative in all cases, however, the short bond distance structure exhibited the least negative overlap. Hay et al. (35) reported ECP calculations on similar heterobimetallic systems.

Hay (172) studied the high-coordinate metal complexes containing lanthanide(III) ions using molecular mechanics calculations. He applied the molecular mechanics technique to 58 known structures of 8- to 12-coordinate aqua- and nitratolanthanide(III) complexes. He also presented various required parameters for the computation of these species, and studied complexes such as $M(OH_2)_9X_3$ and $M(OH_2)_9(BrO_3)_3$ (where X = OTf, $EtSO_4^-$) for M = La,Nd,Gd,Ho,Lu in the former case and M = Pr,Nd,Sm,Yb in the latter case.

Meyer and Hoffmann (173) studied the layered halides Pr_2X_5 (X = Br,I). These authors achieved the unusual composition of M_2X_5 for lanthanide halides (M = La,Ce,Pr; X = Br,I). The Pr_2X_5 layered halide was found to be weakly semiconducting (Pr_2Br_5 3800 Ω cm; Pr_2I_5 1600 Ω cm). Magnetic studies revealed transitions to antiferromagnetic ordering for the paramagnetic phases of Pr_2Br_5 and Pr_2I_5 below T_N (T_N = 50, 37 K). Meyer and Hoffmann (173) made extended Hückel calculations of the bond structure of these compounds, which revealed two narrow low-lying half-occupied *d* bonds with two electrons per Pr_4X_{10} unit cell. Pr_2Br_5 crystallizes monoclinic, forming a layered structure. The metal atoms form zig-zag layers derived from a hexagonal layer.

Kotzian et al. (107) have applied the INDO/S-CI method including spin–orbit coupling based on Rumer spin functions to the hydrated cerium ion. They described an efficient technique to obtain excitation energies, transition moments, and Mulliken populations based on the Rumer-adapted functions method. They obtained the energies of the ground and excited states of $[CeCH_2O)_9]^{3+}$. The $4f \rightarrow 5d$ transitions were found to be split by 10,000 cm^{-1}, due to spin–orbit coupling and ligand field interaction. Kotzian and Rösch are also considering application of this method to other lanthanide hydride ions $[Ln(H_2O)_n]^{3+}$ (n = 8,9).

Kotzian et al. (107) considered several structures of D_{3h}, D_3, C_{3h}, and C_3 symmetries. They noted that the agreement with the experimental spectrum could be obtained for a geometry with low symmetry, (i.e., with the threefold axis removed), although for simplicity, they considered the $[Ce(H_2O)_9]^{3+}$ complex as if it had D_{3h} point-group symmetry. The INDO/S-CI method confirmed the previous Okada et al. (174) assignment of bands in the 37,000–51,000-cm^{-1} region to $4f \rightarrow 5d$ transition. The spin–orbit splitting was computed as 2200 cm^{-1} for the $4f$ state. The Mulliken population analysis revealed a $4f$ occupation of 1.01 electron. It was noted by Kotzian et al. (107) that, because of the compactness of the $4f$ shell, the $4f$ valence electron of Ce^{3+} does not contribute to metal–ligand bonding. This is reminiscent of the $M^{2+}O^{2-}$ bonding discussed before for the diatomic lanthanide oxide. The bonding of $[Ce(H_2O)_9]^{3+}$ is facilitated by donation of lone-pair electrons from H_2O to Ce^{3+}. The σ-type lone pair has the largest overlap with Ce ($6s$, $6p$, and $5d_{z^2}$), and the π-type bonding is composed of H_2O $1b_2$ overlap with Ce($5d_{yz}$) orbital.

4.7.2 Uranocene and Actinocenes

Streitwieser and Müller-Westerhoff (175) have prepared the uranocene–$U(C_8H_8)_2$ sandwich complex. They speculated that the U($5f$) orbitals might play an important role in the metal–ligand bonds in uranocene. The reactivity properties of uranocene have been studied quite extensively by Streitwieser and co-workers (176, 177). Other actinocenes and substituted uranocenes have been experimentally synthesized as well Streitwieser et al. (176) and Karraker (178).

Many actinocenes exhibit beautiful colors with strong absorption in the visible region of the spectrum. Uranocene is green, $NP(C_8H_8)_2$ is yellow to blood red depending on concentration, and $Pu(C_8H_8)_2$ is cherry red (177).

Uranocene, like other actinocenes, is readily prepared by the following reactions:

$$2K + C_8H_8 \xrightarrow{\text{THF}} 2K^+ + C_8H_8^{2-}$$

$$M^{6+} + 2C_8H_8^{2-} \xrightarrow{\text{THF}} M(C_8H_8)_2$$

where M is the actinide. For example, $Np(C_8H_8)_2$ is formed by reacting $NpCl_4$ with $C_8H_8^{2-}$ generated by potassium in tetrahydrofuran (THF) solution. The resulting sandwich complex is crystallized. Uranocene is green and sublimes at 180°C and is very stable in oxygen-free surroundings (175). Several spectroscopic studies (Raman, ^{13}C NMR, H NMR, visible spectra, and photoelectron spectra) of these compounds have been performed.

Chang and Pitzer (9) computed the electronic states of uranocene using *ab initio* RECP, including spin–orbit coupling. Chang and Pitzer employed Gaussian functions, which resulted in a basis set of 208 generally contracted functions. Ninety-four valence electrons of uranocene were explicitly treated. The molecular orbitals were obtained using the SCF method, while virtual orbitals were improved by the improved virtual orbital computation for a better

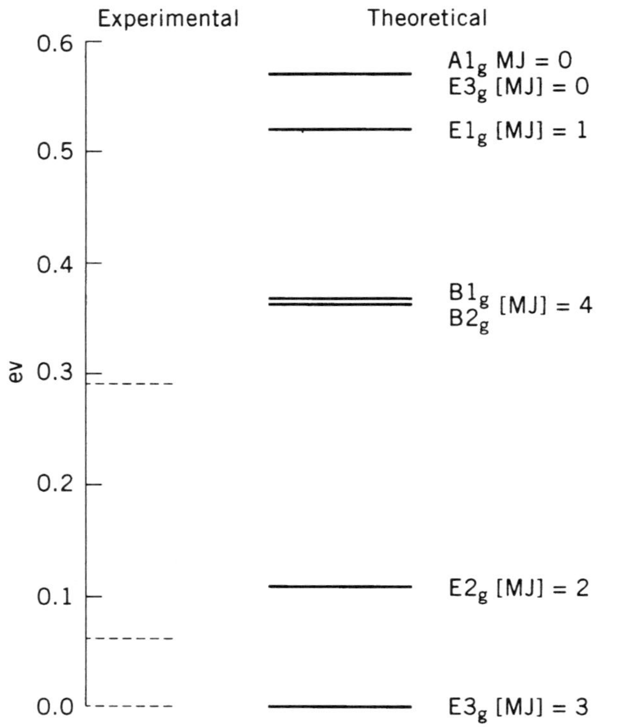

Figure 4.28 The lowest $5f^2$ electronic states of uranocene. (Reproduced from Ref. 9.)

description of excited electronic states. Spin–orbit coupling was included using the double group CI method.

The ground state of uranocene is a E_{3g} state in the D_{8h}^2 double group arising from the $5f^2$ electronic configuration of U. The $f^2\,{}^3H_4$ states, for example, in the D_{8h}^2 double group correlate with A_{1g}, E_{1g}, E_{2g}, E_{3g}, B_{1g}, and B_{2g} ($|M_J| = 4$) representations. The $6d$ orbitals play an important role in bonding in $U(C_8H_8)_2$, but the open-shell $5f^2$ configuration determines the possible electronic states as concluded by Chang and Pitzer.

The Mulliken population analysis of the SCF orbitals obtained by Chang and Pitzer (9) reveals that the charge on U is 0.98+ (rather than the formal 4+ charge). This means that U has donated a significant portion of its positive charge to the ligands. The U population in the ground state is U ($7s^{0.43}7p^{0.11}6d^{1.98}5f^{2.5}$). This is consistent with the primary role of the $6d$ orbitals in the bonding of $U(C_8H_8)_2$.

Figure 4.28 shows the energy-level diagrams for the experimental and the theoretically computed states by Chang and Pitzer of $U(C_8H_8)_2$. As seen from Figure 4.28, the Chang–Pitzer calculation predicts the E_{3g} state as the ground

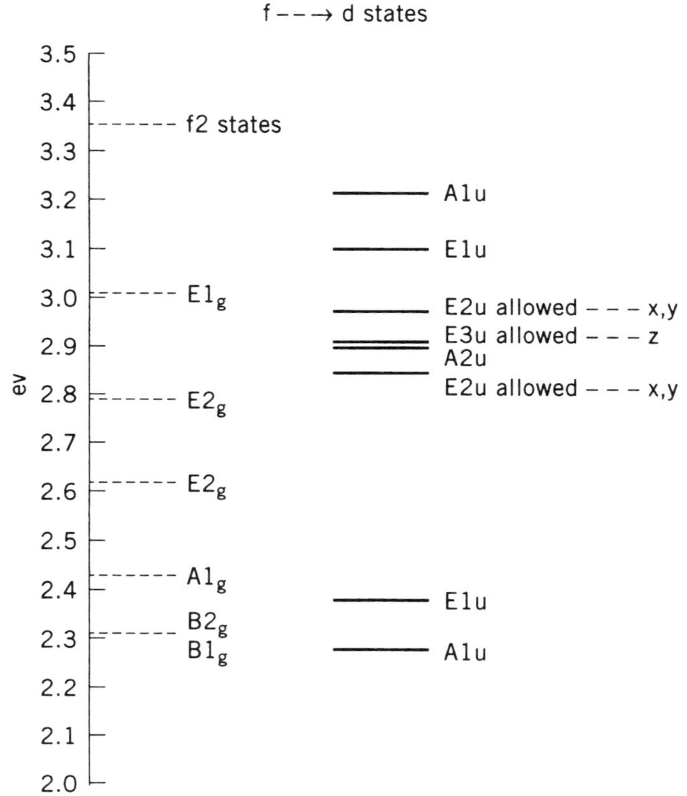

Figure 4.29 The lowest $5f \rightarrow 6d$ states of uranocene. (Reproduced from Ref. 9.)

Table 4.68 First 38 f^2 Electronic States of Uranocene

	State	Energy, eV		State	Energy, eV
38	E_{1g}	2.189	19	E_{1g}	1.234
37	A_{1g}	2.187	18	A_{1g}	1.153
36	E_{3g}	2.122	17	B_{1g}	1.121
35	E_{2g}	1.999	16	B_{2g}	1.118
34	A_{1g}	1.873	15	E_{2g}	1.111
33	E_{2g}	1.869	14	A_{2g}	1.090
32	E_{2g}	1.856	13	E_{3g}	0.999
31	E_{1g}	1.811	12	E_{3g}	0.844
30	E_{3g}	1.741	11	B_{2g}	0.816
29	B_{2g}	1.688	10	B_{1g}	0.811
28	A_{2g}	1.680	9	E_{2g}	0.773
27	B_{1g}	1.679	8	E_{1g}	0.754
26	E_{2g}	1.538	7	A_{1g}	0.565
25	E_{3g}	1.533	6	E_{3g}	0.564
24	E_{1g}	1.504	5	E_{1g}	0.514
23	E_{1g}	1.474	4	B_{1g}	0.364
22	E_{3g}	1.425	3	B_{2g}	0.360
21	E_{2g}	1.388	2	E_{2g}	0.109
20	A_{1g}	1.351	1	E_{3g}	0.000

Source: Reproduced from Reference 9.

state with a low-lying E_{2g} state only ~0.1 eV above the E_{3g} state. The B_{1g} and B_{2g} states corresponding to $|M_J| = 4$ are virtually degenerate. Next, there exists an E_{1g} state and degenerate E_{3g} and A_{1g} states. Figure 4.29 shows the lowest possible electronic states arising from exciting a 5*f* electron into 6*d*.

Table 4.68 shows the computed energy separations of the first 38 electronic states of uranocene obtained by Chang and Pitzer (9). Table 4.69 shows the excitation energies for the electronic states arising from the promotion of a 5*f* electron into 6*d*, together with an indication as to whether an electronic transition into that state is dipole-allowed.

Table 4.69 The 5*f* → 6*d* Excitation Energies for Electronic States of Uranocene[a]

State	Excitation Energy, eV	Electric Dipole Transition	State	Excitation Energy, eV	Electric Dipole transition
A_{1u}	3.211	Forbidden	A_{2u}	2.893	Forbidden
E_{1u}	3.091	Forbidden	E_{2u}	2.837	Allowed (x, y)
E_{2u}	2.961	Allowed (x, y)	E_{1u}	2.385	Forbidden
E_{3u}	2.905	Allowed (z)	A_{1u}	2.280	Forbidden

[a] Six f^2 states are also in this energy range.

Source: Reproduced from Reference 9.

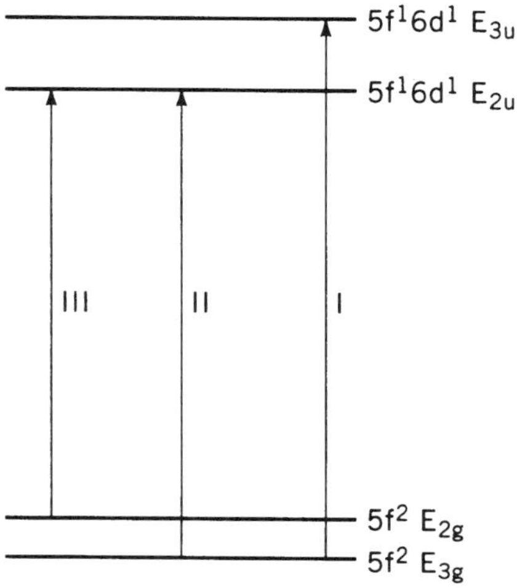

Figure 4.30 Transitions of visible spectrum of uranocene as assigned by Ref. 9.

Figure 4.30 shows three possible electronic transitions that correspond to the visible spectrum is proposed by Chang and Pitzer (9). As evidenced from Figure 4.30, the first allowed transition from the ground state is an $f \rightarrow d$ ($E_{3g} \rightarrow E_{2u}$) transition at 2.84 eV, while the second allowed transition is an $f \rightarrow d$ E_{3u} transition at 2.905 eV (z-polarized). This corresponds to band I (Fig. 4.30) whose experimental transition energy is 2.017 eV and is z-polarized. Band II ($E_{3g} \rightarrow E_{2u}$) has an experimental transition energy of 1.939 eV compared to the calculated value of 2.837 eV and is (x, y)-polarized. Chang and Pitzer sug-

Table 4.70 $U(C_8H_8)_2$ Dipole-Allowed States

State	ΔE, eV	Character	Polarization	State	ΔE, eV	Character	Polarization
B_{2u}	3.814	$\pi \rightarrow d$	x, y	E_{2u}	3.484	$\pi \rightarrow d$	x, y
B_{1u}	3.802	$\pi \rightarrow d$	x, y	E_{3u}	3.477	$f \rightarrow d$	z
E_{3u}	3.796	$\pi \rightarrow d$	z	E_{3u}	3.395	$\pi \rightarrow d$	z
E_{3u}	3.769	$\pi \rightarrow d$	z	B_{1u}	3.377	$f \rightarrow d$	x, y
B_{2u}	3.737	$\pi \rightarrow d$	x, y	B_{2u}	3.377	$f \rightarrow d$	x, y
B_{1u}	3.735	$\pi \rightarrow d$	x, y	E_{3u}	3.349	$\pi \rightarrow d$	z
E_{2u}	3.693	$\pi \rightarrow d$	x, y	B_{2u}	3.327	$\pi \rightarrow d$	x, y
E_{2u}	3.565	$\pi \rightarrow d$	z	B_{1u}	3.322	$\pi \rightarrow d$	x, y
E_{3u}	3.557	$\pi \rightarrow d$	z	E_{2u}	2.961	$f \rightarrow d$	x, y
E_{2u}	3.554	$\pi \rightarrow d$	x, y	E_{3u}	2.905	$f \rightarrow d$	z
E_{2u}	3.522	$\pi \rightarrow d$	x, y	E_{2u}	2.837	$f \rightarrow d$	x, y
B_{2u}	3.491	$\pi \rightarrow d$	x, y	E_{3g}	0.00	f^2	—
B_{1u}	3.490	$\pi \rightarrow d$	x, y				

Source: Reproduced from Reference 9.

gest that the (x, y)-polarized $E_{2g} \rightarrow E_{3u}$ transition (theory 2.796 eV) with an expected experimental energy of 1.959 eV may be buried under band II. Overall, theoretically computed energies of Chang and Pitzer (9) are 0.9 eV too high compared to experimental values. Nevertheless, they reproduced the relative trend of experimental energies. Consequently, Chang and Pitzer (9) have reassigned the observed bands as follows.

The visible bands previously assigned to the ligand to metal $\pi \rightarrow f$ charge-transfer transitions were reassigned to the $5f \rightarrow 6d$ transition. Table 4.70 gives a summary of a few dipole-allowed electronic transitions of $U(C_8H_8)_2$. The characteristic green color of uranocene was thus attributed by Chang and Pitzer (9) to the $5f \rightarrow 6d$ transition ($E_{3g} \rightarrow E_{2u}$, E_{3u}), but this transition also has some admixture of $\pi \rightarrow 6d$ charge-transfer character as inferred from the CI coefficients.

Chang and Pitzer (10) have also investigated $Np(C_8H_8)_2$ as well as $Pu(C_8H_8)_2$. There are other crystal field and ligand field calculations as well on these species. The paper of Chang and Pitzer (9) contains a comprehensive review of the literature. Neptunocene can be best described analogous to uranocene as arising from the $(^4I_{9/2}, 5f^3)$ configuration. The $(^4I_{9/2}, 5f^3)$ yields $E_{(7/2)u}$ ($|M_J| = \frac{7}{2}, \frac{9}{2}$), $E_{(5/2)u}$ ($|M_J| = \frac{5}{2}, \frac{11}{2}$), $E_{(3/2)u}$ ($|M_J| = \frac{3}{2}$), and $E_{(1/2)u}$ ($|M_J| = \frac{1}{2}$) electronic states. Table 4.71 shows the first 11 electronic states of $Np(C_8H_8)_2$ arising from the $5f^3$ electronic configuration. The electronic states shown in Table 4.71 were computed by Chang and Pitzer (10) using the SCF/IVNO/double-group CI procedures analogs to the uranocene. As seen from Table 4.71, the ground state of neptunocene is a $E_{(5/2)u}$ state, although the $E_{(3/2)u}$ and $E_{(7/2)u}$ states are only 0.138 and 0.181 eV, respectively, above the ground state. Hayes and Edelstein (195) have also computed the same ground state and first excited state, but compare $|M_J| = \frac{9}{2}$ for the second excited state using ligand field theory. There is a large energy gap between fifth and sixth states (Table 4.71), and thus Chang and Pitzer (10) suggested that the first five

Table 4.71 The First 11 f^3 States of $Np(C_8H_8)_2$

Number	State	Energy, eV
11	$E_{5/2u}$	1.161
10	$E_{1/2u}$	1.002
9	$E_{3/2u}$	0.946
8	$E_{7/2u}$	0.943
7	$E_{7/2u}$	0.895
6	$E_{5/2u}$	0.848
5	$E_{7/2u}$	0.418
4	$E_{1/2u}$	0.351
3	$E_{7/2u}$	0.181
2	$E_{3/2u}$	0.138
1	$E_{5/2u}$	0.000

Source: Reproduced from Reference 8.

Table 4.72 The First 13 f^3 States of $Pu(C_8H_8)_2$

Number	State	Energy, eV
13	B_{2g}	1.122
12	B_{1g}	1.121
11	E_{3g}	1.050
10	E_{3g}	0.946
9	E_{2g}	0.898
8	A_{2g}	0.829
7	E_{1g}	0.766
6	E_{3g}	0.544
5	E_{2g}	0.445
4	B_{2g}	0.291
3	B_{1g}	0.291
2	E_{1g}	0.138
1	A_{1g}	0.000

Source: Reproduced from Reference 8.

states form the $J = \frac{9}{2}$ complex, while the next six states constitute the $J = \frac{11}{2}$ complex.

Table 4.72 shows the computed energies of 13 electronic states of $Pu(C_8H_8)_2$ arising from the $5f^{14}$ configuration. The ground state of $Pu(C_8H_8)_2$ is thus an $A_{1g}(|M_J| = 0)$ state, while $|M_J| = 1(E_{1g})$ and $|M_J| = 4(B_{1g}, B_{2g})$ are respectively the first and second excited states of $Pu(C_8H_8)_2$. The A_{1g} ground state computed by Chang and Pitzer (10) was fully consistent with the diamagnetic character of $Pu(C_8H_8)_2$ observed by Starks and Streitwieser (179). The results of the Chang–Pitzer computation differed from the Hayes–Edelstein (195) ligand field computation only on the second excited state. Chang and Pitzer (10) assigned the first six states to a $J = 4$ complex and the next seven states to a $J = 5$ complex.

Table 4.73 shows the possible excited u states of $Pu(C_8H_8)_2$ and their computed energy separations (10). The first allowed $(x–y)$ polarized charge transfer $\pi–d$ transition of $Pu(C_8H_8)_2$ from the A_{1g} ground state was computed by Chang and Pitzer to occur at 2.92 eV, while the second transition $(f \rightarrow d)$ is only 0.05 eV higher.

Bursten and Strittmatter (49) reviewed the cyclopentadienyl–acetinide complexes. They presented the development of bonding models for CP–actinide complexes. They considered the photoelectron spectra and bonding of several complexes such as $U(C_5H_5)_3Cl$, $An(C_5H_5)_4$, $U(C_5H_5)_2$, $An(C_5H_5)_3$, $An(C_5H_5)_3L$, $An(C_5H_5)_3X$, $An(C_5H_5)_2Cl_2$, and numerous other complexes of this kind. Since this is a review on the topic, we conclude this section by referring the readers to the excellent review by Bursten and Strittmatter (49) on this topic for additional details.

Dolg et al. (56) have studied the ground state of di-π cyclooctaheraene cerium [cerenocene]. These calculations have revealed that Ce ion is almost

Table 4.73 The *u* Symmetry Excited States of $Pu(C_8H_8)_2$

State	Excitation Energy, eV	Character	Electric Dipole Transition
E_{2u}	3.28	$\pi \rightarrow d$	Forbidden
E_{3u}	3.21	$\pi \rightarrow d$	Forbidden
A_{2u}	3.17	$\pi \rightarrow d$	Allowed (z)
E_{3u}	3.16	$\pi \rightarrow d$	Forbidden
A_{1u}	3.15	$\pi \rightarrow d$	Forbidden
E_{2u}	3.14	$\pi \rightarrow d$	Forbidden
E_{1u}	3.12	$\pi \rightarrow d$	Allowed (x, y)
B_{1u}	3.11	$\pi \rightarrow d$	Forbidden
B_{2u}	3.11	$\pi \rightarrow d$	Forbidden
E_{1u}	3.09	$\pi \rightarrow d$	Allowed (x, y)
A_{2u}	3.08	$\pi \rightarrow d$	Allowed (z)
B_{2u}	2.99	$\pi \rightarrow d$	Forbidden
E_{3u}	2.99	$\pi \rightarrow d$	Forbidden
B_{1u}	2.98	$\pi \rightarrow d$	Forbidden
E_{1u}	2.97	$f \rightarrow d$	Allowed (x, y)
A_{1u}	2.93	$f \rightarrow d$	Forbidden
E_{1u}	2.92	$\pi \rightarrow d$	Allowed (x, y)
E_{2u}	2.91	$\pi \rightarrow d$	Forbidden
E_{3u}	2.84	$f \rightarrow d$	Forbidden
E_{2u}	2.77	$f \rightarrow d$	Forbidden

Source: Reproduced from Reference 8.

entirely in a $4f^1$ electronic configurations corresponding to $[Ce^{3+}(C_8H_8^{1.5-})_2]$. Dolg et al. also accounted for correlation energy using correlation-energy density functionals and by limited coupled electron-pair approximations.

4.7.3 Lanthanides and Actinides inside Carbon Cages ($M@C_n$, $M@C_n^+$)

Kroto et al. (180) showed the existence of an unusually stable C_{60} cluster obtained by the laser vaporization of graphite. The unusually stable C_{60} was called "buckminsterfullerene" or, informally, the "buckyball." Since the appearance of the Kroto et al. (180) paper numerous theoretical and experimental studies have dealt with the buckyball and other carbon cages in the general class of compounds called the *fullerene*. The biography of C_{60} has grown into such an astronomical size.

Heath et al. (181), Cox et al. (182), Kaldor et al. (183), Weiss et al. (184), and O'Brien et al. (185) have shown that metal complexes of the buckyball, especially the complex in which a lanthanum atom at the center of the carbon cage ($La@C_{60}$) can be produced by laser vaporization of a graphite rod impregnated with $LaCl_3$ and subsequent irradiation with and ArF laser. Several metal complexes of the buckyball and other carbon cages such as $M@C_n$

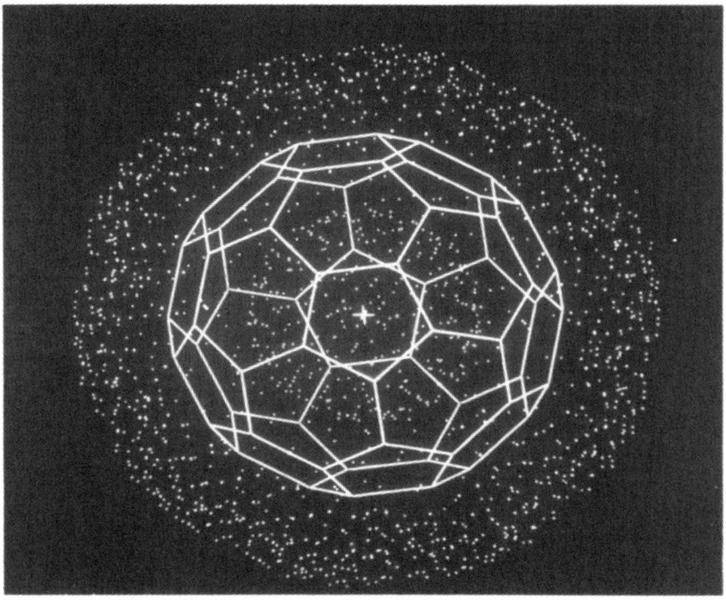

Figure 4.31 The structure of La^+ inside C_{60}. (Reproduced from Ref. 181.)

(M = La,K,Cs,Ca,Ba,Sy) have been produced. Figure 4.31 shows the proposed structure of La^+ inside the C_{60} fullerene cage.

Chai et al. (186) have generated fullerenes with a single and a cluster of La atoms trapped inside of the carbon cages in macroscopic quantities. This was accomplished by Chai et al. (186) by laser vaporization of a lanthanum oxide/graphite composite rod in a flow of Ar gas at 1200°C. Subsequently, when sublimed with C_{60} and C_{70}, they formed an air-stable film containing primarily LaC_{60}, LaC_{70}, LaC_{74}, and LaC_{82}. A solution containing LaC_{82} was prepared by dissolving in toluene. The resulting solution was found to be uniquely stable. Chai et al. (186) also obtained complexes containing up to three La atoms in large carbon cages such as La_3C_{88}. They suggested the use of the @ symbol to represent an atom inside a cage, which we follow in this section. For example, La inside C_{60} will be denoted by La@C_{60}.

Chang et al. (8) have studied the ground states of M@C_{60} and M^+@C_{60} for M = O,F,K,Ca,Mn,Cs,Ba,La,Eu,U. They employed relativistic ECPs for the metal atom and restricted Hartree–Fock (RHF) level of theory to study these complexes. There are other theoretical calculations based on the density functional method by Rosen and Wästberg (187, 188) as well as Ballster et al. (189). The most study of Chang et al. (8) on La@C_{60} showed that the lanthanum atom donates two electrons from the 6*s* orbital to the carbon cage. This, in fact, provided the basis to explain the unusual stability of La@C_{82} observed by Chai et al. (186). In this section we selectively review theoretical and experimental developments so as to gain insight into these unusual carbon cages containing a lanthanide or actinide atom inside.

Chang et al. (8) have performed *ab initio* RHF calculations on C_{60} cages with several metal atoms inside. They used a Gaussian basis set containing up to 556 basis functions and 257 valence electrons. All calculations of Chang et al. (8) took full advantage of the icosahedral symmetry (I_h). In the I_h point group, the central atoms, s, p, d, and f orbitals, correlate into a_g, t_{1u}, h_g, and $t_{2u} + g_u$ representations, respectively.

Table 4.74 shows the electronic states of La@C_{60} and the energy separations computed by Chang et al. (8). The ground state of La@C_{60} is obtained by donating two 6s electrons to the t_{1u} orbital of the C_{60} cage. This results in a $t_{1u}^2h_g^1$ configuration. Note that the single 5d electron of the La atom resides in a h_g orbital. The calculations by Chang et al. (8) have indicated a strong destabilizing interaction between the 6s orbital of La and the C orbitals. This results in a formal charge of 2+ on La, although the Mulliken population analysis of the RHF orbitals obtained by Chang et al. (8) indicated a 0.8 charge as a result of mixing of cage orbitals. The energies computed in Table 4.74 are for the average of $^4T_{1g}$, $^4T_{2g}$, ^4G_g, and ^4H_g states. There is a nearly degenerate electronic state obtained from moving a 6s electron to t_{1u} and one to $5d(t_{1u}h_g^2)$. This corresponds to an average of ^4A_u, $^4T_{1u}$, $^4T_{2u}$, 2^4G_u, and 3^4H_u states.

The ground state of La$^+$@C_{60} was found to be h_g^2, yielding an IP of 9.4.4 eV compared to an upper limit of 9.6.42 eV obtained by Heath et al. (181) and Cox et al. (182). Other electronic states arising from $t_{1u}h_g$ and t_{1u}^2 configurations are 0.3 and 3 eV, respectively, higher in energy.

Table 4.74 Electronic States of La@C_{60}

Configurations		ΔE, eV
$t_{1u}^2h_g^1$	(S = 1.5)	0.0000
$t_{1u}^1h_g^2$	(S = 1.5)	0.0018
h_g^3	(S = 1.5)	0.28
$t_{1g}^1h_g^1t_{1u}^1$	(S = 1.5)	1.39
$t_{1g}^1h_g^2$	(S = 1.5)	1.71
t_{1u}^3	^4A_u	2.22
	^2H_u	2.56
	(Average of configurations)	2.56
$a_g^1h_g^1t_{1u}^1$	(S = 1.5)	2.67
t_{1u}^3	$^2T_{1u}$	2.79
$a_g^1h_g^2$	(S = 1.5)	3.09
$h_g^2t_{2u}^1$	(S = 1.5)	3.30
$h_g^2g_u^1$	(S = 1.5)	4.95

Source: Reproduced from Reference 8.

The density functional theory predicts different ground states for La@C_{60} and La^+@C_{60}. The calculations by Rosen and Wästberg (187) yielded t_{1u}^3 and t_{1u}^2 as the ground states of La@C_{60} and La^+@C_{60}, respectively. The LDF method gave an IP of 9.6.9 eV. However, this is contrasted by Chang et al.'s calculations, which place these states 2.56 and 2.88 eV, respectively, above the ground state. Chang et al. attributed this to the difficulty in the LDF method of distinguishing the total energies of states for which closely spaced orbitals have different occupations.

Table 4.75 shows the electronic configurations, electronic states, and energy separations for Eu@C_{60} computed by Chang et al. Again, the ground state of Eu@C_{60} is obtained by moving two 6*s* electrons from Eu to the C_{60} cage, leaving the $4f^7$ shell essentially an intact nonbonding shell. Since the 4*f* orbital correlates with $t_{2u} + g_u$, this leads to a $t_{1u}^1 t_{2u}^3 g_u^4$ configuration, yielding a $^{10}T_{1u}$ ground state for Eu@C_{60}. Like La@C_{60}, although the formal charge on Eu is 2+, the Mulliken population analysis yields a charge of 0.5+ on Eu.

Table 4.76 shows the electronic states of U@C_{60} obtained by Chang et al. In this case, the ground state is obtained by donating a 7*s* electron to the t_{1u} orbital of C_{60} and one to the 6*d* shell, yielding a $t_{1u}^1 h_g^3 t_{2u}^3$ configuration with ^7A_g, $^7T_{1g}$, $^7T_{2g}$, 2^7G_g, and 3^7H_g states. As seen from Table 4.76, the $h_1^2 t_{1u}^1 t_{2u}^3$, $a_g^1 h_g t_{1u} t_{2u}^3$, and $h_g t_{1u}^2 t_{2u}^3$ configurations are very close to each other. Chang et al. (8) predicted that the spin–orbit effects are likely to be larger for U@C_{60}, especially in states that have increased 6*d* occupancy.

Table 4.77 shows the electronic states of La^+@C_{60}. The ground state of La^+@C_{60} arises from a h_g^2 configuration, which suggests that the ionization from the t_{1u} orbital of $t_{1u}h_g^2$ is more favorable, leading to a h_g^2 configuration. The $t_{1u}h_g$ (triplet) state of La^+@C_{60} is only 0.29 eV higher in energy. However, the $^3T_{1g}$ state arising from t_{1u}^2 is significantly higher.

Tables 4.78 and 4.79 show the electronic states of Eu^+@C_{60} and U^+@C_{60},

Table 4.75 Electronic States of Eu@C_{60}

Configurations		ΔE, eV
$t_{1u}^2(t_{2u}^3 g_u^4)$	$^{10}T_{1u}$	0.00
$a_g^1 t_{1u}^1(t_{2u}^3 g_u^4)$	$^{10}T_{1g}$	0.62
$t_{1g}^1 t_{1u}^1(t_{2u}^3 g_u^4)$	(S = 4.5)	1.39
$h_g^1 t_{1u}^2(t_{2u}^3 g_u^3)$	(S = 4.5)	1.76
$a_g^1 h_g^1(t_{2u}^3 g_u^4)$	$^{10}H_u$	2.47
$a_g^2(t_{2u}^3 g_u^4)$	^8A_u	2.93
$h_g^1 t_{1u}^1(t_{2u}^3 g_u^4)$	(S = 4.5)	3.28
$t_{1u}^3(t_{2u}^3 g_u^3)$	(S = 4.5)	5.91
$t_{1u}^2(t_{2u}^4 g_u^4)$	(S = 3.5)	14.42

Source: Reproduced from Reference 8.

Table 4.76 Electronic States of U@C$_{60}$

Configurations		ΔE, eV
$h_g^2 t_{1u}^1 t_{2u}^3$	$(S = 3)$	0.00
$a_g^1 h_g^1 t_{1u}^1 t_{2u}^3$	$(S = 3)$	0.07
$h_g^1 t_{1u}^2 t_{2u}^3$	$(S = 3)$	0.19
$a_g^1 h_g^2 t_{2u}^3$	$(S = 3)$	1.00
$a_g^1 h_g^1 t_{1u}^1 t_{2u}^2 g_u^1$	$(S = 3)$	1.14
$a_g^1 h_g^2 t_{1u}^1 t_{2u}^2$	$(S = 3)$	1.95
$a_g^2 h_g^1 t_{2u}^3$	^5H_u	2.06
$t_{1u}^3 t_{2u}^3$	^7A_g	2.40
$a_g^2 h_g^1 g_u^3$	$(S = 2)$	2.60
$a_g^2 h_g^1 t_{2u}^2 g_u^1$	$(S = 2)$	3.13
$t_{1u}^2 t_{2u}^4$	$(S = 2)$	3.30
$a_g^2 h_g^1 t_{2u}^1 g_u^2$	$(S = 2)$	3.31
$a_g^1 h_g^1 t_{1u}^2 t_{2u}^2$	$(S = 3)$	3.88
$a_g^1 t_{1u}^1 t_{2u}^4$	$(S = 2)$	4.12
$h_g^1 t_{1u}^1 t_{2u}^4$	$(S = 2)$	4.36
$a_g^1 h_g^1 t_{2u}^4$	$(S = 2)$	6.24
$a_g^2 t_{2u}^4$	$^3T_{2g}$	6.38
$h_g^1 t_{2u}^5$	$(S = 1)$	12.57

Source: Reproduced from Reference 8.

Table 4.77 Electronic States of La$^+$@ C$_{60}$

Configurations		IP, eV[a]
h_g^2	$(S = 1)$	4.43
$t_{1u}^1 h_g^1$	$(S = 1)$	4.72
t_{1u}^2	$^3T_{1g}$	7.20
	(average of configurations)	7.31
	^1H_g	7.43
	^1A_g	7.77

[a]Relative to La@C$_{60}$ ground state.

Source: Reproduced from Reference 8.

Table 4.78 Electronic States of $Eu^+@C_{60}$

Configurations		IP, eV[a]
$t_{1u}^1(t_{2u}^3 g_u^4)$	$^9T_{1g}$	4.74
$h_{1g}^1(t_{2u}^3 g_u^4)$	^9H_u	4.90
$a_g^1(t_{2u}^3 g_u^4)$	^9A_u	5.06
$t_{1g}^1(t_{2u}^3 g_u^4)$	$^9T_{1u}$	6.46
$(t_{2u}^4 g_u^4)$	$^7T_{2g}$	18.86

[a]Relative to $Eu@C_{60}$ ground state.
Source: Reproduced from Reference 8.

respectively. Compared to the $t_{1u}^2(t_{2u}^2 g_u^4)$ ground state of $Eu@C_{60}$, $Eu^+@C_{60}$ has a $t_{1u}(t_{2u}^2 g_u^4)$ electronic configuration. The ionization thus takes place on a t_{1u} cage shell for $Eu@C_{60}$ similar to $La@C_{60}$. The ground state of $U^+@C_{60}$ arises for the $h_g^2 t_{2u}^3$ configuration compared to the $h_g^2 t_{1u} t_{2u}^3$ configuration of $U@C_{60}$. Hence all lanthanide- and actinide-containing C_{60} cages lose an electron from the t_{1u} cage orbital on ionization.

In the most experimental work, Chai et al. (186) extracted, in solution, the $La@C_{82}$ complex. Figure 4.32 shows the FTICR mass spectrum of hot toluene extract of fullerene material produced by laser evaporization of La_2O_3/graphite composite rod. As seen from Figure 4.32, the only La-containing fullerene that survived the exposure to moisture and air was the $La@C_{82}$ cluster. This unusual stability of $La@C_{82}$ and its extractability in toluene solution are rather intriguing. However, the calculation of Chang et al. (8) on $La@C_{60}$ sheds some light on the unusual stability of $La@C_{82}$. As noted before, the ground state of $La@C_{60}$ is formed by promotion of two 6*s* electrons from La to the carbon cage. The magic numbers for large carbon cages are 60, 70, 84, and to some extent 76,

Table 4.79 Electronic States of $U^+@C_{60}$

Configurations		IP, eV[a]
$h_g^2 t_{2u}^3$	$(S = 2.5)$	4.42
$a_g^1 h_g^1 t_{2u}^3$	^6H_u	4.44
$h_g^1 t_{1u}^1 t_{2u}^3$	$(S = 2.5)$	4.94
$a_g^1 t_{1u}^1 t_{2u}^3$	$^6T_{1g}$	5.75
$t_{1u}^2 t_{2u}^3$	$^6T_{1u}$	7.37

[a]Relative to $U@C_{60}$ ground state.
Source: Reproduced from Reference 8.

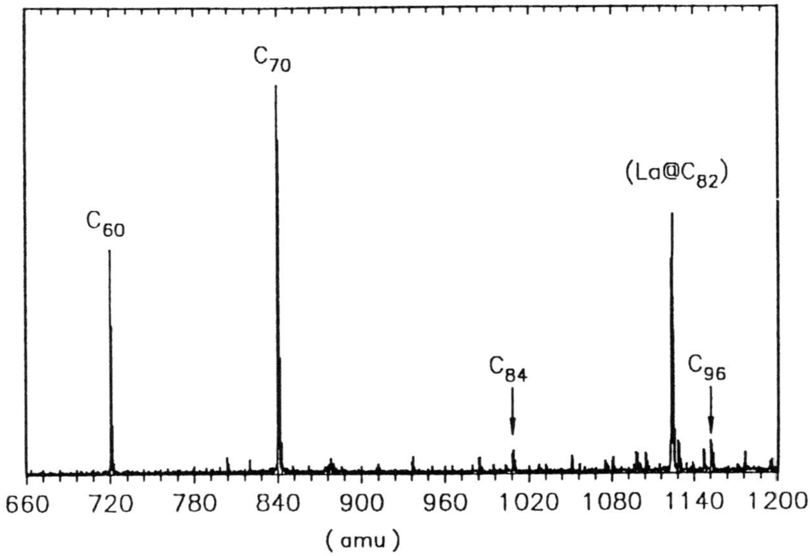

Figure 4.32 FT-ICR mass spectra of not-toluene extract of fullerene material produced by laser vaporization of a 10% La_2O_3/graphite composite rod. This was exposed to air and moisture. (Reproduced from Ref. 186.)

in decreasing order of stability (190). The closest magic number to 82 is 84 for C_n. Since La donates two 6s electrons, the carbon cage accepting this must have two electrons less for the unusual stability. This means that C_{82} can accept two electrons from La, leading to a closed-shell ground state with an unusual stability. This explains why La@C_{82} is unusually stable and is inert to air and moisture and can be extracted in toluene. It is also predicted that the La@C_{82} cluster would have a substantial HOMO-LUMO gap. This also explains why La@C_{74} is the next most stable fullerene containing La.

Coalescence of two carbon cages produces small clusters of up to three atoms present in larger carbon clusters. Strong signals for clusters of composition La_2C_n and even La_3C_n were found by Chai et al. (186). Figure 4.33 shows a portion of the FTICR spectrum obtained after laser-aging of the composite target disk. As seen from Figure 4.33, signals corresponding to La_3@C_{88}, La_3@C_{94}, La_2@C_{106}, and La@C_{118} are quite prominent. Although the exact mechanism of the formation of these species is not clearly established at present, Chai et al. (186) have speculated that they are generated by (La@C_n)–(La@C_n) coalescence reactions. Indeed, the transmission electron microscopy of C_{60} films by Wang and Buseck (191) shows such evidence of C_{60}–C_{60} coalescence to form cylindrical bucky tubes in the solid film. The presence of small clusters of up to three atoms of La in large fullerenes is certainly very interesting.

Several experimental studies have focused on the generation and properties

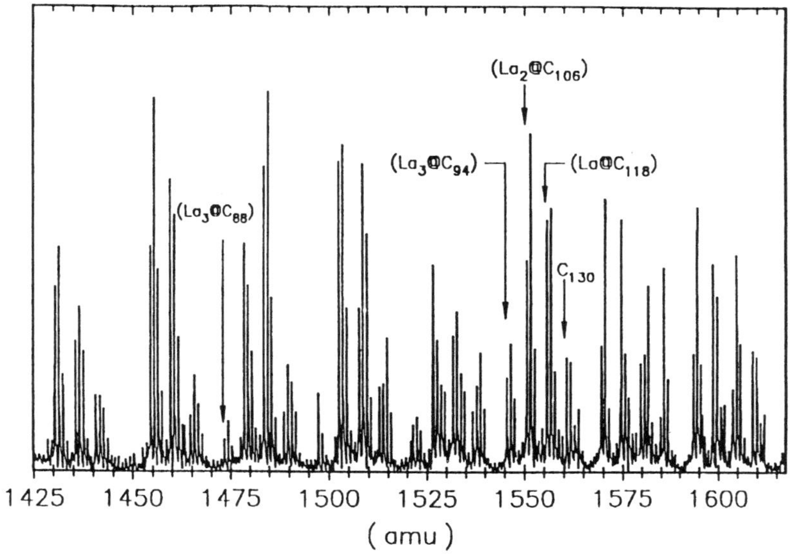

Figure 4.33 FT-T ICR mass spectrum showed the presence of fullerenes containing up to 3 La atoms. (Reproduced from Ref. 186.)

of clusters containing lanthanides. For example, Guiodin et al. (192) generated La_mC_n metal–carbon clusters. Dayane et al. (193) produced clusters of lanthanide atoms through the inert-gas condensation technique. The mass-spectral analysis of Yb_n^+ clusters revealed an overall Gaussian distribution, although Yb_n^+ clusters for $n = 13,19,23,26,29,32,34,37,39,45$ clearly emerge. Likewise, these authors have also generated Sm_n^+ clusters. The distribution of Sm_n^+ clusters exhibited magic numbers for $n = 13,19,26,32,49.5$. Brechignac et al. (194) generated the clusters of Eu and Yb. They also proposed possible structures for Yb_{14}^+ and Yb_{20}^+. It is evident that experimental studies of clusters containing lanthanide atoms are on the increase. There is a clear need for theoretical studies even on dimers of open-shell lanthanide atoms. However, such studies will be complicated by the fact that two 4*f* open-shell atoms will yield numerous possible electronic states with different spin multiplicities. The role of 4*f* orbitals in Ln–Ln bonding and the possibility of any multiple bonding between lanthanide atoms need to be explored.

4.8 CONCLUSION

As seen from this chapter, although there is significant development in both spectroscopic studies and the ligand field model interpretations of lanthanide oxides, there is less understanding of actinide chemistry. At present, relativistic *ab initio* studies on molecules containing actinides are restricted to very few species. While a ligand field model has been applied to almost all diatomic oxides containing Ln atoms, this is not the case as far as *ab initio* methods are concerned. Even in cases in which such calculations have been applied (EuO, GdO, YbO, etc.), very few low-lying electronic states have been studied using such techniques. The INDO/S-CI method of Rösch, Zerner, and coworkers, on the other hand, has enjoyed more success, although typically in this method all calculations are made at single experimental geometry and the integrals are parametrized. Consequently, there is a real need for further developments and application of *ab initio*–based techniques for molecules containing lanthanides and actinides. Of course, the computations of Chang and co-workers (8, 10) on both uranocene and carbon cages containing lanthanides have evidently demonstrated that, indeed, such calculations are doable and play a significant role in our comprehension of the nature of bonding in polyatomics containing lanthanides and actinides. Likewise, the CASSCF computation of U_2 by Pepper and Bursten (46) has shown that relativistic *ab initio* computations are possible on species containing actinide–actinide bonds. Spectroscopic data and their interpretation of actinide compounds need to be addressed more rigorously in the future. It is evident from this review that relativistic effects cannot be ignored for such compounds even to gain a qualitative insight. Furthermore, the open-shell, primarily nonbonding nature of the $5f$ shells, at least in very ionic compounds containing actinides, is likely to result in a large array of electronic states due to the spin–orbit coupling.

REFERENCES

1. D. C. Hoffman, in *Fifty Years with Transuranium Elements*, Robert A. Welch Foundation Conference on Chemical Research XXXIV, Houston, TX, 1990.
2. K. Balasubramanian, in *Handbook on the Physics and Chemistry of Rare Earths*, Vol. 18, *Lanthanides/Actinides: Chemistry*, K. A. Gschneidner, Jr., L. Eyring, G. R. Chapman, and G. H. Lander, eds., Elsevier, Amsterdam, Netherlands, 1994, p. 29.
3. D. Salahub and M. C. Zerner, *The Challenge of d and f Electrons*, American Chemical Society Symposium No. 394, 1989.
4. P. W. Atkins, *Eur. J. Solid State Inorg. Chem.* **28**, 9 (1991).
5. P. S. Bagus, Y. S. Lee, and K. S. Pitzer, *Chem. Phys. Lett.* **33**, 408 (1975).
6. R. W. Field, *Ber. Bunsenges. Phys. Chem.* **86**, 771 (1982).
7. R. W. Field, D. P. Baldwin, E. J. Hill, M. Li, and M. C. McCarthy, "Future Trends in Spectroscopy," *Spectrochim. Acta* Part A, Special Supplement, 75–79 (1989).
8. A. H. H. Chang, W. C. Ermler, and R. M. Pitzer, *J. Chem. Phys.* **94**, 5004 (1991).
9. A. H. H. Chang and R. M. Pitzer, *J. Am. Chem. Soc.* **111**, 2500 (1989).
10. A. H. H. Chang and R. M. Pitzer, unpublished results (1991).
11. R. Nessler and K. S. Pitzer, *J. Phys. Chem.* **91**, 1084 (1987).
12. K. Balasubramanian, *Chem. Phys. Lett.* **135**, 288 (1987).
13. K. Balasubramanian, *Internatl. J. Quantum Chem. Symp.* **22**, 465 (1988).

14. K. Balasubramanian, *J. Phys. Chem.* **93**, 6585 (1989).
15. K. Balasubramanian, *Chem. Rev.* **89**, 1801 (1989).
16. K. Balasubramanian, *Chem. Rev.* **90**, 93 (1990).
17. K. Balasubramanian, *J. Chem. Phys.* **93**, 8061 (1990).
18. K. Balasubramanian and K. S. Pitzer, *J. Chem. Phys.* **78**, 321 (1983).
19. K. Balasubramanian and K. S. Pitzer, *Adv. Chem. Phys.* **67**, 287 (1987).
20. K. Balasubramanian and Ch. Ravimohan, *Chem. Phys. Lett.* **145**, 39 (1988).
21. K. Balasubramanian and J. Z. Wang, *J. Molec. Spectrosc.* **133**, 82 (1989).
22. K. Balasubramanian and J. Z. Wang, *Chem. Phys. Lett.* **154**, 525 (1989).
23. K. Balasubramanian and D. G. Dai, *J. Chem. Phys.* **93**, 7243 (1990).
24. K. Balasubramanian and K. K. Das, *J. Molec. Spectrosc.* **145**, 142 (1991).
25. P. A. Christiansen, W. C. Ermler, and K. S. Pitzer, *Annu. Rev. Phys. Chem.* **36**, 407 (1985).
26. K. K. Das and K. Balasubramanian, *J. Chem. Phys.* **91**, 2433 (1989).
27. K. K. Das and K. Balasubramanian, *Chem. Phys. Lett.* **172**, 372 (1990).
28. K. K. Das and K. Balasubramanian, *J. Phys. Chem.* **95**, 42 (1991).
29. K. K. Das and K. Balasubramanian, *J. Molec. Spectrosc.* **145**, 42 (1991).
30. K. K. Das and K. Balasubramanian, *J. Chem. Phys.* **94**, 3722 (1991).
31. K. K. Das and K. Balasubramanian, *J. Phys. Chem.* **95**, 3979 (1991).
32. P. J. Hay, W. R. Wadt, L. R. Kahn, R. C. Rafenetti, and D. H. Phillips, *J. Chem. Phys.* **71**, 1767 (1979).
33. P. J. Hay and W. R. Wadt, *J. Chem. Phys.* **82**, 270 (1985).
34. P. J. Hay and W. R. Wadt, *J. Chem. Phys.* **82**, 299 (1985).
35. P. J. Hay, R. R. Ryan, K. V. Salazar, D. A. Wrobloski, and A. P. Sattelberger, *J. Am. Chem. Soc.* **108**, 313 (1986).
36. M. Krauss and W. J. Stevens, *J. Comput. Chem.* **4**, 127 (1983).
37. M. Krauss and W. J. Stevens, *Chem. Phys. Lett.* **99**, 417 (1983).
38. M. Krauss and W. J. Stevens, *Ann. Rev. Phys. Chem.* **35**, 357 (1984).
39. Y. S. Lee, W. C. Ermler, and K. S. Pitzer, *J. Chem. Phys.* **75**, 1861 (1977).
40. K. S. Pitzer, *Acc. Chem. Res.* **12**, 271 (1979).
41. K. S. Pitzer, *Interntl. J. Quantum Chem.* **25**, 131 (1984).
42. W. Wadt, *J. Am. Chem. Soc.* **103**, 6053 (1981).
43. W. Wadt and P. J. Hay, *J. Chem. Phys.* **82**, 284 (1985).
44. J. Z. Wang, K. K. Das, and K. Balasubramanian, *Molec. Phys.* **69**, 147 (1990).
45. W. C. Ermler, R. B. Ross, and P. A. Christiansen, *Adv. Quantum Chem.*, **19**, 139 (1988).
46. M. Pepper and D. E. Bursten, *J. Am. Chem. Soc.* **112**, 7803 (1990).
47. M. Pepper and D. E. Bursten, *Chem. Rev.* **91**, 719 (1991).
48. B. E. Bursten and R. J. Strittmatter, *J. Am. Chem. Soc.* **109**, 6606 (1987).
49. B. E. Bursten and R. J. Strittmatter, *Ang. Chemie.* **30**, 1069 (1991).
50. M. Pepper and B. E. Bursten, *Chem. Rev.* **91**, 719 (1991).
51. G. H. Igelmann, H. Stoll, and H. Preuss, *Molec. Phys.* **65**, 1321 (1988).

52. M. Dolg, H. Stoll, A. Savin, and H. Preuss, *Stud. Phys. Theor. Chem.* **62**, 269 (1988).
53. M. Dolg, and H. Stoll, *Theor. Chim. Acta* **75**, 369 (1989).
54. M. Dolg, H. Stoll, and H. Preuss, *Chem. Phys. Lett.* **174**, 208 (1990).
55. M. Dolg, H. Stoll, and H. Preuss, *Chem. Phys.* **148**, 219 (1990).
56. M. Dolg, P. Fulde, W. Kuchle, C.-S. Neumann, and H. Stoll, *J. Chem. Phys.* **94**, 3011 (1991).
57. M. Dolg, H. Stoll, H.-J. Flad, and H. Preuss, *J. Chem. Phys.* **97**, 1162 (1992).
58. M. Dolg, H. Stoll, and H. Preuss, *J. Chem. Phys.* **90**, 1730 (1989).
59. L. V. Szentpaly, P. Fuentelaba, H. Preuss, and H. Stoll, *Chem. Phys. Lett.* **93**, 555 (1982).
60. M. Dolg, H. Stoll, and H. Preuss, in *Handbook on the Physics and Chemistry of Rare Earths*, Vol. 22, K. A. Gschneidner, Jr. and L. Eyring, Elsevier, Amsterdam, Netherlands, 1996, p. 607.
61. P. Pyykkö and J. P. Desclaux, *Chem. Phys.* **34**, 261 (1978).
62. P. Pyykkö and J. P. Desclaux, *Acc. Chem. Res.* **79**, 276 (1979).
63. J. P. Desclaux and P. Pyykkö, *Chem. Phys. Lett.* **29**, 534 (1974).
64. J. P. Desclaux and P. Pyykkö, *Chem. Phys. Lett.* **39**, 300 (1976).
65. J. Pyykkö, *Inorg. Chim. Acta* **139**, 243 (1987).
66. P. Pyykkö, *Relativistic Theory of Atoms and Molecules*, Springer-Verlag, Berlin and New York: Part I (1986); Part II (1992).
67. P. Pyykkö, *Chem. Rev.* **88**, 563 (1988).
68. P. Pyykkö, *Phys. Scripta* **20**, 647 (1979).
69. P. Pyykkö and L. L. Lohr, Jr., *Inorg. Chem.* **20**, 1950 (1981).
70. P. Pyykkö, L. J. Laakonen, and K. Tatsumi, *Inorg. Chem.* **28**, 1801 (1989).
71. P. Pyykkö and J. Jové, *New. J. Chem.* **157**, 17 (1991); P. Pyykkö, J. Li, and N. Runeberg, *J. Phys. Chem.* **98**, 4809 (1994).
72. P. Pyykkö and Y. Zhao, *Inorg. Chem.* **30,** 3787 (1991); *J. Phys. Chem.* **94**, 7753 (1990).
73. Y. S. Lee and A. D. McLean, *J. Chem. Phys.* **76**, 735 (1982).
74. Y. S. Lee and A. D. McLean, *Stud. Phys. Theor. Chem.* **21**, 219 (1982).
75. A. F. Ramos, N. C. Pyper, and G. L. Malli, *Phys. Rev.* **A38**, 2729 (1988).
76. G. L. Malli, ed., *Relativistic Effects in Atoms, Molecules and Solids*, Plenum, New York, 1982.
77. G. L. Malli, *Stud. Phys. Theor. Chem.* **21**, 199 (1982).
78. K. Balasubramanian, P. A. Christiansen, and K. S. Pitzer, *Phys. Rev.* **A43**, 2581 (1991).
79. V. Kellö and A. J. Sadlej, *J. Chem. Phys.* (1991).
80. V. Kellö and A. J. Sadlej, *Chem. Phys. Lett.* **174**, 641 (1991).
81. A. J. Sadlej and M. Urban, *Chem. Phys. Lett.* **176**, 293 (1991).
82. R. Ahrlichs, *J. Chem. Phys.* **82**, 890 (1985).
83. C. W. Bauschlicher and S. R. Walch, *J. Chem. Phys.* **76**, 4560 (1982).
84. C. W. Bauschlicher and S. R. Walch, *J. Chem. Phys.* **78**, 4597 (1983).

85. C. W. Bauschlicher, Jr. and S. R. Langhoff, *Chem. Phys. Lett.* **161**, 383 (1989).
86. C. W. Bauschlicher, Jr. and S. R. Langhoff, *J. Phys. Chem.* **95**, 2278 (1991).
87. S. R. Langhoff, L. G. M. Pettersson, C. W. Bauschlicher, Jr., and H. Partridge, *J. Chem. Phys.* **87**, 268 (1987).
88. S. R. Langhoff, C. W. Bauschlicher, Jr., and H. Partridge, *J. Chem. Phys.* **89**, 396 (1988).
89. D. P. Chong, S. R. Langhoff, C. W. Bauschlicher, Jr., S. P. Walch, and H. Partridge, *J. Chem. Phys.* **85**, 285 (1986).
90. M. Rosi and C. W. Bauschlicher, *Chem. Phys. Lett.* **166**, 189 (1990).
91. R. J. Bartlett, *J. Phys. Chem.* **93**, 1697 (1989).
92. J. Paldus and J. Cizek, *J. Chem. Phys.* **90**, 4356 (1989).
93. F. A. Cotton and G. W. Wilkinson, *Basic Inorganic Chemistry*, Wiley, New York, 1976.
94. B. R. Brooks and H. F. Schaefer, *J. Chem. Phys.* **70**, 5092 (1979).
95. J. Olsen, P. Jorgensen, and J. Simons, *Chem. Phys. Lett.* **169**, 463 (1990).
96. J. Paldus, in *Theoretical Chemistry: Advances and Perspectives*, H. Eyring and D. J. Henderson, eds., Academic Press, New York, 1976.
97. A. D. McLean, M. Yoshimine, B. H. Lengsfield, P. S. Bagus, and B. Liu, in *Modern Techniques in Computational Chemistry*, E. Clementi, ed., Escom Science Publishers, Leiden, Netherlands, 1990.
98. P. Hohenberg and W. Kohn, *Phys. Rev.* **B136**, 864 (1964).
99. W. Kohn and L. J. Sham, *Phys. Rev.* **A140**, 1133 (1965).
100. A. Becke, *Phys. Rev.* **A33**, 2786 (1986).
101. R. G. Parr and W. Yang, *Density Functional Theories of Atoms and Molecules*, Oxford University Press, New York, 1989.
102. D. Salahub, *Adv. Chem. Phys.* **69**, 447 (1987).
103. M. Kotzian, N. Rösch, N. Schröder, and M. C. Zerner, *J. Am. Chem. Soc.* **111**, 7687 (1989).
104. M. Kotzian and N. Rösch, *Eur. J. Solid State Inorg. Chem.* **28**, 127 (1991).
105. M. Kotzian and N. Rösch, *J. Molec. Spectrosc.* **147**, 346 (1991).
106. M. Kotzian, N. Rösch, and M. C. Zerner, *Theor. Chim. Acta.* **81**, 201 (1991).
107. M. Kotzian, N. Rösch, and M. C. Zerner, *Internatl. J. Quantum Chem. Symp.* **25**, 545 (1991).
108. M. Kotzian, N. Rösch, R. M. Pitzer, and M. C. Zerner, *Chem. Phys. Lett.* **160**, 168 (1989).
109. J. Ridley and M. C. Zerner, *Theor. Chim. Acta* **32**, 111 (1973).
110. J. Ridley and M. C. Zerner, *Theor. Chim. Acta* **42**, 223 (1973).
111. R. M. Pitzer and N. W. Winter, *J. Phys. Chem.* **92**, 3061 (1988).
112. A. Bernard and R. Bacis, *Can. J. Phys.* **54**, 1509 (1976).
113. A. Bernard and R. Bacis, *Can. J. Phys.* **55**, 1322 (1977).
114. B. Simard, H. Niki, W. J. Balfour, M. Vassen, and P. H. Hackett, unpublished results.
115. R. F. Barrow, B. W. Bastin, D. L. G. Moore, and C. J. Pott, *Nature* **215**, 1072 (1967).

116. H. Schall, M. Dulick, and R. W. Field, *J. Chem. Phys.* **87**, 2898 (1987).
117. J. L. Elkind and P. B. Armentrout, *J. Phys. Chem.* **90**, 5736 (1986).
118. J. L. Elkind, L. S. Sunderlin, and P. B. Armentrout, *J. Phys. Chem.* **93**, 3159 (1989).
119. J. L. Elkind and P. B. Armentrout, *Inorg. Chem.* **25**, 1078 (1986).
120. P. B. Armentrout and J. L. Beauchamp, *Acc. Chem. Res.* **22**, 315 (1989).
121. J. A. Martinho and J. L. Beauchamp, *Chem. Rev.* **90**, 629 (1990).
122. J. B. Schilling, W. A. Goddard III, and J. L. Beauchamp, *J. Phys. Chem.* **87**, 481 (1987).
123. C. E. Moore, *Table of Atomic Energy Levels*, National Institute of Standards and Technology, Washington, DC, 1971.
124. A. E. Alvarado-Swaisgood and J. F. Harrison, *J. Phys. Chem.* **89**, 5198 (1985).
125. J. Aglada, D. J. Bruna, S. D. Peyerimhoff, and R. J. Buenker, *J. Molec. Spectrosc.* **92**, 299 (1988).
126. Y. M. Chen, D. E. Clemmer, and P. B. Armentrout, *J. Chem. Phys.* **95**, 1228 (1991).
127. J. P. Desclaux, *Atomic Nuclear Data* **12**, 312 (1973).
128. K. P. Huber and G. Herzberg, *Spectroscopic Constants of Diatomic Molecules*, Van Nostrand Reinhold, Princeton, NJ, 1979.
129. I. Gotkis, *J. Phys. Chem.* **95**, 6086 (1991).
130. D. R. Fischell, H. C. Brayman, and T. A. Cool, *J. Chem. Phys.* **73**, 4260 (1980).
131. J. Xin, G. Edvinsson, and L. Klynning, *J. Molec. Spectrosc.* **148**, 59 (1991).
132. J. Shirley, S. C. Scurlock, T. C. Steimle, B. Simard, M. Vasseur, and P. A. Hackett, *J. Chem. Phys.* **93**, 8580 (1990).
133. J. Shirley, W. L. Barclay, Jr., L. M. Ziurys, and T. C. Steimle, *Chem. Phys. Lett.* **183**, 363 (1991).
134. R. J. Van Zee, M. L. Seely, and W. Welter, *J. Chem. Phys.* **67**, 861 (1977).
135. P. Carette and A. Hocquet, *J. Molec. Spectrosc.* **16**, 301; Carette, P., 1990, *J. Molec. Spectrosc.* **140**, 269 (1988).
136. R. D. Suneram, F. J. Lovas, G. T. Fraser, and K. Matsumura, unpublished results (1991).
137. T. C. Steimle and Y. Al-Ramadin, *Chem. Phys. Lett.* **130**, 76 (1989).
138. W. J. Childs, O. Poulsen, and T. C. Steimle, *J. Chem. Phys.* **88**, 598 (1988).
139. J. A. Pople, D. L. Beveridge, and P. A. Dobush, *J. Chem. Phys.* **47**, 2026 (1967).
140. H. Schall, J. A. Gray, M. Dulick, and R. W. Field, *J. Chem. Phys.* **85**, 751 (1986).
141. C. Linton, S. McDonald, S. Rice, M. Dulick, Y. C. Liu, and R. W. Field, *J. Molec. Spectrosc.* **101**, 322 (1983).
142. E. A. Shenyavskaya, I. V. Egorova, and V. N. Lupanov, *J. Molec. Spectrosc.* **47**, 355 (1973).
143. (a) J. M. Delaval, J. Van Heems, and J. C. L. Beaufils, *Can. J. Spectrosc.* **22**, 117 (1977); (b) J. C. L. Beaufils, P. Carette, and J. M. Blondeau, *J. Molec. Spectrosc.* **77**, 1 (1979).
144. M. Dulick and R. W. Field, *J. Molec. Spectrosc.* **113**, 105 (1985).
145. M. Dulick, E. Murad, and R. F. Barrow, *J. Chem. Phys.* **85**, 385 (1986).

146. R. L. Dekock and W. Weltner, Jr., *J. Phys. Chem.* **75**, 514 (1971).
147. M. C. Hannigan, *J. Molec. Spectrosc.* **99**, 235 (1983).
148. C. R. Dickson and R. N. Zare, *Chem. Phys.* **7**, 361 (1975).
149. C. Linton, G. Bujin, R. S. Rana, and J. A. Gray, *J. Molec. Spectrosc.* **126**, 370 (1987).
150. G. Bujin and C. Linton, *J. Molec. Spectrosc.* **137**, 114 (1989).
151. G. Bujin and C. Linton, *J. Molec. Spectrosc.* **147**, 120 (1991).
152. S. D. Gabelnick, G. T. Reedy, and M. G. Chasanov, *J. Chem. Phys.* **60**, 1167 (1984).
153. S. A. McDonald, Ph.D. thesis, MIT, Boston, 1985.
154. P. Carette, A. Hocquet, M. Douay, and B. Pinchemel, *J. Molec. Spectrosc.* **124**, 243 (1987).
155. R. J. Van Zee, R. F. Ferrante, K. J. Zeringue, and W. Weltner, Jr., *J. Chem. Phys.* **75**, 5297 (1981).
156. B. R. Yadav, S. B. Rai, and D. K. Rai, *J. Molec. Spectrosc.* **89**, 1 (1981).
157. C. Linton, D. M. Gaudet, and H. Schall, *J. Molec. Spectrosc.* **115**, 58 (1986).
158. L. A. Kaledin and E. A. Shenyavskaya, *J. Molec. Spectrosc.* **90**, 590 (1981).
159. Y. C. Liu, C. Linton, H. Schall, and R. W. Field, *J. Molec. Spectrosc.* **104**, 72 (1984).
160. C. Linton and Y. C. Liu, *J. Molec. Spectrosc.* **131**, 367 (1988).
161. S. A. McDonald, S. F. Rice, R. W. Field, and C. Linton, *J. Chem. Phys.* **93**, 7676 (1990).
162. C. Linton, M. Dulick, R. W. Field, P. Carette, P. C. Leyland, and R. F. Barrow, *J. Molec. Spectrosc.* **102**, 441 (1983).
163. R. G. Dennig, *Gmelin Handbook*, Supplement A6, pp. 31–79, Springer, Berlin, 1983.
164. A. Cousson, S. Dabos, H. Abazli, F. Nectou, M. Pages, and G. Choppin, *J. Less Common Metals* **99**, 233 (1984).
165. J. J. Katz, G. T. Seaborg, and L. R. Moriss, eds., *The Chemistry of Actinide Elements*, 2nd ed. (2 vols.), Chapman & Hall, London, 1986, p. 1677.
166. *Gmelin Handbook of Inorganic Chemistry* see Ref. 163.
167. C. M. Marian, U. Wahlgren, O. Gropen, and P. Pyykkö, *J. Molec. Struct.* (*Theochem.*) **169**, 339 (1988).
168. M. C. Heaven, J.-P. Nicolai, S. J. Riley, and E. K. Parks, *Chem. Phys. Lett.* **119**, 229 (1985).
169. K. A. Gingerich, *Symp. Faraday Soc.* **14**, 109 (1980).
170. J. V. Ortiz, *J. Am. Chem. Soc.* **108**, 556 (1986).
171. M. A. Makhyoun, B. D. El-Issa, and A. B. Salsa, *J. Molec. Struct.* (*Theochem.*) **103**, 241 (1987).
172. B. P. Hay, *Inorg. Chem.* **30**, 2876 (1991).
173. H. J. Meyer and R. Hoffmann, *J. Solid State Chem.* **95**, 14 (1991).
174. K. Okada, Y. Kaizu, H. Kobayashi, K. Tanara, and F. Marumo, *Molec. Phys.* **54**, 1293 (1984).
175. A. Streitwieser and U. Müller-Westerhoff, *J. Am. Chem. Soc.* **90**, 7364 (1968).

176. A. Streitwieser, U. Müller-Westerhoff, U. G. Sonnichsen, F. Mares, D. G. Morrel, K. O. Hodgson, and C. A. Harmon, *J. Am. Chem. Soc.* **95**, 8644 (1973).

177. A. Streitwieser and S. Kinsley, in *Fundamental and Technological Aspects of Organic f-Elemental Chemistry*, T. J. Marks and I. L. Fragale, eds., Reidel, Dordrecht, 1985, pp. 77–114.

178. D. G. Karraker, *Inorg. Chem.* **12**, 1105 (1972).

179. D. F. Starks and A. Streitwieser, *J. Am. Chem. Soc.* **95**, 3424 (1973).

180. H. W. Kroto, J. R. Heath, S. C. O'Brien, R. F. Curl, and R. E. Smalley, *Nature* (London) **318**, 162 (1985).

181. J. R. Heath, S. C. O'Brien, Q. Zhang, Y. Liu, R. F. Curl, H. W. Kroto, F. K. Tittel, and R. E. Smalley, *J. Am. Chem. Soc.* **107**, 7779 (1985).

182. D. W. Cox, K. C. Reichmann, and A. Kaldor, *J. Chem. Phys.* **88**, 1588 (1988).

183. A. Kaldor, D. M. Cox, D. J. Trevor, and M. R. Zakin, *Z. Phys.* **D3**, 95 (1986).

184. F. D. Weiss, J. L. Elkind, S. C. O'Brien, R. F. Curl, and R. E. Smalley, *J. Am. Chem. Soc.* **110**, 4464 (1988).

185. S. C. O'Brien, J. R. Heath, H. W. Kroto, R. F. Curl, and R. E. Smalley, *Chem. Phys. Lett.* **132**, 99 (1986).

186. Y. Chai, T. Guo, C. Jin, R. E. Haufler, L. P. F. Chibante, J. Fure, L. Wang, J. M. Alforld, and R. E. Smalley, *J. Phys. Chem.* **95**, 7594 (1991).

187. A. Rosen and B. Wästberg, *J. Am. Chem. Soc.* **110**, 8701 (1988).

188. A. Rosen and B. Wästberg, *Z. Phys.* **D12**, 387 (1989).

189. J. L. Ballster, P. R. Antoniewicz, and R. Smoluchowski, *Astrophys. J.* **356**, 507 (1990).

190. F. Diederich, R. Ette, Y. Rubin, R. L. Whetten, R. Bell, M. Alvarez, S. Anz, D. Mesharma, F. Wudl, K. C. Khomani, and A. Koch, *Science*, **252**, 1548 (1991).

191. S. Wang and P. R. Buseck, *Chem. Phys. Lett.* **182**, 1 (1991).

192. A. G. Guiodin, A. Mele, G. Pizzella, and R. Teghie, *Z. Phys.* **D20**, 89 (1991).

193. D. Dayane, A. Benamar, P. Melinon, B. Tribollet, and M. Broyer, Z. *Phys.* **D19**, 191 (1991).

194. C. Brechignac, Ph. Cahuzac, F. Carlier, M. de Fratos, A. Masson, and J. Ph. Roux, *Z. Phys.* **D19**, 195 (1991).

195. R. G. Hayes and N. M. Edelstein, *J. Am. Chem. Soc.* **94**, 8688 (1972).

196. R. B. Ross, J. M. Powers, T. Atashroo, W. C. Ermler, L. A. LaJohn, and P. A. Christiansen, *J. Chem. Phys.* **93**, 6654 (1990).

INDEX